MICROELECTRONIC FAILURE ANALYSIS

Desk Reference, 3rd Edition

Thomas W. Lee, Editor

Dr. Seshu V. Pabbisetty, Editor

The Materials
Information Society

Copyright 1993
by
ASM International®
All rights reserved

No part of this book may be reproduced, stored in a retrieval system, or transmitted, in any form or by any means, electronic, mechanical, photocopying, recording, or otherwise, without the written permission of the copyright owner.

First printing, November 1993
Second printing, October 1994
Third printing, December 1995
Fourth printing, March 1997
Fifth printing, October 1997

This book is a collective effort involving hundreds of technical specialists. It brings together a wealth of information from worldwide sources to help scientists, engineers, and technicians solve current and long-range problems.

Great care is taken in the compilation and production of this Volume, but it should be made clear that NO WARRANTIES, EXPRESS OR IMPLIED, INCLUDING, WITHOUT LIMITATION, WARRANTIES OF MERCHANTABILITY OR FITNESS FOR A PARTICULAR PURPOSE, ARE GIVEN IN CONNECTION WITH THIS PUBLICATION. Although this information is believed to be accurate by ASM, ASM cannot guarantee that favorable results will be obtained from the use of this publication alone. This publication is intended for use by persons having technical skill, at their sole discretion and risk. Since the conditions of product or material use are outside of ASM's control, ASM assumes no liability or obligation in connection with any use of this information. No claim of any kind, whether as to products or information in this publication, and whether or not based on negligence, shall be greater in amount than the purchase price of this product or publication in respect of which damages are claimed. THE REMEDY HEREBY PROVIDED SHALL BE THE EXCLUSIVE AND SOLE REMEDY OF BUYER, AND IN NO EVENT SHALL EITHER PARTY BE LIABLE FOR SPECIAL, INDIRECT OR CONSEQUENTIAL DAMAGES WHETHER OR NOT CAUSED BY OR RESULTING FROM THE NEGLIGENCE OF SUCH PARTY. As with any material, evaluation of the material under end-use conditions prior to specification is essential. Therefore, specific testing under actual conditions is recommended.

Nothing contained in this book shall be construed as a grant of any right of manufacture, sale, use, or reproduction, in connection with any method, process, apparatus, product, composition, or system, whether or not covered by letters patent, copyright, or trademark, and nothing contained in this book shall be construed as a defense against any alleged infringement of letters patent, copyright, or trademark, or as a defense against liability for such infringement.

Comments, criticisms, and suggestions are invited, and should be forwarded to ASM International.

Library of Congress Cataloging-in-Publication data
Microelectronic failure analysis: desk reference/Thomas W. Lee, editor
Seshu V. Pabbisetty, editor.—3rd ed.
Includes index.
1. Electronics—Materials—Testing—Handbooks, manuals, etc.
2. Microelectronics—Materials—Testing—Handbooks, manuals, etc.
3. Microelectronics—Materials—Defects—Handbooks, manuals, etc.
4. Electronic apparatus and appliances—Testing—Handbooks, manuals, etc.
I. Lee, Thomas W.
II. Pabbisetty, Seshu V.

ISBN: 0-87170-479-X

ASM International®
Materials Park, OH 44073-0002

Printed in the United States of America

Acknowledgement

I extend my appreciation to the instructors and authors who have made the long-term commitment to develop and support this demanding program and who met the challenge of making the transition from presenting a technical paper to teaching a basic course; to SEMATECH's support of this book project through Jim Jordan and Dave Manning, who shared my vision of demonstrating that the SEMATECH Partnering Advantage could work in any specialty, and any level, to the benefit of the entire engineering community, while remaining congruent with the larger SEMATECH mission; to colleagues Art Leitten, Don Kortkamp, Ted Guthrie, and Jim Jurgens who provided a positive work environment that fostered creativity, team spirit, productivity and growth; to my family for their understanding about my time commitment; and to my father, Warren Lee, who shared with me the importance of logically determining and correcting the root cause.

FOREWORD

Failure analysis (FA) is key to excellence in electronics; it is a significant contributor to the process of continual quality and reliability improvement, which is critical to manufacturing excellence. It is a part of good customer service.

Failure analysis touches every aspect of microelectronics, from raw materials to field performance. An analyst routinely disassembles and judges the handiwork of entire manufacturing teams. Necessarily, FA professionals become scientific generalists who are competent in a wide spectrum of disciplines.

In today's competitive new market, developments in failure analysis are occurring at a rate that parallels the frenzied pace of the electronics business in general. This results in two universal FA problems: minimizing turnaround times and increasing the percentage of root cause determination, all in the presence of increasing circuit, device, and system complexity.

In this volume, the collective efforts of FA experts are aimed at helping workshop participants address the above issues. This book is intended to be compact, intensive and pragmatic, designed to address many of the varied subjects of concern to failure analysis. The authors have worked hard to provide a compendium of information and techniques, which hopefully can immediately benefit your efforts.

Thomas W. Lee
Editor

WORKSHOP SESSION INSTRUCTORS, CHAPTER AUTHORS, and PROFESSIONAL REVIEWERS

* **Dr. Thomas A. Anderson**
  Chemical and Surface Analysis Lab
  Motorola
  CSPTG
  5005 E. McDowell Rd.
  Mail Drop AZ01/P004
  Phoenix, Arizona 85008
  602-244-3917

* **Mr. Jim Bayless**
  Analytical Chemist
  SEMATECH
  2706 Montopolis Dr.
  Austin, Texas 78737
  512-356-3072

* **Mr. James Beall**
  Manager, Electronic Support Laboratories
  Martin-Marietta Astronautics
  12257 State Highway 121
  Dock 5
  Littleton, Colorado 80127
  303-971-8618
  F 303-977-4739

* **Dr. Jeff Bindell**
  Manager, Analytical Services
  AT&T Corporation
  555 Union Blvd.
  Allentown, Pennsylvania 18103-1285
  213-439-6827
  213-439-6669

* **Mr. Ted Boden**
  TEM Microscopist
  SEMATECH
  2706 Montopolis Dr.
  Austin, Texas 78741
  512-356-3324

Mr. Dave Burgess
  Consultant
  Accelerated Analysis
  470 Laurel Ave.
  Half Moon Bay, California 94019
  415-726-1832

Mr. Dave Burkhard
  Senior Scientist
  Motorola
  ATC Semiconductor Analytical Lab
  Semiconductor Products Sector
  2200 West Broadway Rd.
  Mail Drop M-360
  Mesa, Arizona 85036
  602-962-2890
  F 602-962-2285

* **Mr. Dan Burns**
  U.S. Air Force
  Rome Laboratory RL/ERDR
  Building 3, Room 1080
  525 Brooks Road
  Griffiss AFB, New York 13441-4505
  315-330-2868

Ms. Denise Cantu (Holdman)
  Motorola
  ATC Semiconductor Analytical Lab
  2200 West Broadway
  Mail Drop M-360
  Mesa, Arizona 85202
  602-898-5082

Mr. Joseph Colangelo
  Senior Failure Analysis Engineer
  Texas Instruments, Inc.
  DSEG Failure Analysis Laboratory
  8505 Forest Lane
  M/S 3114
  Dallas, Texas 75266
  214-480-7043
  214-332-6744

Dr. Edward I. Cole, Jr.
  I. C. Failure Analysis
  Division 2142
  Sandia National Laboratories
  Albuquerque, New Mexico 87185-5800
  505-846-1589

Mr. Dan Corum
  Senior Scientist
  Semiconductor Group
  Texas Instruments Inc.
  Houston Device Analysis Laboratory
  Mail Stop 733
  12203 Southwest Freeway
  Stafford, Texas 77477
  713-274-4237
  F 713-274-3555

Mr. William H. Crowell
ESD Control/Data Acquisition
Reliability Analysis Center (RAC)
210 Mill St.
Rome, New York 13440-6916
315-330-4151
F 315-337-9932

Ms. Laurie Dennig
SEM Analyst
SEMATECH
2706 Montopolis Dr.
Austin, Texas 78741
512-356-3027

Mr. John R. DeVaney
President
Hi-Rel Laboratories
East 204 Nora
Spokane, Washington 99207
509-325-5800
F 509-325-9508

Mr. Howard Dicken
President
D M DATA
10170 East Jenan Dr.
Bldg 2
Scottsdale, Arizona 85260-5921
602-451-7449
F 602-451-1890

* Dr. Alain Diebold
Manager, Materials Analysis
SEMATECH/AMD
2706 Montopolis Dr.
Austin, Texas 78741
512-356-3146
F 512-462-4592

* Prof. Henry Domingos
Chairman, ECE Department
Clarkson University
Box 5710
Pottsdam, New York 13699-5710
315-268-7606
F 315-268-7985

* Mr. Edgar A. Doyle, Jr.
Director, Analytical Lab
U.S. Force
Rome Laboratory
Product Evaluation Branch
Microelectronics Reliability Division
525 Brooks Rd.
Griffiss AFB, New York 13441-4505
315-330-4632

* Mr. Dave Dylis
Research Engineer
IIT Research Institute
Reliability Analysis Center
201 Mill St.
Rome, New York 13442-8200
315-339-7055

* Mr. George Ebel
Senior Engineer, Chip Failure Analysis
IIT Research Institute
Reliability Analysis Center
201 Mill St.
Rome, New York 13440
315-337-0900
315-337-7039
F 315-337-9932

Dr. Keenan Evans
Manager, Chemical and Surface Analysis Lab
Motorola
CSPTG
5005 E. McDowell Rd.
Mail Drop AZ01/P004
Phoenix, Arizona 85008
602-244-4608

Mr. Joseph J. Gajda
Senior Engineer
IBM Corporation
Mail Zip 56A
Hopewell Junction, New York 12533
914-894-2558

Mr. Chris Gondran
Materials Engineering
Defect Review and Characterization Lab
SEMATECH
2706 Montopolis Dr.
Austin, Texas 78741
512-356-3548

Mr. Leon Hamiter
President
Components Technology Institute, Inc.
Suite 117
904 Bob Wallace Ave.
Huntsville, Alabama 35801
205-536-1304
F 205-539-8477

* Dr. R. Cotton Hance
Senior Member of Technical Staff, MOS Analytical
 Laboratory
Motorola
MOS Memory Product Division
3501 Ed Bluestein Blvd.
Mail Drop TX11/L4
Austin, Texas 78744
512-928-7199

Note: * indicates reviewer.

Mr. Christopher L. Henderson
  Failure Analysis Department 2275
  Sandia National Laboratories
  P.O. Box 5800
  Albuquerque, New Mexico 87185
  505-844-3216

* Dr. Carolyn F. Hoener
  Materials Analyst
  SEMATECH Materials Lab
  2706 Montopolis Dr.
  Austin, Texas 78741
  512-356-3149

* Mr. Ken Huffman
  Manager, Analytical Lab
  SEMATECH
  2706 Montopolis Dr.
  Austin, Texas 78741

Mr. Martin Johnson
  Failure Analysis Laboratory
  Motorola
  Automotive and Industrial Products Group
  (AIEG)
  Component Engineering Department
  4000 Commercial Ave.
  Mail Drop IL08
  Northbrook, Illinois 60062
  708-480-6699

Mr. James Jordan
  Manager, Yield Improvement
  SEMATECH
  2706 Montopolis Dr.
  Austin, Texas 78741
  512-356-3553

* Mr. Harold F. Jones
  Applications Engineer
  Spin-on Products
  Allied Signal
  17 Timberlake
  Orchard Park, New York 14127
  716-662-2880
  800-247-4519

Mr. Scott Kiefer
  FA Engineer, Product Analysis Laboratory
  Motorola
  Application Specific Integrated Circuits (ASIC)
  1300 N. Alma School Rd.
  Mail Drop AZ50/CH240
  Chandler, Arizona 85224
  602-821-4751

Mr. Shekhar Khandekar
  Manager, Analytical Laboratory
  Level One Corporation
  105 Lake Forest Way
  Folsom, California 94630
  916-985-3670

Mr. Ed Kubel
  Manager, Conference Development
  ASM International
  Materials Park, Ohio 44073-0002
  216-338-5151 ext. 475
  F 216-338-4634

Mr. Thomas Lee
  Chairman, Construction Analysis Group
  SEMATECH
  2706 Montopolis Dr.
  Austin, Texas 78741
  512-356-3799
  F 512-356-7118

Mr. Robert Lowry
  Manager, Analytical Laboratories
  Harris Semiconductor
  Semiconductor Sector
  Palm Bay Rd.
  M/S 62-07
  P.O. Box 883
  Palm Bay (Melbourne), Florida 32901-0101
  407-724-7566
  F 407-729-4053

Mr. J. E. Mann
  Rockwell International Corporation
  Autonetics Group
  3370 Miraloma Ave.
  P.O. Box 4182
  Anaheim, California 92803
  714-762-8111

Mr. Douglas MacCormac
  Manager
  TRW Components International
  19951 Mariner Ave.
  Torrnace, California 90503
  213-214-5500

Dr. J. Thomas May
  Principal Engineer
  Micro-Rel Div. Medtronics, Inc.
  2343 West 10th Place
  Tempe, Arizona 85281
  602-929-5345
  F 602-968-9691

Dr. Tom Moore
  Sr Member Tech Staff, Central Research Laboratory
  Texas Instruments, Inc.
  28815 North Central Expressway
  Dallas, Texas 75243
  M/S 147
  214-995-6117

Note: * indicates reviewer.

Dr. Seshu Pabisetty
Manager, Houston Device Analysis Laboratory
Texas Instruments, Inc.
Semiconductor Group
12203 Southwest Freeway
Mail Stop 7334
Stafford, Texas 77477
713-274-3588
F 713-274-3555

* **Mr. Earl Parks**
Martin-Marietta (GE)
Electronics Lab
Electronics Park Bldg EP 3-15
P.O. Box 4840
Liverpool, New York 13088
315-456-3258

* **Mr. Tom Pinkston**
Chemical/Materials Analysis
SEMATECH
2706 Montopolis Dr.
Austin, Texas 78741
512-356-3030

* **Dr. Pote Pruettiangkura**
Manager, Process Application and Development Department
Component Failure Analysis
13020 Floyd Rd.
P.O. Box 655012 MS-3619
Dallas, Texas 75265
214-917-5268

Dr. Jamie Rose
Senior Scientist/TEM Microscopist
Material & Technology Analysis Lab
Digital Equipment Corporation
30 Forbes Road
NRO5/B4
Northboro, Massachusetts 01532
508-351-4172
F 508-351-4987

* **Mr. Dick Ross**
Manager, Analytical Laboratory
IBM Corporation
1000 River Rd.
Dept. 328, Bldg. 967-2
Essex Junction, Vermont 05452-4299
802-769-2218
F 802-769-1220

Mr. Scott Smith
Motorola
Land Mobile Products Sector
Mail Drop IL02, Room 1506
1301 Algonquin Rd.
Schaumburg, Illinois 60196
708-576-0418

* **Dr. Robert W. Thomas**
104 Cedar St.
Rome, New York 13440
315-336-0286

Mr. Dean Thomasi
Staff FA Engineer
IBM- East Fishkill
General Technology Division
Bldg 502, Zip 95A
Route 52
Hopewell Junction, New York 12533-0999
914-894-5828
F 914-894-6868

Mr. Gene P. Thome
Motorola
Communications, Signal, and Power Technology Group (CSPTG)
Analytical Laboratories, Discrete Devices
5005 E. McDowell Rd.
Mail Drop D-162
Phoenix, Arizona 85008
602-244-6880
602-244-6628
F 602-244-3263

Dr. Lawrence C. Wagner
Manager, Device Analysis Laboratory (Dallas)
Sr Member of Technical Staff
Texas Instruments, Inc.
13532 North Central Expressway
Mail Stop 74
Dallas, Texas 75243
DR copy 5 Dec 90
P.O. Box 655012
Dallas, Texas 75265
214-995-2294
F 214-995-2638

* **Mr. Allen White**
TEM Analytical Scientist
SEMATECH
2706 Montopolis Dr.
Austin, Texas 78741
512-356-3555

Mr. K. Scott Wills
Beam-It Inc.
1610 North I-35 East
Suite 206
Carrollton, Texas 75006
214-446-9203
F 214-446-9303

Note: * indicates reviewer.

Mr. David Wilson
Supervisor, Failure Analysis Lab
Martin-Marietta Astronautics
Receiving Dock 5
12257 State Highway 121
Littleton, Colorado 82107
303-977-4593
303-977-5508
F 303-977-4129

* Mr. Conrad H. Zierdt
Bell Labs
(retired)
4140 Maulfair Dr.
Allentown, Pennsylvania 18103-9663
215-437-2200

TABLE OF CONTENTS

Failure Analysis Procedures and Overview

A Philosophy of Failure Analysis   3
by Howard Dicken

The Start-Up of a Failure Analysis Laboratory   9
by J.E. Mann

Failure Analysis Outline Guide for Semiconductors and Integrated Circuits   15
by James Beall

Electrical and Mechanical Characterization

Curve Tracer Appplications and Hints for Failure Analysis   25
by James Beall and David Wilson

Primer on Device Problems and Curve Trace Characteristics   37
by Douglas McCormac

Inspecting IC Packages with C-Mode Acoustic Microscopy   41
by T.M. Moore

Advanced Radiographic Techniques in Failure Analysis   51
by Joe Colangelo

Specimen Preparation

Mechanical and Chemical Decapsulation   61
by Thomas W. Lee

Plasma Decapsulation Techniques   75
by Dean Thomasi

Particle Verification, Extraction and Identification   79
by Gene P. Thome

Window Etch Procedure for Multi-layer VLSI   87
by David S. Kiefer

Metallurgical and Specimen Decoration Techniques

Cold Sample Preparation Techniques For Low Melting Point Materials   93
by David J. Burkhard and Denise Holdman

Chip and Device Sectioning Techniques   97
by J.J. Gadja

Chemical Etch Formulations and History   111
by Thomas W. Lee

Safety Hints and Ideas   115
by Thomas W. Lee

Plasma Etching   121
by James Beall

Reactive Ion Etch Recipes for Failure Analysis   125
by David S. Kiefer

Dimensional Conversion Chart ... 131

Failure Site Localization Techniques

FALT Probing Bubble Test for Rapid Fail Site Location ... 135
by Gene P. Thome

Liquid Crystal Microscopy ... 141
by Shekhar Khandekar and Kendall Scott Wills

Electron Beam-Induced Current (EBIC) Techniques ... 145
by James R. Beall

Voltage Contrast Techniques and Procedures ... 153
by James R. Beall

Scanning Electron Microscopy Techniques for IC Failure Analysis ... 163
by Edward I. Cole, Jr. and Jerry M. Soden

Emission Microscopy and Liquid Crystal ... 177
by Lawrence W. Wagner, Thomas S. Taylor, and Kendall S. Wills

Photoemission Microscopy ... 181
by Gary Shade and Kendall Scott Wills

Analytical Instrumentation

Instrumental Analysis Techniques ... 199
by Keenan Evans and Thomas A. Anderson

Failure Analysis Acronym List ... 209

Scanning Electron Microscopy and Energy Dispersive X-Ray Analysis ... 211
by John R. Devaney

Microbeam Analysis and Semiconductor Reliability ... 237
by R.K. Lowry

Transmission Electron Microscopy: A Review and a Comparison With High Resolution Scanning Electron Microscopy ... 255
by Jamie H. Rose, Barbara Miner, and Aldo Pelillo

Applications of FTIR Spectroscopy to Semiconductor FA ... 269
by Christopher D. Gondran and Ronald A. Carpio

Failure Modes and Mechanisms

Failure Mechanisms in Integrated Circuits ... 277
by Dan Corum, Swee Yong Kim, and Kendall Scott Wills

Semiconductor Failure Modes and Mechanisms Chart ... 301
by Thomas W. Lee

Failure Modes and Mechanisms of Non-Semiconductor Electronic Components ... 303
by Martin J. Johnson and Scott E. Smith

Limiting Phenomena in Power Transistors and the Interpretation of EOS Damage ... 321
by J. Thomas May

ESD Damage Simulation and Failure Mechanisms ... 329
by Thomas W. Lee

Thermomechanical Effects of EOS ... 335
by Thomas W. Lee

Electrical Overstress in Integrated Circuits ... 353
by J. Thomas May

Reliability for the Failure Analyst ... 357
by David Burgess

Basic Silicon Fractography ... 361
by Dan Pote

Failure Analysis Reporting

Topics in Knowledge-Based Failure Analysis ... 371
by Christopher L. Henderson

Suggestions for a Well-Written FA Report ... 377
by Thomas W. Lee

Appendix

ISTFA 1976-1993 Subject Index ... 381

Resources for Failure Analysis ... 397

Failure Analysis Procedures and Overview

A PHILOSOPHY OF FAILURE ANALYSIS

by Howard Dicken

BASIC CONSIDERATIONS

What is a failure?
What is quality?
What is reliability?
What are the basic causes?
Why is a failure analysis needed?
What are the key steps?
What is a safe sequence?
When is a failure analysis impossible?
What is the goal of a good analysis?

DEFINITIONS

What is a Failure?

If a device has consistently performed the required circuit/system functions under all specified environmental conditions and is no longer doing this, then it has failed!

What is Quality?

Quality is related to the percent of devices that will initially perform the required circuit/system functions under all specified environmental conditions.

Quality, either outgoing or incoming, is measured in PPM (Parts Per Million). This is the percent of units which, after all manufacturing and testing operations, still do not meet the specified performance criteria.

What is Reliability?

Reliability is related to the length of time during which the device will consistently perform the required circuit/system functions under all specified environmental conditions.

Typically, reliability is measured in terms of MTBF (Mean Time Between Failure) or Mean Time to Failure (T_{50}) of the device while in its specified operating environment.

What is a Field Failure?

Usually, a field failure is defined as a failure of the operating system in the user's environment. There are very few component failures in the field. Most component failures occur during incoming test, board assembly and test, and system checkout.

Typical Sequence of Testing

Vendor
Component final test
Component
Incoming test
PC board
Board test
System
System operating test
Field use
Field failure

CAUSES OF FAILURES

What Causes Failures?

A failure is any change of resistance in any electrical path or between any electrical nodes. However, components do not usually fail for electrical reasons. Component failures are caused by either physical, chemical, or mechanical mechanisms. The electrical malfunction is not the cause, but it is a result of the mechanism.

What Causes the Change of Resistance?

1. Electrical overstress, which causes either an increase of resistance (an open) or a decrease (a short).

2. Mechanical/chemical effects, such as high-cycle fatigue or corrosion, which will break an electrical path (open).

3. Component wear out, which causes opens or shorts.

   Note that the change of resistance does not require the extremes (opens or shorts) for failure. For example, a slight de-

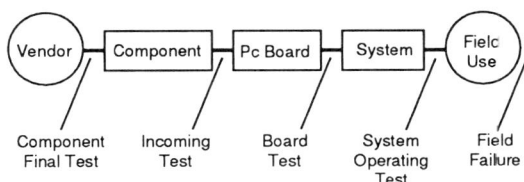

Fig. 1 Typical sequence of testing

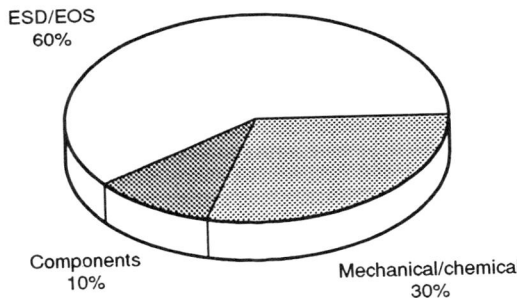

Fig. 2 Initiating factors for field failures in electronic systems

crease of resistance between two nodes, which might be indicated by an increase of leakage current, could cause a failure.

Where are the Failures?

1. In approximately 60% of the cases, electrical over-stress (EOS) will cause a component failure (the failed component is usually a semiconductor device).
2. In about 30% of the cases, mechanical/chemical effects will cause failures at either the box, the board, or the component level.
3. And in approximately 10% of the cases, the failure is initiated by a failure within the component itself.

CATEGORIZING SEMICONDUCTOR FAILURES

Most semiconductor failures can be categorized by both the primary cause and the actual physical mechanisms. By remembering these few basic phenomena, the analysis can minimize much of the routine work. The basic causes that initiate the failure are listed below. For example, the presence of moisture will initiate corrosion or dendritic growth, while a poor device design will allow ESD or EOS damage.

Basic Causes of Semiconductor Failures

- High temperature
- Moisture
- Contamination
- Device design
- Process and process control
- Unsuitable environment/application

The major failure mechanisms are listed below—organized in six basic categories. The following pages illustrate actual examples of most of the mechanisms.

Basic Failure Mechanisms

- Overstress
  - EOS
  - ESD
- Contamination
  - Inversion
  - Corrosion
  - Dendritic growth
- Wear Out
  - Intermetallics
  - Metal migration
  - Oxide
- Mechanical
  - Thermal expansions
  - Shock/vibration
- Electrical Performance
  - Latchup
  - Marginal gain
- Radiation
  - Alpha
  - Total dose

COMPONENT FAILURE CATEGORIES

At the component level, many of the same type of failures still hold, including a large percent for "retest OKs". The chart below shows some typical percentages for the various failure categories of integrated circuits that were returned to the manufacturer.

In this data, the percent of failures caused by ESD/EOS is low, possibly because this represents parts returned to the vendor as compared to all failed components in a system. Other reports indicate that all electrical overstress failures total 25% to 60% of the causes of component failures.

It should be noted that in this data the vendor-related failures are less than 35% of the field returns. Actual physical failures are 11%. This grouping would include the classical temperature-related wear out mechanisms, such as metal migration, intermetallic formation, and oxide wear out. In addition, this group also includes all of the mechanical packaging problems, such as die attach and wire bonds.

IC Field Return Summary

Retest ok	45%
Unknown	3%
Test escape elect.	9%
Die effect	7%
Test escape handling	4%
Ext. mechanical	9%
User ESD/EOS	20%
Assembly defect	4%

FAILURE ANALYSIS PROCESS FLOW

1. Description/History

The description of the device would include the physical/electrical characteristics plus the history of screening and application. This data is necessary to determine the cause of failure and the corrective action.

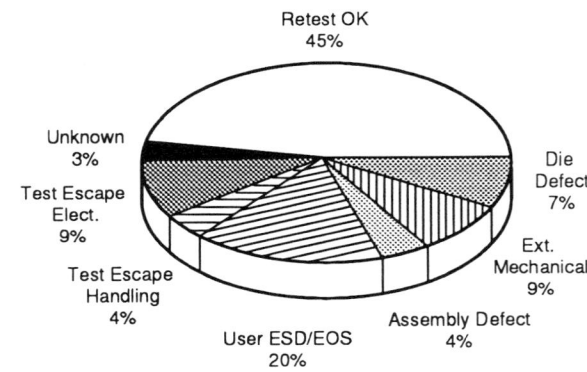

Fig. 3 IC field return summary.

2. Failure Mode

These are the reported characteristics of the failure such as "open", "low gain" etc., as determined by the suggested electrical measurements and the appearance of the curve traces.

3. Safe Analytical Sequence

The is the customized sequence of failure analysis steps which have been determined by the failure mode. This sequence and analytic steps have been selected to provide the maximum information without destroying any evidence.

4. Failure Mechanism

This is the actual physical mechanisms or conditions which have created the observed failure mode.

5. Cause

These are the external conditions or environment which triggered the failure mechanism.

6. Corrective Action

This consists of a list of one or more actions or changes to eliminate, avoid or minimize the cause or failure mechanism. Some examples are:

- Reduction of environmental temperature
- Monitor moisture content of package
- Specify different device
- Monitor assembly procedures
- Require "spit" shields in assembly process
- Establish vendor SPC
- Eliminate ultrasonic cleaning
- Add input protection network for ESD
- Minimize temperature extremes
- Add mechanical protection

The information for each of these categories is developed by performing the failure analysis, using the listed equipment and following the safe sequence listing.

DESCRIPTION/HISTORY

One of the key requirements for any successful failure analysis is an extensive listing covering the description and history of the failed part. This description should not only include details on the part, including schematics, pinouts etc., but

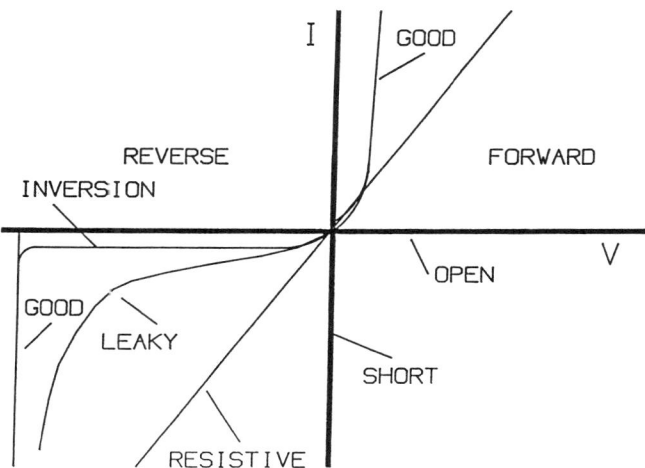

Fig. 4 Junction characteristics.

should also include information of the history of the part including testing, screening, and time of use. In addition, this data should include details as to the time and place of the failure. A detailed listing of required data is given on the following pages.

The more data that is available, the better the chance of an accurate and fast analysis. Some of the benefits are:

- SPEED—Valuable personnel time is not wasted trying to gather data or recreate information such as pinouts, etc. In addition, data on the type of failure and environment of the failure allows the analysis to safely skip some initial steps and go directly to the key analytical steps.
- ACCURACY—Detailed background data helps to avoid some potentially wrong assumptions.
- ROOT CAUSES—Usually the applications data will allow the analysis to more accurately determine the root cause of the failure and in addition, suggest a realistic corrective action.

The background data should be a necessary part of any analysis procedure and should be part of any communication with the customer.

HISTORY/DATA

1. Part Number _____
2. Manufacture _____
3. Data Code _____
4. Process Technology _____
5. Age of Technology _____ years
6. Feature Size _____ microns
7. Die Area _____ Sq mils
8. Time off Failure (check one)
   - Component Final Test _____
   - Component Incoming Test _____
   - Board Level Test _____
   - System Test _____
   - Field Operation _____
9. Application Environment (check one)
   - Office _____
   - Industrial _____
   - Military-Ground _____
   - Military-Space _____
   - Military-Mobile _____
   - Navy-Ship _____
   - Navy-Submarine _____
   - Airborne-Fighter _____
   - Airborne-Helicopter _____
   - Missle _____
   - Cannon _____
10. Operating Environmental Temperature _____ deg C
11. Level off Screening/Testing (check one)
    - None _____
    - Burn-In _____
    - MIL-STO-883 _____
    - Class S _____
    - ESS _____
12. Package type (check one)
    - Hermetic _____
    - Plastic _____
13. Package Material (check one)
    - Metal _____
    - Epoxy _____
    - Ceramic _____
14. Package Seal (check one)
    - Molded _____
    - Soldered _____
    - Welded _____
    - Frit _____
    - Epoxy _____
15. Repeated Failure Mode _____ Y _____ N
16. Failures Let Related _____ Y _____ N
17. Percent off Failures by Lot _____ %
18. Spice modeling parameters available _____ Y _____ N

A Philosophy of Failure Analysis

Failure Analysis Procedures for Semiconductors

There are many different approaches and procedures to determine the cause of a microcircuit failure. Each failure analysis laboratory usually develops its own checklist. The steps and sequence can vary. However, they must be organized to get the maximum information in a sequence of tests without affecting or destroying any evidence.

To minimize the chance of destroying evidence, the following analytic sequence of steps is suggested:

Benign
1. Description/History/Application
2. External Visual
3. X-Ray
4. Fine Leak (He)

Non-Benign
5. V/I Electrical Test (Curve Tracer)
6. Hermeticity
7. PIND - V/I Electrical Test
8. External Clean - V/I Electrical Test
9. Bake - V/I Electrical Test
10. Temperature Cycle - V/I Electrical Test
11. External SEM

Possible Destructive
12. RGA
13. Delid
14. Internal Visual

Possible Destructive (contd)
15. V/I Electrical Test
16. Clean - V/I Electrical Test
17. Internal SEM and EDX
18. EBIC
19. Voltage Contract
20. Bake - V/I Electrical Test
21. Liquid Crystal
22. IR Test
23. Passivation Glass Strip
24. Probe/Isolate

Destructive
25. Oxide Strip
26. Metal Strip
27. Decoration
28. Cross Section

Safe Sequence of Failure Analysis for Integrated Circuits

Steps	Open	Hard short	Resistive short	Leaky	Inversion	Intermittent	Unstable	Functional failure	Out of spec
1. History	X	X	X	X	X	X	X	X	X
2. Photo	X	X	X	X	X	X	X	X	X
3. Ext Visual	X	X	X	X	X	X	X	X	X
4. X-Ray	X	X	X	X	X	X	X	X	X
5. Curve Trace	X	X	X	X	X	X	X	X	X
6. Functional El.	X	X	X	X	X	X	X	X	X
7. He Leak	X		X	X	X		X		X
8. Gross Leak			X				X	X	
9. PIND/Elect		X				X			
10. Ext SEM			X	X					
11. Bias Bake/El.			X	X	X		X	X	X
12. Temp Cy/EL					X	X			
13. Ex Clean/El.			X	X			X	X	X
14. RGA			X	X	X		X	X	X
15. DELID	X	X	X	X	X	X	X	X	X
16. Int Visual	X	X	X	X	X	X	X	X	X
17. Electrical	X	X	X	X	X	X	X	X	X
18. Bake/El.			X	X	X		X	X	
19. SEM/EDX	X	X	X	X	X		X	X	
20. EBIC			X	X	X		X	X	
21. Volt. Contrast	X	X				X			
22. Hot Spot		X	X	X		X	X	X	
23. Glass Strip			X	X	X	X	X	X	
24. Probe	X	X	X	X		X	X	X	X
25. Metal Strip		X	X	X		X	X	X	X
26. Oxide Strip		X	X	X	X	X		X	
27. Decorate		X	X	X		X			
28. Cross Section		X							
29. Review/Correl.	X	X	X	X	X	X	X	X	X
30. Report	X	X	X	X	X	X	X	X	X

THE START-UP OF A FAILURE ANALYSIS LABORATORY

by J. E. Mann

Starting a failure analysis laboratory? A reasonable request until you study the requirements. The questions now start to flow and require carefully thought-out answers. What talents are necessary for adequate staffing? What and how large an equipment budget do you propose? What is the company or corporate backing, and what level of analysis is expected? Is supplier interfacing needed? Must the analysis results meet internal system design, or will they be used for corrective action at suppliers or subcontractors?

The extent of reporting and record keeping must be determined. How long are the records kept? Reporting, record keeping and the inventory of failed components are a single task as important as the analysis work itself.

Now the failure analysis laboratory is set up, and it appears successful. The round starts again. Equipment must be updated as technology goes from bipolar to MOS to C-MOS to MIS and microprocessor. How and when does the laboratory decide to bring in special analysis equipment or seek special labs equipped with the required analytical tools? An attempt will be made to answer these questions.

FAILURE ANALYSIS REQUIREMENTS

The primary objective of failure analysis is to identify the cause of failure in the quickest and most economical way. This does not mean a "quick and dirty" shot, but rather the most efficient and thorough level of analysis required by the situation. Following these procedures, a cost-effective operation will be visible to all management.

Level of analysis is a critical factor in costing the job and depends upon the circuit information supplied, length of time the part has been in operation, and a knowledge of the failed component, i.e., application of part using critical electrical parameters where cost of redesign is prohibitive. Many times the only necessary information needed is: "Was the failed part removed or should further troubleshooting be continued?"

In critical application, you may be asked to check the part down to the starting material for pinpointing the failure causes. Therefore, you should define levels of analysis and set a priority system into action. The priority system aids both the laboratory and the operators in obtaining answers in the required length of time. An example of definition of three levels of analysis is described by the following:

Minimum level. Electrical testing of the part to supplied specifications will be performed and a report of defective parameters will be made if the part is defective electrically. The results of specific parameter values within the specified limits may also be supplied if required.

Medium level. Electrical test is a minimum level. X-radiographic work, hermeticity tests and/or pin-to-pin curve trace evaluations when applicable or appropriate will be performed.

Depending on results of these tests, the part may be vibrated under power and while being electrically monitored to indicate possible intermittency.

Any defective part package would be opened when applicable and an internal microscopic visual inspection made. Such things as electrical overstress, loose bonds, diffusion anomalies, and/or surface contamination/foreign materials may be indicated at this level.

In-depth level. This level includes both of the above sections and additional work to isolate and identify the failure mechanism(s) and relate it (them) to the part usage, testing, and/or manufacturing processes and procedures.

Such techniques as internal electrical probing of circuitry and voltage nodes, isolation and testing of circuit components, electron microprobing of particles or contamination observed, wet chemical analyses of contaminants (if applicable), bond/lead pull testing, scanning electron microscope analyses, metallographic analysis of circuitry and/or packaging, etc., will be employed when appropriate to achieve the objectives of isolating the failure mechanism(s) and identifying it (them).

Talents for excellent failure analysts come from many disciplines: chemical, electrical, physical, mechanical, or any other. Usually, the most successful analyst is one with a diversified background using more than one discipline. This person must be one who likes to work in difficult situations. He must be observant of every minute detail and remember all the past analyses of which he has knowledge. He must under no circumstances become jaded with the job or act all-knowing in this work. In other words, he is an electronic pathologist dealing in "cause of failure." Depending upon the background, allow six months to one year for a complete analyst to develop. Otherwise, hire a specialist from another corporation or semiconductor manufacturer.

Setting up a laboratory is dependent upon the levels of analysis needed to accomplish the identified task. The essential items for any failure analysis laboratory are:

1. Transistor curve tracer
2. Set of micromanipulators
3. Stereomicroscope 10-60×
4. Metallurgical microscope 50-600×
5. Camera for microscopes
6. Small polishing equipment

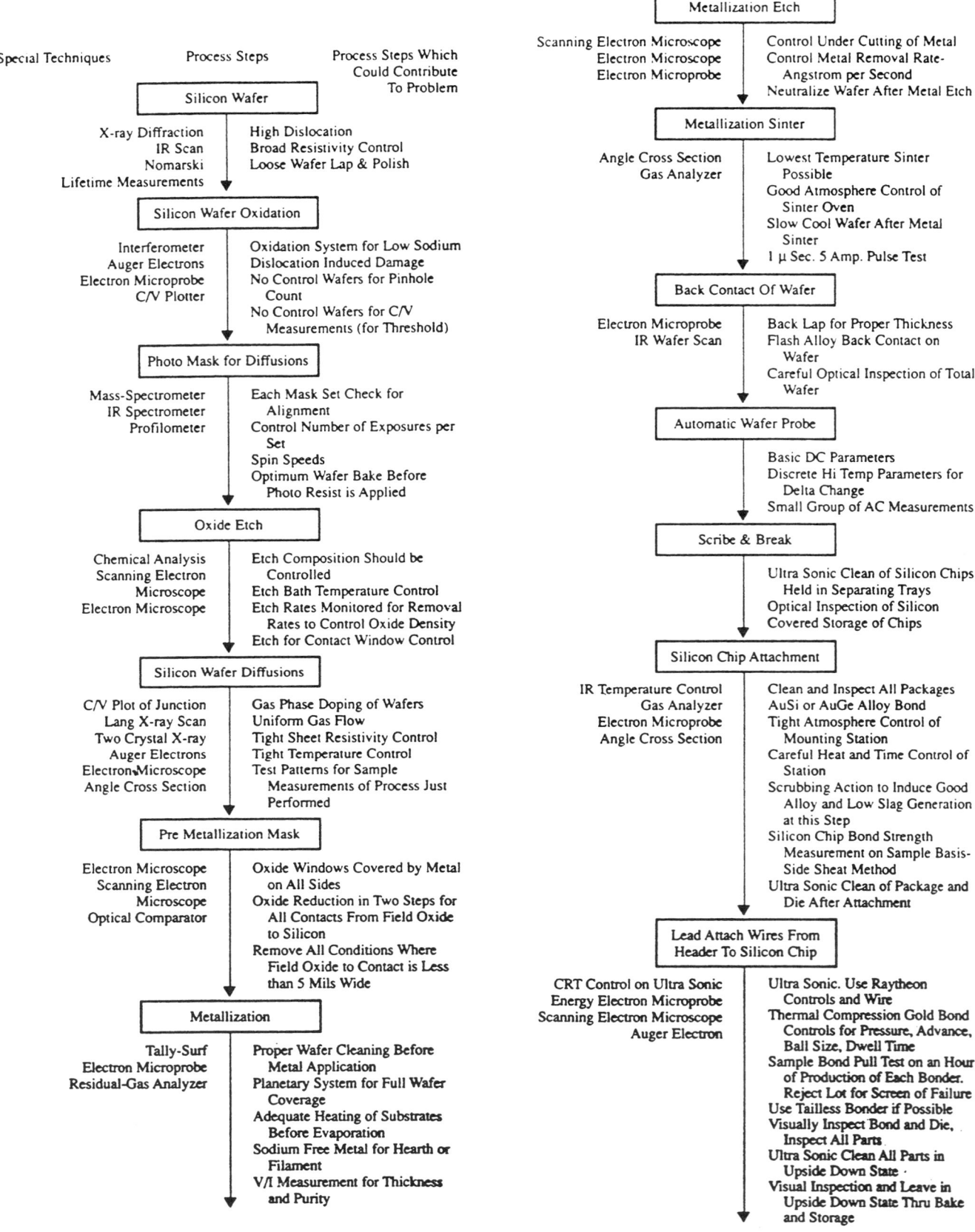

Fig. 1a Types of analysis needed for identified problem area in semiconductor processing. (continued)

The Start-Up of a Failure Analysis Laboratory

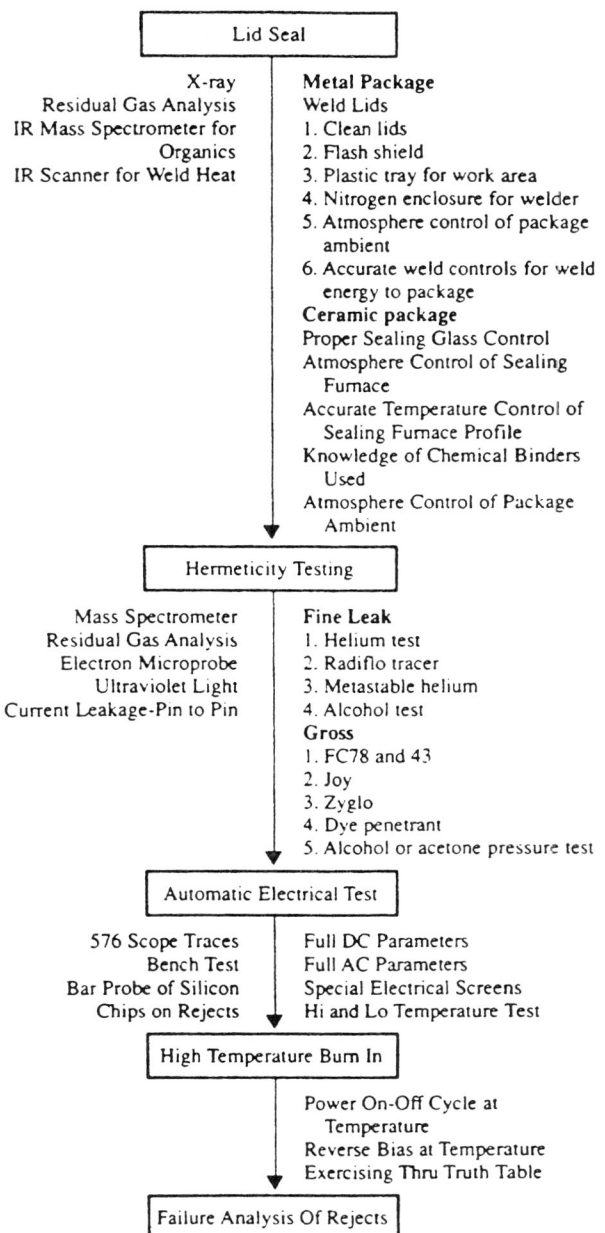

Fig. 1b Types of analysis needed for identified problem area in semiconductor processing.

7. Linear plug-in for curve tracer

The above listed equipment would complete about 80% of the analysis.

With the above equipment, one requirement is to develop a pin-to-pin understanding of all ICs used in a specific system. For example, a complete set of curve trace characteristics for the Minuteman system could be on file. RADC and NASA also use this system. Even LSI can be checked for failures by a careful circuit analysis using curve tracer equipment along with the micromanipulator probes for node voltages. Transistor elements in integrated circuits or LSI can be tested for electrical parameters by using the same microprobes. The initial laboratory described will define the failure electrically and, by using the microscope with camera attachment, photograph evidence which can be used to show surface type failure. Using the electrical data and location, the analyst can cross-section the part for failures in the subsurface of the device.

Any number of good epoxy or acrylic encapsulating materials can be used for the cross-section. There are a number of quick set materials available commercially. A simple plastic cup with a taper on the walls can be used for holding the potting material.

One of the most valuable tools for a small failure analysis group is a library. Reference materials give guidance and support to the laboratory in substantiating their findings. These items are not expensive and give guidance and support to the laboratory in substantiating their findings.

It is important to use the phone and discuss your findings with the manufacturers and any large laboratories with which you can develop a communication link.

Small failure analysis operations must, by charter, operate on a maximum of innovation due to the lack of special equipment for analysis. There are times when a request for checking a highly specialized part is made. In this case, the innovation of the laboratory can be made evident. A simple example would be to use a circuit board that has the failed or suspect component on the board. In this fashion, the parts can be tested, and the problem as to pin location can be resolved. If one considers the cost of special programs for the automatic testers, 160 to 200 hours can be saved in program cost alone.

When the failure analysis laboratory has become successful, management requests an equipment list to expand the area for more complete analysis to increase cost effective work for corrective action by improving design or getting corrective action, taken at the component supplier. The equipment list grows and suggested items are:

1. X-ray unit for radiography of packaged components or multi-layer boards
2. Hermetic seal tester
3. Test console with power supplies, oscilloscopes, current meters, digital voltmeter, and resistor matrix for circuit loading
4. Circuit tester for maximum number of different circuit types, e.g., T^2L, P-MOS, N-MOS, and C-MOS
5. Bond pull tester-silicon die shear tester
6. Three station grinding and polishing table
7. Metallograph capable of 1000X or better with oil lens
8. Scanning electron microscope

The laboratory can now complete the analysis on almost any device or circuit. I say "almost" because a unique RF device, I-R sensor, or phase lock loop can create some problems. Now you should consider a talented circuits engineer for the opening you've created. By proper use and training, he/she will become an asset as valuable as your capital equipment.

Considering the equipment now available, the next step is to find an independent laboratory to augment your own equipment. The need for FTIR scanning of plastic, surface analysis, mass spectroscopy, or residual gas analysis can be handled on an as-needed basis. Therefore, your capital is lower and space is not a problem. This assumes, of course, that such equipment is not available in your company or corporation—if it is, then use that equipment.

The ultimate scheme for a failure analysis laboratory is found in major corporations. The total capability is represented by the equipment and the specialist working together for an analysis task. An example is:

1. Electrical test for total AC and DC test with special plots for acquisition of failure data.
2. Individual test equipment dedicated to special device types: memories, CPUs, and microprocessors.
3. Multiprobe stations capable of exercising functioning wafers or chips, with capability to put probes on discrete lines for voltage nodes.
4. Hermetic and gas analysis equipment for checking both leak rate and absorbed gases confined to the interior of the package.
5. IR equipment for thermal scanning or looking through the silicon for damage or checking organic materials.
6. Scanning electron microscopes, some with EDAX attachments, electron microprobes, ion microprobes, Auger, ESCA, and SIMS. A combination of these instruments can detail any surface problem; from semiconductors to unique weld failures.
7. Complete metallography laboratory with specialist in metal-semiconductor-thin film. These people can help direct the region to be analyzed or improve the sample for analysis by special angle sections.

The list could go on and include a microchemist capable of quoting in ppm the water extractable elements found in milligrams of a sample.

A complete list of the analytical equipment available plus the relative sensitivity of each technique follows:

DETECTION LIMITS OF INSTRUMENTAL METHODS[2]

Method	Sensitivity (g)
Gas chromatography	10^{-8} to 10^{-12}

See details in Tables 2 and 3 of Reference 2.

Thin layer chromatography
   Color reaction 10^{-6}
   Fluorescence 10^{-9}

Nondestructive detection of pesticides and indoles via color reaction of several electron acceptors. Use of fluorgenic labeling and fluorescence detection increases sensitivity.

Mass Spectrometry
   Electron impact 10^{-12}
   Spark source 10^{-13}
   Ion probe 10^{-15}
   Chemical ionization 10^{-10}
   GC/MS/with computer 10^{-11}

Detects all elements and most inorganic compounds. Instrumentation is generally complex and expensive. The GC/MS computer system is most versatile and universally useful for analytical purposes.

Liquid chromatography
   Refractive index detector 10^{-6}
   Ultraviolet/visible detector 10^{-9}
   Fluorescence detector 10^{-10}
   SID and FAD detectors 10^{-11}

Attainable sensitivity depends on detector used. Ultraviolet/visible limited to those compounds such as FAD and SID are more universal and sensitive.

Neutron activation analysis 10^{-12}
   Responds to all elements.
Atomic absorption spectroscopy
   Flame 10^{-9}
   Flameless 10^{-12}

Over 60 elements detectable with different sources.

Atomic emission spectroscopy 10^{-9}

High sensitivity and multi-element analysis via inductively coupled plasma.

Infrared spectroscopy
   Standard techniques (pure samples) 10^{-6}
   Gases and solids (special techniques)   (ppb)
   Fourier transform infrared 10^{-9}

Responds to all samples with infrared absorption; special techniques enhance sensitivity.

X-ray fluorescence 10^{-7}

Used for elements with atomic numbers above 11.

Optical microscopy 10^{-12}

Simple, rapid method for particulate analysis.

Anodic stripping voltametry 10^{-8}

Can analyze 10 to 20 elements; best for Cu, Pb, Zn, Cd Surface analysis

ESCA	10^{-10}
ISS	10^{-10}
Auger	10^{-10}
SIMS	10^{-15}

Detects and identifies atoms in first several atomic layers of a surface; among the most sensitive methods known.

Polarography
I/C polarography	10^{-8}
Pulsed polarography	10^{-10}
Stripping voltametry	10^{-11}

Detects organics and most metallic elements and compounds.

Ion-selective electrodes	10^{-15}

Sensitivity shown is for detection of Cu; otherwise sensitivities vary, depending on element and electrode.

GC/ultraviolet-photoelectron spectroscopy	10^{-5}

Using a direct-coupled GC, spectra can be obtained in less than 1 min on 10— g quantities.

Wet chemical methods
Nitrate by enzyme reagent	(ppb)
Hg in water	(ppb)

Many methods involving chemical reaction, color-metric detection give ppm and ppb or finer measurements in large sample sources such as air, water.

Refractive index	(ppm)

Simple, rapid method, only useful for binary mixtures whose components have a wide difference of refractive index.

Proton NMR
Continuous wave (single scan)	10^{-4}
Fourier transform (approx. 10,000 scans)	10^{-6}

Detects all organic and diamagnetic organometallic compounds that contain hydrogen atoms; instrument costs range from moderate to expensive; NMR offers additional ability to give structure and identity for compounds.

Special Analysis

The decision to go for special analysis is dependent upon many factors—none of which should be, "Let's see what we find."

Special analysis on any surface analytical equipment must be carefully planned in order to develop meaningful results:

1. Plan the number and type of samples to be analyzed.
2. Develop detailed information on the samples.
3. Know as much as is possible of the processes used to manufacture the part.
4. Be free and open with the technical analyst on what you're looking for and how certain processes may contribute to the problem. Keep an open mind to other possibilities which may develop as the information becomes available.
5. Have open discussion with the manufacturer, technical analyst, and engineers in your company.

With these basic guidelines, a successful effort should result, thereby determining the failure cause and identifying corrective actions necessary.

Special analysis can lead the failure analysis group into many strange fields of effort; as an example, our labs have developed many special screen test methods to ensure reliability. Each special test was developed because failure analysis identified problems and came forth with a screen test to protect the system user from those identified hazards.

The test method becomes effective if each laboratory communicates with other laboratories by direct-line communications or publishes reports on their own findings. This is another key to awareness of the problems found throughout the community.

REPORTS AND RECORD KEEPING

For the initial reporting, a simple letter or letter form will more than handle the reporting of the failure and its apparent cause. Setting up a spreadsheet of database computer log for recording the incoming work with completion dates also logged will ensure good record keeping. A laboratory notebook for each analyst is a simple method of keeping a file of failure modes. When the program becomes larger, data plotting is extremely helpful. Following this method, trends of failure and peak failure modes are self evident.

On major programs, either commercial or government, where large amounts of components are used, a computer program is recommended that handles all removed parts. The major type programs have part removals that for specified engineering reasons are not analyzed. Examples are parameter margin and noise. These units are a stockpile to occasionally check for other failure modes or as retest to compare with a ghost-type failure—i.e., temperature sensitive, pressure sensitive, output high or low.

The proper computer database gives rapid access to past data on failure histories and trends, thereby aiding the analyst in doing a comprehensive job.

GOVERNMENT VERSUS OEM FAILURE ANALYSIS

At the start of increased awareness of reliability, the government was a forerunner in developing the guidelines and spending the necessary millions to achieve those reliability goals. The suppliers of electronic equipment used the funding and knowledge to improve their process and assembly in the commercial market.

There is very little difference in the requirements now for military and large OEM manufacturing. Both now use either AQL or LTPD systems for maintaining control of the quality of products produced.

The military set the stage for the present role of complete failure analysis of electronic components removed from systems or subsystems. It was proven that a consistent effort at this level could lead to corrective action that showed increases in the reliability of systems out in the field. Military's major problem is to keep abreast of new systems produced by other countries. A state of constant change in design and improvement is going on. Many times the other factor is small volume production that places the system on a learning curve which doesn't always reach maturity.

Major OEM manufacturers now entering the electronic field are going through the same pitfalls and learning curve found by the military; i.e., automobile manufacturers, computer, machine controls, and communications are deeply engrossed in failure analysis and corrective actions to improve their products.

Many people now feel the OEM requirements can be more of a challenge than the military requirements. This is based upon the large quantities required for OEM, and the lower price per unit is a key factor in profit and loss. The OEM is a far less forgiving group when the time comes to discuss charges for corrective action or redesign. Due to his volume, he attracts a large group of competing companies for his business. It becomes critical for OEM suppliers to have fast economical corrections for field or test problems.

FAILURE ANALYSIS OUTLINE GUIDE FOR SEMICONDUCTORS AND INTEGRATED CIRCUITS

by James Beall

STRATEGY

Establish present electrical parametric condition and verify the reported failure mode prior to beginning the analysis. Get as much information as possible from the personnel that used and tested the device, to learn the circumstances of the failure and application of the device. If the reported failure mode cannot be verified, a test sequence should be generated that will most likely reproduce the reported condition. If necessary, consider placing the failed device in the same circuit configuration it was in when the reported failure occurred. Also consider other environmental conditions which may result in a repeat of the reported failure i.e., temperature cycle, operation at temperature, PIND test, HTRB, etc. Remember that the device that presently tests good may have a latent failure and you have not evaluated it sufficiently to expose it. If the device you have tests good in all test conditions, it could signify that the wrong device was identified in the next assembly failure isolation. This again may suggest that this device be placed back into the original application circuit to determine if a correct isolation of the failed device was made.

PROCESS

Using the outline guide, lay out a failure analysis plan that would provide a high confidence level for determining the root cause of failure. The failure analysis sequence must be performed in an order that will not effect or preempt a subsequent step. The failure analysis outline guide has been structured with this in mind but, in some cases it may be appropriate to modify the sequence of these steps. It is important that the failure analyst understand the operation of this device in the next assembly application and the circumstances in which it failed. List what condition(s) could produce the reported fail mode. The failure analyst must think through the analysis steps and visualize what will be done to the failed device and what is expected to result from each step in the analysis. If it is uncertain whether a given step is needed or important, the safest position is to include it. These decisions will be easier to make as the analyst becomes more experienced in failure analysis.

Throughout the failure analysis procedure it is important to determine the physical condition/appearance of the failed device. This is identified at various points in the outline guide. These points are included to establish the present condition of the device for reference later in the analysis should a relevant factor develop. It is recommended that all discernible deviations from normal be described in writing or if appropriate documented by photograph. This data can be invaluable later in the analysis when you are relating the condition of the device at receipt.

Curve tracer evaluation is the most important technique for assessing the electrical condition and integrity of a device. This should be the first electrical evaluation used on a failed device for establishing its present condition. For example low current VI (voltage/current) response can provide a measurement of junction breakdown quality by examining the breakdown knee at very low currents (pA, nA, & uA) and provide an indication of electrical overstress by performing a pin-to-pin evaluation on integrated circuits at low currents (uA & low mA). This is a non-degrading technique that can be referenced to a known good device for comparison. When a parameter abnormality is identified it is important for the analyst to determine what degradation or failure mechanism would cause this parameter response. This is important because it is that degradation or failure mechanism that you are pursuing to determine the failure cause.

USING THE OUTLINE GUIDE

At the beginning of the Outline Guide it is important to record all pertinent information relative to the device to be analyzed. This is important historical information and it may not be available from the device later in the analysis e.g., part number (including dash number if applicable), lot date code, manufacturer, serial number. Writing down in your words your understanding of the reported device discrepancy is important for developing the appropriate failure analysis plan as well as for later reference should you realize that you have encountered a different device discrepancy.

KEY

Items identified in the Key are items that are repeatedly required during the course of the analysis to establish present condition, change in condition, or document evidence relative to the analysis. These items can be included at any point in the analysis at the analysts discretion but are identified in the Outline Guide at points where they are expected to be required by use of the asterisks (*). A lab data sheet or lab notebook should be used to record supplemental and pertinent data and information related to an analysis. These data can become extremely important later in the analysis when you need to verify when a particular condition was first noted or when a change in device performance began. It is important because the condition or change may be the result of deprocessing in the analysis. Your primary objective as you pursue the root cause of failure is to preserve the "as received" state of the failure mechanism.

EXTERNAL ANALYSIS

External analysis encompasses for the most part all analysis of the device prior to opening the device. This is the point where the present condition of the device, the "as received" condition, is established. There are three primary factors to address at this point: 1) the reported condition of the device, 2) the measured condition of the device i.e., failed condition and verification of reported failure, and 3) the reference condition for comparison purposes for the remaining conduct of the analysis. This is your last opportunity to perform these analytical steps for this analysis. However, you must consider the consequences of performing or not performing each step. For example, if you see evidence of a package crack during external microscopic exam, you would elect not to perform external rinse or backfill gross leak testing. Also you would not perform an external rinse with DI water or dilute HF followed by DI water if you observed contamination or corrosion on the package until you had analyzed/identified it. And you cannot attempt to capture particles in PIND testing if you are planning on performing gas ambient analysis; in fact, it may not be practical to attempt to capture particles after gas analysis. At times it will appear to be a toss-up on which way to go, and it is likely your decision will be incorrect at times. The important thing to strive for is to learn from these experiences and use this to help in decision making in the future. Some optional analysis steps that are not listed in the external analysis section are DC/AC parametric or functional testing, testing at different supply voltages, schmoo plotting of data, acoustic microscopy, package thermal resistance, and high temperature reverse bias testing. These and others should be considered by the analyst to determine if they are applicable to the specific analysis and may be of benefit.

The notes that are included at the end of each outline guide section, "Consider consultation before continuing", applies throughout the course of the analysis. From experience it has been shown that it is common for the analyst to get so involved in the analysis that the obvious is missed. The purpose of this statement is a reminder that if at any time in the analysis something doesn't seem right or doesn't make sense, take time at that point to review with someone what data you have acquired, your interpretation and what you are uncomfortable about. The person you discuss this with does not have to be an expert in failure analysis. Many times what we overlook is the obvious, and sometimes just formulating the information in your mind can clarify or point out a key factor.

INITIAL INTERNAL ANALYSIS

Decapping the device is a critical step in the analysis process. The device is extremely vulnerable to damage and contamination and important evidence contained in the package may be lost. If you encounter a new package configuration consider opening another device other than the failure to get some experience. As pointed out earlier, most steps in this outline guide are optional; the analyst should periodically review the analysis course set earlier to determine if it is still appropriate. Generally the course of an analysis will change one or more times through the analysis because the analyst becomes more familiar with the failure mechanism and directs the analysis towards identifying the root cause of the failure.

Special attention should be given to microscopic examination of the die surface. This is the first opportunity to examine the die surface and it is in as close to an "as received" condition as practical. It will never be any closer. If it is a complex die it is not practical to examine the complete die at 200× magnification. In this situation you would give particular attention to those areas that electrical testing has shown to be the most likely areas where the failure may reside. Documentation of what is measured and observed through the analysis is very important and this includes photographs. Photographs should be used to document any abnormalities that are disclosed in the analysis.

Vacuum bake and surface rinse are related to devices exhibiting electrical leakage or surface inversion characteristics. They would not be performed until an analysis of suspect areas was performed e.g., Auger or FTIR. Also vacuum bake or surface rinse with dilute HF may not be adequate because they do not identify the cause, but only the effect of what was on the surface. It should be mentioned here that the analyst should continue to look for improved analytical methods and techniques and incorporate them where appropriate to maintain an up to date failure analysis process.

DIE SURFACE ANALYSIS

This section of the outline guide addresses more specific analysis steps. This section will not be applicable to some device analyses as the analysis will have been completed before this point. These are reminders of what steps may be relevant to your present analysis. Electrical measurements continue to be important to maintain an awareness of any influence the failure analysis process has had on the device.

Infrared or liquid crystal analysis would be used to locate a high current leakage or shorted site. One must be careful not to overstress the failure site when attempting to locate a failure site with one of these techniques. The die acts as an excellent heat sink and large temperature deltas are not normally found. Also the temperature sensitivity for these two techniques are not as high as we would like them to be. If the near vicinity can be identified, voltage probing may help in pinpointing the site. Refer to the Micromanipulator Probing paragraph in this section.

Die glassivation removal is beneficial to mechanical voltage probing and usually necessary for SEM voltage contrast examination. Glassivation removal presents the risk of possible damage to metallization and die passivation oxide. Evaluation of etchant performance on like die or selective area of removal can reduce the level of risk. Also consider dry (plasma) etch glassivation removal processes.

SEM voltage contrast analysis provides the advantages of not burdening the circuit, not producing physical or performance degradation, and ease of probe placement. It also presents some disadvantages and it becomes the decision of the analyst to select the appropriate measurement technique.

SEM EBIC analysis provides the ability to evaluate the diffused junction regions with good sensitivity and resolution for locating and identifying defect and degradation sites. The disadvantages presented by EBIC are generally outweighed by the benefits. The primary disadvantage of EBIC is possible irradiation damage and residual trapped charge introduced into the oxide by the electron beam. This is particularly a problem with surface sensitive device technologies e.g., low collector current bipolar circuits (analog and low power digital) and MOS circuits. This disadvantage is accepted where electron beam damage is not a factor as all relevant electrical measurements have been completed and only the location and identification of the junction defect remains.

In some cases it may be necessary to mechanically probe the circuit nodes in order to isolate the failure site. This may also require circuit conductors to be cut for measuring individual device electrical parameters or pinpointing the location of a short or leakage path. Circuit conductors can be opened by cutting or scraping over the metal with the probe point, using photo resist and wet etch through a small window cut in the resist, using an ultrasonic modulated probe or a laser probe. A technique that has been effective in locating pinhole shorts under metallization is to probe the voltage along the length of the shorted conductor. This typically requires a voltmeter with six or more digits of voltage resolution to avoid unusually high currents through the failure site. The principal is to locate the point of lowest voltage. This identifies the immediate area of the short by the lowest IR drop along the conductor. By taking care in probing one can minimize the damage to the conductor surface and allow high magnification examination by light microscopy or SEM to look for evidence of localized metallization topography indentation or melting at the failure site. By etching the metal off, the actual failure site can be located and examined.

Metallization, thermal oxide, nitride, and polysilicon layers can be preferentially removed to examine and locate defects. For example, metallization or polysilicon discontinuities at vias, oxide pinholes or masking/etching defects can be located using these delayering techniques.

SUBSTRATE/SURFACE CHEMICAL OR ELEMENTAL ANALYSIS

The same is true for this section as the previous section. As appropriate, any of these steps can be performed earlier in the analysis sequence. It would be an unusual case that would require the use of all steps identified in this section. The sequence or order of the steps would depend upon the location to which the failure has been isolated.

Subsurface failure sites in dielectrics, vias, conductors, and diffusions can be evaluated by metallurgical microsection and metallographic examination. Due to its destructive nature microsectioning is normally performed near the end of an analysis. Microsectioning may be surface lapping (parallel polishing from top down to delayer the die surface), backside lapping (parallel polishing from bottom up to approach the die surface structure from the die side), angle lapping (parallel polishing from the top down at a small angle to the surface to increase the apparent size of physical features along the angle e.g., oxide thickness, diffusion depth), and cross-sectioning (parallel polishing at an angle of 90 degrees from the surface to obtain a side view of the sample). Etching and staining may be used to remove polishing artifacts/smearing of material and to delineate areas of interest. Examination is typically performed using a metallurgical light microscope or SEM. Microsectioning is very dependent on the skill of the operator. Automatic sectioning equipment is an alternative but is a compromise in obtainable quality.

Organic chemical analysis is very important to failure analysis. Currently there have been 105 basic elements identified. There are over a million organic compounds. Typically contamination and residues that result in failure are present in microscopic quantities. Organic chemical analysis of small volumes (micron to sub-micron) has not been available until the arrival of electron spectroscopy for chemical analysis (ESCA) and fourier transform infrared (FTIR) microscopy. These instruments provide the ability to analyze nano to picogram sample quantities. Because organic chemicals are compounds predominantly made up of hydrogen, oxygen, nitrogen, and carbon molecules, the analytical spectra primarily relate atomic bonding and molecular characteristics. Therefore, these instruments are not well suited for "witch hunt" analyses, where you have an unknown sample that you would like to identify. This is typically difficult to impractical to accomplish. If an organic chemical spectra data base were available it would be too time consuming to perform a total comparison search. To improve the odds of identification it is necessary to develop a list of likely and possible chemicals for comparison to the unknown. Reference spectra for these chemicals are stored in a library for computer comparison testing. These organic chemical analysis instruments will continue to become more valuable as application experience and data bases increase.

The scanning electron microscope (SEM) is one of the more valuable instruments available to the failure analyst. Its high resolution/high magnification and large depth of field serve the semiconductor failure analyst well in the world of micron to sub-micron feature sizes. The SEM is the most likely instrument to find application prior to this point in an analysis sequence. With its diversity in capability it can find application in nearly every step of a failure analysis. Having this degree of importance it is mandatory that every analyst be very proficient in the application of the SEM. This includes voltage contrast, EBIC, as well as SEM accessories such as energy dispersive x-ray (EDX) analysis.

The electron microprobe x-ray analyzer is similar to the SEM with EDX. The electron microprobe is designed to meet the performance requirements of wavelength dispersive x-ray analysis (WDX) in addition to EDX. EDX and WDX analyses compliment each other in the way they detect the emitted x-rays. In EDX, x-rays are detected using a solid state lithium-doped silicon detector. This detector will detect x-rays for the majority of the basic elements. At low energy, it is limited by the energy filter effect of the detector window material, usually at sodium for a beryllium window, or boron for a polypropylene window. The window material blocks the lower energy x-rays and thereby excludes the detection of the lighter elements. Another limitation of the EDX detector is that only a single x-ray event can be handled within the detector response time. The typical detector response time is in the 1-10 microsecond range. If two or more x-ray photons arrive within a single detection period the detected signal is discarded and the new detection period is started. With higher x-ray photon flux, the number of x-ray photon coincidences at the detector is greater. This is referred to as <u>pulse pileup</u> and is measured as detector <u>deadtime</u> (percent of single x-ray photon detection events to multiple detection events). The key advantage of the EDX system is that it detects the x-ray photons over the total elemental range. This provides the total elemental spectrum within one acquisition period. In comparison, WDX employs a crystal spectrometer, which easures the angle of refraction for an x-ray photon having a specific wavelength. Therefore a given spectrometer can detect only a single x-ray wavelength at a time. To measure a different wavelength, the angle between the crystal spectrometer and the detector must be mechanically changed. Typically individual elements produce multiple orders of wavelength points and it may be necessary to acquire more than one wavelength measurement per element identified.

A comparison of EDX and WDX will best show how they compliment each other. EDX acquires the overall elemental analysis at essentially the same point in time and is faster. WDX has the higher resolution, 2-5eV versus 130-150eV for EDX. Where peak overlap is a problem for EDX it may be resolved or reduced with WDX. This is important in quantitative analysis and elemental mapping. Typically WDX provides a higher detection sensitivity and wider elemental detection range than EDX. Another benefit of WDX is the improved ability to map the presence of minor constituents. Generally minor constituents produce lower x-ray photon emission (count) rates. In EDX if the electron beam current is increased to increase the x-ray count rate it will at some point result in increased EDX detector deadtime. To reiterate, this occurs because it increases the x-ray photon flux for all elements and increases the probability of pulse pileup or deadtime. This does not produce the same effect in WDX because the spectrometer passes only the x-ray photons for a single wavelength point. Together EDX and WDX represent a powerful combination. EDX provides a quick analysis of the overall elemental composition. WDX can then be used to resolve special cases or provide a more detailed analysis.

For the most part the analytical techniques described to this point require a cubic micron or more of sample volume. In the semiconductor failure analysis field a micron can be referred to as a bulk analysis. For most analyses this volume is way too large. This problem has been partially solved by the arrival of surface analyzers, Auger and ion microprobe. They are only partial solutions because of the limited detection sensitivity of Auger and limited depth resolution, quantification, and in some applications limited detection sensitivity of ion microprobe. Surface analysis still remains a powerful and valuable analytical tool. A few of the many applications are oxide layers on conductive films, localized areas of thin contaminated films, Ion implant profiles, contaminates in oxides, trapped contaminants beneath plated films, and the list goes on and on.

FAILURE MECHANISM/ROOT CAUSE OF FAILURE

It is difficult to determine when the root cause of a failure has been identified. In some cases it is not necessary to determine the root cause e.g., when there is no interest in identifying what corrective action should be taken to eliminate future failures for the same cause or, identifying what screens may be used to remove additional devices that contain the same defect or, determining what impact it will have on built hardware containing the same devices; but where there is only a need to determine whether a failure is inherent to the next assembly application or to the device.

One approach to determining whether the root cause has been identified is to test your finding with the question of "Are there any other factors that could effect the identified root cause?". For example, a wire bond lift-off failure. The root cause would not be the wire bond lift-off because there are a number of other factors that would determine why this bond lifted. Some of the possibilities would include: contamination on the bond pad, wrong wire bond bonding schedule, marginal wire bond schedule parameter, wrong wire bond wire composition, oxidation on wire bond pad metal, etc. Another question that needs to be answered is, "Considering the application of this finding (end use of this analysis finding), what effect would these additional factors have?". If the application of this finding was to identify a wire bond problem to the device supplier, it would be inappropriate to continue the analysis. However, if it were the intention to inform the device supplier what wire bond problem he had, determine the extent of the problem in existing product and ensure that subsequent lots he delivers after correcting the problem do not have this problem, then it would be important to determine what other factors were responsible.

Another technique that can be used to "test" your analysis conclusion or root cause finding is proof by contradiction. That is, if this fact is true then something cannot be true, or if this fact is not true then something must be true. There is added risk in this approach. If you disprove your conclusion you must be prepared to review your analysis and correct your finding. The most appropriate time to "test" your opinion using proof by contradiction is during the course of the analysis.

A final recommendation is to perform a "post failure analysis review and assessment" to critique your failure analysis procedure and identify recommended improvements to be considered for future failure analyses. The intent is to continue to question your analysis procedures and look for ways to improve them. This includes minor as well as major improvements.

In most cases the failure analysis must be documented in a report. Following are some recommendations for report writing. Write the report as if you are telling someone what you did through the course of the analysis. Many times the analyst will describe only the high points of an analysis and this is difficult for the unfamiliar reader to appreciate what considerations were given to the analysis. Without this rationale it is impossible to understand why a specific analysis step was, or was not, performed. Describe all experimental and analytical procedures in detail, and include a list of equipment used. This will be invaluable if an experiment needs to be reproduced at a later time. Include interpretation of all test results; don't leave it to the reader. Photos and diagrams should follow the text in a logical sequence. Identify all photos and diagrams with captions to explain what is shown, and magnification if applicable. Finally, identify any literature used for reference.

Failure Analysis Outline Guide

SEMICONDUCTOR/INTEGRATED CIRCUIT
FAILURE ANALYSIS OUTLINE GUIDE

DATE: _____ PROGRAM: _____ I.D. NO: _____

PART NO: _____ PART NAME: _____ S/N: _____

LOT NO: _____ TEST NO: _____ MANUFACTURER: _____

REPORTED FAILURE/DEGRADATION MODE: _____

_____

_____

_____

_____

_____

_____

_____

KEY: * Electrical Parameter(s) measurement required to detect change. Curve Tracer photographs required if anomalous. Lab Data Sheet required.

           ** Photomicrographs required in these steps. Identify what is shown and the magnification.

I. EXTERNAL ANALYSIS

        ○   Electrical parameters (record on Lab Data Sheet) _____

        ○   External microscopic exam _____

**   ○   X-Ray _____

        ○   Hermetic seal check (fine/gross) _____

*   ○   Environmental tests (specify) _____

*   ○   PIND test/particulate capture _____

        ○   Gas ambient analysis _____

NOTE: Consider consultation before continuing.

Fig. 1a Semiconductor/integrated circuit failure analysis outline guide

II. INITIAL INTERNAL ANALYSIS

* ○ Decap device _____
  _____
  _____

** ○ Surface microscopic exam _____
  _____
  _____

* ○ Vacuum bake _____
  _____
  _____

NOTE: Consider consultation before continuing.

III. DIE SURFACE ANALYSIS

○ Infra-Red topographical plot/liquid crystal _____
  _____
  _____

** ○ Die glassivation removal _____
  _____
  _____

** ○ SEM voltage contrast analysis _____
  _____
  _____

** ○ SEM EBIC analysis _____
  _____
  _____

** ○ Micromanipulator voltage probing/Isolate metal runs _____
  _____
  _____

** ○ Metalization, thermal oxide preferential removal (Delayering) _____
  _____
  _____

NOTE: Consider consultation before continuing.

Fig. 1b Semiconductor/integrated circuit failure analysis outline guide

IV. SUBSTRATE/SURFACE CHEMICAL OR ELEMENTAL ANALYSIS

** ○ Microsection and metallographic exam _____

** ○ Chemical analysis/FTIR microscope/ESCA _____

** ○ Scanning Electron Microscopy – Energy dispersive X-Ray analysis _____

** ○ Electron microprobe analysis _____

** ○ Auger, ion microprobe _____

V. FAILURE MECHANISM/ROOT CAUSE OF FAILURE

_____

Analyst

Fig. 1c Semiconductor/integrated circuit failure analysis outline guide

Electrical and Mechanical Characterization

CURVE TRACER APPLICATIONS AND HINTS FOR FAILURE ANALYSIS

by James Beall and David Wilson

ABSTRACT

A curve tracer provides a powerful and unique capability for visual examination of the continuous voltage versus current response for a wide range of electronic and electrical devices. The failure analyst will find that the curve tracer is one of the most commonly used instruments during semiconductor device analysis. A thorough understanding of the instrument and the information to be gained from the I-V curves is required to maximize the benefit to an analysis. This paper will describe the operation of the curve tracer and the curve interpretation.

Figures 24 through 29 show typical equipment examples.

INTRODUCTION

Device acceptance testing generally is based on single point data for a given parameter. This assumes that the parameter response apart from these data points is characteristic of a typical device. When evaluating a suspect or failed device, this assumption should not be made. A common habit that one can fall into is that of only evaluating or examining a suspect device within the manufacturer's specified parameters on the published data sheet. One should routinely examine device performance above and below these specified parameter ranges or points. The only limitation for this practice is that whatever measurements are made, they must not overstress or degrade the device.

To understand the cause for the present condition of a device it is important to determine how that device's characteristics appear and how they compare to a good device. For example, if one is analyzing a device with low junction breakdown it would be important to evaluate the reverse leakage current response over a voltage range, the breakdown knee response, and the breakdown voltage response with current. This is key to determining why the breakdown for example, is below the specified voltage at the 10 µA test current. It is also important to look for other parameters that may have been affected when the device failed. For example, low or high collector current beta performance, leakage current response, forward biased junction response, pulsed parameter test performance, junction capacitance versus voltage characteristics, parameter response over temperature, etc. These interrelationships can be very valuable in determining the failure/degradation cause or scenario.

Curve tracer evaluation of integrated circuits is beneficial for assessing the present status of the interface or peripheral circuitry. The procedure involves bipolar-voltage curve tracing on a pin-to-pin basis at low (nA to low µA) currents. Keeping the test current at a low level will prevent further degradation of a device. It is helpful to compare the pin-to-pin I-V responses of the suspect device with the responses of a known-good device. This procedure quickly determines two things:

1. It shows apparent areas of degradation and possible paths of circuit overstress.
2. It determines whether further ATE testing may cause added circuit degradation.

Keep in mind that ATE test programs are intended to force specified parametric current and voltage levels. This could lead to further degradation of an input or output circuit and mislead or complicate the analysis of a damaged device. Therefore, a curve tracer evaluation should be the first step in electrical testing to determine the present condition of a device to be failure analyzed, and the specified current and voltage levels should be approached from lower values while monitoring the device response.

SEMICONDUCTOR JUNCTION CHARACTERIZATION

An understanding of the information available from a curve tracer requires a fundamental understanding of semiconductor operation and basic parameter characteristics. There are a number of good reference textbooks that address these topics [1-5]. This section contains information on semiconductor junction characterization and the conditions that the failure analysts may encounter.

FORWARD-BIAS

The typical junction should have a uniform smooth exponential current response with increasing voltage as shown in Fig. 1. On the linear scale displayed, the curve for a silicon de-

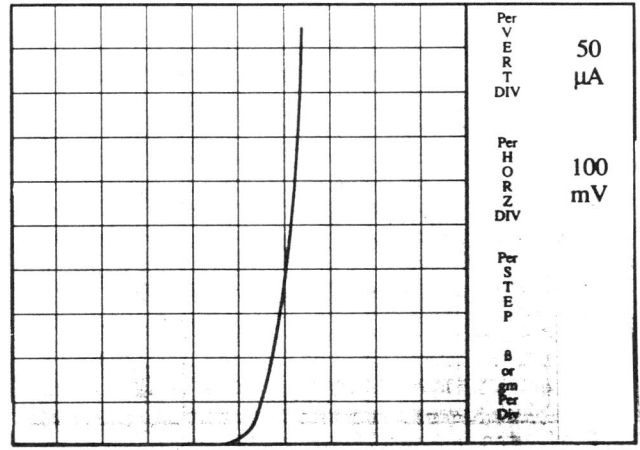

Fig. 1 Typical forward-bias I-V characteristic.

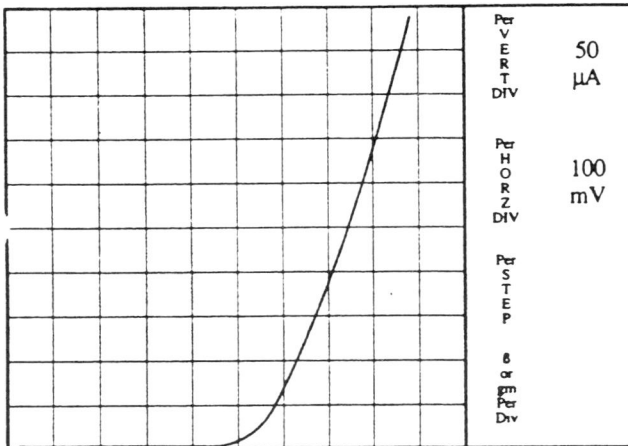

Fig. 2 Abnormal forward-bias I-V characteristic due to the presence of a series resistance of 500 ohms.

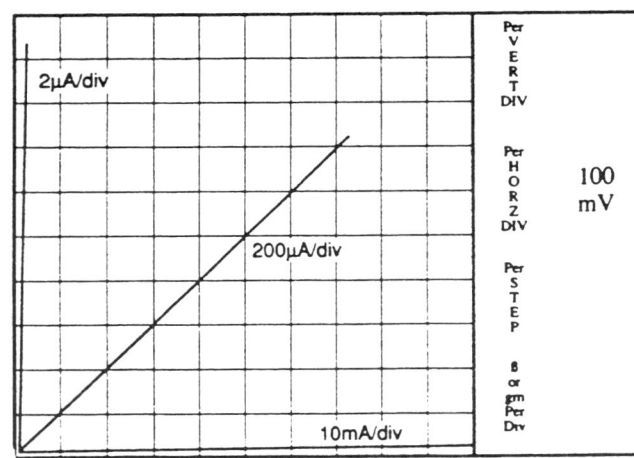

Fig. 3 500 ohm resistor display using 3 different current scales.

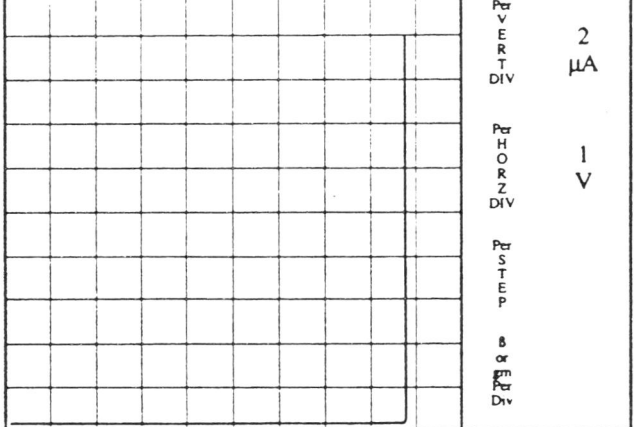

Fig. 4 Avalanche breakdown of a reverse-biased base-emitter junction.

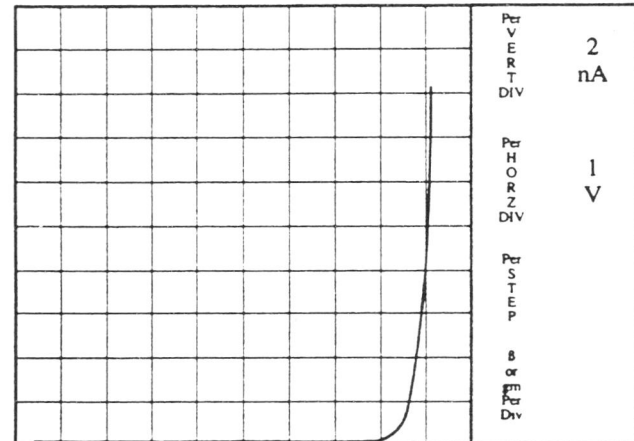

Fig. 5 Same junction as shown in Fig. 4 using leakage current mode. Note 3-order decrease in vertical current per division range.

vice becomes near asymptotic to the voltage axis in the 0.6-0.7 volt range. If this was displayed as the log of the current versus voltage it would be a straight line at these low current levels. The resistance of the junction is approximately 26/I, where resistance is in ohms and I (current) is in milliamps. For instance, at 1 milliamp the resistance would be 26 ohms and at 26 milliamps the resistance would decrease to 1 ohm. In addition to the junction resistance there is the ohmic body resistance and interconnect or contact resistances. The forward-biased junction voltage has a negative temperature coefficient shifting approximately -2.0 to -2.5 mV/°C.

Abnormal Characteristic: It is unusual to find an abnormal forward biased junction I-V response curve. One possible anomaly is the presence of a high series resistance, for example at a bond interface. The I-V response curve would have additional slope after the knee as shown in Fig. 2. In this case there is an additional resistance of approximately 500 ohms. A note of caution is appropriate at this point. When using a curve tracer to view a resistive slope it is important to pay attention to the current and voltage scales. Figure 3 displays a 500 ohm resistor using three different current scales. The resistor can have the appearance of a short, a resistor, or an open simply by changing the current range on the curve tracer. This underlines the importance of specifying just what a "short" means in report verbage. At one setting, it may appear to be excess leakage; at another, an extremely low resistance, perhaps milliohms.

Reverse-Bias

Under reverse-bias conditions, the typical junction will have a constant low level of current (reverse saturation current) until the voltage reaches the point where breakdown occurs [6]. The two mechanisms responsible for junction breakdown are *zener breakdown* and *avalanche breakdown*.

Curve Tracer Applications and Hints

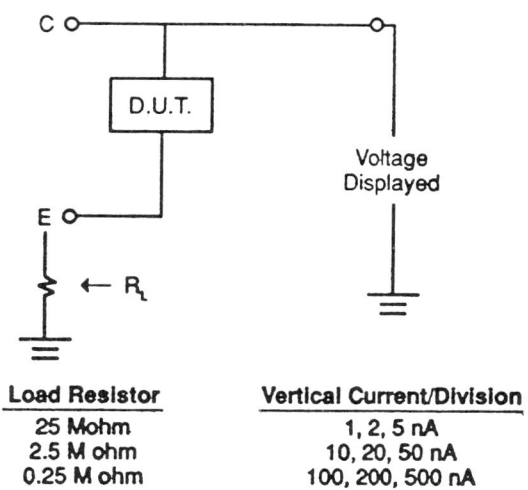

Load Resistor   **Vertical Current/Division**
25 Mohm   1, 2, 5 nA
2.5 M ohm   10, 20, 50 nA
0.25 M ohm   100, 200, 500 nA

Figure 6. Curve tracer leakage current measurement configuration.

Fig. 6 Curve tracer leakage current measurement configuration.

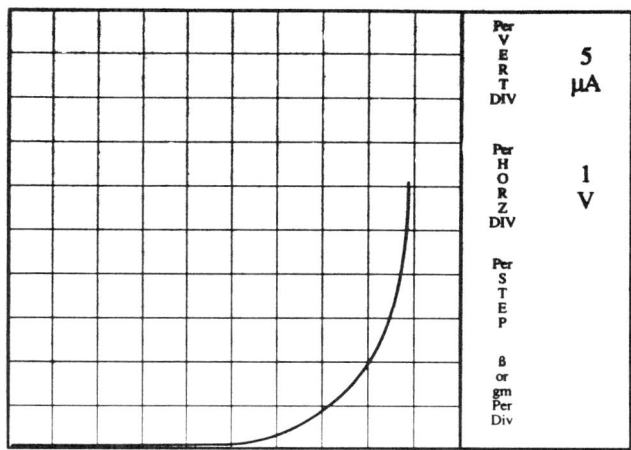

Fig. 7 Same junction as shown in Fig. 4 following reverse-bias DC stress.

Zener breakdown occurs in heavily doped semiconductor junctions due to the large electric field produced across the narrow depletion region. This field causes separation of valence electrons from their respective nuclei and a large current results at the breakdown voltage level.

Avalanche breakdown occurs in semiconductor junctions doped at lower concentrations. The reverse bias produces movement of holes and electrons across the depletion region. The acceleration of these charges increases with increased bias until the point is reached that they have enough energy to free valence electrons during collisions. These electrons produce additional valence electrons in a process referred to as avalanche multiplication. Again, as in the case of zener breakdown, a large current results at the breakdown voltage level. Zener breakdown predominates for breakdown levels up to 5 volts, between 5 and 8 volts both mechanisms are present, and above 8 volts avalanche breakdown predominates.

Due to the differences in breakdown mechanisms, the temperature coefficient of the two processes are opposite. <u>The zener breakdown voltage decreases with temperature and the avalanche breakdown voltage increases with temperature</u>. This can be a factor where devices are operated over large temperature ranges. For example, diodes with breakdown voltages above 8 volts can be temperature compensated by connecting a forward-biased and reverse-biased junction in series, and the coefficients effectively cancel.

A typical avalanche reverse-bias breakdown curve is shown in Fig. 4. This junction has a sharp breakdown just below 9 volts. Breakdown is normally specified at 10 microamps; this device measures approximately 8.8 volts at this point. This same junction is shown in Fig. 5 with the current scale decreased by 3 orders of magnitude (the mode switch is set to leakage current). A note of caution on leakage current measurements: Comparing Fig. 4 and 5 it is noted that at 10 nanoamps the voltage level is 9. This should be below 8.8 volts at this low current level indicating that the voltage displayed is higher than the voltage actually across the device. This higher level is due to the instrument configuration used for the leakage measurements (Tektronix 576 Curve Tracer). A resistor to ground is present when the mode switch is set to leakage. The effect in Fig. 5 is a resistive slope of 25 megohms after the breakdown knee. The increase is 10 nanoamps through 25 megohms, or 0.25 volts. The equipment configuration is illustrated in Fig. 6.

The resistor to ground varies depending on the current/division so the resistive slope will vary accordingly. It is possible to overstress and damage this resistor to ground so the leakage setting should be used with caution.

Reverse breakdown voltage measurement should not be performed on PIN diodes. The parameter specified is the leakage at a specific voltage rather than breakdown voltage at a given current. During avalanche breakdown, a junction with the intrinsic region is apparently damaged by injection of carriers in the reverse-bias direction. This may be aggravated by current concentration due to a negative resistance characteristic [7]. Reverse breakdown voltage measurement on PIN diodes can result in permanent damage to the breakdown level.

Abnormal Characteristics: A number of abnormal characteristics can occur for different reasons on a reverse-biased junction. These characteristics include:

1. Reduced breakdown voltage indication.
2. Soft breakdown knee.
3. Channelled curve.
4. Walkout of the breakdown voltage.
5. Unstable curve.

Conditions that produce these abnormal characteristics are electrical overstress (EOS), surface contamination, and manufacturing defects.

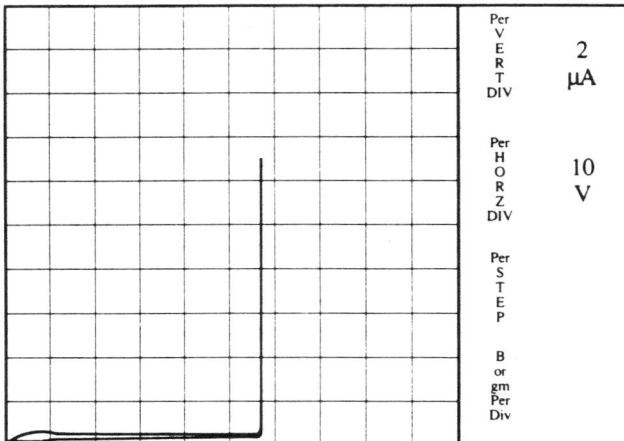

Fig. 8 Reverse junction breakdown I-V response prior to ESD-stress.

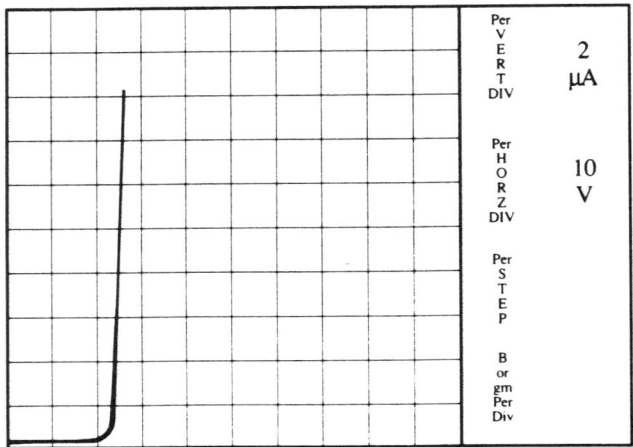

Fig. 9 Same reverse junction I-V response after ESD degradation. In this case the apparent BV was degraded by more than 25 volts.

Electrical overstress is a common cause of degraded junctions. The overstress can be forward or reverse bias and DC, AC, or transients. Forward-bias electrical overstress has been reported to cause a soft breakdown knee characteristic [8]. The phenomena is believed to be due to the formation of a positive charge sheet in the oxide at the oxide to silicon interface. Degradation due to forward-bias electrical overstress is not as likely a degradation due to reverse-bias overstress because of the power limitations in the former.

Current is often limited by the interconnect wire size and the voltage is limited by intrinsic conduction of the diode. In many cases when forward-bias electrical overstress occurs the wire will be fused open but the junction characteristics will not be degraded.

Reverse-bias electrical overstress can cause both an apparent reduction in breakdown voltage levels and soft breakdown knees. The effect that occurs is related to the type and location of the damage. Active trap sites are generated and metal atoms migrate into the silicon. DC electrical stress causes softening of the knee as shown in Fig. 7. This characteristic was produced by forcing 600 mA through a reverse-biased base-emitter junction. The avalanche breakdown voltage is still at 8.8 volts as it was prior to the stress. This indicates that damage sites have been produced near the junction and these sites contribute leakage current as they are incorporated into the depletion region with increasing voltage.

Very fast transients, such as an electrostatic discharge (ESD), can cause reduction of the apparent breakdown level. Figure 8 shows the reverse breakdown characteristic on a junction field effect transistor (JFET) prior to a simulated ESD stress. Following this stress the displayed breakdown voltage has been reduced as shown in Fig. 9. There is slight softening of the knee and more resistance after the breakdown. If a high enough current were forced through the junction to follow this resistive path out to the original breakdown voltage level, avalanche breakdown would occur. The curve displayed in Fig. 9 is not true avalanche breakdown since the electric field is not high

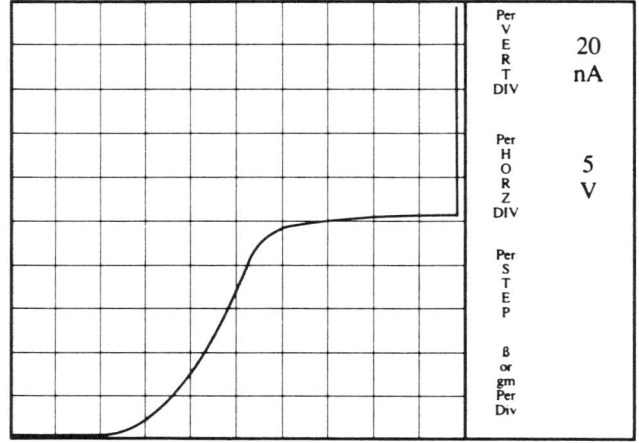

Fig. 10 Inversion characteristic associated with ionic contamination.

enough to support multiplication. To have a high current flow below the level where the depletion region field can support avalanche multiplication indicates that a source of electrons has been reached by the depletion region. This may be due to aluminum from the contact spiking into the silicon and at the voltage level for the depletion region to reach this aluminum, high current flow is produced.

Surface inversion due to ionic contamination of the oxide can cause reduction of the breakdown voltage and the formation of inversion channels [9, 10]. Positive ions such as sodium are readily transported to the silicon-silicon dioxide interface and can invert the underlying silicon. The inversion layer will then modify the depletion region characteristics. The reverse-biased junction has little current flow until the reverse-bias is high enough to incorporate a carrier generation site. When the carrier generation site is incorporated current flows through the inversion layer. The resistance associated with this inversion layer leads to the saturation characteristic as shown in Fig. 10.

One situation has been encountered on a bipolar device where a definite saturated channel characteristic has been observed due to electrical overstress. This is the case of the JFET with electrical overstress damage to either the source to gate junction or the drain to gate junction. The channel is seen when displaying the reverse-bias characteristic of the opposite junction. This can be explained in the following manner. The damage is near the source contact. When the depletion region reaches this point, current will be generated. This current has to travel through the pinched-off channel in the n-type source/drain diffusion, to the drain contact. Analogous to the inversion layer resistance, the channel resistance produces the limited current characteristic.

Walkout of the breakdown voltage is commonly observed on devices due to positive charge in the oxide either as a result of ionic contamination or following long periods of forward-bias stressing [8]. This is depicted in Fig. 11. The appearance of these curves is dependent upon the current limiting resistance on the curve tracer and on the adjustments made to the collector supply as the voltage is increasing. Often more of a sawtooth appearance results. Walkout will also be observed as a thermal effect since breakdown voltage increases with junction temperature.

Manufacturing defects related to the silicon die can produce reduced breakdown voltage levels, soft breakdown knees or unstable curves. These characteristics will occur with defects in the silicon within the depletion region.

An example of a defect that will produce an unstable curve is a cracked die. High current flow may occur before breakdown or at breakdown, the curve may jump between different characteristics. This is depicted in Fig. 12.

Another characteristic that can readily be viewed on a curve tracer is noise generation at voltage levels close to the breakdown point. In this situation, current will flow at one voltage level and then the curve will jump to a second stable condition. The curves will alternate between these two characteristics in a bistable manner. This generally occurs at low current levels near the onset of avalanche breakdown, as shown in Fig. 12. One explanation for the switching between the two curves is that as current flows to one breakdown site an appreciable voltage drop develops across the internal series resistance. The field strength at the depletion region will decrease below that necessary to maintain avalanche and so the current flow stops in this area and begins in another location that has a higher avalanche breakdown voltage. When the initial site cools down, breakdown at that location is reinitiated. Oscillation between these two locations then results. Additional semiconductor noise sources include burst noise, zener noise, and microplasma noise [8, 11].

TRANSISTOR CHARACTERIZATION

The information in the previous section on the characterization of semiconductor junctions is relevant for transistor base-emitter and base-collector junction characterization. Additional transistor parameters that are commonly measured on the curve tracer include:

1. Common-emitter current gain, beta or h_{FE}.
2. Collector-emitter characteristics, breakdown and leakage.
3. Base-emitter and base-collector saturation characteristics.

COMMON-EMITTER CURRENT GAIN

The low frequency common-emitter current gain, beta or h_{FE}, is defined as the ratio of the collector current to the base current (I_C/I_B). This can be measured directly on the curve tracer. The small-signal current gain, h_{fe}, is the change in the collector current divided by the change in base current at a given operating point. This is also usually defined at a given frequency and cannot normally be performed on the curve tracer. The current gain of a transistor is controlled by the manufacturing process with effects produced by the quality of silicon and silicon dioxide, the diffusion profile and geometry, and

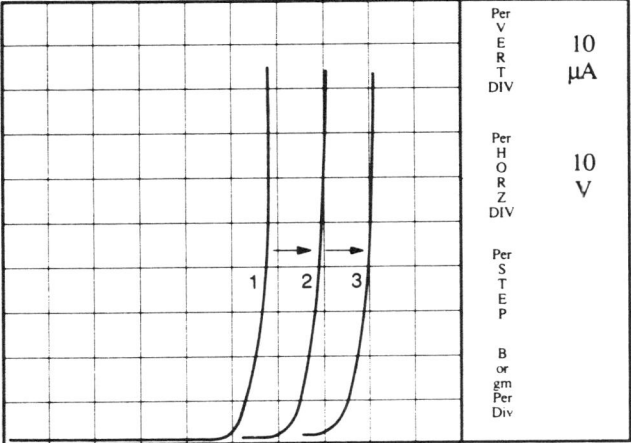

Fig. 11 Depiction of reverse breakdown "walkout" where the I-V response drifts from #1 to #3 as the test current is increased.

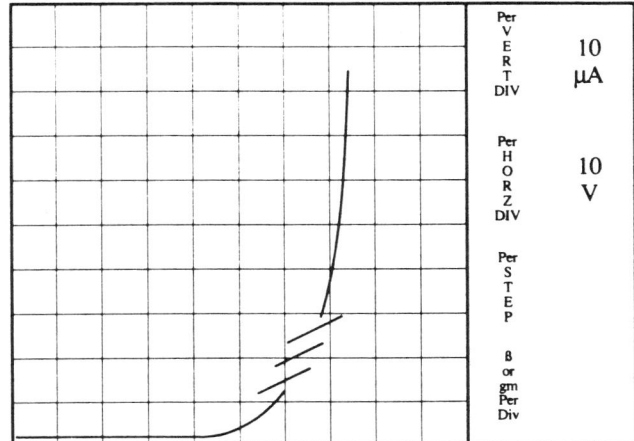

Fig. 12 Depiction of reverse-bias breakdown I-V characteristic for a cracked die. This instability is just one of many characteristic aberrations that a cracked die may show, including soft breakdown, jittering of the entire curve, and a short.

dopant concentration. The gain will vary with temperature, current level, and frequency due to the interaction of a variety of effects [12].

For the failure analyst, the low-current gain is typically the most important gain measurement [13]. This gain can be degraded by reverse-bias base-emitter junction current flow. The characteristics of this include:

1. Low gain at low collector currents.
2. Normal gain at higher collector current levels.
3. Increase of collector to emitter breakdown voltage in the low-current region.
4. Little or no damage to the base-emitter reverse-bias characteristics.
5. Nearly full recovery of the gain with high temperature bake.
6. Little or no degradation with a high temperature reverse bias bake (as would be expected with ionic contamination).

These six items provide the basis for assessing a device during a failure analysis. The low current gain is dominated by the recombination of electrons and holes in the emitter depletion layer. This recombination current is composed of both bulk and surface components. Electrical stress likely increases the surface recombination component and thus reduces the low current gain. The increase seen in the collector to emitter breakdown voltage is simply an effect of this gain degradation since BV_{CEO} is indirectly proportional to gain and the low current gain is the most severely degraded [12]. The recovery characteristics, (item 5), are also consistent with surface effects.

COLLECTOR-EMITTER CHARACTERISTICS

Measurement of breakdown voltage and leakage can be made in both polarities between the collector and emitter. These parameters are usually specified for collector positive on NPN transistors and collector negative on PNP transistors. The opposite polarity measurement has a breakdown voltage approximately equal to the base-emitter breakdown voltage. A common usage for this latter measurement is to verify continuity between the collector and emitter in the case when the base wire is open. A forward-bias overstress involving the base can open the base wire. This measurement will verify that the connections to the collector and emitter are still intact and also will show the condition of the base-emitter junction.

The breakdown voltage measurement between the collector and emitter is an important parameter for power transistors particularly in switching applications. Breakdown voltage can be measured with:

- The base open, BV_{CEO}.
- The base-emitter connected by a resistor, BV_{CER}.
- The base-emitter biased, BV_{CEX}.
- The base-emitter shorted, BV_{CES}.

The voltage versus current curves for these measurements are different, with voltage increasing as the measurements go down the list. Caution must be exercised during the measurement of these parameters on the curve tracer due to the negative resistance characteristic encountered after current flow begins [14, 15]. The most difficult to measure because of this instability is BV_{CEO}. Figure 13 depicts the negative resistance characteristics of BV_{CEO}. When power is left on for several seconds, the curve shifts as shown in Fig. 14. The die heats up during the time the voltage level is increasing (right-hand trace) and this produces a higher current level during the decreasing voltage time period (left-hand trace). Monitoring the shift of the retrace voltage is a good indicator of the die temperature. As a rule BV_{CEO} should not be measured during a failure analysis unless there is a specific need. Where the measurement is considered pertinent to the analysis, it is recommended that the device be tested using a system or circuit specifically made to test this parameter. A note of caution is in order at this time. $BV_{CEO(sus)}$ is often specified at a relatively high current level and is to be measured under pulsed conditions. A common mistake is to perform this measurement with the base drive set to pulsed operation. This does not apply a pulse to the collector supply but rather forces it to be a DC supply. This is the worst case power dissipation situation and can easily result in the destruction of the de-

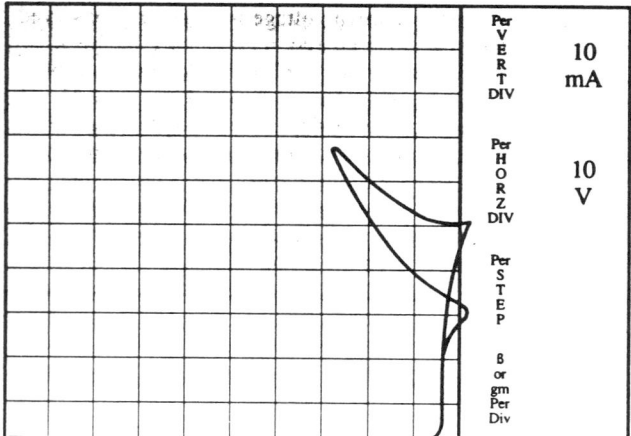

Fig. 13 BV_{CEO} negative resistance characteristic.

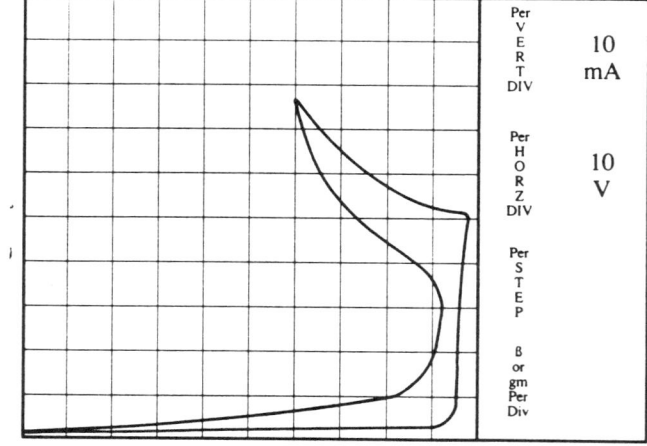

Fig. 14 Same display as Fig. 13 following several seconds of heating due to the power dissipation.

vice, a high-power plug-in does, however, pulse the collector supply. Another caution in the measurement of $BV_{CEO(SUS)}$, where a given current must be sustained. The curve tracer maximum power setting must be adjusted so that the voltage adjustment control advances the current on the vertical portion of the breakdown characteristic in a smooth and controllable manner. A common mistake is to perform this measurement with the power setting at the maximum 220 watts. The transistor will break down, and short due to excessive current.

Measurement of leakage current between the collector and emitter can be made with the same connections as the breakdown voltage measurements. Two abnormal conditions encountered are a resistive short or a channel. The resistive short can be produced by a variety of overstress conditions [16]. The cause of this type of failure is normally due to second breakdown. First breakdown is avalanche breakdown as previously discussed. Second breakdown occurs when the temperature of the silicon becomes sufficiently high to thermally produce carriers in excess of the background concentration. At this point the base current no longer controls the collector current. A large current flows and thermal runaway occurs [4]. The dopant in the emitter region diffuses through this very hot region and forms a pipe of opposite dopant material in the base region [14]. Electrically, the base-emitter breakdown may show little or no effect while the collector-emitter path becomes resistive.

A channel characteristic can occur between the collector and emitter that has the appearance of a low-current gain curve. This effect can be due to a surface inversion in the base or collector regions. As with other inversion problems a high temperature unbiased bake will cause the current level to decrease and a high temperature biased bake will cause recurrence.

BASE-EMITTER AND COLLECTOR-EMITTER SATURATION CHARACTERISTICS

Transistor base-emitter and collector-emitter saturation voltages are the minimum voltage levels that occur at collector saturation. Collector saturation is defined as the point where an increase in base current produces no significant increase in collector current. Saturation parameters usually provide a margin of safety to allow for beta degradation over operating life. This safety margin typically assumes a forced beta of 10, or for low gain power devices a forced beta of 5 is sometimes used. Base-emitter and collector-emitter saturation voltages are measured using the lower volts/cm or the expanded voltage feature to provide maximum sensitivity and readability. Saturation voltage margins can be measured by observing either the base-emitter or collector-emitter voltage with sufficient sensitivity, as the base current is gradually decreased from the forced beta level. When the base or collector voltage exceeds the maximum specified saturation voltage, this base current divided into the forced beta base current will provide the saturation safety margin. An abnormal saturation voltage characteristic indicates possible wrong die (no epitaxial layer), poor die backside metallization, poor solder wetting, bad wire bonds, or excessive resistance in the lead crimp or weld terminals.

TRANSISTOR OPERATING PARAMETERS

An application of the curve tracer is ballparking or approximating the typical operating parameters for an unknown transistor. This procedure is useful to the failure analyst because it can confirm expected parameter performance when evaluating a suspect or degraded transistor. The following discussion describes the procedure to follow, assuming a silicon bipolar transistor, not an alloy germanium transistor.

First the emitter, base and collector are identified by examining the forward and reverse biased I-V characteristics. The emitter to base will have the lower reverse breakdown voltage (BV), the base to collector will have the higher reverse BV, and the collector to emitter will not have a low forward biased voltage. This identifies the emitter, base and collector terminals. The voltage polarity for the forward-bias condition differentiates the PNP from the NPN transistor. A silicon transistor typically has current flow starting at a forward bias voltage of about 0.5 volts (cut-in), a mid-range for use of about 0.6 volts (active),

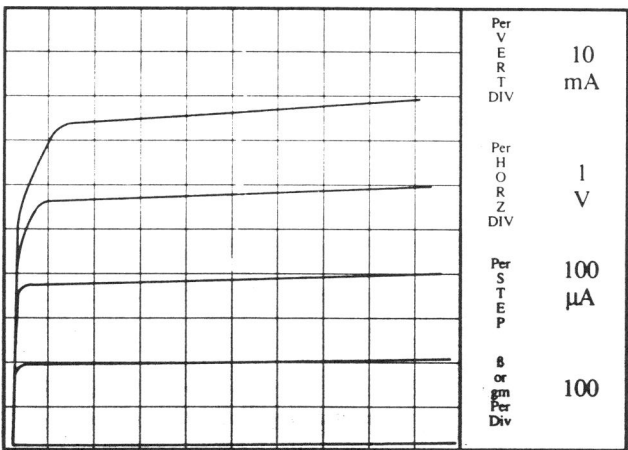

Fig. 15 Family of curves for NPN transistor.

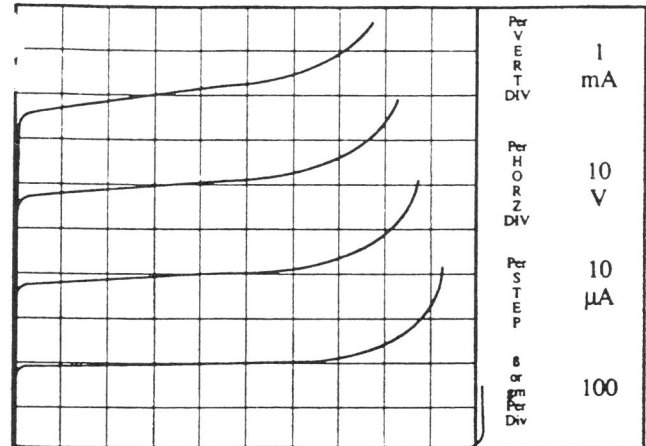

Fig. 16 Family of curves for NPN transistor with collector to emitter breakdown occurring.

and an upper use level of about 0.7 volts (saturation) [1, pg. 256]. Forward-bias the emitter to base junction with the curve tracer to obtain a 0.6 volt drop across the junction. Measure the current that produces this 0.6 volt drop. This should be a reasonable base current level expected for operation in the active region. A typical general purpose low-power NPN transistor was used to produce the figures for this example and it measured 200 microamps at 0.6 volts V_{BE}. Configure the curve tracer in the following manner:

- Connect the emitter, base and collector terminals of the device to the corresponding terminals of the curve tracer.
- Set the polarity for the device type, NPN.
- Set the base step generator for a current per step of .1 mA.
- Set the number of steps to at least 4.
- Set appropriate current limiting resistance in the collector supply (eg. 6.5 ohms).

Increase the collector supply to 5-30% of the measured collector to base breakdown voltage. Adjust horizontal volts/division and vertical current/division to keep the curves displayed on the CRT. Figure 15 displays a family of curves produced with this configuration. If there is an indication of collector to emitter breakdown, the collector voltage should be reduced to a point below this breakdown. For this transistor the collector to base breakdown was measured to be 190 volts and the collector to emitter breakdown occurs in the 70 to 100 volt range as shown in Fig. 16.

The display can be adjusted by increasing or decreasing the base drive current. Check the device package temperature to keep from exceeding the ambient heat sink capacity. The display can be expanded to allow direct measurement of $V_{CE/(SAT)}$ by decreasing the horizontal volts/division. The horizontal volts/division can also be switched from collector to base to allow direct measurement of $V_{BE(SAT)}$. Compilation of these parametric data will provide a good summary of the device operating capability. Select some devices at random or, better yet, set up some friendly competition between yourself and a fellow worker and practice measuring device parameters. Collect the parametric data and then compare them with the manufacturer's data sheet. To increase the challenge have someone install the devices in black or opaque boxes and bring out their three leads. This experience will sharpen your skill in using the curve tracer and interpreting device parametric responses.

INTEGRATED CIRCUIT PIN-TO-PIN EVALUATION

As mentioned earlier, curve trace evaluation of integrated circuits is beneficial for assessing the present status of the interface or peripheral circuitry on a given device. The procedure involves bipolarity voltage curve tracing on a pin to pin basis at low (nA to low µA) currents. Keeping the test current at a low level will prevent further degradation of a damaged device. It is helpful to compare the pin to pin I-V responses of an unknown device with responses of a good device. This procedure quickly determines two things:

1) It shows apparent areas of degradation and possible paths of circuit overstress.

2) It determines whether further ATE testing may cause added circuit degradation.

Keep in mind that ATE test programs are intended to force specified parametric current and voltage levels. This could lead to further degradation of an input or output circuit and mislead or complicate the analysis of a suspect or failed device. Therefore a curve tracer evaluation should be the first step in electrical testing or evaluation to determine the present condition of a device to be failure analyzed.

To start, select two pins and the evaluation configuration desired e.g., input pin to power pin or ground pin or another device pin. It is helpful to select a like function pin on the failed or suspect device to compare or the same pin on a comparison device. Sometimes, good devices are unavailable, cannot be obtained within a reasonable time, or are prohibitively expensive. In this case, pins with similar function, such as inputs and outputs, can be compared on the same device. This method generally characterizes only the first few components within the circuit, usually, the input protection networks and output drivers. The same connections are made between each side of the curve tracer and the two pin combinations selected. This allows the curve tracer to be switched between the two pin combinations to look for differences. Another method of comparison is to use photographs. Figures 18 and 19 show the typical response curves from an input pin to the power pin, pin 14. Figures 20 and 21 show responses between an input pin and the ground pin, pin 7. As is shown, Fig. 18 is very similar to Fig. 19 and Fig. 20 is very similar to Fig. 21. In comparison Fig. 22 and 23 show a shorted and a degraded response curve from an input to the power pin, pin 14. The test current levels should typically be kept below 10 microamps. It is paramount that the failed or suspect device not be overstressed or degraded by this evaluation. Besides the test current level the analyst must also be aware of the maximum voltages impressed on the device circuit nodes. These node voltages may result in rupture of capacitor oxides or MOS gate oxides.

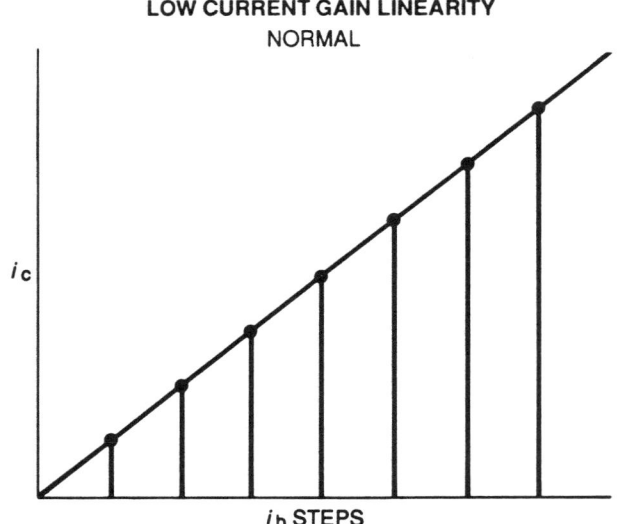

Fig. 17 Illustration of low current gain linearity

Curve Tracer Applications and Hints

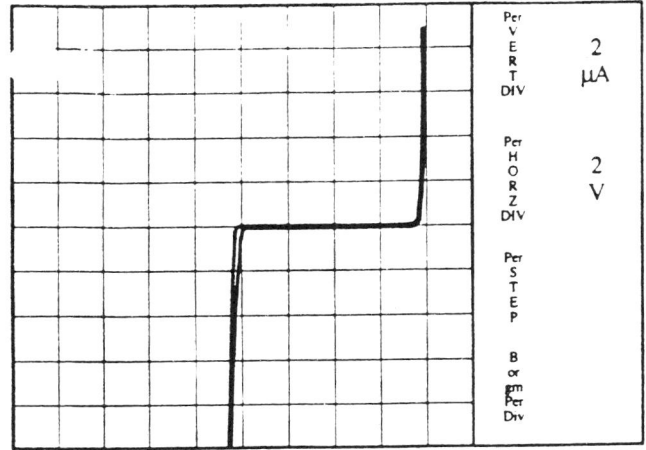

Fig. 18 Positive and negative polarity response curve for pin 10 (input) to pin 14 (VCC).

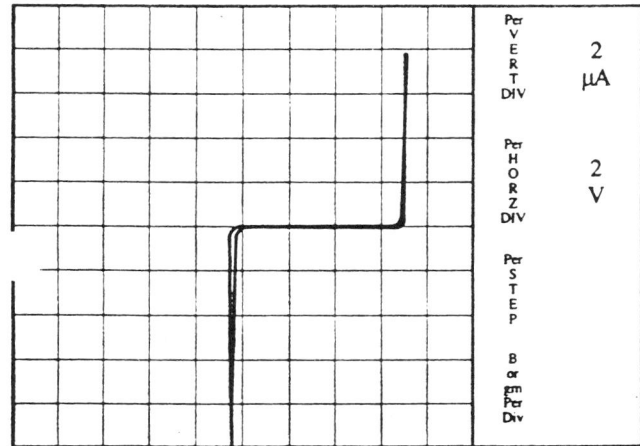

Fig. 19 Response curve for pin 12 (input) to pin 14 (VCC).

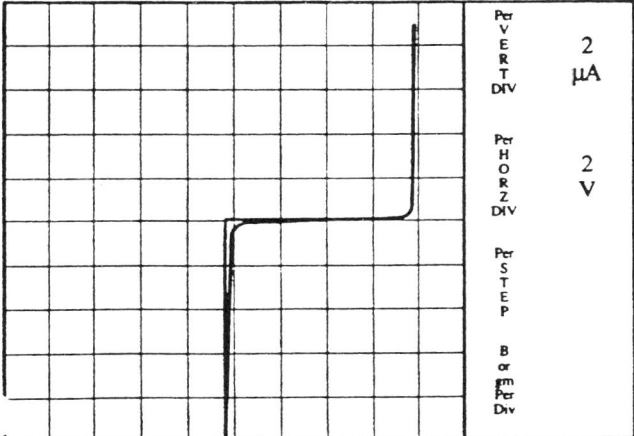

Fig. 20 Positive and negative polarity response curve for pin 10 (input) to pin 7 (gnd).

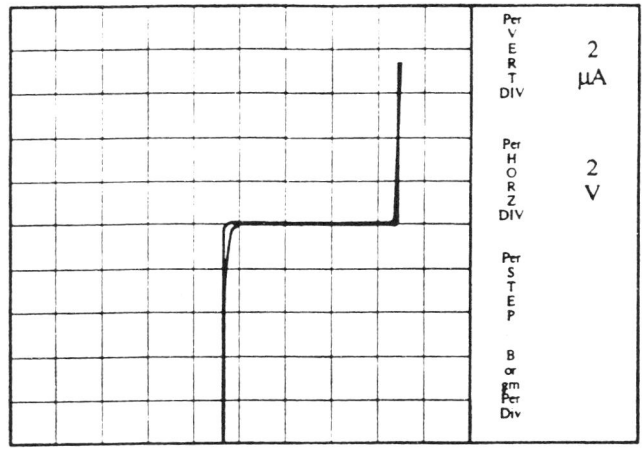

Fig. 21 Response curve for pin 12 (input) and pin 7 (gnd).

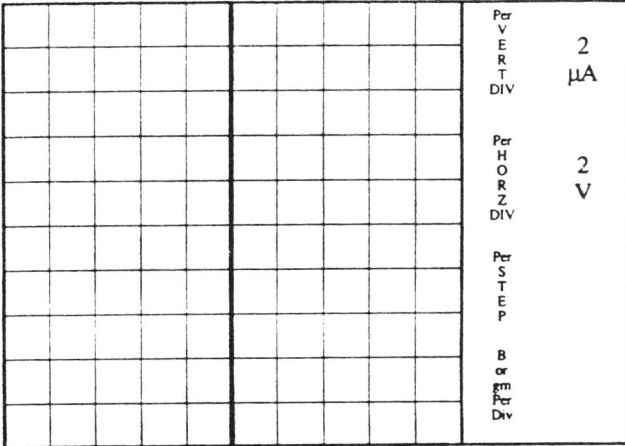

Fig. 22 Example of a shorted input response curve for pin 9 (input) and pin 14 (VCC).

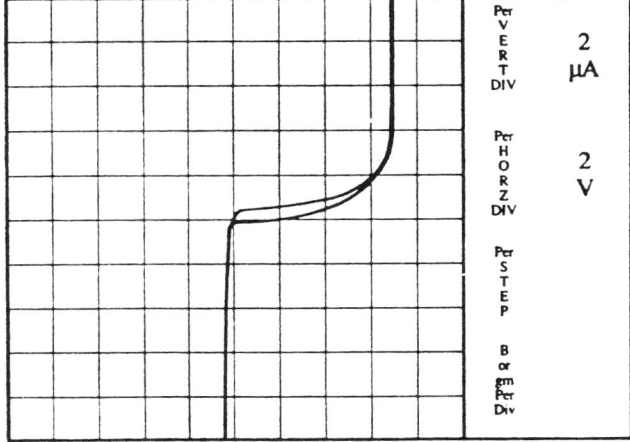

Fig. 23 Example of a degraded input response curve for pin 1 (input) and pin 14 (VCC).

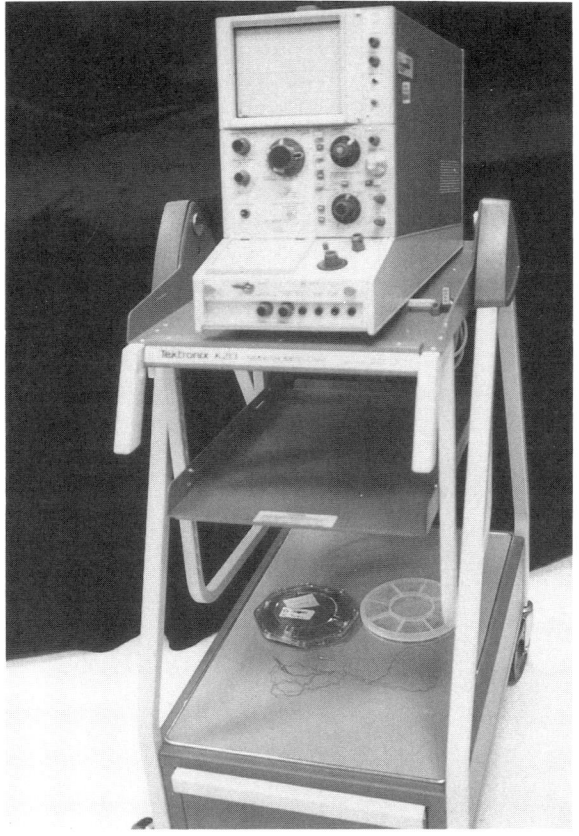

Fig. 24 Tektronix 577 curve tracer.

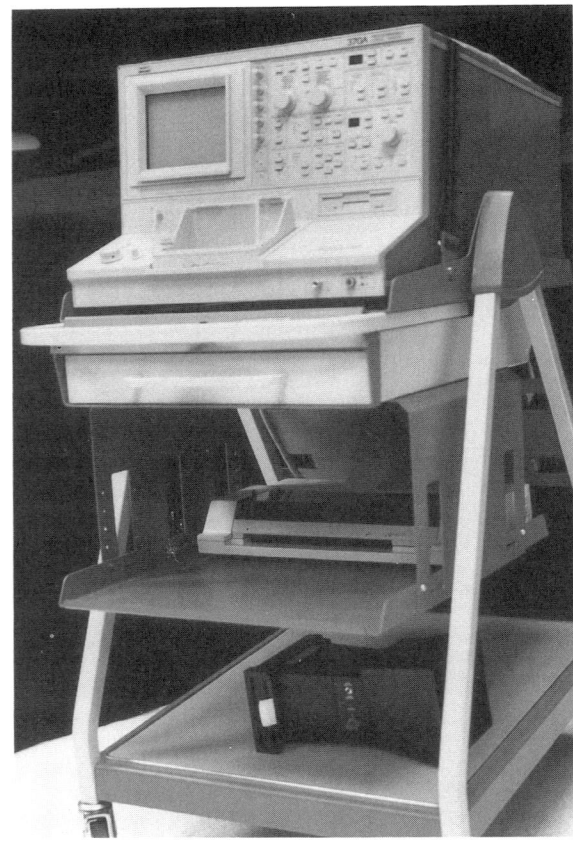

Fig. 25 Tektronix 370 curve tracer.

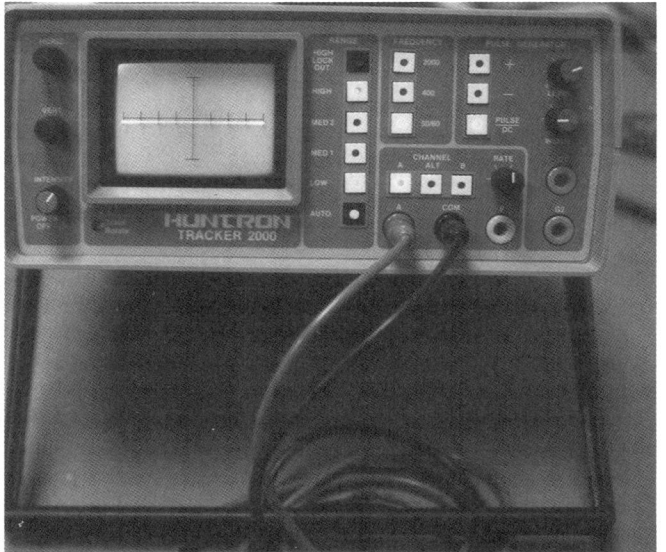

Fig. 26 Huntron Tracker® economy continuity checker and curve tracer.

Fig. 27 The Hewlett-Packard parametric tester.

Each pin combination should be evaluated using both polarities. The I-V response curve is examined for variations in resistive slopes and differences in forward and reverse biased voltage transition points. With experience and the related pin circuitry one can predict/interpret the I-V response curve. When an abnormality occurs in an I-V curve the experienced

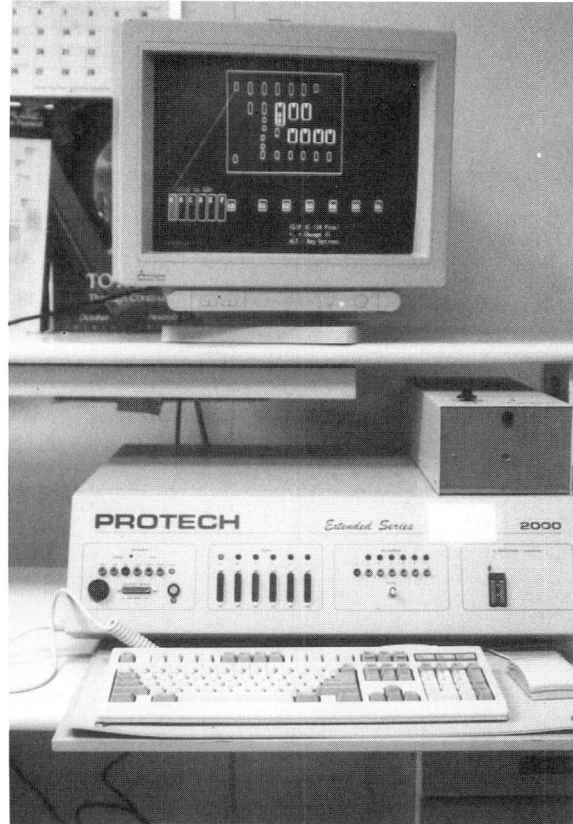

Fig. 28 Protech in-circuit IC tester and identifier.

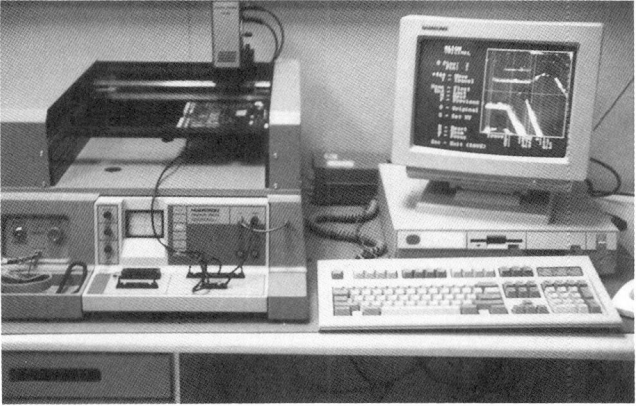

Fig. 29 Huntron board curve tracer test system.

analyst can identify the responsible component or structure within the I.C.

SUMMARY

Many curve tracer applications have been described. These applications show the power and versatility of this instrument. The I-V response curves provide an unequaled capability to the analyst for evaluation of parametric and circuit characteristics. These applications should be considered as a starting point that can be expanded and built on.

ACKNOWLEDGEMENTS

The authors would like to thank Joe Abbott and Chuck Reed for review of this text and their many helpful comments.

REFERENCES

1. *Electronic Devices and Circuits,* J. Millman, C. Halkias, Published by McGraw-Hill Book Company, Inc., 1967.
2. *Fundamentals of Semiconductor Devices,* J. Lindmayer, C. Wrigley, Published by D. Van Nostrand Company, Inc., 1965.
3. *Physics and Technology of Semiconductor Devices,* by A. Grove, Published by John Wiley and Sons, Inc., 1967.
4. *Semiconductor Power Devices,* S. Ghandi, Published by John Wiley and Sons, Inc., 1977.
5. *Reliability and Degradation* - Semiconductor Devices and Circuits, M. Howes, D. Morgan, Published by John Wiley and Sons, Inc., 1981.
6. *Motorola Zener Diode Manual,* Published by Motorola, Inc. 1980.
7. *Avalanche Characteristics and Failure Mechanisms of High Voltage Diodes,* by H. Egawa, in IEEE Transactions on Electron Devices, Vol. ED-13, No. 11, 1966.
8. *Burst Noise and Walkout in Degraded Silicon Devices,* by J. Schenck, in Proceedings of 6th Annual Reliability Symposium, 1967, pp. 31-39.
9. *Failure Analysis of Surface Inversion,* by W. Schroen, in Proceedings of the Reliability Physics Symposium, 1973, pp. 117-123.
10. *Mechanisms of Channel Current Formation in Silicon P-N Junctions,* by D. Fitzgerald and A. Grove, in Proceedings of the Physics of Failure in Electronics, 1966, Volume 4, pp. 315-332.
11. *An Investigation of Microplasma Noise in Zener Diodes With the SEM,* by C. Varker, in Proceedings of the Reliability Physics Symposium, 1971, pp. 155-162.
12. *Transistor Circuit Design,* by J. Walston, J. Miller, Published by McGraw-Hill Book Company, Inc., 1963.
13. *Diagnosis and Analysis of Emitter-Base Junction Overstress Damage,* by M. Jensen, R. Milburn, in Proceedings of the Electrical Overstress Electrostatic Discharge Symposium, 1981, pp. 101-105.
14. *Avoiding Second Breakdown,* by W. Roehr, Motorola Semiconductor Products Inc. Application Note 415A.
15. *How to Safely Check Sustaining Voltage on Power Transistors,* Unitrode Corporation Design Note 5
16. *Limiting Phenomena in Power Transistors and the Interpretation of EOS Damage,* by J. T. May, in the Microelectronics Failure Analysis Desk Reference, Published by ASM International, 1990.
17. *Modeling the Bipolar Transistor,* by Ian Getreu, published by Tektronix, Inc, Beaverton, Oregon, 3rd impression Nov, 1979.
18. *The XYZs of Using A Scope,* available by writing Oscilloscope Primer, DS 47-824, Tektronix, Inc, P.O. Box 300, Beaverton, Oregon 97077.
19. *Semiconductor Device Measurements,* by John Mulvey, Tektronix Measurement Concept Series, Beaverton, Oregon, 1968.

20. *Testing the Bipolar Transistor With the 370 Curve Tracer,* Tektronix Application Note 48W-6756.

21. *Testing the Power MOSFET With the 370 Curve Tracer,* Tektronix Application Note 48W-6757.

22. *Novel Technique For Measurement of MOSFET Threshold Voltage at Very Low Channel Currents,* by G.C. Holmes of British Telecom, Apr, 1985.

A PRIMER ON SIMPLE DEVICE PROBLEMS AND CURVE TRACER CHARACTERISTICS

by Douglas McCormac

These curve trace drawings are to complement the previous article. Note that Figs. 1-9 and 12 are in quadrants 1-4 (A.C. mode, center zero)

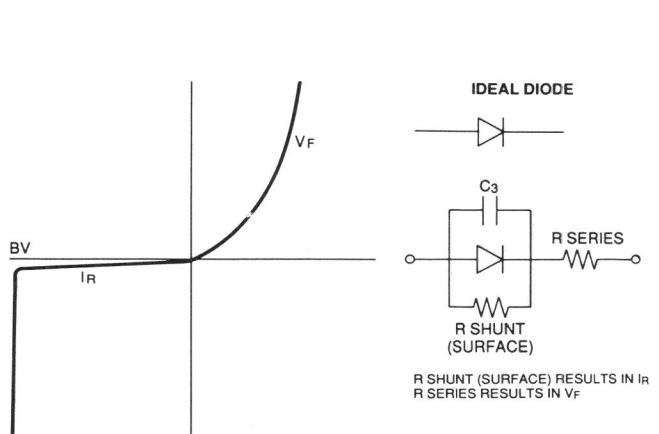

Fig. 1 Illustration of parasitic elements in real vs. ideal diodes, and effect on characteristic

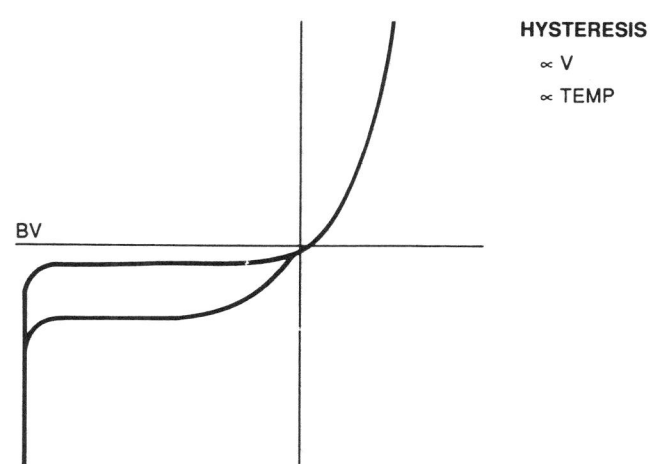

Fig. 2 Illustration of hysteresis, which can be proportional to voltage or temperature.

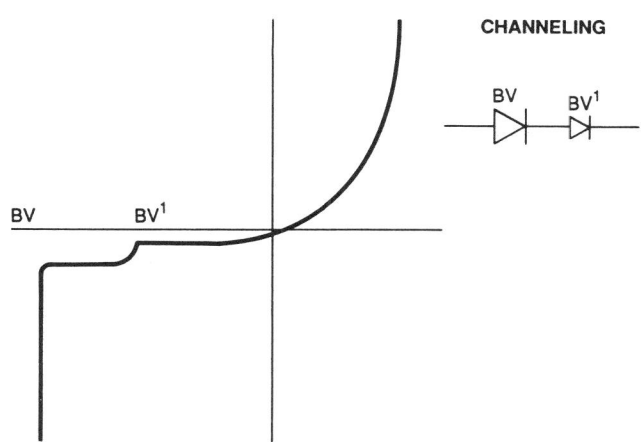

Fig. 3 During channeling, inversion of the surface results in abnormally high leakage.

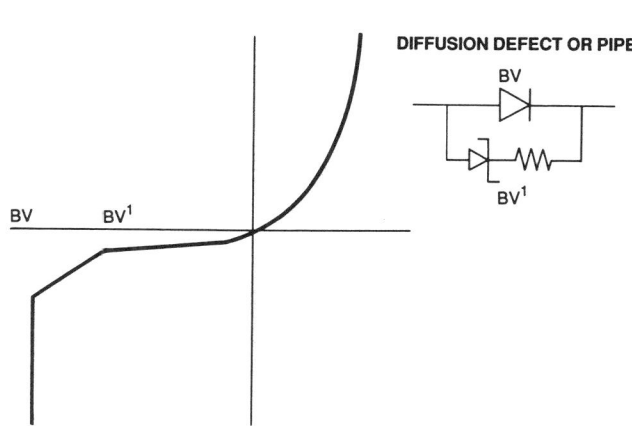

Fig. 4 A diffusion defect or pipe effectively parallels the function with a smaller device with a lower breakdown voltage.

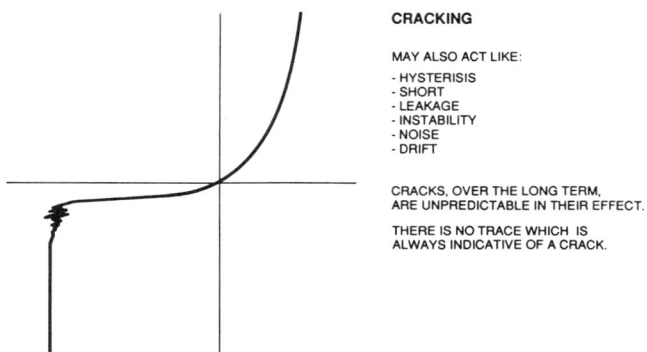

Fig. 5 A cracked die may imitate hysteresis, a short, leakage, instability, noise, or drift; while the above trace is typical, there is no trace which is always indicative of a cracked die.

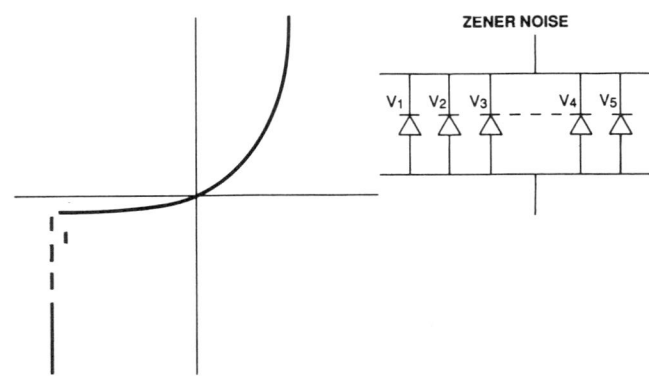

Fig. 6 Zener noise can be represented as many small diodes in series, each with a slightly variable reverse breakdown voltage.

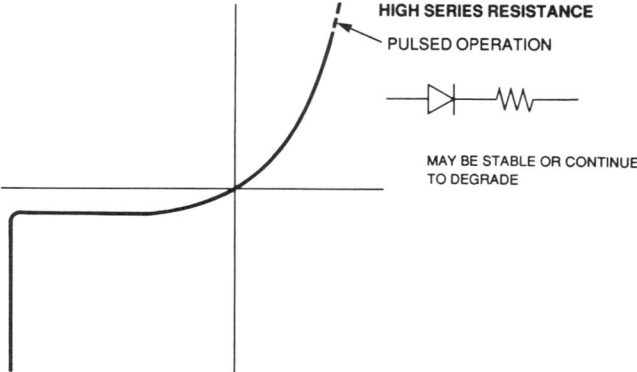

Fig. 7 High series resistance can be stable or unstable

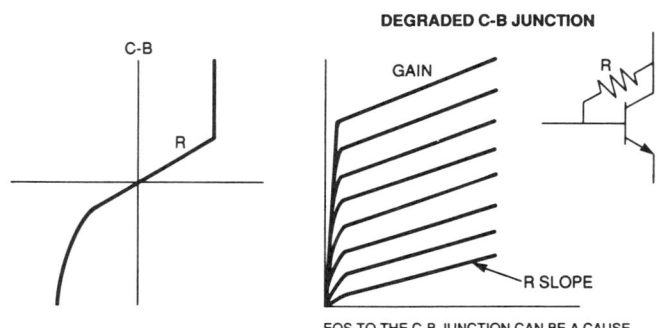

Fig. 8 A degraded C-B junction will offset the characteristic family by the amount of leakage

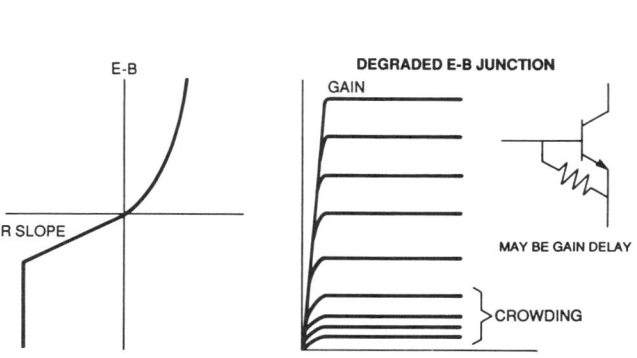

Fig. 9 A degraded E-B junction will decrease h_{FE}, result in poor carrier injection, current crowding, and high E-B reverse leakage.

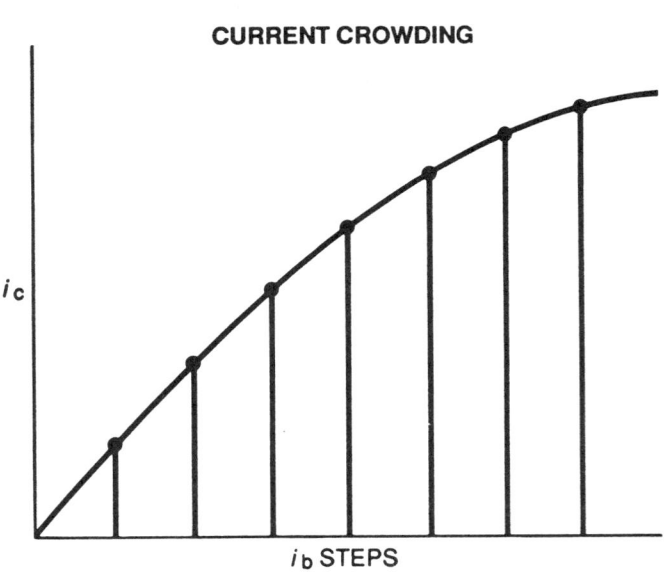

Fig. 10 Detailed illustration of current crowding

A Primer on Simple Device Problems

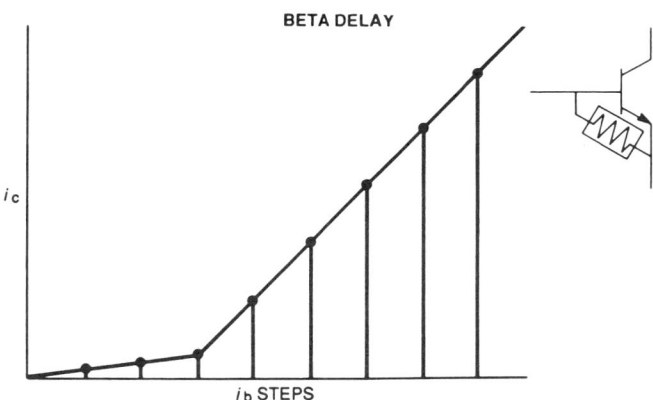

Fig. 11 Detailed illustration of beta or h_{FE} delay

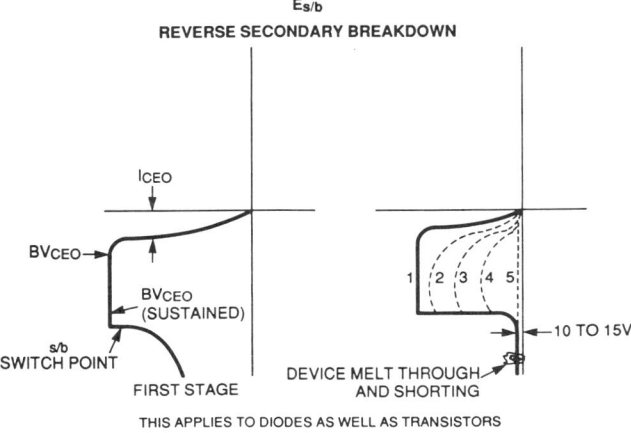

Fig. 12 Illustration of second breakdown progressing from stable BV_{CEO} (sus) condition

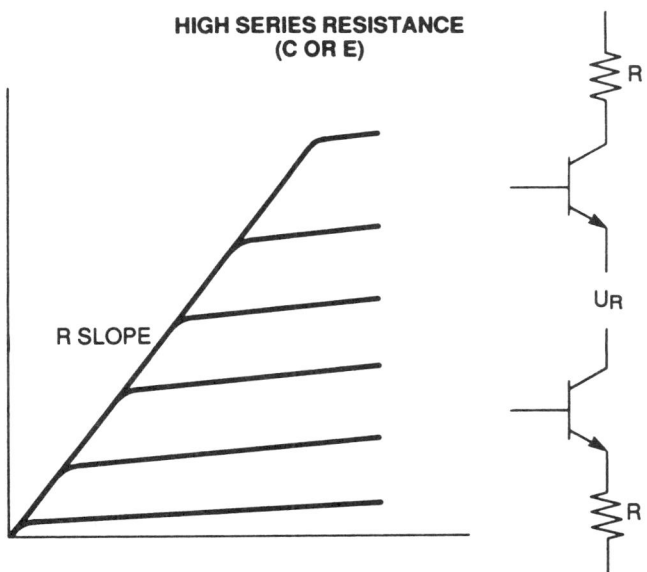

Fig. 13 Series resistance within a transistor collector or emitter can decrease the slope of the saturation curve.

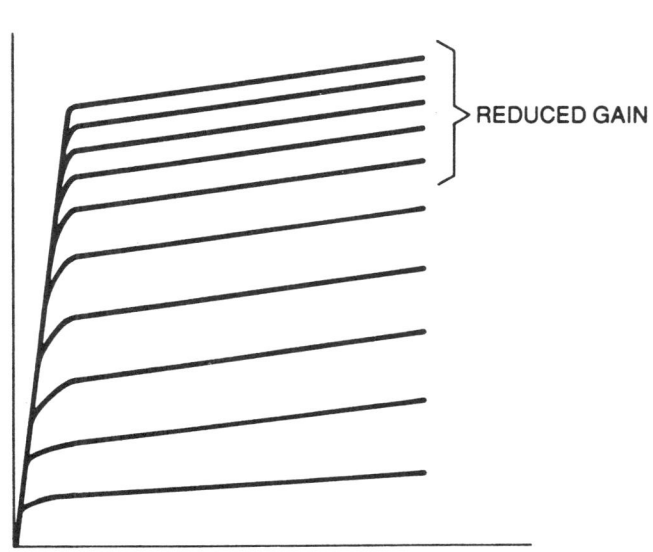

Fig. 14 Series resistance in the base region can compress the characteristic family at high current.

INSPECTING IC PACKAGES WITH C-MODE ACOUSTIC MICROSCOPY

by T.M. Moore

INTRODUCTION

C-mode acoustic microscopy, or C-AM, is rapidly being adopted by packaging researchers and IC failure analysis labs because it provides nondestructive imaging of moisture/thermal-induced damage, such as package cracks and delaminations. C-AM has become an important tool in the development of improved molded packages. Automatic polarity analysis of the echo pulse is used to assist in the identification of delaminated interfaces. Images of the three-dimensional internal structure of the package are produced that are often useful in recognizing package defects and in determining the mechanism for a package failure.[1]

The C-AM is a hybrid instrument with characteristics of both the scanning acoustic microscope (SAM) developed at Stanford in the early 1970s, and the C-scan which has been a part of the nondestructive test (NDT) industry since the 1950s.[2] The characteristics of each of these methods is briefly reviewed.

The term C-scan comes from early NDT nomenclature. The C-scan image is an image of a planar region at a constant depth within the sample. The C-scan image is formed by mechanically scanning a piezoelectric transducer above the specimen and electronically gating the signal in time. The broad-band C-scan transducer has a small numerical aperture (NA) lens for sub-surface imaging. C-scan imaging has played a major role in the macroscopic imaging of sub-surface flaws in industrial components with center frequencies in the range of 1-10MHz. Although modern C-scan instruments provide RF signals, the echo signal has traditionally been rectified for easier interpretation, and hence all phase information was discarded. C-scan inspection takes advantage of the ability of ultrasound to penetrate optically opaque solids.[3,4]

The SAM employs a large NA lens in order to excite surface waves on the sample. Instead of the large water baths of C-scan, a tiny water droplet acoustically couples the transducer and sample. The lens is formed by grinding a hemispherical cavity into the tip of a sapphire rod. The large difference in acoustic velocities between the sapphire and the water droplet produces excellent focusing with a single surface lens. Image contrast is formed by the combined interference of longitudinal and shear waves reflected from the surface and from beneath the surface, and leaky surface waves. Narrow-band RF pulses with frequencies in the range of 100MHz-3GHz are used. Precision mechanical scanning is employed for microscopic imaging. Both the amplitude and phase of the reflected pulses are measured and used to produce images of the mechanical properties of the near-surface region. Sub-micron lateral resolution and penetration are achieved at the highest frequencies. In the C-scan method, lateral resolution is typically limited by absorption in the sample. In contrast, lateral resolution with SAM is limited by high frequency attenuation in the coupling water droplet.[5]

The SAM was first demonstrated by Lemmons and Quate in 1973.[6,7] Its development was strongly encouraged by biomedical researchers who anticipated improved contrast in images of tissue samples. Contrast in the SAM is determined by the large variation in elastic properties in these samples compared to a relatively small variation in dielectric properties, which determine optical contrast. Tissue samples normally required complicated chemical staining for optical microscopy. In the years following its discovery, the frequency of the SAM was continuously increased until the lateral resolution was comparable to that of optical microscopes.[8]

The application of reflection acoustic microscopy to the inspection of IC packages represents the convergence of the capabilities focused C-scan and high frequency SAM. The modern C-AM is a hybrid instrument that combines precision mechanical scanning for microscopic inspection of small samples, sophisticated RF signal analysis and display, and a broad-band transducer with a small NA lens for subsurface imaging.[9,10] C-AM center frequencies in the range of 15-200MHz are intermediate between the frequencies commonly used for C-scan and SAM. C-AMs are now available from several instrument manufacturers.

Scanning laser acoustic microscopy (SLAM) and x-ray radiography are other commonly used methods for nondestructive inspection of packaged ICs. Although these methods have overlapping capabilities with C-AM for this application, the individual advantages of C-AM, SLAM and x-ray imaging make them complementary techniques. The capabilities of these techniques will be briefly compared with those of C-AM.

SAM APPLICATIONS IN THE IC INDUSTRY

SAMs operating at frequencies of 1GHz and higher became commercially available after 1982. The application of SAM in the semiconductor industry was initially the high frequency inspection of thin dielectric layers and conductors on the device surface. Optional broad-band transducers in the intermediate frequency range of 30-100MHz were available for sub-surface studies such as die attach inspections. Die attach inspection through ceramic or metal packages was already demonstrated using high resolution C-scan equipment. Prior to acoustic inspection, die attach inspection was performed primarily with x-ray radiography. It was soon learned that reflected sound indicates the true percent area bonded while x-ray inspection shows only voids in the die attach material. Both C-scan instruments and SAMs equipped with broad-band transducers with center frequencies in the range of 30-100MHz have been used for die attach inspection in metal and ceramic packages.[11-22]

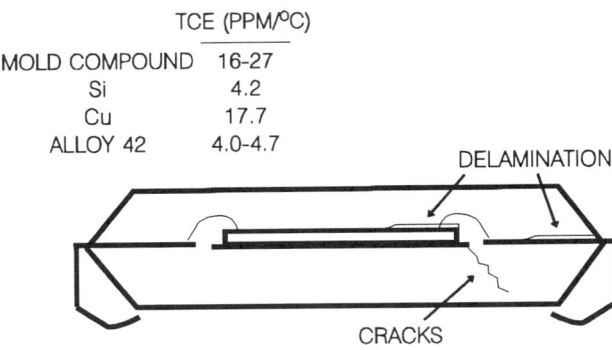

Fig. 1 Thermal coefficient of expansion (TCE) mismatches in plastic IC packages.

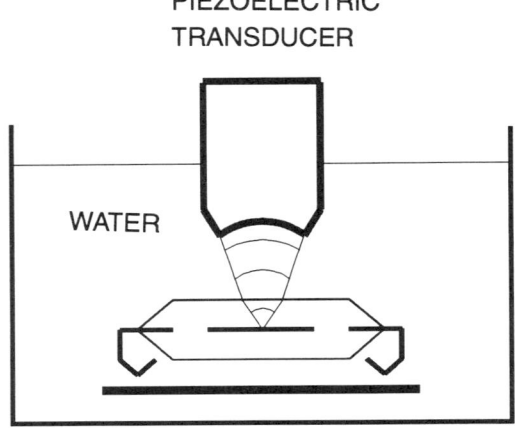

Fig. 2 The inspection of IC packages with the C-mode acoustic microscope (C-AM).

Studies using reflection acoustic microscopy to plastic package inspection began appearing in the US around 1985. These early applications were being performed mostly by Japanese IC manufacturers. The echo signals in the studies typically were rectified and delamination detection was based solely on amplitude imaging. However, these instruments incorporated precision microscope-type scanning and advanced data analysis and presentation features. These early studies were instrumental in correlating the amount of damage detected in plastic packages after reflow soldering to the moisture content in the package.[23-28]

After 1988, the limitations of the detection of delamination by amplitude alone in plastic-packaged ICs were recognized. Reports using even more sophisticated instruments with the ability to detect phase inversion and other features dedicated to IC package inspection began to appear.[1,29-32] With these new C-mode acoustic microscopes, studies have been performed to correlate electrical testing and destructive physical analysis with the results of acoustic inspection in order to understand the moisture sensitivity of modern surface mount packages during board mounting.[33-36]

The integrated circuit (IC) packaging industry is involved in the rapid conversion from the packaging of small dies in conventional through-hole dual in-line packages (DIPs) to the packaging of large high-functionality dies in space-efficient surface mount packages. This conversion has attached increased importance to a basic design problem with the molded IC package. The design of the molded package involves the joining of materials with very different thermal coefficients of expansion (TCEs) (Fig. 1). The packaging of larger and larger dice in thin surface mount molded packages can result in the development of internal stresses sufficient to exceed the mechanical strength of the package during temperature cycle reliability testing.[37]

The internal stress situation is further exacerbated by the fact that the plastic mold compound tends to absorb moisture from the air during shipping and storage. This becomes a problem during mounting of the device to the circuit board. Wave soldering of DIPs delivers a comparatively lesser thermal stress to the body of the package than that experienced by surface mount packages. When surface mount parts are mounted, the entire package body is exposed to soldering temperatures. Absorbed moisture expands during the mounting operation. This greatly increases internal stresses and promotes delamination and package cracking. This "moisture sensitivity" is a problem primarily with surface mount package designs.

Devices designated as being moisture sensitive are currently being shipped in special dry bags. Limits have been recommended for the maximum duration of exposure to air before mounting. These limits are based on a recommended moisture level threshold for the appearance of moisture/thermal-induced damage during mounting.[38] Dry packing does not increase the manufacturing cost of the devices. However, the possibility of mechanical damage and production delays associated with moisture control (such as in the baking of overexposed devices) represent a risk of assembly cost increase. IC manufacturers and mold compound producers are improving package designs and mold compound characteristics in order to provide moisture insensitive packages.

THE C-AM FOR IC PACKAGE INSPECTION

For IC package inspection, center frequencies in the range of 15-100MHz are used. Broad-band acoustic pulses are focused to a point within the IC package (Fig. 2). The pulse repetition rate is typically limited to 10KHz due to decay of the reverberations that occur between the transducer and sample. The transducer is precisely scanned in a plane parallel to the plane of the package for microscopic imaging. At internal interfaces, a fraction of the incident acoustic energy is reflected and detected by the same piezoelectric transducer and converted back into an electrical signal. The amount of incident acoustic energy that is reflected depends on many factors including the materials in contact at the interface, the mechanical properties of the interface, absorption, and the size and orientation of the interface. The echo signal is analyzed and characteristics of the signal are used to form images of internal structures and defects. Sophisticated signal analysis techniques are used to extract characteristics from the echo signal such as amplitude, phase and depth. Because sound is a matter wave, the technique is sensitive to cracks that are invisible to x-ray radiography. The sample and transducer are acoustically coupled by a water bath.[1]

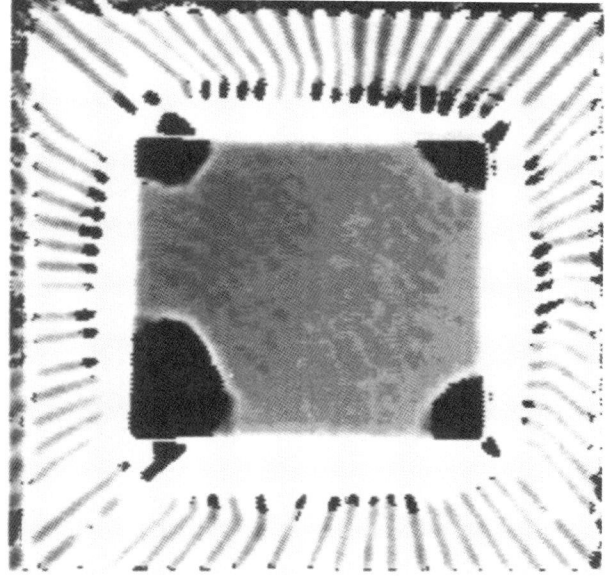

Fig. 3 SAM image of the die area of a 68PLCC. The die is 3.6mm by 3.1mm. The reflected intensity image is presented as a grey scale image on a white background. The delaminated areas are indicated as black areas superimposed over the intensity image.

SIGNAL ANALYSIS FOR IC PACKAGE INSPECTION

Polarity analysis of acoustic echo signals provides information about delamination at internal interfaces in plastic-packaged ICs. Figure 3 shows a C-AM micrograph (top view) of the die area of a 68-pin plastic leaded chip carrier (68PLCC) that has been subjected to 168 hours of exposure to 85 °C/85%RH (relative humidity) followed by vapor phase reflow (VPR) soldering. In Fig. 3 the image of the reflected intensity of the primary sub-surface echo is displayed as a grey scale image on a white background. Those areas identified as delaminated by polarity analysis are indicated in black superimposed over the intensity image. The device in Fig. 3 demonstrates the typical pattern of delamination initiation at the die corners with the subsequent spread of this delamination toward the die center. Delamination on the tips of the leads is also evident.

Figure 4 shows typical examples of acoustic echo signals from an area with good adhesion, and a delaminated area, on the die of a 68PLCC such as the one shown in Fig. 3. In each of the two echo signals in Fig. 4, a reflection from the top surface of the package, and a later sub-surface reflection from the interface between the mold compound and the die can be seen. Note the 180° inversion of the single reflection at the delamination, and the deeper partially resolved reflections in the signal from the die surface area with good adhesion to the package.

For an explanation of the phase inversion phenomenon, consider the simplified example of plane wave reflection, at nor-

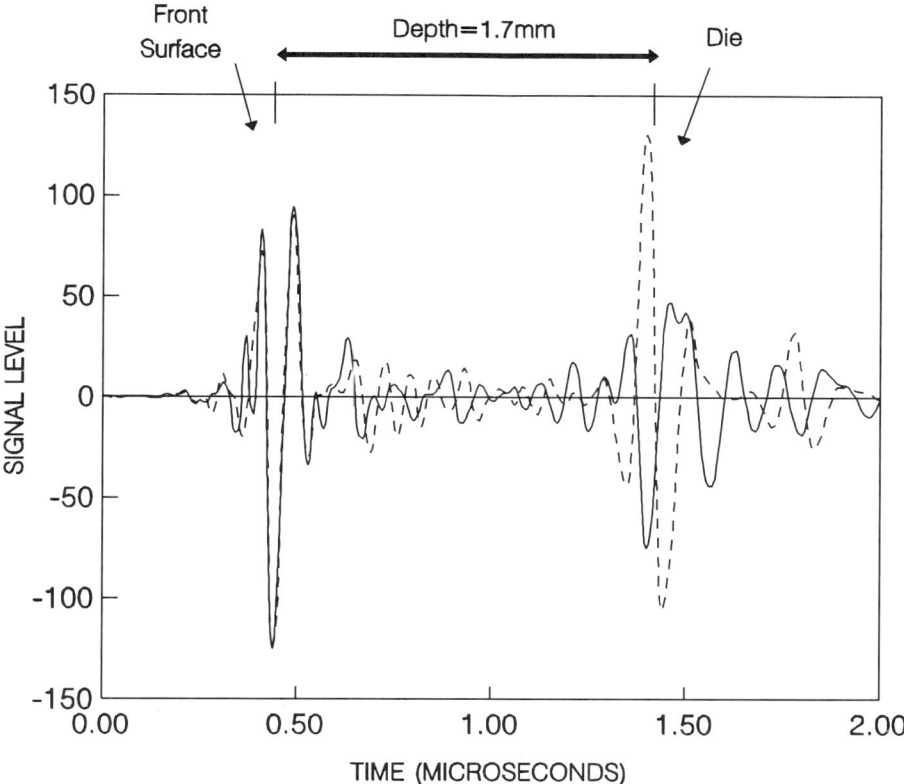

Fig. 4 Typical acoustic echo signals (15MHz) from an area of good adhesion (solid) and a delaminated area (dashed) at the mold compound/die interface in a 68PLCC.

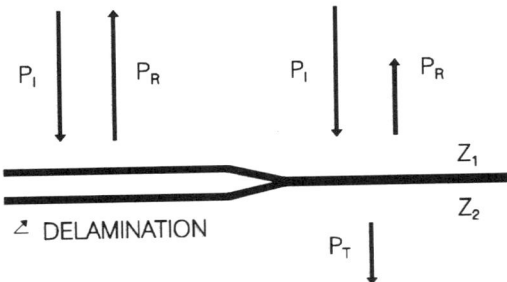

Fig. 5 Reflection at normal incidence of a plane wave at a delamination (left) and at a bonded interface (right).

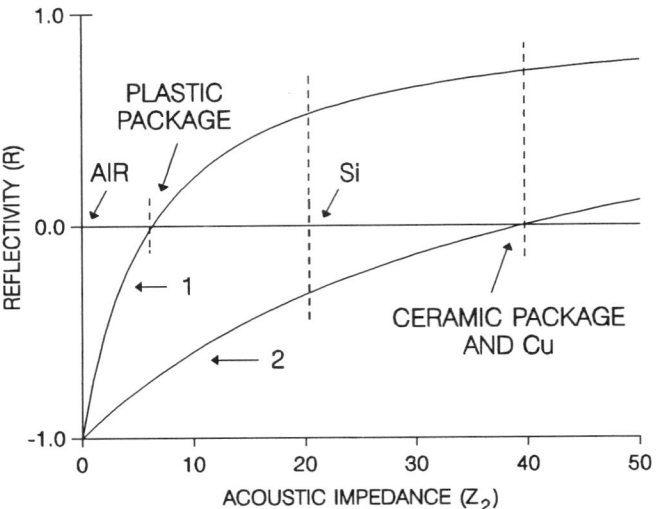

Fig. 6 Ideal acoustic reflectivity (R) versus acoustic impedance of the second layer (Z_2) for plastic packages (Curve 1) and for ceramic packages (Curve 2). (Z units: $10^5 g/cm^2 sec$)

Fig. 7 Backside intensity image of a 68PLCC with a moisture/thermal-induced package crack. The circular features in the four corners of the image are due to scattering of the acoustic probe at surface depressions produced by the mold ejector pins. The image frame is 20mm square.

mal incidence, at an ideal interface (Fig. 5).[39] The incident plane wave has the sinusoidal acoustic pressure amplitude P_I and reflected and transmitted pressures amplitudes P_R and P_T, respectively. As a result of the boundary conditions that the acoustic pressure and particle velocity in both materials must be equal at the interface, the frequency remains unchanged across the interface, and the reflected and transmitted pressure amplitudes can be described as functions of the acoustic impedances, Z_i, of the two materials.

$$P_R = (Z_2 - Z_1)/(Z_2 + Z_1) \quad \text{(Eq. 1)}$$

$$P_T = 2Z_2/(Z_2 + Z_1) \quad \text{(Eq. 2)}$$

The acoustic impedance is the ratio of the acoustic pressure to the particle velocity per unit area and is defined as the product of the density (ρ_i) and the speed of sound (v_i) in layer i.

$$Z_i = \rho_i v_i \quad \text{(Eq. 3)}$$

Equation 1 is plotted in Fig. 6 for two values of Z_1. Curve 1 is calculated for Z_1 equal to the impedance of mold compound (plastic package), and Curve 2 for Z_1 equal to the impedance of Al_2O_3 (ceramic package). Note that each curve passes through the horizontal axis at the value of Z_2 for the appropriate package material. This indicates that there should be no reflection at an ideal interface between identical materials, as expected. In Fig. 6, reflectivities less than zero refer to reflected pulses with inverted phase relative to the incident pulse.

Curve 1 in Fig. 6 shows that at bonded interfaces between the plastic mold compound and the die (MC/die), and between the mold compound and a Cu lead (MC/Cu), the transition is from lower to higher acoustic impedance. Therefore, P_R is positive at these interfaces and there is no phase inversion. However, at a delamination or a package crack, which is represented by an interface between mold compound and air (MC/air), ideally 100% of the energy is reflected, and the phase of the reflected pulse is inverted relative to the incident pulse. This model does not include the effects of attenuation losses. Attenuation losses in plastic packages can often obscure the increase in the amplitude of signals reflected at delaminations, especially on the lead frame. Phase inversion is very important for reliable detection of delamination and cracks in plastic packages.

The simplified plane wave model (Eq.1) is useful in describing phase inversion of reflected acoustic pulses at delaminations and cracks. In practice, apparent phase shifts at similar interfaces due to multi-layer interference effects, frequency dependent attenuation and spatial resolution limitations can also be encountered. In addition, an interface at a constant depth can produce reflections with a continuous variation in phase shift between bonded and delaminated areas that are not explained by the simplified model. These intermediate phase shifts may provide additional information on the condition of the adhesive bond.[48] However, in spite of these practical limitations, the detection of phase inversion in reflected acoustic pulses has proven to be extremely useful in the detection of delaminations in plastic IC packages, and is a distinct advantage of inspection with reflected sound.

Curve 2 in Fig. 6 describes the ideal reflectivities in ceramic (Al_2O_3) packages. In typical ceramic package applications, the C-AM is used to inspect the die attach layer. C-AM offers an advantage over x-ray inspection of die attach quality in that C-AM images indicate only voids in the die attach layer.

As shown in Table 1, the acoustic impedances of Al_2O_3 and a Cu are very similar. The acoustic impedance of the ceramic package is so high that phase inversion detection is not applicable in ceramic package inspection. However, this is compensated for by the fact that the amplitude contrast in an image is typically very high. In a ceramic package with Cu leads, for example, the ceramic/Cu reflections are weak compared to a reflection from a crack or a disbonded lead. These defects provide almost 100% contrast in C-AM inspection. Inspection of eutectic die attach quality also typically shows dramatic contrast between bonds and disbonds. Polymeric die attach adhesives provide lower, but sufficient, contrast. The speed of sound in materials is typically less than 13km/sec. This is roughly four orders of magnitude less than the speed of light. The time delay between returning echoes can be easily measured and images with three-dimensional information can be displayed. This is a unique advantage of a reflection acoustic technique and is often useful in determining the mechanism for package crack formation, for example. Figure 7 shows the reflected intensity image for a 68PLCC that has cracked during board soldering (backside view). Figure 8 is an isometric plot of the information in the time-of-flight image that is acquired simultaneously with the image in Fig. 7. The shape of the crack system, which resembles the stands at a sports arena, can be clearly seen. From similar time-of-flight images it has been learned that this type of crack initiates at the center of the die pad edges and propagates to the back surface.

Table 1: Acoustic Parameters

Material	v, m/sec	ρ, g/cc	Z, 10^5 g/cm² sec
Al_2O_3	10400	3.8	40
Cu	4400	8.9	39
Si	8430	2.4	20
Mold Compound	~ 3500	1.8	6.3
Water	1480	1.0	1.5
Air	343.	1.2×10^{-3}	4.1×10^{-4}

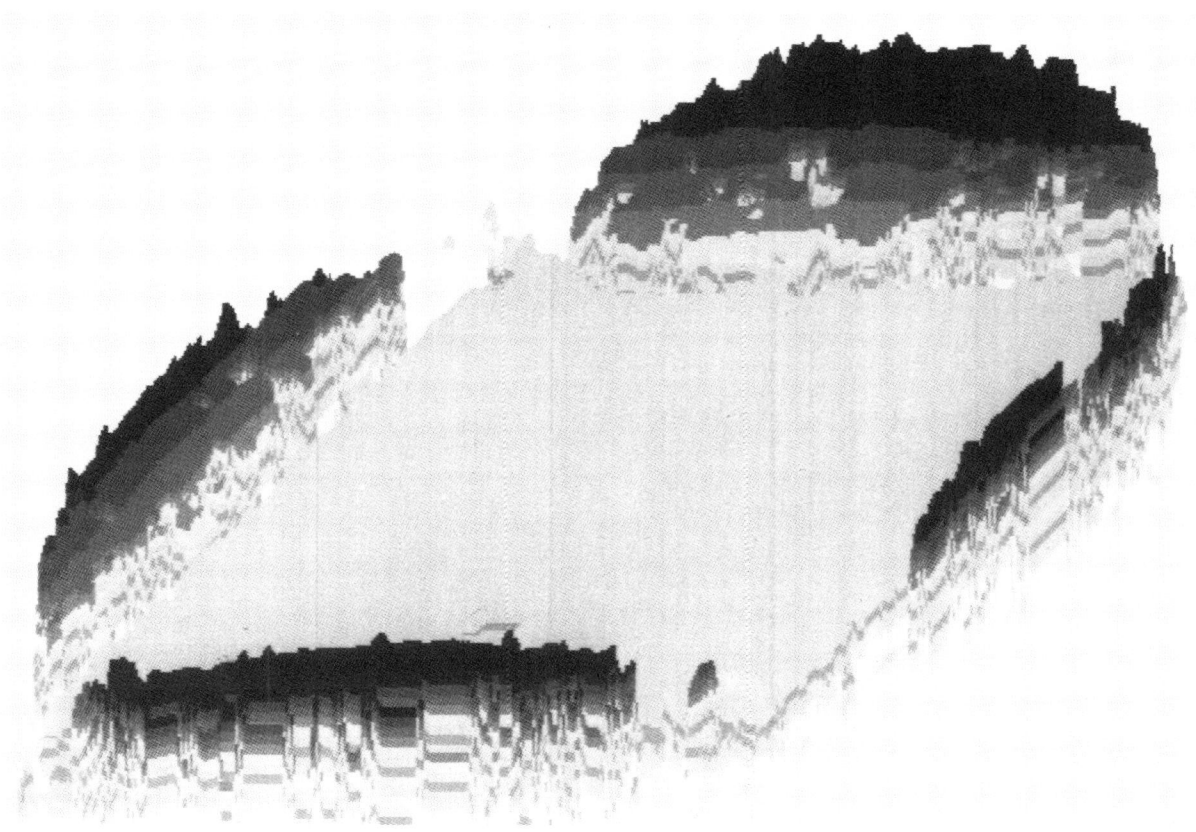

Fig. 8 Isometric plot of the time-of-flight image acquired simultaneously with the intensity image shown in Fig. 7.

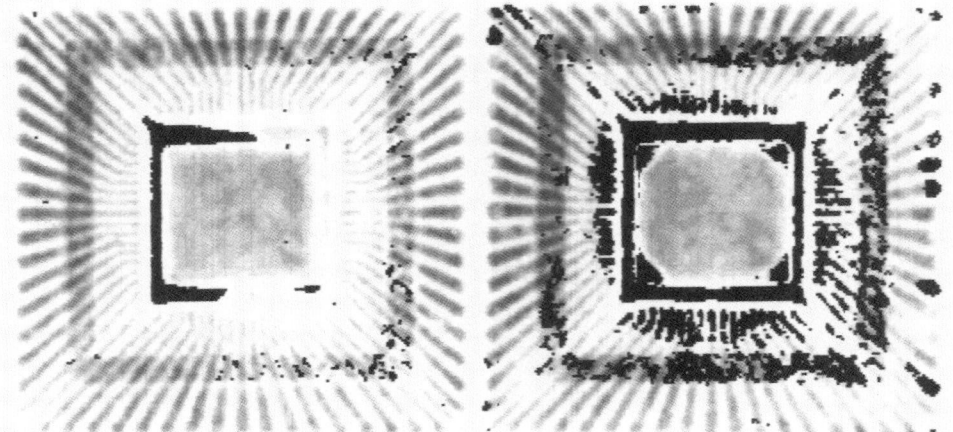

Fig. 9 Delamination images of the same 68PLCC taken initially (**a**) and then after VPR exposure (**b**). The reflected intensity image is presented as a grey scale image on a white background. The delaminated areas are indicated as black areas superimposed over the intensity image. 68PLCCs are 24mm square.

PLASTIC PACKAGE INSPECTION

Figure 9 shows two acoustic micrographs of the same 68PLCC at different times. The initial C-AM image of this device appears in Fig. 9(a). This device was subsequently saturated with moisture (0.32 wt.%) during 168 hours of exposure to 85 °C/85%RH, subjected to VPR soldering, and imaged again. This post-VPR image is shown in Fig. 9(b). The micrographs in Fig. 9 include both amplitude and phase information. The image of the amplitude of the sub-surface reflection is displayed as a grey scale image on a white background. Delaminated interfaces are designated by total black superimposed over the amplitude image. The delaminated areas were identified by phase analysis of the reflected acoustic pulse.

Figure 9 demonstrates moisture/thermal-induced damage in plastic surface mount packages. In the initial C-AM image, the only significant delamination appears on the die pad periphery surrounding the die.

After VPR, delamination has appeared on the entire die pad periphery, the corners of the die, on the leads (predominantly at the internal terminations), and at scattered locations on the lead tape. Studies have indicated that the primary reliability threat is delamination at the die surface. It is likely that this delamination will spread from the shear stress maxima at the die corners toward the die center during subsequent temperature cycle reliability testing. Typically, as the delamination spreads, stress-induced damage will occur at the die surface within the shrinking boundary of good adhesion. And in the delaminated corners, shear displacement between package and die will damage wire bonds. If delamination at the die surface is initiated during board mounting, it spreads during subsequent temperature cycling and leads to early device failure due to stress-induced damage to the device and wire bond degradation, package cracks and delamination failure of the device. [33,36]

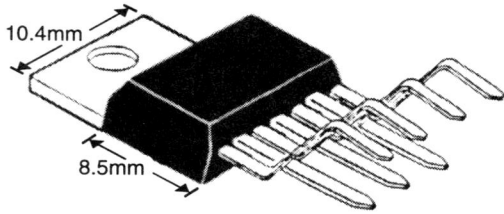

Fig. 10 The TO-220 power IC package with Cu heat sink.

INSPECTION THROUGH HIGH IMPEDANCE SUBSTRATES

Since the acoustic impedances of Cu and Al_2O_3 are so similar, Curve 2 in Fig. 6 can be used to predict reflectivities for die attach inspections in ceramic packages, power heat sink packages or other high impedance substrates such as metal assemblies for microwave components.

For example, Fig. 10 is a sketch of a standard TO-220 molded power IC package. The package is characterized by its thick (1.2mm) Cu heat sink which is necessary to manage the heat produced by high power devices. The die is attached directly to the inside surface of the heat sink with an experimental Pb/Sn alloy solder. Figure 11 shows two 50MHz C-AM images of the die attach in the same device at different times. Figure 11(a) is an initial image, while Fig. 11(b) was recorded after 200 cycles of temperature cycle reliability testing. The dark areas denote good adhesion and are areas where most of the energy in the acoustic pulse was transmitted into the die. The bright areas in the die attach indicate almost total reflection due to disbonding. Note the reduction of the area of good adhesion after temperature cycling. A real-time x-ray image was taken after temperature cycling and is shown in Fig. 12. Due to x-ray attenuation in the thick Cu heat sink, the x-ray image required a significant amount of image processing to reveal contrast produced by the die attach layer. The die attach voids observed in the x-ray image taken after temperature cycling agree well with the features in the initial C-AM image. The C-AM image at 200

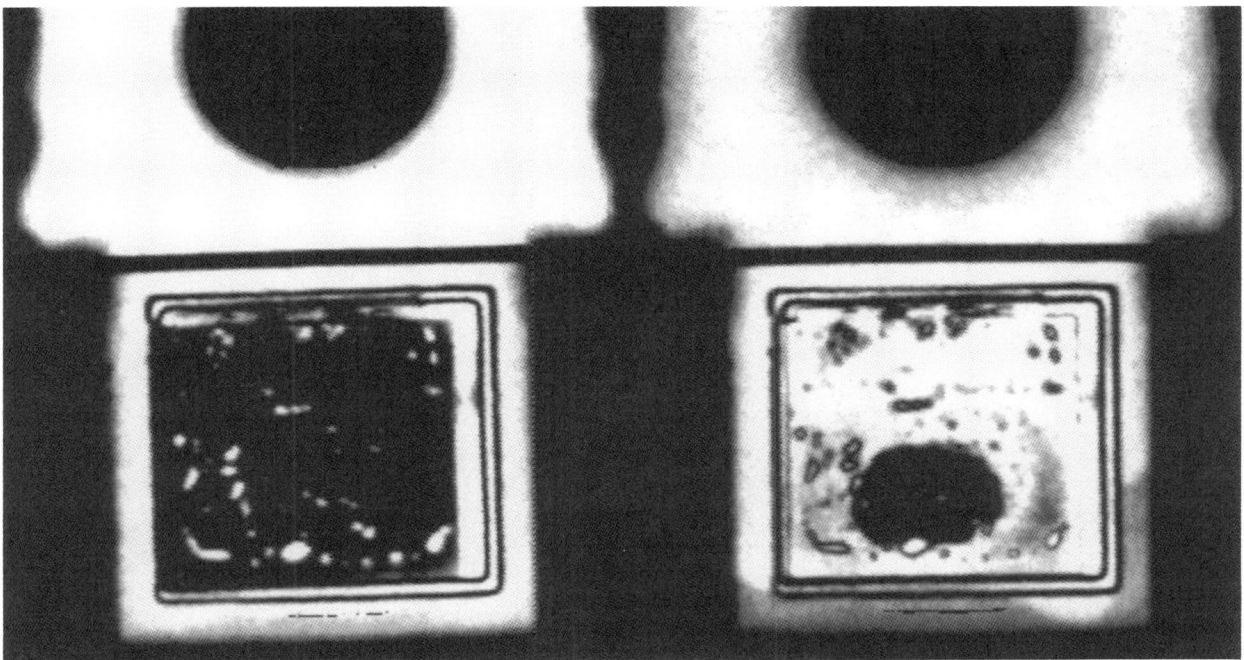

Fig. 11 Die attach images in a TO-220 package before testing (**a**) and after 200 temperature cycles (**b**). Bright areas indicate a high reflected intensity. Dark areas in the die attach region indicate good adhesion. The heat sink is 7mm wide at the die attach region.

Fig. 12 Real-time X-ray image of the device shown in Fig. 11 taken after temperature cycling.

cycles indicates a reduction in the total area of adhesion that was not detectable by x-ray radiography.

DETAILS OF C-AM INSPECTION

The lateral resolution (d) of a spherically focused probe for sub-surface inspection can be estimated by:[40]

$$d \approx \lambda F/D \qquad (Eq. 4)$$

Here, λ is the acoustic wavelength, F is the focal length, and D is the diameter of the lens. Typically the effective F/D ranges from 1 to 1.6 for plastic package inspection. At a center frequency of 20MHz the wavelength is approximately 170 μm and the calculated resolution is 270 μm. This is an ideal resolution that does not account for attenuation. The observed resolution is roughly 400m at a depth of 1.6mm (or 3.2mm round trip) in a standard PLCC package. Mold compound attenuation varies considerably from one formulation to the next and this dramatically affects the lateral resolution. Both the penetration and resolution are noticeably degraded by temperature shock damage to the mold compound.

The acoustic attenuation in mold compound has been reported to be 40dB/cm at 15MHz and to increase rapidly with increasing frequency.[41] The package preferentially attenuates the higher frequencies. As a result, acoustic inspections of standard PLCC packages are limited to roughly 20MHz. However, thin plastic packages (mm) are often inspected at frequencies up to 90MHz.

Attenuation also affects the depth resolution, especially in plastic package inspection. The decay period for a broad-band echo pulse is roughly 1.5λ which creates a "dead zone" after each reflection in which later reflections cannot be temporarily resolved without the use of frequency domain analysis methods. This dead one becomes thinner as the center frequency is increased.

In ceramic packages, the attenuation is much less (<.05dB/cm), but the speed of sound is almost three times greater than that of mold compound. Hence, one wavelength at 50MHz in Al_2O_3 is 200 μm and roughly equivalent to one wavelength in mold compound at 20MHz. A lateral resolution approaching one wavelength is observed at a center frequency of 50MHz and a depth of 1.6mm in Al_2O_3.

The sensitivity of reflection acoustic imaging is superior to the lateral resolution of the technique. For example, 25 μm (1

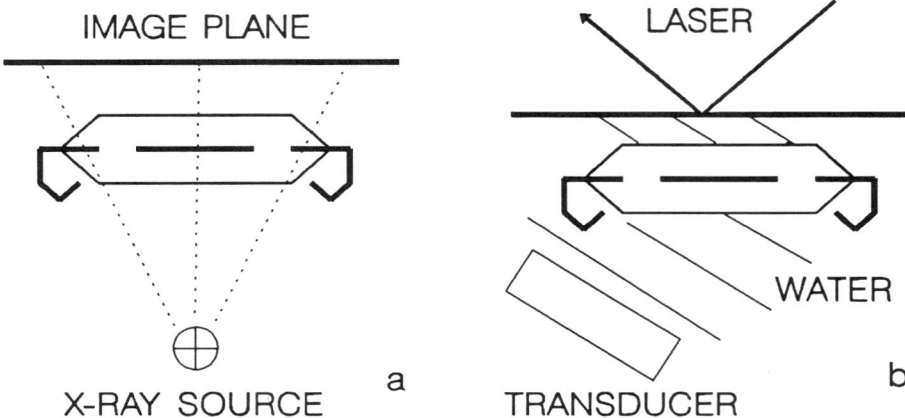

Fig. 13 X-ray radiographic inspection (a) and scanning laser acoustic microscopy (SLAM) (b) of IC packages.

mil) bond wires are often seen in 20MHz C-AM at 1.6mm in plastic package inspections in which the acoustic wavelength is 170 μm. A weakly reflecting object with lateral dimensions smaller than a wavelength that is easily detected in reflection may produce only a negligibly small contrast effect in transmission. However, the apparent size of a small, non-specular reflector in the image will be determined by the lateral resolution available.

The sensitivity of reflection acoustic imaging to thin cracks with lateral dimensions much larger than a wavelength is an important aspect of C-AM inspection. Sound is a matter wave and depends on molecular vibrations for propagation. Theory predicts that crack openings greater than the particle displacement amplitude produced by the interrogating sound wave should be detectable. Experiments with steels, for example, indicate that air-filled cracks on the order of 10nm are detectable.[4,42]

COMPLEMENTARY NONDESTRUCTIVE TECHNIQUES

C-AM is one of three complementary techniques commonly used in the nondestructive inspection of packaged integrated circuits (ICs). A transmission acoustic technique, scanning laser acoustic microscopy (SLAM), and x-ray radiography are also widely used for this application (Fig. 13). These techniques can each produce images of packaged ICs, and therefore have partially overlapping capabilities. However, due to certain unique capabilities of each technique, they are complementary for the nondestructive inspection of IC packages.

Real-time x-ray inspection offers the best lateral resolution and is unsurpassed for nondestructive wire sweep inspection. Images are produced at TV rates and manipulation of the sample is simple and uncomplicated. Although commonly used for die attach inspection, x-ray imaging shows only the location of voids in the die attach layer, and not the total area of attachment. X-ray inspection often exhibits limited contrast for low mass thickness die attach layers on metal substrates with high mass thickness. Similarly, detection of package voids can be limited by lead frame shadowing.[32]

A recent development in x-ray inspection in the semiconductor industry is x-ray laminography. This is a tomographic technique that produces an image of x-ray attenuation at a specific plane within the sample. At this time, x-ray laminography has been applied almost exclusively to the inspection of solder joint quality for surface mount process control.[43,44]

SLAM is a transmission acoustic technique developed in the early 1970s at Zenith and later transferred to Sonoscan in 1974.[45] SLAM incorporates concepts similar to those presented by Sokolov in 1936 for the first proposed acoustic microscope.[46] In SLAM, a planar sound wave is transmitted through the sample onto a mirrored coverslip. The disturbances produced in the coverslip by the transmitted wave are imaged with a scanned laser beam. The technique produces either an acoustic shadow image or an interference pattern by comparison of the signal to a phase reference. SLAM offers the advantage of real time imaging which is more suitable for process control than a technique involving mechanical scanning. SLAM has been successfully applied to situations where real time imaging is critical, and in applications involving samples with irregular shapes and very thin multi-layer construction that cannot be imaged well by reflected sound due to the "dead zone" effect described earlier. SLAM applications have included the inspection of tape automated bonding (TAB), thin film ceramic chip capacitors, and real-time die attach inspection.[9]

SLAM operates at frequencies typically up to 500MHz and potentially offers better lateral resolution than C-AM for certain applications. However, in practical application the lateral resolution of C-AM images of IC packages is typically superior to that of SLAM images at the same frequency.[47] SLAM cannot be applied to die attach inspection in packages with air cavities, such as ceramic DIPs and pin grid arrays, unless the lid is removed.

SUMMARY

The C-AM is a hybrid instrument with characteristics of both the high frequency SAM and the C-scan recorder.[9,10] For the inspection of delaminations in layered IC packages both amplitude and phase are measured and used to produce images of the internal structure and locations of interfacial delamination. Precise scanning is used for optimum lateral resolution.

Small numerical aperture transducers provide sub-surface imaging of internal package interfaces. In addition, the depth of the primary sub-surface reflection is recorded for pseudo-three-dimensional representations of package crack systems and void positions.

The C-AM has proven to be well suited to plastic package inspection. Although the plastic mold compound strongly attenuates sound and thereby limits resolution, the planar, layered design of IC packages is ideal for C-AM inspection. The typically featureless and planar front surface facilitates sub-surface imaging.

The C-AM has been a key factor in the understanding of the moisture sensitivity problem in surface mount plastic IC packages. Polarity analysis of reflected acoustic pulses has proven to be extremely useful in the detection of delaminations in plastic IC packages, and is an important advantage of inspection with reflected sound. Work is currently underway to provide quantitative limits on the amount of delamination that is acceptable from a reliability standpoint so that C-AM inspection can be incorporated into standard test methods for reliability assurance. Current standard test methods specify destructive cross sectioning methods.

The nondestructive nature of C-AM means that in reliability evaluations involving a dense matrix of variables, fewer devices are required. Previously, a fraction of the remaining devices would have to be sacrificed for destructive analysis at each inspection interval. Also, the data of reliability evaluations using C-AM are more valuable because the initiation and propagation of damage can be tracked on individual ICs through the test.

The development of one-dimensional, or two-dimensional scanned array technology for acoustic imaging would have a dramatic impact on the effectiveness of C-AM for real-time applications such as process control by eliminating the need for mechanical scanning.

REFERENCES

1. T. Moore, Proc. Int. Symp. Testing and Failure Analysis, 1989, pp. 61-67.
2. Y. Bar-Cohen and A.K. Mal, in Metals Handbook, 9th edn., ASM International, 17, 1989, pp. 231-277.
3. R.C. McMaster, Nondestructive Testing Handbook, Vol. 2, Ronald Press, 1959, p. 43.
4. J. Szilard, in Ultrasonic Testing, J. Szilard (ed.), John Wiley and Sons, 1982, pp. 1-23.
5. C.F. Quate, A. Atalar, H.K. Wickramasinghe, Proc. IEEE, 67, 1979, pp. 1092-1114.
6. R.A. Lemmons and C.F. Quate, App. Phys. Lett., 25, 1974, pp. 251-253.
7. R.A. Lemmons and C.F. Quate, Proc. 1973 IEEE Ultrasonic Symp., 1973, pp. 18-21.
8. C.F. Quate, IEEE Trans. Sonics and Ultrason., SU-32 (2), 1985, pp. 132-135.
9. L.W. Kessler and S.R. Martell, Proc. Int. Soc. for Testing and Failure Anal., 1991, pp. 491-504.
10. B.T. Khuri-Yakub, in New Technology in Electronic Packaging, B.R. Livesay and M.D. Nagarkar, ASM International, 1990, pp. 311-315.
11. J.L. Rose and P.A. Meyer, Mater. Evaluation, 31 (6), 1973, p. 109.
12. G.J. Curtis, in Ultrasonic Testing, J. Szilard (ed.), John Wiley and Sons, 1982, pp. 495-555.
13. N.J. Burton and D.M. Thacker, Proc. Int. Soc. for Testing and Failure Analysis, 1985, pp. 187-192.
14. K. Shirai, K Kobayashi, T. Noguchi and T. Goka, Proc. Int. Soc. for Testing and Failure Analysis, 1988, pp. 47-52.
15. M.J. Mirasole, Proc. Int. Soc. for Testing and Failure Analysis, 1988, pp. 77-88.
16. R.A. Lemmons and C.F. Quate, Appl. Phys. Lett., 25 (5), 1974, pp. 251-253.
17. R.G. Wilson, R.D. Weglein and D.M. Bonnell, Semiconductor Silicon 1977, 1977, pp. 431-435.
18. C.F. Quate, Semiconductor Silicon 1977, 1977, pp. 422-430.
19. A.J. Miller, Inst. Phys. Conf. Ser. No. 67: Section 8, 1983, pp. 393-398.
20. A.J. Miller, Acoust. Imaging, 12, 1982, pp. 67-78.
21. H.K. Wickramasinghe, J. Micros., 129, 1983, pp. 63-67.
22. H.R. Vetters, et al., Scanning Electron Microscopy/III, 1985, pp. 981-989.
23. M. Sakimoto, et al., Proc. Int. Symp. for Testing and Failure Anal., 1985, pp. 173-177.
24. A. Kitayama, H. Tabata and H. Suziki, Proc. IMC, 1986, pp. 462-469.
25. S. Ito, et al., Proc. Elect. Comp. Conf., 1986, pp. 360-365.
26. S. Okikawa, et al., Proc. Int. Symp. for Testing and Failure Anal., 1987, pp. 75-81.
27. S. Kuroki and K. Oota, Proc. Elect. Comp. Conf., 1989, pp. 885-890.
28. A. Nishimura, S. Kawai and G. Murakami, Proc. Elect. Comp. Conf., 1989, pp. 524-530.
29. T.M. Moore, Texas Instruments Technical Report TR-088778, Feb. 1988.
30. J.E. Semmens and L.W. Kessler, Proc. Int. Symp. for Testing and Failure Anal., 1988, pp. 211-215.
31. R. Birudavolu, Proc. Surface Mount 1989, 1989, pp. 751-766.
32. A. van der Wijk, K. van Doorselaer, Proc. Int. Symp. for Testing and Failure Anal., 1989, pp. 69-74.
33. K. van Doorselaer and K. de Zeeuw, Proc. Elect. Comp. Conf., 1990, pp. B49-B53.
34. T. Moore, R. McKenna, S.J. Kelsall, Proc. Int. Symp. Testing and Failure Analysis, 1990, pp. 251-258.
35. T. Moore, R. McKenna, S.J. Kelsall, J. Surface Mount Tech., 4 (3), 1990, pp. 31-38.
36. T. Moore, R. McKenna, S.J. Kelsall, IEEE Int. Reliability Physics Symp., 1991, pp. 160-166.
37. K.R. Kinsman, J. Metals, 40 (6), 1988, pp. 8-13.
38. IPC-SM-786, "Impact of Moisture on Plastic Package Cracking," and IPC-Test Method 650-2.6.20, "Plastic Surface Mount Component Cracking," Institute for Interconnecting and Packaging Electronic Circuits (IPC), Lincolnwood, IL, 1991.

39. L.A. Kinsler, et al., Fundamentals of Acoustics, John Wiley and Sons, 1982, p. 125.
40. R.S. Gilmore, R.A. Hewes, L.J. Thomas, and J.D. Young, in Acoustical Imaging, 17, H. Shimizu, N. Chubachi, and J. Kushibiki (eds.), Plenum Press, 1989, pp. 97-109.
41. B.T. Khuri-Yakub, in New Technology in Electronic Packaging, B.R. Livesay and M.D. Nagarkar (eds.), ASM International, 1990, pp. 311-315.
42. J. Krautkramer and H. Krautkramer, Ultrasonic Testing of Materials, Springer-Verlag, 3rd edn., 1983, p. 28.
43. B. Baker, Electronic Manufacturing, Feb. 1989, pp. 20-22.
44. C. McBee, Circuits Manufacturing, Jan. 1989, pp. 67-69.
45. L.W. Kessler, Proc. IEEE, 67, 1979, pp. 526-536.
46. S. Sokolov, USSR Patent No. 49, Aug. 31, 1936.
47. R.K. Mueller and R.L. Rylander, IEEE Spectrum, 1982, pp. 28-33.
48. T.M. Moore, Reliable Delamination Detection by Polarity Analysis of Reflected Acoustic Pulses, Proc. Int. Symp. for Testing and Failure Anal., 1991, pp. 49-54.

ADVANCED RADIOGRAPHIC TECHNIQUES IN FAILURE ANALYSIS

by Joe Colangelo

The use of x-ray energy for nondestructive evaluation has been common practice for years. The use of radiographic techniques reveals many defects, and internal features, in parts and materials.

In the past, the primary method for capturing an image was to place the sample on a sheet of film and expose it to x-ray energy for a predetermined time. The amount of x-ray energy that passes through the sample to reach the film is determined by the sample's density and atomic structure. The areas of higher absorption will allow less energy to pass, resulting in darker features on the film (assuming a positive imaging medium such as Polaroid film). This creates the familiar "shadowgraph."

The acquisition of an x-ray system which provides magnified images in real time (see Fig. 1) eliminates the trial-and-error method: expose sample, develop film, vent frustration, adjust position, adjust power, and repeat. Historically, this has been the typical method of x-ray evaluation.

A comprehensive failure analysis can only proceed in one direction since it involves destructive tests. Accurately capturing information pertaining to the failure in each phase of the analysis is critical because steps cannot be retraced. Clearly, the addition of this real-time capability provides a wealth of information at the front end of the analysis that was previously unattainable. X-ray analysis is as simple as mounting the sample on the five-axis manipulator and adjusting for the optimum viewing angle. The power level is continuously adjustable from 30 to 160 kV and up to 1 mA for penetration of a wide range of samples. Table 1 summarizes the performance features of a typical real time x-ray system:

Table 1 Performance Summary of the Feinfocus FXS 160.52

Magnification	Up to 200×
Resolution	4 to 6 µm
Maximum power	160 kV, 1 mA
Cycle time	5 minutes
Manipulation	Five axes, 0.01-mm resolution, joystick or CNC-controlled
Image processing	Pseudo color, frame integration, background subtraction, intensity profile/histogram, distance measurement, superimposed text
Data storage	Photograph, videotape
Customer interaction	During analysis

Instead of exposing film, the x-ray energy is converted to a video signal by an image intensifier detector tube. The x-ray tubehead is classified as a micro focus type, because the electron beam is focused to a very small spot on the target, approximately 5 microns in size. Because the x-rays are emitted in a conical pattern from a point source to the target, geometric magnification of the image in real time is achieved. This is done by projecting the image onto the image intensifier (see Fig. 2). By changing the positions of the image intensifier and sample with respect to the x-ray source, magnification of up to 200X is possible with very low distortion.

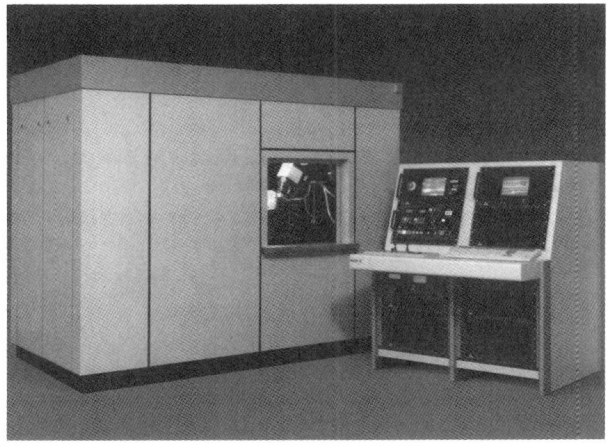

Fig. 1. External appearance of the Feinfocus FXS160.52 machine.

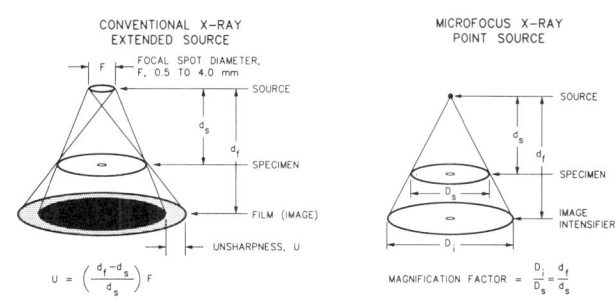

Fig. 2. Illustration of x-ray magnifications and sharpness.

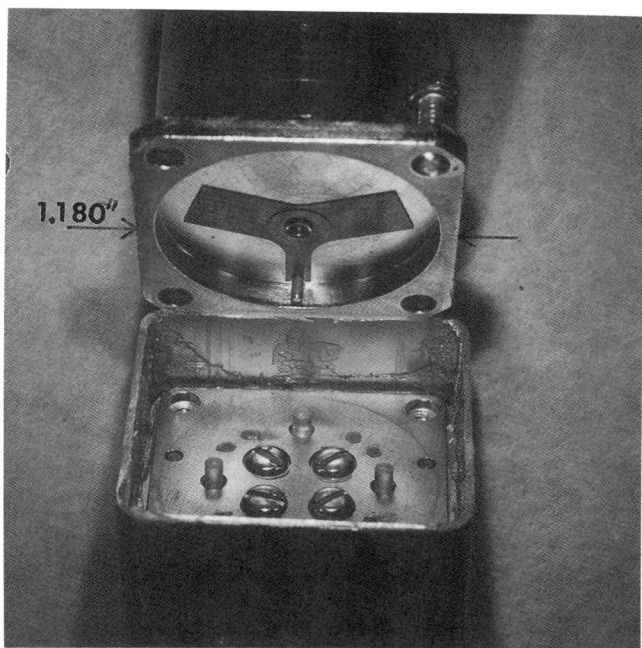

Fig. 3. RF shutter partially disassembled.

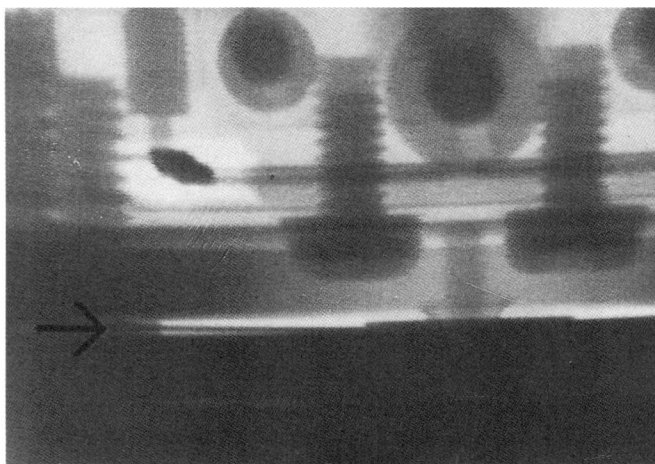

Fig. 4. X-ray image side view, with shutter de-energized.

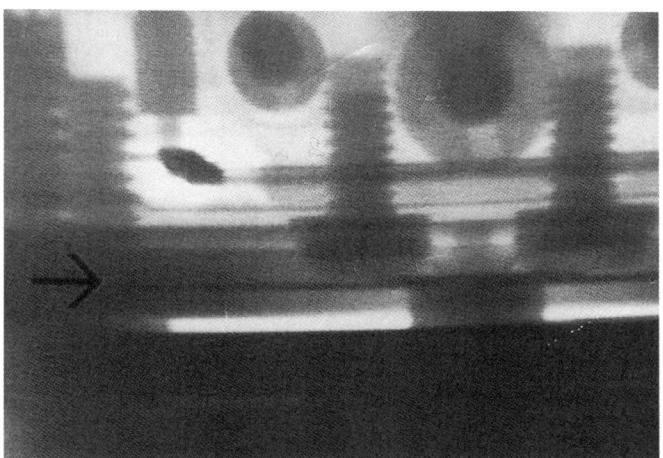

Fig. 5. Failed unit, energized.

The purpose of this section is to demonstrate the capabilities of this system and to illustrate its direct impact on productivity. This will be done by describing actual case histories, although it is difficult to accurately present the benefits of real time imaging with photographs.

APPLICATION EXAMPLES

1. One program was experiencing repeated field failures on an RF shutter assembly. Figure 3 shows the unit cut open to expose the Y-shaped 0.007 inch-thick beryllium copper (BeCu) switch plate. This plate was driven by a solenoid, which moved the three clear plastic plungers in the lower section, when it was energized.

   The photos show, from a side view, the plate in the rest position (Fig. 4), and the failed unit with the coil energized (Fig. 5). Figure 6 shows the operation of a good unit for comparison. The plate has flexed under the pressure of the switch return springs and the switch contacts have changed position. The failure was due to a fatigue fracture in the small arm of the plate which rides on the locating pin. This allowed the plate to rotate out of position. The internal action of the switch was evaluated in real time, with no destructive analysis. The benefit of this became apparent when customer-owned material was returned from the field, and destructive analysis was not authorized. Subsequent failures were quickly identified.

2. In this example, an outside vendor was performing a hand-soldering operation of connectors and plated-through holes on a motherboard with inconsistent and unacceptable results. Because of the high thermal conductivity of the internal copper plane layers, an adequate temperature was not achieved for the solder to fill the barrels. The unsupported barrels would crack and become intermittent during environmental stress screening. Conventional radiographic inspection would have been difficult and time consuming, even if a suitable large chamber was available. Using five-axis manipulation, the optimum viewing angle was quickly established.

   Figure 7 is an overall view of the motherboard, with a 6 inch scale for reference. It was a 20 layer, 160 mil-thick polyimide PWB with 2 oz. (2.8 mil copper) plane layers. Figure 8 reveals a partially filled plated-through hole near one of the round connectors. Figure 9 is a close-up, and Fig. 10 is the same image processed to enhance the defect. The photograph also shows the annular ring of each layer, which gives an indication of layer registration. With a videotape of the x-ray results, the vendor initiated training sessions for the operators to eliminate the problem.

3. This example illustrates the importance of proper viewing angle to highlight failures with x-ray. Figure 11 is a conventional x-ray image using wet film (negative) processing. Many of these RF filter assemblies were failing in system burn-in. Conventional x-ray could not detect any anomalies; but,

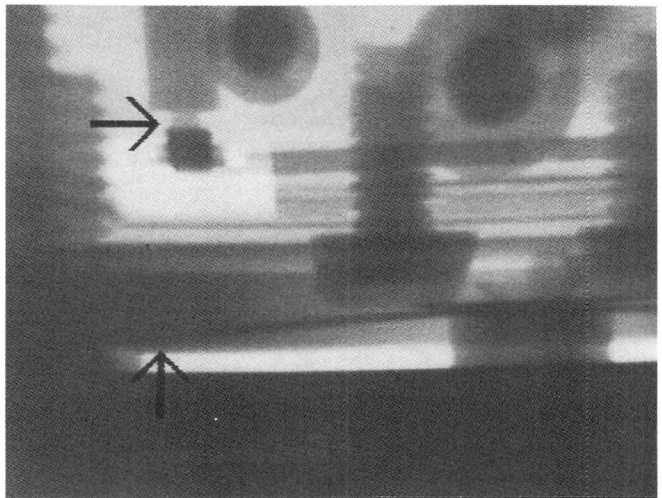

Fig. 6. Good unit, energized.

Fig. 7. Motherboard, 20 layer, polyimide.

Fig. 8. X-ray image of filled, plated-through holes.

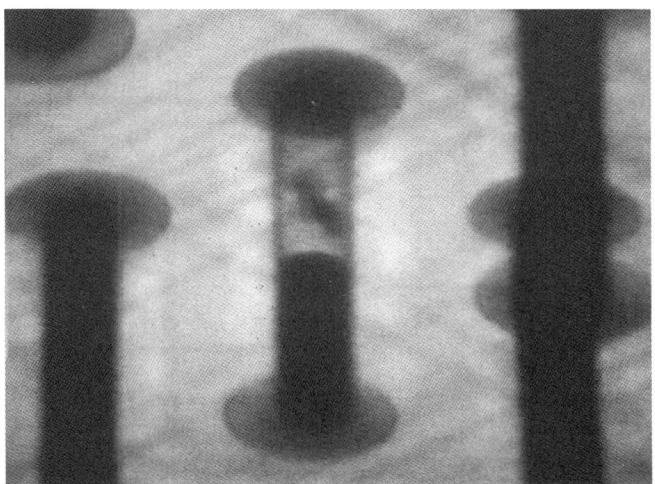

Fig. 9. Higher magnification of unacceptable solder fill.

when rotated while examining with real time x-ray (Fig. 12), the crack became evident when properly aligned with the x-ray source.

Figure 13 shows the crack from a different angle: looking down through the barrel of the SMA connector. Figure 14 is a scanning electron microscope image of the crack taken during the course of destructive analysis, to confirm the x-ray results. This problem was quickly resolved because the argument that the crack was caused by stresses induced during destructive analysis was no longer valid.

4. The all-ceramic hybrid in Fig. 15 failed electrically and also failed subsequent particle impact noise detection (PIND) testing. The construction of the package made lid removal nearly impossible without introducing contamination internally. X-ray examination (Fig. 16) quickly revealed a conductive epoxy moisture "getter" that had separated from the substrate and shorted the internal circuitry. In the processed image (Fig. 17), the getter and 1 mil gold wire bonds, a few of which are fused, are clearly visible. Some minor voiding was also detectable in the eutectic under the large I.C.

5. Figure 18 shows a power hybrid which was evaluated for die attach integrity. Each of the four cells contained a power Schottky rectifier die that required a low thermal impedance to the 75 mil copper substrate. Using x-ray analysis, the process could be modified to reduce voiding to an acceptable level. Figures 19, 20, and 21 demonstrate some image processor functions that can assist in the inspection.

In Fig. 19, the intensity profile across the bottom of the screen represents the relative intensity of the area traversed by the horizontal cursor. Figure 20 shows a 7.09 mil diameter void which was measured using the cursors. Figure 21 shows the image processed to enhance the boundaries of the voided areas. The die size is 180 × 180 mils. This quick turn-around evaluation can reduce the time required for process optimization.

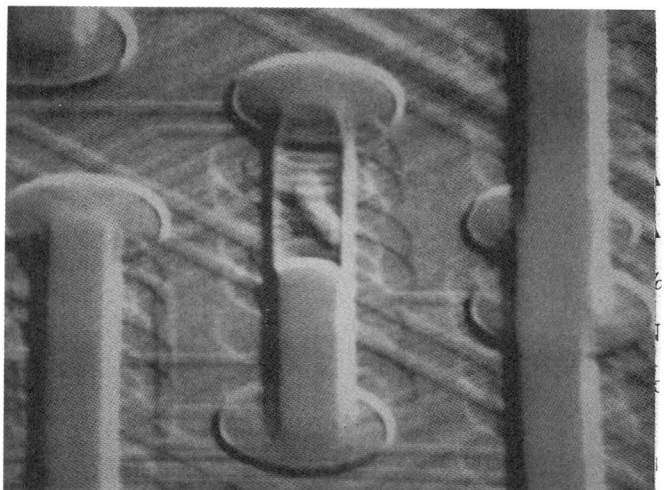

Fig. 10. Processed image revealing layer registration.

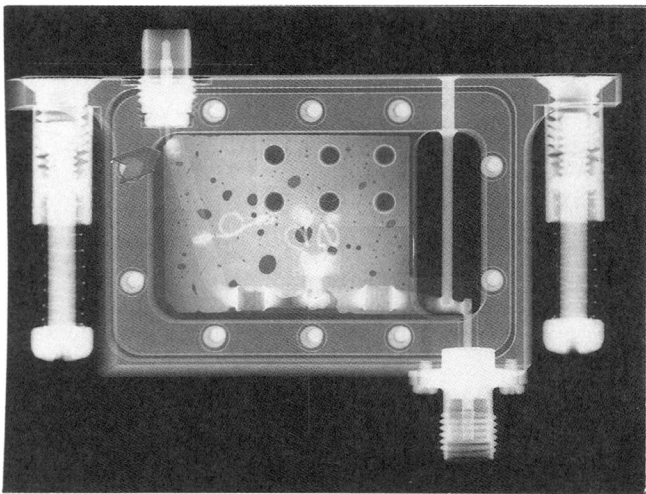

Fig. 11. "Wet-film" x-ray shadowgraph of filter module. (A) denotes problem solder joint.

Fig. 12. With manipulation in real time, crack becomes apparent.

Fig. 13. View of same joint by looking through SMA connector.

6. One critical (and mystifying) failure was resolved with a confidence level that would have been impossible with any other analysis technique, including standard x-ray. An encapsulated connector on a cable assembly failed during system burn-in (Fig. 22). The failure mechanism was the lack of solder where the wire terminated in the solder cup. Other connectors examined showed varying degrees of the same anomaly. The problem always occurred on an odd-numbered pin (i.e., on the same side as the marking ink). During assembly, each solder joint was visually inspected for a fillet prior to encapsulation, so it appeared that the solder may have reflowed either during or after the encapsulation process. The hypothesis was that the curing process for the marking ink on the outside of the connector involved a heat gun, and this uncontrolled process may have reflowed the solder. This did not seem likely because there was no evidence of heat damage on the connector body.

Using the real-time x-ray system, a test was set up to verify the hypothesis. A heat gun was placed 3 inches from the connector while it was examined. At approximately 90 seconds, the solder became molten and was completely wicked out of the odd-numbered solder cups by the silver-plated stranded wire (Fig. 23). The results were recorded on videotape and presented to the project. Other physical evidence was collected to support the x-ray data. Because of the convincing evidence, the problem was quickly resolved and corrective action was implemented.

7. Figure 24 demonstrates the usable magnification levels attainable in real time. This is a temperature-sensing device with a 0.4 mil Nichrome® wire sensing element. Decapsulation to expose this extremely fine wire was practically impossible. X-ray analysis revealed damage to some of the windings in the coil. This most likely occurred prior to the encapsulation process, while the exposed winding was susceptible to damage.

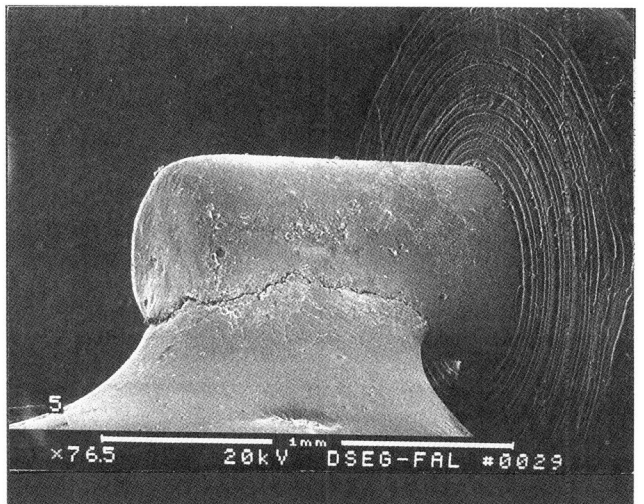

Fig. 14. SEM image of the crack confirmed x-ray results.

Fig. 15. All-ceramic hybrid assembly.

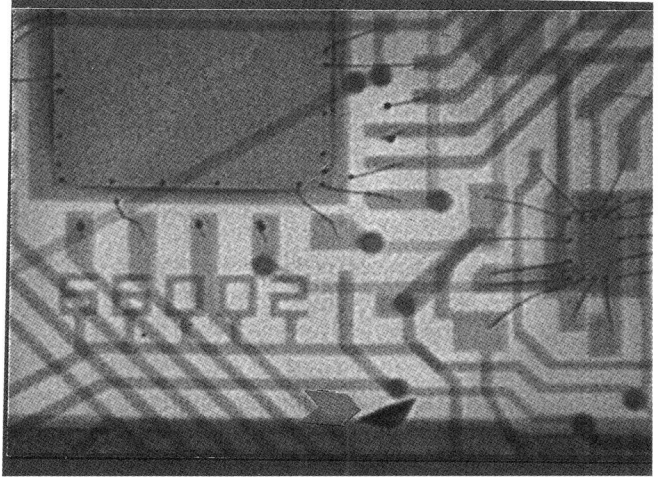

Fig. 16. X-ray image revealing particle, a conductive epoxy "getter."

Fig. 17. Processed image, enhancing detail. Fused 1-mil bond wires are visible.

COMPONENT DEGRADATION

Using the x-ray analysis technique to examine solder joints on populated PWBs has been extremely successful. Voids in surface mount solder and unfilled plated-through holes are clearly visible. Because the inspection time is decreased to minutes, much less than conventional techniques, inspection of high risk production board samples is feasible. This system has also successfully been used to determine defects flagged with other inspection systems actually did not exist.

When examining active components, it becomes important to limit the total exposure so that the absorbed dose of radiation does not exceed amounts which might cause damage. Table II lists the maximum allowable dose ranges for various device technologies.

With most real-time x-ray systems, many factors determine the dose rate: accelerating voltage and current are continuously

Fig. 18. Quad power rectifier hybrid.

adjustable, distance to the source (inverse square relationship), duration of exposure, and the amount of inherent filtration in the packaging, if any. Preliminary tests were conducted to determine the actual dose rates under various conditions so that

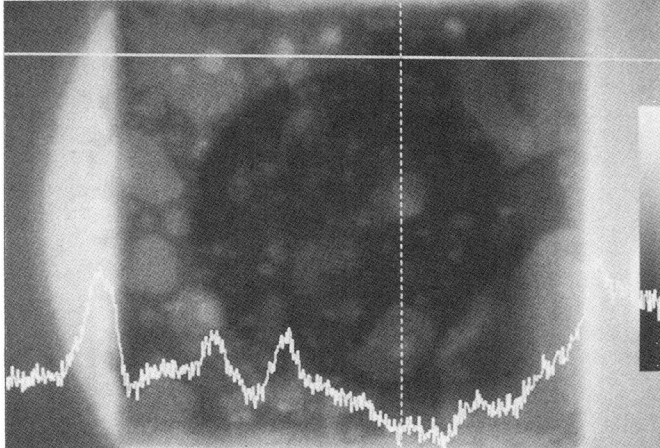

Fig. 19. Intensity profile of voiding along horizontal cursor.

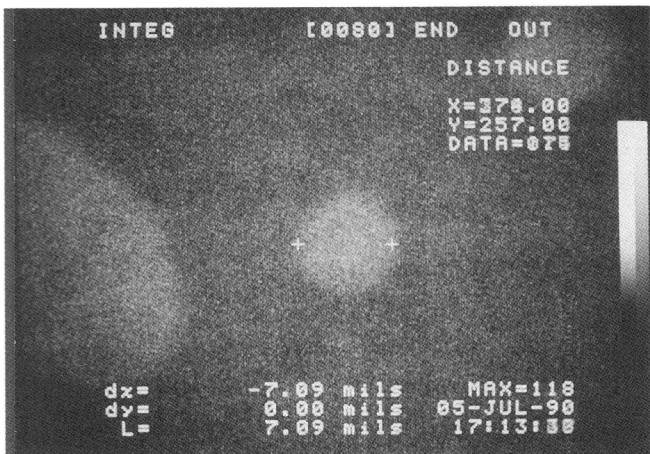

Fig. 20. Measurement of 7.09 mil diameter solder void.

Fig. 21. Processed image, enhancing detail. Die size is 180 x 180 mils.

Fig. 22. Connector with reference designator on the encapsulation material.

"live" boards can be examined with some confidence that reliability will not be affected.

The effectiveness of an x-ray filter material depends primarily upon thickness, density, atomic structure and accelerating voltage. The x-rays are produced by a stream of electrons striking a tungsten target. The energy is emitted in the form of a band of frequencies in the x-ray spectrum. Higher accelerating volt-

Table 2 Radiation Damage Threshold for Common Device Technologies

Generic Technology	Damage Threshold, RAD (Si)
ECL, TTL, GaAs	200 k to 10 meg
Linear, I^{2}L	15 k to 1 meg
MNOS, PMOS	8 k to 100 k
CMOS, VMOS, NMOS	0.8 k to 10 k
Quartz crystals (natural)(a)	100 to 1 k

(a) Crystals will show part per million frequency shifts, although the effect is temporary and will anneal out.[3]

ages not only increase the radiation intensity but also cause a shift in the envelope towards the shorter wavelengths, as shown in Fig. 25. This short wavelength energy provides greater penetration.

The information necessary for image generation is contained in the x-rays of various intensity which pass through the sample. Most of the lower frequency, or longer wavelength, "soft x-rays," are absorbed by the sample and do not contribute to image quality. The purpose of filtering, therefore, is to block these soft x-rays so as to minimize the dose absorbed by the sample. At an accelerating voltage of 100 kV, a .030 inch-thick layer of aluminum absorbs approximately half of the x-ray energy. Hence, it is called the "first half value layer"(4). An equivalent absorber at this voltage is approximately 1.7 mils of copper. This is convenient when estimating the filtering characteristics of copper plane layers in a PWB.

Table 3 illustrates equivalence factors of metals at various voltages.

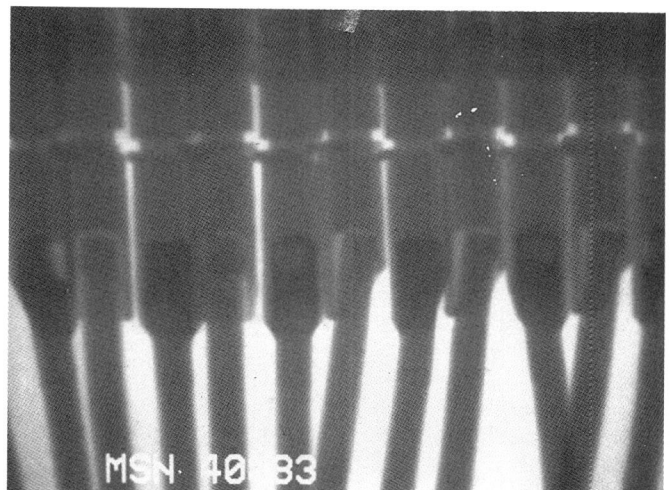

Fig. 23. Alternate cups without solder correspond to the side with the marking ink.

Fig. 24. Processed image of an encapsulated temperature sensing device with damaged 0.4-mil (10.2 μm) diameter Nichrome® wire.

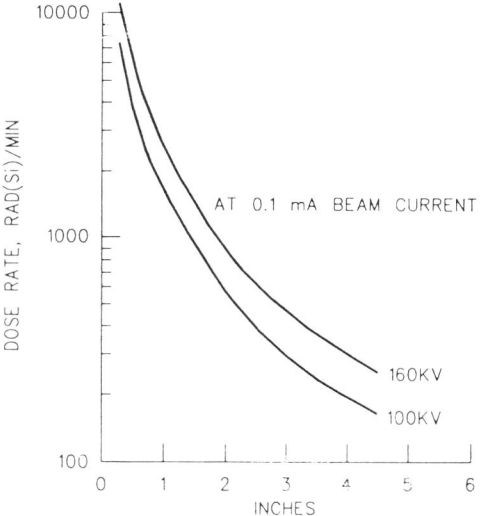

Fig. 25. Typical x-ray envelopes, as a function of accelerating voltage and beam current.

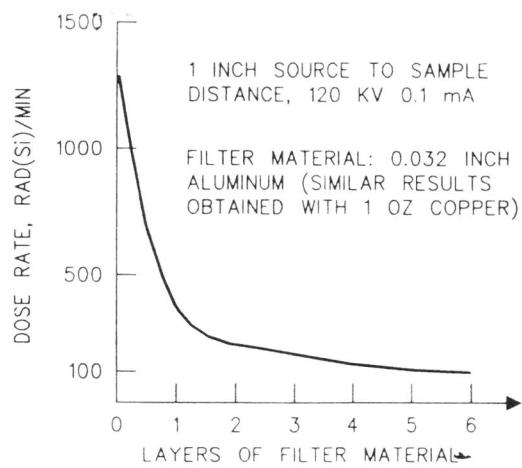

Fig. 26. Alternate cups without solder correspond to the side with the marking ink.

Table 3 Approximate equivalence factors of metals

Metal	Density, gm/cm³	Atomic Number	50 kV	100 kV	150 kV	220 kV
Magnesium	1.74	12	0.6	0.6	0.05	0.08
Aluminum	2.70	13	1.0	1.0	0.12	0.18
2024 Al alloy	(2.99)b	(14)	1.4	1.2	0.13	0.14
Steel	(7.81)	(26)		12.0	1.0	1.0
18-8 Steel	(7.82)	(26)		12.0	1.0	1.0
Copper	8.96	29		18.0	1.6	1.4
Zinc	7.14	30			1.4	1.3
Brass(a)	(8.32)	(29)			1.4	1.3
Lead	11.4	82			14.0	12.0

Note: Aluminum is taken as the standard at 50 kV and 100 kV, and steel at the higher voltages.[5] **(a)** Tin or lead alloyed in the brass will increase these factors. **(b)** Numbers in parenthesis are estimates based on representative alloys.

It is apparent from these equivalence values that the x-ray attenuation properties of the materials aren't directly proportional to either density or atomic number. The relative effectiveness of these metals as x-ray filters will also vary as a function of accelerating voltage.

The testing performed on the Feinfocus FXS-160 involved measuring the total dose at different voltages, source to object distances, and with various layers of filtering. Thermoluminescent dosimeters (TLDs) were used to measure the ionizing total dose. They were exposed at various dose rates for a fixed amount of time. Since the effects of ionizing radiation are a function of the absorbing material, the dose rate must be converted to units of RADs (radiation absorbed dose) silicon for correlation to the limits in Table 2. A RAD is defined as "the absorbed dose of any ionizing radiation which is accompanied by the liberation of 0.01 joule of energy per kilogram of absorbing material"(4).

The data from some of these tests was plotted and is shown in Fig. 26. With this information, and by compensating for the filtration inherent in the packaging materials (or adding more), the dose rate in RADs silicon can be estimated. With the rate information and total dose ranges in Table 2, inspection time limits can be established which would ensure only insignificant levels of degradation.

SUMMARY

An advanced radiographic capability has been demonstrated for effective nondestructive evaluation of components and assemblies. Empirical data support the premise that examination of active components is feasible without sustaining permanent damage. Although many studies exist pertaining to the characteristics of x-rays and the effects of radiation on electronic components, many questions persist. We are currently attempting to quantify the true impact on components in the unique environment that exists with the Feinfocus system.

The data contained in this report are preliminary and the values presented should not be construed as absolute. Future emphasis will focus on the questions of: (1) the accuracy of the various measuring devices at the energy levels of interest, (2) the absorption characteristics of silicon as a function of accelerating voltage, and (3) assessing the true impact on semiconductors. This question arises because the data reflect absorption in bulk silicon. In most cases the damage occurs in the oxide layers, which represent a small fraction of the mass of the die, and may also possess different absorption characteristics than doped silicon.

ACKNOWLEDGEMENTS

The author would like to thank Frank Poblenz of the DSEG Radiation Effects Laboratory for his technical assistance with the dose rate measurements.

REFERENCES

1. *Fundamentals of Microfocus Radiography,* Application Note, Feinfocus USA Inc., Agoura Hills, CA, pp. 5-7.

2. MIL-HDBK-728/5, Radiographic Testing, 1985, p. 45.

3. A similar chart is available from IRT Corporation, San Diego, CA.

4. *Ionizing Radiation and Its Effects Upon Electronic Components,* Application Note, Nicolet Test Instruments Division, Madison, WI.

5. *Radiography in Modern Industry,* 2nd Edition, Eastman Kodak Company, 1957, p. 18.

Specimen Preparation

MECHANICAL AND CHEMICAL DECAPSULATION

by Thomas W. Lee

WARNING! The procedures outlined in this paper involve the use of materials which are among the most powerful and poisonous known. Persons employing any techniques discussed in this paper should do so only after they have been fully trained in these and related procedures, employ and use the complete range of FULL protective clothing, face shields, booties, etc., and only after their involvement in these procedures has been approved by the relevant Safety Personnel in their organizations. The procedures described in this paper represent the author's personal understanding of each of the subject matters discussed, but neither the author nor Motorola, Inc. makes any representation with respect to the safety or appropriateness of any of the techniques described.

INTRODUCTION

To most analysts who have a fair amount of experience with semiconductor devices, mechanical and chemical decapsulation is a mundane and familiar task, often regarded as simple and repetitious. However, experience quickly shows that subtle variations in general procedures can result in widely varying quality. If proven methods and logical operations are not followed, valuable specimens can be ruined. Although almost everyone associated with failure analysis has some general knowledge about chemical decapsulation, procedural guidelines are often sketchy. Procedures used may be the result of opinion, limited experimentation, or narrow focusing on a few experiences. Much has been written on decapsulation, and increasing product sophistication will require continuing development.

For example, the general procedures of mechanical decapsulation are familiar, but fast methods of material removal, selection of proper tools, and unencapsulated sectioning methods may not be known. In chemical decapsulation, it is important to understand the deterioration of the acid with time, the importance of correct temperature, and the quenching procedures which do not result in metallization attack. The chemistry involved is seldom a part of engineering undergraduate curriculums, such as general inorganic chemistry. For example, discussion of sulfuric acid is common; discussion of oleum is not.

Device designers obtain improved thermal, electrical, and reliability performance through the use of new and tenacious encapsulants. These present special decapsulation problems. The availability of the acids, at any cost, is a growing concern. Finally, the disposal of these toxic or carcinogenic decapsulants can present procedural and legal problems.

The quality of chemical decapsulation can make or break an analysis!

Once a specimen is damaged by improper decapsulation, the most elaborate analytical machines and procedures cannot restore its value to the analysis.

MECHANICAL PREPARATION

The most important consideration in mechanical preparation is that the material removal reduces the time that the internal structures are exposed to acids and other solvents. Using an X-ray shadowgraph as a guide, selective removal of the material can be performed such that the area of interest can be quickly exposed when acid is applied. This avoids overexposure of the device internal structures to the acid. The mechanical removal of as much material as possible is important because it has a major effect on total specimen preparation time, since mechanical removal is quick, and chemical removal is slow.

Plastic encapsulants are "filled" with fiberglass, quartz sand or crushed fused quartz. Filling adjusts the properties of the plastic. Quartz is used because of its low coefficient of expansion and high thermal conductivity. Figure 1 shows an example of the quartz grains, and Fig. 2 shows fiberglass rods in typical TO-92 devices which have been partially decapsulated. Quartz is much harder than the fiberglass, and generally harder than the steel alloy in metal cutting tools. Examples of the comparative hardnesses of electronic materials are given in Fig. 18. A hard-

Fig. 1. Although some decapping tasks will be straightforward, most jobs involve the removal of tenacious, filled plastic encapsulants.

Fig. 2. SEM view of fiberglass rods used for filler, particularly in devices encapsulated with silicone. (100×)

Fig. 3. Optical photomicrograph of appearance of top of wire loop in a TO-220 device being ground. (30×)

ened steel drill bit can be dulled on a single specimen. As a result, carbide bits should be used for milling on packages with fiberglass filler, and diamond bits should be used for those filled with quartz.

Both carbide and diamond bits are available for hand-held motor-tools, and their relatively high cost ($10 or more each) is offset by their lifetime in service. See the article on *Particle Verification* (Gene Thome) for a photograph of a typical hand-held motor-tool. Quartz is also nearly as hard as the silicon carbide (SiC) in metallurgical sanding discs, resulting in slow progress with sandpaper disc material removal. Diamond grinding wheels should be used for surface grinding. Their cost is quickly offset by their speed and convenience. The relatively high cost of metallurgical grade diamond wheels can be offset by using standard 8 in. wheels from a lapidary supply, or "rock shop."

Using these grinding wheels, the plastic should be removed from IC or transistor packages until the top of the highest wire loop is barely intersected by the wheel. This will leave an ellipse of aluminum exposed on the ground surface. Chemical treatment should be initiated at this point. Figure 3 shows a typical example of an intersected wire.

Sometimes, the plastic can be very difficult to dissolve, especially in products where the composition of the plastic is not known. Because of the longer exposure to the acid, the aluminum may be dissolved from the pad windows, and all pins may be open. In this case, the plastic can be ground even further than described above, until the wire bonds are intersected at the die. At this point, the plastic will be approximately 3 to 5 mils thick, and you may be surprised to find that you can see the die surface *through* this thin layer of normally black plastic. In ICs, this will leave a square or rectangular pattern of sliced wire bonds in the ground surface of the plastic package, outlining the shape of the die. Then, the time of application of the acid can be very short, and there will be little damage to the metallization on the die.

Fig. 4. SEM view of gold ball bonds sliced close to the device surface during plastic grinding. (50×)

Figure 4 shows the surface of a die prepared in this manner. The sacrifice that is made, of course, is that the die must be probed for post-decapsulation electrical data, since the bond wires would then be open.

Another use for the diamond grinding wheel is the removal of the caps from cerdip packages. The ceramic cap and base of the package is harder than the SiC in sanding discs, or the tungsten carbide (WC) in carbide tool bits. The ceramic package halves are cemented together with a type of lead glass (the exact

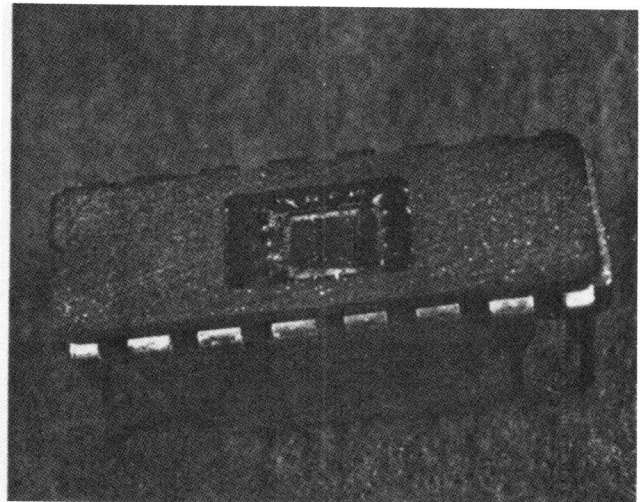

Fig. 5. Optical micrograph of cerdip package opened by grinding away top ceramic cap with diamond wheel. (3×)

composition is usually proprietary) with a low melting point and relatively low strength. Most opening methods use a pliers or blade arrangement to apply lateral pressure to the glass seal and "pop" the lid away. This method is fast, but will sometimes cause a pin to be broken out of the glass, and be airborne on the wire. As a result, the delidded device cannot be retested in a socket.

Alternately, the diamond grinding wheel can be used to thin the hard ceramic lid until the depression in the backside of the lid over the die is intersected, and the package cavity is breached. Then, the flap of ceramic remaining is tweezered away, leaving a clean cavity opening. Since both the "pop" and the grind methods generate particles, the samples must be ultrasonically cleaned following either procedure.

Figure 5 shows a typical example of a cerdip IC opened by this method. The diamond wheel can also be used to gain entrance to glass hermetic packages, such as the diode shown in Fig. 6, by carefully grinding the glass away.

The metallurgical "slow saw," using a diamond-tipped blade, is also a useful tool. This saw can remove materials from large encapsulated devices, such as power modules, very easily and accurately with a minimum of damage. When using this saw, you must be very careful to never contact the die with the blade. The diamond grit is coarse enough to cause fracturing of the die to the extent that it will be ruined for section. Figure 7 shows an example of a slow saw cut into a glass hermetic diode package, resulting in exposure of the cavity and die.

Another useful tool is the micro sand blaster. Using aluminum oxide, silicon carbide, or sodium bicarbonate as abrasives, the plastic can be selectively removed in an area that has already been prepared by coarse grinding. This greatly shortens decapsulation time. It is important to proceed with great care. The air and particle stream can form a vortex if the abrasive contacts the silicon die, severe damage occurs immediately. Figure 9 shows a typical example.

Fig. 6. SEM view of glass diode with cavity exposed by grinding sides with diamond lap. (30×)

Fig. 7. SEM view of glass diode with cavity exposed by slicing with diamond blade slow saw. (70×)

In discrete devices, the slow and careful removal of all encapsulant, with the exception of that around the base of the leads of the package, followed by careful decapsulation by the immersion method, will result in the exposure of the die and internal interconnections while leaving a support for the leads themselves. Figures 10 and 13 show examples. This will make

(a)

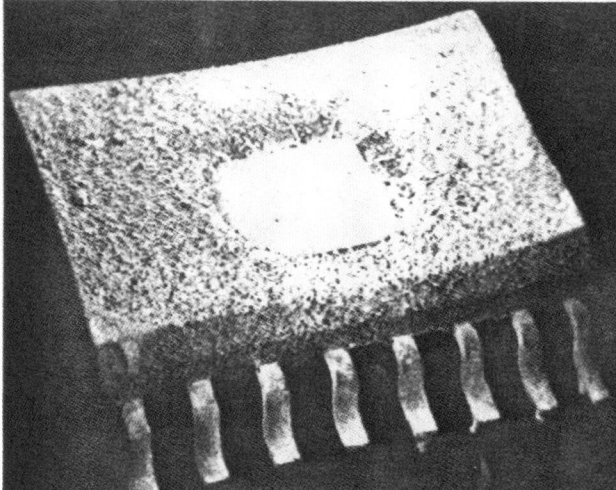

(b)

Fig. 8a and b. Views of phase locked loop frequency synthesizer in 16-pin 501C package before and after chemical decapsulation by the dropped method. (8×). In the above photos, the importance of sandblast preparation of the package is evident. The acid will unavoidably capillary over the package surface, and the top of the package will be dissolved in areas other than those desired. It is advantageous to become expert at the removals of as much material as possible in the area of interest before applying acid.

probing unnecessary, which would be difficult because of the uneven bottom surfaces. Retesting for data such as Vce(sat), or Io in power ICs, may be easily performed without probes. Biased reliability testing, such as HTRB, can be performed on the opened device.

When performing this type of decapsulation, it is important to remember the great affinity of nitric acid for copper. If the leads in the package have been bent close to their egress from the plastic, the plating will crack. The acid can react with the copper through these cracks, and the leads can be eaten off or undercut close to the device body, as shown in Fig. 11. Be sure to tin the leads with solder to minimize this, or the advantage of

Fig. 9. Skill is required with the sandblast method. Here is the effect of contacting the die with the abrasive particle stream, catastrophic damage is immediate (arrow). (20×)

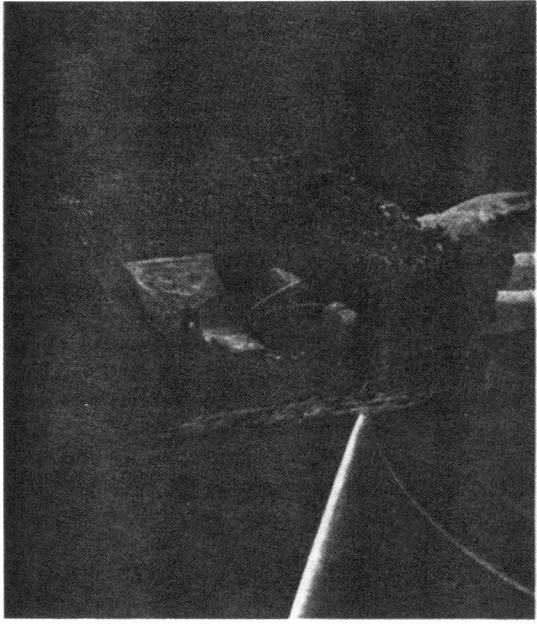

Fig. 10. TO-92 which has been sandblast prepared and opened by the dropper method. (20×)

the method will be lost. Even with tinning, the leads will be undercut. Careful handling, accompanied by the soldering of the part to a supporting substrate, can protect it mechanically so that it can be handled in an ordinary manner without breaking off the now-delicate wires. Figure 12 shows a typical device prepared in this manner.

If a quality decapsulation has been performed, the device will still be operational.

"PLASTICS 101"

Plastics used for encapsulation are *polymers,* or long-chained organic molecules. Examples of polymers occurring in nature are cellulose, protein, and DNA. Plastic polymers are synthesized by two processes; addition and condensation,

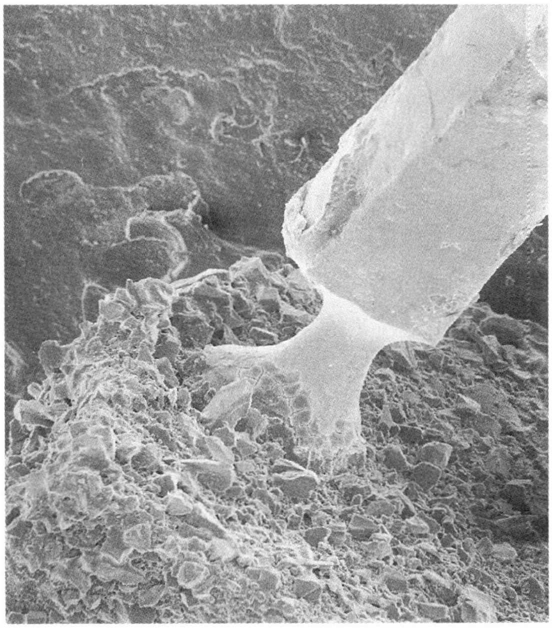

Fig. 11. SEM detail of the leadwires in the device in Fig. 10. Necking down makes them delicate. Notice filler grains. (50×)

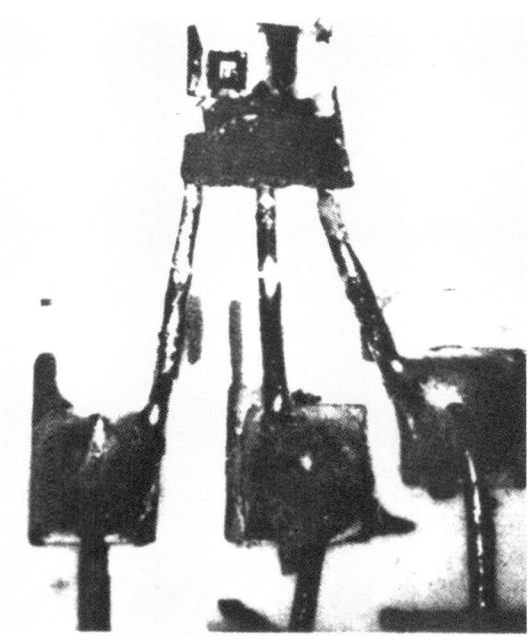

Fig. 12. Here, opened TO-92 has been mounted to a substrate for mechanical support of delicate leads. (8×)

Fig. 13. This plastic thyristor has been selectively decapsulated using sandblast and immersion. (12×)

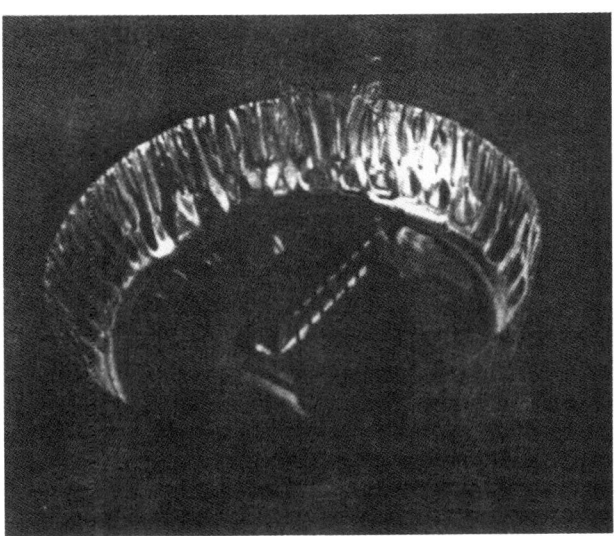

Fig. 14. This plastic integrated circuit is supported on a copper block for heat conduction.

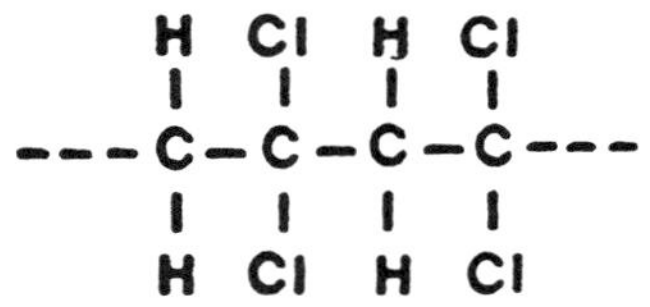

Fig. 15. Diagram of polyvinylidene chloride, a linear polymer, or polyolefin.

where molecular weights into the millions are obtained by sticking many monomers together into long, linear chains.

An example of an addition polymer is polyvinyl chloride, or ordinary PVC pipe, and Saran Wrap®. Figure 15 is a diagram of PVC, and its shape is typical.

Thermoplastics are long-chained polymers which melt upon exposure to high temperatures, and freeze or solidify at lower temperatures. **Thermosetting plastics** cannot be melted; they thermally decompose when heated beyond a certain temperature. They are obtained when crosslinks are established between chains of linear polymer molecules, like the rungs of a ladder. *Benzene rings* are used to form the *crosslinks,* and lengthen and complicate the chains. This forms a dense and strong structure, much like bracing on a bridge.

Thermosetting plastics are mechanically strong and chemically unreactive. The first such material to be manufactured which capitalized on this structure was Bakelite®. Chemist Leo Bakeland was exploring reactions between phenol and formaldehyde when he found that some of the material he had reacted in a flask had formed a black substance that he could not remove with any known solvent. He named his discovery Bakelite®, shown in the diagram in Fig. 16. This mechanism is also called *addition polymerization*.

Bakelite® quickly found its way into the electrical industry, being used for switchboards, transformer insulation, terminal strips, tube sockets, carbon composition resistors, and molded mica capacitors. Bakelite® was "filled" with mica, sand, paper, cloth, asbestos, etc., to adjust its mechanical and electrical properties. Bakelite®, despite its age, is one of the most difficult electronic plastics to dissolve without damage to the component. Another example of a crosslinked polymer is Vulcanized® rubber.

Another crosslinking mechanism available is *condensation polymerization*. Some monomers have protruding small structures such as H^+ or OH^- which, in the presence of a catalyst, can attach to protruding molecules in other parts of the monomer to form strong and stable molecules, such as H_2O, CH_2, or NH_2. The resulting materials are polyethers, polyesters, polyamides, etc., and typical tradenames are Mylar® and Nylon®.

Condensation polymerization is also used to form the epoxy plastics. Epoxies are the reaction product of an epoxide such as epichlorohydrin, with biphenol. The presence of a catalyst and heat causes these materials to form long chains and crosslinks. Even more complex and stronger plastics can be formed by reacting novolac and epoxy resins together. This structure contains chains of phenol molecules braced with epoxy structures, in the form of branches and crosslinks. These are the *epoxidized novolacs,* or *novolac epoxies*.

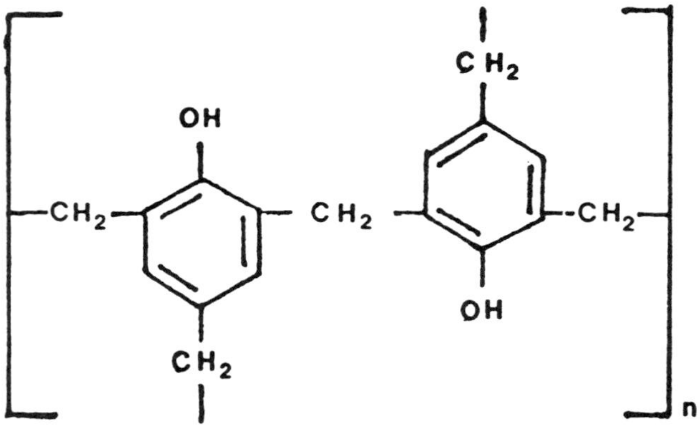

Fig. 16. Structure of Bakelite®, showing chain of interlocking benzene rings.

Curing of the plastic is even more complete when the water resulting from the condensation process above, is reacted with hexamethylene tetramine, yielding formaldehyde and ammonia. The formaldehyde restarts the phenol-formaldehyde reaction, consuming any unreacted resin. The end product is then ammonia gas, which is easily removed. The resulting plastic is an *anhydride novolac,* or *anhydride epoxy*.

FUMING NITRIC ACID

Acids

The acids used to dissolve plastic encapsulants are fuming acids, which are very powerful oxidizers. These acids are dangerous to handle. Basic knowledge of their properties is essential to avoid injury. Their use has been found to involve some basic ground rules that can usually guarantee a good decapsulation result.

Fuming nitric acid is a specific solvent for the *phenolic epoxies*. Nitric acid is produced by the *Ostwald Process,* where ammonia, synthesized by the *Born-Haber Process,* is oxidized to form nitric oxide, NO, then nitrogen dioxide, NO_2. The NO_2 reacts directly with water to form nitric acid, HNO_3. Fuming nitric acid results when an excess of NO_2 is dissolved in the HNO_3 under pressure.

The physical characteristics of fuming nitric acid are unmistakable. It is red in color, and the high vapor pressure causes red fumes under low pressure to collect in the bottle above the acid. It is important to note that concentrated nitric acid (90%) is white, or slightly yellow, in color. Yellow nitric acid will also fume when unstoppered, but the fumes will be white. It will make red fumes when reacting vigorously with metals. This acid may be called "fuming nitric acid" in chemical supply catalogs, and confused with red fuming nitric acid. It may not be suitable for decapsulation purposes, dissolving the metallization on the die.

Red fuming nitric acid is unstable at both room and elevated temperatures. The acid has little effect on plastic at room ambient, but elevating the temperature to approximately 100 °C will cause it to decapsulate the device in a few minutes. Higher temperatures only decompose the acid. It is important to note that

Fig. 17. SEM view of edge of silicon die attacked by oleum because of excessive time or temperature. (100×)

the acid is ready for use immediately upon boiling. Any further boiling will evaporate the dissolved NO_2, causing it to revert to ordinary concentrated nitric acid, dissolving the metallization. Any copper present will cause the acid to become a clear green color, giving an indication that the acid is spent, and must be renewed.

Red fuming nitric acid is becoming increasingly expensive and difficult to procure. One commonly stated reason is its hazardous nature and its effect on the carrier's insurance risk and DOT regulations. It was, until very recently, used as the oxidizer in liquid-fueled rockets. The Titan missile was an example. Kerosene or dimethylhydrazine was used as the fuel. The rocket motor was simple, because the acid and fuel were *hypergolic;* they ignite on contact. The danger of handling the acid, and the presence of nitric oxides in the rocket exhaust gases caused the abandonment of these liquid propellants. The cryogenic fuels liquid oxygen and hydrogen were substituted. If you visit a missile museum you can witness the care with which the armed services handles fuming nitric acid. (See references.) The abandonment of use for rocketry has almost certainly affected the price and availability of the acid. It is available from only a few distributors, listed in the resources chapter of this book.

The hypergolic properties of the acid and fuel underlines the necessity for separation of acids and solvents when designing a lab. Mixing more than about 5 drops of fuming nitric and any alcohol will result in an explosion with amazing violence. This property is in direct conflict with the neutralization methods outlined below, but the reaction of the two is highly advantageous over the use of water. It is critically important for safety reasons to keep the volume of reactants low. Please refer again to the "Warning" section at the beginning of this article.

The mechanisms by which acids dissolve these plastics are not well understood, but oxidization certainly is the dominant mechanism. Oxygen plasmas are known to be very effective on these plastics, verifying the hypothesis.

diamond = 7000

silicon = 2480

alumina = 2100

tungsten carbide = 1880

beryllium oxide = 1250!!!

quartz = 820

glass = 530

steel = 500 to 800

!!! = NEVER grind beryllium oxide (BeO) ceramics unless there is a filtered exhaust, or it is held beneath water. BeO dust is deadly, causing *Pneumoconiosis* or *Berylliosis* (TLV = 0.002 mg/m^3).

Fig. 18. Comparative hardnesses of common electronic materials on the Knoop Indentor scale. Tool steel is relatively soft.

THE DROPPER METHOD

Devices can be opened by the immersion method, the dropper methods, or the machine method. With ICs, the dropper or machine methods are preferable because the device leads are not exposed to the acid. It is very important to keep the specimen hot, and exposure to reacted or spent acid very short. The aluminum weighing dish shown in Fig. 14 has an important difference from the usual decapsulation method employed. Notice that there is a block of copper beneath the specimen. This provides heat directly to the bottom of the device, instead of depending on radiation, convection, or conduction up the leads, and minimizes exposure of the specimen to cold diluted acid. Acid is then applied to the device with a disposable dropper. Quench the device by removing it from the block and rinsing in a spray of isopropanol. Do not fill the dish with a spray of acetone. It will make a mess of insoluble goo and dissolve the metallization on the device.

THE IMMERSION METHOD

If the plastic material is partially removed, and the device is properly suspended in a beaker of acid, the plastic can be selectively removed. This immersion method works well with both nitric and sulfuric acids. Figure 13 shows an example of a device opened by this method.

CLEANUP

Figure 22 illustrates one effective method of stopping the action of nitric acid and cleaning the part. Notice that the device is dipped in acid at room temperature, for a few seconds maximum. This tends to remove any remaining sludge. This sludge is usually a greenish mass that is unremovable with any known solvent except cold fuming nitric. Following the acid dip, the device is rinsed with isopropanol.

Note that this procedure involves the mixing of *hypergolic* reactants. Again, refer to the warning statement. However, it is

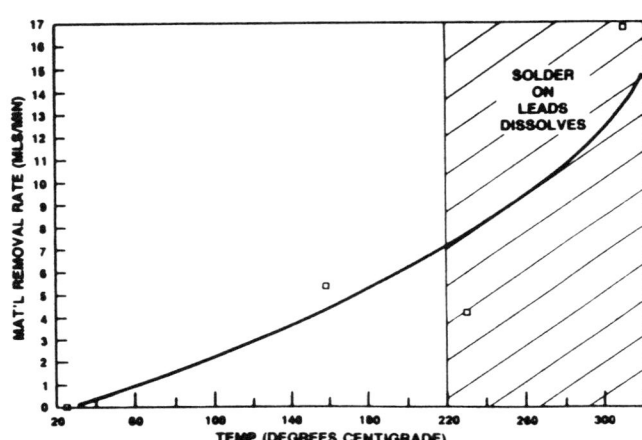

Fig. 19. Graph showing the rate of material removal using oleum at various temperatures. Note that at higher temperatures, faster decapsulation is accompanied by danger of attacking silicon die.

well known that neutralization of the acid with water, regardless of the care exercised, will almost always result in dissolving of the metal, especially in the bond pads of ICs. Using isopropanol avoids this problem. The final clean may be performed in many solvents, and the best choice is often a subject of debate among analysts. The alcohols are *hygroscopic;* they "getter," or absorb water and evaporate, leaving a dry specimen.

FUMING SULFURIC ACID

Sulfuric acid is a solvent specific to anhydride epoxies. Sulfuric acid, H_2SO_4, is the most important chemical in the industrialized world. It is used in such vital applications as the coagulant bath in the production of synthetic fibers, pickling of steel, ammonium sulfate fertilizer production, digestion of titanium ore for white paint pigment, and (as oleum) in the manufacture of detergents. It has greasy texture, explaining its archaic name, "oil of vitriol."

It is manufactured by the contact or older chamber process. Finely divided sulfur, "flowers of sulfur," is burned in air to produce sulfur dioxide. SO_2 is further oxidized to form sulfur trioxide, SO_3, a dense white gas. Dissolving SO_3 in water will make sulfuric acid, but it is not a very efficient process. The SO_3 is instead dissolved in concentrated (98.3%) H_2SO_4 resulting from an earlier step in the process, as shown in Fig. 20. This feedback loop enriches the concentration of the acid with SO_3 until there is no longer any water in solution. The resulting product is fuming sulfuric acid (oleum), or SO_3. SO_3 is an odd material, easily frozen to a white solid which has three allotropic forms α, β, and γ. The beta form decomposes explosively. Unlike SO_2, it cannot be compressed, because it also solidifies with pressure.

The oleum or SO_3 is shipped to major customers in 6,000 to 12,000 gallon insulated railway tank cars or tank trucks. The reason for the insulation is that SO_3 freezes at 114 °F, forming what is known as *unmeltable polymer*. In the wintertime, the

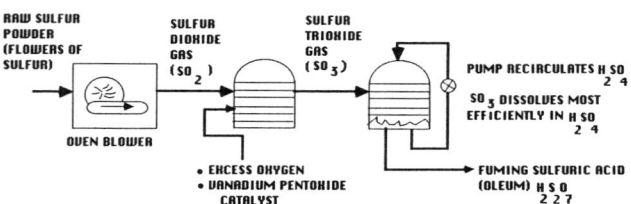

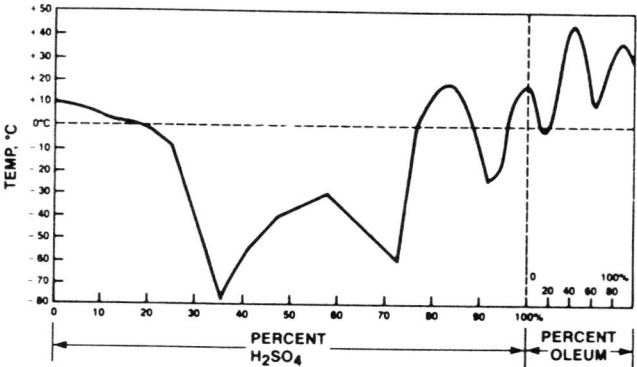

Fig. 20. Diagram of the process for manufacturing sulfuric acid. Note the feedback loop indicated by the arrow. This results in the enrichment of the acid concentration, and eventually yields fuming sulfuric, or OLEUM.

Fig. 21. Graph showing the complexity of the physical behavior of sulfuric acid, sulfur trioxide, and oleum. Although only the melting point is shown, other physical properties are equally complex.

tank car contents may freeze, and the contents are thawed with steam or electric heaters before unloading through hoses. Oleum is stored in heated tanks until used as is, or diluted to H_2SO_4. Shipping oleum avoids the uneconomical transportation of water.

Sulfur trioxide is the decapsulant in oleum. You can prove this by freezing oleum, which, if done carefully, will cause crystals of SO_3 to drop out of solution. If you remove one of these crystals with tongs, you will find that melting it onto a device will dissolve the plastic. Handling solid SO_3 (Sulfan®) is dangerous. It is a difficult material to buy in small quantities because of bottling, packaging, and transportation problems. See the appendix for sources.

Ordinary concentrated sulfuric acid will also decapsulate epoxy-encased devices, but the metals in the device will be blackened and attacked. The acid is heated to drive away the moisture because sulfuric absorbs water from the atmosphere. Sulfuric is so strongly hygroscopic that a beaker, filled to the rim and left open to the air, will overflow in a day or two. However, this process has a limit. Sulfuric will only absorb water until its vapor pressure equals that of the ambient air.

You may have been advised to "boil all the water out of the sulfuric" before use. You cannot boil ALL the water out of sul-

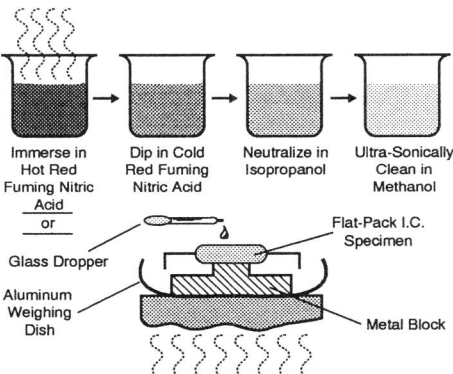

Fig. 22. Diagram showing the recommended method for decapsulating epoxy packages with fuming nitric acid and neutralizing procedure.

furic acid, regardless of how long you leave it on the hot plate. Concentrated H_2SO_4 will always contain some water. The maximum concentration that can be achieved by boiling is 98.3%, a number so predictable that it has been employed as a laboratory standard. The reason for this behavior is that water and sulfuric acid form an *azeotrope* with a depressed boiling point at 98.3%. An *azeotrope* is the liquid analog of a *eutectic*. In oleum, applying extra SO_3 ties up this 1.7% excess water. Alternately, boiling of oleum by leaving it on the hot plate too long will ruin it by driving off the excess SO_3, and it will revert to ordinary 98.3% concentrated sulfuric acid. The graph in Fig. 21 shows how complex the composition and properties of sulfuric acid and oleum are.

Oleum at room temperature has practically no effect on epoxy, and it must be heated to about 150 °C to begin to react. Heating it to an excessive temperature will dissolve the solder used to prepare the leadframe for decapsulation, and attack the silicon, particularly where heavily doped, as shown in Fig. 17. There is a range of temperatures at which oleum will decapsulate a device in a sensible amount of time without causing this undesirable damage, and this is represented by the graph of material rate in Fig. 19.

CLEANUP

The chart in Fig. 23 shows the recommended method for cleaning a device decapsulated with oleum. First of all, the quenching is performed in the opposite manner from that with fuming nitric; *deionized water (D.I.) is used first* for quenching. Dilute sulfuric acid reacts with the metals used in devices at a much slower rate than dilute nitric acid. An ultrasonic cleaner should be used. The D.I. should be changed whenever it becomes cloudy or grayish, and the process continued until it becomes clear. Then, the device is immersed in methanol to complete cleaning and leave the specimen completely dry. Quenching first in methanol will result in the formation of the poisonous ester *dimethylene sulfate*, a gas to which you may be exposed while you are inspecting your specimen beneath the microscope. So poisonous is this material that the chemical references sometimes refer to it as "WAR GAS." This is unaccept-

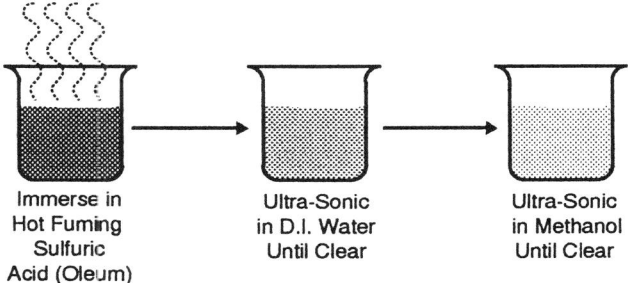

Fig. 23. Diagram showing recommended method for chemically decapsulating anhydride epoxy packages with fuming sulfuric acid, with neutralization procedure.

able. **Always quench and clean specimens decapsulated with fuming sulfuric acid in water first.**

CAUTION! DO NOT CONFUSE dimethyl sulfoxide [$(CH_3)_2So$] (DMSO) with the poisonous dimethyl sulfate (or dimethylene sulfate) [DMS, $(CH_3)_2SO_4$]. Diethyl sulfate (DES) is equally poisonous.

It is often convenient to allow decapsulation to proceed while performing another task, but even in oleum this will blacken the metals if decapsulation has been completed and the device is exposed to the acid for an excessive time. Watch your specimens closely. Also, it is advisable to check the device while decapsulation proceeds, to avoid overexposure to the acid. When removeing the device to check for decapsulation progress, do not quench the acid in D.I. It will blacken metals. Instead, tap the specimen lightly, observe the progress while holding it in the airflow of a fume hood, and re-immerse it without contact with water, if it needs additional exposure. *Notice that this seemingly contradictory advice applies only to checking progress, not to the final neutralization.*

SAFETY

The fumes given off by these acids consist of vapors, mists, and *anhydrous* gases from which the acids are originally made, and they revert to acid upon contact with the moisture in your lungs. This does not necessarily cause immediate reaction or sensation, but may result in delayed serious injury. Exposure to low concentrations over a period of time can result in a *chemical pneumonia*, recovery from which is extremely slow, perhaps months. Fortunately, NO_2 has a sharp and irritating metallic odor in low concentrations. The sensitivity of the human nose, generally in the PPM range for many gases, may detect its presence but cannot, of course, be solely relied upon because sensitivity varies among individuals, and the olfactory organ can fatigue. SO_3 does not seem to have a particular odor, and is thus more treacherous. It will go into your chest, and symptoms may be delayed. **NEVER USE FUMING ACIDS WITHOUT A FUME HOOD** and other safety precautions your Safety Department recommends.

Burns from these acids are painful and slow to heal. Contact with fuming nitric acid is immediately painful, and you will experience an unmistakable stinging even from the smallest spatter. The skin turns green (*necrosis*), and the acid will continue to react to depth. It has an affinity for protein. Neutralize the area with a paste of sodium bicarbonate, or baking soda, or a soak in the solution. If you are burned, get medical attention. Contact your safety department, a hospital poison control center, or paramedics. Post their phone numbers prominently in your laboratory.

BE FAMILIAR WITH YOUR FUME HOOD. Eddy currents caused by obstructions can cause dead air spaces. Unusual winds outdoors can stop airflow in some systems, or cause fumes to be drawn back into the building. Know the location of any switches or circuit breakers that can shut off the vent motors without your knowledge. Have the circuit alarmed in case of a power failure. Check periodically for leaks that can admit fumes back into the workspace in a low concentration; this is the source of many cases of chemical pneumonia.

Hydrofluoric acid, HF, occasionally used to clean glass or quartz fill residue from the specimens, is an insidious hazard. The records of poisoning with this acid are most plentiful in the etched glass industry. HF can sit on your skin for a few minutes unnoticed. **Contact with HF results in no immediate sensation.** Meanwhile, however, it penetrates downward into the body. Symptoms include burning and decomposition of the skin, and a peculiar *prickling sensation*. One of the most com-

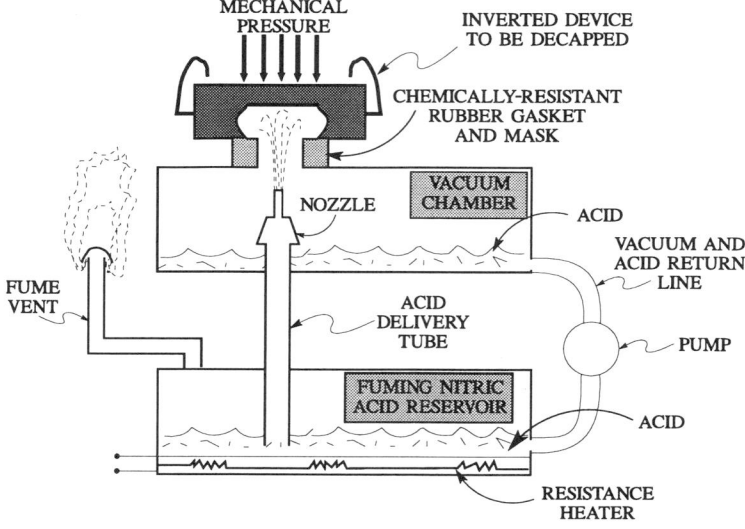

Fig. 24. Block diagram of a typical chemical decapsulation machine.

SOLVENT → Xs = dissolves slowly Xc = cold	Fuming Nitric	Fuming Sulfuric	Con H_2SO_4	Dynasolve 100 ®	Dynasolve 160 ®	Dynasolve 210 ®	Decap ®	Uresolve Plus ®	Acetone	Methanol	Isopropanol	Chlorothene VG	Methylene Chloride	Perchlorethylene	Methylethylketone	Xylene	Toluene	Allied A-20 ®	Freon TTF
Material to be dissolved	1	2	3	4	5	6	7	8	9	10	11	12	13	14	15	16	17	18	19
asphalt fill ("tar")				X								X	X						
coating, film resistor								X											
epoxy, anhydride		X																	
epoxy, Novolac	X																		
epoxy, cast, two-part				X															
Nylon				X														X	
oil (polyolefins)												X	X			X	X	X	X
paint, enamel									Xs			X	X	Xs	Xs				
Plexiglas and acrylics			X			X										Xs	Xs		
polyvinyl chloride (PVC)			X										Xs						
silicone device packages								X											
silicone die coating			Xc															X	
silicone dielectric gel					X			X										X	
silicone dielectric grease								X					X						
silicone rubber																		X	
solder flux (rosin-type)									X		X	X							
wax (paraffin)												X	X			X	X		
wax (sealing)									X	X									

Fig. 25. Matrix showing best solvents for a variety of materials commonly encountered during the performance of failure analysis on electronic components.

mon burn locations with HF is *beneath the fingernails*. The real danger lies in deep penetration, generally associated with larger quantities or longer exposure. When the HF penetrates to the bone, it may perform an ion exchange reaction, replacing the calcium carbonate $CaCO_3$, or aragonite, with calcium fluoride CaF_2. The fluoride is very brittle. This is a highly desirable process in the enamel on the surface of the teeth; this familiar *fluoridation* hardens the tooth surface and prevents decay. But you can readily understand why the *amount* of fluoride that you receive is critical. Additionally, HF FUMES ARE CLASSIFIED AS AMONG THE MOST DANGEROUS ACID FUMES KNOWN.

COMMERCIAL DECAPSULATION MACHINES

Much of the variability in decapsulation is skill-related, and more uniform results can be obtained, especially with unskilled operators, through the use of commercial chemical decapsulation equipment. Some of the safety problems are also eliminated, and the purchase of such equipment may be much more acceptable to your Safety Officers than the use of open chemical decapsulation techniques. These machines are manufactured for use with both sulfuric and fuming nitric acids.

The operating principles of chemical decapsulation machines are relatively straightforward. The machines offer strong advantages over the manual methods for the reasons itemized in the discussions that follow. A block diagram of a typical machine appears in Fig. 24.

The acid is heated in the machines by a thermostatically-controlled hot plate, and in the case of nitric acid, the temperature is held to about 70 °C. The temperature of a beaker of acid on a hot plate is difficult to control because the air draft of the required fume hood causes the beaker temperature to differ from the hot plate surface temperature considerably. The thermal coupling between a beaker and the hot plate surface is also unpredictable, contributing to temperature variations, and poor process control. Decapsulation machines correct this problem.

The cold acid is drawn directly from the bottle by a small-diameter hose. This minimizes the handling of an open bottle of acid, and the risk of an accident associated with the pouring of the acid into a small beaker.

A capillary tube immersed in the hot acid reservoir is used in conjunction with a vacuum on the device to draw the acid stream into the reaction chamber. Should sidewall breakthrough occur, acid will not spray out because the vacuum seal will have been broken.

During manual decapsulation, the acid can flow out over the device surface, removing the plastic in a much larger area than originally desired. Fluoric rubber or similar soft, flexible, chemically resistant material is used to make a set of masks through which the acid is applied to devices of various sizes and shapes. This results in dissolved areas that are relatively uniform from device to device, and steep-walled to allow the leads to remain intact in the decapped device.

Automatic controls are used. Where a large number of devices of the same type must be opened with uniform results, operator skill becomes unimportant.

The machine cabinet offers a double-containment system for the acid, minimizing the risk of accidents.

A plastic mold decapsulation system for fuming nitric acid is manufactured by Nippon Scientific Co., represented by the ICOM Corporation, 734 Silver Spur Road, Suite 201, Rolling Hills Estates, Ca., 90274. A decapsulation system for use with concentrated sulfuric acid, called the "Jet-Etch," is manufactured by B & G Enterprises, 34C Hanger Way, Watsonville, Ca. There are other manufacturers of these systems, and the user should make his choice only after studying literature from all available manufacturers.

PREFERRED ORGANIC SOLVENTS

If you are analyzing devices from a number of manufacturers, various types of components, and potted assemblies, each will be found to respond best to a specific organic solvent. The plastics involved may be such materials as epoxies, urethanes, silicones, polysulfides, etc.

Many solvents and solvent formulations are useful for the removal of plastic materials. Some of these solvents are discussed below, and are discussed briefly to aid your familiarity with them.

You will find that you will eventually assemble a large collection of decapsulants for convenience. Over a period of time, a systematized list can help to aid your memory, as the variety of both plastics and solvents can be large. In Fig. 25, a chart of solvents and encapsulants has been constructed for this purpose.

Blank lines have been left in the chart so that it can be expanded and tailored for your particular operation. Although this paper is primarily concerned with mineral acids, organic solvents and commercially prepared solvent mixtures are important. Most chemical suppliers have technical customer service representatives that are very helpful.

- **acetone**—$(CH_3)_2CO$, C_3H_6O, CH_3COCH_3, dimethyl ketone, was originally one of the materials obtained from the destructive distillation of wood. It is now synthesized. Acetone is a solvent for oil, wax, lacquer, varnish, rubber, and some resins. Nail polish remover is a common use. Acetone has an odor threshold of 20 to 140 ppm, and the TLV is 750 ppm. Acetone is a popular degreaser and cleaner, miscible with water and alcohol.
- **acetic acid**—CH_3COOH, "HAc", ethanoic acid, or glacial acetic acid, is present in vinegar and distilled vinegar in a concentration of about 3%. Acetaldehyde, C_2H_4O, or ethanal (as opposed to ethanol), is the starting material for the manufacture of acetic acid. The odor threshold is 0.2 to 1 ppm, and the TLV is 10 ppm. Splashing of acetic acid in the eyes can cause an unnerving temporary blindness that can persist for days. Acetic acid is miscible with water and alcohol, and is a common buffer in etch formulations. Acetic acid is easily evaporated from these solutions, however, and excessive boiling will remove its buffering action.
- **amyl acetate**—$C_7H_{14}O_2$, isoamyl acetate, banana oil, pear oil, is a solvent which boils at 142 °C. It is soluble

in alcohol and ether, and slightly soluble in water. Amyl acetate is used in applications where the vapors are likely to be inhaled, such as in thinners for wax, paint, and dye, including silver paint for SEM specimens. Amyl acetate is also used for artificial flavoring and perfumes.

- **benzesulfonic acid**—$C_6H_5SO_3H$, is soluble in water, and is a component of stripper formulations.
- **dimethyl hydrazine**—UDMH, $(CH_3)_2NNH_2$, unsymmetrical dimethylhydrazine, uns-dimethylhydrazine, is a rocket fuel, used with oxidizers such as fuming nitric acid. UDMH is used for decapsulating plastic devices.
- **dimethyl sulfoxide**—$(CH_3)_2SO$, or C_2H_6OS, or DMSO, is a byproduct of the production of wood paper pulp manufacture, where cellulose is digested by reaction with sulfuric acid. DMSO is odorless, colorless, and hygroscopic. DMSO may be present in hydraulic fluid, antifreeze, and paint and varnish removers. DMSO is a solvent for Orlon®. DMSO is a well-known analgesic and anti-inflammatory. DMSO is best known, however, for its unique penetrant carrier properties. Although DMSO is not toxic by itself, it may carry toxic materials into the body. DMSO is a component of some organic decapsulant formulations, and may be used hot to dissolve epoxies.
- **ethanol**—C_2H_5OH, or grain alcohol, boils at 78.3 °C. Ethanol is used as a solvent for medicines, perfumes, and flavorings which are intended for human consumption or contact. It is present in alcoholic beverages in a percentage equal to half the label "proof" number. Industrial ethanol is rendered unfit for human consumption by adding a denaturant, usually acetone, benzene, ether, gasoline, kerosene, or methanol.
- **ethyl silicate**—is used as a dehydrating agent in organic chemical etch solutions.
- **ethylene glycol monomethyl ether**—is $C_4H_{10}O_2$, is a cellusolve, also known as carbitol, EGME, DIGLYME, ethyl cellusolve, ethylene glycol monomethyl ether, diethylene glycol monomethyl ether, 2-ethoxy ethanol, "cellosolve," "cellosolve solvent," or glycol ether. Other cellusolves in this family are methylene glycol monomethyl ether, propylene glycol monomethyl ether, and diethylene glycol monobutyl ether or 2-butoxyethanol (butyl cellusolve). Cellusolves are solvents for nitrocellulose, and are used in lacquer, varnish remover, cleaning solutions, leather dye, and pigments. Ethylene glycol monomethyl ether acetate (EGMEA) has a sweet odor, and is used in automotive lacquer to impart a high gloss, and it is a solvent in some photoresists. Ethylene glycol monobutyl ether acetate has a fruity odor, and is used for similar purposes. All of the cellusolves have odors, detectable at about 1 ppm or less, and TLV values of 5 ppm, but the IDLH value is 2500 to 6000 ppm. Cellusolves are reproductive hazards; more hazardous when heated. EGMEA is used in photoresists, along with ethyl and butyl acetates and xylene. EGMEA appears in commercial products, such as paint, wherever the manufacturer needs the solvent to have a pleasant odor, and long evaporation time. Butyl, ethyl, and methyl cellusolves are broad-spectrum solvents for silicone varnishes, RTV silicon rubbers, urethane coatings, elastomers, and foams.
- **ethylenediamine**—$C_2H_8N_2$, is also called 1,2-ethylenediamine, 1,2-ethanediamine, and dimethylenediamine, is a fuming, hygroscopic liquid that smells like ammonia. When heated, it will dissolve polyimide without attacking metals.
- **formic acid**—H_2CO_2, HCOOH, or methanoic acid, is the poison in ant and bee stings. It is used in some organic decapsulation formulations. Formic acid is damaging to skin. Formic acid is the compound formulated by the body from ingested methanol, which results in permanent blindness.
- **glycol ethers (general)**—have recently been subject to much negative attention in the semiconductor industry. The glycol ethers are: diethylene glycol monobutyl ether, diethylene glycol monobutyl ether acetate, diethylene glycol monoethyl ether, diethylene glycol monobutyl ether acetate, ethylene glycol monoethyl ether, ethylene glycol monoethyl ether acetate, ethylene glycol monomethyl ether, ethylene glycol monomethyl ether acetate, propylene glycol monomethyl ether, propylene glycol phenyl ether, triethylene glycol dimethyl ether.

In addition to semiconductor processing, the glycol ethers are used for, or in: brake fluid, cleaning solutions, cleansing cream, cloth, coatings, dry cleaning fluid, dye, enamel, feed additive, ink, lacquer, leather, perfume fixative, photographic film, plasticizers, soap, sun tan lotion, textile conditioners and printing, varnish and varnish remover, wax, wood stain, and yarn.

- **isopropyl alcohol**—C_3H_8O, isopropanol, or 2-propanol, or IPA, is used in familiar rubbing alcohol, a topical refrigerant and relaxant, in concentrations of 70% to 90%. Isopropanol is also used as a solvent in shellac, creosote (a tarry mixture which contains cresol), gums, and resins. Isopropanol has a fragrant odor, with a TLV of 400 ppm and a high IDLH value of 20,000 ppm, somewhat less than ethyl alcohol. IPA evaporates more slowly than ethanol or methanol. Although ingestion of 100 ml of IPA can be fatal, it is a comparatively safe solvent. Like methanol and ethanol, IPA forms azeotropes with other organics, but it is insoluble in salt solution, so adding salt to an azeotropic mixture of solvents containing IPA will cause it to separate out.
- **methanol**—CH_3OH, CH_4O, wood alcohol, was originally obtained, along with phenol, creosol, acetone and other products, by the destructive distillation of wood during the making of charcoal. Methanol boils at 65 °C. It is now synthesized from petroleum. Methanol is a common solvent, present in paint strippers, lacquer thinner, and shellac, and in gasoline as an antifreeze and octane booster. Ingestion of methanol liq-

uid or vapors in quantity results in physiological effects similar to grain alcohol, except that blindness results, because the body reacts methanol to yield formic acid. Ingestion of 30 ml of methanol can result in death. Methanol has a flash point of 12 °C (54 °F), and an ignition temperature of 470°C, and is explosive in air in concentrations from 6 to 35%. Methanol is miscible with most organic solvents, forming azeotropes. Methanol is one of the denaturants added to ethyl alcohol to render it unfit for drinking, and exempt from tax. It is hygroscopic, carrying away water as it is removed from the specimen. Methanol is used for removing urethane coatings, polyesters, and silicone oils, and contaminants such as animal oils.

- **methylene chloride**—CH_2Cl_2, methyl chloride, chloromethane, or dichloromethane, is a quick-evaporating, nonflammable wide-spectrum solvent which has been used as a refrigerant and local anesthetic. It is impossible to heat methylene chloride unless it is mixed with another slow-evaporating solvent. It is useful for dissolving methyacrylate and other thermoplastics.

- **n-methyl pyrrolidone**—C_5HONO NMP or m-pyrol n-methyl-z-pyrrolidone, is used hot to deflash and decap epoxy-encapsulated components, and to remove, or develop, positive photoresist. m-pyrol is combustible and has a TLV of 100 ppm.

- **n,n dimethyl formamide**—C_3H_7NO, DMF, DMFA, will dissolve polyether urethane foams, Orlon®, and other polyacrylics. DMF has a slow evaporation rate. DMF can be absorbed through the skin.

- **ortho-dichlorobenzene**—$C_6H_4Cl_2$, ortho-dichlorobenzene, and para-dichlorobenzene, are strongly aromatic, and are used for familiar mothballs. Its odor threshold is 2 to 4 ppm. The TLV of o-dichlorobenzene is 50 ppm, where that of p-dichlorobenzene is 1000 ppm. o-dichlorobenzene has a vapor pressure of 1.2 mm, and p-dichlorobenzene has a vapor pressure of 0.4 mm. Both compounds can be absorbed through the skin, and are carcinogenic. The dichlorobenzenes may be constituents of strippers, and they may be used in failure analysis for dissolving tough plastics such as Nylon®.

- **perchloroethylene**—CCl_2CCl_2, "perc," boils at 121 °C, has low toxicity, and is nonexplosive. It is used in pharmaceuticals, medicines, and as a solvent, spot remover component, and cleaner.

- **phenol**—C_6H_5OH, C_6H_6O, or carbolic acid (as opposed to carbonic acid, H_2CO_3), is manufactured from benzesulfonic acid. Phenol is miscible with water, ether, glycerol, oil, and hydroxides, and boils at 182 °C. Phenol is a component of many strippers, useful in failure analysis for dissolving plastics, including the polyimides. Phenol, as "carbolic acid," was used for years as an antiseptic and disinfectant in hospitals, then disapproved because it was identified as a suspected carcinogen. A fatal dose may be as little as 1 gram. Interestingly, phenol is present in one popular mouthwash at 1.4% concentration. TLV = 5 ppm, IDLH = 100 ppm; its odor is detectable at 0.005 to 0.5 ppm.

- **potassium hydroxide**—KOH, is the electrolyte for large nickel-cadmium batteries (nicads), used primarily in aircraft. KOH is commonly sold in sticks or pellets. It is used as an etchant for polysilicon and silicon nitride. Hot KOH will dissolve Pb, Cd, Zn, Ta, Nb, Al, Mg, and Be. KOH is used to adjust the pH of decapsulant solutions so that metals will not be attacked.

- **trichloroethylene**—TCE, is a strong and effective solvent and degreaser which will dissolve tar and wax. Unfortunately, it has a rather agreeable odor, and it is an unquestioned carcinogen. It is a prohibited material in industry.

Most chemical suppliers have technical customer service representatives that are very helpful in finding the proper solvent for your particular application. Dynaloy, Inc., of Hanover, New Jersey, manufacturer of the popular Uresolve and related solvents, has made a similar chart, focusing on the use of their specialized solvents for dissolving various plastics.

The MSDS

All vendors of chemicals are now required to include a copy of the Material Safety Data Sheet by Federal Law. This sheet tells the composition of the chemical, the toxicity, and its dangerous properties, neutralization, disposal, etc. Read and collect these sheets, and make them readily available to your analysts. Find a MSDS guidebook which contains terms and definitions, and have it available nearby. Your laboratory personnel have the right to know all the hazards associated with the materials with which they work. In the descriptions above, TLV = "threshold limit value," a long-term exposure concentration limit. IDLH = "immediately dangerous to life and health."

DISPOSAL

All disposal procedures should be specifically approved by your Environmental and Safety Departments before they are employed.

The sophistication of your operation will depend on the size of your Company and its resources, but in all cases, **DISPOSAL PROCESSES MUST BE LEGAL.** Individuals may now be **personally liable** for damage to health and the environment occurring from the use of improper or illegal disposal methods. You may, in some cases, be liable even for the improper actions of disposal companies who you contract with to handle your waste. This liability can include fines and imprisonment. KNOW THE LAW, and keep in constant communication with your Environmental, Health, and Safety experts.

NEVER store waste in buried tanks unless they are in concrete bunkers. Take a course in hazardous waste management.

Mineral acids

Under no circumstances pour water into fuming acids or fuming acids into water. The familiar rule of thumb, "always pour acid into water, never reverse," offers no protection with

FUMING acids. Spattering, an explosion, and copious fumes, may result from either case.

The spent acids may be *aspirated* to acid drains, where they are safely diluted with large volumes of water. In general, mineral acids are relatively easy to legally dispose. They will react with calcium in the soil to form salts that are stable, and similar in composition to naturally-occurring minerals. One neutralization technique consists of reacting the acid with crushed limestone. These acids may release fumes during neutralization, so dilution beforehand is wise.

Organics

In the case of organics, such as chlorinated hydrocarbons, the problem of disposal is much more complex than with mineral acids. **Usually, no chlorinated or fluorinated solvents, or oil, should be dumped into a solvent drain.**

Your city may have a list available, which names prohibited and toxic substances, which you must handle with specialized procedures for compliance with local statutes, Examples of common representative substances may be: carbon tetrachloride, methylene chloride, perchloroethylene, phenol, trichloroethane, and trichloroethylene. Chlorinated solvents are sometimes used bottle-to-bottle, disposed of in a closed container, and identified with a special waste label. All disposal procedures should be specifically approved by your Environmental and Safety Departments before they are employed.

Prohibited substances

Substances such as cyanide, copper, and cadmium may not be disposed into industrial waste drains, or solvent drains, in most cities.

REFERENCES

The following articles are good general information to have on file in your lab:

1. *The Chemistry of Failure Analysis,* by Mike Jacques, 17th Annual Proceedings of the International Reliability Physics Symposium, 1979, pp. 197-208.
2. *Basic Integrated Circuit Failure Analysis Techniques,* by D. Platteter, 14th Annual Proceedings of the Reliability Physics Symposium, 1976, pp. 248-255.
3. *Decapsulation of Epoxy Devices Using Oxygen Plasma,* by D.D. Wilson and J.R. Beall, 15th Annual Proceedings of the Reliability Physics Symposium, 1977, pp. 82-84.
4. *Three Decapsulation Methods for Epoxy Novolac Type Packages,* Proceedings of the Reliability Physics Symposium, 1980, pp. 107-109.
5. *Tough Analysis Problems That Have a Solution,* by Mike Jacques, Advanced Techniques in Failure Analysis Symposium, 1978, pp. 124-136.
6. *Improved Technique for Decapsulation of Epoxy-Packaged Semiconductor Devices and Microcircuits,* by B.L. Wensink, Solid State Technology Magazine, Oct. 1979, pp. 107-111.
7. *Chemical Decapsulation Revisited,* by T.W. Lee, Proceedings of ISTFA-1987.
8. *TO-92 Transistor Decapping Techniques,* by Robert W. Cook, Motorola R&QA Analytical Laboratories, Phoenix, Arizona, Oct. 1980.
9. *Stripping of Polyset Molding Compounds,* a technical information bulletin from the Morton Chemical Company, 1275 W. Lake Avenue, Woodstock, Illinois, 60098, Mar. 1976.
10. *Modern Chemistry,* by Charles E. Dull et. al., Holt, Rinehart and Winston, Unit 8, "The Halogens and Sulfur," pp. 414-417.
11. *Inorganic Chemistry; An Advanced Textbook,* by T. Moeller, Wiley, Chapter 14, "Oxides of Sulfur," notice Table 14-10 and Fig. 14.6.
12. *The Chemistry of Familiar Inorganic Substances,* by R.T. Sanderson, see Chapter 12: "Sulfur-Oxygen Chemistry," Arizona State University Press.
13. Gable, Betz, and Maron, J.A.C.S., 72, (Freezing points of oleum), 1950, pp. 1445-1448.
14. *H_2SO_4,* product information brochure available from General Chemical (formerly Allied Chemical), 6519 E. Quaker Road, Orchard Park, New York, 14127, 716-662-2880 and 800-247-4519.
15. *Sulfur Trioxide and Oleum, storage and handling,* product brochure available from E.I. DuPont De Nemours & Company, Inc., Wilmington, Delaware, 19898.
16. *Plasma Etching, An Introduction,* edited by Dennis M. Manos and Daniel L. Flamm, Academic Press, 1989.
17. *Silicones,* by R.N. Meals and F.M. Lewis, Reinhold Plastics Application Series, N.Y., 1959, pp. 6-131.
18. *Textbook of Polymer Science,* "Hetrocyclic, Ladder, and Inorganic Polymers," 2nd edition, by Fred W. Billmeyer, Jr., Wiley-Interscience, pp. 457-461.
19. *Preparative Methods of Polymer Chemistry,* by W.R. Sorensen and T.W. Campbell, Wiley-Interscience, pp. 73, 87 and 171.
20. *Epoxy Molding Compounds,* Motorola Materials Technology Laboratory Technical Report MTL-82-10, by Bill Hunter and Ron Thomas, 15 Sept., 1982.
21. *Industrial Toxicology,* by R. Irving Sax and Richard J. Lewis, Van Nostrand-Reinhold, 1987.
22. *Principles of Hazardous Materials Management,* by R.D. Griffin, Lewis Publishers, 1988.
23. *MSDS sheets*—from chemical manufacturers, including E-12275 (12/88) and E-91997 (2/87) on SO_3 and H_2SO_4 from DuPont.
24. *Resource Conservation and Recovery Act,* Environmental Protection Agency.
25. *Destruction of Hazardous Chemicals in the Laboratory,* by G. Lunn and E.B. Sansone, Wiley-Interscience, 1990.

PLASMA DECAPSULATION TECHNIQUES

by Dean Thomasi

This chapter illustrates and discusses the plasma removal of packaging materials. Two specifically selective material removal techniques and the required plasma processing parameters are detailed.

As integrated circuit semiconductor devices become more complex, the task of decapsulating devices, while retaining electrical functionality and package structural integrity, has become essential in isolating defects. Use of plasma etching, with only the top electrode powered (non-RIE) as opposed to reactive ion etch (RIE) produces good results. This allows the failure analyst to remove packaging materials while retaining full device functionality. The contamination problems of wet chemical processes are avoided. The removal of packaging materials allows visual analysis and the use of fault isolation techniques to determine the root cause of device failure.

The two techniques which have been found to be highly effective at retaining device functionality and package integrity are milling and masking. The milling involves preparing the sample by milling before exposure to the plasma. The masking involves mechanically or electrically shielding specific areas of the device through the use of fixturing, which allows only specific areas to be plasma etched.

Figure 1 schematically illustrates a parallel-plate plasma system. The etch gases flow between the parallel plates of the system, where they are ionized by RF energy. The ionized gas (plasma) chemically etches the packaging material from a device located on the lower non-powered electrode. The gaseous by-products are removed from the system by the vacuum pump. By contrast, non-RIE processes and wet chemical techniques (Fig. 2) remove packaging materials isotropically.

Material selectivity can be controlled by etch gas selection. For the purposes of this article, oxygen (O_2) and Freon® (CF_4)

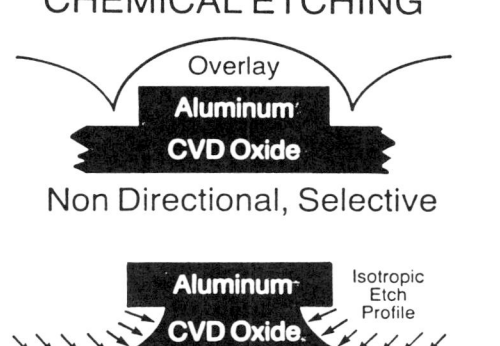

Fig. 1. Schematic of parallel-plate system in plasma mode, without RIE effects.

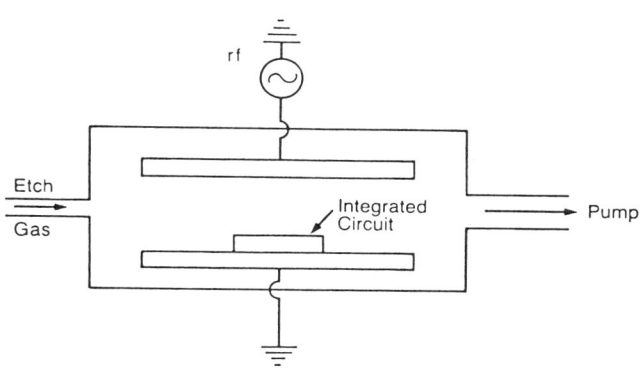

Fig. 2. Etch characteristics for non-RIE plasma, or wet-chemical etching, are isotropic and undercutting results.

are found to quickly etch packaging materials while providing selectivity to chip passivations.

Temperature and power control are also important parameters. These parameters determine both the etch rate and the amount of material which will sputter from the device leads and plate walls and redeposit on the device surface. At higher temperatures and RF power exposures, a greater amount of sputtering takes place. This sputtering will mask removal, leaving packaging material on the device surface. Lower temperatures and power levels will decrease the etch rate. Excessive heat or power can also overheat and damage the device. In general, maintaining the device at room temperature and limiting power exposure during processing are the most important variables for retaining full functionality.

SYSTEM DESCRIPTION

The plasma system used to develop these processes is a DryTek® Model DRIE-100, with its chamber modified to contain a single pair of parallel plate water-cooled electrodes. The system allows RF power, gas type and flow rate, chamber pressure, substrate temperature, and electrode spacing to be varied and controlled independently. The particular system processes in either plasma or RIE mode. For the processes which follow, the parallel plate plasma system is set in the plasma (or non-RIE) mode with the top electrode driven by the RF generator). In the RIE mode, the bottom electrode is driven.

The first step involves removal of the packaging material by milling. Preparing a device in this fashion reduces the time required for plasma etching and ensures that the chip surface is exposed before package integrity is compromised, i.e., before the leads break off. As an example, a plastic dual in-line package (PDIP) is prepared by milling as illustrated in Fig. 3 before

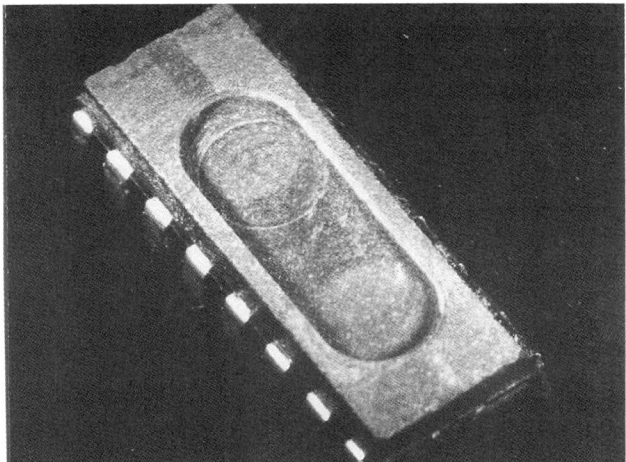

Fig. 3. View of a device in the PDIP package after preparatory milling.

Fig. 4. Device housed in masking fixture, showing exposed die in center.

Fig. 5. View of the masking fixture top and bottom parts, with a prepared SMT device in the center for size comparison.

it is exposed to the plasma. To obtain this result, top and side view x-rays are taken to determine the chip placement and the depth of the bond wires. This information is critical for determining the correct depth, width, and length of the trench to be milled in the package surface. To ensure that the bond wires and lead frame remain intact after processing, it is recommended that approximately 0.25 millimeters of plastic be allowed to remain over the bond wires to protect them from the milling operation. Also, the trench milled should be slightly longer and wider than the integrated circuit die to allow easy access for mechanical probing, should it become necessary.

The second approach involves fixturing which masks the device such that only the area to be etched is exposed to the plasma. An example of the fixturing used to house and mask a surface mount (SMT) device is shown in Fig. 4. In this example, the dimensions of both are 25 × 65 mm. The cover and bottom pieces are 1.5 and 6 mm thick, respectively. In the cover, a 5 × 16 mm mask hole is created by milling and filing. The mask hole should be of the same dimensions as the die and should be located directly over the die when both fixturing halves are joined (Fig. 5). For this example, the bottom cavity is 10 × 17 mm, and 4.5 mm deep. The cavity should be sized and located to accommodate the SMT device such that with the top cover in place, the entire device is enclosed. Only the area to be etched should be exposed to the plasma. A small spring is located beneath the SMT device to keep its top surface flush and in contact with the inside of the cover. This prevents the plasma from entering the fixturing and removing packaging material around the leads. Unchecked, this would eventually compromise package integrity.

NON-RIE PLASMA PROCESS

After milling and housing the device in fixturing, it is exposed to the non-RIE plasma process using the following parameters:

RF power —200 watts

Electrode temperature —25 °C

Electrode spacing —3.2 cm

Chamber pressure —800 millitorr

Gases/flows — O_2 at 100 sccm

— CF_4 at 10 sccm

Etch time 6 hours, composed of 6, one-hour runs to remove 25 mils of plastic packaging material (sccm = standard cubic centimeters per minute)

The plasma reacts with and consumes only the epoxy part of the plastic packaging material. Therefore, the inorganic filler material must be removed with a stream of clean compressed air or nitrogen after each one-hour run.

Increasing the chamber pressure to 1300 millitorr and increasing the CF_4 flow to 15 sccm will increase the process speed greatly, because much more ionized gas is available for etching. Unfortunately, this increase in etch rate is obtained at the expense of selectivity. Passivation attack and loss of device functionality will occur very quickly if vital areas of the chip surface are exposed before complete decapsulation has been achieved.

It is recommended that this faster process is used to remove material quickly to get the reacting surface near the chip surface. Then use of the normal or reduced CF_4 process (discussed below) will ensure that full circuit functionality is retained.

If the packaging material etches unevenly, plasma etching of the device surface can be limited by reducing the CF_4 flow rate to 3.0 sccm. This same compensating adjustment should be made when etching photoresist and polyimide.

In Fig. 6, the appearance of a typical device subjected to milling, followed by plasma etching with the recommended processes, is shown.

Figure 7 presents a detailed view of the die from a device following plasma etching which used the masking fixture. Figures 8 and 9 show the appearance of a die before and after plasma polyimide removal, respectively.

RECOMMENDATIONS

The processes discussed in this paper should be considered a starting point for the development of various plasma processes; non-essential functional devices are recommended as test samples for the development of all plasma processes. Never risk devices under analysis for the development or optimization of a process.

CONCLUSIONS

Non-RIE plasma processing for the removal of packaging materials has been shown to be useful, particularly when wet chemical methods would cause contaminants or materials to be dissolved or redeposited on the interior surfaces of the device.

Proper sample preparation and device masking enhance the speed with which the plasma penetrates to, and exposes, the die surface. With this method, decapsulation can be accomplished without compromising package integrity.

The obvious sacrifice of this method is speed. Where chemical methods can potentially gain access to the die in a plastic package in seconds, this method requires hours.

As integrated circuits become more complex, with smaller dimensions, new materials, and more layers, the use of these plasma techniques will become a necessity for effective failure analysis. Plasma methods will become necessary for electrically isolating fail sites and determining the root cause of device failure.

ACKNOWLEDGMENTS

The authors wish to thank the past and present members of the Physical Analysis project, whose input made this paper possible, specifically Paul Marchetti and John Powlis. Thanks also

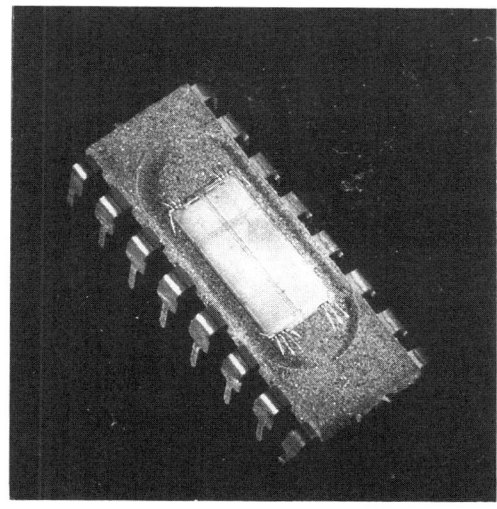

Fig. 6. Oblique view of device following plasma decapsulation.

Fig. 7. Optical micrograph of DRAM die following plasma decapsulation.

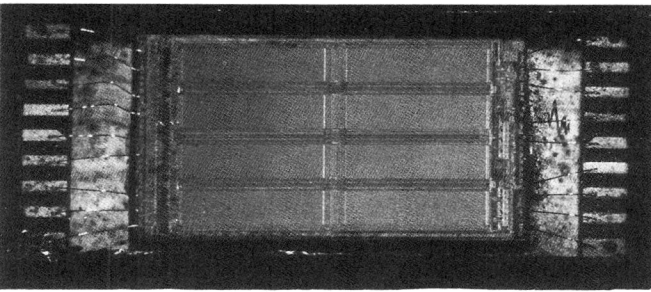

Fig. 9. Optical micrograph of die surface following plasma removal of polyimide passivation.

Fig. 8. Optical micrograph of packaged device with polyimide die coating before plasma etching.

to Jim Boyd of Drytek for his many helpful discussions. We also wish to thank Norma Hedelund for her contributions during the preparation of this manuscript.

REFERENCE

1. D. Thomasi, P. Marchetti, J. Powlis, *Failure Analysis Applications of Plasma,* ISTFA Symposium Proceedings, 1987.

PARTICLE VERIFICATION, EXTRACTION, AND IDENTIFICATION

by Gene P. Thome

Semiconductor devices are subjected to particle impact noise detection (PIND) testing because there is a chance that a loose metallic particle could cause a short, resulting in device or circuit failure. PIND testing has been used since at least 1965,[1] and has been proven to be very capable of detecting particles within package cavities. A number of analyses have been documented and published in which critical failures were found to be caused by particle shorts.[5,6]

EXPLANATION OF PIND TESTING

During PIND testing, semiconductor devices in metal packages are shocked and vibrated according to various programs specified in MIL-STD-750C, method 2052. To perform this test, the device is fastened to an acoustic transducer head with a couplant adhesive and subjected to pre-shocks and co-shocks of 50 to 1800 g with a duration of 100 microseconds. Alternating with the shocks are vibrations with a peak amplitude of 10 or 20 g, during which the metal package is acoustically monitored by feeding the signal from the transducer to an audio amplifier. The amplifier output is monitored for clicks and pops, and the time-domain waveform is observed with an oscilloscope.

The shocks are designed to ensure that any particle present is dislodged, and is free to move about within the package. Repeated collisions of the particle with the inside of the package during vibration periods generate the audio clicks, and result in rejection of the device.

REASON FOR PARTICLE RETRIEVAL

PIND rejects may not be retested. Procedures prohibit retesting because the particle, moving rapidly around within the package during the vibration phase of testing, may lodge or "lock up," somewhere in the interior of the device. The device may then pass subsequent testing; experience shows that even multiple retests can be passed with a locked or lodged particle. This particle could, however, dislodge at a later time and cause a failure. Consequently, devices must be discarded upon the failure of the first test sequence. This discarding requirement, the reputability and absolute accuracy of PIND tests, and the hazard represented by various particle types, are often enthusiastically contested. PIND testing, therefore, has been subjected to many thorough and critical reviews,[2] Adolphsen's[8] in particular.

Because the devices being tested are frequently of military quality, and comparatively expensive, the occurrence of PIND failures must be prevented. Consequently, failure analysis using methods which build high confidence in the identification of particle size, shape, and composition is extremely important in operations where PIND testing is required. The data not only validate or refute the PIND test result, but define the location of the particle source (if any), formulation of corrective actions, and prevention of further component yield loss.

Because x-ray inspection methods may not yield acceptable results, the retrieval and identification of particles from the internal cavity of metal-can semiconductor devices is very delicate and labor-intensive, requiring extensive skill and practice. There is no known tool available which is specifically intended for the particle-free decapping of metal-can devices. Conventional package openers are intended for high productivity, not for cleanliness. During cap removal, these machines may generate significantly more particles than the failed device may already contain, hopelessly confusing identification efforts. The reporting of inaccurate data can frustrate problem solving efforts. It is the purpose of this section to illustrate a method of particle retrieval which can identify particles without generating additional ones.

PROCEDURE

Over the years, experience has proven the following illustrated step-by-step procedure to be the most valuable for the verification of the presence of particulates in devices. Experience was gained with discrete devices in packages such as the TO-18, TO-5, and TO-3, but it applies to virtually any metal-can part, regardless of the semiconductor technology within. Outline drawings for these packages appear in the Appendix of this Desk Reference.

1. Using a Dremel® Moto-tool®, with an abrasive stone, shown in Fig. 1, or with an abrasive disk, shown in Fig. 2a, grind the top perimeter of the can as shown in Fig. 4, 5, and 6. Be careful not to grind all the way through the metal. If the can is ground through, the sample must be discarded, because the interior will quickly become contaminated with metal shavings and abrasive particles shed by the grinding stone, hopelessly disguising the original package contents.

2. Next, using the fine abrasive stone shown in Fig. 2b (right), polish the coarsely ground surface. The metal in the perimeter of the package should become thin enough to be flexible and it should depress, or distort, when the pressure of the polishing wheel is applied. This flexible *"paper-thin"* metal should form a noticeable groove around the perimeter of the can. The groove is the most significant indicator of the satisfactory completion of grinding and polishing. Figures 7 and 8 detail the appearance of this groove.

3. When the entire circumference of the metal package has been polished, remove all grinding debris with a cotton swab and alcohol. Blow the device dry with compressed air or dry nitrogen.

4. Using machinist's blue-line marking liquid, or an indelible marker, apply coloring to the thin metal in the groove created in the above operation.

This procedure will make it possible to positively identify any particles sourced from the package exterior during the opening operation, should they inadvertently enter the package cavity.

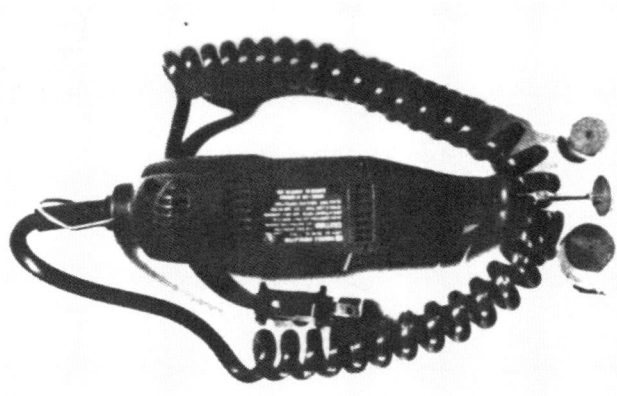

Fig. 1. Dremel® Moto-Tool® with grinding and polishing stones.

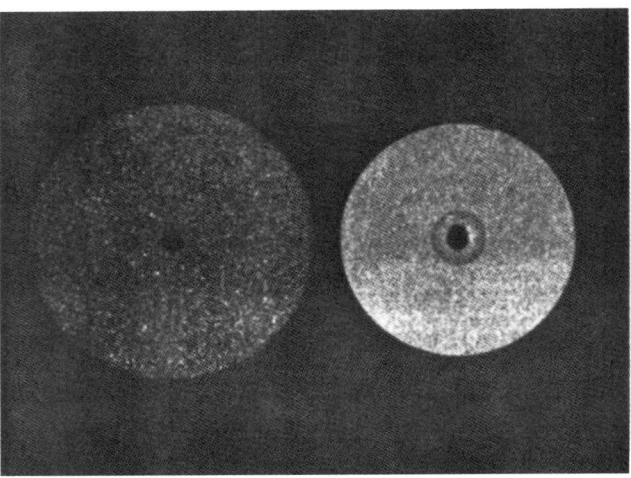

Fig. 2. Detail of abrasive disks, used for both grinding (left) and polishing (right). (1.5×)

Fig. 3. Exterior view of typical TO-3 device at the beginning of the operation. (2×)

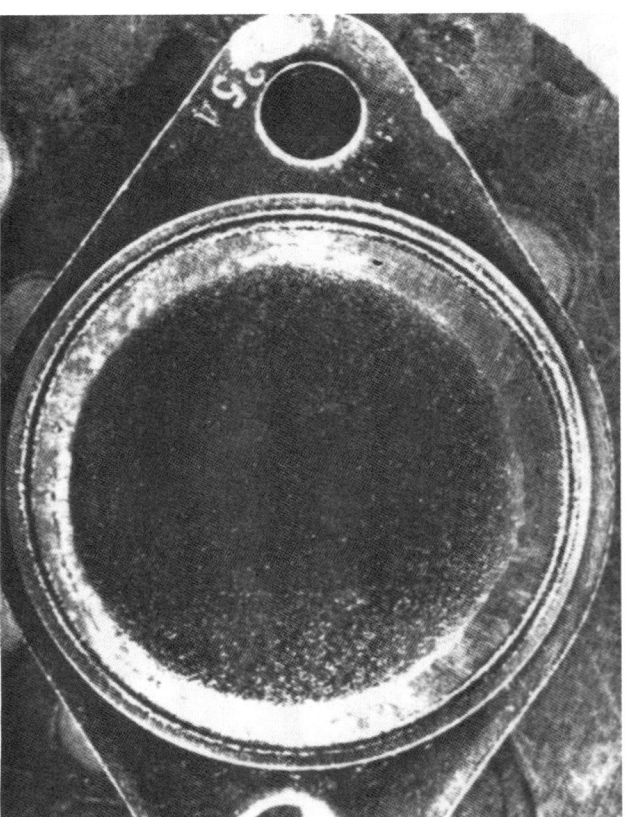

Fig. 4. Top view of TO-3 package exterior following first coarse grind operation. (2×)

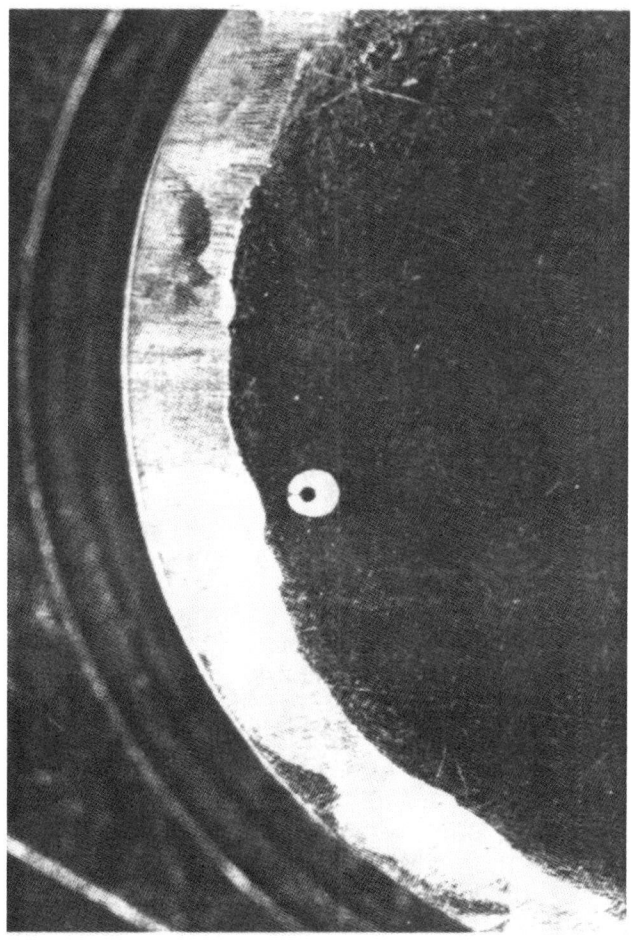

Fig. 5. Detailed view of top perimeter of device after coarse grinding. (4×)

Fig. 6. Profile view of package perimeter surface after coarse grinding. (9×)

5. Puncture the thin metal in the polished groove with a scalpel having a *new* blade. Ensure that the blade is particle-free by first checking its surfaces beneath a binocular microscope. Insert the blade through the thin metal in the polished groove. Using an outward cutting motion similar to that used with the old manual blade-type can openers of years ago, open the package. *Never* use a sawing motion, because the downward stroke may introduce particles into the package. Figure 9 details this cutting operation, and Fig. 10 shows the appearance of a typical opened device.

The analyst is advised to experiment and practice with several devices each, of as many package styles and sizes as available. In this manner, proficiency in the grinding techniques above, and the extraction techniques following, can be further developed and mastered. This technique is not simple to master, and several breached cavities can be expected before proficiency is accomplished. The cost of the devices, and the importance of the positive identification of the particles within the package, should justify extensive practice.

6. With the cap or top of the package removed, obvious particles consisting of weld splash, solder balls, etc., can be removed, photographed, and analyzed. In most cases, static charge or remanent induced magnetism of the tip of the scalpel blade is sufficient to cause the adherence of particles. The particle can then be transferred to an adhesive "tack strip" for further handling. Although it is acknowledged that tack strips have been in use for years,[4,5] the uniqueness of the method presented here bears repeat coverage.

7. To locate and remove particles which are hidden from view, or which may have become trapped during test shocking and vibration, center a tack strip over the top of the device decapped by the procedure outlined above. Seal the tack strip to the device by applying pressure along its perimeter. Figure 11 shows the result of this process. Here, a strip of lead (Pb) adhesive-backed tape is being used as a tack strip.

8. Invert the unit, and use a scalpel handle or other metallic object to tap the bottom periphery of the package header to dislodge any particles which may have "locked-up." They will then fall onto the tack strip and be held in place for subsequent handling and examination without the risk of loss. Remove the tape carefully. Figure 12 shows the appearance of a typical tack strip after removal from an opened package, with particles adhering to its surface.

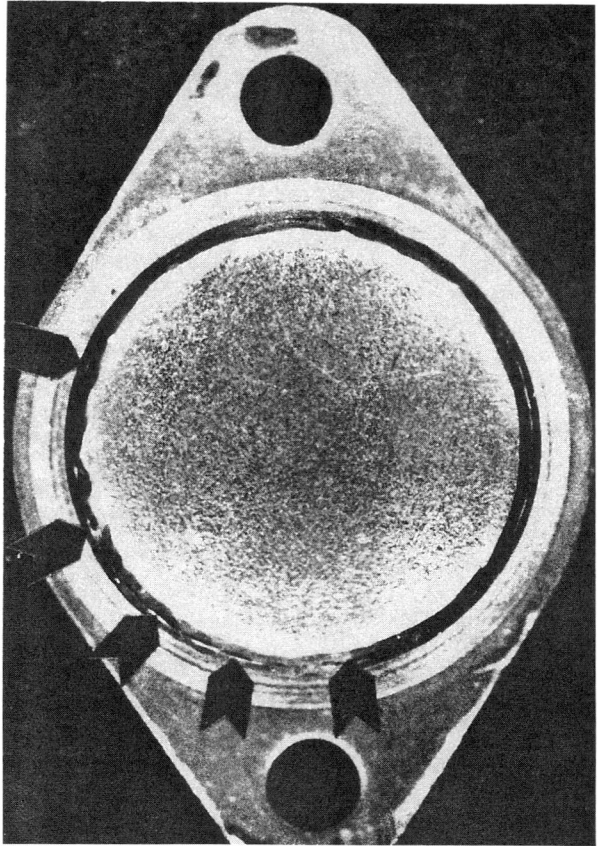

Fig. 7. View of top of device after polishing with fine abrasive. Arrows indicate groove.

PARTICLE IDENTIFICATION

Physical Appearance

9. Figures 13 through 17 will help in the identification of the physical appearance of particles which have been injected into the package during opening. For example, metal scrapings can be identified by folds from compression and striations from abrasives. Glass shards have a telltale conchoidal, or shell-like, fracture surface. Shards frequently have iron (Fe), lead (Pb), and other elements present, along with small bubbles and other physical imperfections, that can distinguish them from chips of silicon. Pieces of wire and wire bonds are unmistakable. Flakes of plating can be identified by comparing their texture to other plating within the package. Typical examples are shown by SEM photos in Fig. 13 through 15. Figures 16 and 17 show the appearance of particles originating, or shed, from the grinding or polishing stones.

General contaminants of a size insufficient for optical methods can again be identified by their shape, and the superior depth of field of the SEM makes it the ideal instrument choice for this purpose. Murphy's paper[2] shows several examples.

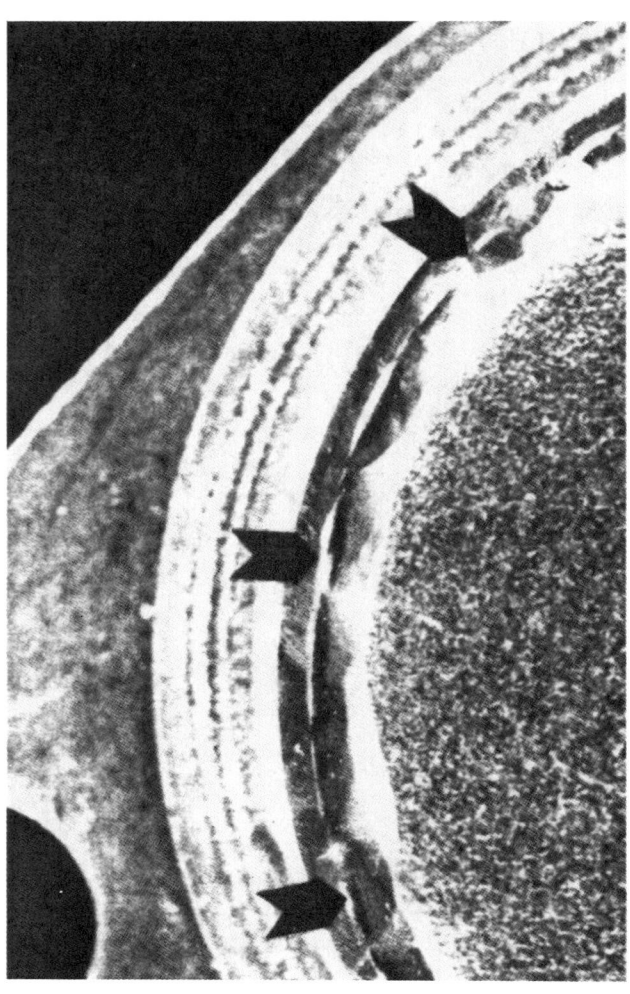

Fig. 8. Detail of groove after polishing. Metal is paper-thin. (4×)

10. Optical methods, such as a low-power binocular microscopy, are applicable for large particles. For example, Fig. 20 and 21 show a typical weld splash particle. Its spherical shape implies a former molten state, and more detailed analysis is unnecessary.

Chemical and/or Elemental Composition

General contaminants for which appearance alone is insufficient for identification, can be characterized most effectively by a SEM with energy dispersive x-ray analysis (EDXA). The resulting spectra may reveal elements which are foreign to the semiconductor processing industry, allowing their probable source to be determined. Examples are calcium (Ca) and potassium (K) in Fig. 18. Calcium was also identified in Fig. 19. Many general references are available in the identification of unusual or foreign elements in semiconductor processing and failed devices, and these should be reviewed for references to particles.[3]

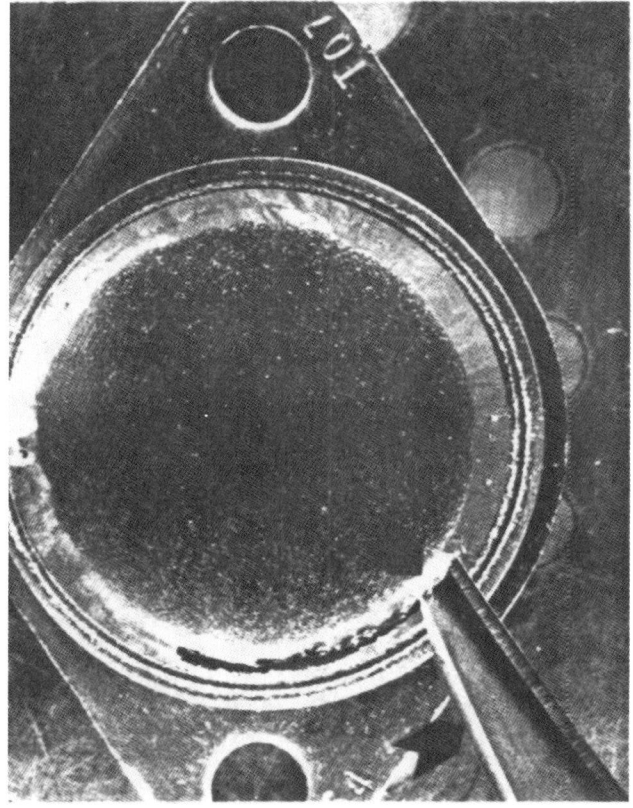

Fig. 9. Scalpel blade shown during cutting operation. Cutting strokes should be outward only.

Fig. 10. View of interior of device after complete removal of cap by method shown in Fig. 9.

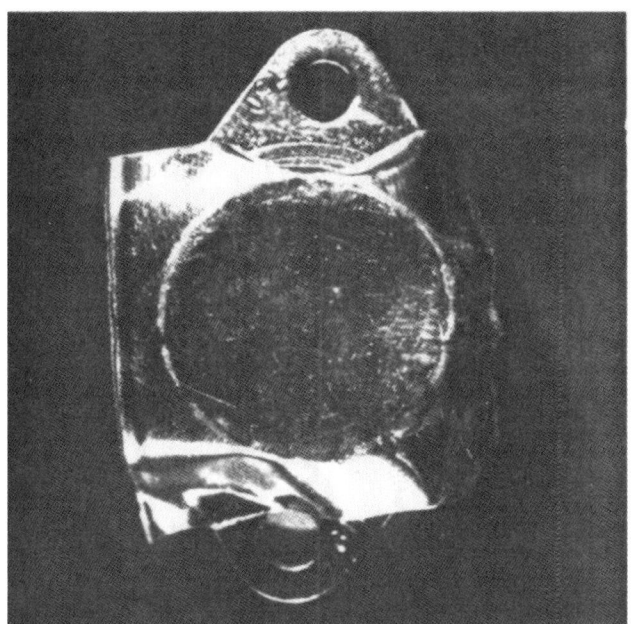

Fig. 11. Lead (Pb) tape tack strip after pressing onto top of decapped device.

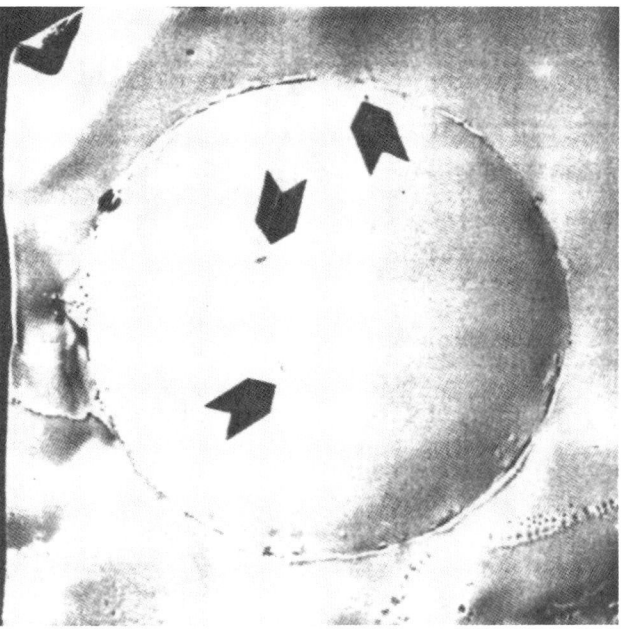

Fig. 12. Adhesive side of tack strip after removal from device. Arrows show location of trapped particles.

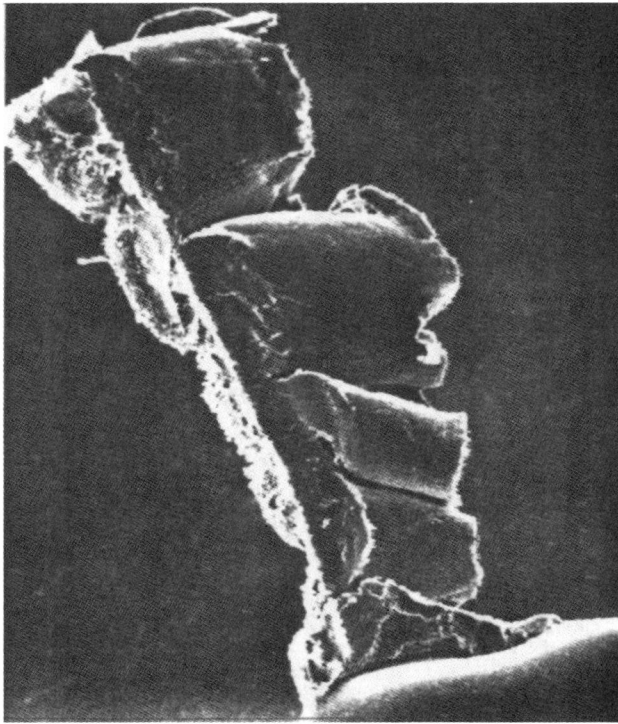

Fig. 13. SEM view of particles injected into the package cavity during opening procedure, showing obvious compressive folding. It is composed mostly of nickel. (200×)

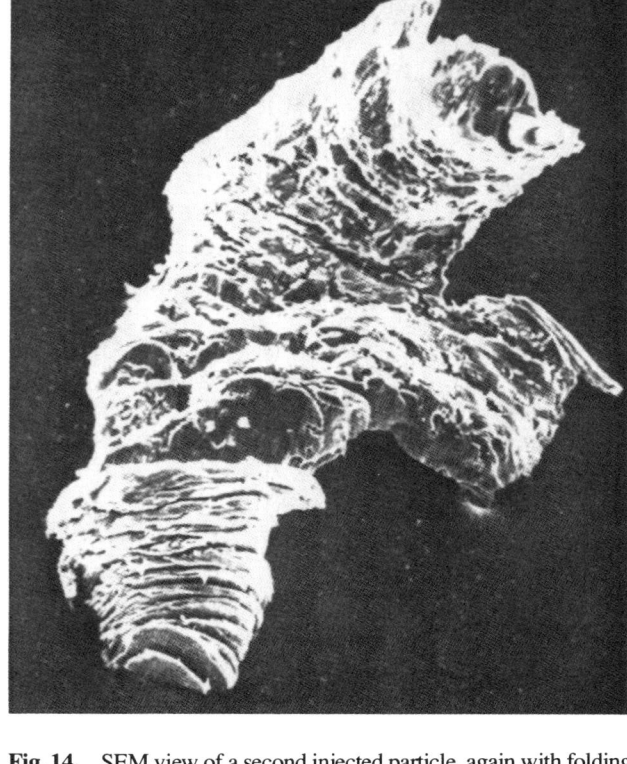

Fig. 14. SEM view of a second injected particle, again with folding or wrinkling of surface indicative of compression. (300×)

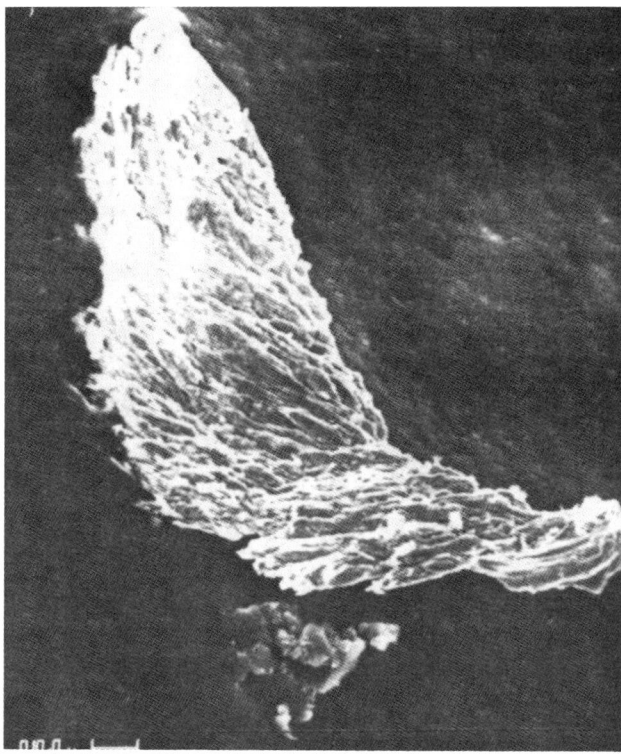

Fig. 15. Another example of an injected particle. (500×)

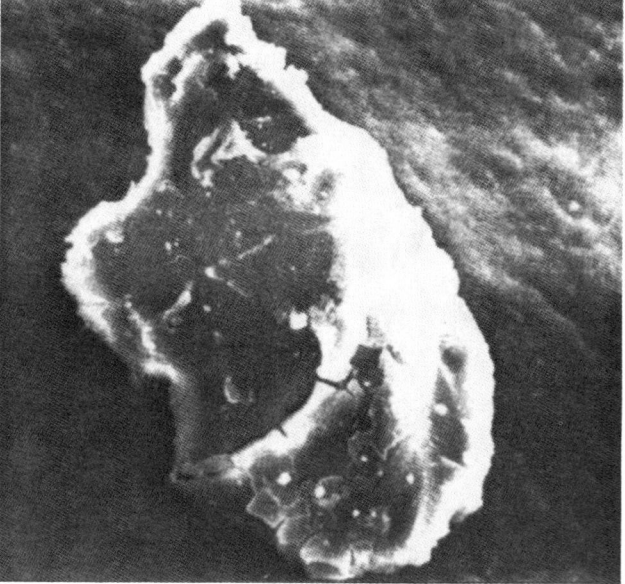

Fig. 16. SEM view of silicon carbide (SiC) particle injected into the package by inadvertently grinding through metal cap. (700×)

Particle Verification, Extraction, and Identification

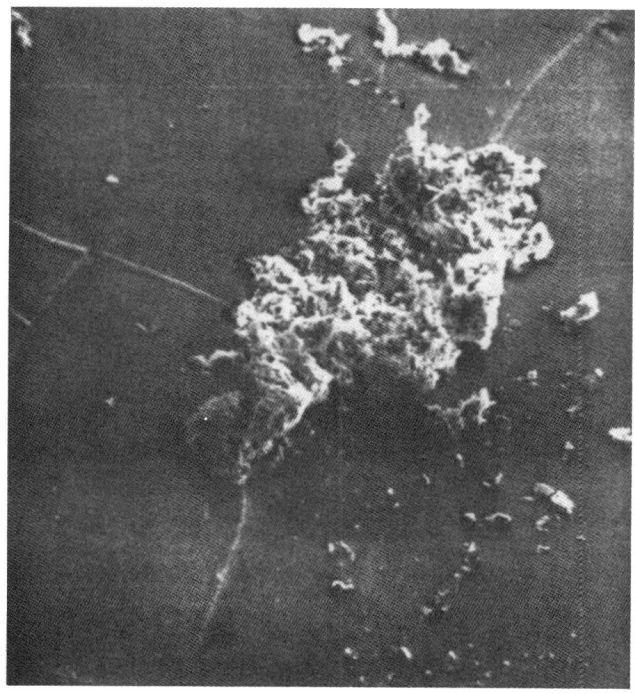

Fig. 17. SEM view of typical "dust" particle injected during polishing procedure when package was breached. (200×)

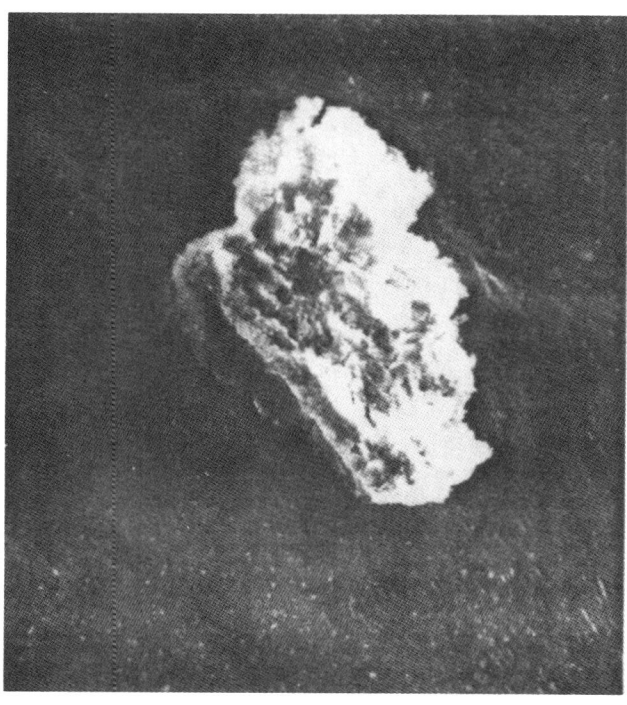

Fig. 18. SEM view of particle measuring 3.8 × 2.3 mils. It consists of Si, Ca, Cl, and O, and is almost certainly human debris, such as skin flake. (500×)

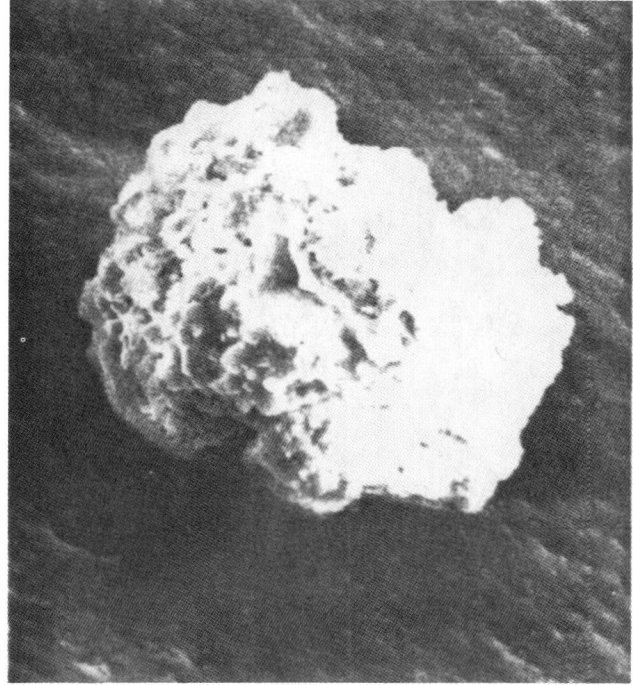

Fig. 19. Particle measured to be 2.1 × 2.8 mils, containing Ca and Cl, probably a water evaporite deposit. (700×)

Fig. 20. Optical photo of header of opened part. Arrows indicate position of weld splash particle; composition is nickel (Ni), and package plating is Ni. (10×)

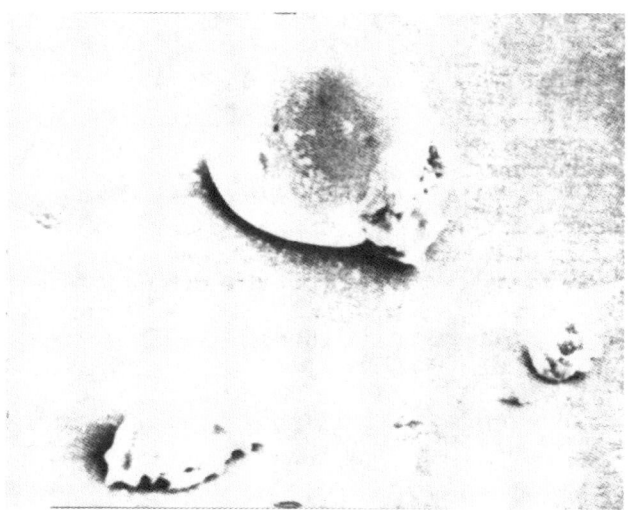

Fig. 21. SEM view of solder sphere having a diameter of 2.5 mils, probably sourced during die bonding. (400×)

REFERENCES

1. Paper in the *Institution of Environmental Sciences Annual Technical Meeting Proceedings,* 1965.

2. *The Particle Atlas-Edition Two,* by W.C. McCrone and J.G. Dilly, Ann-Arbor Science Publishers, Inc.

3. *Screening techniques for Intermittent Shorts,* in the Proceedings of the 10 Annual International Reliability Physics Symposium, 1972, pp. 19-22.

4. *Hybrid Technology Loose Particles and Coating Materials,* by Charles Murphy, in the Proceedings of the 13th Annual International Reliability Physics Symposium, 1975, pp. 248-252.

5. *Failure Analysis Techniques Applied in Resolving Hybrid Microcircuit Reliability Problems,* by Geroge H. Ebel, in the Proceedings of the 15th Annual International Reliability Physics Symposium, 1977, pp. 70-81. (See Fig. 15, p. 76)

6. *Developing An Approach To Semiconductor Failure Analysis And Curve Tracer Interpretation,* by Joe Patterson, in the Proceedings of the 16 Annual International Reliability Physics Symposium, 1978, pp. 93-100. (See photo 5b, page 96)

7. *PIND's Role As A Failure Analysis Tool,* Ralph E. McCullough, J.C. Burru, and Willie R. Reynolds, Proceeding of ATFA, 1979, pp. 23-26.

8. *The Effectivity of PIND Testing,* by John W. Adolphsen, in the Proceedings of the Advanced Techniques of Failure Analysis Symposium 1979 (ATFA-79), p. 27.

9. *The Identification and Elimination of Human Contaminants in the Manufacture of ICs,* by Robert W. Thomas and Donald W. Calabrese, Proceedings of ISTFA, 1985, pp. 169-176.

WINDOW ETCH PROCEDURE FOR MULTI-LAYER VLSI

by David S. Kiefer

Many state-of-the-art VLSI technologies utilize multi-layer metallization schemes to increase circuit density. In failure analysis and design debugging, it is often necessary to probe one of the bottom layers of metallization without disturbing the top layers. The procedure outlined in this paper presents a method that allows a small window to be cut down through the top layers, exposing the desired area without altering the circuit's performance. The procedure is relatively simple and most failure analysis labs should have all the necessary equipment.

By the use of photoresist techniques and selective etching, the desired underlying structure is exposed. With the information supplied, the technique could easily be applied to other technologies. This paper identifies specific equipment, materials, and recipes used to perform the procedure.

Failure analysis and design debug of multi-layer VLSI devices is a difficult task even under the best circumstances. Understanding the complex design is a formidable task in itself. However, sometimes just getting access to the suspected malfunctioning transistor or node is also difficult. With the introduction of a three layer metal, polyimide passivated bipolar technology, a need became apparent to expose the bottom layer of metallization for electrical probing while maintaining the electrical integrity of the top two-layers. A technique called selective window etching was developed to accomplish this task. The procedure uses photoresist to mask the die surface while selective etch techniques are used to bore a small hole through the top metallization and dielectric layers, thus exposing the area of interest. This paper outlines specific equipment and chemicals, as well as the detailed recipes used to accomplish the window etch. Most well-equipped failure analysis labs should have access to the necessary materials and equipment. The specific procedures described herein are for three-layer metal polyimide passivated bipolar gate arrays and two-layer metal polyimide passivated CMOS gate arrays. Figure 1 shows a SEM micrograph of a completed window cut through a three-layer metal ECL SRAM which was used to evaluate bit errors on the device.

EQUIPMENT AND MATERIALS

While most labs are equipped with some form of the following equipment and materials, brand names and models are given on the most critical, since the process parameters presented in the procedure are specific to these. With minimum effort, the procedure could be adapted to other equipment and/or materials.

- Metallurgical microscope with adjustable iris and UV filter NIKON OPTIPHOT
- Two-speed cross-sectioning wheel   LECO GP60
- Plasma etcher—   Technics PEII
- Curing oven capable of   +125°C
- Fume hood
- Positive photoresist—   AZ P4620
- Photoresist developer—   AZ 421K
- Photoresist stripper—   AZ 300T
- Metal etch—phosphoric-nitric-acetic mixture
- Oxide etch—buffered HF 6%
- Apiezon® wax
- Xylenes

PROCEDURE

1. Identifying the Area of Interest

When dealing with multi-layer VLSI, the area of interest is often buried beneath one or more layers of metallization and dielectric and may be difficult to visually locate through a metallurgical microscope. A Calma® plot is very useful in locating the area of interest. Using X and Y coordinates and easy to find landmarks (such as bonding pads, vias, etc.) the area can easily be located.

2. Initial Device Preparation

The etching procedure used to expose the underlying structure often damages the bonding wires. The recommended material used to protect the bonding wires is Apiezon wax. The wax has a melting point of 150 °C and once cooled stays hard and impermeable during all subsequent etching procedures.

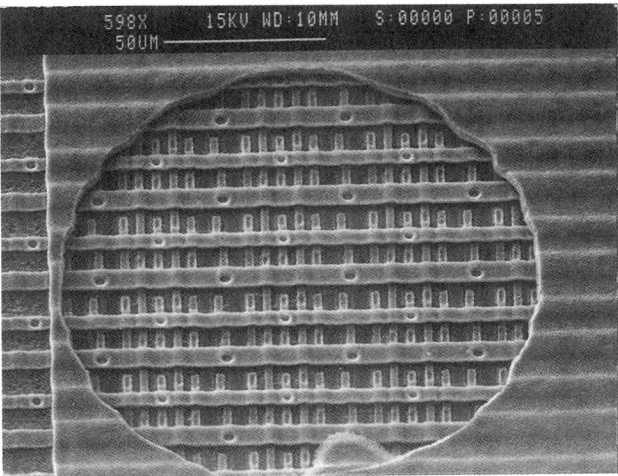

Fig. 1. SEM micrograph of window cut in three-layer metal ECL SRAM to evaluate bit errors.

The device and a Petri dish of wax are heated on a hot plate to 175 °C. When the wax becomes soft, a Q-tip® is used to deposit the wax around the bonding wires. Figure 2 shows a pin grid array (PGA) package with wax applied to protect the bonding wires. After completion of the window etch, the wax is easily removed with xylenes.

3. Photoresist Spin-On

A high speed cross-sectioning wheel is used for a photoresist spinner. A carrier suitable for holding the device at high RPM must be mounted very close to the center of the disc. After securing the device in the carrier, it is necessary that the ambient lighting be dimmed so the resist is not exposed. It should be noted that from this point until the completion of the hard bake, all ambient lighting should be minimized to avoid inadvertent exposure of the resist. The package cavity is flooded with AZP4620 photoresist. The wheel is then brought to 1000 RPM and maintained at that speed 10 seconds. A uniformly thin layer of photoresist should now be apparent. Figure 3 shows a high speed cross sectioning wheel used for a photoresist spinner. A PGA carrier can be seen mounted to the center of the wheel.

4. Photoresist Soft Bake

In order to stabilize the resist for exposure a soft bake is performed in a small oven at 95 °C for 15 minutes.

5. Photoresist Exposure

The device is placed under a metallurgical microscope with an adjustable iris. In order to locate the area of interest under the microscope, a UV filter must be placed in the light source to prevent inadvertent exposure. The light source should also be dimmed to the minimum amount of light required by the analyst's eyes. Once the area of interest has been located, the iris is adjusted to expose the desired size window. The UV filter is removed and the light source is increased to maximum intensity. All neutral filters and polarizers should be removed during the exposure. Using a Nikon Optiphot® with a 12V, 100 watt light source, the exposure time is two minutes.

6. Photoresist Develop

Developing the photoresist is accomplished by immersing the device in potassium hydroxide (KOH) developer for 60 seconds. This is followed immediately by a thorough DI water rinse and a nitrogen blow dry.

7. Photoresist Hard Bake

The photoresist hard bake will desensitize the coating to further light exposure, and toughen it for the selective etching procedure. The hard-bake temperature is 125 °C and the time is 25 minutes. Even though the coating will be desensitized to light, care should still be taken to avoid exposure to high-intensity UV light sources.

After completion of the hard bake, the device is ready for selective etching to expose the underlying area of interest. The exposed window should be inspected to ensure all the resist has been cleared and the window is of the desired size and in the correct location. Figure 4 shows a completed window in the photoresist prior to any selective etching.

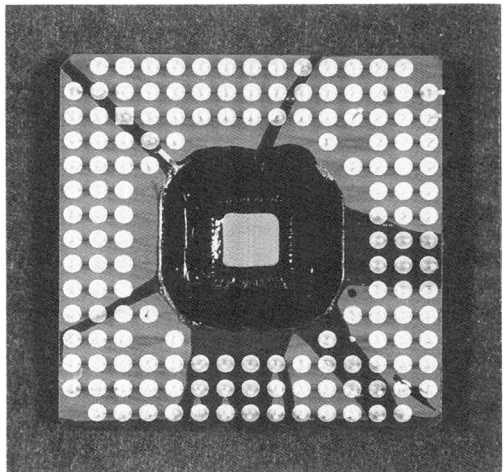

Fig. 2. 149 PGA package with Apiezon wax applied to protect the bonding wires.

Fig. 3. High-speed cross-sectioning wheel modified for use as a photoresist spinner.

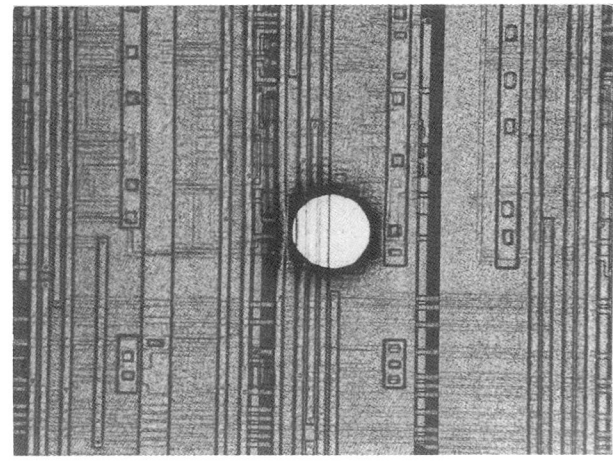

Fig. 4. Completely cleared window in photoresist after hard bake.

Two examples of selective etching are presented. One for a three-layer metal, three-layer polyimide technology used to fabricate bipolar gate arrays. The other is for a two-layer metal, two-layer polyimide technology used to fabricate CMOS gate arrays. In general, the photoresist is resistant to most wet chemicals and plasmas used in the selective etching procedure. Close attention should be paid to the integrity of the photoresist coating during the etch process. If it appears to be wearing thin, cracking, or bubbling, the resist should be stripped and re-coated, and a new window exposed in the same location. Two etches that cause accelerated wear-out of the resist coating are CF_4 plasmas and chemicals which have a very high pH (basic). Further, the temperature of the resist coating should not be elevated higher than 70 °C during any of the selective etching steps.

SELECTIVE ETCH FOR THREE-LAYER METAL POLYIMIDE PASSIVATED BIPOLAR TECHNOLOGY

A cross-sectional view of this technology is shown as Fig. 5. The top layer of metallization covers about 95% of the die surface and is used for power and ground distribution. The first and second layers of metallization are used as transistor and cell interconnect and are generally the layers of interest for electrical probing and E-beam analysis.

First, the top polyimide (final passivation) is removed using an oxygen plasma. With the Technics PEII system the pressure is 250 millitorr and the RF power is 300 watts. About 10 minutes of etching time is required to fully expose the top layer of metallization in the window area.

The top layer of metallization is removed using a buffered aluminum etch at 70 °C. The device is placed on a hot plate and observed with a stereo zoom microscope. After the metal etch is placed on the die surface, the window area is observed until the top metal clears, generally in about 2 minutes. Immediately after clearing, the device is placed in a beaker of DI water to stop the reaction.

After rinsing and drying, the window area is inspected under a metallurgical microscope. If the top metal has been completely cleared, the first and second layers of interconnect should now be visible. Figure 6 shows an example of a completely cleared window in the top layer of metal.

The remaining polyimide dielectric is removed using an oxygen plasma under the same conditions. The etch time to expose both layers of interconnect is about 15 minutes. The win-

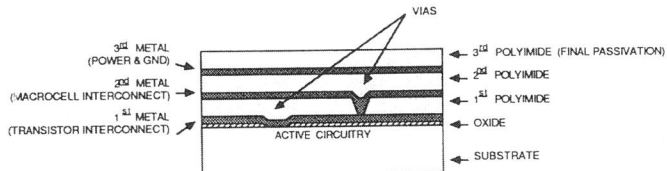

Fig. 5. Cross-section drawing of three-layer metal bipolar technology.

Fig. 6. Window in bipolar array after etching top metal. Bottom two-layers of interconnect are now visible.

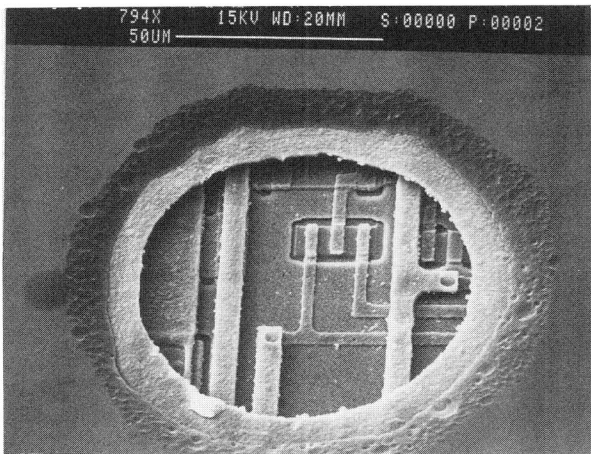

Fig. 7. SEM micrograph of completed window in a three-layer metal bipolar gate array.

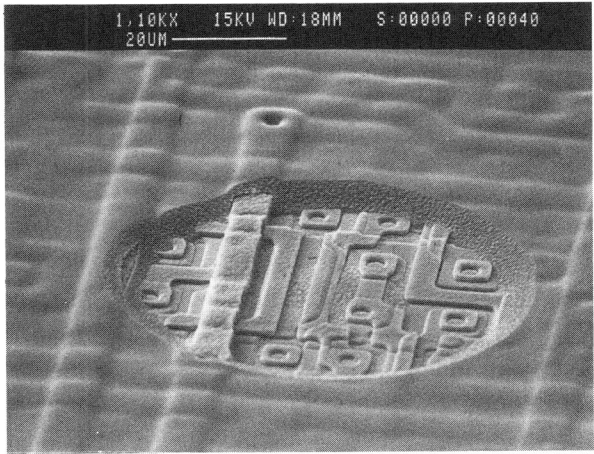

Fig. 8. SEM micrograph of a completed window in a DLM CMOS gate array.

dow is now complete and is ready for electrical probing or E-beam analysis. If desired, the Apiezon® wax can be removed with xylene and the coating of resist can be removed with the stripper at 70 °C.

Figure 7 shows a SEM micrograph of a completed window in a three-layer metal bipolar array. The aluminum interconnect has been completely exposed and the area of interest is ready for analysis.

SELECTIVE ETCH FOR TWO-LAYER METAL POLYIMIDE PASSIVATED CMOS TECHNOLOGY

Since there are few large power buses covering the active areas with CMOS technology, a much simpler procedure is used to expose the area of interest. The top and interlevel polyimide are removed in one step. Both layers are etched in an oxygen plasma with a pressure of 250 millitorr and 250 watts of power. With these conditions, the etch time is about twelve minutes. Figure 8 shows a SEM micrograph of a completed window in a DLM CMOS gate array. Both layers of metallization are exposed and the device is ready for electrical probing or E-beam analysis.

A relatively simple procedure for exposing underlying structures in multi-layer VLSI has been presented. The procedure is non-destructive and does not alter electrical performance, thereby allowing electrical characterization of an underlying structure. The procedure is useful in both failure analysis and design debug of multi-layer circuits. The equipment used to accomplish the window etch procedure is common to most failure analysis laboratories. The procedure could easily be adapted to any multi-layer VLSI technology.

REFERENCES

1. *Procedures for Probing Underlying Structures in Multi-layered-I Devices While Maintaining Electrical Integrity,* by C. Baynes, D. Kiefer, R. Milburn and C. Nakata, Proceedings of ISTFA-88, pp. 127-132.

2. *Advanced Failure Analysis Techniques for Polyimide Passivated Bipolar Gate Arrays,* by C. Baynes and R. Milburn, Micro Electronic Packaging Technology Proceedings, 1989, pp. 265-271.

Metallurgical and Specimen Decoration Techniques

COLD SAMPLE PREPARATION TECHNIQUE FOR LOW MELTING POINT MATERIALS

by David J. Burkhard and Denise Holdman

A new, three-step, cross-sectioning technique that maintains flatness across materials of differing hardnesses has been developed especially for the examination of low melting point solders. There are several advantages to this technique: reproducibility of results, ease of operation, planarity across materials of different hardnesses, and reduction/elimination of embedded polishing compound into soft materials. This polishing technique is especially well suited for the examination of Pb-Sn solders in die attach, solder coatings on leads, and in flip-chip applications. Although initially developed for solder applications, this technique is useful for most materials. A review of the procedure used and a brief overview of the polishing process mechanics leading to the development of this procedure will be presented.

The procedures used for polishing soft materials like Pb-Sn solders have been well documented.[1,2,3] Problems occur when solders are adjacent to harder materials such as a copper lead on a semiconductor package or as a solder die attach between a silicon die and a heatsink. Some polishing methods consist of using slurries or using diamond paste on a napless cloth with an oil lubricant. Slurries have been observed to embed into the solder when sandwiched between harder materials. Friction effects cause heating of the sample during diamond polishing. Polishing can be considered a form of fine grinding. For grinding, it has been established that there is a strong dependence on the surface temperature to the energy per unit surface area that is ground. Temperatures of 1650 °C or higher can be generated in a ground surface.[4] Although polishing does not generate this amount of heat, samples have been observed to become quite warm and even hot to the touch. These temperature increases can lead to surface melting, metallurgical changes, or edge rounding. The recrystallization temperature of the material being examined must also be considered to determine if the polishing procedure altered the microstructure. For example, eutectic PbSn solder has a recrystallization temperature around room temperature so any heating of the sample will change the microstructure over time.[5] Also, plastic deformation and flow can easily occur with a moderate temperature increase at the sample surface. These factors influence the amount that the solder will be recessed below the level of the adjacent materials.

The use of slurries to compensate for the heating effects is inadequate on these types of samples because the polishing compound has a tendency to embed into the solder. One way to overcome the problems associated with rough polishing is to make the solder harder by reducing any impact of heating caused by friction. One method to accomplish this is to cool the sample or platen during polishing. This has the result of keeping all of the surfaces planar because the solder becomes harder and does not as easily deform. Another benefit is that the polishing compound is less likely to become embedded into the solder surface because the solder is harder. As a result, a planer final polished surface can be easily obtained with an appropriate final polishing cloth. A general rule that can be followed is to use the grinding or polishing technique(s) for the most sensitive material present in the sample to be cross-sectioned.

TECHNIQUE

This sample preparation technique is performed on a single specimen polisher that has a stationary platen and moves the sample over the grit-coated papers and polishing cloths (Minimet® polisher from Buehler). The platens are placed in bowls in order to contain any liquid that may be placed onto the grit or onto the polishing cloth surface. Metal platens need to be fabricated to fit into the bowl (three inch diameter, 1/4 inch thick). This is to ensure adequate heat transfer during the rough polishing step. The platen can be externally cooled by pouring liquid nitrogen down the sides of the bowl along the edge of the platen. A platen could be fabricated that would be internally chilled. References to speed, time, and pressure are based upon the control units on the polisher.

The key to producing good polishes in samples where soft materials are adjacent to harder materials is to avoid any embedding of the grit, polishing compound, or pieces of the sample that may be removed into the soft material. Each of the steps have been optimized to greatly reduce the possibility of any material becoming embedded into the solder. A paraffin type wax is rubbed onto the grit paper to limit the amount of SiC that can break off and freely roll around on the grit paper. A kerosene-based lubricant is used to slowly break down the wax and provide lubrication. The second step, the key to the whole procedure, involves cooling of the platen. This has the effect of removing any temperature increase that may cause any of the problems previously mentioned. Chilling of the platen maintains excellent planarity. This second step is performed using diamond paste on a chemotextile polishing cloth. The last step uses a cloth that is a grown-in-place poromeric material with 0.05 μm colloidal silica as the polishing compound. Occasionally the solder may be tarnished after cleaning with DI water and soap. If excessive tarnish is present, the sample can be repolished and rewashed. A small amount of tarnish (oxidation) will tend to highlight the grains in the solder and has not been observed to interfere with any SEM analysis.

Grinding

Although any size grit may be used, better results can be achieved with a medium grade grit. Photograph 1 shows a typical finish of PbSn solder between a silicon die and a copper

Fig. 1. Typical sample after grinding step. PbSn solder between Si (top) and Cu (below).

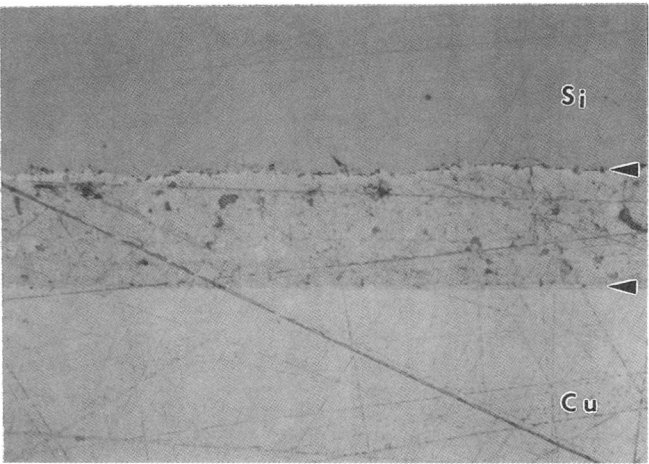

Fig. 2. Same sample after rough polishing without platen chilling. Note amount of surface relief.

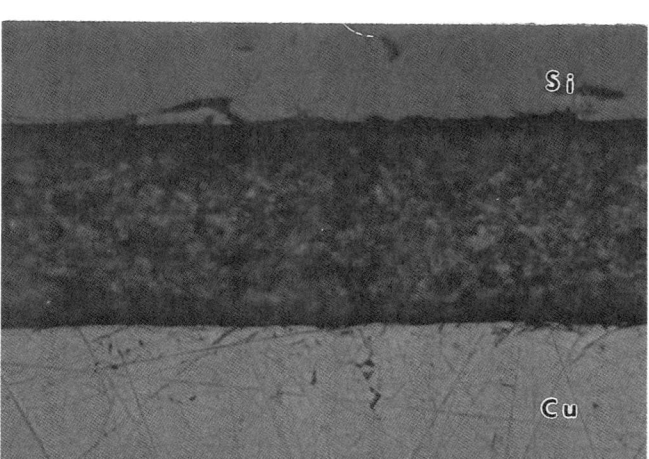

Fig. 3. Same sample after rough polishing with platen chilling. Planarity between layers is excellent.

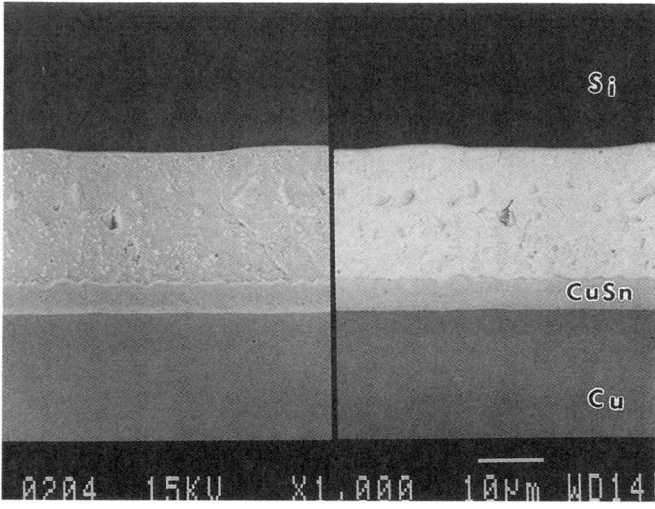

Fig. 4. Secondary and backscattered images of PbSn die attach after final polishing. Excellent planarity between layers is achieved.

heatsink after grinding. A medium pressure and low speed are used for a short period of time. Excessive grinding on a piece of grit paper wears the paper too much and may allow the grit to become embedded into the solder. Insufficient grinding may not expose the grit if too much wax is rubbed onto the grit paper.

Rough Polishing

The key to the procedure is the chilling of the platen. Liquid nitrogen is used since it is generally readily available in a laboratory environment and leaves no residue on the polishing cloth. A thermocouple to monitor the platen temperature could be placed in the platen. Temperatures less than –25 °C to –30 °C should be avoided since the lubricant starts to solidify. A good temperature range to use is somewhere between 0 °C and –20 °C. The liquid nitrogen should be slowly poured down the side of the bowl between the platen and the bowl until the platen is sufficiently cooled. A diamond grit size of approximately 3 μm is used with high speed and high pressure. This rough polishing step is repeated until all of the grinding damage has been removed. Photograph 2 is a typical image of surface relief observed during rough polishing without chilling the platen. Photograph 3 is a typical image with the same diamond size with chilling the platen. Chilling the platen maintains excellent planarity across the silicon die, solder, and copper heatsink.

Final Polishing

Final polishing is performed on a grown-in-place poromeric material with 0.05 μm colloidal silica as the polishing compound. This type of cloth is dense, has virtually no nap, and produces an excellent finish on most materials. The pressure used is approximately half of the full pressure and the speed is nearly maximum. The time for final polishing can be as long as 30 minutes with 15 minutes being typical. Photograph 4 is a typical secondary electron and backscattered electron image of a polished structure.

CONCLUSION

A readily available method has been proposed that produces excellent polishes in solders when adjacent to harder materials. This technique has the ability to maintain sample flatness across all of the materials while ensuring the microstructure of the solder is maintained. This technique is not limited to solders but can be applied to any material where improved edge retention is desired.

REFERENCES

1. *Metals Handbook, Volume 8 Metallography, Structures and Phase Diagrams,* American Society for Metals, Metals Park, OH, 1973, pp. 132-134, 138-140.

2. *Metallography Principles and Procedures,* St. Joseph, MI: Leco Corporation, 1977, p. 21.

3. Gunter Petzow, *Metallographic Etching,* American Society for Metals, Metals Park, OH, 1976, pp. 719-723.

4. George E. Dieter, *Mechanical Metallurgy,* New York: McGraw-HIll Book Co., 1978, pp 80, 84.

5. Dev Gupta, private communication.

CHIP AND DEVICE SECTIONING TECHNIQUES

by J. J. Gajda

1. ENCAPSULATION MOUNTED CHIPS

The technique described facilitates easy mounting of chips for either angle or perpendicular sections. The technique is used on chips with or without full metallization layers and bonding pads. When polished sections of the highest quality are required, this procedure is used. Several metals with different hardnesses can be prepared without damage. Molds are fabricated to produce 3X, 5X and 10X magnification angle plastic preforms as well as perpendicular preforms. Once the molds are available, several thousand preforms are made. Figure 1 shows the easy mounting of chips on angle and 90 °C preforms. Prior to mounting the chips, the preform angle mounts are put into the polishing fixture (Fig. 2) and the top of the mount is polished away leaving a planar edge for perfect mounting. Two final polishing steps will be described: one for junction delineation and the other for interfacial film identification on Al, and Al-Cu metallization.

Precise Chip Mounting Procedure

A silicon chip can be bonded to the angled surface with a suitable adhesive. Super-glue (polymethyl-methacrylate) is often used. A thin, uniform layer of adhesive is spread on the surface to keep the sample at the desired angle. A low-power microscope (40X) is needed in mounting chips. The chip must be set directly parallel with the slope of the angle. The polished line is used as a guide during mounting. The mounting step is of extreme importance. When polishing through the active portion of the transistor, the polished plane must be perpendicular to the examined area. A diagonal polishing plane is undesirable. Any size transistor, diode or integrated circuit can be also mounted in this manner. The bottoms of packaged devices are ground flat before being mounted. Some semiconductor chips, when bonded to the header, might be slightly tipped; therefore, the angle is unknown. The angles on finished samples can be measured precisely to minutes by the use of an optical goniometer. A focused light beam is reflected from the edge of the beveled surface, thus splitting the light beam. The part of the focused light beam that is reflected off the unpolished surface is projected onto a referenced comparator. The goniometer is then rotated, so the projected light beam from the beveled plan is lined up with the initial, referenced point. The angle is measured by the goniometer as the difference between the two reflected points.

After the specimen is mounted, the entire cap is placed into a larger diameter, thin-walled plastic container. The epoxy "Araldite" is then poured over the entire cap, covering the specimen completely. Araldite® was found to be most desirable. It exhibits low viscosity, enabling it to flow into cavities and adhere to the edge of the sample. In angle-polished sections, the Araldite® prevents rounding off, and a sharp angle is kept throughout the polishing cycle. The Araldite® is sufficiently hard and has a high resistance to the acids used in junction delineations. Normal curing temperature is 55 °C for 10 minutes.

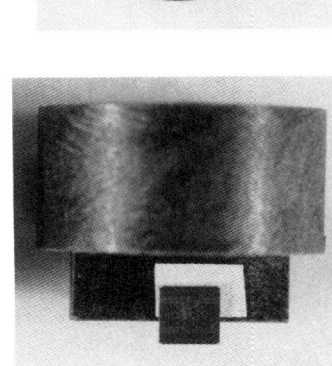

Fig. 1 Preformed plastic mounts.

Fig. 2 Polishing fixture accommodating mounts with chips.

Polishing Fixture

To retain the desired angle during rough and fine polishing, the specimen must be kept perpendicular to the surface. A suitable fixture was fabricated to accommodate the specimen mounted tightly into a holder. Figure 2 shows a metallurgical cap with a mounted sample held securely in the polishing fixture.

The entire fixture is then placed on the runners shown in the background. Grinding is performed by moving the fixture in a back-and-forth motion across the grinding surface. Movement of the sample to or away from the grinding plane is controlled by a micrometer. A micrometer was built into the fixtures so the sample could be easily moved against or away from the grinding surface. This fixture also controls the amount of material that will be ground away. Settings can be made to control the amount of material that is ground away, down to 0.5 mil. This feature is important when grinding near the transistor's active area, and it decreases changes of grinding through the area. The entire fixture, including the sample, is placed on a set of runners to facilitate a reciprocating motion across the grinding papers.

Rough Grinding

Preliminary rough grinding is started on 320-grit silicon carbide paper, lubricated with water. When the chip has a thin coating of Araldite® still around it, it should be abraded on successive papers of 440 and 600-grit silicon carbide. The edge of the semiconductor chip should be exposed with 600-grit paper. It is desirable to lap the semiconductor material on a used 600-grit silicon carbide paper with a glass plate backing. This procedure minimizes excessive chipping and reduces deep gouges. The material is also being lapped away at a faster rate than on medium polishing cloths. Extreme caution should be exercised here as some devices have active areas that are only 2 μm^2 and can be easily polished away. The progress of polishing should always be viewed under a microscope. When the wall of the diffused base junction is reached (usually recognized by steps in the SiO_2 layer), or before it is actually reached (depending on the size of the transistors active area), the 600-grit polishing should terminated.

Final Polishing

The samples should be thoroughly washed before proceeding to the final polishing. Portions of the semiconductor surface actually are smooth after the 600-grit polished step. With the surface in this condition, less time is spent in final polishing.

During final polishing, diamond paste on PAN-K® pellon paper was found most satisfactory for the suitable preparation of silicon surfaces. The hardness of diamond is such that uniform abrasion takes place, producing a flat surface practically free from relief effects. A clean fast-cutting action is obtained, along with assurance of an extremely long life of the polishing laps. A medium grade of diamond is initially used (4 μm-5 μm). The coarse grades tend to chip and gouge semiconductor materials. The PAN-K paper acts as a soft cushion pad during final polishing and holds the diamond particles securely. The low nap feature of Pellon® paper also keeps the sample flat and the angle sharp. Polishing with 4 μm-5 μm diamond paste is continued until a clear surface is obtained. High-speed wheels promote clean and fast cutting surfaces. The area of interest is usually reached at this step. Subsequent steps will remove scratches but only minute amounts of material will be polished away. The specimen is transferred to a wheel containing 1 μm-2 μm diamond on PAN-K® pellon paper. The depth of the damaged layer is further reduced by this step. Any remaining scratches are then removed by using 0.25mm diamond slurry on "Struers MOL" cloth along with a few seconds of touch up, using 0.1 μm aluminum oxide powder on "Buehler Microcloth®". Both of these last, fine polishing steps must be kept to a minimum to prevent any rounding. The polished semiconductor should be scratch-free. Reproducible results are obtainable in junction delineation work with proper surface preparation.

Examples in Junction Delineation

In performing a construction analysis of vendor devices, the hot process steps within the silicon should be evaluated. Once the silicon surface is prepared with the outlined procedure, junction delineation etches can be applied. A versatile etch consists of 20 parts nitric acid and 1 part hydrofluoric applied for 1 second. The results of the polishing procedure and junction delineation etch are shown in Figure 3. The silicon surface is scratch free after etching. The various structures, such as the trench isolation, N+ subcollector P region, and emitter implant junctions are readily discernible. From one section the following observations can be made; barrier layer undercut, barrier layer not covering entire contact opening, "mouseholes" in silicon on the uncovered side, and voids in the glass between closely spaced stripes (aspect ratio). Structure enhancement and SEM imaging will be discussed later. However, in the backscattered mode, various insulator layers can be observed. In this example, from a single plane cross section, several inherent problems with the quality of this product are revealed. The backscatter image also reveals the chip surface was planarized by an RIE etchback process, preceded by an application of spun on glass (SOG).

2. POLISHING Al, Al-Cu ON Al-Cu-Si

The following microsectioning technique utilizes the encapsulation method previously described and differs only in the final polishing step. Surface damage was minimized in the previous sectioning technique through successively fine polishing: using a 0.5 μm diamond paste, 0.1 μm Al_2O_3 slurry, and chemical polishing.

A successful alternate method of eliminating surface damage is to use a colloidal silica slurry for final fine polishing in place of the 0.5 μm diamond paste and 0.1 μm Al_2O_3 slurry. This procedure has proven to be a reliable sample-preparation technique for subsequent viewing of interlevel via hole microstructures, including resistive interfacial films and microcracks. The only additional items needed are colloidal silica, a ring stand, a 500 ml separatory funnel with stopcock plug, Tygon® tubing to fit the lower stem of the funnel, and a hosecock clamp.

Setup of the apparatus requires a dedicated polishing wheel because colloidal silica slurry is incompatible with diamond paste. The separatory funnel, with the use of the ring stand, ty-

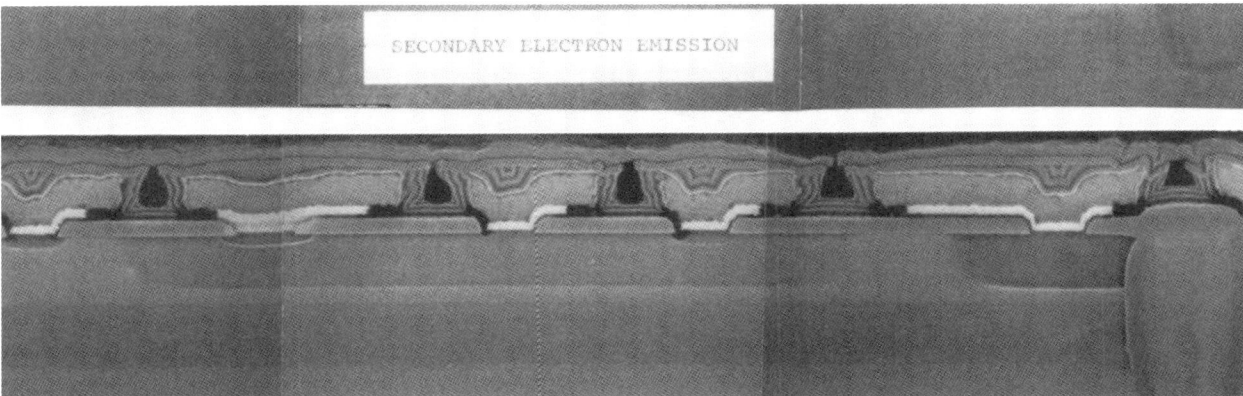

Fig. 3 Junction delineation of a state of-the-art high speed bipolar structure or device.

Fig. 4 Test site via chain.

gon® tubing and hose-cock clamp, is situated so that the colloidal silica solution can continuously drip onto a polishing wheel fitted with a low-nap polishing cloth.

An automatic polishing device holds an encapsulated chip (sample) and moves across a rotating wheel in a periodic fashion. Less than 100 grams of weight is placed on the sample during the polishing operation. It is important to check the polishing cloth for any silica particles that may have separated from the slurry prior to placing the sample on it. This is accomplished by lightly fingering the cloth as the wheel rotates at a slow to medium speed, as colloidal silica continuously drips on it. Silica grit can be removed by flushing the polishing wheel momentarily with water. The polishing wheel should always be examined before placing the sample on it.

By visually inspecting the sample periodically during the fine polishing, it can be determined when the desired cross section is attained. The next step is preparing the specimen for SEM viewing by cleaning the sample with soap and DI water, etching it in an aluminum etch, cleaning it a second time with soap and DI water, and drying it with an air gun. An etch, identified as the "slow aluminum etch", is used to delineate interfacial films between first and second level via holes. The etch has the following ingredients:

Slow Al Etch	
180 ml	—DI Water
360 ml	—Phosphoric Acid
60 ml	—Nitric Acid
72 ml	—Acetic Acid

The etch acts as a chemical polish and removes approximately 50nm of the surface after one minute. A 90-second etch time is recommended.

When cleaning interlevel via holes for interfacial films, avoid scrubbing the etched surface of the sample with a cotton swab. Scrubbing may break up the protruding interfacial film and destroy all evidence of it. Instead, dab the etched surface lightly with a soapy cotton tip while holding the mount under running DI water. Standard SEM mounting and preparation techniques apply from this point on and prior to viewing the sample in the SEM.

A Cambridge Stereoscan 250 SEM, which is equipped with a removable upper rail, easily accommodates the large, encapsulated microsectioned samples.

Examples of Metallization Fails

Via chain test sites are incorporated in chips that are subjected to process steps identical to those of the product. The

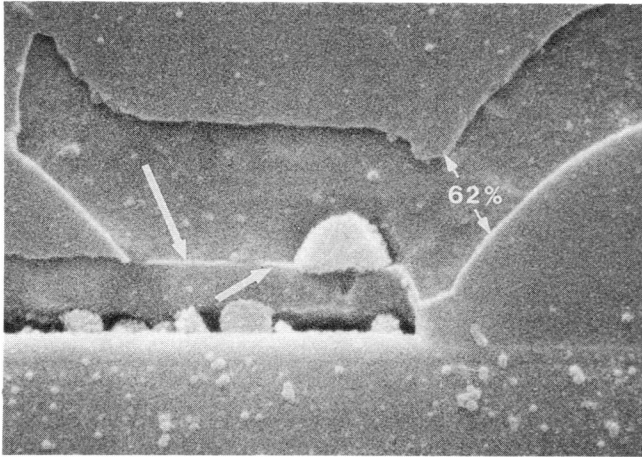

Fig. 5 Al$_2$O$_3$ film between first and second level lands.

chains are tested at each level, and problems can be observed on wafers going through the manufacturing process. The specification for a 1,200 via string of 1 μm vias is 120 ohms. In this example, a problem job exhibited a uniformly high resistance across individual rows of vias. Rough polishing was terminated in the region between rows of vias (Fig. 4). The individual row of metallization was only subjected to the silica slurry polishing procedure. Surface damage was eliminated and after employing the "slow aluminum etch", an interfacial film of Al$_2$O$_3$ can be observed to be the cause of the uniform high resistance (Fig. 5).

Another example of this particular polishing technique is shown in Fig. 6. Alloy penetration occurs when the barrier layer between metal and platinum silicide (PEAL$_2$) deteriorates. In this example, the barrier layer exhibits marginal peripheral coverage. The aluminum has come into contact with the PtSi to form PtAl$_2$ with subsequent silicon dissolution under the PtAl$_2$.

3. NON-ENCAPSULATION—BLOCK (OLD)

Description

This technique evolved from IBM's Solid Logic Transistor (SLT) line, where junction depths were measured from beveled sections. Chips without metallization were waxed down on metal blocks with given angles, polished on a fitted glass wheel and stained for junction control. In the early 1970s, with the advent of 1K bit MOS chips and bipolar integrated circuits, the technique was adapted to 90° sections. The chips were simply attached to the top of a metal block with glycol phthalate wax. The final polish was performed on a glass wheel with a finish corresponding to a 15.0 μm mesh. Obvious benefits from this technique include; no chip encasing, simultaneous optical inspection during polishing, high degree of spatial precision on dense chips, and ease of handling for further analysis. One of the objectionable features of this technique is the mechanical damage and smearing of the soft metallization. A protective glass layer had to be deposited over unprotected metal. Even with a glass layer present, the metal was not amenable to micro-

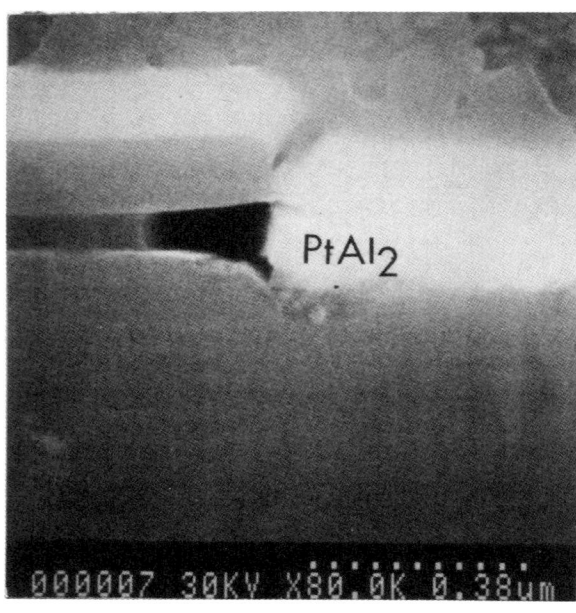

Fig. 6 Metal penetration into bulk Si.

structure etching, interfacial film determination, barrier layer integrity, and prevention of film delamination. In the 1980s with very large scale integrated circuits (VLSI) being routinely fabricated, a large portion of the chip area was allocated for interlevel metallization and various interconnections utilizing vias and vertical studs. Interconnection reliability has become the dominant factor limiting VLSI chip reliability. The present failure modes being experienced are (a) electromigration, (b) corrosion, and (c) stress induced deformation. With three and four levels of interconnections encompassing different metals and alloys on present products, new polishing procedures were developed with the non-encapsulated method.

4. NON-ENCAPSULATION— BLOCK (New)

All problems and defects associated with metal interconnections were addressed with the Araldite and successive fine grit polishing procedures. The problem with the non-encapsulation technique and glass wheel polishing was the distortion of the soft metallization patterns. To succeed in producing a damage free surface, after the glass wheel, another step was added. A bronze wheel containing a microcloth was impregnated with a Syton® silica gel (SiO$_2$- 0.5 μm). The defect region is reached using the glass wheel. The final step using the Syton takes about 30 seconds of touch up polishing, and renders the surface free of any mechanical damage. Just by changing the final procedure, results were obtained that rivaled the Araldite encasement technique previously described. The fine polish step removes about 200 μm of material; therefore, the defect region is retained for further analysis utilizing structure enhancements, techniques and SEM observation.

Other refinements to the final Syton polishing include Glanzox (Al$_2$O$_3$). Instead of glass wheels, the rough polishing can

Chip and Device Sectioning Techniques

Table 1 A Comparison of FE Gun Operation with Tungsten Filament

	Thermal Electron Gun		Fe Electron Gun
	Tungsten Filament	LaB$_6$ Pointed Cathode	
Brightness (A/cm^{-2}·sr)	10×10^6	10×10^7	10×10^9
Electron Source Diam (nm)	10×10^4	10×10^4	10
Electron Energy Range (eV)	2	2	0.2
Life (hr)	40	100	1000
Vacuum Gun Chamber (Pa)	10×10^{-3}	10×10^{-3}	10×10^{-7}

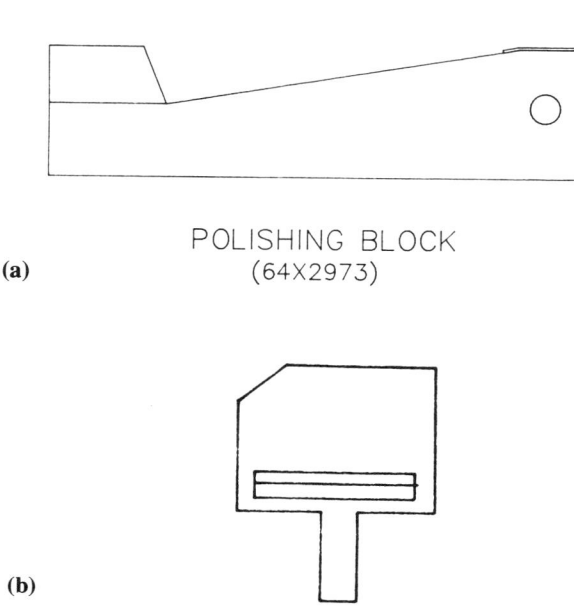

Fig. 7 Polishing block and stud insert.

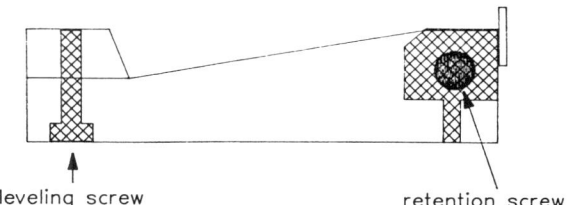

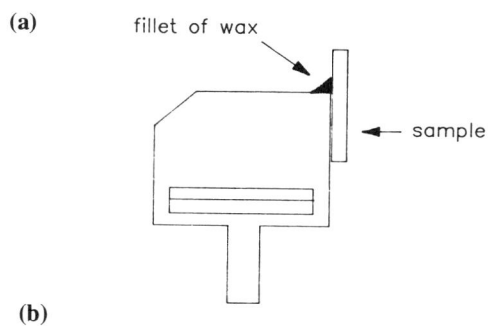

Fig. 8 Chip attachment and polishing block.

be performed on polishing pads with diamond grits from 30 mm down to 0.1 μm. Another innovation is the design of a common polishing and SEM fixture that eliminates chip removal and handling. A stud insert is placed into the polishing block (Fig. 7 and 8). The chip is placed on this insert for polishing. After all polishing and surface etching is performed, the entire stud with the chip is placed into the SEM chamber. Further polishing can be performed after SEM analysis without removing the sample from the stud.

Equipment

Polishing Table/8" Wheels	Buehler Polimet 1
Diamond Polishing Pads	3M
Final Polishing Pad	Buehler
Final Polishing Slurry	Syton HT50 or Glanzox
Polishing Block and Stud	IBM
Inverted Microscope	Unitron
Hot Plate	Corning
Stereo-Microscope	Bausch & Lomb

Samples

The area of interest should be recorded with photos and it may be helpful to mark the area with a laser, to make the area easy to locate when cross-sectioning.

Mounting

Obtain a Polishing Block (64 × 2973) and a Stud Insert (64 × 2976). See Fig. 7 and 8.

Mount the sample on the stud insert with glycol phthalate wax, using a hot plate set at approximately 130 °C. The sample should be aligned with the area of interest protruding out past the edge of the stud approximately 1/8". Alignment is easily done using a stereo-microscope, although any method that accomplishes the alignment can be used. **IMPORTANT:** remove all excess glycol wax with acetone to make the sample as clean as possible.

NOTE: It's important to remove the fillet of wax between the sample and the stud (see Fig. 8).

Leveling

Leveling of the sample should be accomplished as soon as possible and should be checked and corrected with each different grit pad as you section into the sample. When viewed on an inverted microscope it can be judged which leveling screw has to be turned. The amount the screw is turned will be learned with experience (see Fig. 8).

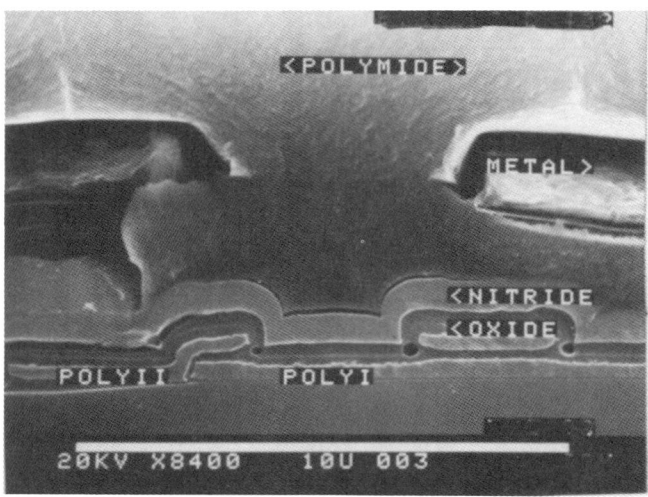

Fig. 9 Double polysilicon layers in MOS memory devices. (8400×)

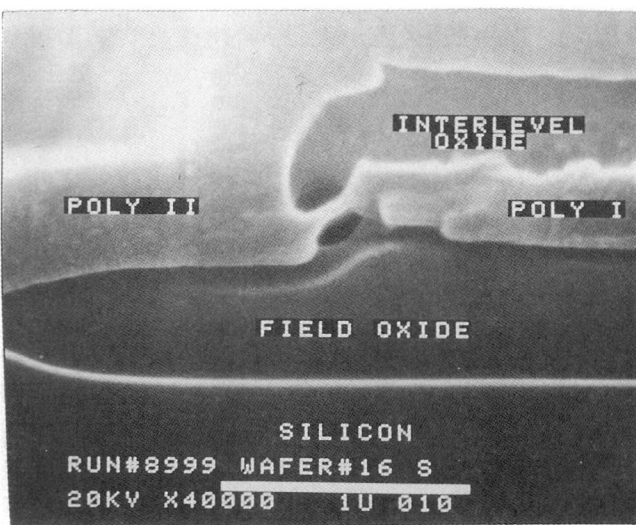

Fig. 10 Polysilicon grain slip.

Sectioning (Soft Metal)

Insert the stud into the polishing block and tighten the retention screws evenly. Leveling screws should be bottomed out.

Polishing is accomplished by using various grit Diamond Polishing Pads and Syton HT50® or Glanzox® polishing slurry. Diamond pad grits are 30, 15, 6, 3, 1, .5 and .1 micron. The grit to be used depends on the distance to be polished and the desired finish, i.e., the greater the distance the larger the grit that should be used. **EXAMPLE** - if the distance to be polished is 6 or 7 mm, you might start with a 30-micron pad and work down using various grit pads to approximately .3 mm with a .5 micron pad and finish polish into the area of concern with Glanzox. There is no set rule to the polishing steps; each person should develop a method that he is comfortable with.

Sectioning of Tungsten Studs

Identify the failed area on the sample and cut out to the proper size with a laser. Mount the sample on the stud for 90° section. Start sectioning the sample with a 15 micron (Al-Ox) grid paper or 5 micron silicon carbide grid paper, depending how close you are to the area of concern. Use water as a lubricant. When you get to within 2 microns of the defect area, switch to 1 micron (aluminum-oxide) grid paper. Stay on the 1 micron (Al-Ox) grid paper until you are 1/4 of the way into the defect area. Continue sectioning the sample with 1/2 micron diamond paste on a cloth wheel. Use a mixture of glycerine, soap, and water as a lubricant. When you have polished about 1/2 way into the defect area, switch to a 1/4 micron diamond paste on a cloth wheel using the same lubricant. Stay on this wheel for about 15-20 seconds. Finish off the sample by sliding the sample from side to side on a static Compol-80 wheel 3 or 4 times. Clean immediately after each step with detergent (e.g., Mr. Clean) and water. Blow dry with air and inspect.

5. LASER AND FRACTURE CLEAVE

The laser and fracture cleave method of cross section is an excellent choice for in-process characterization. It is simple, not time consuming, and retains the integrity of the device structure. The rapid turnaround time and SEM examination permits real-time, in-line physical characterization. This technique uses a microscope to focus the light from a laser, to scribe lines that bracket a specific structure or area in a semiconductor wafer segment or chip surface. This facilitates cleaving through that structure or area of concern. The line of fracturing is defined by the laser scribe lines and will yield a microscopically rough fracture surface which may be viewed directly.

The resulting cleavage will project through the structure or area of interest and yield a 90° (vertical) section. This cross section can be treated, if required, using preferential etching techniques to delineate specific layers or microstructures for SEM/EDX or SEM topographical analysis.

When an entire wafer, or large pieces of it, are available, a diamond tip tool can be used as a scribe. The line is scribed perpendicular to the sectioned cleave. This fast method is applicable to a MOS device with repetitive cell structures. It cannot be used in a "one of a kind" situation such as one in which a single contact has been identified as an electrical fail. The fracture cleave works well on double polysilicon structures often used in MOS memories. The polysilicon separated by polyoxide is relatively hard, and the structure is not altered. Topography and layer thicknesses can be measured very easily on an MOS device using this technique.

Laser Cleave Section

1. Laser scribe sample.
2. Cleave sample.
3. Etch, if required, to delineate structure.

Figure 9 depicts a typical SEM micrograph of a double polysilicon test device vertical structure sectioned by laser cleaving. This technique is recommended in the observation of double polysilicon layer (DPS) vertical structures above the

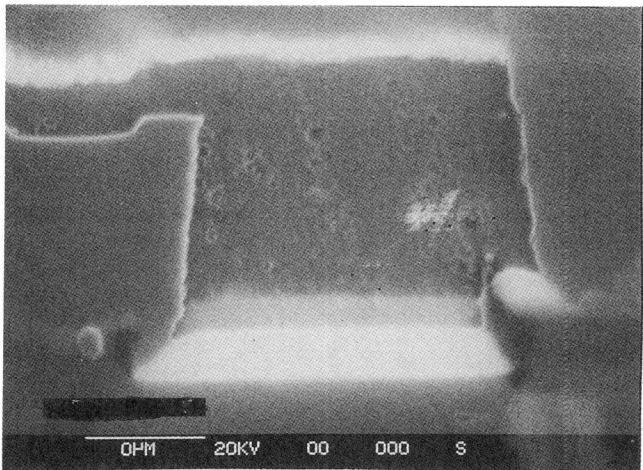

Fig. 11 Al-Cu intermetallic buildup (below interface film).

substrate surface (passivation layers, metallurgy and polyimide). The cleaving of some samples (those with layers that are ductile or flexible) may have to be done under liquid nitrogen to prevent distortion or extrusion of the layers. With this technique, many samples can be prepared for analysis in a short time.

The technique was also applied for low breakdown between polysilicon layers. The analysis revealed that polysilicon grains along poly 1 edges slip vertically when pushed up during gate 2 oxidation especially over the field oxide (see Fig. 10). The cleaved section was etched with a glycerated buffered HF solution.

Chemical Structure Enhancements

1. Al, Al-Cu, Al-Cu-Si

The previous section described sectioning techniques for arriving at the precise defect location. Initial examination using a field emission SEM can be used on unetched and uncoated samples. Once this information is obtained, chemical etching with dilute solutions can be applied to the specimen. Metallographically prepared sections usually show little distinction as conventional SEM images in the as-polished condition. Present stripe metallization conductors utilized in the semiconductor industry include Al, Al-Cu, and Al-Cu Si. With Al-Cu and Al-Cu-Si interconnections, both Cu and Si precipitates can be observed after a 90-second etch on the polished samples. The etch, known as the slow aluminum etch, was discussed in Section 2 (Polishing Al).

The etch is very forgiving with an etch time of 90 to 180 seconds. Figure 11 shows a pileup of Al_2Cu intermetallic at the lower portion of the stud after 120 seconds of etch time. After feedback to Manufacturing Engineering, it was determined the evaporation process was stopped and then started, resulting in an Al_2O_3 film barrier isolating copper-rich Al in the lower portion of the film.

The etch can also be used to evaluate step coverage in lift-off aluminum-based metallizations. During the evaporation process, the lower portion of the film in a cavity or contact hole does

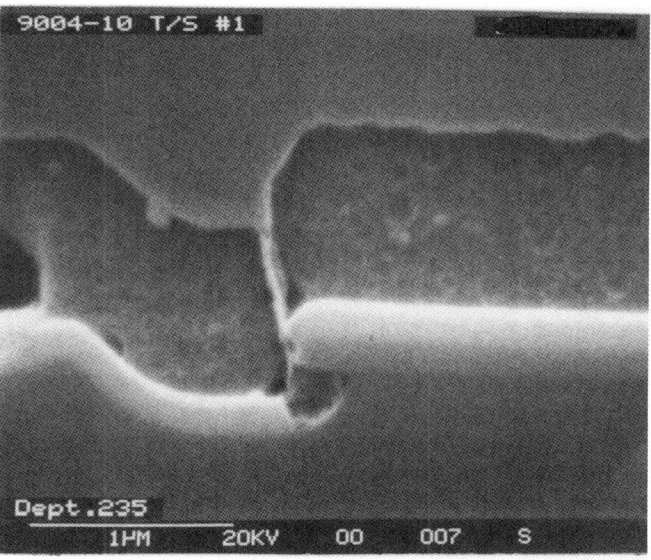

Fig. 12 Aluminum oxide seam at vertical step.

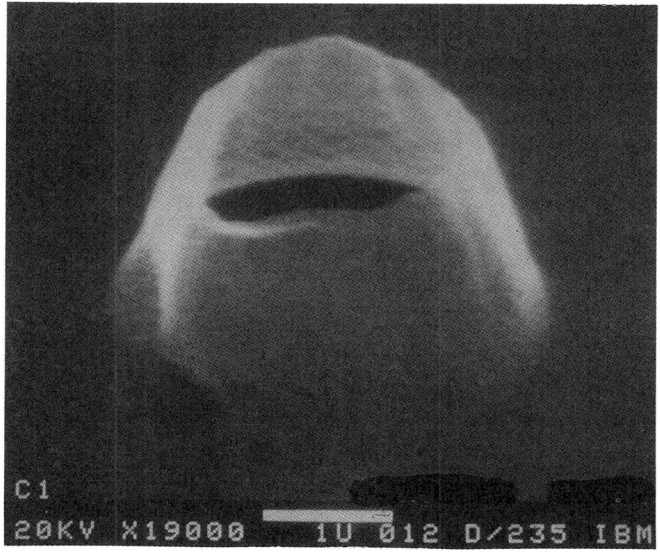

Fig. 13 Aluminum oxide shell after etch.

not connect with the elevated portion until the late stages of evaporation. This results in a sidewall oxidation and vertical oxide seam that is quite discernable after slow aluminum etch (see Fig. 12).

Another etch used primarily to evaluate the integrity of Al_2O_3 as a corrosion protective layer in reactive ion etching experiments is as follows:

Aluminum Etch

1 Part Hydrogen Peroxide

2 Parts Hydrochloric Acid

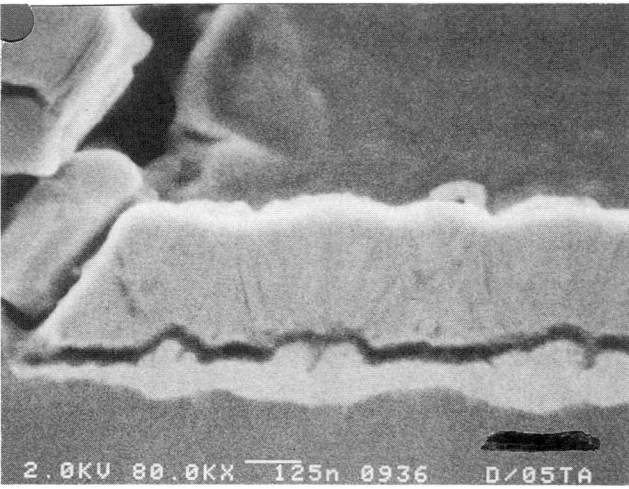

Fig. 14 Interface shown above the PtSi.

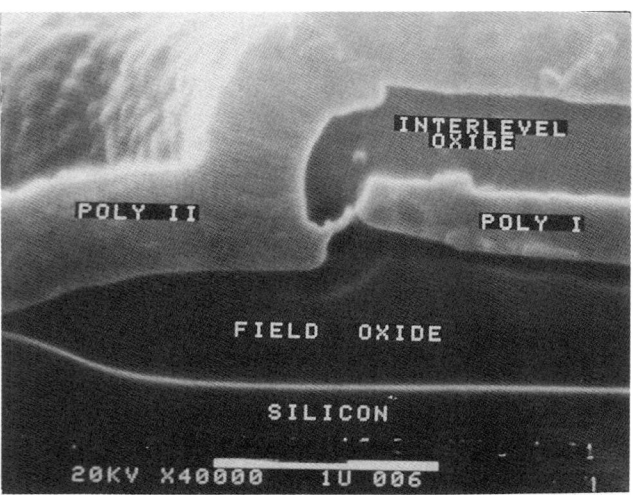

Fig. 15 Oxide etch delineating a short between polysilicon layers.

This etch is used for 120 seconds. It etches away the metallization completely, leaving an envelope of Al_2O_3 surrounding the entire metallization indicating a good quality film. In corrosion-susceptible metal, the oxide film is discontinuous and of poor quality (see Fig. 13).

6. BARRIER LAYERS

Many high performance bipolar circuits use Schottky diodes as collector clamps, and as diodes in memory cells. The barrier height of the silicide has to be maintained through the several high temperature fabrication steps used in the process. Barrier layers of Ti-N, Ti, Ti-W or $Cr-Cr_2O_3$ are often used to prevent a metallurgical reaction between aluminum-based metallization and the silicide. The barrier height will otherwise decrease upon reaction, resulting in a lower voltage needed to turn on the cell. These layers can be delineated from the silicides and metallization by the following etch:

Barrier/Silicide Delineation Etch

6gms Sodium Hydroxide

20gms Potassium Ferrocyanide

50ml Distilled Water

The solution is heated to 80 °C and the specimen is dipped into the solution for 10 seconds. The thin-film details must be analyzed prior to junction delineation because the barrier etches usually obliterate the conductor stripes.

Figure 14 depicts a Schottky diode contact with nodular Pt silicide and a composite barrier layer of Ti and $Cr-Cr_2O_3$. The nodular platinum silicide*, PtSi, caused an increase in resistance at the Ti-Pt silicide interface, which resulted in a high V_F.

---

*Platinum silicide is represented throughout this section by the chemical formula "PtSi". There are, in fact, three platinum silicides; PtSi, Pt_2Si, and Pt_3Si.

7. POLYSILICON—OXIDE LAYERS

The analysis of state-of-the-art bipolar and metal oxide semiconductors (MOS) requires an etching demand in which either the poly or bulk silicon needs enhancement or the various doped thermal oxides and deposited glass need to be highlighted. If a short is suspected between double polysilicon layers, an oxide etch can be used. An etch of the following composition can enhance shorted regions between polysilicon or metallization patterns.

Polysilicon/Oxide/Metallization Etch

80ml Glycerine

400ml Ammonium Fluoride

20ml Hydrofluoric Acid

An example is shown in Fig. 15 denoting a short condition. Etching the polysilicon can develop the grains or show surface topography problems. An etch of the following composition is used successfully.

Polysilicon Etch

400ml Acetic Acid

400ml Nitric Acid

10ml HF

8. SILICON JUNCTION DELINEATION

Etching of silicon is usually performed after all the subtle metallization etches are completed. The solutions used to show various junctions destroy the metallization integrity. Junction etches can be applied to samples prepared by the encapsulated, non-encapsulated and fracture cleave techniques. One of the most important physical parameters measured in silicon device processing is the diffusion or implanted junction depth. This measurement can be performed by electrical probing, such as spreading resistance (SRP) or a capacitance voltage (C-V)

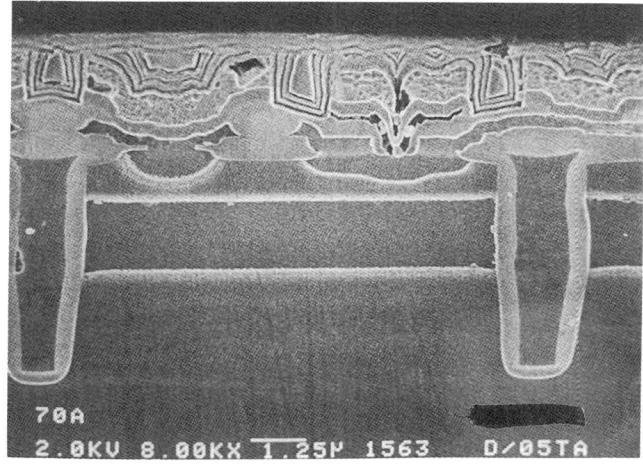

Fig. 16 Example of polysilicon base and emitter N-P-N transistor.

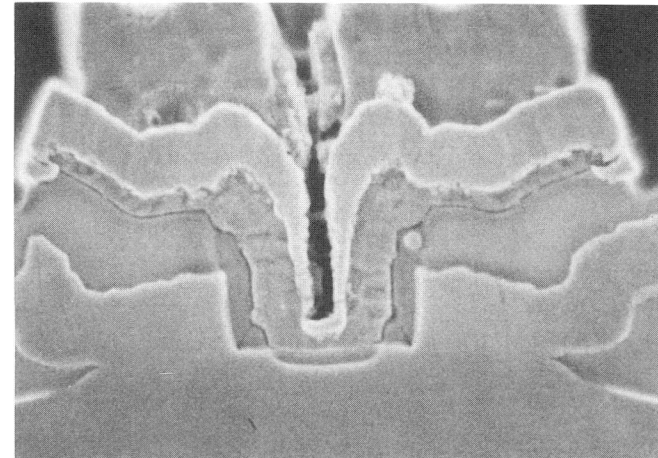

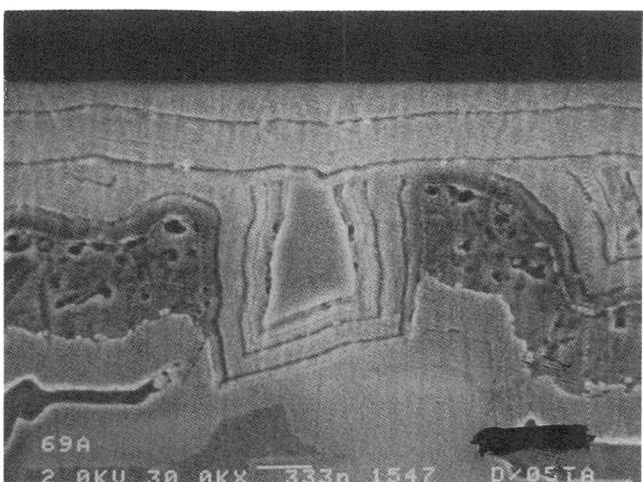

Fig. 17 Layers of doped CVD glass, silicon nitride, Si_3N_4.

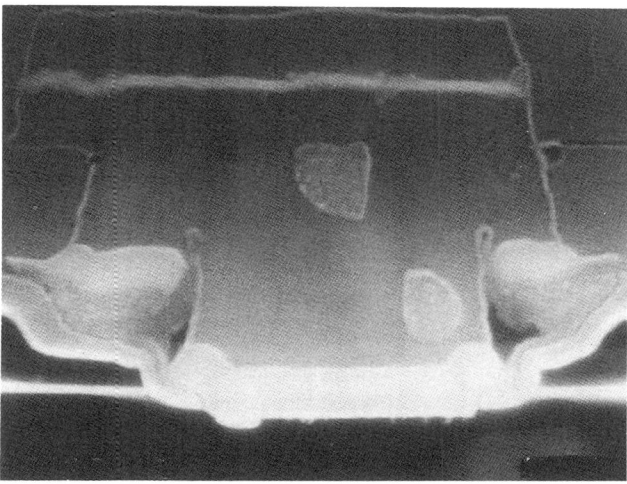

Fig. 18 Inside and outside oxide seams are delineated with RIE.

method; however, sectioned chips containing all junctions can be simply stained or etched and observed on the SEM for simple measurement. With the present resolution and magnifications on the SEM, all delineation is performed on perpendicular sections. The etching procedures utilize cotton swabs saturated with the etching solution and administered over the polished silicon surfaces. The procedure delineates silicon material of different conductivity type and dopant concentration gradient. In today's devices this includes epitaxial layer, N+ subcollector, polysilicon trench, intrinsic and extrinsic base, and emitter implant (see Fig. 3). As can be observed, this technique is useful down to the smallest device geometry and the shallowest depth.

The junction solutions commonly used are HF-based. The etching mechanism is the oxidation-reduction (REDOX) reaction, combined with differential etching rates. Mixtures of HF and HNO_3 give excellent results. One of our versatile etches is a combination of 1 part HF and 20 parts HNO_3. A wide range of HF:HNO_3 ratios has been used in failure analysis laboratories. An example of the 20:1 etching is depicted in Fig. 16; all of the junctions are delineated in this advanced bipolar structure. The same etch can be used on MOS devices with similar excellent results. In some cases it is desirable to etch samples with a 7:1 buffered HF for 10 seconds prior to junction etches. This sequence enhances the various insulator layers as shown in Fig. 17.

Once the polishing and etching techniques are operational, opportunities arise in process characterization, process fails and complete construction evaluations. Once the expertise is measured, one of a kind field fails can also be successfully analyzed.

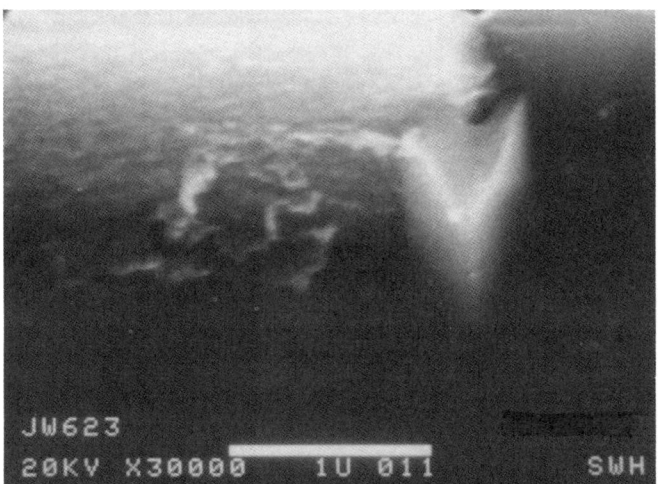

Fig. 19 Contact penetration after RIE delineation.

9. REACTIVE ION ETCHING ENHANCEMENTS

Insulator Layers

In failure analysis, plasma and reactive ion etching (RIE) were discussed in the context of selective insulation removal to expose multiple layers of interconnect metal. In this section, perpendicular cross sections will be structurally enhanced as was shown with chemical etching. In some cases, combinations of both dry and wet etching will be shown. In advanced bipolar devices, emitter and base polysilicon, silicon nitride, thermal and various deposited glass layers, metallization, contact silicide, and barrier layers have to be delineated. Chemical etching by itself is not sufficient to distinguish all layers.

Fluorocarbon gases commonly available in analysis laboratories (e.g., CF_4 and C_2F_6) combined with oxygen, are used as structure delineators or contrast enhancers. Mixtures of 10% to 20% oxygen with CF_4 or C_2F_6 were shown to give satisfactory results on metallized patterns. Silicon nitride layers are etched faster than thermal oxides. The chemically vapor deposited (CVD) glass layers can be differentiated according to the level of their doping. They etch twice as fast as undoped oxides. The polysilicon and bulk silicon have the higher etch rate at the 10% to 20% oxygen levels. In going to the 53% oxygen range, the oxide etch rate is higher than the polysilicon etch rate.

In failure analysis, many problems are encountered in the interconnection metallurgy in which aluminum oxide seams at steps cause field or reliability problems. In these problems, the combination of chemical etching and plasma etching becomes an important tool. Figure 18 shows an oxide seam after a slow aluminum etch and a DE-l00® plasma etch (10% oxygen - 90% CF_4). The same etch can be used to remove the silicon from beneath contacts on polished sections. By tilting the specimen the entire contact can be observed for evidence of barrier breakdown (see Fig.19). The samples depicted were all sectioned with the advanced encapsulated technique previously described. All dry etching is performed with the chip still on the polishing fixture, with the same fixture compatible for the SEM examination.

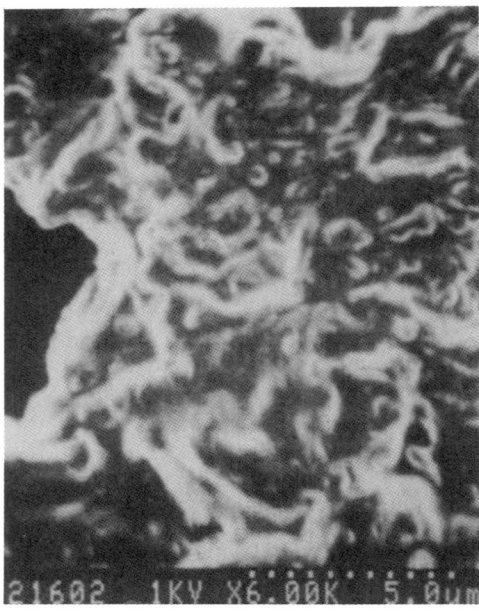

Fig. 20 Ceramic substrate at 1 kv. No visible grains.

10. SCANNING ELECTRON MICROSCOPE (SEM)

Thermal Electron Gun (SEM)

Today's increased circuit complexities and decreased device geometries dictate the need for a high-resolution scanning electron microscope (SEM). One SEM used in the work described here was a Cambridge Stereoscan® 250 equipped with a turbo-pumped, 10 port chamber, lanthanum hexaboride (LaB_6) electron source, a series 100 high-resolution stage and a series 200 microanalysis stage. The large chamber is designed with a removable upper rail in which the specimen stage rides. By removing the upper rail, an additional 25mm of working distance is achieved to facilitate viewing of large specimen such as encapsulated micro-sectioned samples and ceramic substrates. Interfaced with the SEM is a Princeton Gamma-Tech System III® Microanalysis System, allowing analysis of integrated circuits. The other SEM's used are an AMR 1400, with an ion pumped chamber and a LaB_6 electron source; and a Hitachi S-500, with a diffusion pumped chamber and a tungsten electron source.

Before any sample is prepared for examination in the SEM, variables such as size, shape and conductivity properties of the specimen must be taken into consideration.

The size and shape of a specimen to be analyzed are determined by two characteristics of the SEM: the size of the SEM chamber; and the degrees of freedom afforded by the SEM stage. Larger chambers, such as the one previously mentioned, will allow for longer working distances and X, Y, Z movements. This makes it possible to examine large samples such as portions of wafers and modules, with minimal sample preparation. In addition to viewing these large specimens, small microcircuits must also be examined. This requires large distances to be

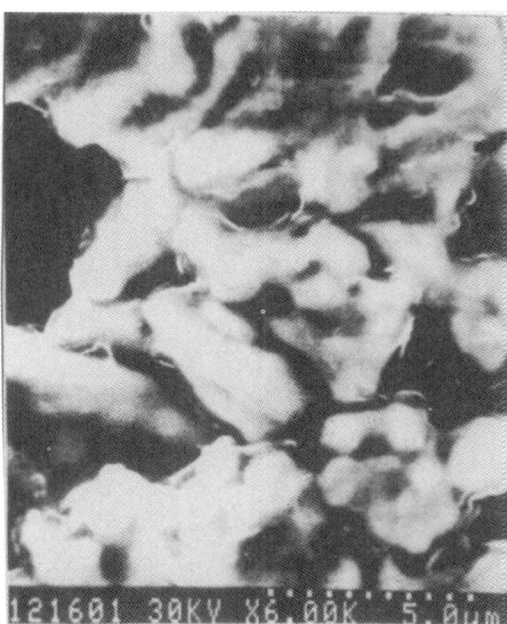

Fig. 21 Ceramic substrate at 30 kv, showing grains.

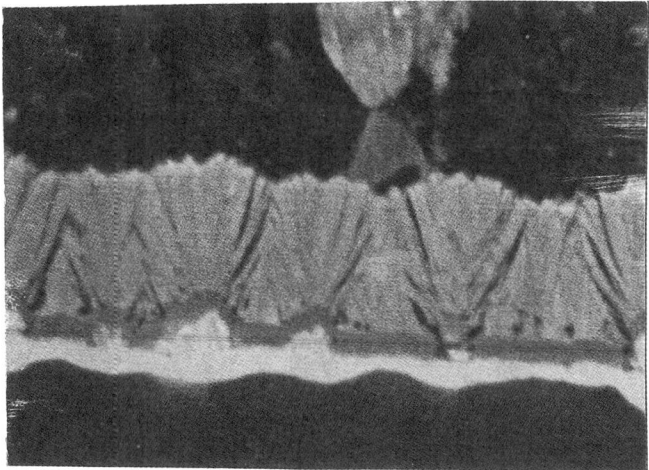

Fig. 22 Interface observed above Cr-Cr_2O_3 layer. (100× magnification)

covered along the Z-axis in small working distances for certain energy dispersive x-ray (EDX) analyses. In short, high resolution operation of the SEM requires small working distances to 10 mm, while EDX analysis may require working distances to 40mm. This establishes the need for a larger specimen chamber on a SEM dedicated to failure analysis of integrated circuits.

The vast majority of scanning electron microscopes use a tungsten hairpin cathode as an electron source. The tungsten hairpin is the easiest electron source to use because it is simple in construction and works well at pressures up to 10^{-4} Torr.

On a SEM with a lanthanum hexaboride (LaB_6) cathode system, the following advantages can be realized: 1) higher resolution, 2) longer emitter life, 3) increased beam brightness (approximately ten times that achieved by tungsten hairpin filaments) and 4) greater emitter stability. The stability of a LaB_6 emitter makes it extremely efficient in obtaining accurate quantitative microanalysis results. The increased brightness with LaB_6 allows the use of TV imaging at high magnifications with excellent signal-to-noise ratios. This enables a sample to be viewed more rapidly as it is manipulated in the chamber, with increased throughput for quality control applications. Because of the smaller spot sizes obtained, and the higher brightness available with a LaB_6 emitter, higher resolution at low accelerating voltages can be achieved. An operating pressure of 10^{-6} Torr or better is essential for optimum LaB_6 operation. This is in contrast to the 10^{-4} Torr required for tungsten filaments.

SEM Sample Coatings

With the exception of highly conductive materials such as metals, most sectioned samples going into the scanning electron microscope require a conductive coating before they can be examined adequately in the SEM. Two methods that have proved to be satisfactory are sputter coating and high vacuum evaporation. Sputter coating can be accomplished using a Technics Hummer V® Sputtering System with a Au/Pd cathode. Use of a quartz deposition thickness monitor ensures consistent and accurate film thickness from sample to sample. A thickness of 15.0 to 20.0 nm proves adequate in preventing charge buildup during examination in the SEM. Au/Pd is used because of its low granularity as compared with gold. This yields a very thin continuous film which is more polishable than gold.

One advantage of sputter coating is that it provides a continuous coating layer on parts of the specimen that are not in line of sight of the target. This means that highly textured structures or complex surfaces will be adequately coated. The ability of the target atoms to "go around corners" is particularly important in coating non-conductive passivation layers, ceramic substrates and sharp edges created by 90° cross sections on unencapsulated microcircuits.

High-vacuum evaporation of carbon is used to prepare specimens for x-ray analysis. However, upon occasion; such as when an insulator is being studied for topography; carbon and metal can be used as a coating. This can reduce specimen charging effects.

Field Emission SEMs (FESEMs)

There are many advantages of FESEMs over tungsten filament or lanthanum hexaboride (LaB_6) pointed cathode guns. Three primary ones are: beam brightness, resolution and low voltage operation capability.

Table 1 (10) compares FE gun operation with tungsten filament and LaB_6 guns. Equation 1 represents the spatial resolution of SEMs in general. Since the energy spread is so low in FE guns (0.2 eV), the image spread is 1/10th that of a tungsten filament gun, for example. This, coupled with the increased beam brightness, allows higher resolution at low operating voltages.

Low operating voltages and higher resolution mean less beam penetration, making the FESEM the most suitable instrument for producing secondary electrons near the sample sur-

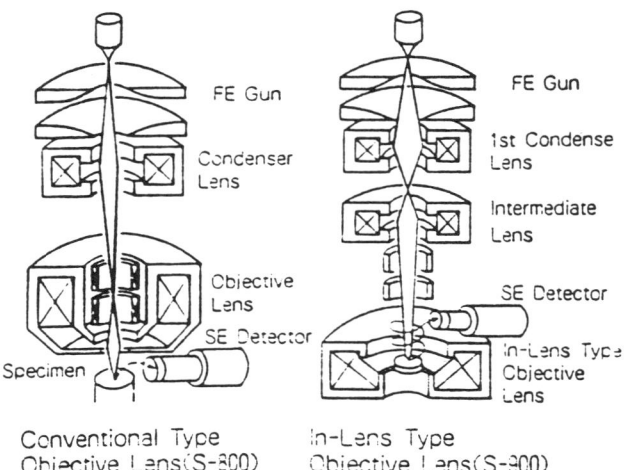

Fig. 23 Schematic of conventional and "in-lens" system.

face without having to coat the sample.

$$DC = \frac{\Delta V}{V} \cdot Cs \cdot a \qquad (Eq\ 1)$$

Where: V = Electron energy spread
V = Accelerating voltage
Cs = Aberation coefficient of objective lens
a = Aperture angle

The advantage of low voltage operation can be seen in Fig. 20. Figure 21 shows a ceramic substrate viewed at 30 kv. The grains of the ceramic material are apparent. However, when the precise area of the same sample is viewed using 1kv, the grains are no longer visible.

The grains were coated with an organic bonding material that was transparent at 30 kv. Here is an example of the need for alternate operating parameters to differentiate varying conditions.

The main trade-off that must be considered in FESEM vs. other SEM types is vacuum. The required four orders of magnitude is achieved by ion gun pumping in the column, but at the cost of longer pump down time/lower throughput.

Thermal Field Emission

The present state-of-the-art thermally assisted source SEMs deliver high resolution with material contrast needed for polished cross sections. Resolution of 5 nm at 1 kv can be attained on the International Scientific Instruments (ISI) DS-130F® scanning electron microscope. At high kv operation the resolution capability is 2 nm at 30 kv. Improvements to the system were made using a newly developed high-brightness thermal field emission gun (TFEG) and an improved condenser lens system. These improvements, along with the unique "dual stage" configuration, provide ultra high resolution in the top stage and excellent specimen handling capability, combined with exceptional resolution in the bottom stage.

The benefits derived from these systems include cross sec-

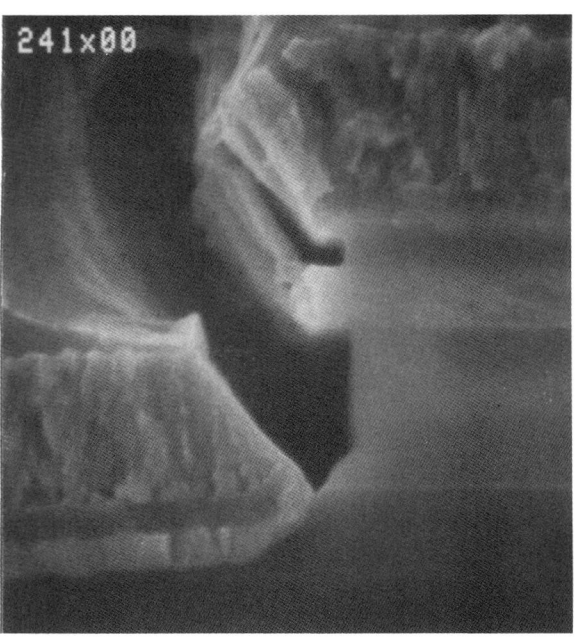

Fig. 24 Buffered HF undercut of the oxide (80× magnification).

tioned devices, which do not have to be coated, and usually don't have to be etched, for further enhancement. This improvement becomes very important in the examination of barrier layers such as Ti, Ti N, Pt Si, Cr, Cr_2O_3, TiSi, TaSi, and associated interface interactions. As these films are being reduced in thickness both coverage and interface resistance problems cause yield related concerns. In one problem set on bipolar chips with Schottky barrier diode, high V_F readings on test sites were observed. Several of the individual diode structures were sectioned using the previous outlined techniques. Examination of the interface using the thermal field emission SEM on uncoated or unetched revealed an interface problem above the Cr-Cr_2O_3 barrier layer (see Fig. 22). The PtSi Schottky Contact was also highly irregular indicating the original Si surface contained some contamination on an oxide film. This information was given to photo process engineers and to the metals group for quick problem solving.

Cold Cathode Field Emission

A cold cathode field scanning electron microscope (CFE-SEM), with an ultra high resolution "immersion lens" has been introduced to observe semiconductor devices. Combinations of a high-brightness, field emission electron source and a highly excited objective lens (to minimize aberrations) was shown to achieve 1nm resolution at 30 kv and 3 nm at 1 kv. A schematic of the electrical optical system of a conventional FE-SEM and "in-lens" system appears in Fig. 23. In the "in-lens" objective the probe diameter can be made smaller and signal collection improved by a through lens configuration. This system gives high resolution imaging and becomes important as devices are designed with submicron design rules. Low accelerating voltages are needed to increase compositional contrast.

This CFESEM, introduced by Hitachi, is known as the S-

900. In recent papers it has been described as obtaining resolutions equal to that of a 300 kv transmission electron microscope (TEM) when used on semiconductor cross sections. This utilization of this SEM over a TEM provides many advantages. Sample preparation time is greatly reduced, considerably more observation area has been increased, and TEM artifact interpretations are eliminated. Cross-sectioned chips can be polished further to image many planes and reviewed sequentially in the SEM. By contrast, the TEM preparation allows only one sectional plane after complex and time consuming preparation.

An example utilizing the CFESEM is depicted in Fig. 24. Both insulator and metal layers can be observed at low kV operation. In these experiments, the extent of undercut oxide can be readily observed in over etched contact holes.

The disadvantage in using both the thermal field and cold cathode SEM is the incompatibility of the polishing stud inserts. After fine polishing, the chip has to be removed from the polishing insert and put on a special fixture designed precisely for field emission SEM. During this transfer, cross sectioned chips can easily be damaged or contaminated if extreme caution is not used. Waxed-down chips should not be heated for removal. Organic films deposited on the chip surface during heating can be observed with low kv analysis. Compensations made during the cross section operation, such as extending the chip a longer overlap distance, helps later in chip transfer. The edge of the chip is now scribed with a diamond tip and ready to place in to the special fixture for field emission SEM examination.

Special thanks to...

Terry Irish for his contribution on the equipment and mounting procedures using the nonencapsulation block polishing method.

Francis Trudeau for his introduction of thermally assisted field emission "immersion lens" SEM into the IBM East Fishkill failure analysis laboratories.

B. Averill, D. Betterton, S. Aliotta, G. Hawks, V. Tom and K. Maloney for their dedication and excellent work on failure analysis. And to E. Hearn and S. Boettcher for contributing examples from the cold cathode field emission SEM.

APPENDIX SUPPLIES LIST

Item and Manufacturer

Syton
Monsanto Inorganic Chemicals Division
800 N. Lindenburgh Blvd.
St. Louis, MO 63166

Glanzox— 3250
Compol-80
Fujimi Corporation
Compol— 80
747 Church Road
Elmhurst, IL 60126-1420

Araldite
Ciba Products Corp.
Robinson Bros. Chemicals, Inc.
Brooklyn, NY

Glass Discs
Grayco Optical Corp.
91-4 Colin Drive
Holbrook, NY 11741

Diamond Discs
3M Co.

Mr. Clean
Proctor and Gamble
Cincinnati, OH 45202

Shipley AZ1350
Shipley Company

MOL Cloth
DP Cloth
Pellon Paper

Struers Scientific Inst.
Copenhagen, Denmark

Microcloth
Carbimet Paper
Buehler Ltd
41 Waukegan Rd.
Lake Bluff, IL 60044

Diamond Compounds
Paste 0.5 um
Industrial Diamond Laboratories, Inc.
528 Tiffany St.
New York, NY 10474

REFERENCES:

1. AMR Technical Bulletin 1105-174, *Specimen Charging and the Effect of Acceleration Potential in the SEM*.
2. Dansky, A. H., 1981. "Bipolar Circuit Design for a 5000-Circuit VLSI Gate Array," IBM Journal of Research and Development: 25, 116.
3. Danyew, R. R. and Hammond, B. R., 1989, "A Microfinishing Technique For Semiconductor Failure Analysis". International Sympo-

sium For Testing and Failure Analysis Proceedings, pp. 161-165.

4. Danyew, R. R. and Fitzgerald, W. F., 1980, Polysilicon Sectioning and Delineation Techniques. IBM Internal Technical Report 19.0495.

5. Favaron, J., 1990. "Plasma Delineation of Silicon Chip Cross Sections". International Symposium For Testing and Failure Analysis Proceedings, pp. 117-120.

6. Ferris, S. D., Joy, D. C., Leamy, H. J., and Crawford, C. K., 1975. "A Directly Heated Lab Electron Source". Eighth Annual Scanning Electron Microscopy Symposium, pp. 12-18.

7. Gajda, J. J. and Trudeau, F. G., 1991, "Utilization of High Resolution SEM in the Failure Analysis of Bipolar Chips". International Symposium on Testing and Failure Analysis. American Society for Metals.

8. Gajda, J. J., 1990. "Failure Analysis of Semiconductor Devices". VLSI Reliability Seminar, Seoul, Korea.

9. Gajda, J. J., 1987. Failure Analysis Tutorial, International Reliability Physics Symposium.

10. Gajda, J. J., 1987. Component Reliability Through Test Site Analysis, In *Reliability Key to Industrial Success,* ed. S. Sunderesan and D. Keccecrough, pp. 165-173, American Society for Metals.

11. Gajda, J. J. and Wilson, H. R., 1985. Electrochemical Chemical Society Fall Meeting.

12. Gajda, J. J., Irish, T. H., and Trudeau, F. G., 1983. "Novel Techniques for Analyzing Device High Via Resistance".International Symposium on Testing and Failure Analysis.

13. Gajda, J. J., Grose, D. A., Longo, R. A., and Wildman, H. S., 1982. A Simple Evaporation Process for Producing Improved Interlevel Via Resistance. IEEE Transactions on Components, Hybrids and Manufacturing Technology, Vol. No. 4, December.

14. Gajda, J. J., 1982. Failure Analysis Instructional Teaching Manual. IBM Internal Course.

15. Gajda, J. J., Lindstrom, G. J., and DeLorenzo, D. J., 1981. Interlevel Insulation Reliability Evaluation. IEEE Transactions on Components, Hybrids and Manufacturing Technology, Vol. 4, No. 4, December.

16. Gajda, J. J., Trudeau, F. G., and Wade, J. A., 1981. Semiconductor Structure Enhancement for SEM Analysis. International Symposium for Testing and Failure Analysis, pp. 11-21.

17. Gajda, J. J., DeLorenzo, D. J., and Wade, J. W., 1979. "Failure Analysis Techniques and Failure Mechanism Identification Utilizing a Plasma Etcher". International Reliability Physics Symposium, April.

18. Gajda, J. J., 1974. Techniques in Failure Analysis. Proceedings of International Reliability Physics Symposium, pp. 30-37.

19. Gajda, J. J., 1964. Evaluation of Semiconductors Through Angle Sectioning and Junction Delineation. IBM Internal Technical Report, 22.111.

20. Ghate, P. B., 1986. Reliability of VLSI Interconnections. American Institute of Physics.

21. Kehl, G. L., 1949. *Principles of Laboratory Practice,* McGraw-Hill, p. 424.

22. Morrissey, J. M., Wade, J. A., Caraballo, R. A. and LaColla, R., 1981. Silicon Dioxide Defect Location and Analysis on VLSI DPS Structures. International Symposium on Testing and Failure Analysis, October.

23. Negatini, T. and Saito, S., 1989, "Development of a High Resolution SEM and Comparative TEM/SEM Observation of Fine Metal Particles and Thin Films". Instrument Physics Conference Proceedings, Chapter 12, pp. 519-522.

24. Reilley, T., Pelillo, A., and Miner, B., 1990. Comparison of Ultra High Resolution Immersion Lens CFESEM and TEM For Imaging of Semiconductor Devices. 12th Proceedings of International Congress for Electron Microscopy, pp. 622-623.

25. Thomas, S., 1983. Scanning Electron Microscope Analysis of Fracture Cross-Sections In Integrated Circuits Process Characterization, *Scanning Electron Microscopy,* IV, pp. 1585-1593.

26. Yamada, M., Suzuki, K., Watanabe, and T., Nagatani, T., 1986. *Hitachi Instrument News,* Ninth Edition, p 24.

CHEMICAL ETCH FORMULATIONS AND HISTORY

by Thomas W. Lee

Aluminum etch is a formulation which usually consists of a mixture of 65 to 75% phosphoric acid (H_3PO_4), 1 to 5% nitric acid (HNO_3), with 0 to 15% glacial acetic acid (CH_3COOH) and water as buffers. Surfactants should be used with this etch to detach bubbles of hydrogen which form and passivate the active metal surface from the action of the etch, producing artifacts. Recirculation of the solution is also necessary.

Aqua regia, "water of kings," was named in Latin in the days of alchemy because it is one of the few liquids which will dissolve gold. The only other gold solvents are the alkali cyanides, NaCN and KCN, and a solution of KI and iodine. Aqua regia is a mixture of nitric (HNO_3), and hydrochloric (HCl) acids. Controversy exists over its classical composition, usually taken to be any variation along a scale from 1:3 to 3:1 nitric/hydrochloric, and in a range of concentrations. The composition can be varied for particular metals. Aqua regia dissolves all common metals and gold (Au), osmium (Os), platinum (Pt), and palladium (Pd). It does not dissolve tantalum (Ta), niobium (Nb), titanium (Ti), ruthenium (Ru), rhodium (Rh), or iridium (Ir). Consequently, aqua regia has been important in the early separation process steps in the refining of complex metal ores for years. Aqua regia is a good broad-spectrum etch for removing metals from silicon dice. It will also attack photoresist and other organics.

Buffered oxide etch (BOE) consists of a mixture of ammonium fluoride (NH_4F) and hydrofluoric acid (HF) in various ratios. The HF is buffered with NH_4F because it etches more uniformly when buffered. Also, unbuffered HF will attack photoresist materials. The NH_4F:HF ratio is important. Ratios greater than 4:1 will result in crystals of NH_4F dropping out of solution. The etch rate is dependent upon this ratio. For example, 5:1 etches at about 1200 Å/min, 10:1 at about 550 Å/min, 25:1 at about 250 Å/min, and 50:1 at about 150 Å/min. BOE can also be known by a variety of names, such as; "Bell's No. 2," "COE" (common oxide etch, "7:1," etc, etc). BOE is a good passivation and oxide stripper, and it will also delineate titanium nitride. BOE is used to prepare silicon surfaces for hot processing, and to etch openings in SiO_2 layers. BOE is unstable when exposed to air, and will evaporate to leave colorless crystals of NH_4F behind.

Dash etch is commonly thought to have been named because it is applied to the sample quickly, or "dashed-on." This is not true. The etch was developed by Mr. Dash, and originally for *germanium*. The development of Dash Etch in 1956[1] precedes Sirtl Etch, and Dash's work is referenced in Sirtl's paper.[4] Dash etch consists of HF and nitric acid buffered with acetic acid. Dash etch is used for junction delineation. When the specimen is held beneath a strong light for a few 10s of seconds, the junctions will be stained in order of doping concentration. This etch does not accentuate polishing damage. Dash Etch has also been called "white etch," "Bell's #1," "silicon etch," "10:1," "dislocation etch," etc, etc. When buffered with acetic acid, Dash Etch is generally named for the specific formulation, such as "1-3-10." See "Rosetta Stone," Fig. 1.

Pinhole etch usually consists of diluted HF-HNO_3. Pinhole etch delineates voids in thin films by selectively etching the silicon beneath. The selectivity of HF-HNO_3 to silicon over oxide is poor, so the etch must be used carefully. Pinholes expose the silicon layer beneath, and result in shorts if they are large enough to allow deposited metal to penetrate. Pinholes can be decorated by plating with silver or copper, and revealed by etching the layer beneath. Pinholes can also be revealed electrolytically by immersing the specimen in methanol and applying a large D.C. bias, which causes bubbles of hydrogen to be formed.

Pirahna etch consists of sulfuric acid, H_2SO_4 and hydrogen peroxide, H_2O_2. Pirahna etch is a powerful organic stripper, and its name is said to be derived from this property; it "eats everything." A very similar material is peroxydisulfuric acid (PDSA), which is manufactured by electrolyzing sulfuric acid.

Schimmel etch consists of Sirtl Etch buffered with acetic acid. For more information, see reference 2. The formula consists of 75 gms of CrO_3 dissolved in 1 liter of D.I. water. The quantity of 49% HF that is added to the solution is varied with the resistivity; for 0.2 Ω-cm and above, it is mixed with two parts of HF. For resistivities below 0.2 Ω-cm, the former solution is cut with 0.5L more water. Delineation occurs in 5 to 15 minutes, with the lower resistivities etching more quickly. Schimmel etch will delineate dislocations, slip planes, swirl defects, stacking faults, and oxygen clusters in <100> material.

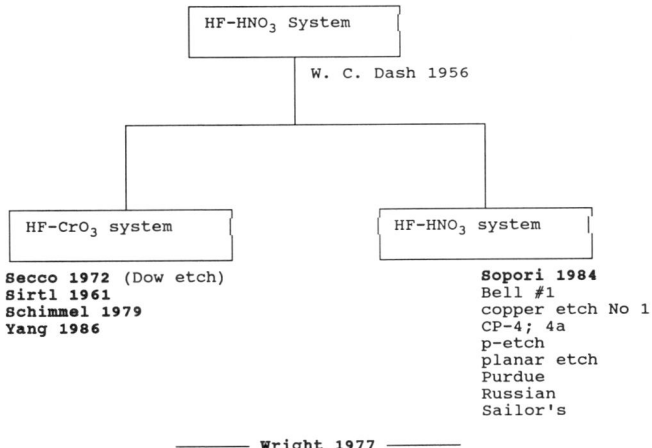

Fig. 1 The Rosetta Stone of Silicon Etch Families

Secco Etch consists of one part by volume of 0.15 molar potassium dichromate ($K_2Cr_2O_7$) and two parts of hydrofluoric acid (HF); 44 gms of dichromate to 1 liter of water. This stock is then added to two parts 49% HF. Secco Etch is slow-acting, requiring about 20 minutes to produce decoration in material with resistivities from 1 to 10,000 Ω-cm. Ultrasonic agitation reduces this time to 5 minutes. For more information, see reference 3. Secco Etch produces elliptical features at dislocations in both <100> and <111> silicon, and delineates swirl defects, etch pits, stacking faults, and oxygen clusters. Secco etch does not delineate slip planes. Secco etch is sometimes called "Dow etch," because Secco did this work at Dow-Corning.

Sirtl Etch. In their 1961 paper, Von Erhard Sirtl and Annemarie Adler of Siemens in Munich, Germany describe a chromic acid etch which decorates defects, such as dislocations and stacking faults, in silicon. The mix specified in the paper was 46

Liquid Etchant Reaction Chart for Microelectronic Materials

	HF	HCl D	HCl C	HNO_3 D	HNO_3 C	$HClO_4$ D	$HClO_4$ C	H_2SO_4 D	H_2SO_4 C	H_3PO_4 D	H_3PO_4 C	$H_2C_2O_4$	HNO_3/HCL	HNO_3/HF	CrO_3/H_2SO_4	KOH D	KOH C	KI/I_2
Ag			h	+	+	+	+		+		h		+	+	+			+
Al	+	+	+	+	h	h	h	h	+	+	h	+	+	+	+	+	+	+
Au																		h
Be	h	+	+	+	h	+	h	+	+	+	+	+	+	+	+	+	+	h
Cd	+	+	+	+	+	+	+	+	+	+	+	+	+	+	+	+	+	h
Co		+	+	+		+		+	+	+	+		+	+				h
Cr	+	+	+							+			h	+				h
Cu			h	+	+	+	+		+		h	+	+	+	+		+	+
Fe	+	+	+	+		+		+	+	+	+	+	+	+	+		h	h
Ge			h		h		h	h					+	h	+			+
Hf	+		h						+		h		+	+				
Ir																		
Mg	h	+	+	+	h	+	h	+	+	+	+	+	+	+	+		+	+
Mn	+	+	+	+	+	+	+	+	+	+	+	+	+	+	+			+
Mo				+	h	+	h		h				+	+				
Nb	+									h				+		h	+	
Ni		h	+	+		+		+	+	+	+	+	+	+				h
Os		h	h										h					
Pb		+	+	+		+				h	h		+	+		+	+	
Pd			h	+	h	+		h	h		h		+	+	h			
Pt									h		h		h	h				
Re		+	+	+	+	+	+	h	h				+	+	h		h	
Rh									h						h		h	
Ru																		
Si														+		+	+	
Sn	+	h	+	+		+		h	+	+	+		+	+	+	h	h	+
Ta	+								h				+				+	
Ti	+	h	+					h	+	h	+	h	+					
V	+		h	+	+	+	+		+		h		+	+	+			
W									h				h	+				
Zn	+	+	+	+	+	+	+	+	+	+	+	+	+	+	+	+	+	+
Zr	+		h						+	h			+	+				
Al_2O_3	+														+			
BeO	+		h		h								h	+				
BPSG																		
GaAs																		
$MoSi_2$																		
SiO_2	+														+			
Si_3N_4	h											+	+		h	h	h	
Ta_2O_5																		
$TaSi_2$																		
TiN																		
$TiSi_2$																		
WSi_2																		

Note: D = dilute; C = concentrated; + = reaction at 25 °C, h = reaction at 100 °C or above.
Ag, Au, and Cu are also etched by KCN and NaCN.
Cr is also etched by HCl and $SnCl_2$.
Ge is also etched by KOH and H_2O_2.
Pd is also etched by HCl and Cl_2.
Re is also etched by Br_2-H_2O_2.
Ru and Os are also etched by Na_2O_2, or KOH and $KClO_3$.
Ti is also etched by hot concentrated $AlCl_3$ and dry Cl_2.
W is also etched by fused KOH and KNO_3.

gms CrO_3 in 100 gms of 40% HF, and was reported to be stable with use, and useful over a wide range of concentrations. Sirtl etch should not be stored for long periods in plastic bottles, of materials such as Nalgene®, because the bottles will slowly decompose, turn green, and become brittle enough to break.

Sopori Etch is a mixture of 36 parts of 49% HF, 20 parts of CH_3COOH, and x parts of HNO_3, where x is varied from 1 to 2. This etch was originally discussed in reference 5.

Superoxol etch consists of dilute HF and hydrogen peroxide, H_2O_2, in various formulations.

White Etch. HNO_3 and HF form "white etch," the basis for most silicon etch formulations, including "pinhole etch." White etch is generally formulated at 1 part HF to 4 parts HNO_3. When mixed with process-concentration acids; 49% HF and 70% HNO_3, white etch is fast. It is extremely corrosive and dangerous. Its reaction with silicon is exothermic and effervescent with the release of red fumes. The silicon removal rate is nonlinear with time as the solution warms. White etch is used for thinning wafers and as a chemical polish. White etch can also be known as "CP-4," etc, etc. White etch also dissolves tantalum, chromium, tungsten, and titanium.

Wright etch is a defect etch for silicon and an improved version of Sirtl etch developed by Margaret Wright Jenkins of Motorola and described in her paper, reference 6. The Wright etch consists of a mixture of 30 ml of 5 molal chromic (CrO_3), 60 ml of hydrofluoric (HF), and 30 ml of nitric (HNO_3), buffered with 60 ml of glacial acetic acid and 60 ml of water. Two grams of copper nitrate ($Cu(NO_3)_2$) or copper sulfate ($CuSO_4$) are also added to this etch to decorate the junction. The Wright Etch decorates defects on both the <100> and <111> planes in n- and p-type material, over a wide range of resistivities. The Wright etch preferentially delineates defects resulting from high-temperature processing steps. The Wright etch is slow; defects begin to show in 10 to 15 minutes.

Yang Etch is a defect etch formulation based on the CrO_3-HF system developed by K.H. Yang, and described in reference 7. Yang etch is made by dissolving 150 gms of CrO_3 in 1 liter of H_2O, and diluting the resulting "stock" solution with an equal amount of D.I. water. Yang etch will delineate all common defects, with hot processing-induced defects showing after 2-3 minutes, and defects in the starting material showing after 10 - 15 minutes.

References

1. W.C. Dash, Journal of Applied Physics, 27, 1956, p 1193.
2. *An Examination of the Chemical Staining of Silicon*, by D.G. Schimmel and M.J. Elkind of Bell Labs, in the Journal of the Electrochemical Society, January, 1978, pp 152-155.
3. *Dislocation Etch for (100) planes in Silicon*, by F. Secco d'Aragona, Journal of the Electrochemical Society, Vol 119, No. 7, July 1972, pp -950-, and Phys Status Solidi (a), 7, 1971, p 577.
4. *Chromsäure-FluRsäure als spezifisches System zur Ätzgrubenentwicklung auf Silizium, (Chromic Acid-Hydrofloric Acid as Specific Reagents for the Development of Etching Pits in Silicon)*, by Von Erhard Sirtl and Annemarie Adler, Z. Metallik, 52, 1961, pg 529-531.
5. *A New Defect Etch for Polycrystalline Silicon*, by B.L. Sopori, Journal of the Electrochemical Society, Solid-State Science and Technology, Technical Notes, March 1984, pp 667-672.
6. *A New Preferential Etch for Defects In Silicon Crystals*, by Margaret Wright Jenkins, Journal of the Electrochemical Society, 124, 1977, p 757.
7. *An Etch for Delineation of Defects in Silicon*, by K.H. Yang, Journal of the Electrochemical Society, 131, 1984, page 1140.
8. *Metallographic Etching*, by Günter Petzow, Max Planck Institute for Metals Research, Institute for Materials Science, Stuttgart, West Germany, 1976. Available from the American Society for Metals (ASM), Materials Park, Ohio, 44073, 216-338-5151.
9. *Metallographic Principles and Procedures*, 1977, Edited by Cornelius A Johnson, Leco Corporation, 3000 Lakeview Ave, St.Joseph, Missouri, 49085, 616-983-5531.
10. *Hazardous Chemicals Desk Reference*, N. I. Sax and R. J. Lewis, Van Nostrand Reinhold Co.
11. *ASTM Standard Definitions, Guides, and Practices for Surface Analysis*, an AVS Monograph containing a reprint of ASTM Committee E-42 on Surface Analysis, *Standard Terminology Relating to Surface Analysis*, publication E673-79a, from the Annual Book of ASTM Standards, ASTM, 1916 Race Street, Philadelphia, Pa, 19103, July 1989.
12. *Standard Test Method for Crystallographic Perfection of Silicon by Preferential Etch Techniques, ASTM Standard F47-88*, available from the American Society for Testing and Materials, (ASTM), 1916 Race Street, Philadelphia, Pa, 19103.
13. *Cross Sectioning Technique for Metallization Step Coverage Analysis*, by Martha Brandt, Mary Jones, and Jon Roberts, Eaton Corporation Application Note #90.2
14. *Stacking Fault Nucleation in Epitaxial Silicon on Variously Oriented Silicon Substrates*, by S. Mendelson, Journal of Applied Physics, May, 1964, Vol 35, No. 5, pp 1570-1581.
15. *Techniques for Structural Evaluation and Delineation of Shallow Layers in Si VLSI Substrates*, by W. Maszara, C. Carter, and G.A. Rozgonyi
16. *SEMATECH Chemical Safety Handbook*, 3rd Edition, by Corporate Safety Department, National Semiconductor Corporation, Santa Clara, California, 1986
17. *Quick Reference Manual For Silicon Integrated Circuit Technology*, Section 5, "Chemical Recipes" by W.E. Beadle, J.C.C. Tsai, and R.D. Plummer, Wiley Interscience.
18. Research Triangle Institute *Report ASD-TQR-63-316*, Vol x.

SAFETY HINTS AND IDEAS

by Thomas W. Lee

DISCLAIMER

The following information is believed to be reliable, and is based on experience in failure analysis. It should not be considered as a recommendation for the establishment of policies, procedures, and practices in your work area. All such policies, procedures and practices should be submitted to, and approved by, your safety and environmental departments before implementation. The author, ASM, SEMATECH, and Motorola, Inc., cannot be held responsible for any actions resulting from the interpretation of the below information, whether or not approval has been obtained from appropriate corporate officials within your company.

IT IS IMPORTANT TO CONSIDER FAILURE ANALYSIS, AND THE ELECTRONICS INDUSTRY IN GENERAL, A CHEMICAL PROCESSING INDUSTRY. FAILURE ANALYSIS, BECAUSE IT MUST REVERSE-PROCESS ELECTRONIC DEVICES, IS AN ACTIVITY WHERE MANY OF THE CHEMICALS USED IN THE ELECTRONICS INDUSTRY ARE COLLECTED INTO A SINGLE AREA. THIS INCREASES AND COMPLICATES THE POTENTIAL HAZARDS OVER ANY SINGLE PROCESSING STEP UTILIZED IN THE ORIGINAL MANUFACTURING OF THE DEVICES.

Use protective clothing and equipment. Use gloves and a face shield. Wear a lab coat made from a material that will not react with the acid. Cotton, for example, if not treated, will become full of holes from contact with sulfuric acid. If you handle large bottles of acid, wear a rubber apron and booties. Split and cracked bottles do occur; handles occasionally fall off; a previous user may have left the cap loose or cross-threaded. You may be tempted to hurry, and take shortcuts. Don't do it. Take the time to gown up. No job is so important that you should endanger your health, or risk injury, or disablement.

Under no circumstances pour fuming acids into water, or water into fuming acids. The familiar rule of thumb, "always pour acid into water, never the reverse," offers no protection with FUMING acids. Spattering, an explosion, and copious fumes, may result from either case.

Disposal of spent mineral acids may be accomplished by aspiration to acid drains. The water flow in the aspirator safely dilutes them. Beware that a thin film of viscous sulfuric acid will adhere to the walls of the pickup tube in the aspirator until the vacuum (water) is turned off. Then, a few drops of undiluted acid may then drip back out of the pipe, and possibly onto your hand. Passing clear water through the aspirator removes this hazard. Locate the aspirator at the rear of the hood, so that fumes and splashing occurring during the dilution cannot emerge from the hood. In general, mineral acids will react with calcium in the soil to form salts that are stable, and similar in composition to naturally-occurring minerals. One neutralization technique consists of reacting the acid with crushed limestone. These acids release fumes during neutralization, so dilution beforehand is wise.

Be alert to fumes. The human nose is an unusually reliable and sensitive indicator of the presence of fumes, usually in the PPM range. The human olfactory is sensitive, but not discriminatory. For example, ammonia is quickly identifiable by most people, but most acids smell about the same. Fumes from HF can do you a lot more damage, a lot more quickly, than HCl, although they smell similar.

CHEMICAL ODORS

The following chemicals can be detected by their odor before their concentration reaches the threshold limit value (TLV):

- *acetic acid
- *acetone
- *ammonia
- *butyl alcohol
- *chlorine
- *chlorobenzene
- *p-dichlorobenzene
- *dimethyl formamide
- dioxane
- ethylene glycol monomethyl ether (cellusolves)
- *ethyl acetate
- *ethyl alcohol
- ethylenediamine
- ethylene glycol
- hydrogen cyanide
- *isopropyl alcohol (isopropanol)
- methyl alcohol (methanol)
- methylene chloride
- *methy ethyl ketone (MEK)
- *napthalene
- *nitric oxide
- *nitrogen dioxide
- *ozone
- *phenol
- phosphine
- *silane

Note: (*) chemicals can be detected by the nose in concentrations an order of magnitude or greater **below** the threshold limit value (TLV).

- *mineral spirits
- *toluene
- 1,1,1 trichloroethane
- *xylene

It is also important to realize that the nose can build a tolerance to certain odors (fatigue), and lose sensitivity as a result. Additionally, it is possible that very long exposures to undetectably low levels of chemicals can cause damage.

Never work around chemicals if you have a cold; your nose will be much less sensitive, and you may inhale fumes in dangerous concentrations.

It is important to know which fumes you can smell, and which are insidious. The following chemicals can be present in concentrations **greater than the TLV** before the nose can detect them. **Watch Out** for these fumes with detectors other than your nose!! They have weak odors.

- **arsine
- carbon tetrachloride
- **chloroform
- cyanogen chloride
- **diborane
- diethyl acetamide
- ethanolamine
- hydrazine
- **hydrogen bromide (hydrobromic acid)
- **phosphine
- **trichlorobenzene

Notice that many of the chemicals in this list are gaseous dopants used in ion implantation and plasma processing.

CARCINOGENS

The following chemicals are capable of causing cancer, or are suspected of being able to cause cancer. Be informed that much of the data has been generated from rat and other mammal studies, and that human susceptibility may be vastly different. Cancer develops over periods of time which may be measured in years, sometimes from a single exposure (crysotile asbestos). Cancer is frequently progressive and fatal, cures are difficult, and the disease may reappear after years of dormancy. On the positive side, cases of spontaneous remission have been documented.

- asbestos
- arsenic
- arsenic trioxide
- arsine

Note: (*) chemicals can be detected by the nose in concentrations an order of magnitude or greater **below** the threshold limit value (TLV). (**) chemicals can be present in concentrations **an order of magnitude above the TLV** before they can be detected by the human nose.

- benzene
- beryllium
- beryllium oxide dust
- carbon tetrachloride
- chromic acid
- chromium
- ethylene oxide
- gallium arsenide
- hexachlorobutadiene
- hexachloroethane
- methylene chloride
- nickel sulfamate
- potassium dichromate (or -bichromate)
- tetrachloroethylene
- trichloroethylene
- vinyl chloride

Beware of symptoms of poisoning in yourself or others. The below list was compiled from the descriptions of human reactions caused by mild chemical poisoning in several references. Note that there could be myriad other causes of these symptoms. If you notice these symptoms in yourself or coworkers, low-level chemical poisoning could be taking place. Consult your physician.

- Lassitude, drowsiness, weakness, anemia.
- Headache, dizziness, dilation of pupils, incoordination, blurred vision, impairment of reaction time, impaired judgment, hearing loss, excessive perspiration, weak pulse.
- Irritability, nervousness, insomnia, mental depression.
- Coughing, chest pains, sneezing, inflamed nose and throat, nose bleeds, difficult breathing, rapid breathing.
- Nausea, loss of appetite, abdominal pains, diarrhea, dryness in the mouth, excessive saliva secretion, garlic-like halitosis.
- Persistent cold, bronchitis, flu, or asthma symptoms.
- Skin rashes, bloodshot or irritated eyes, discolored skin, flushed complexion, drying and cracking of the skin.

DO NOT CONFUSE FEAR WITH SAFETY. INSTEAD, ASSURE THAT LAB MEMBERS ARE CAUTIOUS, ALERT, AND EDUCATED ABOUT CHEMICALS AND THEIR EFFECTS.

Never use fuming acids without a fume hood. The fumes given off by these acids consist partially of the gases from which the acids are originally made; they are acid **anhydrides.** If you breathe these fumes, they form acid upon contact with the moisture in your lungs. You may also breathe acid mists, or acid **aerosols** if the acids are heated or sprayed. Depending on the fumes and the concentration, reaction may be immediate or delayed. The acid may scar or destroy the **alveoli,** the tiny sacs

Safety Hints and Ideas

in your lungs which exchange oxygen and carbon dioxide from your blood. Inhalation of low concentrations of acid fumes over a period of time can result in a chemical pneumonia, the recovery from which is extremely slow (months). **Burns** from these acids are painful and slow to heal. Contact with fuming nitric acid is immediately painful, even from the smallest spatter. The skin turns green, and the acid will continue to react to depth.

Hydrofluoric acid, HF, is an insidious danger. The most common burn resulting from this acid is beneath the fingernails. Contact results in no immediate sensation. The acid quickly penetrates the skin to depth. Sensation is not apparent until the acid has been present for minutes. At the least, the burn will cause an aggravating prickling sensation that persists for days. If the acid is present in sufficient amounts, and for a sufficient time to penetrate to depth, it may perform an ion exchange in the bones, altering the calcium carbonate ($CaCO_3$), or aragonite, to calcium fluoride (CaF_2). The fluoride is very brittle. This reaction is highly desirable with the enamel on the surface of the teeth; it prevents decay. But you can readily understand why dentists are very careful about the **amount** of fluoride (as stannous fluoride) that you receive. HF is treacherous. HF fumes are classed with the **most dangerous acid fumes known!** Consider all etches containing HF as being potentially as dangerous as HF itself.

Keep water running in the fume hood during all procedures. It is cheap insurance against injury, because almost all emergency treatments begin with the words, "Immediately flush the affected area with copious amounts of water." If the water is already running, you save precious time finding the handle and turning it on. An eyewash or emergency shower should also be immediately available.

For small burns, keep a bottle of soda, or soda-water solution, handy at all times. You can help to arrest the action of acid before you take time to walk to, or call, your medical or safety department.

Be familiar with your fume hood. Eddy currents caused by obstructions can cause dead air spaces. Unusual winds outdoors can stop airflow in some systems, or cause fumes to be drawn back into the building. Know the location of any switches or circuit breakers that can shut off the vent motors without your knowledge. Have the circuit alarmed in case of a power failure. Check periodically for leaks that can admit fumes back into the workspace in a low concentration; this is the source of many cases of chemical pneumonia.

Do not allow overcrowding in a chemical lab. For example, a bump to the elbow of a person holding a beaker of acid could result in a burn.

Organics are complicated and numerous. They are usually present in the FA lab as reagents for specific purposes, or as a part of formulations for decapsulation or other processes. Some solvents, such as isopropanol, or common rubbing alcohol, are relatively safe, while others can be carcinogens, a mutagens, or a teratogens. The health effects may be delayed and can become extremely serious. Reaction time of the body can vary with the compound; benzene causes almost immediate poisoning while the effects of other organics can take place over years at very low concentrations. Consult your MSDS file for specific information about each of the organics you use. Look the chemical up in any of the toxicologic references.

Label all beakers and containers. At the least, the label should identify the contents. It is also advisable to include information such as name of the analyst, time, project number, etc.

Use dispensers because they eliminate the handling of large (usually gallon) bottles of acid, and pouring. Pouring can result in inadvertent mixing, bottles with incorrect contents, spills, and dropped bottles. Dispensers can be set to deliver small quantities of acid very reliably. Common sense dictates that there is less danger if accidents involve small quantities than large ones.

Keep a copy of the MSDS for each chemical in your lab within easy reach. Your analysts have the right to know the properties and dangers of the materials they are working with by law, and the knowledge can help raise safety awareness levels in the lab.

Prohibit cloth handtowels from the lab. It is inevitable that a towel will be used to clean up a small spill, then left for an unsuspecting analyst to wipe his hands on. Eliminate this source of burns. If a handtowel is designated specifically to a sink in a non-chemical area near where chemicals are handled, buy or rent the type which are loaded with a chemical color indicator.

Conduct audits. Look for violations of rules, note them, and discuss the results with lab members. Insist on verification of corrective measures.

Wear spats if you handle gallon bottles of chemicals. The handles of bottles can break off, the bottle can split, or the previous user could have left the cap loose or cross-threaded. A shoe full of acid ensures a debilitated analyst.

Replace rubber gloves regularly. They deteriorate, and may leak due to holes or cracks. **Never** put on a rubber glove that is wet inside. Inflate them to check for leaks. Do not share gloves. Make sure they are durable for the chemicals to which they will be exposed. Do not immerse your gloved hand in chemicals if at all avoidable. When you remove your gloves, rinse and wipe the gloves, then your hands. Do not store them wet.

Store chemicals in vented and exhausted cabinets. Storing chemicals in unvented acid cabinets will ensure a face full of fumes when you open the door. If the acid cabinet is located outdoors, has gasketed doors, and becomes heated from sunlight, pressures can build up. Follow the manufacturer's recommendations for fume exhause and pressure relief.

Use temperature-limited, explosion-proof hotplates. They may be proportionally controlled, have a remote thermostatic controller, or be hermetically sealed. A spark from a hot plate contact in a solvent hood is unacceptable. The flash point of liquids can be exceeded unless maximum temperatures are limited.

Do not allow open flames. Propane torches, Bunsen burners, and blowpipes are trouble.

Use raised rubber mats on the floor. They decrease the probability that a dropped bottle will shatter or burst.

Separate your trash. Mixing of contaminated paper with waste paper can increase the quantity of trash which must be treated as contaminated waste.

Never drink DI water. It tastes like chalk anyway.

Practice good housekeeping. Excessive trash invites a fire.

Do not allow clutter on fume hood decks. It invites a spill or other accident.

Use only as much reagent as you really need.

Never pipette by mouth. Use a rubber bulb. If you think that this rule is unnecessary, examine the condition of the inside of some used bulbs. Remember that lifting a pipette out of the liquid while the bulb still has a vacuum will cause the liquid to burst up into the bulb. Air is a lot less viscous than any liquid.

Wear eye protection with side shields. Droplets of acid seem to have the ability to find their way behind ordinary glasses.

Encourage lab members to volunteer for hazard or fire training.

Try not to work alone in a hood.

Ensure that only approved chemicals can be purchased or stored in the lab. Materials such as fuming acids, petroleum distillates, cyanide, heavy metals, poisonous gases, carcinogens, etc., should be present only if demonstrably necessary.

Collect solvents without water. Disposal services can distill solvents, or reuse them in applications requiring a broad spectrum solvent, such as paint strippers or thinners. Mix them with water, and you make a mess with a larger volume which can't be distilled.

Acid and solvent bottles should be stored in acid and solvent cabinets. The mixing of solvents and acids, in liquid or vapor form, may generate a fire, an explosion, or poisonous products. The following are examples of poisonous esters that can result from the mixing of solvents and acids:

- Sulfuric acid and methanol combine to make dimethylene sulfate, so poisonous by inhalation it is classed as a war gas.
- Nitric acid and methanol combine to make nitromethane, a contact poison. Nitromethane is an explosive (race car fuel).
- Any acid will combine with cyanide stripper salts to yield Prussic acid, HCN, a poison by inhalation (it smells like almonds).
- Alcohols will combine with phosphoric acid to yield POCL, phosphine, or other gases, classed as nerve poisons.
- There are no second chances with some of these materials.

Don't create strange etch concoctions without good reason. Recipes are in existence which call for mixing nitric, sulfuric, and glycerine, which can make nitroglycerine. One advises the mixing of potassium cyanide and nitric. Some formulas call for picric acid, an explosive used in World War I. Some ask for a few drops of liquid bromine, which has a high vapor pressure and will puff into your face as you loosen the stopper. Some ask for perchloric acid, so dangerous it is a prohibited substance in many facilities. Call the developer of the etch and ask the necessity of these chemicals. Only use them if you are sure nothing else will work. Try to use standard etches whenever possible.

Be careful with lab plasma etchers. We all know that opening the door of a plasma etcher which has just been deenergized and returned to atmospheric results in a horrible smell. Believe your nose. Plasma etchers can make, along with fluorine and chlorine: ozone, nitrogen oxides, aluminum chloride, cyanogen, cyanogen chloride, hexachlorobutadiene, hexachloroethane, and tetrachlorobutadiene. Ensure that the chamber is thoroughly vented before opening. Have a small funnel exhaust installed over the door of the chamber. If you have trouble obtaining a response, ask your safety department for help.

Be cognizant of the following additional hazard possibilities:

- R.F. fields—from dielectric heaters, plasma reactors, ICP spectrometers
- Ultraviolet radiation—plasma reactors, mask aligners
- Laser radiation—various analytical and processing equipment
- X-rays—XRF, ion implanters, radiographic equipment
- Cryogenic fluids
- Bottled gases

REFERENCES

1. G. Luch and E.B. Sansone, *Destruction of Hazardous Chemicals in the Laboratory,* Wiley-Interscience, 1990.

2. N. I. Sax and R.J. Lewis, Sr, *Hazardous Chemicals Desk Reference,* Van Nostrand Reinhold Company, 1990.

3. N. I. Sax and Richard J. Lewis, *Dangerous Properties of Industrial Materials,* Van Nostrand, 1989. (3-volume set)

4. *CRC Handbook of Laboratory Safety,* several editions, CRC Press, Inc, Boca Raton, Florida, revised every year.

5. *Prudent Practices for Handling Hazardous Chemicals in Laboratories,* Committee on Hazardous Substances in the Laboratory, National Research Council, Washington, D.C., 1981.

6. *MT&S Laboratory Safety, Health, and Environmental Manual,* document No. 91010434A-OPS, April 10, 1991, available from SEMATECH, 2706 Montopolis Drive, Austin, Texas, 78741, 512-356-3500.

7. *The MSDS Pocket Dictionary,* edited by J.O. Accrocco, Genium Publishing Corp. 1145 Catalyn Street, Schenectady, NY, 12303-1836, 518-377-8854, 1988.

8. *Occupational Safety and Health Administration,* 200 Constitution Ave, N.W., Washington, D.C. 20210, 202-523-7894.

9. *Compact School and College Administrator's Guide for Compliance with Federal and State Right-to-Know Regulations,* The Forum for Scientific Excellence, Inc., Philadelphia, PA, J.B. Lippincott Co., 1971.

10. *Safety and Health Guide for the Chemical Industry,* OSHA # 3091, U.S. Dept. of Labor, Occupational Safety and Health Administration, 1986.

11. Sybil P. Parker, et al, *McGraw-Hill Dictionary of Chemical Terms,* McGraw-Hill, 1985.

12. *The Merck Index,* Merck and Company, Rahway, New Jersey, 1989.

13. Marshall Sittig, *Handbook of Toxic and Hazardous Chemicals and Carcinogens,* 3rd Edition, (in 2 volumes), Noyes, 1991.

14. L. Bretherick, *Bretherick's Handbook of Reactive Chemical Hazards,* Butterworths, 1990.

15. G. Weiss, *Hazardous Chemicals Data Book,* 2nd Edition, by Noyes Data, 1986.

16. R.J. Lewis Sr., *Rapid Guide the Hazardous Chemicals in the Workplace,* 2nd Edition, Van Nostrand-Reinhold, 1980. (Handbook format)

17. *Chemical Safety Handbook*, by the Corporate Safety Department of National Semiconductor Corporation, 2900 Semiconductor Drive, Santa Clara, Ca, 95052-8090, 1986.

18. *Storage and Treatment of Hazardous Wastes in Tank Systems,* US Environmental Protection Agency, Noyes Data, 1987.

19. R.D. Griffin, *Principles of Hazardous Materials Management,* Lewis Publishers, 1991.

PLASMA ETCHING

by James Beall

This review discusses the techniques and methodologies for using a plasma etcher/asher and is tailored for the failure analyst. The decapsulation of plastic packages, removal of organics, and etching of chip passivation are included. The benefits of plasma deprocessing include minimal risk of physical damage to the device and elimination of chemical residues encountered with wet chemical etches.

TERMINOLOGY

Plasma—RF power is applied to a gas in a chamber at a reduced pressure producing ions, atoms, free radicals, and free electrons. These activated gaseous species produce etching by physical or chemical reactions with solid surfaces. Much research has been performed on this subject. The most common frequency, 13.56 MHz, is a band allotment for medical diathermy machines.

Plasma etching—A chemical process in which the plasma combines with the solid surface to form products, some of which are volatile. It should be noted that during normal plasma etching operations, physical effects such as ion bombardment of the surface are also occurring. These effects cause reactive etching (RIE).

Plasma ashing—The chemical reaction of oxygen plasma with an organic. The term ashing is likely a carryover from high temperature ashing of samples. Atomic oxygen reacts with the organic to produce a fine ash which can be blown away.

Barrel reactor—Barrel reactors are commonly used for failure analysis work due to their availability, ease of operation, and low cost. The term *barrel* refers to the shape of the sample chamber. Surrounding the barrel is an electrode that supplies the RF power.

Parallel-plate reactor—Parallel plate plasma etching systems are configured with an upper electrode (RF signal) and a lower electrode (ground). This arrangement produces more of a physical effect on the sample due to ion bombardment. These systems are more expensive and require grounding of the sample. They do offer additional controllability of the system variables [4].

PLASMA CHEMISTRY

Plasma chemistry is enormously complicated. Some of the chemical reactions related to the removal of passivation from integrated circuits are shown below [5,9,10,12,13]:

1. $CF_4 + \text{energy} \rightarrow CF_3 + F$
2. $Si + 4F \rightarrow SiF_4$
3. $SiO_2(s) + CF_x(g) \rightarrow SiF_y(g) + CO, CO_2(g)$
4. $Si_3N_4(s) + CF_x(g) \rightarrow SiF_y(g) + CFN(g)$

Note: Only gases are produced in reactions 3 and 4.

5. Addition of H_2 to CF_x [9,10]:
   - Free fluorine atoms are efficient etchers of silicon
   - CF_3 radicals preferentially etch SiO_2 versus Si because of polymer formation on Si and not on SiO_2.
   - Hydrogen scavenges free fluorine, so from Equation 1 there will be an excess of CF_3.
   - Selectivity of the removal of SiO_2 versus Si improves.
   - If too much H_2 is added, then carbon-bearing radicals (polymers) collect on all surfaces and etching stops.
   - Generally less than 40% H_2 is used.

6. Addition of O_2 to CF_4 [10,11]:
   - Oxygen combines with carbon to form CO, CO_2, or F_2CO.
   - Effect is opposite of adding H_2, i.e., Si etch rate is increased.
   - Generally less than 10% O_2 is used.

Note: Mixtures of O_2 and CF_4 are available. Certain mixtures are patented by LFE and are considered industry standards. The mixtures produce a higher etch rate than CF_4 alone, without excessive loss of SiO_2 to Si selectivity.

7. Gold + fluorine
   - Exposed gold surfaces within an IC package can volatize and redeposit on surfaces as gold fluoride.

LITERATURE REVIEW

Articles were reviewed to obtain operating conditions reported for plasma stripping operations related to failure analysis. The table accompanying this article provides a brief summary of the data. This summary should only be used as a guide to the location of the information. Review of the articles is required to properly understand the processes, as the number of variables is large, the machines differ in operation, and all desired variables are seldom present in each reference.

PLASMA STRIPPER OPERATION

The operation of a typical plasma stripper is as follows:

1. Place parts in chamber.
2. Evacuate chamber with a roughing pump.
3. Adjust inlet gas flow rate or pressure.
4. Turn on and adjust RF input power.
5. Adjust and minimize reflected RF power. Some systems do this automatically.
6. When stripping is complete, shut off power, gas, and vacuum.
7. Remove part and inspect results.

Additional steps which may be required include:

Table 1 Operating Conditions Reported for Plasma Stripping Operations

MATERIAL TO BE ETCHED	GAS	POWER	PRESSURE	FLOW RATE	TIME	SYSTEM	REF
SiO_2 & Si_3N_4	O_2/C_2F_6	200 W	—	100 SCCM/ 100 SCCM	10-30 min.	DRYTEK DRIE-100 (parallel plate)	[2]
Si_3N_4	O_2/CF_4 1/20	100-300 W	—	400 SCCM	40-300 sec.	LFE LTA-302 (barrel)	[3]
SiO_2 & Si_3N_4	O_2/CF_4	200 W	53 Pa	100 SCCM/ 10 SCCM	45 min.	DRYTEK DRIE-100 (parallel plate)	[4]
SiO_2	O_2/CF_4	—	.7 torr	1 5 ml/min	—	Plasmod (barrel)	[5]
SiO_2	CF_4	25 W	—	—	—	Plasmod (barrel)	[6]
SiO_2 & Si_3N_4	O_2/CF_4 10/90	100 W	.6 torr	2.5 psi (f.r. unk.)	—	(barrel)	[7]
SiO_2 & Si_3N_4	O_2/CF_4 8/92	200-350 W	.2 torr	—	6-15 min.	LFE 502 (barrel)	[8]
SiO_2 & Si_3N_4	H_2/CF_4 30/70	100 W	.3 torr	—	4-8 hrs.	LFE PDS/PDE 301 (barrel)	[9]
P-doped SiO_2	O_2/CF_4 8/92	100 W	.8-1.3 torr	—	8-15 min.	Plasmod (barrel)	[14]
undoped SiO_2	Ar/SF_6 20/80	100 W	.8-1.3 torr	—	45-120 min.	Plasmod (barrel)	[14]
Epoxy Package	O_2	20 W	—	150 cc/min	3-5 days	Plasmod (barrel)	[1]
Epoxy Package	O_2/CF_4	200 W	.8 torr	100 SCCM 10 SCCM	6 hrs.	DRYTEK DRIE-100 (parallel plate)	[2]
Epoxy Package	O_2/CF_4 90/10	300 W	.2 torr	—	2 hrs.	LFE 502 (barrel)	[8]
Epoxy Package	O_2/CF_4 90/10	100 W	1 torr	25 SCCM	2-3 hrs.	Plasmod (barrel)	[14]
Polyimide/ Photoresist	O_2/CF_4	200 W	.4 torr	100 SCCM/ 3 SCCM	15 min.	DRYTEK DRIE-100 (parallel plate)	[2]
Photoresist	Argon	75 W	.175 torr	—	500A/min	Plasmod (barrel)	[14]

1. Mechanically remove portion of epoxy package prior to package ashing (oxygen) [1,2].
2. Remove package and blow off ashed epoxy and loose grains of filler material [1,2].
3. Coat portions of the package with silicone grease to eliminate ashing [1].
4. Coat gold with photoresist or use a lower RF power level to avoid contaminating die surface with gold deposit (CF_4 plasma) [8].
5. Etch in Techni-strip® at 60 °C to remove gold deposition (CF_4 plasma) [8].

The roughing pump used on the plasma stripper should be vented to the outside, as some readicals and gaseous reaction products are toxic. Special note: Consideration should be given to the new laws regarding the use of CFCs and the destruction of atmospheric ozone:

1. Use only approved gases
2. Use a scrubbed or neutralized exhaust or a burn-box for on-site destruction.

The pump oil should be changed regularly. The oil should be disposed of in the proper manner as hazardous waste [15]. A special oil is required for the pumps when oxygen is used as the gas. Oils such as super-refined hydrocarbons, silicones, and Fomblin® are expensive, and should be chosen carefully.

The process of using a plasma stripper for failure analysis is relatively simple; however, there are a number of process and device variables that will affect the results. To establish the correct parameters to use for a given device it is advisable to use a "practice" part. Plasma parameters can then be maximized to obtain the desired results.

REFERENCES

1. "Decapsulation of Epoxy Devices Using Oxygen Plasma," D. Wilson and J. Beall, in *15th Annual Proceedings Reliability Physics*, Las Vegas, Nevada, April 12-14, 1977.
2. "Failure Analysis Applications of Plasma," D. Tomasi, P. Marchetti, and J. Powlis, in *Proceedings of the International Symposium for Testing and Failure Analysis*, Los Angeles, California, Nov. 9-13, 1987.
3. "The Use of Plasma Chemistry in Failure Analysis," M. Pfarr and A. Hart, in *18th Annual Proceedings Reliability Physics*, Las Vegas, Nevada, April 8-10, 1980.
4. "Integrated Circuit Overlay Removal by Planar Plasma Etching," by P. Marchetti, J. Powlis, in *Proceedings of the International Symposium for Testing and Failure Analysis*, Los Angeles, California, Oct. 22-24, 1984.
5. "Plasma Etching as Applied to Failure Analysis," W. Jones, in *12th Annual Proceedings Reliability Physics*, Las Vegas, Nevada, April 2-4, 1974.
6. "Plasma Etching Proms and Other Problems," J. Devaney and A. Sheble, in *12th Annual Proceedings Reliability Physics*, Las Vegas, Nevada, April 2-4, 1974.
7. "Failure Analysis Techniques and Failure Mechanisms Utilizing a Plasma Stripper," J. Gajda, D. DeLorenzo, and J. Wade, in *17th Annual Proceedings Reliability Physics*, San Francisco, California, April 24-26, 1979.
8. "Plasma Etching Techniques for Failure Analysis," G. Thome, in *Proceedings of the International Symposium for Testing and Failure Analysis*, Los Angeles, California, October 19-23, 1981.
9. "A Procedure for the Nondestructive Removal of Glassivation From Integrated Circuits," J. Abbott, in *Proceedings of the International Symposium for Testing and Failure Analysis*, Los Angeles, California, October 20-24, 1986.
10. "Some Aspects of the Fluorocarbon Plasma Etching of Silicon and Its Compounds," by J. Coburn and E. Kay, *IBM Journal of Research and Development*, Vol 23, No. 1, January 1979.
11. "Choosing Gases for Plasma Dry Etching," J. Webber, *Microelectronic Manufacturing and Testing*, January 1985.
12. "Plasma Etching of Oxides and Nitrides," A. Weiss, *Semiconductor International*, February 1983.
13. "Dry Etching of SiO_2 and Si_3N_4," P. Singer, *Semiconductor International*, May 1986.
14. Application Notes from March Instruments Incorporated.
15. "Dry Etch Chemical Safety," J. Ohlson, *Solid State Technology*, July 1986.

REACTIVE ION ETCH RECIPES FOR FAILURE ANALYSIS

by David S. Kiefer

Increased circuit densities, smaller feature sizes, and the use of multi-layer technologies are an ever-increasing challenge to today's failure analyst. Often, uncovering the underlying defect in an intact form can be difficult. Wet etch techniques no longer do an acceptable job of deprocessing today's integrated circuits. The introduction of triple-layer metal polyimide passivated bipolar gate arrays and triple-layer metal oxide passivated CMOS gate arrays has led to the development of dry etch techniques for failure analysis applications. This paper outlines detailed recipes for reactive ion etching (RIE) of critical layers on these new technologies. Etch recipes are included for most of the common layers and materials used in today's VLSI devices, i.e., aluminum, titanium tungsten, silicon nitride, silicon dioxide, polyimide, and polysilicon. The recipes are specifically for failure analysis purposes, the goal being to remove the desired layer while causing minimal damage or disturbance to neighboring layers.

MACHINE DESCRIPTION

The RIE etcher used is a commercially available table-top machine costing less than $75K. The low price and small footprint make it well suited for most failure analysis laboratories. A photograph of the machine is shown as Fig. l. The system was purchased with its own table, which houses the etcher, vacuum pump, RF generator, and toxic gas cabinet, all in one unit. The etcher has four regular mass flow controllers (MFCs) and two corrosion-resistant controllers. The machine utilizes a 600 watt 13.56 MHz RF generator and a 21 CFM corrosive series vacuum pump. The chamber is configured with a four-inch diameter bottom electrode spaced 1.5 inches from the ten-inch top electrode. Necessary external facilities are scrubbed exhaust for pump, hood and toxic gas cabinet; compressed air (60 psi); for pneumatics; nitrogen for chamber bleeds and line purges; and a water recirculator to heat and cool the chamber. The etcher was purchased with a small toxic gas cabinet suitable for two lecture-bottle-size gas tanks with the associated regulators and mass flow controllers MFCs. The non-corrosive gasses are supplied remotely via poly flow tubing. Because of the dangers and restrictions involving chlorine based gases, it is advisable to consult with safety engineering as early as possible in the procurement cycle. The machine has memory for storing ten unique process recipes. Each recipe has programmable power levels, gas flows, etch time, and prerun pump down pressure.

RIE ETCH BASICS FOR FAILURE ANALYSIS

Reactive ion etching (RIE) is a close physiochemical analog of sandblasting the surface of an integrated circuit. Depending on the gas used, the operating pressure, and the power density,

Fig. 1. Photos showing RIE system used for this paper.

 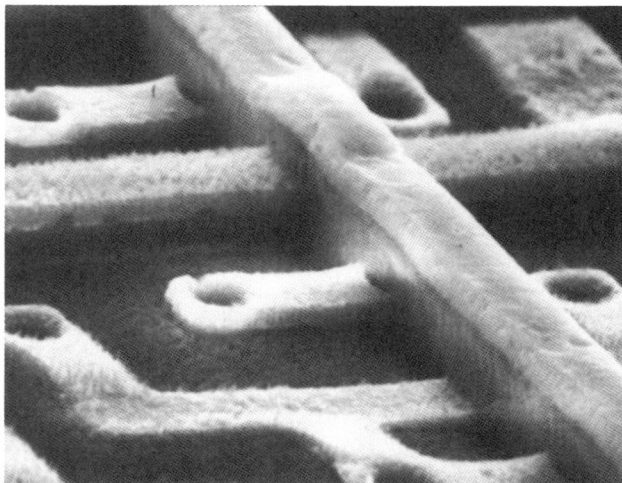

Fig. 2. SEM photos showing etch profile of TEOS oxide on three-layer metal CMOS gate array.

radically different outcomes can be achieved in an RIE etcher. The two basic modes are anisotropic etching, for the "vertical sidewall" etching of the desired layer, and isotropic etching for blanket removal of a layer at any angle. Generally, the lower the chamber pressure and power density, the more anisotropic the etch becomes. For failure analysis purposes, the lowest possible power should be used which will yield the desired results. High power densities can cause underlying damage, such as junction leakage and gate rupture. For the recipes included in this paper, chlorines are used to etch metal layers, fluorines are used to etch silicon-containing layers, and oxygen is used to etch organics such as polyimide. For failure analysis deprocessing, the most important factor in selecting an etch gas is how it will affect the underlying layers.

Special considerations need to be taken when etching packaged integrated circuits in an RIE etcher. For plastic packages using gold bonding wire it is advisable to remove the gold wire from the package; gold tends to redistribute either on the chamber walls or the IC surface (or both). If the bonding wires must be kept intact, they can be coated with photoresist to inhibit gold redistribution. Also, some epoxy molding compounds and die attach materials can act as chamber/die surface contaminants. Another condition to avoid is sharp objects pointing upward from the bottom electrode. When etching pin grid array (PGA) packages, it is advisable to either shield the external pins with a stainless steel fixture, or remove them entirely. Chamber cleanliness is also a factor. For best etch performance and repeatability, a thorough alcohol wipedown of the reactor, followed by an oxygen plasma, should be used before a sample is etched.

ETCH RECIPES

Silicon Dioxide

An anisotropic etch of plasma TEOS oxide was developed for the purpose of exposing all three layers of metallization on high-density CMOS arrays. Because of the anisotropic nature of the etch, the electrical integrity of the device is maintained so electrical probing or E-beam voltage contrast analysis can be performed on any layer. The following parameters are used for the etch:

POWER = 200 watts
PRESSURE = 120 millitorr
GAS = CF_4 +8%O_2
FLOW = 15 sccm
TEMP = 25 °C
TIME = 10 min.

With these conditions, the etch rate is approximately 2400 Å/min

SEM photomicrographs showing the etch quality are included as Fig. 2. In this photo, 2.4 microns of plasma TEOS oxide has been anisotropically etched exposing all three layers of metallization on a CMOS gate array.

Other gases can also be utilized for oxide etching. CHF_3 has been widely used for wafer fabrication oxide etching. However, in a single chamber system used for failure analysis, small amounts of organic contaminants are usually present in the chamber or in the IC package itself. CHF_3 is more prone to organic polymer build-up, which can inhibit the etch. Experience has shown that CF_4 is generally the gas of choice for failure analysis oxide removal. However, when higher oxide to silicon selectivity is needed, such as etching an oxide layer above polysilicon, CHF_3 is a good choice.

Silicon Nitride

Silicon nitride is commonly used as a final passivation and as a contaminant barrier above bipolar and MOS structures. Wet etch solutions for silicon nitride generally react slowly and cause damage to underlying layers. RIE of silicon nitride is faster and has a better selectivity to oxide. The etch recipe for removal of a 7500 Å layer of plasma-enhanced silicon nitride is as follows:

Reactive Ion Etch Recipes

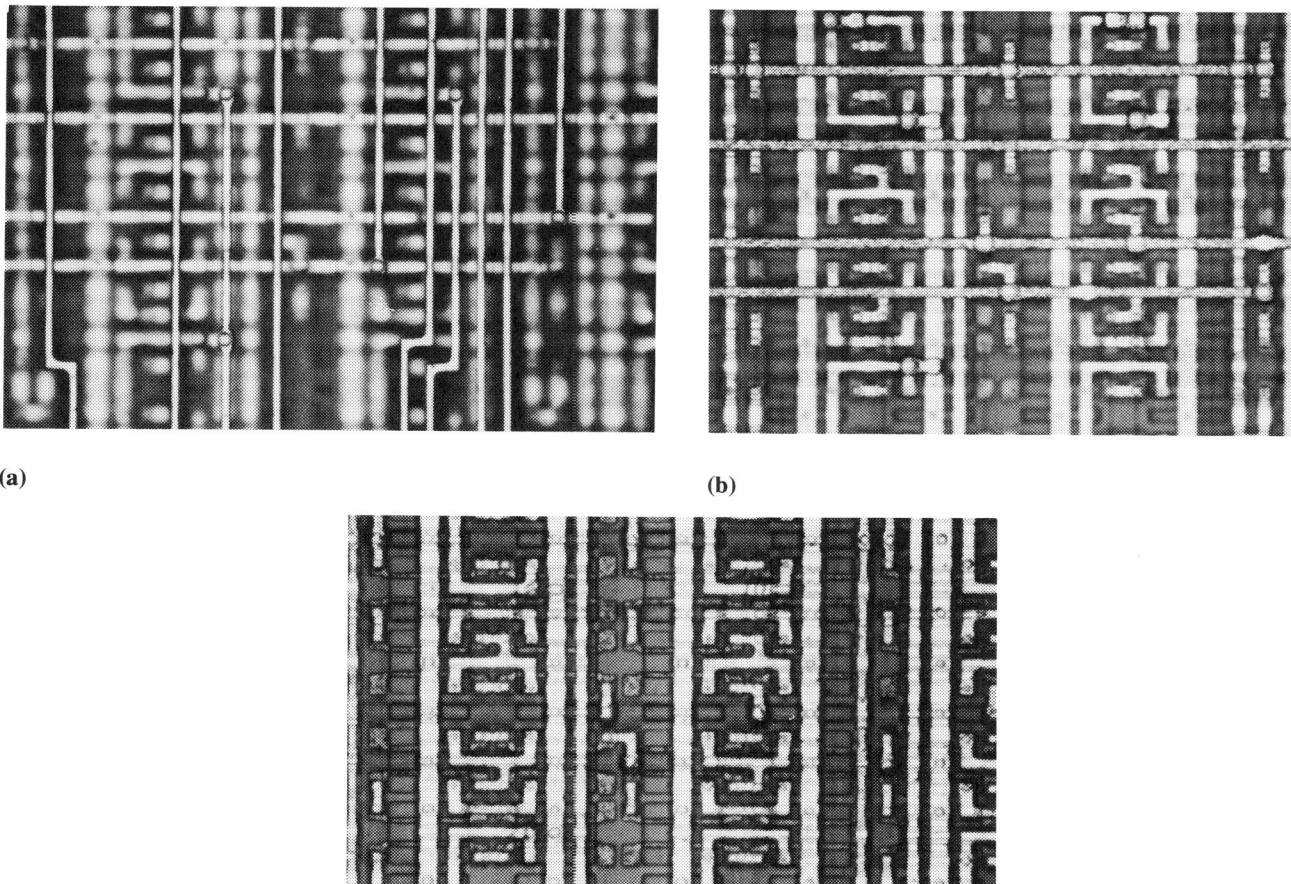

Fig. 3. Optical photos showing metal etch sequence on three-layer metal CMOS gate array.

POWER = 150 watts
PRESSURE = 200 millitorr
GAS = CF_4
FLOW = 15 sccm
TEMP = 250 °C
TIME = 100 sec.

With this formula, the etch rate is about 4500 Å/min.

Aluminum

Often, it is desirable to etch a top layer of aluminum interconnect without damaging the underlying layer of metallization. Wet etch techniques generally undercut in the via areas causing considerable damage to the underlying layer. An aluminum etch was developed for the removal of interconnects on high-density CMOS gate arrays, which leaves the underlying metal layers, even the vias, intact. The etch recipe is actually three process steps consisting of a punch-through etch which removes the native aluminum oxide, a main etch which removes the aluminum layer, and a corrosion prevent-etch which removes any residual chlorine left from the main etch. The following parameters are used.

PUNCH-THROUGH ETCH
POWER = 150 watts
PRESSURE = 180 millitorr
GAS = BCl_3
FLOW = 40 sccm
TEMP = 400 °C
TIME = 25 sec.

The etch rate for aluminum is approximately 4000 Å/Min.

MAIN ETCH
POWER = 150 watts
PRESSURE = 140 millitorr
GAS = BCl_3 and Cl_2
FLOW = 25 sccm(BCl_3); 5 SCCM(Cl_2)
TEMP = 40 °C
TIME = 40 sec.

The etch rate is approximately 12,000 Å/min.

CORROSION-PREVENT ETCH

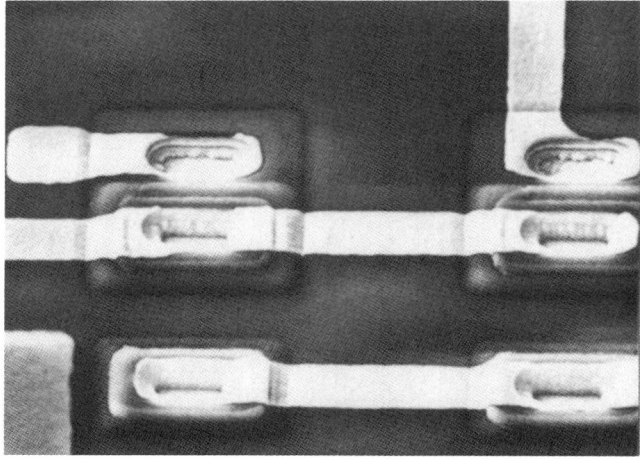
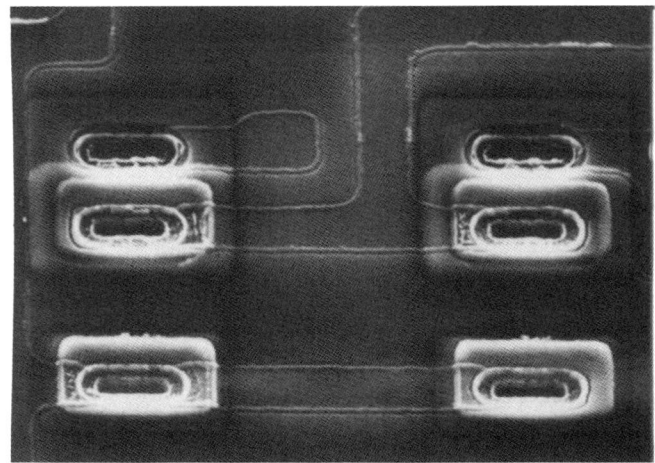

(a) (b)

Fig. 4. SEM photos showing Ti-W etch quality used as barrier metal on bipolar gate array.

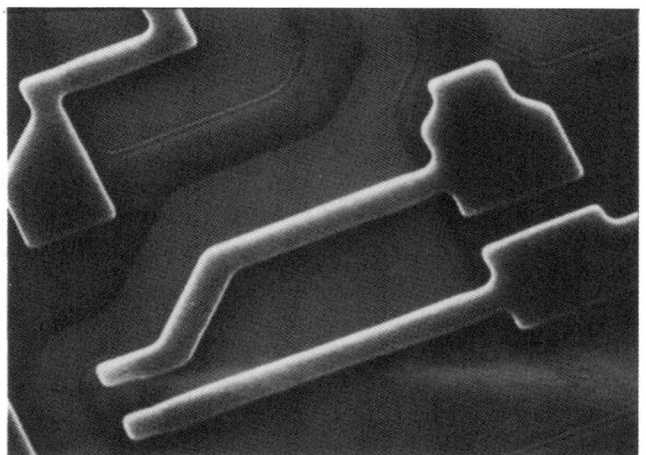
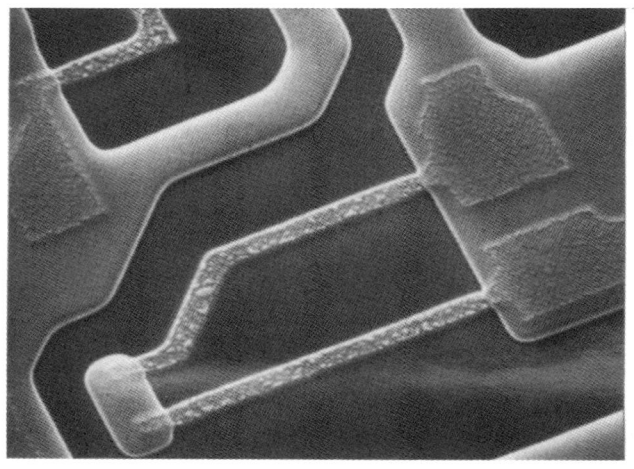

(a) (b)

Fig. 5. SEM photos showing polysilicon etch quality on CMOS gate array.

POWER = 200 watts

PRESSURE = 500 millitorr

GAS = O_2+CF_4

FLOW = 40 sccm(O_2); 20 sccm (CF_4)

TEMP: 40 °C

TIME = 40 sec.

Optical photographs showing the layer by layer etching sequence on a three-layer metal CMOS gate array are shown as Fig. 3. Note that the vias are left intact after removal of the previous metal layer.

Titanium-Tungsten

A Ti-W etch was developed for removal of Ti-W barrier metal used beneath bottom aluminum on the MCA III bipolar arrays. Wet etch techniques using H_2O_2, or aqua regia, have difficulty removing the Ti-W from the contact areas. A chlorine-based RIE etch yields improved results without causing significant damage to the contact areas. The etch is run at a high pressure, yielding an isotropic etch which clears the steep sidewalls into the contacts. No corrosion-prevent etch is performed, so a thorough flushing with DI water is necessary immediately after removal from the etcher. The recipe to etch a 1.5KA layer is as follows:

POWER = 250 watts

PRESSURE = 400 millitorr

GAS = Cl_3+Cl_2

FLOW = 60 sccm(BCl_3); 25 sccm (Cl_2)

TEMP = 40 °C

TIME = 200 sec.

The etch rate is approximately 500 Å/min.

SEM photos showing Ti-W etch results appear as Fig. 4.

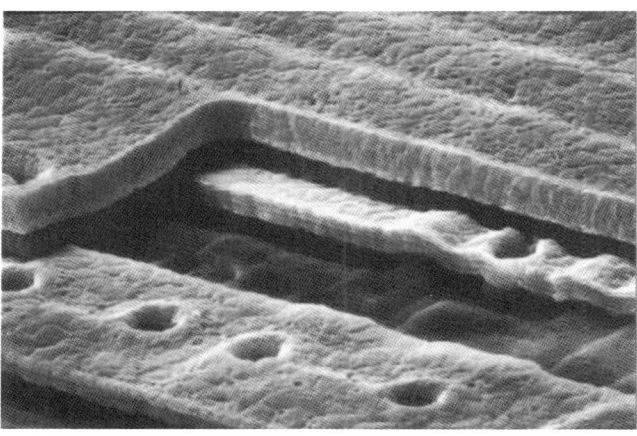
(a)

(b)

Fig. 6. SEM photos showing polyimide etch quality on three-layer metal bipolar gate arrays.

Polysilicon

The selective removal of polysilicon gates which leaves the underlying gate oxide intact is often required during CMOS failure analysis. Very few etches provide the necessary selectivity to remove a 3,000 Å polysilicon layer while leaving a 220 Å gate oxide layer intact. The best selectivity for polysilicon oxide has been achieved using pure chlorine. $CF_4 + O_2$ has also been used; however, the results are not as desirable. The following parameters are used for a chlorine chemistry etch of a 3,000 Å thick phosphorous doped polysilicon layer:

POWER = 200 watts

PRESSURE = 200 millitorr

GAS = Cl_2

FLOW = 60 sccm

TEMP = 40 °C

TIME = 25 sec.

Etch rate is approximately 8KA/min.

SEM photos showing a gate area before and after etching are included as Fig. 5.

Polyimide

Polyimide is used as an interlevel dielectric on three layer metal bipolar gate arrays. The polyimide can either be blanket etched, layer by layer, or anisotropically etched to expose all three layers of metal while maintaining electrical integrity. The recipe used to etch through 4.5 microns of polyimide is as follows:

POWER = 200 watts

PRESSURE = 150 millitorr

GAS = O_2

FLOW = 10 sccm

TEMP = 25 °C

TIME = 300 sec.

The etch rate is 10,000 Å/min.

SEM photos showing the resulting etch profile on a three-layer metal polyimide passivated bipolar array are shown in Fig. 6.

CONCLUSION

Reactive ion etching will undoubtedly play an increasing role in multi-layer VLSI failure analysis. With a properly maximized recipe, RIE achieves superior results compared to wet chemical deprocessing. Specific etch parameters and supporting photographic documentation of the etch quality and profiles have been presented. The recipes presented in this paper are unique to the machine described herein. However, the etch concepts can be adapted to any table top RIE machine.

REFERENCES

1. Addison R. (Randy) Crocket, private communication, especially with respect to polysilicon.

2. D. Tomasi, P. Marchetti, and J. Powlis, "Failure Analysis Applications of Plasma," in *Proceedings of the International Symposium for Testing and Failure Analysis,* Los Angeles, California, Nov. 9-13, 1987.

3. M. Pfart and A. Hart, "The Use of Plasma Chemistry in Failure Analysis," , in *18th Annual Proceedings Reliability Physics,* Las Vegas, Nevada, April 8-10, 1980.

4. W. Jones, "Plasma Etching as Applied to Failure Analysis," in *12th Annual Proceedings Reliability Physics,* Las Vegas, Nevada, April 2-4, 1974.

5. J. Coburn and E. Kay, *"Some Aspects of the Fluorocarbon Plasma Etching of Silicon and Its Compounds,"* IBM Journal of Research and Development, Vol 23, No. 1, January 1979.

6. J. Webber, "Choosing Gases for Plasma Dry Etching," in *Microelectronic Manufacturing and Testing,* January 1985.

7. A. Weiss, "Plasma Etching of Oxides and Nitrides," *Semiconductor International,* February 1983.

8. P. Singer, "Dry Etching of SiO_2 and Si_3N_4," *Semiconductor International,* May 1986.

9. G. Matusiewicz, "RIE for Failure Analysis," *Proceedings of the International Symposium for Testing and Failure Analysis,* Los Angeles, California, November 6-10, 1989.

10. S. Mortis and E. Widener, "Anisotropic Etching For Failure Analysis Applications," *Proceedings of the International Symposium for Testing and Failure Analysis,* Los Angeles, California, November 6-10, 1989.

Dimensional Conversion Chart

DIMENSIONAL CONVERSIONS							
	Inch (")	Mil	Microinch	Centimeter (cm)	Millimeter (mm)	Micron (μ or μM)	Angstrom (A)
Inch (")	1	10^3	10^6	2.54	25.4	2.54×10^4	2.54×10^8
Mil	10^{-3}	1	10^3	2.54×10^{-3}	2.54×10^{-2}	25.4	2.54×10^5
Microinch	10^{-6}	10^{-3}	1	2.54×10^{-6}	2.54×10^{-5}	2.54×10^{-2}	2.54×10^2
Centimeter (cm)	0.3937	3.937×10^2	3.937×10^5	1	10	10^4	10^8
Millimeter (mm)	3.937×10^{-2}	39.37	3.937×10^4	0.1	1	10^3	10^7
Micron (μ or μM)	3.937×10^{-5}	3.937×10^{-2}	39.37	10^{-4}	10^{-3}	1	10^{-4}
Angstrom (A)	3.937×10^{-9}	3.937×10^{-6}	3.937×10^{-3}	10^{-8}	10^{-7}	10^{-4}	1

1. To convert from units along left edge to units along top edge, **multiply** by the number in the box.
2. To convert from units along top edge to units along left edge, **divide** by the number in the box.

IMPORTANT NUMBERS
Avogadro's number: 6.03×10^{23} molecules/g-mole
Boltzmann's constant $k = 8.62 \times 10^{-5}$ ev/°K
Mass of an electron in free space $m = 0.911 \times 10^{-27}$ g
1 ev = 1.60×10^{-12} erg = 1.60×10^{-19} joule

```
Velocity of light in free space  =   2.998x10^10 cm/sec
Mobility of electrons in GaAs    =       8600 cm^2/V-sec
Mobility of electrons in Ge      =       3900 cm^2/V-sec
Mobility of holes in Ge          =       1900 cm^2/V-sec
Mobility of electrons in Si      =       1500 cm^2/V-sec
Mobility of holes in Si          =        475 cm^2/V-sec
Mobility of holes in GaAs        =        250 cm^2/V-sec
Energy gap of SiO2               =        8.0 eV
Energy gap of GaAs               =        1.4 eV
Energy gap of Si                 =        1.1 eV
Melting point of SiO2            =       1700°C
Melting point of Si              =       1412°C
Melting point of GaAs            =       1238°C
Melting point of Ge              =        937°C
```

Failure Site Localization Techniques

FALT PROBING METHOD FOR RAPID EOS/ESD FAILURE SITE LOCATION

by Gene P. Thome

Failure analysis of a MOS power transistor frequently becomes a deeply involved, time-consuming task because each chip consists of thousands of paralleled individual transistor cells. Aluminum source metallization obscures much of the device surface, concealing many structural features (Photo 1). Isolating a single damaged cell from this sea of identical cells by conventional SEM methods is consequently very time consuming.

STUDY OF ALTERNATIVE METHODS

Various hot-spot isolation techniques using thermal methods have been in existence since the earliest days of failure analysis. They were studied as alternatives, and the chart below compares their physics and attributes. These methods are all satisfactory to varying degrees for the location of suspected damage or failure sites on power chips, when these sites are manifested only as hot spots:
- infrared thermal imaging
- liquid crystals
- pressurized Freon® bubble
- thermally sensitive fluorescent phosphors
- thermal indicator paints or coatings

When applied to the problem of hot spot failure location on power MOS dice, however, these above techniques were too expensive, too slow, or unnecessarily elaborate for the purpose. For example, infrared thermal imaging, thermally sensitive fluorescent phosphors, and optical emission (EMMI) have high resolution and sensitivity, but they are expensive and cannot illustrate dynamics. Some liquid crystals, such as MBBA, may be carcinogenic. Liquid crystals may require a hot chuck with very accurate controls to thermally bias the specimen in the center of the crystals' dynamic range. Some of the particular chemicals required for the above techniques, such as liquid crystals and europium thenoyltrifluoroacetonate (EuTTA) may be difficult to obtain. Many of the above competing methods were determined to be unacceptable when speed, cost, and simplicity were important factors.

Detailed experimentation with the sensitivity of some of these techniques showed that the conventional bubble test, which is implemented by powering a device immersed in a low-boiling point liquid indicator, usually a Freon®, lacked the sensitivity necessary to detect small thermal gradients. Liquid crystals were a problem because the failure sites in proximity to the source bond were occluded by a large meniscus around the source wire, owing to the viscosity of the crystal.

After this study of the options, it appeared that an investment of time in the maximizing of the sensitivity of the old bubble test would be the most expedient approach to developing a technique for the rapid and economical location of hot spots.

The first sensitivity improvement was based on the assumption that the thermal mass of the liquid was responsible for the low sensitivity of the bubble test. It was reasoned that thinning of the film, resulting in the reduction of its volume in the vicinity of the fail site, would increase the sensitivity of the test. Experimentation with a variety of available solvents verified that this was true.

Next, it was reasoned that the most effective indicator would be a liquid with a boiling point just above room ambient.

CHOICE OF LIQUID INDICATOR

The use of methanol, or methyl alcohol, CH_3OH, has been based on experimentation with a limited menu of organic liquids which were readily available in the laboratory at the time. Surprisingly, a deliberate search for a liquid organic thermal indicator material, utilizing tables of chemical properties, did not necessarily produce a better candidate.

The properties of the candidate organic liquid indicator were to be: low boiling point in a range around room ambient (27 °C), low heat of vaporization, poor thermal conductivity, high electrical volume resistivity, and non-hygroscopic. The liquid also needed to be of low toxicity, non-carcinogenic, nonflammable, and of low environmental impact. The database search for organic liquids with boiling points between the arbitrarily chosen limits of 20 °C to 35 °C produced a list of 65 compounds. However, the above constraints quickly narrowed the choices because many of the compounds were totally unacceptable for use in the lab electrical test environment[3]. Examples are: carcinogenic (vinyl alcohol), explosive (diiodoacetylene and silane), environmentally unsafe (Freon® 11, Freon® 152), flammable (dimethyl ether), odoriferous (acetaldehyde, butyric acid, and ethyl mercaptan), or poisonous (arsine, borane, hydrogen cyanide, and phosphine). Obviously, none of these materials are acceptable for use in the intended electrical test environment. Table 1 contains a few examples of the candidate liquids within the boiling point range selected, regardless of their suitability. Ultimately, methanol was selected because its evaporation rate was compatible with the heating rate of the chips, it left no residue, and its fumes were the least annoying in the absence of ventilation.

Notice that common methanol remains one of the few reasonable choices. Other common solvents such as acetone (56.48 °C(, ethyl alcohol (ethanol) (78.32 °C), and isopropyl alcohol (1-propanol) (82.5 °C) have boiling points that are too high.

This derivative of the bubble test is called failure area location technique, or FALT. FALT has proven to be a viable ana-

Table 1 Example candidate liquids for use as thermal indicators[1,2]

BP, °C	Chemical name	Chemical formula
20.6	1, 1 dimethylcyclopropane	$1, 1, -(CH_3)_2C_3H_4$
20.8	acetaldehyde	CH_3CHO
21.6	difluoroiodomethane	F_2CHI
21.1	1, 2 dichloro-1,2 difluoroethylene	$CFCl=CFCl$
22.6	isopropenyl chloride	$CH_3CCl=CH_2$
25-31	dibromodifluoromethane (freon)	Br_2CF_2
27	**ROOM AMBIENT TEMPERATURE**	
27.5	n-methylaziridine	
27.8	isopentane	$CH_3CH_2CH(CH_3)_2$
28	vinyl ether	
28	1, 2-dichloro-1, 1, 2-trifluoroethane	(CFC123a)
30	dodecafluoropentane	
30	iodotrifluoroethylene	
30	propylethylene	
30.7	1, 2-difluroethane (Freon 152)	
30.7	ethylene difluoride	$CH_2FCH_2F^®$
31.2	isoamylene	
31.36	furan	C_4H_4O
31.5	methyl formate	HCO_2CH_3
32.04	**methanol**	CH_3OH
32	iodoacetylene	
32.3	dichlorofluoroacetonitrile	
32.5	butyl fluoride	
32.5	methyl isopropyl ether	
32-33	dichloroacetylene	
34	dimethylketene	
34	isoprene	$CH_2=C(CH_3)CH=CH_2$
35	1-fluoro-1, 2-dichloroethylene	
35-36	ethoxyethylene	

lytical method. FALT probing reduces the search time for locating a failed cell on a power MOSFET die by approximately 90%. This reduction in time is compared to the time that would be normally expended for examination of the die, cell by cell, to find a melt site. If no melt site were apparent, then the location of the damaged area would require examinations alternating with selective layer removal. The economy of this method derives from the elimination of the thermally controlled stage, as with liquid crystals, Further, it eliminates the wait for computer image processing, as with infrared thermal imaging. (Photos 2,3,4).

Although common methanol is one of the most desirable organic indicator liquids present in the search and in Table 1, the analyst is reminded that methanol is definitely poisonous. Inhalation of excessive fumes can result in symptoms similar to those resulting from ingestion. Methanol ingestion can result in blindness. Although the quantity of methanol used is about one drop, it is recommended that ventilation be used when performing this test. Consult with, and follow the recommendations of, your company's safety officers.

EQUIPMENT AND PROCEDURE

The equipment items used to perform the FALT procedure were:

- Tektronix, 576 curve tracer
- Alessi probe, model REL-4100A
- Polaroid camera

Fig. 1. Photo 1. Bubble Test—Device is submersed in indicator liquid and powered via clips and flying leads. Depending on the nature of the short, bubbles may emanate from the damage site and mark its location.

- Perkin-Elmer ETEC SEM
- Additional items: syringe, methanol, aluminum block, 3-inch glass Petri dish, and double-stick tape.

FALT Probing Method

Fig. 2. Photo 2. Typical FALT probe set-up. The chip is powered and submersed in methanol. Bubbles emanate from the damage site and tend to spiral upward within the liquid film. Petri dish and block are not in use.

TEST SEQUENCE

1. Secure the device to a 1/4-inch thick metal block to hold it away from the pool of alcohol.
2. Secure the metal block to the center of the Petri dish to retain any excess alcohol.
3. Secure the Petri dish to the probe base for stability.
4. Attach probe station to the power source and adjust the probe needles to contact the chip. In this work, the probe needles were placed on the gate and source contact areas.
5. With the syringe, apply a drop of alcohol to the die and observe the time required for evaporation to dryness, while monitoring any changes in the appearance of the die surface. Apply a second drop of alcohol and start applying voltage as in step 6.
6. While pulsing the curve tracer power, gradually increase the voltage until the alcohol film on the chip's surface begins to evaporate.

The experimental damaged dice were powered by a Tektronix 576, 577, 370, or 371 curve tracer, and contacts were made with a probe station, in this case manufactured by Alessi. **It is important to note that the current levels required to**

Fig. 3. Photo 3. SEM view of typical MOSFET die with metallization removed. This particular geometry contains more than 6500 cells.

produce the desired results in this work ranged from 50 to 200 mA. As the alcohol approaches its vaporization point, a circular dry area will appear and enlarge as the pulsing continues. The time required for the manifestation of the circular dry area is directly dependent upon the severity of the hidden damage conditions; more specifically, the physical size and resistance of the damaged region and the area of the device which is turned on as a result of the damage. This process is repeated until the same dry area is observed to appear regularly. It was found that a thin film of methanol will evaporate most quickly in a 5 to 10 mil (0.005 to 0.010 inch) diameter area centering on the damage site because it has the highest temperature of any area in the die. Its location is photographically recorded for attention during the remainder of the analysis.

Failure Site Interpretation—Once the failure location is identified, delayering is then required to reveal the exact cause of failure. The procedure is as follows:

1. Photograph the general area for orientation, then photograph suspect cell(s) at higher magnification, 2—3 KX.
2. Chemically remove metal and glass layers one at a time and photograph the suspected cell area after each operation.
3. The delayering process is generally continued until sufficient evidence is available to determine whether the failure cause is mechanical or electrical (Photos 6—10).
4. Should there be no manifestations of a failure mechanism at the failure site location after delayering, the chip's surface is etched with a delineating agent—usually 5:1:1 HNO_3, HF, HAc silicon etch (Photos 11 and 12).

Fig. 4. Photo 4. Failure site revealed through progressive delayering process. At the top left, passivation glass and Al have been removed. Detail of the site appears at the top right. At the bottom left, the TEOS glass layer has been removed. In the bottom right, the polysilicon layer has been stripped. This is ESD damage.

Occasionally, no conclusive evidence can be found at the suspected site or elsewhere. Current data show that false positive indications have been found to occur about 5% of the time. However, the author feels a 95% success rate, coupled with the enormous time and cost savings over the alternative methods, is more than acceptable.

LIMITATIONS AND AREAS FOR FUTURE WORK

Use of this method, and interpretation of the action of the liquid on the surface of the die, is complicated by the properties of the device under test and the alcohol itself. It should be noted that when the FALT test was applied to good units, the area where the alcohol evaporated to dryness, first always centered on the source bond (Photo 3). The reason for this effect is that alcohol is hygroscopic, and absorbs water from the atmosphere, forming an azeotrope, with decreased volume and sheet resistance. Additionally, alcohol is a polar liquid, inherently conductive. The alcohol film across the surface of the die tends to form a voltage divider between the drain and source, which may bias the gate positive and turn the device "on" if the gate is terminated with a high impedance. The source bonding wire is therefore a location on the die surface which may run hotter than its surroundings under normal forward bias conditions, and the alcohol would evaporate more quickly at this location.

AZEOTROPES

The tendency for organic liquids to form azeotropes can be used to advantage. Automotive antifreeze, an azeotrope of ethylene glycol and water, is familiar. A few thousand azeotropes are known. If a safe liquid with a boiling point below room ambient were mixed with another liquid, the boiling point of the resulting azeotrope is determined by its composition, "tailoring" it to the desired temperature range, and boiling would stop. For example, a minimum boiling point azeotrope consists of a mixture of chloroform and methanol. Two more interesting azeotropes are listed in the CRC Handbook: methyl formate with n-pentane (21.8 °C) and trichlorofluoromethane with 2-methylbutane (23.16 °C). There is room for extensive experimentation in this area.

RESULTS

Efficient failure site location is a necessary talent in power MOSFET failure analysis. Mastery of this technique was found to eliminate much of the search time required by other methods, including conventional SEM inspection damage location methods.

By following the steps listed in this article, the properly equipped analyst can quickly become proficient. Improved turnaround time, productivity, and report quality were the result of the application of this simple test when damaged power MOS devices were the subject of the analysis.

REFERENCES

1. *The Merck Index,* 11th Edition, Merck and Company, Rahway, New Jersey, 1989.

2. *Handbook of Chemistry and Physics,* 72nd Edition, David R. Linde, CRC Press, 1991.

3. *Hazardous Chemicals Desk Reference,* by N.I. Sax and R.J. Lewis, Van Nostrand-Reinhold, New York, 1987.

4. "Thermal Analysis of Microelectronic Devices Using a Hot Spot Detection System Based Upon Vapor Bubble Technology," by K. Hollmna et al, *Proceedings of ISTFA-82,* pp. 43-52.

5. "Condensation Thermography—A Novel Approach For Locating Short Circuits and Determine Surface Temperatures in Semiconductor Die," by Aaron DerMarderosian et al, *Proceedings of IRPS,* 1981, pp. 276-281.

6. "A Method of Detecting Hot Spots on Semiconductors Using Liquid Crystals," by John Hiatt, in *19th Annual Proceedings of the Reliability Physics Society,* 1981, pp. 130-133.

7. "Remote Thermal Imaging With 0.7 µm Spatial Resolution Using Temperature-Dependant Fluorescent Thin Films," by P. Kolodner and J.A. Tyson, in *Applied Physics Letters* 42(1), 1 January 1983, pp. 117-119.

8. "Microscopic Fluorescent Imaging of Surface Temperature Profiles with 0.01 °C Resolution," by P. Kolodner and J.A. Tyson, in *Applied Physics Letters* 40(9), 1 May, 1982, pp. 782-784.

9. *Hawley's Condensed Chemical Dictionary, 12th Edition,* Revised by Richard J. Lewis, Van Nostrand-Reinhold, 1993.

10. "Toxic and Acutely Toxic Chemicals," Appendix C, from *Safety and Health Guide for the Chemical Industry,* U.S. Department of Labor, Occupational Safety and Health Administration, 1986 OSHA 3091, Appendix I and II.

11. "Liquid Crystals," by J.L. Fergason, in *Scientific American,* 211, Aug. 1964, pp. 76-85.

12. *CRC Handbook of Chemistry and Physics,* Section 6, "Fluid Properties," CRC Press, 1991, pp 6-121 to 6-153.

13. *The Chemist's Ready Reference Handbook,* by G.J. Shugar and John A. Dean, McGraw-Hill, 1990, pp27.5 -27.8, "Azeotropic Distillation.

LIQUID CRYSTAL MICROSCOPY

by Shekhar Khandekar and Kendall Scott Wills

ABSTRACT

To perform successful failure analysis on a semiconductor device, precise knowledge of the exact fail site location is important. Several techniques are used by failure analysts to determine the failing location. Photo emission microscopy, voltage contrast and liquid crystal are among these techniques. Liquid crystal microscopy is the most economical and simplest to set up. The technique is based on the change in electro-optical properties of the liquid crystals as a function of voltage, current or temperature caused by the defect.

1. INTRODUCTION

The most critical problem concerning failure analysis of integrated circuits lies in determination of the precise location of electrical defects. Locating the defect site in current devices is complicated by the small geometries, 1 micron or less, of the transistors used to form the integrated circuit and the large number of transistors which are combined to generate the required logic function. Internal probing, once the main stay of failure analysis, has become more difficult. The ability to sharpen the needles to less than 0.5 µm and then place them repeatedly on a metal line without damage to the metal has not developed as fast as the increment in integration.

Several new tools have been developed to test the current generation of semiconductors. The new techniques are electron beam (E-beam) and laser probing (1), Emission Microscopy, electron beam induced current (EBIC) and optical beam induced current (OBIC). A complimentary technique to these techniques is liquid crystal microscopy (LCM). The change in the electro-optical properties of the crystal as a function of voltage, current or temperature is used to visualize the action of the failing circuit (2). Liquid crystal microscopy (LCM) is important, easy and economical as it can be set up for a small fraction of the cost as is required for E-Beam or Emission Microscopy.

2. LIQUID CRYSTALS

2.1. Principle Of The Technique

The technique works on the principle of liquid crystal's inheritant property of "visible transition from anisotropic state to isotropic state".

2.2. Types Of Liquid Crystals

Two types of liquid crystals are typically used in failure analysis: Cholesteric and Nematic. There are magnetostrictive and voltage sensitive liquid crystals available on the market. Some market research is required to find the correct product for one's needs. The demand for liquid crystals is low in comparison to other chemicals. The low demand has set up a rather volatile industry. One company, BDH, however, has continued to be a constant supplier of liquid crystals to the failure analysis.

Cholesteric LCs display color changes in response to localized temperature differences by monochromatically reflecting incident light. Some of the Cholesteric LCs are not sensitive enough to detect low power dissipation shunts.

Nematic LCs possess optical birefringence properties. The crystals are rotated by electric fields, changing the material's property of reflection of polarized light. Defects like polysilicon to diffusion shorts, typical of ESD failures, are easily isolated with high spatial resolution, down to 1 micron, when nematic liquid crystals are used.

2.3. Selection Of Liquid Crystal Type

Various liquid crystals are commercially available. For example, 4-cyano-4'hexyl-biphenyl, commercially known as K-18 by BDH, is available from various sources. K-18 is popular LC since its state transition temperature (STT) is 29.9 °C which is very close to the room temperature. Picart *et al.* (3) described several of the LCs used for the LCM. Bahr *et al.* (4) reported the properties of the smectic 'C' phase, and smectic 'A' phase of compounds with high spontaneous polarization.

LC K-21 (BDH) has a STT of 42.83 °C. Other LCs are available with their STT ranging from –30 °C to 60 °C. Merck T74 and T75 display color bands as a function of temperature. LC for special application are commercially available with STT of 83.5 °C, 60 °C, as well as that of 107 °C.

Several vendors sell LC kits in which they label these LCs as LC1, LC2 etc.

Fig. 1 A typical setup for liquid crystal testing contains A) probe station with polarizers, B) temperature controller to adjust the temperature of the device under test to the transition temperature of the liquid crystal, C) some visual monitor - in this case a video camera.

3. APPLICATION

The liquid crystal microscopy is very useful in locating failures related to excessive leakage current. The technique has been effectively used on CMOS, MOS, Bipolar and BiCMOS technologies. Many types of current conducting phenomena can be detected by this technique. Care should be taken when selecting a LC to ensure the temperature range of the transition is compatible with the fail mechanism.

4. EQUIPMENT REQUIREMENTS

Failure analysis laboratories already possess almost all the equipment required for this technique, a power supply or some form of a stimulus (a tester or a pattern generator), a temperature controller and a microscope with two polarizers. The first polarizer is in the initial light path to the device under test. The second polarizer, analyzer, is in the reflected light path. The polarizer near the light source should be rotatable to align the line of polarization with the optic axis of the liquid crystal. As the liquid crystal changes state, the optical axis of polarization of the liquid crystal changes from in line with the polarizers, to a polarization out of phase, or in some cases, random with respect to the analyzer.

Areas of the test device which do not change state remain clear. The light passes through the polarizers unattenuated. Regions of the test device which do change state cause the polarization to shift. The analyzer then attenuates the light causing enhanced contrast between the two regions of the test die.

The power supply or the stimulus is used to power-up the device and to run the failed pattern. To increase the viewability of the failure, the power to the device is pulsed. The power pulse causes the fail-site to wink or flicker. The power pulse also helps control the temperature of the device. On devices where pulsing the power is not acceptable, the suspected failing circuit must be activated at a frequency which can be seen by the human eye, and which causes the temperature of the fail-site to increase sufficiently to cause the liquid crystal to change states.

The temperature controller is used to vary the temperature of the device around the STT. The closer to the STT without going over the more sensitive to state changes the liquid crystal becomes. One equipment supplier adds a temperature jogger to the temperature controller to allow the temperature to slowly fluctuate across the transition temperature. During the time the temperature is infinitely close to the transition temperature the ultimate sensitivity is reached. The ultimate sensitivity is only supported for a short time, typically 2 to 10 seconds. Pictures of the fail-site must be taken at this time. If the work is to be done on a wafer or a device which has lost its bond wires, a probe station can be set-up to do liquid crystal microscopy.

Other accessories required, other than the liquid crystal bottle are: small bottles for mixing the liquid crystal, syringe, a glass beaker, latex gloves, filter paper, a small soft brush, methyl alcohol (methanol) or ether. Acetone can also be used to dilute the LC. Acetone is required to clean the device after the LCM. Test the solvent with a small test sample before use. Some solvents react with the liquid crystal causing the crystal to gel.

Fig. 2 A video image of liquid crystal testing where the defect is shown by the black spot at the arrow. Note no black spot appears on the structures which are identical next to the defective structure.

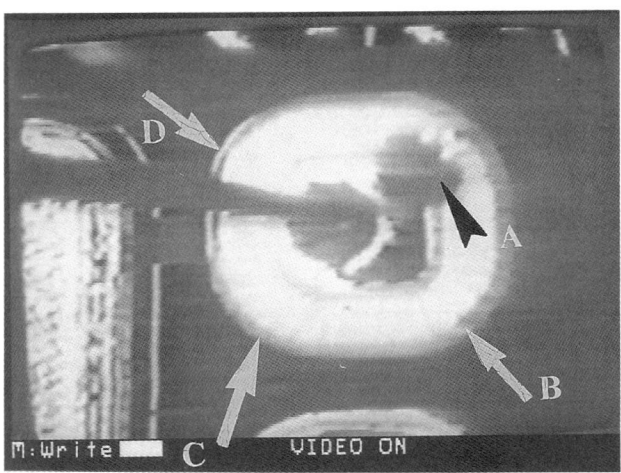

Fig. 3 A video image of liquid crystal testing of a diode. The black spot at arrow A is the location of the defect. Compare the black spot to the other corners of the diode, arrows B, C and D.

5. LIQUID CRYSTAL MICROSCOPY

5.1. Liquid Crystal Preparation

Take about 1 ml of liquid crystal with the syringe and pour it into the bottle for mixing. Take 10 ml of methanol and pour it into the bottle with liquid crystal thus giving you a 1:10 mixture of LC and methanol. This ratio works well for applying a uniform and thin coat of LC on the surface of the die. Pure LC is very viscous and does not spread uniformly on the surface of the die.

Some liquid crystals are supplied as a powder. In such cases place a small piece of crystal on the sample. Heating the sample above the transition temperature will cause the crystal to melt and flow over the sample.

Liquid Crystal Microscopy

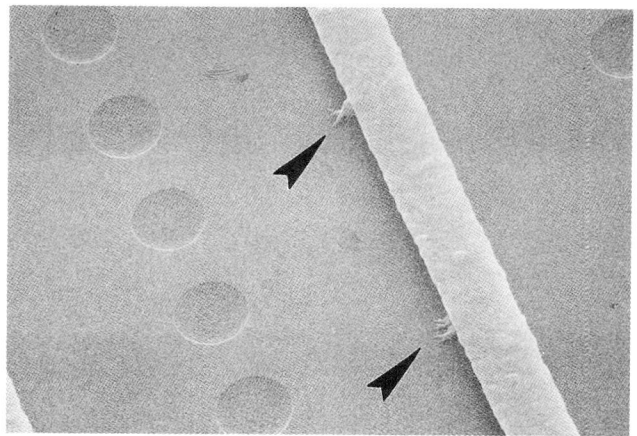

Fig. 4 A SEM image of poly filaments caused by electrical over stress. The defects are pointed to by the arrows.

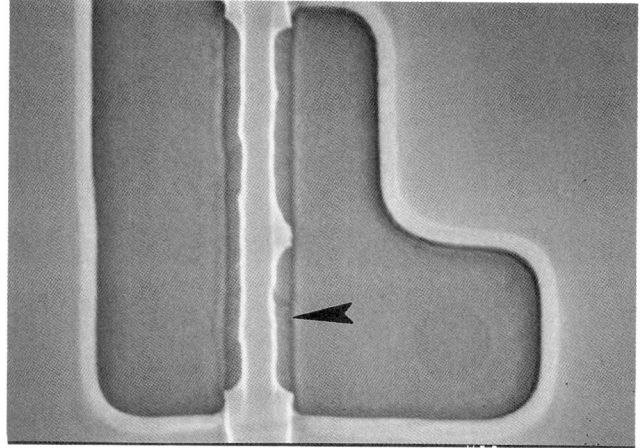

Fig. 5 A SEM image of a gate oxide damaged by electro-static discharge.

5.2. Preparing The Device For CM

The device to be analyzed is decapsulated by using an appropriate technique for the device which maintains electrical integrity. The device is then mounted on a socket under the microscope which has the polarizers. The device is connected to the test stimulus and readied for testing.

The LC is then applied to the die with the help of syringe. This is done by applying a few drops of diluted LC on the die surface and spreading the crystal with a small brush or filter paper. The brush or filter paper will eliminate any additional LC from the die. The die can be spun to get a uniform coating if the coating is applied before insertion into the electrical test fixtures.

The device is then powered up as per test conditions. The device is heated beyond the STT of the LC being used. At this time make sure that the polarizers are in place. Let the device cool down. As the device cools down to near the STT of the LC, a hot spot will be detected. Try to rotate the polarizer to get the best viewing. Remember, the temperature reading on the controller may be different than the temperature at the surface of the device.

The device is then deprocessed by selectively etching the various layers which form the integrated circuit to identify the defect.

6. SENSITIVITY OF LIQUID CRYSTALS

Good results are obtained at a power dissipation of 1 uW or better. A resolution of 1 micron square, and temperature sensitivity of 0.1 °C can be obtained. For smaller structures, the sensitivity is much lower. For buried layers, a loss in resolution has been experienced.

7. LIMITATIONS

Only current-conducting failure mechanisms can be detected. Failures which occur at temperatures other than those

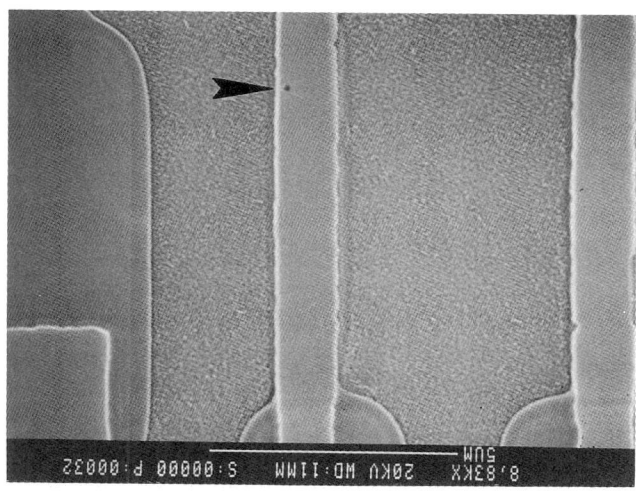

Fig. 6 A SEM image of a gate oxide pinhole. The pinhole is shown at the arrow.

close to the liquid crystal's state transition temperature will be difficult to detect with this technique.

Limited work has been performed on the magnetostrictive and voltage sensitive liquid crystals. The primary reason for the limited work is the rather low device operating frequency with which these crystals work, less than 1 MHz. The rather low operating frequency and the difficulty of obtaining a consistent supply of these liquid crystals has prevented them from becoming prominent in the failure analysis laboratory.

8. SAFETY

Although no evidence of any major effects on the human body have been documented, irritation has been observed on

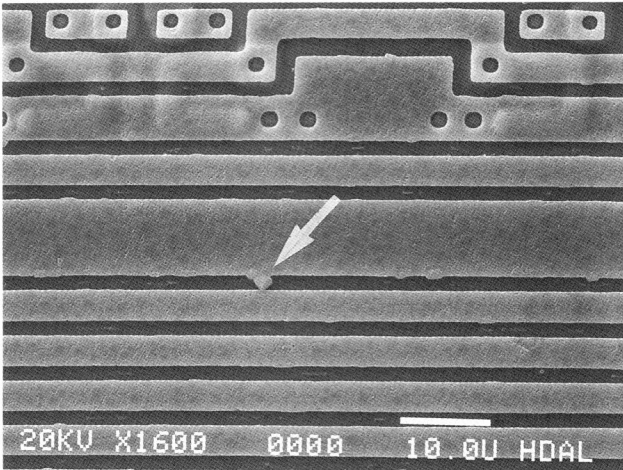

Fig. 7 A SEM image of a metal short.

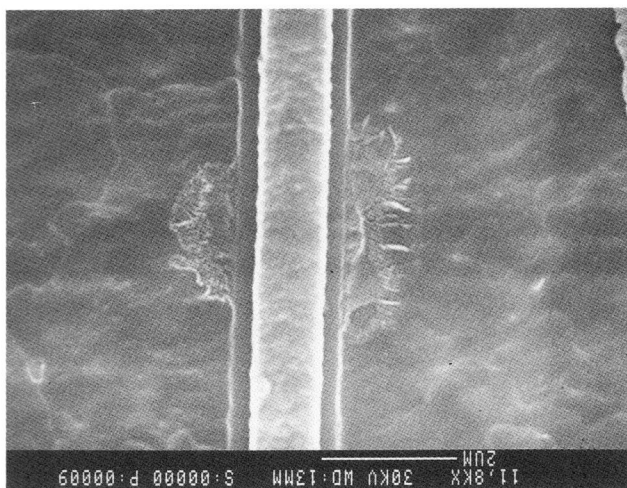

Fig. 8 A SEM image of the residue left from a contaminate. Residue shown at arrow.

the skin of laboratory rabbits. LC is not known to be toxic material. Nevertheless, care should be taken not to expose the body to liquid crystal. If accidental contact does occur, the LC should be washed off. The proper MSDS (Material Safety Data Sheet) should be obtained from the vendor of the LC before use. When using LC's at elevated temperatures some of them produce a foul odor. They need to be used where the area around the test die can be vented.

9. CONCLUSIONS

An inexpensive method of liquid crystal microscopy for fail-site isolation has been presented. This technique is fast and easy to set up but must be customized by the user per the requirements of each device to be analyzed. The technique is limited to defects which will generate enough heat to change the state of the liquid crystal being used. For defects which generate heat the power dissipation of the failure must be greater than 1 uW. Liquid crystals which are magnito constrictive and voltage sensitive are available with some research into potential suppliers.

10. EXAMPLES OF PROBLEMS FOUND WITH LIQUID CRYSTAL

See Fig. 1-8.

Acknowledgments

Thanks is extended to the various sources of liquid crystal listed in the following section. Without their help this paper would not be possible. Special thanks to Temptronic Corp. for providing video tape of equipment setup and applications.

Bibliography

1. Proceedings of the 1st European Conference on Electron and Optical Beam Testing of Integrated Circuits, Grenoble, France, 1987. Microelectronics Engineering, 7, 1987.
2. Bauman et al., 5th International Conference on Reliability and Maintainability, Biarritz, France, 1986.
3. Picart et al., "Visualization if VLSI Integrated Circuits by means of Ferro electric Liquid Crystals," SPIE Vol. 1080, Liquid Crystal Chemistry, Physics and Applications (1989), p 131-139.
4. Bahr Ch. et al., Discussion Meeting on Physics and Chemistry of Unconv. Org. Material, Wiesbaden-Naurod, West Germany, 29 Apr-1 May 1987, p 925-927.
5. Kubalek et al., Microelectronics Engineering, Volume 10, Numbers 1-4, May 1990.
6. Strnad R., "Sensing of Cold Spots on Integrated Circuits by Using Liquid Crystals," Sensors Expo Proceedings, 1987, p 369-374.
7. Doane et al., "Field Controlled Light Scattering from Nematic Micro Droplets," *Applied Physics Letter*, Vol. 48, No. 4, 27 Jan. 1986, p 269-271.
8. Fleuren E. M., "A Very Sensitive, Simple Analysis Technique using Nematic Liquid Crystals," IRPS proceedings, 1983, p 148-149.
9. West G. J., "A Simple Technique for Analysis of ESD Failures of Dynamic RAMS using Liquid Crystals," IRPS Proceedings, 1982, p 185-187.
10. Crow et al., "A New Liquid Crystal for Field-Effect Viewing of 5V Vcc CMOS Logic Families," IRPS Proceedings, 1982, p 179-184.
11. Goel et al., "Liquid Crystal Technique as a Failure Analysis Tool," IRPS Proceedings, 1980, p 115.
12. Hiatt J., "A Method of Detecting Hot Spots on Semiconductor Using Liquid Crystals," IRPS Proceedings, 1981, p 130-133.
13. Burgess et al., "Improved Sensitivity for Hot Spot Detection Using Liquid Crystals," IRPS Proceedings, 1984, p 119-121.
14. Taylor et al., "Leakage Detection Techniques: A Comparative Study," ISTFA Proceedings, 1989, p 5-13.
15. Batchman et al., "Liquid Crystals in Failure Analysis of Analog and Digital IC Chips," IRPS Proceedings, 1986, p 133-136.
16. Batchman T.E., "The Use of Liquid Crystal Materials for Fault Detection in VLSI Circuits".
17. Fleuren G., "Liquid Crystal Microthermography State of the Art".

ELECTRON BEAM-INDUCED CURRENT APPLICATION TECHNIQUES

by James R. Beall

The electron beam-induced current (EBIC) mode of SEM operation is a powerful tool for analyzing and evaluating bipolar semiconductor devices. The EBIC mode could very well surpass the secondary electron or surface imaging mode in value for bipolar semiconductor applications. EBIC provides the capability for measuring subsurface electrical and physical parameters. These measurements are performed without mechanical damage to the surface.

In the SEM, electrons are emitted from the cathode and are accelerated within the electron gun. The acceleration of the electrons is determined by the potential difference between the cathode and the anode in the gun. The energy of the electron is measured in electron volts (eV). An electron that has been accelerated through a potential of 20 kV has an energy of 20 KeV. As the electrons travel through the column, they alternately diverge and converge along the optical axis as they pass through the magnetic lenses. This is the process of demagnifying the electron crossover point from the electron gun to a finite point on the specimen. During their travel down the column, electrons having greater or lesser energy levels than nominal are removed by the apertures. Therefore, the primary electrons arriving at the specimen surface are considered to be nearly monoenergetic.

The electron interactions with the specimen surface are very complex. However, the two principal mechanisms are elastic and inelastic electron scattering. It is basically through these mechanisms that the primary electron energy is dissipated. Elastic scattering can be described as a change in electron direction with little energy loss. An elastic collision results from an interaction between an electron and the nucleus of an atom. Inelastic scattering can be described as an energy loss with little change in electron direction. The elastic collisions produce the backscattered electrons. An inelastic collision consists of two mechanisms: the interaction between an electron and the nucleus of an atom and the interaction between an electron and an electron in an atom. The first interaction produces Bremsstrahlung x-rays, and the second produces secondary electrons and elemental characteristic x-rays. Those electrons that travel into the specimen and are absorbed produce specimen current.

As electrons penetrate the specimen surface, the direction of travel becomes random. It is determined by the number of collisions and the angle of deflection for each. Also, the depth of penetration depends on the number of collisions and the initial energy of the primary electron. Therefore, the amount of energy dissipated in a solid depends on the density of the material, the energy of the primary electron, the rate of primary electrons incident on the specimen surface, or exposure time, and the area scanned.

Two basic factors that determine the electron range in solids are the acceleration voltage and the atomic density of the solid. Therefore, a desired electron range can be preselected based on the material density. If an electron beam strikes the surface of a given thickness of material and the range is sufficient for partial electron penetration, the majority of these electrons can be collected and measured. Figure 1 shows how the number of collected electrons is related to the film thickness. This figure assumes a constant acceleration voltage and a similar atomic density on each side of the detector gate. Electrons that cross the detector gate are measured. The detected electron signal is inversely proportional to the film thickness and depicts the complement of the surface topography. At position "A," no absorbed electrons are measured, at "C" a majority are measured and at "B" and "D" a portion are measured. A visual interpretation of the detected signal will provide a qualitative assessment of the surface topography.

RANGE OF ELECTRON PENETRATION IN SOLIDS

The electron interaction and penetration in solids is a subject that has been studied in detail. Additional information is contained in References 1-7.

As expressed earlier, the interactions between an electron beam and a solid surface are complex. A wealth of energy spectra are emitted from a specimen surface. Aside from the back scattered electron (BSE) energy, the energy emitted from the surface has little effect on the magnitude of absorbed electron current. The primary electron energy is dissipated in two basic mechanisms, elastic and inelastic electron scattering. Elastic scattering is the primary mechanism for reflecting energy from the specimen surface, and inelastic scattering is the primary mechanism for absorbed current and electron-hole pair current.

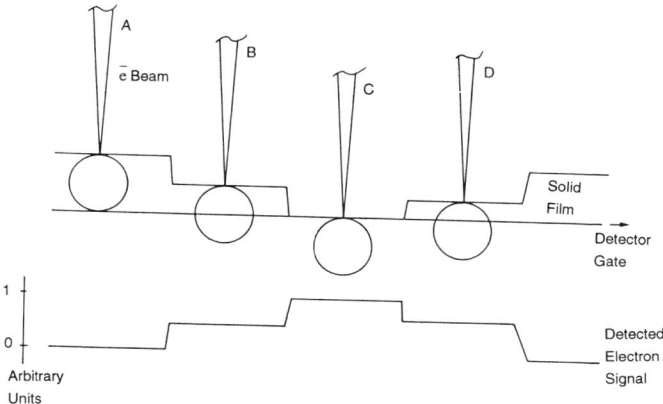

Fig. 1. Electron range as a function of film thickness.

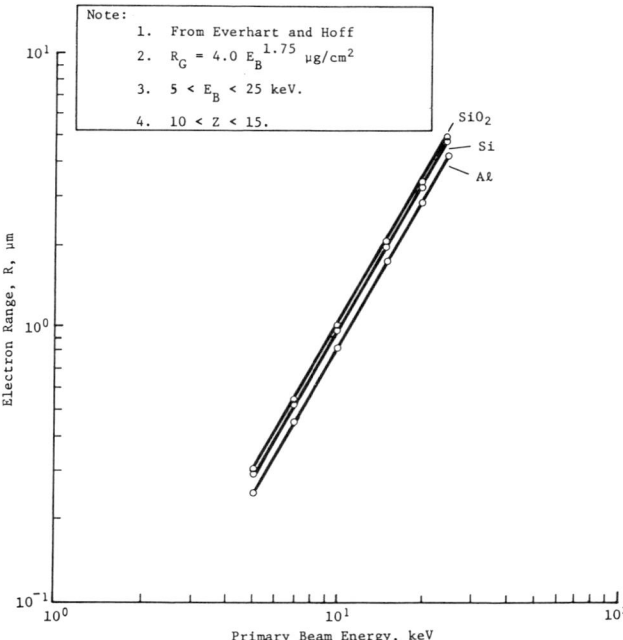

Fig. 2. Electron range (R) versus primary beam energy.

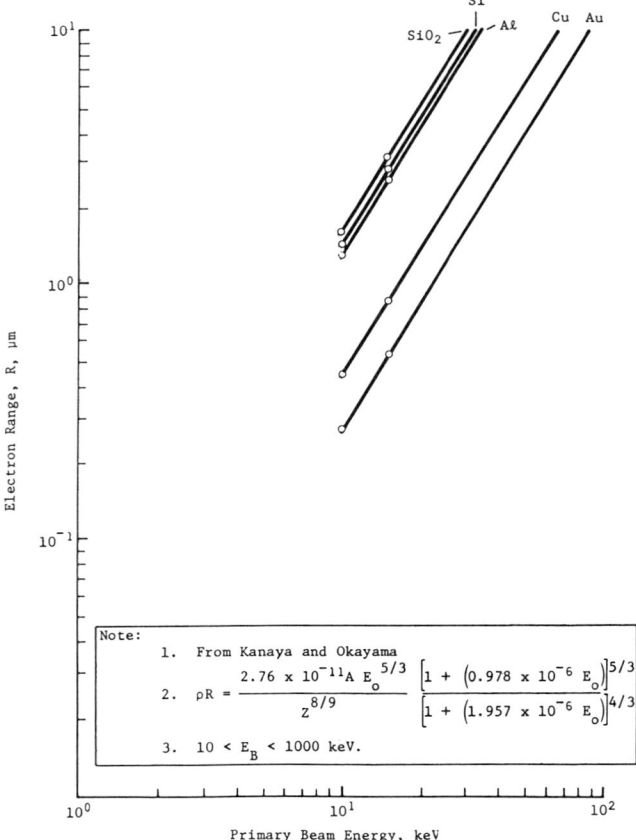

Fig. 3. Electron range (R) versus primary beam energy.

The absorbed current in a specimen is the difference between the electron beam current and the specimen reflected current. This current, commonly referred to as sample or specimen current, must have a return to instrument common to prevent sample charge accumulation. A second current induced in the specimen is an electron-hole pair current. Electron-hole pairs are produced in the specimen by inelastic interaction between a primary electron and the valence electron of an atom. An ionizing energy frees the valence electron leaving a hole. A single primary electron can produce 10^4 electron-hole pairs. The following equation can be used to calculate the energy required to generate e-h pairs in a material using electron beam excitation:

Generation Energy = Material Bandgap Energy $\times$ 2.1 + 1.3 eV

Electron-hole multiplication is the primary mechanism for absorbed electron energy dissipation. For example, in silicon, electron-hole generation requires approximately 3.6 eV. Therefore, a 15 KeV electron can generate 4.2×10^3 electron-hole pairs.

In most semiconductors containing N and P diffusions, the applied or generated fields cause the electrons and holes to flow in opposite directions. Most electron-hole (EH) pairs will recombine spontaneously unless they are formed in, or near, the depletion region, where the fields are strong enough to separate them. Metal atoms in semiconductors produce traps which can severely reduce EH pairs through recombination. The electrons or holes that diffuse across a junction will produce an external current. The number of electrons or holes that diffuse across the junction depend on the distance from the point of generation to the junction depletion region and the diffusion length for the diffused region. Diffusion length is the average distance of carrier travel from the point of generation to recombination (distance traveled during lifetime). The resultant external current flow is typically 10^3 times greater than the primary electron current. It is this electron-hole multiplication current that is used in EBIC applications.

Many theoretical studies and experiments have been conducted to determine the electron range for various elemental materials. Some of these studies evaluated the probability for collision, extent of interaction, resultant energy loss and change in direction of travel. This is a random interaction mechanism and must be evaluated by probability approximations. Other studies have used experiments in which the degree of electron penetration through various elemental thin films could be measured. Through the evolution of these studies came expressions for deriving the electron range based on these theoretical and experimental data.

The electron ranges derived from these expressions do not generally yield the same result. This is due in part to the definition of electron range and the basis from which the expression evolved. Electron range may imply the measurement of the average path length, or the practical/extrapolated range as measured from the point of beam incidence on the specimen surface. The measurement of the average path length would yield a greater distance than the average range from beam incidence. (The electron does not travel in a straight line from the point of

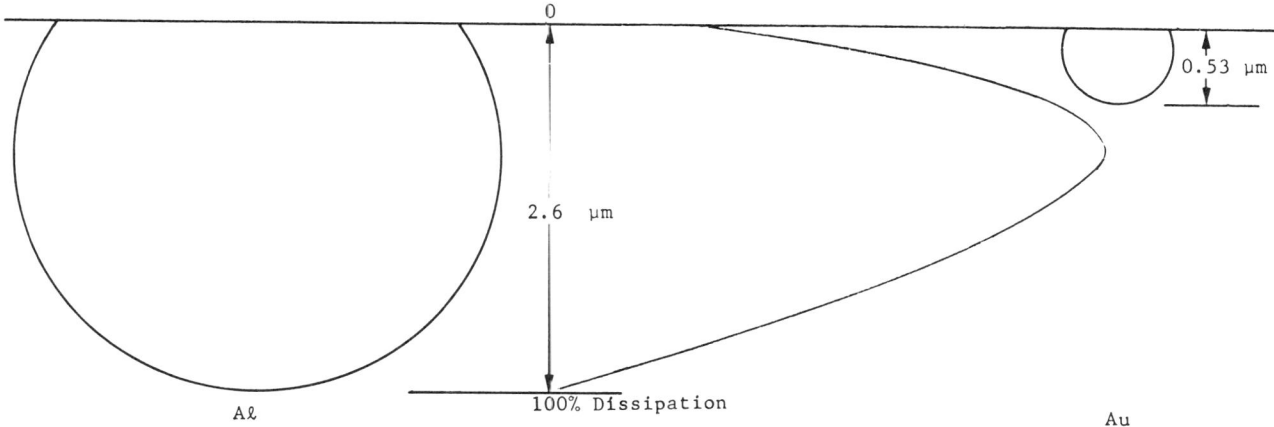

Fig. 4. Electron range for aluminum and gold at 15 Kev (from Kanaya and Okayama Ref 6) and electron energy dissipation for aluminum (Everhart and Hoff Ref 5).

incidence). For this case, the electron range as measured from the point of beam incidence is used.

Two equations were selected for calculating the electron ranges for EBIC applications. One equation was derived by Everhart and Hoff (Ref 5), which is based on energy-dissipation range. This range equation is valid for acceleration voltages of 5 to 25 KeV and atomic elements 10 through 15. This would include semiconductors employing aluminum, silicon dioxide, and silicon systems. The equation is:

$$R_G = 4.0\, E_B 1.75$$

where E is the beam energy in KeV and R is the range $\mu g/cm^2$

(ρR). The equation was plotted showing the relationship of electron range in microns and acceleration voltage in KeV for Al, Si, and SiO_2. Ref. Fig. 2.

The second equation was derived by Kanaya and Okayama (Ref 6), which is based on a modified diffusion model. This range equation is valid for acceleration voltages of 10 to 1000 KeV for all elements. This would include all semiconductor material and metallization systems. The equation is:

$$\rho R = \frac{2.76 \times 10^{-11} A\, E_O^{5/3}}{Z^{8/9}} \frac{[1 + (0.978 \times 10_{E_O}^{-6})]^{5/3}}{[1 + (1.957 \times 10_{E_O}^{-6})]^{4/3}}$$

where E is the beam energy in eV, A is atomic weight in g, Z is the atomic number, ρ is the density in g/cm^3 and R is the range in cm. This equation was plotted showing the relationship of electron range in microns and acceleration voltage in KeV for Au, Cu, Al, Si, SiO_2. Ref. Fig. 3.

These graphs of electron range versus acceleration voltage show the significance of film density on electron range. A comparison (from Kanaya and Okayama) (Ref. 6) of electron range and energy dissipation profiles for Al and Au are shown in Fig. 4. This represents the respective ranges for a 15 kV beam voltage. The energy dissipation profile is important in determining average electron range.

The application of range data in conjunction with semiconductors provides a quantitative and qualitative analysis capability. This capability is related to surface and junction characteristics in conjunction with the underlying diffusions. The diffusions act as an electron energy analyzer. As discussed earlier, the beam-induced electron-hole multiplication current is the major current for these applications, and the semiconductor provides a unique capability for measuring this current. Figure 5 shows a representative cross section of a PNP transistor. In this example the beam induced current is measured from the base. The holes (minority carriers) generated in the base, which diffused across the base-emitter or base-collector junction, produce a net increase in electrons. The increase in electrons results in a current flow through the current amplifier input and ground to the respective emitter or collector junction where they recombine with excess holes. The multiplication current flows only when an electron beam generated bias or external reverse bias is applied to at least one junction of the transistor. For example if the emitter and collector were disconnected from ground, the current flow measured at the base would be the absorbed electron beam current. In the example in Fig. 5, if the current amplifier were connected to the emitter or collector with the base at ground, the amplifier input polarity would be inverted and the current level would be indicative of the recombination current for the respective junction.

TEST DEVICE INTERCONNECT CIRCUITRY

Basic requirements for the interconnect circuitry are 1) low noise pickup, 2) low leakage current, 3) package adaptation flexibility, 4) operation flexibility, and 5) minimal electron beam interference. To reduce noise pickup and leakage current, it is necessary to minimize lead lengths and circuitry external to the specimen chamber; to obtain operational flexibility, it is necessary to make interconnect circuit changes external to the specimen chamber; therefore, the best compromise is to locate the interconnect programming switches at the socket interface.

The specimen stage contained an 18-pin electrical feedthrough. To increase the number of feedthroughs, a high density 61-pin vacuum rated connector was installed. These two connectors were wired to connectors internal to the specimen chamber (Fig. 6 and 7). The internal connectors provide a package adapter disconnect. Mating connectors, interconnect

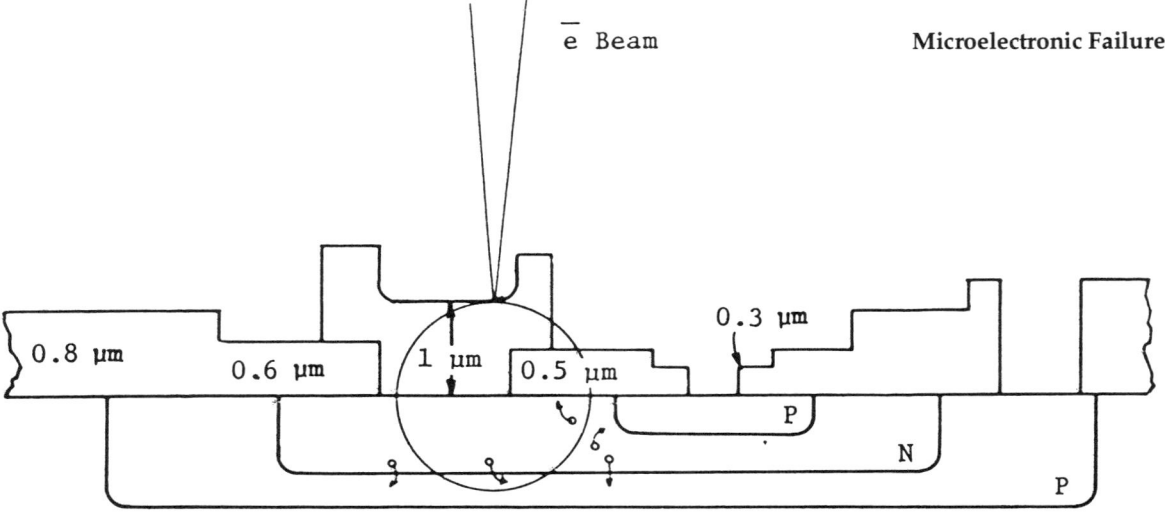

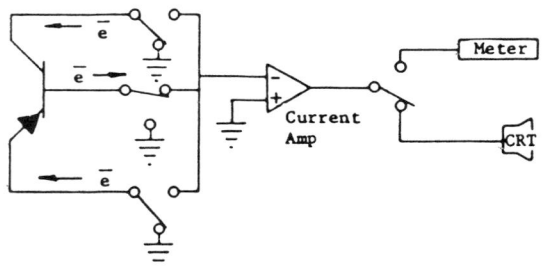

Fig. 5. Cross section showing a typical PNP transistor and current amplifier interconnect circuit.

Fig. 6. Dual-in-line adaptor installed on the specimen stage.

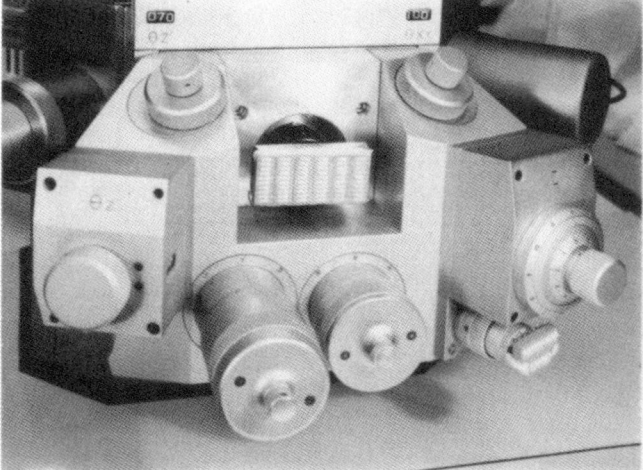

Fig. 7. Two switch matrixes installed on the specimen stage.

wires, and a package test socket provide the different adapters needed for various device package styles. An effort was made to use plug-in sockets for all device packages; however, a suitable mechanical holder could not be found for flatpack style packages. Until a better technique becomes available, the flatpack devices are tack-soldered to a printed circuit card. Metal shields were used to minimize direct exposure of nonconductive package surfaces to the electron beam. Also, nonconductive surfaces on the adapters were coated with conductive paint. Note: This was more important to voltage contrast interference than EBIC interference. All interconnect wiring between the test adapter and the disconnect socket was shielded with a grounded wire sleeve.

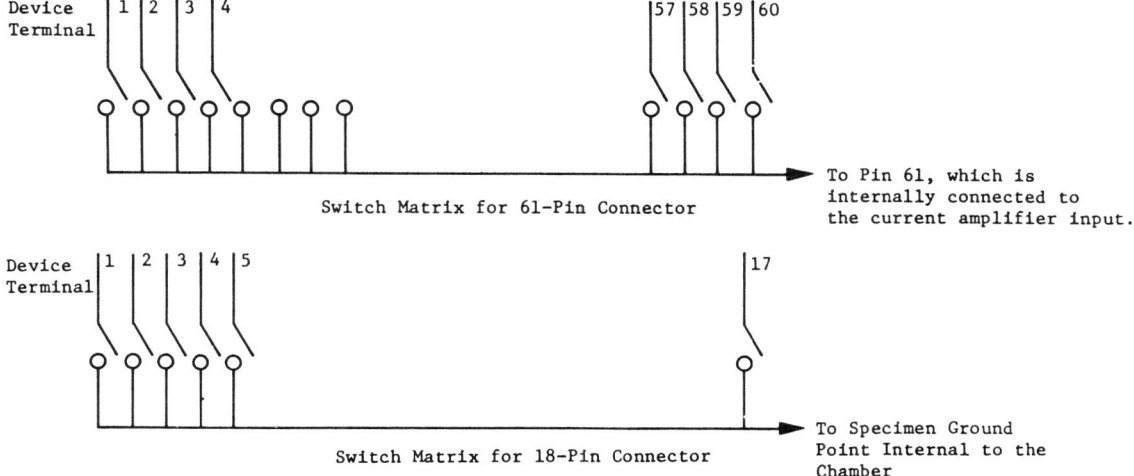

Fig. 8. Schematic circuit for switch matrices.

A switch matrix was constructed and installed external to the chamber and immediately next to the interface connectors. Mechanical DIP switches were used to reduce the switch matrix size and lend inductance. Single-pole single-throw switches were used. The 18-pin connector was used for ground termination switching and the 61-pin connector was used for current amplifier switching (Fig. 7). If single-pole double-throw DIP switches are available with satisfactory isolation ratings, a complete switch matrix could be constructed on a single connector. The isolation obtained for these switch matrices after assembly, cleaning, and vacuum bake was less than 100 pA at 20 volts per switch channel. Subsequent vacuum baking has not been necessary to maintain this isolation; however, it is recommended that it be examined for each case; and especially in high humidity environments. A circuit schematic for the switch matrices is shown in Fig. 8.

The same adapters and interface connectors were used for voltage contrast studies. A 60×10 crosspoint patch board was used to supply the required signal and supply voltages. All interconnecting wires were shielded. Special care should be taken to prevent vibrations from being transmitted along the interconnecting cable to the specimen chamber and electron column. Blowers in equipment mounted in the same cabinet are typically the major sources of vibration.

Measuring the beam-induced currents in a discrete transistor is straightforward. When measuring EBIC in integrated circuits the measurement points are not as accessible. The transistor junctions that have a resistor or resistors in parallel with the junction produce poor EBIC levels. This is due to the majority of recombination current flowing through the parallel resistance, therefore all junctions are not accessible. When necessary it may be possible to open the parallel resistance circuit by scribing open metallization. Current amplifier input impedances are generally greater than 10 megohms.

Although this paper illustrates the use of mechanical switches and basic bipolar structures the principles of EBIC presented apply to bipolar semiconductor structures of any scale, including submicron dimensions.

REFERENCES

1. P.R. Thornton, Chapman and Hall, "Scanning Electron Microscopy," 1968.
2. O.C. Wells, *Scanning Electron Microscopy,* McGraw-Hill, 1974.
3. *Quantitative Scanning Electron Microscopy,* by D.B. Holt, M.D. Muir, P.R. Grant, and I.M. Boswarva, (Ed), Academic Press, 1975.
4. *Practical Scanning Electron Microscopy,* by J.I. Goldstein and H. Yakowitz, (Ed), Plenum Press, 1975.
5. T.E. Everhart and P.H. Hoff, "Determination of Kilovolt Electron Energy Dissipation vs Penetration Distance in Solid Materials," in *J. Applied Physics,* Vol 42, 1971, pp. 5837-5846.
6. K. Kanaya and S. Okayama, "Penetration and Energy-Loss Theory of Electrons in Solid Targets." in *J. Physics D,* Vol 5, 1972, p. 43.
7. T.E. Everhart, et al., "A Novel Method of Semiconductor Device Measurements," in *Proc IEEE,* Vol 52, 1964, pp. 1642-1647.

EBIC References Related to Failure Analysis Applications

J.R. Beal and L. Hamiter, Jr., "A Valuable Tool for Semiconductor Evaluation and Failure Analysis," in *Proceedings of the Reliability Physics Symposium,* 1977, pp. 61-69.

J.R. Beall, "Quantitative Measurement of Metallization," in *Proceedings of the Advanced Techniques in Failure Analysis,* 1978, pp. 175-186.

R.H. Sorenson, I. Thomson, and L. Adams, "Application of Scanning Electron Microscope, EBIC Mode to Semiconductor Evaluation and Failure Analysis," in *Proceedings of the Advanced Techniques in Failure Analysis,* 1979, pp. 136-142.

J.R. Beall, D.D. Wilson, W.E. Echols, and Lt. M. Walters, "SEM Techniques for the Isolation of Failures in Memory Circuits," in *Proceedings of the International Symposium for Testing and Failure Analysis,* 1980, pp. 1-8.

J.M. Patterson, "Semiconductor Junction Temperature Measurement Using the Electron Beam Induced Current Mode in the Scanning Electron Microscope," in *Proceedings of the Reliability Physics Symposium,* 1984, pp. 109-112.

EBIC References Related to General Applications

J.F. Bresse, in *Proceedings of the 5th Annual SEM Symposium,* IITRI, Chicago, IL, 1972, p. 106.

H.C. Casey and R.H. Kaiser, "Analysis of n-Type Gas With Electron Beam Excited Radiative Recombination," by in *J. Electrochem. Soc.*, Vol 114, 1967, pp. 149-153.

T.H.P. Chang and W.C. Nixon, "Electron Beam Induced Potential Contrast on Unbiased Planar Transistors," in *Solid-State Electron*, Vol 10, 1967, pp. 701-704.

W. Czaja, "Detection of Partial Dislocations in Silicon with the Scanning Electron Beam Technique," in *J. Appl. Phys.*, Vol 37, 1966, pp. 918-919.

W. Czaja and J.R. Patel, "Observations of Individual Dislocations and Oxygen Precipitates in Silicon with a Scanning Electron Beam Method," in *J. Appl. Phys.*, Vol 36, 1965, p. 1476.

"A Method for Detecting Imperfections in tHin Insulating Films," in the *Trans. of the Eighth National Vacuum Symp. (American Vacuum Society)*, 1961, pp. 830-835.

"Use of a Scanning Electron Microprobe Device to Investigate p-n Junctions in Semiconductors," by S.A. Ditsman, et al., in *Bull. Acad. Sci. USSR, Phys. Ser. (USA)*, Vol 32, 1968, pp. 885-886.

"The Direct Observation of Electrical Leakage Paths due to Crystal Defects By Use Of The Scanning Electron Microscope," by I.G. Davies, et al., in *Solid-State Electron*, Vol 9, 1966, pages 275-279.

"Evaluation of passivated integrated circuits using the scanning electron microscope," by T.E. Everhart, in *Extended Abstracts (Electrochem. Soc.). Electronics Division*, Vol 12, 2, New York Meeting, Oct. 1963, pp. 2-4.

T.E. Everhart, et al., "Evaluation of Passivated Integrated Circuits Using The Scanning Electron Microscope," in *J. Electrochem. Soc.*, Vol 111, 1964, pages 929-936.

T.E. Everhart, et al., "A Novel Method of Semiconductor Device Measurements," in *Proc. IEEE*, Vol 52, 1964, pp. 1642-1647.

T.E. Everhart, "Certain Semiconductor Applications of the sCanning Electron Microscope," in *The Electron Microprobe*, Proc. Symp. held in Washington D.C., Oct. 1964, pp. 665-676.

J.W. Gaylord, "Microplasma observations in silicon junctions using a scanning electron beam," in *J. Electrochem. Soc.*, Vol 113, 1966, pp. 753-754.

A. Gopinath, "On Scanning-electron-microscope Conduction-mode Signals in Bulk Semiconductor Devices: Linear Geometry," in *J. Phys.D: Appl. Phys.*, Vol 3, 1970, pp. 467-472.

D. Green and H.C. Nathanson, "Observation of Inversion Layers under Insulated-Gate Electrodes Using A Scanning Electron Microscope," in *Proc. IEEE*, Vol 53, 1965, pp. 183-184.

D. Green, "A Method of Examination of Semiconductor Oxides with a Scanning Electron Microscope," in Proc. Eighth Annual Electron and Laser Beam Symp., Univ. of Michigan, April 1966, pp. 375-384.

W.H. Hackett, "Electron-Beam Excited Minority-carrier Diffusion Profiles in Semiconductors," in *J. Appl. Phys.*, Vol 43, 1972, pp. 1649-1654.

F.H. Harris, "Low Velocity Electron Probe Investigation of Thin Dielectric Films," in *Proc. National Electronics Conf.*, Chicago, 1964, pp. 227-232

M. Patzakis, "New Method of Observing Electron Penetration Profiles in Solids," in *Appl. Phys. Lett.*, Vol 18, 1971, pp. 7-10.

H. Higuchi, et al., "Amplification of Bombarding Electron Beam Current in Transistors," in *Extended Abstracts (Electrochem. Soc.). Electronics Division*, Vol 13, 1, 1964, pp. 140-141. (Toronto Meeting, May 1964)

H. Higuchi and M. Maki, "Observations of Channels of MOS Field-Effect Transistors Using a Scanning Electron Microscope," in *Japan. J. Appl. Phys.*, Vol 4, 1965, pp. 1021-1022.

H. Higuchi and H. Tamura, "Measurement of the Lifetime of Minority Carriers in Semiconductors with a Scanning Electron Microscope," in *Japan. J. Appl. Phys.*, Vol 4, 1965, pp. 316-317.

B.H. Hill, "Simple Device for Measuring Beam Current in a Scanning Electron Microscope," in *Rev. Sci. Instrum.*, Vol 39, 1968, p. 1369.

P.H. Hoff and T.E. Everhart, "Carrier Profiles and Collection Efficiency in Gaussian p-n Junctions Under Electron Beam Bombardment," in *IEEE Trans. Electron Dev.*, ED-17, 1970, pp. 458-465.

K. Kanaya and S. Okayama, "Penetration and Energy-Loss Theory of Electrons in Solid Targets," in *J. Phys. D*, Vol 5, 1972 pp. 43.

H. Kanter and E.J. Sternglass, "Interpretation of Range Measurements for Kilovolt Electrons in Solids," in *Phys. Rev.*, Vol 126, 1962, pp. 620-626.

J.J. Lander, et al., "Microscopy of Internal Crystal Imperfections in Si p-n Junction Diodes by Use of Electron Beams," in *Appl. Phys. Lett.*, Vol 3, 1963, pp. 206-207.

G.V. Lukianoff, "Electrical Junction Delineation by Scanning Electron Beam Technique," in *Solid-State Technol.*, March 1971, pp. 39-43.

N.C. MacDonald and T.E. Everhart, "Direct Measurement of the Depletion Layer Width Variation vs. Applied Bias for a p-n Junction," in *Appl. Phys. Lett.*, Vol 7, 1965, pp. 267-269.

N.C. MacDonald and T.E. Everhart, "Scanning Electron Microscope Investigation of Planar Diffused p-n Junction Profiles Near The Edge of a Diffusion Mask," in *J. Appl. Phys.*, Vol 38. pp. 3685-3692.

H.F. Matare and C.W. Laakso, "Scanning Electron Beam Display of Dislocation Space Charge," in *Appl. Phys. Lett.*, Vol 13, 1968, pp. 216-218.

C. Munakata, "Measurement of the Homogeneity of a Semiconductor with an Electron Beam," in *Japan. J. Appl. Phys.*, Vol 4, 1965, p. 815.

C. Munakata, "On the Voltage Induced by an Electron Beam in a Bulk Semiconductor Crystal," in *Japan. J. Appl. Phys.*, Vol 5, 1966, pp. 756-763.

C. Munakata, "Applications of the Barrier Electron Voltaic Effect," in *Proc. Eighth Annual Electron and Laser Beam Symp.*, Univ. of Michigan, April 1966, pages 357-374.

C. Munakata, "Measurement of Minority Carrier Lifetime with a Non-Ohmic Contract and an Electron Beam," in *Microelectron. Rel.*, Vol 5, 1966, pp. 267-270.

C. Munakata and H. Todokoro, "A Method of Measuring Lifetime for Minority Carriers Induced by an Electron Beam in Germanium," in *Japan. J. Appl. Phys.*, Vol 5, p. 249.

C. Munakata and H. Watanabe, "Measurement of Resistance by Means of Electron Beam: II," in *Japan. J. Appl. Phys.*, Vol 5, 1966, pp. 1157-1160.

C. Munakata, "Detection of Resistivity Variation in a Semiconductor Pellet with an Electron Beam," in *Microelectron. Rel.*, Vol 6, 1967, pp. 27-33.

C. Munakata, "An Electron Beam Method of Measuring Diffusion Voltage in Semiconductors," in *Japan. J. Appl. Phys.*, Vol 6, 1967, p. 274.

N.F.B. Neve and P.R. Thornton, "Electrical Effects of Crystal Imperfections Studied by Scanning Electron Microscopy," in *Solid-State Electron.*, Vol 9, 1966, pp. 900-901.

R.F.W. Pease, "The Scanning Electron Microscope," in *IEEE Spectrum*, Vol 4, 1967, pp. 96-l02.

L. Pensak, "Conductivity Induced by Electron Bombardment in Thin Insulating Films," in *Phys. Rev.*, Vol 75, 1949, pp. 472-478.

H.R. Potts, "Secondary-Electron Analysis of Electronic MicroCircuits," in *Microelectronics and Reliability*, Vol 6, 1967, pp. 173-175.

K.V. Ravi, et al., "Electrically Active Stacking Faults in Silicon," by in *J. Electrochem. Society*, Vol 120, 1973, pp. 533-541.

B.W. Schumacher and S.S. Mitra, "Measuring Thickness and Composition of Thin Surface Films by Means of an Electron Probe," in *Proc. Electron. Components Conf.*, Washington, D.C., 1962, pp. 152-161.

M.B. Shaw and G.R. Booker "A New Method for Obtaining SEM Beam-Induced Conductivity Images," in *IITRI69*, 1969, pp. 459-464.

S. Shirai, et al., "Atomic Number Analysis by Specimen Current in Scanning Electron Microscope," in *Sixth Int. Cong. Electron Microscopy*, Kyoto, Aug. Sept. 1966, pp. 199-200.

G.V. Spivak, et al., "Contrast of Images of a p-n Junction in the Scanning Electron Microscope," by (1965), in *Electron Microscopy. Proc. Third European Regional Conf.*, Prague, Sept. 1964, pp. 285-286, (in German).

G.V. Spivak and G.V. Saparin, "Investigation of p-n Junctions by Means of Scanning Electron Microscopy with Electron and Ion Bombardment of the Semiconductor Specimen," in *Bull. Acad. Sci. USSR, Phys. Ser. (USA)*, Vol 30, 1966, pp. 788-791.

G.V. Spivak, et al, "Determination of the Depth of a p-n Junction Using a Scanning Electron Microscope," in *Sov. Phys. Semicond.*, Vol 3, 1970, pp. 1304-1306.

G.V. Spivak, et al, "Development of the Theory of Contrast Formation in the Scanning Electron Microscope Image of a p-n Junction," in *Bull. Acad. Sci. USSR, Phys. Ser. (USA)*, Vol 32, 1968, pp. 1046-1051.

R.M. Stern, et al., "Dislocation Images in the High Resolution Scanning Electron Microscope," in *Phil. Mag.*, Vol 26, 1972, pp. 1495-1499.

D.V. Sulway, et al., "Direct Observation of Electrical Faults in Planar Transistors Made in Epitaxially Grown Silicon," in *Solid-State Electron*, Vol 11, 1968, pp. 567-568.

J.W. Thornhill and I.M. Mackintosh, "Application of the Scanning Electron Microscope to Semiconductor Device Structures," in *Microelectron. Rel.* Vol 4, 1965, pp. 97-100.

P.R. Thronton, et al., "Quantitative Measurements by Scanning Electron Microscopy—1, The Use of Conductivity Maps," in *Microelectron. Rel.*, Vol 5, 1966, pp. 291-298.

P.R. Thornton, et al., "Failure Analysis of Microcircuitry by Scanning Electron Microscopy," in *Microelectronics and Reliability*, Vol 6, 1967, pp. 9-16.

P.R. Thornton, "Scanning Electron Microscopy in Device Diagnostics and Reliability Physics," in *IEEE Trans. Electron Dev.*, ED-16, 1969, pp. 360-371.

C.J. Varker and E.M. Juleff, "Electron Beam Recording in Silicon Dioxide with Direct Read-Out using the Electron Beam Induced Current at a p-n Junction," in *Proc. IEEE*, Vol 55, 1967, pp. 728-729.

H. Yakowitz, et al., *Proceedings of the 6th Annual SEM Symp.*, IIRTI, Chicago, IL (1973), p. 173.

T. Yanagowa, "Influence of Epitaxial Mounds on the Yield of Integrated Circuits," in *Proc. IEEE*, Vol 57, 1969, pp. 1621-1628, Errata 58, 1960-1961, 1970.

VOLTAGE CONTRAST TECHNIQUES AND PROCEDURES

by James R. Beall

INTRODUCTION

Integrated Circuit Failure Analysis

The demand for increased electronic circuit capability has driven the complexity of integrated circuits to higher densities and smaller feature sizes. This in turn has required a major change in the tools used to analyze these increasingly complex and dense integrated circuits.

Early in this growth period the benefits of qualitative voltage contrast were realized in applications for small and medium scale integrated circuits. Although voltage contrast was not dictated by the complexity and density of these circuits, it provided significant benefits over existing techniques such as mechanical probing. Qualitative voltage contrast provided the analyst the ability to observe the functional operation of integrated circuits. Analyses could be performed more quickly and with less risk of mechanical damage. Also these circuits were more forgiving; they were more tolerant to electron beam radiation damage. They did not require numerous signal and power interconnections. Early circuits were not glassivated and metal gate MOS was more radiation tolerant than silicon gate MOS that was developed later. These less complex, less dense, and more tolerant circuits provided the early application development ground for voltage contrast.

As the LSI, VLSI and VHSIC technologies were developed, the requirements for failure analyzing these circuits became more stringent. The need for qualitative voltage contrast continued but was complicated by a number of factors. They were much more susceptible to electron beam radiation damage. Circuit operating frequencies increased, making timing margins smaller. They were glassivated and typically contained multilayered interconnects. Feature sizes continually decreased in size. Lower power circuits were more susceptible to upset due to injected node current. And signal and power interconnects to the device multiplied. VHSIC circuits add other factors. They contain very high equivalent gate counts. Feature sizes have been reduced to one micron or less. Interconnections require up to three conductor layers with the uppermost layer being ground and power planes that covered 80% or more of the chip surface. These factors established the requirement for quantitative voltage contrast as well as special preparation and operation techniques to allow the application of voltage contrast. The continuous growth in technology presses for improved applications and new developments in electron beam voltage contrast for it to continue to serve the needs of the failure analyst.

Benefits of Voltage Contrast

Electron beam voltage contrast represents a significant improvement over previous techniques when measuring node voltages and circuit timing or observing functional operation. Mechanical probing is tedious, risks damage, and may burden the circuit. Liquid crystals are more frequency response limited, and increase the risk of contamination. Electron beam voltage contrast is presently the only practical method of observing complex circuit operation and measuring node voltages and circuit timing response. There are two key factors that limit these analyses for complex integrated circuits. They are small feature size of circuit conductors and circuit susceptibility to upset due to loading by the measuring instrument. Electron beam voltage contrast can support present 1.0 micron circuit conductors and up to low gigaHertz operating frequencies. Present electron beam instrument performance will require additional improvements to keep pace with tomorrow's technology.

Alternatives

For qualitative voltage contrast there are limited alternatives. Viewing the circuit function can be performed on a limited basis using liquid crystals. A thin layer of liquid crystals is spread over the die surface and a field plate is floated on the surface (Ref 1). This technique is limited to operating frequencies up to several hundred Hertz.[1] A major concern with using this technique is the effect it may have on the circuit under test. There is the risk of introducing contamination to the die surface or dissipating trapped surface charge which could result in recovery of the failure.

Another application that may help in isolating the area of functional failures are circuit current signatures.[2] By monitoring the supply current response in conjunction with the signal and clock stimuli the defective circuit cell(s) can be identified by abnormal current signatures.

Comparison of Qualitative vs Quantitative Voltage Contrast

There are significant differences between qualitative and quantitative voltage contrast. The reference of either one as QVC has added to the confusion between these two voltage contrast modes.

Qualitative voltage contrast provides a relative indication of the voltage levels present on the sample (Fig. 1). The relation between sample node voltage and image contrast deltas is nonlinear. In addition, image contrast or secondary electron yield is affected by other factors. These issues are addressed later in fundamentals on qualitative voltage contrast.

Quantitative voltage contrast provides the ability to measure the actual voltage level present on the circuit node. At best these voltage levels can be resolved to about 10 mV. Typically these voltage levels are sampled points plotted along a time base (Fig. 2). The time resolution in a time versus voltage plot is directly related to the electron beam on-time. For example, if the beam blanker turns on the electron beam for a period of one nanosecond, each sampled point on the plot is representative of a one nanosecond increment of time. This process will be discussed in more detail in the section on fundamentals of quantitative voltage contrast. It should be apparent that in order to measure a circuit node's rise time or propagation delay time, one should have 5 to 10 sample points within the measurement period. The more sample points, the higher the waveform fidelity.

The basic difference between qualitative and quantitative voltage contrast is that qualitative provides a relative indication of the voltage level present, and quantitative provides a real measure of the voltage level and response time. Both qualitative and quantitative voltage contrast provide valuable benefit to failure analysis applications.

Role of VC in F/A

Qualitative voltage contrast can serve as a "bird dog" to locate functional failure sites. It provides the ability to observe circuit function and trace signals through functional circuit blocks and from block to block. The result is a very expedient method of locating the area, signal line, or point of a failure. When a circuit fails due to an intermittent or sporadic condition, qualitative voltage contrast may still work but it becomes more difficult to locate or identify the failed area. Qualitative voltage contrast can isolate failures in reduced time. This is achieved by the visual display of the circuit in operation, enabling a rapid understanding of the failed circuit function. It also presents minimal risk of damage to a circuit under examination. <u>Damage is minimized by operating with an electron beam voltage below 5 Kv.</u> Qualitative voltage contrast can be performed on a typical SEM with little or no instrument modification. Most SEM instruments contain interconnect feedthroughs to allow circuit operation in the SEM.

In isolating the failed area using qualitative voltage contrast, if the cause for circuit failure is not visible, further analysis would be performed using quantitative voltage contrast. Further assessment of related circuit node voltage and timing waveforms should pinpoint the reason for failure. In this manner qualitative and quantitative voltage contrast complement each other in the analysis of integrated circuits.

FUNDAMENTALS

Qualitative Voltage Contrast

Voltage Contrast Process

Voltage contrast is the result of a potential on the specimen surface influencing the secondary electron yield or detection level. Secondary electrons are low energy electrons and for aluminum their kinetic energy ranges from 1 to 15 eV.[3] With these energy levels, secondary electrons are easily influenced by nearby electric fields. A positive potential present on the speci-

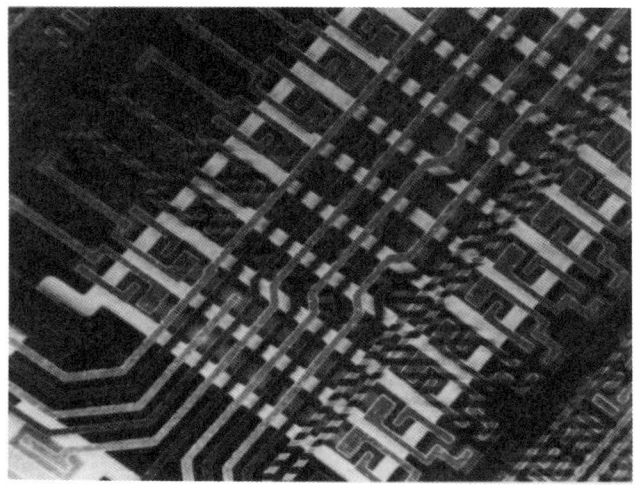

Fig. 1. Showing Voltage Contrast (candy stripes) from a Memory Circuit. Magnification 250×.

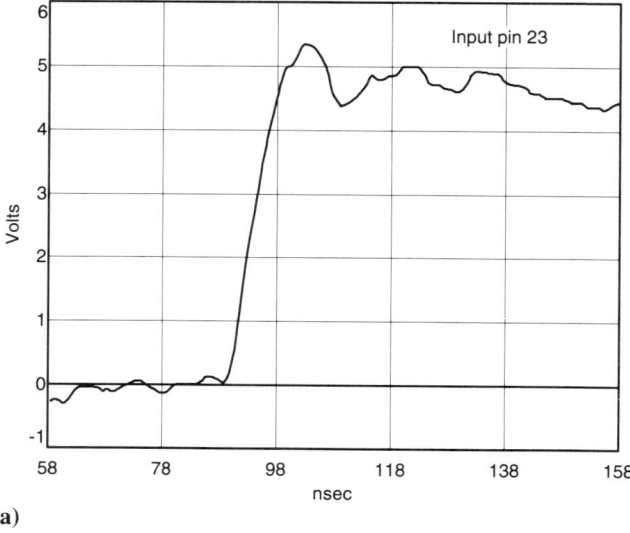
(a)

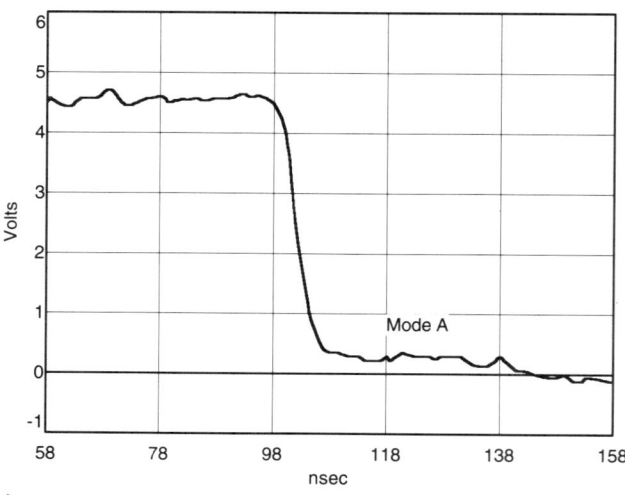
(b)

Fig. 2. Example of Sampled Circuit Node Voltage, (a) Waveform of the Input Signal, (b) Waveform of the Internal Circuit Node.

Voltage Contrast Techniques and Procedures

men at the point of electron beam incidence reduces the number of electrons that escape from the specimen surface for detection. The converse is also true in that if a negative potential is present, an increased number of secondary electrons will escape. The difference in secondary electron levels is related to the surface potential and provides a qualitative relation of voltage present.

The sensitivity of secondary electrons to surface potentials can be appreciated by the degree of image contrast that is produced by circuit node levels in the 0-5 Volt range. Therefore it is easy to see how electron-beam-induced surface charging can interfere with node voltage contrast.

In addition to secondary electrons being influenced by circuit node voltages, they are also influenced by trapped surface charges. Therefore it is important to minimize the introduction of trapped charges produced through irradiation of the specimen by the electron beam. This is accomplished by reducing the beam acceleration voltage. Reduced beam voltage decreases the penetration depth of primary electrons, thereby reducing the depth of trapped charge. Also the beam current is reduced to decrease the number of primary electrons deposited on the specimen surface.

Another method of reducing surface charging is to operate with the acceleration voltage at or near the crossover point. Operating at a voltage above the crossover point (2-3 Kv and higher) results in a net negative surface charge because of the accumulation of electrons. As the crossover point is approached the number of reflected electrons approaches the number of incident electrons and the net surface charge decreases to zero. As the beam voltage decreases below crossover, the number of reflected electrons exceeds the number of incident electrons and the net surface charge becomes positive. By operating near the crossover point the beam-induced surface charge is minimized.

Application Modes

The ability to observe a circuit in operation provides for a number of applications. Obvious is the isolation of circuit failure. In addition, there is support in the development of circuit schematics and block diagrams, comparison of normal and suspect circuit operation, evaluation of functional threshold for signal amplitude and frequency and power supply levels, memory bit map generation, and propagation and upset levels of EMI/RFI signals in circuits.

It is helpful to observe the total or major portion of a device in operation with a low frequency (5 Hz) cyclic address pattern applied to the circuit. Observing the circuit image at a TV scan rate allows quick visual recognition of circuit functional blocks.

The signal excitation frequency is an important factor affecting the object circuit visibility. For example when the frequency is too low, the voltage contrast stripe pattern tends to segment the conductor strip. This occurs when the contrast stripe width is much greater than the conductor width. When the contrast stripe width is equal to or smaller than the conductor width, the conductor is accentuated. The excitation frequency is selected in relation to the SEM photography frame scan speed. This relationship is identified by the following equation

$$\text{Excitation Frequency} = \frac{\text{Cycles per Frame}}{\text{Frame Period (Sec)}} \quad \text{(Eq. 1)}$$

The number of cycles desired per frame is dependent on the image magnification and conductor stripe widths. A typical value would be 30 to 40 cycles per frame. Also, by orienting the majority of conductor stripes at a 45° angle with the top and side of the SEM monitor, a voltage contrast pattern is superimposed at a 45° angle on the conductor stripe. This also reduces the segmented appearance from the superimposed voltage contrast pattern.

To determine the best acceleration voltage, start at the lower extreme (2 Kv or below) and gradually increase it to obtain the optimum image contrast. If the optimum contrast point is exceeded, it may be necessary to reduce the acceleration voltage to the optimum voltage, shut down the instrument, and vent the chamber to atmosphere to neutralize specimen charging. Return the SEM to operation and proceed with voltage contrast examination.

Circuit failure isolation can range from routine to very challenging. Hard failures in static circuits are typically the easier to locate. Intermittent failures and failures in dynamic circuits are typically more difficult to locate. These can require innovative methods. As an example, refresh times in dynamic memories can be extended considerably beyond device specifications. Device specifications include margins for temperature, device performance variation, and data retention guarantee.

Also data strobes and clock trains can be modified by applying setup pulses at shorter periods (higher frequencies) and key transition event pulses at longer periods (lower frequencies). This produces wider voltage contrast stripes for the key event pulses and makes them more visible. Typically these strategies are first demonstrated using known good devices. These types of innovations can resolve what can appear as insurmountable obstacles to voltage contrast applications.

QUANTITATIVE VOLTAGE CONTRAST

Voltage Contrast Process

The voltage contrast process for quantitative measurement is the same as for qualitative imaging. Numerous technical papers have been published on the subject of quantitative voltage contrast. Some key papers are found in references 3-7. The primary difference is that the shift in the secondary electron energy distribution is used to produce quantitative voltage measurement. This is accomplished through the application of an energy spectrometer that will be described later.

A quantitative voltage contrast system requires a number of functions. Figure 3 shows the block diagram for a typical system. In this example the data and waveform display, system setup and control, and digital averaging software are contained in and provided by the master computer.

In quantitative voltage contrast the data obtained from the system establish a reconstructed waveform. This waveform depicts the voltage-time response from a measured circuit node. The process of reconstructing a waveform is as follows.

The beam blanker performs two functions. First, it is used to determine the time coincidence point of measurement for the

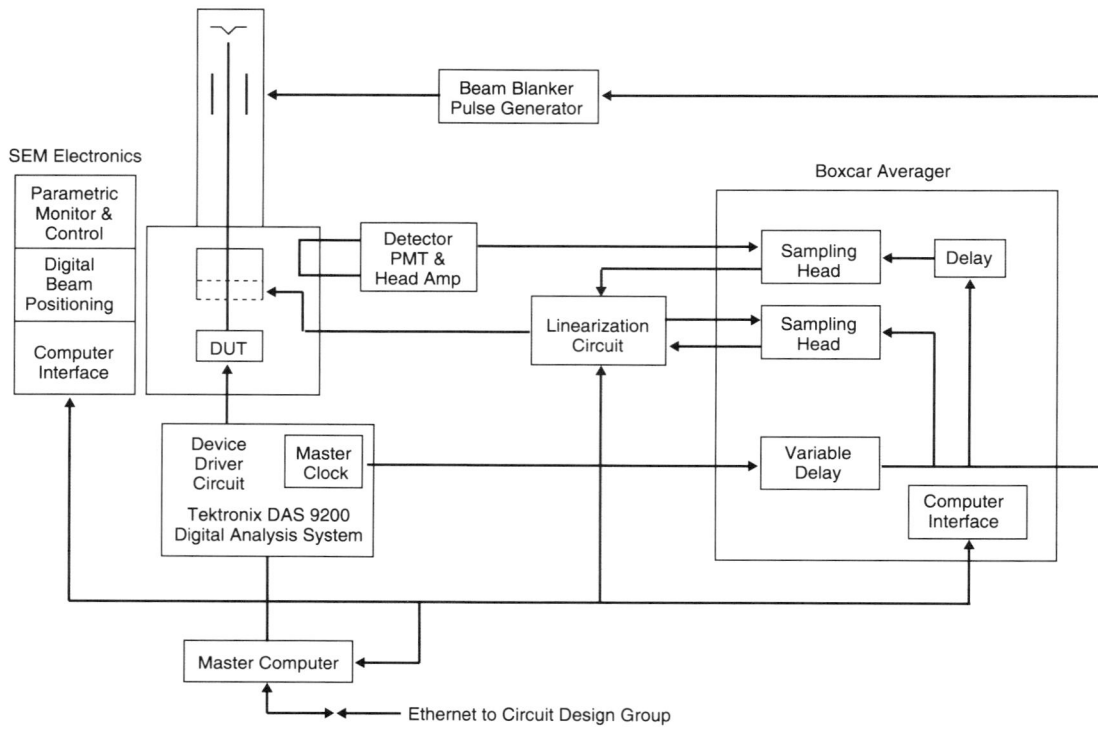

Fig. 3. Block Diagram of a Typical Quantitative Voltage Contrast System.

circuit node being measured. Second, it controls the electron beam-on or measurement time. The beam on time determines the minimum time period that can be resolved on the measured waveform.

A more detailed description is needed to explain this process. The measured waveform depicts a node voltage response with reference to time using a sampling process. Multiple node voltage points are sampled along a time base to form the point by point plot of the voltage levels. Figure 4 shows an example of the sampling process. Each point represents an increment of time that is the period of time the electron beam is on. Each point represents a number of sample measurements that are averaged to improve signal-to-noise margin. The point measured is determined by the time coincidence with electron beam turn on. Determining the time coincidence to reduce searching for a specific circuit node event or transition may require some calculation. This calculation would consider the expected circuit response with respect to the circuit clock pulse. Time synchronization is maintained between the circuit clock and the beam blanker strobe circuit.

Waveform sampling provides a significant reduction in the required measurement system bandpass frequency response. This results in lower system noise and cost.

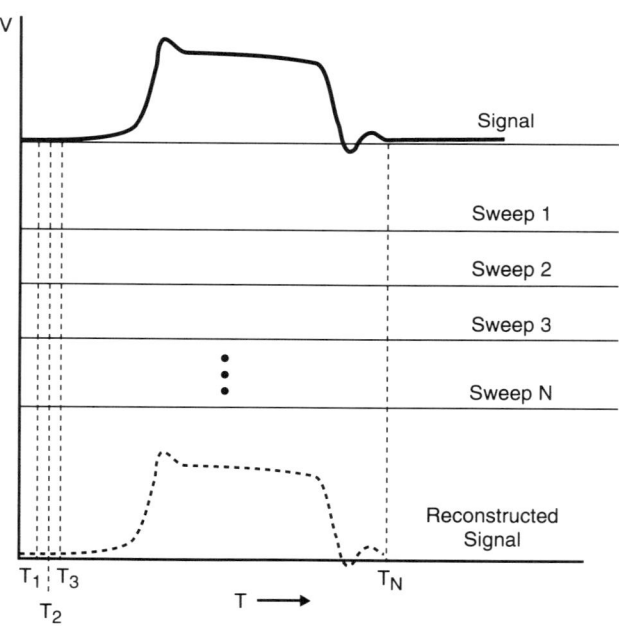

Fig. 4. Depiction of How a Sampled Voltage Contrast Waveform is Generated.

Voltage Contrast Techniques and Procedures

Voltage quantitation is performed by the energy spectrometer. The energy spectrometer is located in or below the final lens in the electron column. Primary beam electrons pass through the energy spectrometer and impact the specimen surface. The energy spectrometer provides the ability to sense the change in secondary electron energy distribution produced by a voltage on the specimen surface.

A more in-depth description of the quantitative voltage contrast measurement process is required, however. Primary electrons incident on the specimen surface produce multiple interactions. Two interactions key to voltage contrast produce backscattered electrons and secondary electrons. Secondary electrons are instrumental in voltage contrast measurements. Secondary electron emission and energy distribution are dependent on surface work function, topography, and fields present at the measured surface.

Several examples illustrating this follow:

1. If the primary electrons produce a carbon deposit on the specimen surface, it would produce a change in the surface work function and result in reducing the emitted secondary electrons, i.e., if the surface changed (> 50 Å of carbon thickness) from aluminum to carbon;
2. If the primary electron beam landing site is on a flat surface and the beam drifts to a high point, e.g., onto a nodule or hillock, the secondary electron emission would increase (the reverse would occur if the beam drifts into a recess or hole);
3. If the surface potential increased in a positive direction, the secondary electron emission and energy distribution would decrease; and
4. If the potential changed on a nearby surface (beam charging, close-by circuit node, or nearby interconnect bond wire), it may also influence the detected level and energy distribution of the secondary electrons.

All of these effects on secondary electron emission and energy distribution, with exception of number 3, would produce circuit node measurement error. Therefore one needs to minimize the opportunity for these interferences. Major effects on quantitative voltage contrast measurement are due to changes in secondary electron energy distribution. Although secondary electron emission can influence quantitative voltage contrast measurements, it has a lesser effect.

Secondary electrons emitted from the specimen surface are directed into the energy spectrometer by the extraction grid. An example of a typical cylindrical energy spectrometer is shown in Fig. 5.

To appreciate the sensitivity of the energy spectrometer one needs to have an idea of the signal magnitude. As shown in equation 2, a 100 nanoamp electron beam current and a 200 picosecond beam-only period would result in only 125 primary electrons incident on the specimen surface.

$$(100 \exp -9 \text{ amps} \times 200 \exp -12 \text{ seconds}) \, 6.242 \exp 18 \text{ electrons/amperes-second} = 125 \text{ electrons} \quad (\text{Eq. 2})$$

With a proportionately low secondary electron emission, the energy spectrometer transmission efficiency must be high. As the primary beam current increases, the signal to noise ratio improves, but the beam spot size increases and resolution decreases. The predominant path of the secondary electrons from the specimen surface is through the spectrometer extraction, retarding, and collector grids to the secondary electron detector and head amplifier. The extraction grid provides a high normal field gradient between the specimen circuit node and the spectrometer. This provides increased isolation from surrounding potential fields. The field gradient is typically 900-1000 volts/mm. The retarding grid(s) regulates the level of detected secondary electron energy distribution so that it is constant. The function of the retarding grid will be discussed in more detail later. The collector grid directs the secondary electrons into the detector. The suppression grid keeps extraneous tertiary electrons from entering the detector.

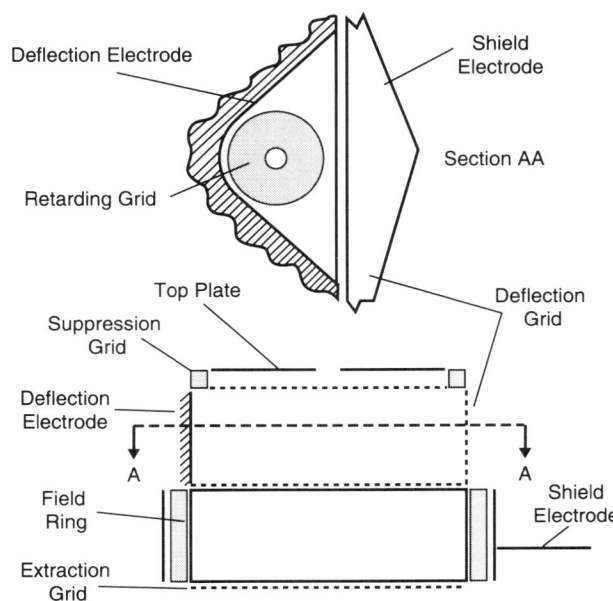

Fig. 5. A Side View and a Cross Sectional View of a Typical Cylindrical Energy Analyzer.

These tertiary electrons are produced by backscattered electrons impacting the top of the spectrometer or generated in the electron column by primary electron impact. The source of backscattered electrons is primary electron impact on the specimen surface. Additional extraneous tertiary electrons are generated by backscattered electron impact on wall surfaces and grid wires of the spectrometer. These extraneous electrons can contribute to voltage measurement error. Therefore, the spectrometer walls and grid wires are coated with carbon to reduce tertiary electron emission levels.

Secondary electrons are directed from the energy spectrometer to the secondary electron detector or scintillator. The secondary electron detector, photomultiplier tube (PMT), and head amplifier are similar to those in a standard SEM, except that they typically have higher frequency response and improved noise level.

As in the SEM, the scintillator converts the secondary electrons into light impulses (photons). These photons are transmit-

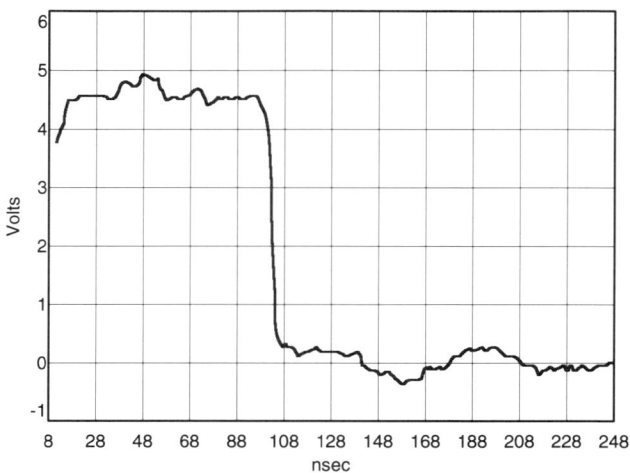

Fig. 6. Input and Internal Nodal Voltage Contrast Waveforms.

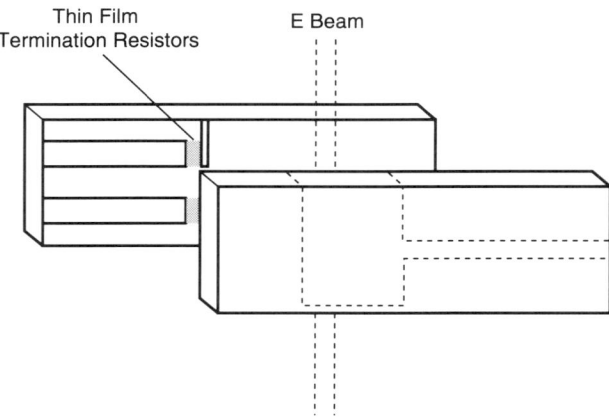

Fig. 7. Diagram of a Two Plate Beam Blanker.

ted to the PMT. The PMT amplifies the photons through electron avalanche multiplication and converts them to an electrical signal. This electrical signal is further amplified by the video amplifier. The output of the video amplifier is directed to the signal processing and feedback control circuitry.

The signal processing and feedback control circuit includes the boxcar averager, linearization, comparator and variable delay functions. The boxcar averager is a gated signal averager that provides the ability to average and store multiple measurements successively sampled at the same point on a time base, that is, synchronized sampling. Signal averaging is required to improve the signal-to-noise ratio. The number of samples averaged typically ranges from 20 to 100 samples per sampled point.

The comparator provides an output in comparison to a stored reference related to a zero potential level on the specimen surface. For example, as the specimen surface potential goes positive, the output of the comparator, would go positive. The comparator output is applied to the linearization circuit to compensate and linearize the voltage over the full response range. The typical response range is 10 to +10 volts. The linearization circuit output is applied to the retarding grid of the energy spectrometer. This voltage represents the equivalent of the voltage present on the specimen circuit node being measured. The variable delay circuit controls the time increment and number of samples averaged per point along the time axis.

A description of the overall quantitative voltage contrast system closed loop operation is as follows. The beam blanker turns on the electron beam at a selected point on the time axis. The primary electrons illuminate the selected specimen circuit node at the precise point in time. The retarding grid is driven by the comparator and linearization circuits to the voltage necessary to null the comparator.

For example, assume the specimen circuit node voltage is +2.5 volts. With some other voltage applied to the retarding grid, assume zero volts, the comparator would find the detected secondary electron energy level was too low. The comparator would produce a correction voltage level equivalent to +2.5 volts. This null voltage is equivalent (within the voltage accuracy) to the voltage level on the specimen circuit node. This assumes that all error sources are below the maximum limits.

Each sample point on the time base is analyzed (measured), averaged over 20 to 100 samples, then stored in the boxcar averager. Then, the sample point is incremented to the next point on the time base. This sequence is continued until a complete voltage waveform is produced as a plot along the time axis (Fig. 4). The required voltage and time resolution determines: the number of sample points that need to be averaged per waveform point; the number of points sampled along a waveform; and the electron beam "on" period. An example of a measured waveform is shown in Fig. 6.

APPLICATION MODES

In applying quantitative voltage contrast one would follow the same principles as circuit troubleshooting and failure isolation. As discussed previously, qualitative voltage contrast can be used as a "bird-dog" to locate the failure area. Once located, quantitative voltage contrast can be used to further analyze the failed area to evaluate timing and voltage level compliance.

Selective etching is a technique that plays an important part in gaining access to internal circuit nodes. Complex circuits contain multi-layer conductor and oxide films. Every year, operating frequencies are pushed upward. To gain access to buried circuit nodes it becomes necessary to selectively remove overlaying films without interrupting conductor paths. When working with an unknown chip layout, the use of other test devices is required to develop an access strategy. Windows are chemically etched in power and ground planes to gain access to underlying circuit nodes. Underlying oxide layers are plasma etched to expose conductors without undercutting conductors.

An important factor for high frequency circuits is to minimize the volume of conductor and oxide material that is removed. These adjacent materials play an important role in determining the characteristic impedance of circuit interconnects. Their removal can significantly alter the circuit node rise, fall, and propagation delay times.

INSTRUMENT REQUIREMENTS

Qualitative Voltage Contrast

The basic SEM provides the capability to perform qualitative voltage contrast with minor modification. It requires power and signal line access to the device in the sample chamber. This by itself allows static and low frequency voltage contrast application. Modification may also include adding a beam blanker to support stroboscopic observation. A beam blanker provides the ability to synchronize the electron beam with the circuit operating frequency. This allows observation of circuit operation above a few hundred Hertz.

Electron Column

The standard SEM electron column normally meets the needs of qualitative voltage contrast. Key column capabilities are acceleration voltage range and selectivity, beam strobe control, and high secondary electron detection sensitivity. The low voltage range of the acceleration voltage is important in reducing beam-induced trapped surface charge and active device irradiation damage. As a minimum, the acceleration voltage should be adjustable down to 5 Kv. Lower voltage is desirable to reduce the risk of irradiation damage and surface charging. Variable voltage adjustment in this range provides the option of selecting an acceleration voltage at or near the surface charge crossover point.

A beam blanker switches the electron beam on and off by deflecting the beam on and off the column axis. The beam blanker is typically located between the electron gun and the first electron lens. The blanker consists of 2 deflection plates. Figure 7 shows a typical configuration. With zero potential on the plates the electron beam travels unobstructed down the column. Application of a potential to the plates deflects the beam off the column axis and this prevents primary beam electrons from traveling down the column to the specimen. A beam blanker allows strobing of circuit nodes to display higher frequency voltage contrast images than would be possible in real time.

The sensitivity of the secondary electron detector is important as typical voltage contrast applications produce low secondary electron levels. Scintillator, photomultiplier tube (PMT), and head amplifier sensitivity should be measured and periodically checked to ensure high performance.

Instrument Sample Chamber

The sample chamber must provide the capability for interconnecting power and signals from outside the sample chamber to the device in the chamber. The interconnecting leads must be adequately shielded to prevent interference with secondary electron trajectories. Interconnecting leads pose two possible interference sources: insulation charging and electrical fields. Shielding over these leads will contain stray fields and prevent them from disturbing the secondary electron trajectories.

Sample Excitation

Sample excitation is dependent on the complexity of the circuit being analyzed. For the less complex circuits (up to LSI), square wave or word generators are practical for excitation. More complex circuits (VLSI and up) require sophisticated address and timing patterns. Depending on the circuit requirements, an excitation circuit can be constructed, the actual circuit application can be operated, or a commercially available system can be utilized. Systems such as the Tektronix DAS 9200 or equivalent are very flexible in producing complex patterns by software control and are practical to program on an individual part basis.

QUANTITATIVE VOLTAGE CONTRAST

Electron Column

The electron source in a quantitative voltage contrast system can be a standard tungsten filament, but is typically an LaB_6 or field emission source to increase beam current density for comparable spot sizes.

Much like in qualitative voltage contrast, a beam blanker switches the electron beam on and off by not deflecting the beam or moving it away from the column axis. In quantitative voltage contrast the blanker may consist of 2 or 4 deflection plates. A 4 plate blanker is used to reduce time differences inherent in 2 plate blankers. In a 4 plate blanker, the electron beam travels in an elliptical path and always passes over the column axis in the same direction. This eliminates time non-linearities in 2 plate blankers where the beam travels back and forth across the column axis.

Quantitative voltage contrast beam blankers must turn a beam on and off within 100 to 500 picoseconds. This performance requires stringent deflection plate positioning and impedance matching. Complementary pulses are typically applied to opposite deflection plates to increase the rate of beam deflection. Refer again to the drawing of the two plate beam blanker assembly in Fig. 7. To drive the electron beam off axis a negative pulse is applied to one plate while a positive pulse applied to the opposite plate. The negative plate repels the beam as the positive plate attracts the beam in a push-pull fashion. These pulse rise and fall times are extremely short so the beam blanker must be matched to the impedance of the pulse driver. This eliminates reflections and ringing which could compromise blanker performance.

There are two basic types of energy spectrometers in use today. One is the Feuerbaum or cylindrical spectrometer that is located immediately below the final lens. The second is an in-the-lens spectrometer and the spectrometer grids are located in the final lens. Some benefits of the in-the-lens spectrometer are shorter working distance with resultant decreased electron beam spot size, and greater separation of the extraction grid and retarding grid, providing improved voltage resolution. Recent systems typically use the in-the-lens spectrometer.

Instrument Sample Chamber

The sample chamber for quantitative voltage contrast has the same basic requirements as qualitative voltage contrast. It must provide a socket or adaptor to hold the specimen, along with interconnections for power and signals to the specimen. However, devices are typically more complex and therefore require larger packages with a greater number of terminals, and they typically operate at higher frequencies, which requires interconnection impedance matching. A larger number of interconnect cables present the possibility of specimen stage drift. Specimen position drift cannot be tolerated when working at the magnification required by feature sizes found in complex circuits.

Sample Excitation

As circuits grow in complexity the stimulus required for device excitation becomes more challenging. Using VLSI/VHSIC automatic test equipment is impractical for excitation in failure analysis except where test software already exists. This is typically not the case. An alternative solution is an easily programmed system that provides versatile timing and frequency by software control. A system similar to the Tektronix model DAS 9200 may meet these requirements. The trend with new quantitative voltage contrast systems is to incorporate the excitation capability in the system.

Quantitative Voltage Contrast System

The quantitative voltage contrast system has made significant progress over the last few years. Systems available today provide badly needed help to the failure analyst who is faced with analyzing complex one-of-a-kind circuits. In today's situation there may be no circuit schematics, die maps, or internal chip layout descriptions available. This makes analysis of these circuits extremely difficult.

Today's quantitative voltage contrast systems provide the ability to store normal circuit operation using a good device and compare these voltage contrast signatures to signatures of the failed device to identify the location of the failure. Without this capability it becomes very costly, even practical, to perform failure analysis. It is encouraging that the cost of these systems is continuing to decrease.

REFERENCES

1. "Liquid Crystals Display Techniques for Analyzing Microprocessors," by D.J. Burns, G.E. Jacobcic, and M.L. Wangler, in *Proceedings of the Advanced Techniques in Failure Analysis*, 1979, pp. 194-198.
2. "Application of Quiescent Current Signature Analysis to Failure Analysis of Integrated Circuits," by S.C. Anderson, in *Proceedings of the International Symposium for Testing and Failure Analysis*, 1989, pp. 27-36.
3. "Fundamentals of Electron Beam Testing of Integrated Circuits," by E. Menzel and E. Kubalek, in *Scanning*, 1983, 5, pp. 103-122. (Reference has an extensive list of Voltage Contrast references.)
4. "Electron Beam Test Techniques for Integrated Circuits," by E. Menze and, E. Kubalek, in *Scanning Electron Microscopy*, 1981, I, pp. 305-322.
5. "Voltage Contrast: A Review," by A. Gopinath, K. Gopinathan, and P. Thomas in *Scanning Electron Microscopy*, 1981, I, pp. 375-380.
6. "Secondary Electron Analyzers for Voltage Measurements," by E. Menzel and M. Brunner, in *Scanning Electron Microscopy*, 1983, I, pp. 65-75.
7. "Secondary Electron Detection Systems for Quantitative Voltage Measurements," by E. Menzel and E. Kubalek, in *Scanning*, 1983, 5, pp. 151-171.
8. "SEM Techniques for the Isolation of Failures in Memory Circuits," J.R. Beall, D.D. Wilson, W.E. Echols and Lt. M. Walter, in *Proceedings of the International Symposium for Testing and Failure Analysis*, 1980, pp. 1-8.
9. "SEM Stroboscopic Techniques—Their Application to Failure Analysis of LSI's," by H. Yuasa, M. Fujita, and N. Manabe, in *Proceedings of the International Symposium for Testing and Failure Analysis*, 1980, pp. 9-14.
10. "Failure Analysis and Fault Isolation of a 1Kbit Schottky RAM by SEM Voltage Contrast," by M. Walter, G. Bernhardt, and J. Carroll, in *Proceedings of the International Symposium for Testing and Failure Analysis*, 1980, pp. 131-135.
11. "Voltage Contrast SEM Observations with Microprocessor Controlled Device Timing," by J.B. Bindell and J.N. McGinn, in *Proceedings of the Reliability Physics Symposium*, 1980, pp. 55-58.
12. "SEM Techniques for the Analysis of Memory Circuits," by J.R. Beall, D.D. Wilson, W.E. Echols, and M.J. Walter, in *Proceedings of the Reliability Physics Symposium*, 1980, pp. 65-72.
13. "Phase Dependent Voltage Contrast—An Inexpensive SEM Addition for LSI Failure Analysis," by D. Younkin, in *Proceedings of the Reliability Physics Symposium*, 1981, pp. 264-265.
14. "Automated Contactless SEM Testing for VLSI Development and Failure Analysis," by M. Macari, K. Thangamuthu, and S. Cohen, in *Proceedings of the Reliability Physics Symposium*, 1982, pp. 163-165.
15. "The Practical Implementation of Voltage Contrast as a Diagnostic Tool," by K.J. Bertsche, and H.K. Charles, Jr., in *Proceedings of the Reliability Physics Symposium*, 1982, pp. 167-178.
16. "Electron Beam Testing for Verification of Voltage Distribution on VLSI Circuits," by J.S. Wolcott, in *Proceedings of the International Symposium for Testing and Failure Analysis*, 1982, pp. 149-155.
17. "Internal Node Testing by Tester Aided Voltage Contrast," by M.J. Walter, C.A. Eldering, K.M. Krevis, and J.R. Haberer, in *Proceedings of the International Symposium for Testing and Failure Analysis*, 1982, pp. 156-161.
18. "Identification of Ceramic Capacitor Shorts by Voltage Contrast in Scanning Electron Microscope," by Dr. D.T.Y. Wei, in *Proceedings of the International Symposium for Testing and Failure Analysis*, 1984, pp. 203-208.
19. "Observation of Latch-Up Phenomena in CMOS ICs by Means of Digital Differential Voltage Contrast," in *Proceedings of the International Symposium for Testing and Failure Analysis*, 1984, pp. 265-266.
20. "Digital Processing and Color Coding of Voltage Contrast Images," by E.L. Miller, in *Proceedings of the International Symposium for Testing and Failure Analysis*, 1985, pp. 63-71.
21. "LSI Failure Analysis Using an Electron-Beam Tester Directly Combined to an LSI Tester," by H. Hosoi, H. Yuasa, M. Kudoh, K. Nikawa, and T. Ohiwa, in *Proceedings of the International Symposium for Testing and Failure Analysis*, 1985, pp. 78-84.
22. "E. Beam Testing Image and Signal Processing for the Failure Analysis of VLSI Components," by J.P. Collin and T. Viacroze, in *Proceedings of the International Symposium for Testing and Failure Analysis*, 1985, pp. 89-97.
23. "Fault Contrast: A New Voltage Contrast VLSI Diagnosis Technique," by A.R. Stivers and D.C. Ferguson, in *Proceedings of the Reliability Physics Symposium*, 1986, pp. 109-114.
24. "Failure Analysis by Dynamic Voltage Contrast Development of a Semi-Automatic System," by N. Giraud-Liria, G. Perez, and A. Cuquel, in *Proceedings of the International Symposium for Testing and Failure Analysis*, 1987, pp. 67-74.

25. "Gated-Pulse Stroboscopy for Passivated Device Imaging," by E.L. Miller, in *Proceedings of the Reliability Physics Symposium*, 1987, pp. 118-125.

26. "Overview of Voltage Contrast Techniques and Applications," by J.R. Beall, in *Proceedings of the International Symposium on Testing and Failure Analysis*, 1988, pp. 1-8.

27. "Voltage Contrast Imaging of Passivated Devices," by E.L. Miller, in *Proceedings of the International Symposium on Testing and Failure Analysis*, 1988, pp. 9-14.

28. "Voltage Contrast Instrumentation and Analysis Methods," by D.S. Koellen, in *Proceedings of the International Symposium on Testing and Failure Analysis*, 1988.

29. "Dynamic Voltage Contrast SEM Failure Analysis of a 32K ROM," by M. Foltz, in *Proceedings of the International Symposium on Testing and Failure Analysis*, 1988, pp. 15-20.

30. "Electron Beam Testing of VLSI Circuits," by E.B. Sziklas and J.S. Wolcott, in *Proceedings of the International Symposium on Testing and Failure Analysis*, 1988, pp. 21-30.

31. "Failure Analysis of DRAMs," by R. Lemme, M. Gentsch, and R. Kutzner, in *Proceedings of the International Symposium on Testing and Failure Analysis*, 1988, pp. 31-40.

32. "Nodal Waveform Analysis of Recoverable Gate Array Functional Failures," by J.E. Zeferjahn, in *Proceedings of the International Symposium on Testing and Failure Analysis*, 1988, pp. 41-46.

33. "Electron Beam Probing of VLSI Circuits Through IEEE-488 Interface Control," by I.S. Kim and J.F. Polcari, in *Proceedings of the International Symposium for Testing and Failure Analysis*, 1989, pp. 81-86.

34. "Failure Analysis of Vendor Produced CMOS Modules," by T.E. Rothwell and J.R. McLean, in *Proceedings of the International Symposium for Testing and Failure Analysis*, 1989, pp. 99-108.

35. "Practical Applications of Dynamic Fault Imaging," by P.A. Cundall, in *Proceedings of the International Symposium for Testing and Failure Analysis*, 1989, pp. 277-284.

SCANNING ELECTRON MICROSCOPY TECHNIQUES FOR IC FAILURE ANALYSIS

by Edward I. Cole, Jr. and Jerry M. Soden

The scanning electron microscope (SEM) has become as standard a tool for IC failure analysis as the optical microscope. The SEM's advantages over light microscopy include greatly increased depth of field, much higher magnification, increased working distance, and improved imaging of surface topography. In addition, the interaction of the electron beam with the IC enable unique imaging and analytical capabilities. The strengths, limitations, and effects upon the device being analyzed must be understood to use the SEM effectively. SEM settings that are appropriate for one analytical technique may be unsuccessful, may give misleading results, or may damage the device being analyzed, if used for a different technique.

This section reviews conventional and new SEM techniques for IC analysis. All of these techniques can be performed on a standard SEM. The use of advanced electron beam test systems is also discussed. The workshop is designed to provide beneficial information to both novice and experienced failure analysts. Topics to be covered are (1) standard techniques: secondary electron imaging for surface topology, voltage contrast, capacitive coupling voltage contrast, backscattered electron imaging, electron beam induced current imaging, and x-ray microanalysis and (2) new SEM techniques: novel voltage contrast applications, resistive contrast imaging, biased resistive contrast imaging, and charge-induced voltage alteration. Each technique will be described in terms of the information yielded, the physics behind technique use, any special equipment and/or instrumentation required to implement the technique, possible damage to the IC as a result of using the technique, and examples of using the technique for failure analysis.

STANDARD SEM TECHNIQUES

Secondary Electron (SE) Imaging for Surface Morphology

Information Technique Yields: SE imaging generates a high resolution, large depth of field image (compared to optical microscopy) depicting the surface morphology of the sample examined. This is the most commonly used imaging mode of the scanning electron microscope.

Physics Behind Technique Use: SE image contrast is generated by differences in SE emission efficiency with topography. Figure 1 illustrates the electron beam interaction products from a passivated integrated circuit generated by a 10 keV primary electron beam. A primary electron beam incident on a solid will create many excited electrons in the target material. These excited electrons are scattered isotropically, some having enough energy to escape the solid at the surface. SEs are those electrons emitted from the surface with energy ≤ 50 eV. A primary electron beam with energy > 100 eV will yield an SE energy distribution whose shape is determined by the work function, Fermi level, and other material parameters. Below 100 eV numerous other factors determine SE emission. The SEs originate only from the top 30 nm of passivated integrated circuits. Scattered electrons generated deeper than this do not have the energy required to escape to the surface. Figure 2 displays primary electron beam scenarios with the surface. This "edge effect" results in a larger number of SEs being generated with a "brighter" image. It is the change in SE emission efficiency with primary beam/sample angle that produces the SE image. Other considerations that can reduce or increase image

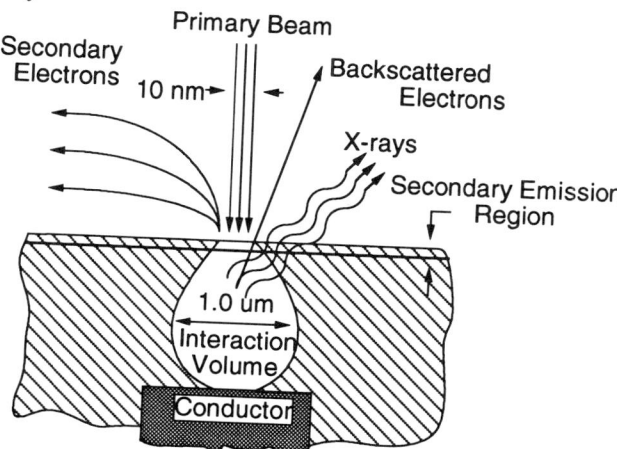

Fig. 1. Primary electron beam interaction products.

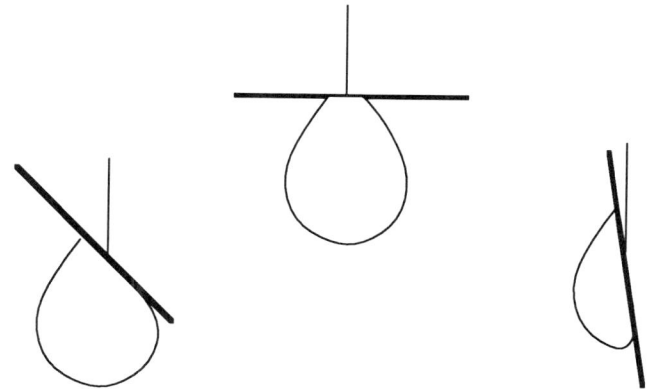

Fig. 2. Electron beam interaction volume variation with angle.

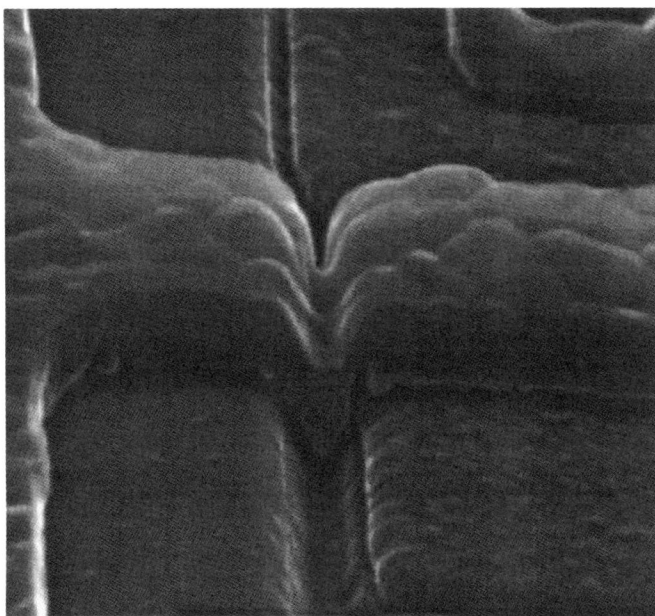

Fig. 3. Example of SE morphology imaging.

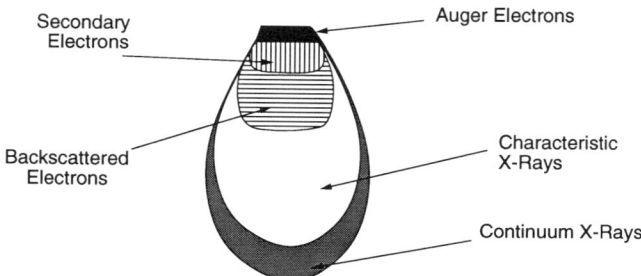

Fig. 4. Distribution of electron beam interaction products.

quality are the relative detector location, electron beam energy and current, and the target material.

Difficulty of Implementation: The only equipment necessary to acquire SE images is a scanning electron microscope. Whether or not any sample preparation is performed depends upon the sample type and information desired. Any packaged integrated circuit SE observation requires removal of enough of the package to expose the IC surface to be examined (metal lid removal, plastic grinding or etching, etc.). Sample charging by the electron beam is the most common problem encountered during SE imaging. A net charge will be produced at the integrated circuit surface if the number of electrons absorbed from the primary electron beam is not equal to the number exiting the sample. This generates a surface charge which increases in magnitude until equilibrium is reached. The charging effect is beneficial in other electron beam imaging techniques, but its effect in topography imaging is to reduce spatial resolution. To eliminate surface charging a thin conductive coating (~20 angstroms of C or Au/Pd) may be applied to the surface by vacuum evaporation or D.C. sputtering. Careful selection of the primary electron beam energy will also reduce charging effects. For example, by grounding the substrate of an integrated circuit and using a high enough primary electron beam energy such that the surface is shorted to the substrate through the interaction volume, charging can be greatly reduced and an adequate image produced. This technique was used to produce the SE images in Fig. 3. The surface passivation was removed.

Possible IC Damage: No direct "physical" damage occurs with electron beam testing. Alteration of the threshold voltage on MOS transistors is possible. This change in threshold voltage results from irradiation damage to the gate oxides of MOS transistors. The damage is generated by direct primary electron collisions and by x-rays generated through interactions in the surface layer(s). The primary electrons quickly alter the interface trap density and occupancy as well as the fixed charge levels in the gate oxide. If desired, low temperature annealing (100-150 °C) may be used to restore the threshold voltage to near its original value.

If surface layer removal is performed, there is the possibility of device damage during the removal process. Conductive coating to eliminate charging will prevent/hinder microelectronic operation and should be removed to restore device operation. Note that removal of any coating may damage the IC under examination. However, removal of carbon coatings with an oxygen plasma is relatively safe.

Backscattered Electron (BSE) Imaging

Information Technique Yields: BSE imaging detects differences in elemental atomic number on and below the surface. It is primarily used to detect sharp atomic number gradients under the passivation, such as those caused by impurities and metal voids.

Physics Behind Technique Use: BSEs are electrons with energies E such that $50 \text{ eV} < E \leq E^*$, where E^* is the primary beam energy. Unlike SEs, the BSEs escape from much deeper in the surface, most coming from about 1/3 the maximum depth of the electron beam/device interaction volume (Fig. 4). This greatly reduces any of the surface effects that hamper SE imaging. The major factor determining BSE contrast in images is the atomic number of the target, indicating that *nuclear scattering* is the principal electron interaction. Because of the complex elastic and inelastic scattering processes that create BSEs, no exact theory is possible, but two general statements about imaging expectations can be made. First, the spatial resolution is limited by the size of the interaction volume at the BSE source, therefore the spatial resolution will always be less than or equal to the SE image resolution. Second, experimentation with different beam energies can improve spatial resolution depending upon the BSE source and its depth. Figure 5 is a BSE image revealing electromigration voids in a metal conductor under a passivation layer.

Difficulty of Implementation: The equipment necessary to perform BSE imaging is a scanning electron microscope and a BSE detector. The detector is normally a large area semiconductor, positioned directly over the sample. A digital imaging system is advantageous to image acquisition/manipulation, but not necessary.

Possible IC Damage: Same as with SE imaging.

Fig. 5. Example of BSE imaging showing electromigration voids. (30KeV)

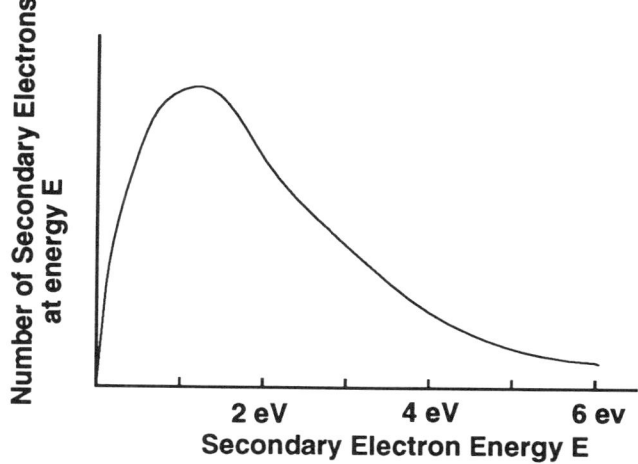

Fig. 6. A typical secondary electron energy distribution.

Voltage Contrast (VC) Imaging

Information Technique Yields: VC imaging creates an image in which the intensity is largely determined by the static voltages on an integrated circuit. By analyzing the variations in brightness, the logic levels of a digital integrated circuit can be determined and the voltage on internal circuit nodes can be measured.

Physics Behind Technique Use: VC imaging takes advantage of the variation in secondary electron (SE) emission efficiencies with applied bias on an integrated circuit. Figure 1 displays the electron beam interaction products between a passivated IC and a 10-keV primary electron beam. SEs are defined as those electrons emitted from the surface with energies of 50 eV or less. The physics of SE emission is complicated, involving SE generation, transport to the surface, and escape from the surface. The typical shape of the SE energy distribution is shown in Fig. 6. This shape is largely independent of primary electron beam current and energy when using primary electron beam energies above 100 eV. Because of their low energies, SEs are generated only from about 30 nm or less from the surface. The ratio of the electron flux escaping the surface of an exposed device to the total electron flux injected into a device varies with the material examined and the primary electron beam energy used. For microelectronic materials at relatively low beam energies (<2 keV) more electrons escape the surface than are injected. At higher energies and extremely low primary electron beam energies (<100 eV) fewer electrons escape than are injected. Under both conditions the surface has a net charge. For the case of low beam energies a positive charge builds up on the surface. This charge will prevent lower energy SEs from escaping the surface and a positive *equilibrium voltage* (about 0.5 V with a 1.0 keV beam) will be established at which the escaping and incident electron flux are equal.

If the surface of the target is forced by an external power supply to a different voltage, then the SE image will have increased or reduced signal depending upon the difference between the forced voltage and the equilibrium voltage. The conductors of a depassivated integrated circuit may be forced in such a manner, as shown in the image of the CMOS static RAM cell in Fig. 7a. The lowest voltage region (V_{SS}) are bright and the highest voltage regions (V_{DD}) are dark. If the applied voltage is altered during the image scan, then the changes in applied voltage can be seen as "candy-stripes" on the IC conductors (Fig. 7b).

Passivated devices may also be observed using static VC. If the primary electron beam energy is increased to the point where more electrons are injected than escape, a negative charge develops on the surface. The voltage can equal that of the primary beam for the insulating glass, obscuring all image information. However, if the primary beam energy is increased to the point where the primary electron interaction volume reaches the substrate of the integrated circuit, a charge leakage path through the substrate is generated. Because of hole-electron pair production in the interaction volume, this pear-shaped region (Fig. 1) is a conductor. This conducting volume will force the surface to the voltage of the buried conductors, thereby affecting the SE emission in a manner similar to the low beam energy situation described above. As a rule of thumb for typical IC materials (Si, Al, SiO_2) the following expression may be used as a rough estimate of primary electron beam penetration,

$$R = 0.022\, E^{1.65} \qquad \text{(Eq. 1)}$$

where R is the primary electron beam penetration depth in microns and E is the primary electron beam energy in keV.

The techniques described here generate images which yield qualitative voltage information. Quantitative information may be obtained in two different ways. The first is the simple comparison of image intensity as a function of voltage. This method has an accuracy of about 0.5 V. The second, preferred method

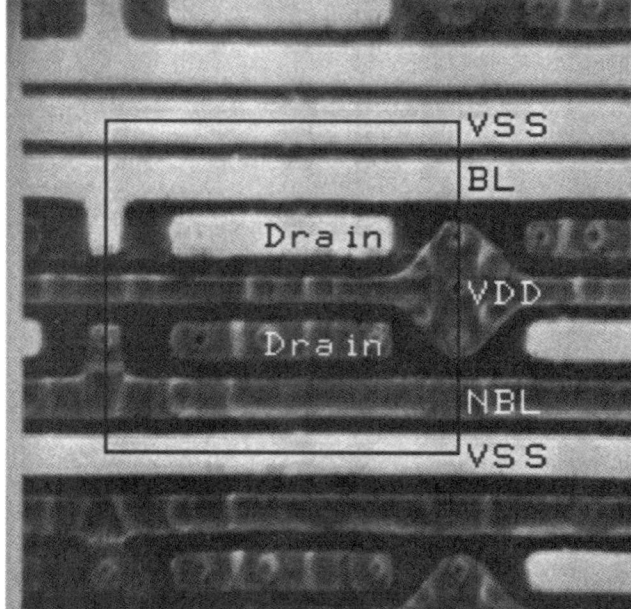 

Fig. 7. (a) VC of a memory cell and (b) "candy stripe" VC showing the effect of changing the applied voltage during the image scan.

for voltage measurement, is to employ an energy spectrometer on the emitted SEs. This technique has a voltage accuracy of 10 mV.

Difficulty of Implementation: The equipment necessary to perform static VC are a scanning electron microscope, an electrical vacuum feed-through, and a voltage supply. By applying the physics described above, static VC images and voltage measurements are routinely acquired throughout the microelectronics industry. For device and state comparison, a digital image acquisition system is advantageous for image acquisition/manipulation, but not necessary. Factors which limit the voltage and spatial resolution include surface charging from incomplete beam penetration on passivated surfaces, surface contaminants, and local electric fields of structures from neighboring regions. These must be overcome/minimized to optimize performance.

Possible IC Damage: No direct "physical" damage occurs with electron beam testing. Examination of devices at beam energies of 1 keV or less is nondestructive. The same considerations for MOS devices described for SE imaging apply here as well.

Capacitive Coupling Voltage Contrast (CCVC) Imaging

Information Technique Yields: The CCVC mechanism permits imaging and measurement of dynamic voltages beneath passivation layers. Through primary beam blanking, CCVC facilitates buried conductor as well as diffusion imaging and depth measurement with no applied bias. The major advantage of CCVC over static voltage contrast is the absence of any irradiation damage.

Physics Behind Technique Use: This description builds on the physics of secondary electron generation (SE) described in the static voltage contrast (VC) section. CCVC imaging uses

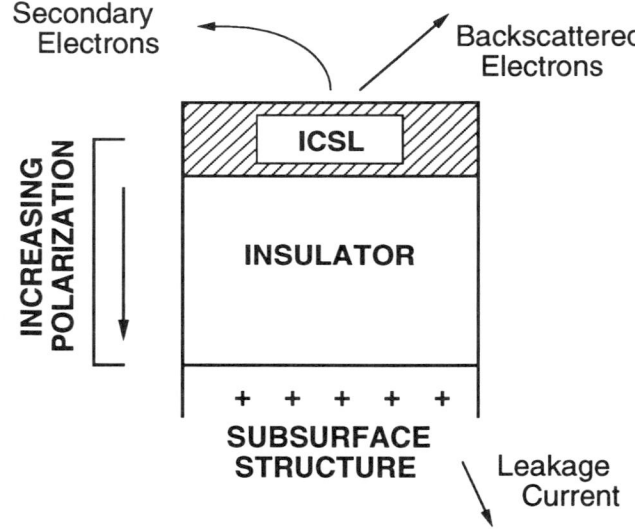

Fig. 8. Physics of CCVC generation.

the passivation layer as a discharging capacitor to generate a dynamic image of changing subsurface voltages. CCVC is performed at fast scan rates to increase the time resolution of the dynamic signal. As described in the VC physics section, at low primary electron beam energies (< 2 keV) more electrons escape than are injected by the primary electron beam. A net positive charge will build up on the surface, preventing lower energy SE's from escaping the surface and decreases the SE image intensity. An equilibrium voltage is reached when the net charge accumulated on the device does not change with time. A bound surface charge will be produced at the Induced Conduc-

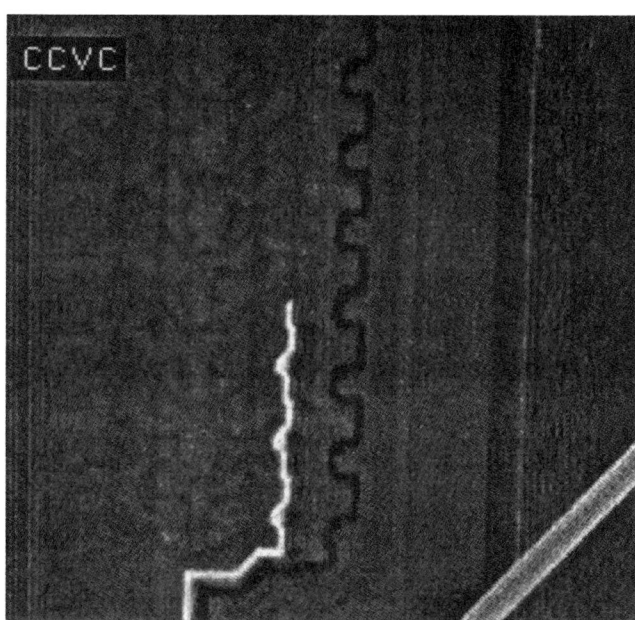

Fig. 9. Example of CCVC imaging on an open conductor.

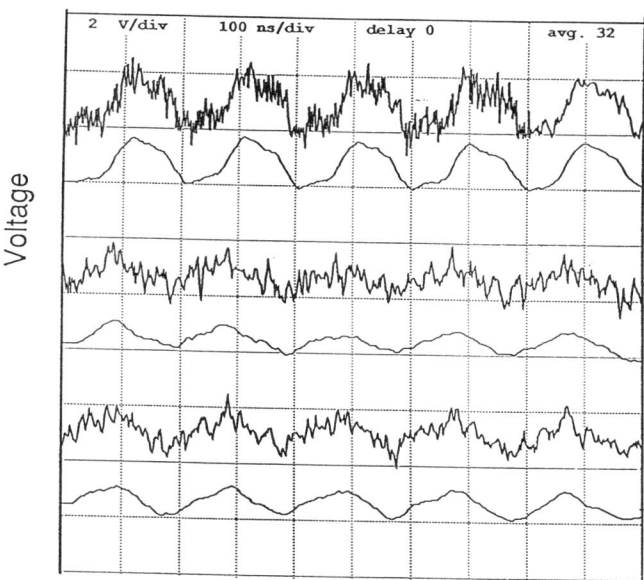

Fig. 10. Example of an EBT waveform.

tive Surface Layer (ICSL) when structures below the maximum beam penetration depth change potential, and the material between them and the surface becomes polarized (Fig. 8). The CCVC signal is the change in the number of SEs caused by this bound charge potential. As the primary electron beam scans across the surface the bound change is dissipated via differences in SE emission, and equilibrium is reestablished. This differs from static VC imaging where a conducting path to the biased structure exists. The time for the bound charge to decay to the equilibrium potential depends on the passivation, the primary beam energy, the depth of the structure examined, and the primary electron flux. The smaller the incident electron flux, the longer the CCVC decay times. Unfortunately, the signal-to-noise ratio (SNR) decreases with reduction in incident electron flux. As with VC, image comparison and energy spectrometry may be used for voltage measurement. Figure 9 displays a CCVC image of a failing bit line (bright-negative transition) and a functional bit-not line (dark-positive transition). The time for the CCVC signal to decay to equilibrium may also be used to measure voltage.

An additional use for CCVC is the imaging of buried conductors with no applied bias. This is achieved by blanking the primary electron beam and allowing the passivation surface to "leak" to the stage potential, ground. This normally takes about 30 seconds. The beam is then scanned across the surface, which will go to a positive equilibrium voltage as described above. The time it takes the surface to reach the equilibrium voltage depends inversely on the depth of buried conductors under the passivation, permitting layer identification through different equilibrium times.

Difficulty of Implementation: The equipment necessary to perform CCVC is a scanning electron microscope, an electrical vacuum feed-through, a voltage supply, and a switching mechanism to generate dynamic voltages. The scanning electron microscope must scan at or near video rates for adequate time resolution to observe CCVC. Because of the SNR-time resolution trade off, a frame grabbing digital image acquisition system is necessary to obtain adequate signal resolution via averaging. This same imaging system should be synchronized to a beam blanking unit to examine buried conductors with no applied bias. The factors that limit static VC imaging also must be dealt with when using CCVC.

Possible IC Damage: There is no damage associated with CCVC, other than package lid/plastic removal.

Advanced Electron Beam Test (EBT) Systems

Information Technique Yields; Advanced electron beam test (EBT) systems generate voltage waveforms for conductors on ICs under test. These waveforms can have voltage and timing resolutions of 10 mV and 10 ps respectively. The advanced EBT systems also permit comparison of control and failing devices through image subtraction as well as comparison of CAD layout data with SEM images.

Physics Behind Technique Use: The data acquisition physics of advanced EBT systems is automated VC and CCVC. Time resolution is increased beyond that of standard SEM scan rates by employing a beam blanker. The beam blanker consists of electrostatic plates that can be biased at a fast rate to move the scanning or "spotted" electron beam away from and back to the IC under test. A VC image of the IC in a given state may be generated by triggering the beam blanking plates to permit electron beam interaction with the sample only when the IC is in the state of interest. By repeated cycling of the vectors applied to

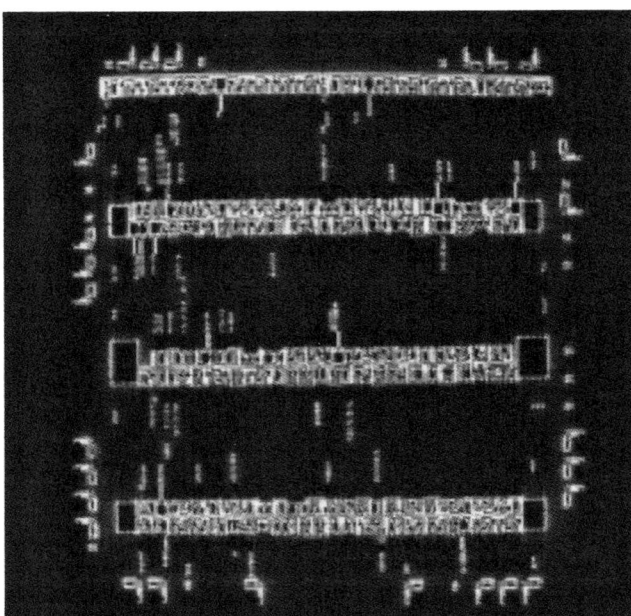

Fig. 11. Example of EBIC imaging of an entire die.

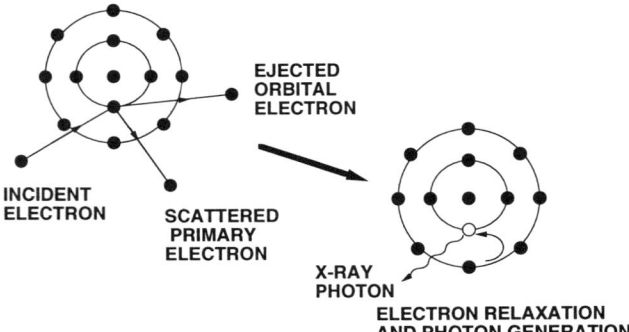

Fig. 12. Characteristic x-ray generation.

the IC, a VC image of a given logic state, showing the complete field of view, may be acquired. A digital image storage system is required for strobed VC image acquisition and display.

The waveforms generated by EBT systems employ a secondary electron energy spectrometer to obtain quantitative voltage and timing information. The energy spectrometer consists of a retarding grid in front of a secondary electron detector. The change in the number of secondary electrons reaching the detector with an applied retarding grid voltage may be used to determine the voltage at the surface of the IC. A waveform is generated using the beam blanking plates described above with the difference that the electron beam is not scanning when unblanked, but in spot mode over the conductor being measured. The waveforms in Fig. 10 display single and 20 averages of three different conductors.

An additional feature of EBT systems is that CAD information on layout, schematic, and circuit modeling may be incorporated into the EBT system so that the CAD data may be used during IC analysis. The features of the CAD interface and its utilities depend on the vendor.

Difficulty of Implementation: The equipment required to implement advanced electron beam testing of ICs is readily available from several vendors. The equipment cost ranges from ~$200K to ~$700K depending upon the vendor and options desired. Full utilization of EBT systems normally requires the application of sophisticated stimuli to the IC under test, which adds to a system's cost. The compatibility of existing CAD databases with the EBT system is another factor that must be considered when evaluating EBT systems. One last consideration is that EBT is that EBT systems normally operate only at low electron beam energies (≤ 1.25 keV). While this is ideal for most VC and CCVC observations, certain other SEM techniques cannot be used under these conditions.

Possible IC Damage: Because of the low primary electron beam energies used in EBT systems, there is no damage to ICs other than package lid/plastic removal.

Electron Beam Induced Current (EBIC) Imaging

Information Technique Yields: EBIC imaging localizes regions of Fermi level transition. EBIC is primarily used to identify buried diffusions and Si defects.

Physics Behind Technique Use: When the primary electron beam is scanned across a sample, collisions in the target material from electron-hole pairs within the bulk of the sample. The relatively low ionization energies (less than 10 eV) of materials used in integrated circuit manufacturing allow a single 10 keV primary electron to produce as many as 500 to 1000 free electron-hole pairs. These pairs usually recombine randomly in the material; however, if production occurs in a space-charge (depletion) region, the charge carriers will be separated by the junction potential before recombination. The large number of pairs per primary electron generates an EBIC signal much larger than the incident beam current. The magnitude and direction of the induced current is used to generate an image localizing where junction potentials occur. In contrast to secondary electron and backscattered electron scanning modes, *the EBIC signal detector is the device itself.* By controlling the primary electron beam energy, the depth of the diffusions examined can be differentiated. Various device pins may be sampled to observe different EBIC signals.

Difficulty of Implementation: The equipment necessary to perform EBIC imaging is a scanning electron microscope, current to voltage amplifier, and an electrical vacuum feedthrough. The only sample preparation needed is lid/plastic package removal. The passivation may or may not be removed for EBIC analysis. No electrical driving equipment is necessary since the integrated circuit is driven only by the electron beam. Because of the small signal generated, a digital image acquisition system would be advantageous to image acquisition/manipulation, but not necessary. Electrical testing to determine proper node selection is also desirable. An example of an EBIC image is shown in Fig. 11. Diffusions across the entire IC are visible in Fig. 11.

Possible IC Damage: The same considerations for IC damage described for SE imaging apply to EBIC imaging.

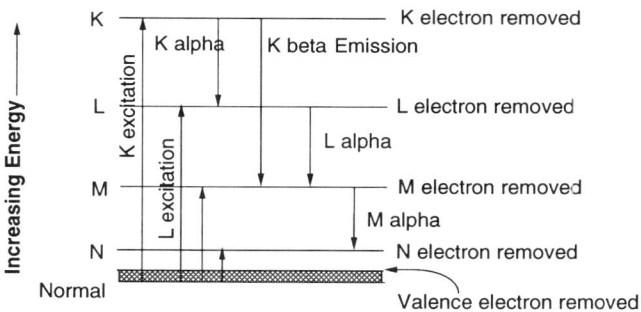

Fig. 13. Atomic electron energy level transitions.

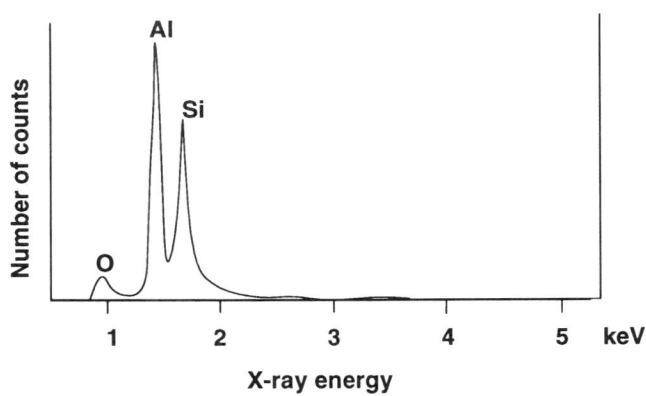

Fig. 14. An example x-ray spectra.

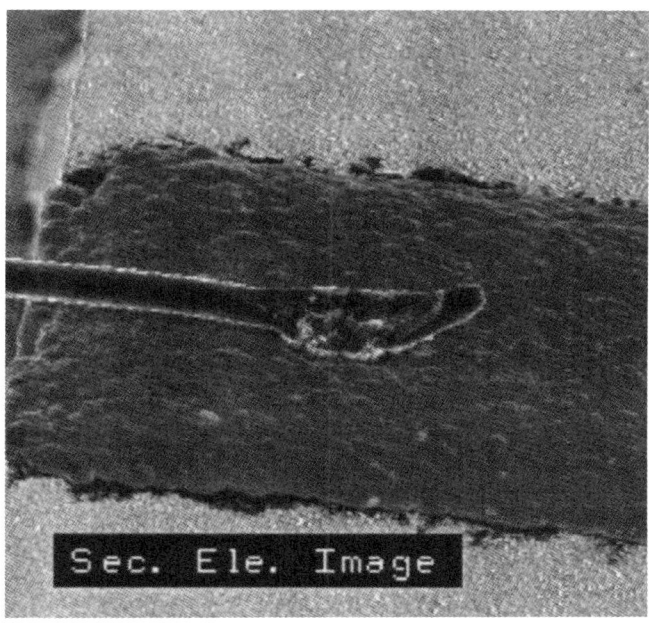
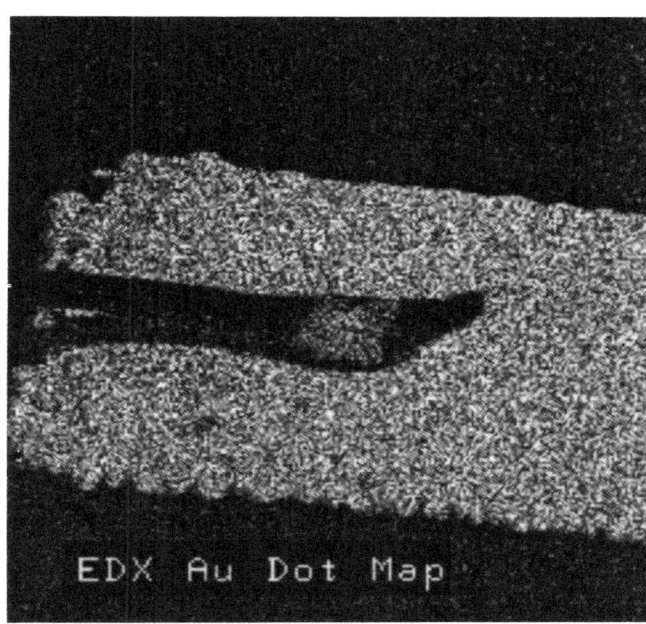

Fig. 15. An example of a SE and a Au x-ray dot map with the same field of view.

X-Ray Microanalysis

Information Technique Yields: X-ray microanalysis permits identification and localization of elemental constituents of a sample, with resolution limits of 0.5 to 0.1 weight percent.

Physics Behind Technique Use: Two types of x-ray generation occur during primary electron beam interactions with target atoms. *Bremsstrahlung* or continuum x-rays are generated as the electrons are decelerated through interactions with nuclear cores of atoms. These x-rays are generally considered the "noise" component of the total x-ray signal. The continuum x-ray have an intensity that varies as $(E_o-E)/E$, where E_o is the electron beam energy and E is the x-ray energy.

Characteristic x-rays are the second type of x-ray generation from electron beam interaction. When the electron beam displaces an inner core electron, an electron from a higher energy level will decay to replace the missing, inner electron (Fig. 12). The transition from the higher energy level to the lower level will be accompanied by an energy release, whose magnitude is the difference between the two electron energy levels. A photon (x-ray) with this "characteristic" energy is one form of energy release. Because the inner core electrons of different atoms are separated by known, discrete energy levels, elemental identification is possible from observing the energies of the characteristic x-rays produced. Figure 13 summarizes some of the permitted energy level transitions for atomic electrons. The K, L, M, and N refer to different energy levels.

X-ray energy spectra of a sample may be acquired with the beam scanning over a sample area or localized to a particular point. A sample spectra is shown in Fig. 14. These spectra identify the elements in that region. A scanned image or dot-map in which the contrast is modulated by detection of a certain x-ray

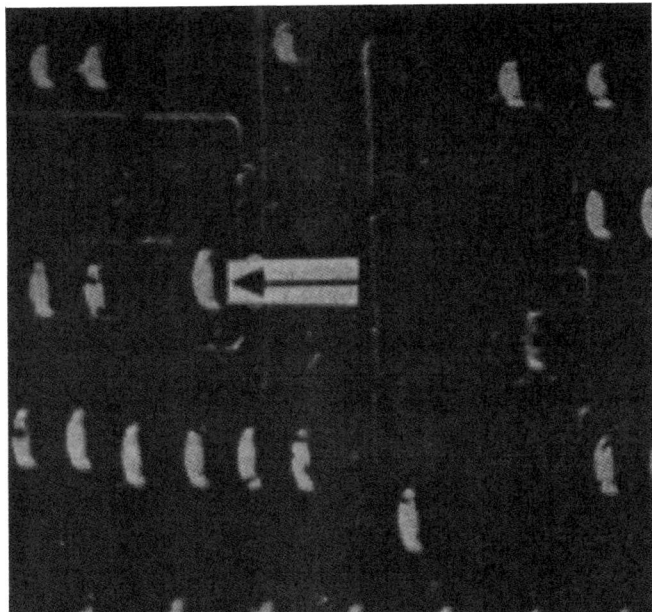

Fig. 16. Example of passive voltage contrast.

energy displays the spatial distribution of that x-ray energy (element) source. A secondary electron image and dot-map for Au characteristic x-rays is shown in Fig. 15.

Difficulty of Implementation: The equipment necessary to perform x-ray microanalysis is an electron beam source, such as a scanning electron microscope or an electron-microprobe, and an x-ray detector. Two types of x-ray detectors are available. First, a wavelength dispersive x-ray (WDX) detector uses Bragg interference to generate an x-ray wavelength spectra. The WDX detector takes advantage of an interference crystal and the Bragg equation:

$$n\lambda = 2d\sin\Theta \qquad (Eq.\ 2)$$

where λ is the x-ray wave length, Θ is the x-ray angle of incidence, d is the interplane spacing of the reflection crystal, and n is a quantum number (1, 2, 3,...). By changing the angle of incidence different wavelengths (energies) may be observed. (For photons, energy and wavelength differ by a constant). The actual geometry for WDX detectors is quite complicated. More details are available in the reference section. The typical energy resolution of WDX detector is < 10 eV.

The second type of detector is an energy dispersive x-ray (EDX) detector. The EDX detector employs a semiconducting medium, usually lithium-drifted silicon, to convert x-ray energy into charge. The x-rays deposited into the semiconductor will create electron-hole pairs in the semiconductor. These hole pairs are collected by biasing the semiconductor and are converted into a voltage pulse. The larger the energy of the x-ray the greater the number of hole pairs and the larger the voltage pulse. The magnitude of the voltage pulse is then used to determine the incident x-ray energy. The typical energy resolution of EDX detectors is < 150 eV, but it is faster than WDX analysis.

Quantitative x-ray analysis is possible with both detection systems but requires implementing a complicated methodology which considers electron beam current, x-ray production efficiency, sample absorption of x-rays, and other factors. Information on quantitative analysis may be found in the references section.

Possible IC Damage: Same as for SE imaging.

ADVANCED SEM TECHNIQUES

Passive Voltage Contrast Imaging at Various Angles

Information Technique Yields: Conductor continuity and gate oxide ruptures can be located using the variation in VC with primary electron beam angle. This technique is used on depassivated ICs.

Physics Behind Technique Use: Passive VC imaging employs the physics described in the SE emission and VC sections along with the variation in secondary electron emission efficiency with tilt. The secondary electron emission efficiency increases with tilt angle as 1/cos(tilt angle). Low primary electron beam energies (~1 keV) are used for passive voltage contrast to avoid large amounts of deposited negative charge and maximize the difference in image contrast between floating and grounded conductors. Grounded conductors will not change voltage with tilt because of a charge path to ground. Floating conductors will increase in voltage as a result of increased secondary electron emission, until an equilibrium voltage is achieved where the number of primary electrons injected equals the number of secondary electrons emitted. The change in the equilibrium voltage of floating conductors with tilt can be seen in the VC image to differentiate grounded and floating regions.

Figure 16 is an example of using passive voltage contrast to locate a faulty gate contact. The passivation and metal have been removed from the IC. The polysilicon contact highlighted by the arrow is bright, indicating a path to ground similar to most of the other contacts to the silicon substrate in Fig. 16. The path to ground of the polysilicon contact is the result of a gate oxide rupture between the polysilicon and the silicon substrate. The dark contact to the right is electrically floating and charged positive because no ground path defect exists.

Difficulty of Implementation: The equipment necessary to use passive voltage contrast is an SEM with low primary bean energy capability and the capacity to ground structures on an IC. Almost all SEMs have the capability to tilt the sample and ground connections in the vacuum chamber. Sample preparation and access to regions on the IC that need to be grounded are the major difficulties with this technique.

Possible IC Damage: Same as for CCVC imaging.

Using the Electron Beam to Charge and Test Nodes

Information Technique Yields: Electron beam charging of nodes is used on hybrid interconnect networks to verify conductor continuity on substrates.

Physics Behind Technique Use: Electron beam charging of nodes is a combination of SE, VC, and (if desired) EBT systems. The electron beam is used in spot or raster mode to inject charge at one node of a hybrid conductor network. The injected charge will generate an equilibrium voltage on the conductor as

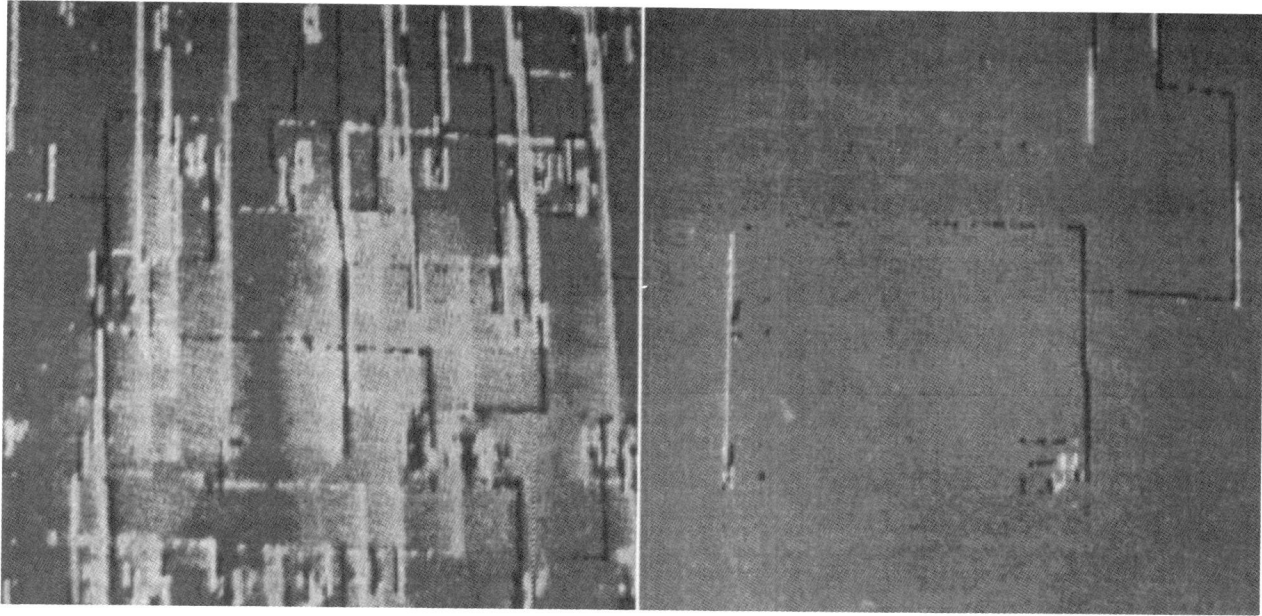

(a) (b)

Fig. 17. Example images of initial and processed images of CCVC coupled with I_{DDQ} testing.

discussed in the SE and VC sections. The electron beam is then quickly moved to another node on the hybrid network that should have electrical continuity with the first node, and therefore should be charged to the same equilibrium voltage. Depending upon the amplitude of the electron beam induced equilibrium voltage and the electron beam system used, either the difference in image contrast or the actual voltage of the second test node is recorded and compared. The process is then repeated for all other nodes on the hybrid interconnect network to verify continuity.

Electron beam charging of nodes can also be used on ICs. Techniques employing this are described in later sections.

Difficulty of Implementation An SEM with a computer controlled electron beam is required to use electron beam charging of nodes on hybrid interconnect networks. No vacuum feed through or other sample connections are required. No sample preparation is required. The technique works best when the area of interest may be seen in one field of view so that only electron beam positioning is required. If the sample must be moved then the data acquisition time will increase. Obviously the data acquisition time increases with the number of nodes to be tested. The test node locations must be entered into the computer driving the electron beam position for charging and testing.

Possible IC Damage: For hybrid interconnect testing there is no damage using electron beam charging.

CCVC Coupled with Quiescent Current Drain (I_{DDQ}) Testing

Information Techniques Yields: CCVC imaging coupled with I_{DDQ} testing permits localization of nodes generating anomalous I_{DDQ} values on a CMOS IC.

Physics Behind Technique Use: I_{DDQ} testing of CMOS ICs examines the quiescent current values (I_{DDQ}) after all switching transients. Large values of I_{DDQ} (above ~50 µA for most static CMOS) may indicate defects or a reliability risk. The I_{DDQ} testing technique is powerful because only one parameter needs to be examined for analysis but localization of the nodes producing high I_{DDQ} values can be difficult. The CCVC technique has been applied to localize the nodes generating high I_{DDQ} values. When performing I_{DDQ} testing the vectors that generate high I_{DDQ} are recorded. It is assumed that the same defect will produce the same or similar I_{DDQ} value when activated by different applied vectors. CCVC images are then acquired of the IC at the different vectors causing the elevated I_{DDQ} value. The CCVC images are then compared and processed to retain only the conductors that have the same potential shift. The left-hand side of Fig. 17 is a single CCVC image with a high I_{DDQ} applied vector. The right-hand side of Fig. 17 is the result of comparison with three other CCVC images with different vectors applied, but the same I_{DDQ} value. Only the information common to all four images is retained. The eventual outcome is that only the node(s) responsible for the high I_{DDQ} is present in the image.

Difficulty of Implementation: The same hardware necessary for CCVC imaging is required. In addition, an IC tester capable of performing I_{DDQ} testing and vector recording as well as a digital image processing system for image comparison and processing is needed. No image alignment software is needed since all images for comparison will be from the same IC. The data acquisition time will be reduced if the entire IC die or area of interest can be put into the same field of view. If the IC must be moved the data acquisition time will increase accordingly.

Possible IC Damage: Same as with CCVC imaging.

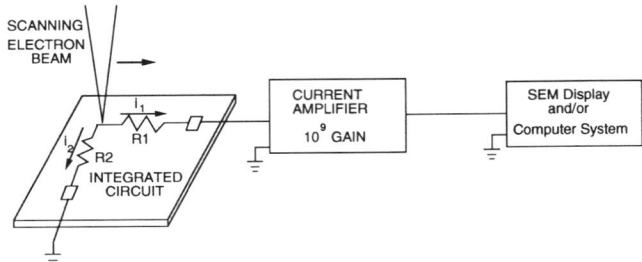

Fig. 18. Block diagram of an RCI imaging system

Resistive Contrast Imaging (RCI)

Information Technique Yields: RCI generates a relative resistance map between two test nodes of a passivated integrated circuit. The map generated will display buried conductors on an integrated circuit and may be used to localize open conductors.

Physics Behind Technique Use: RCI obtains resistance information by using the integrated circuit as a complex current divider. Figure 1 displays the electron beam interaction products between a passivated integrated circuit and a 10-keV primary electron beam. To obtain RCI information the primary electron beam energy is increased until the tip of the interaction volume intersects the buried conductor of interest. A portion of the primary electron beam current will be injected into the conductor. Using an amplification configuration as shown in Fig. 18, the currents induced by electron beam exposure will have a path out of the integrated circuit. The relative resistance between the electron beam position on the circuit and the test nodes determines the direction and amplitude of current flow. The current, on the order of nanoamps, is amplified and used to make a resistance map of the conductors. Usually the power and ground inputs are used as test nodes because of their global nature across the integrated circuit. However, other node combinations may be used if desirable/indicated. If a resistance change occurs along a conductor relative to the test node combination selected, such as an open conductor, the RCI image will display an abrupt contrast change at the open site. The RCI image in Fig. 19 localizes an electromigration open along a clock line.

Difficulty of Implementation: The equipment items necessary to implement RCI are: a scanning electron microscope, current to voltage amplifier, and an electrical vacuum feedthrough. The only sample preparation needed is lid/plastic package removal. The passivation is not removed for RCI. By increasing the primary electron beam energy, multilevel conductors under metal may be observed. No electrical driving equipment is necessary since the integrated circuit is driven only by the electron beam. Because of the small signal generated, a digital image acquisition system would be advantageous to image acquisition/manipulation, but not necessary. Electrical testing to determine proper node selection is also desirable. Another induced current effect, Electron Beam Induced Current (EBIC) should be avoided. EBIC generates currents several orders of magnitude greater than RCI which can mask the

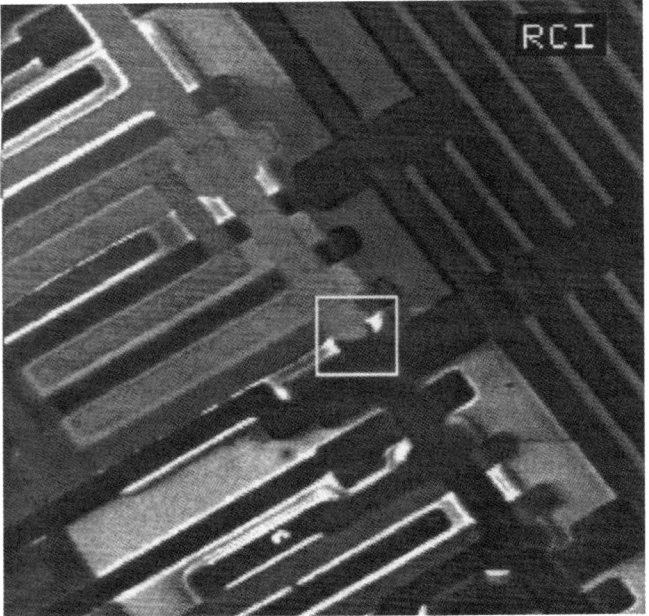

Fig. 19. Example of RCI imaging locating an open conductor.

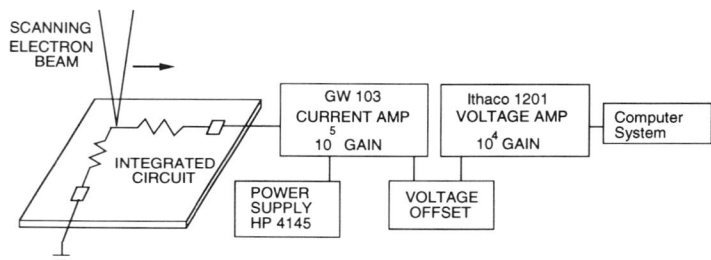

Fig. 20. BRCI imaging system.

RCI signal. EBIC signals are generated primarily from buried diffusions and can be avoided by using lower primary electron beam energies. With practice in selecting the proper primary electron beam energy and current, RCI data may be acquired readily. Unfortunately, not all internal conductors with defects will be identified using RCI. Escape from detection occurs when the current paths from the defect-containing conductor to the IC pins are too convoluted and no difference in resistance occurs across the open site relative to the IC pins.

Possible IC Damage: No direct "physical" damage occurs with electron beam testing. A possible alteration of the threshold voltages on MOS transistors is possible. This change in threshold voltage results from irradiation damage to the gate oxides of MOS transistors. At the primary electron beam energies used for RCI, no direct primary electron/gate oxide interactions occur. However, some Bremsstrahlung x-rays generated by interactions in the passivation layer will deposit energy in the gate oxide. This x-ray dose will alter the interface trap density and occupancy as well as the fixed charge levels in

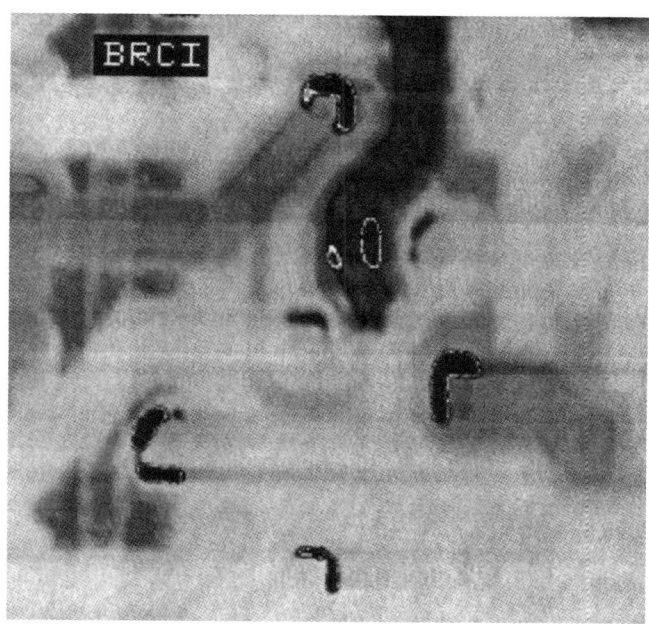

Fig. 21. Example of BRCI locating an open metal conductor.

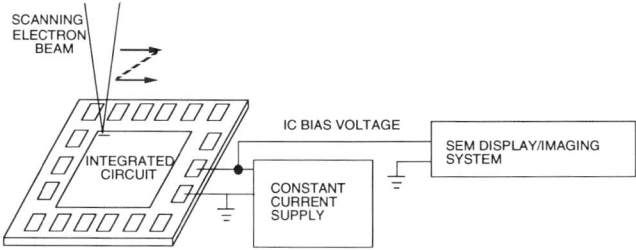

Fig. 22. CIVA imaging system.

the gate oxide. Experiments on 3 micron, commercial grade, MOS transistors indicate that 40 images with 1 micron spatial resolution may be acquired before the threshold voltage shifts by 5%. If desired, low temperature annealing (100-150 °C) may be used to restore the threshold voltage near or to its original value.

Biased Resistive Contrast Imaging (BRCI)

Information Technique Yields: BRCI generates a logic map of a biased integrated circuit. The image generated resembles a static voltage contrast map with the passivation removed and the topography subtracted. The BRCI technique keeps the passivation intact.

Physics Behind Technique Use: This description builds on the physics of resistive contrast imaging (RCI) generation described in the RCI section. As indicated in the RCI description, the major limitation to RCI is that only resistance differences relative to the test nodes are identified. To overcome this limitation the basic RCI technique has been modified to examine biased, passivated CMOS devices. When biased, all conductors of a properly functioning, fully static CMOS integrated circuit are connected to V_{DD} or V_{SS}, either directly or through one or more transistors. Under these conditions, the RCI data generated using V_{DD} and V_{SS} as the test nodes is binary for the device. This BRCI image is generated by monitoring the subtle shifts in the power supply current of the integrated circuit as an electron beam is scanned over the device surface. These shifts occur as electrons are injected into the conductors of the device. The resultant image yields the information of a static voltage contrast image, but with the lower radiation dose associated with RCI. The experimental setup for performing BRCI at Sandia and an image example are given in Fig. 20 and 21. Figure 21 shows an open bit line in a memory. The open site was not observed in an RCI image. The amplification system is in two stages. First the power supply current is amplified and then offset to eliminate the DC component. The offset signal is then amplified to observe the fluctuations in I_{DD} with electron beam position.

Difficulty of Implementation: The equipment set for BRCI is identical to that for RCI except for the addition of a second voltage amplifier. The voltage offset is an integral part of the current amplifier used. The small BRCI signals generated, which comprise only a fraction of the beam current, and the additional system noise from biasing the test device make a digital image processing system very desirable. To avoid the built-in current limitation (~ 250 µA) of the current amplifier while biasing, the bias voltage should be as low as possible.

Possible IC Damage: Same as RCI.

Charge-Induced Voltage Alteration (CIVA) Imaging

Information Technique Yields: CIVA was developed to localize open conductors on both passivated and depassivated CMOS ICs. CIVA facilitates localization of all open interconnections on an entire IC in a single, unprocessed image. CIVA has been applied to an analog bipolar technology with similar results.

Physics Behind Technique Use: CMOS ICs with open conductor lines may function at low to moderate (< 50 kHz) frequencies. The reason for this functionality is that significant quantum mechanical electron tunneling across the open can transport enough charge at low frequencies to maintain functionality. The maximum operating speed depends upon the nature of the open.

Even though the ICs may be functional with open conductors, charge injection into the floating portion of the conductor may cause significant loading that can overwhelm the open's tunneling capacity. CIVA takes advantage of this tunneling capacity to create an image of "loaded" areas. The CIVA image is generated by monitoring the voltage shifts in a constant current power supply as the electron beam is scanned over a biased integrated circuit. As electrons are injected into non-failing conductors the additional current, on the order of nanoamps, is readily absorbed and produces little change in the supply voltage. When charge is injected into an electrically floating conductor, the voltage of the negative conductor becomes more negative. This abrupt change in voltage on the floating conduc-

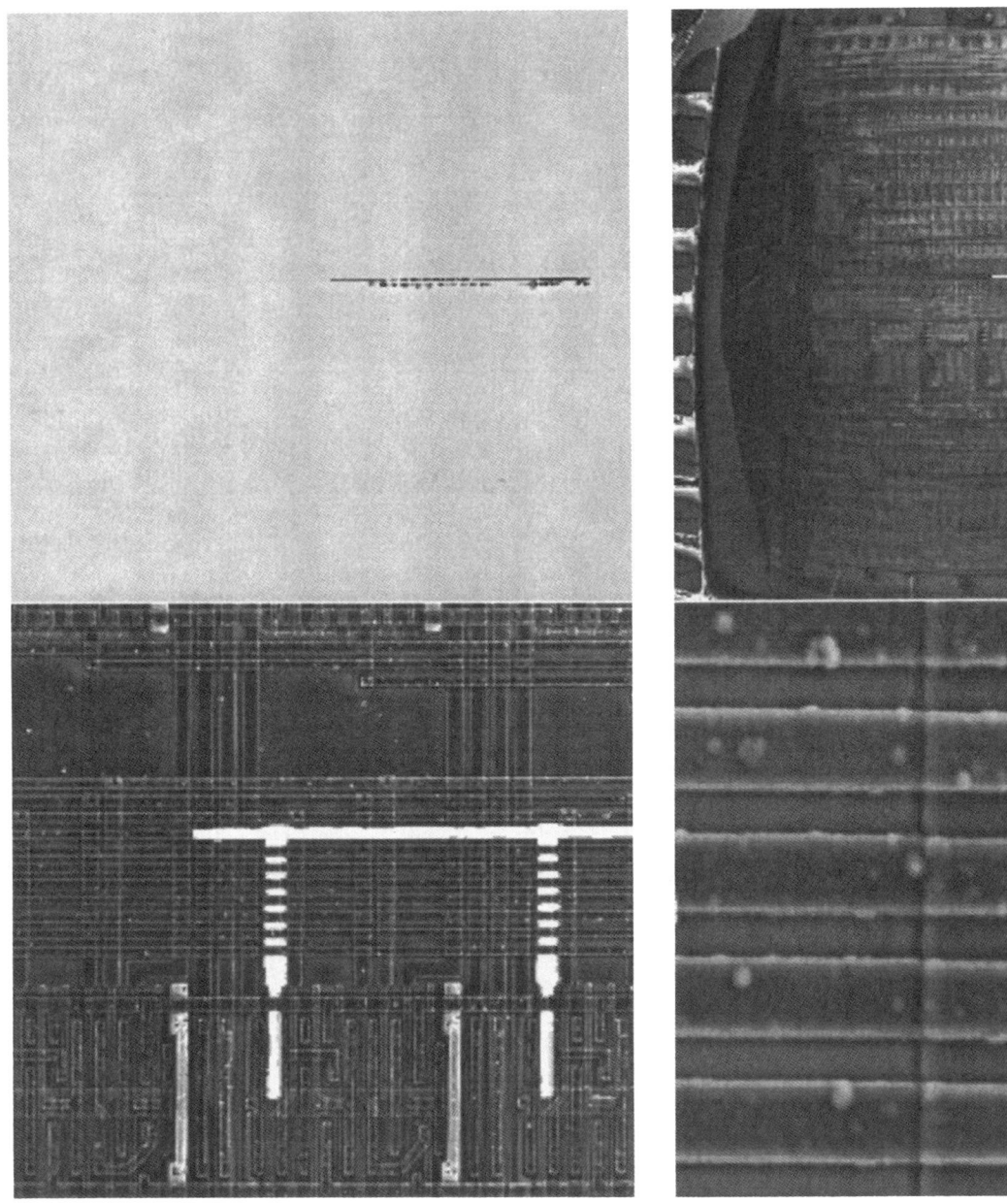

Fig. 23. Image examples of CIVA at low magnification (upper left), overlaid with an SE image (upper right), and higher magnification CIVA/SE images localizing an open conductor (lower images).

tor generates a temporary shift in the voltage demand of the constant current source supplying bias to the integrated circuit. The shift in power supply voltage can be either positive or negative depending on the proper state of the floating conductor. As the electron beam moves away from the floating conductor the previous equilibrium is quickly (~ 100 μsec depending upon the bandwidth of the current source) reestablished. The shifts observed in the power supply voltage, even for opens that exhibit significant tunneling, are on the order of 100 mV with a 5 V supply voltage. These relatively large shifts produce images in which the contrast is dominated by the open conductors.

Transistors with "weak" drive capacity have also been identified using CIVA.

Figure 22 displays a block diagram of the experimental setup used to generate CIVA images. Figure 23 shows an example of CIVA imaging on a passivated IC with an open conductor. The top left image in Fig. 23 shows a CIVA image with no processing. The other three images show an overlay of the CIVA signal and secondary electron images at different magnifications. The highest magnification shows the open conductor at a polysilicon step.

Difficulty of Implementation: The equipment necessary to perform CIVA imaging are a scanning electron microscope, a constant current source, and an electrical vacuum feedthrough. The only sample preparation needed is lid/plastic package removal. A digital image acquisition would be advantageous to image acquisition/manipulation but is not necessary. The passivation may or may not be removed for CIVA analysis. The IC must be biased into a non-contention state, but no complicated vector set is required. By increasing the primary electron beam energy, multilevel conductors under metal may be detected. Like RCI and BRCI, the selection of proper primary electron beam energy includes using energies that just reach the buried conductors and avoid EBIC signal generation. See the sections describing RCI and BRCI for a complete description of proper beam energy selection.

Possible IC Damage: Same as with RCI imaging.

ACKNOWLEDGMENTS

The authors would also like to thank Richard E. Anderson, Ann N. Campbell, and Daniel L. Barton for their careful review of and valuable contributions to this workshop.

The authors would like to recognize the contributions of James R. Beall, Martin Marietta Astronautics Group, to previous ISTFA workshops on electron beam testing (particularly voltage contrast and EBIC techniques).

This work was performed at Sandia National Laboratories supported by the U.S. Department of Energy under contract # DE-AC04-7DP00789.

REFERENCES FOR ALL SECTIONS

For SE, VC, EBIC, BSE, and X-ray:

1. L. Reimer, *Scanning Electron Microscopy*, Berlin, Springer-Verlag, 1985.

For VC, CCVC, EBIC, RCI, AND BSE:

2. J.R. Beall, "Voltage Contrast Techniques and Procedures" and "Electron Beam-Induced Current Application Techniques," *Microelectronic Failure Analysis Desk Reference*, Supplement Two, November 1991.

3. E.I. Cole Jr., et al., "Advanced Scanning Electron Microscopy Methods and Applications to Integrated Circuit Failure Analysis," *Scanning Microscopy*, Vol 2, No.1, pp. 133-150, 1988.

For CCVC:

4. E.I. Cole Jr., "A Novel Method for Depth profiling and Imaging of Semiconductors Devices Using Capacitive Coupling Voltage Contrast," *Journal of Applied Physics*, Vol 62(12), p 4909-4915, 1987.

For RCI:

5. E.I. Cole Jr., et al., "Resistive Contrasting Applied to Multilevel Interconnection Failure Analysis," *Proceedings of IEEE VLSI Multilevel Interconnection Conference*, p 176-182, 1989.

6. C.A. Smith et al., "Resistive Contrast Imaging: A New SEM Mode for Failure Analysis," *IEEE Transactions on Electron Devices*, ED-33, No. 2, p 282-285, 1986.

For RCI and BRCI:

7. E.I. Cole Jr., "A New Technique for Imaging the Logic State of Passivated Conductors: Biased Resistive Contrast Imaging," *Proceedings of IEEE International Reliability Physics Symposium*, p 45-50, 1990.

For X-ray:

8. J.I. Goldstein et al., *Scanning Electron Microscopy and X-Ray Microanalysis*, New York, Plenum Press, 1984.

For CIVA:

9. E.I. Cole Jr., and R.E. Anderson, "Rapid Localization of IC Open Conductors Using Charge-Induced Voltage Alteration (CIVA)," *Proceedings of IEEE International Reliability Physics Symposium*, pp. 288-298, 1992.

For Passive Voltage Contrast:

10. J. Colvin, "A New Technique to Rapidly Identify Gate Oxide Leakage in Field Effect Semiconductors Using a Scanning Electron Microscope," *Proceedings of the International Symposium for Testing and Failure Analysis*, p 331-336, 1990.

For CCVC combined with I_{DDQ}:

11. R. Bottini, et al., "Failure Analysis of CMOS Devices with Anomalous IDD Currents," *Proceedings of the International Symposium for Testing and Failure Analysis*, p 381-388, 1991.

For electron beam charging of hybrid test nodes:

12. C. Hilbert and C. Rathmell, "Design and Testing of High Density Interconnection Substrates," *Proceedings of NEPCON West*, pp. 567-579, 1990.

13. IRPS Tutorial, "Scanning Electron Microscopy Techniques," March 30, 1992.

EMISSION MICROSCOPY AND LIQUID CRYSTAL

by Lawrence C. Wagner, Thomas S. Taylor, and Kendall S. Wills

INTRODUCTION

Leakage related failures comprise a major portion of the failures observed for all integrated circuit technologies. This includes not only leakage parameter failure but also functional failures which occur because of leakage at an internal circuit node. Two dominant techniques for the precise isolation of leakage and leakage induced failures have been developed in the last decade: liquid crystal and emission microscopy.

From a failure analysis perspective, both liquid crystal and emission microscopy techniques have advantages and disadvantages. Both techniques share a common advantage: the precise identification of the location of the failure. While both techniques are fairly easy to use, emission microscopy has a significant advantage in this area. Problems with control of device temperature, liquid crystal thickness and thermal backgrounds on high power devices are eliminated. The sensitivity of these techniques is difficult to evaluate. The sensitivity of the liquid crystal approach is limited only by the power which can be dissipated through the leakage site. Emission microscopy, on the other hand, can be sensitive to very low leakage levels, independent of power dissipation. However, if the failure mechanism is of a non-emitting type, the failure site cannot be identified regardless of the power dissipated through the failure site.

The two techniques will be presented. Analysis of leakage failures by both techniques will be used to illustrate the differences between them. A combination of the two techniques has been found to be extremely effective for the analysis of leakage related failures.

HISTORY OF LEAKAGE ISOLATION TECHNIQUES

Parametric leakage failures comprise a major segment of integrated circuit failures which must be analyzed. A large number of techniques to localize the site of leakage failures have been developed. These can be generally categorized as thermal detection techniques (frequently referred to as "hot spot" techniques) or non-thermal techniques.

True thermal hot spot detection techniques included boiling freon®, IR thermal mapping and liquid crystal. IR thermal mapping has been primarily limited by the spatial resolution of the thermal mapping systems, as well as the sensitivity of the systems to small changes in temperature. Boiling freon® (other solvents can be used) approaches consist of immersing the device under test in the solvent and microscopically observing localized boiling of the solvent at the "hot spot". Boiling freon® approaches have the potential for high spatial resolution and fair sensitivity when the temperature of the freon® applied to the device can be controlled to a temperature just under the boiling point of the freon. However, control of the freon® evaporation in an enclosed system has resulted in a lack of convenience in setting up the test. Thus commercial systems for this "hot spot" technique have been expensive. Liquid crystal has emerged as the most popular of these techniques. It has the significant advantages of: 1) ease of use, 2) good sensitivity to thermal heating (achieved by control of the device temperature to a point just below the liquid crystal transition temperature, 3) good rate of success in isolation of failures and 4) low cost.

Non-thermal techniques have been proposed and used on a rather limited basis until the emergence of emission microscopy as introduced by Intel. EBIC (electron beam induced current) and OBIC (optical beam induced current) techniques have been shown to be workable but the difficulty of set-up and lack of consistent positive results have resulted in a lack of general utilization. Both techniques involve the detection of current changes in a device under test when a rastered beam is used to create electron-hole pairs in the device under test. In EBIC, an electron beam in a scanning electron microscope is used while in OBIC, a laser beam is typically employed. The rapid evolution of confocal laser scanning microscopes may allow OBIC to emerge as a next generation leakage/defect isolation approach for more general application. Another proposed approach has been the use of magnetic particles to trace the current path on a device under bias. A suspension of magnetic tape particles in a solvent is applied to the device. Current paths are decorated due to the magnetic field generated by current flow. It has similar advantages to liquid crystal in terms of its ease of use, sensitivity and success rate but at a higher cost. This technique has apparently not been widely used.

Thus liquid crystal and emission microscopy have emerged as the most popular thermal detection and non-thermal detection techniques, respectively. This has been driven primarily by the ease of use of the techniques and relatively high success rates in the isolation of failures. Liquid crystal has also had the advantage of low equipment cost. While both techniques exhibit high success rates, neither technique is universally successful. This article will examine the success of each technique as applied to a variety of CMOS and bipolar failures in an effort to understand the relative success rates of each. In the final analysis, it becomes apparent that these techniques are complementary rather than competitive. The high success rates of each technique independently evolve into a very excellent success rate when the two techniques are used in tandem.

EQUIPMENT AND MATERIALS

The liquid crystal station includes standard bench test equipment, an AC function (pulse) generator with DC offset, and an optical microscope with adjustable polarizers after the light source and after the objective. Device temperature is controlled

Fig. 1 Typical liquid crystal hot spot.

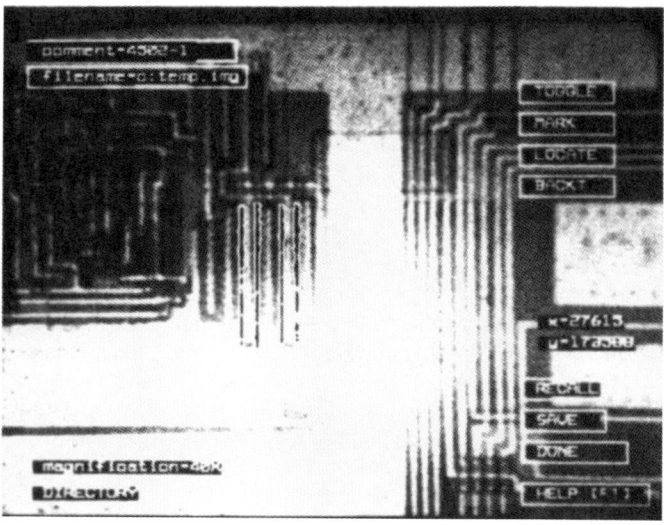

Fig. 2. Emission microscope output typical of transistor turn-on.

with a Temptronic TPO315A temperature controller with thermal chucks for different package types. A heat gun or freeze-mist is used for less precise control of device temperature. The liquid crystal used was type K18 from EM Industries (4-cyano-4'-n-alkylbiphenyl). Although many types of liquid crystal are available, this type was chosen because its nematic-to-isotropic phase change occurs at 29 °C, slightly above room temperature. In some applications, higher temperature transition may be desirable and some work has been done with such crystals.

A number of emission microscopes are currently available. The present work was performed on an EMMI (KLA Industries) equipped with objective lenses up to 100 ×. The power supply is connected through an ammeter so that the leakage current in the failing state can be monitored.

For devices which require functioning to reproduce the failure, an IMS2000 (Integrated Measurement Systems) tester is used. The test head is fitted to a 21 × 21 zero-insertion-force PGA socket through a custom load board. Non-PGA packages are tested using an adaptor socket to the PGA outline. A B&L Microzoom microscope with polarizing lens is mounted above it, allowing liquid crystal analysis to be performed directly on the test head. For emission microscopy, a cable is available which plugs into the socket on the test head and goes through a side port of the emission microscope to an equivalent socket mounted on the emission microscope stage.

LIQUID CRYSTAL TECHNIQUES

Two liquid crystal techniques were used: 1) "standard" liquid crystal techniques, and 2) a "heat and cool" method. The standard liquid crystal technique employs polarized light to enhance the liquid crystal transition of a powered device. The liquid crystal applied to the device is relatively thin and clear below the transition temperature of the liquid crystal. A dark spot (see Fig. 1) results from localized heating associated with the "hot spot". To further enhance sensitivity, the temperature of the device can be controlled using a hot chuck or a heat gun. If the temperature is controlled near the transition point for the liquid crystal, the detectable heat generated from a leakage site can be smaller.

The "heat and cool" method involves the application of a thicker layer of liquid crystal so that the liquid crystal has a "bubbly" appearance. The device is then heated with a heat gun (or the heat of the device) past the liquid crystal transition point and sprayed with freeze mist or allowed to cool. When the liquid crystal is sufficiently heated, it appears clear. As the device cools, the "bubbly" appearance will return and the last clear spot will identify the leakage site. Once identified, the hot spot can be localized by reducing voltage and/or stabilizing temperatures. This method is simple, quick and easy to set up. Polarized light and a slow rate of cooling may be used to enhance the sensitivity of the technique.

For state dependent leakages, the device must be placed in the failing logical state. Often, functional failures have higher I_{CC} (leakage between the supply voltage and ground) levels associated with a particular test vector. Identifying the logical states with high I_{CC} allows the analyst to maximize current flow through the defect site, typically the optimum condition for the defect isolation. A useful technique is to drive V_{CC} (power supply voltage on device) with an low frequency AC signal (1V P-P @ 0.1 to 10 Hz) with a 5 volt DC offset. This "toggles" the hot spot without changing the state of the device. This may also be accomplished by increasing and decreasing the voltage on a power supply and thus retaining device functionality. At times, leakage may be light sensitive rendering liquid crystal ineffective.

For devices with high standby currents, the device must be cooled or placed in a lower temperature mode. The device can be cooled using freeze mist or a cold chuck. To place the device in a lower temperature mode, toggling V_{CC} on and off at a relatively high frequency works fairly well unless the leakage is state dependent or the device has a power up clear circuit. Liquid crystal is generally more difficult to perform on this type of device. Changing to a liquid crystal with a higher transition temperature is also possible.

PHOTOEMISSION TECHNIQUES

Standard emission microscopy techniques with commercially available equipment have in general been applied. Some of the techniques used to bias devices for liquid crystal are also applicable for emission microscopy. For example, use of a tester to drive state dependent leakages is required when performing emission microscopy.

All input pins must be tied to a high or low voltage level. Failure to do so will result in emission sites at some transistors whose gates are floating. An example of this is shown in Fig. 2. After tying the pin low, the emission site was no longer observable.

On bipolar devices, the amount of V_{CC} which can be applied will depend on the device type and the type of internal components since emission sites can result from conduction of current on large transistors normally "on". High currents on forward biased emitter base (EB) bipolar junctions are detected as emissive sites.

The voltage applied to V_{CC} should be as high as possible without jeopardizing device functionality. On CMOS devices, V_{CC} should not generally exceed 7-8 volts. Increased voltages cause higher fields on defective junctions and gate oxides, increasing the probability of electron transitions to the conduction band where they can eventually recombine with holes to generate photons. Thus, photoemission at some defects is very voltage dependent.

The acquisition time of the emission microscope is controlled by two parameters. The first, "analog", determines the amount of time that the camera acquires an image. The second, "digital," is the number of times the image is acquired and added by the image processor. Due to various factors such as camera heating during analog acquisition, detector non-linearities, and the subtraction of a background for the same acquisition time, there is not a simple relationship between the acquisition parameters and the intensity of the acquired emission site. While greater intensity is generally obtained with greater acquisition times, there are fluctuations which show that some analog/digital combinations are more efficient than others. Although the total imaging time is equally dependent on analog and digital settings, the defect intensity is much more dependent on the analog time. In fact, there appears to be a minimum analog time necessary to detect the defect.

Liquid crystal analysis can be made more sensitive by careful control of the device temperature. Liquid crystal analysis of state dependent leakages can be performed by superimposing an AC signal onto V_{CC}. Emission microscopy sensitivity is improved when the voltage is set as high as possible. Knowledge of a system's analog/digital characteristics can also make emission collection more efficient. For both techniques, placing a device in the proper state to apply proper voltage levels at the defect location was critical in resolving many failures.

CONCLUSIONS

Overall, liquid crystal was found to have a better success rate at detecting leakages than emission microscopy. However, emission microscopy generally pinpointed the defect more accurately and took less time. Emission microscopy also located some leakages which were too low for liquid crystal analysis. Emission microscopy performed slightly better on gate oxide defects while liquid crystal excelled with conductor related defects. Neither emission microscopy nor liquid crystal was able to isolate failures due to inversion mechanisms and overetched contacts. It was found that emission microscopy is less likely to find gate oxide defects on PMOS transistors than on NMOS transistors, apparently since the gate and substrate doping of PMOS transistors are the same type. Both techniques fared better on CMOS technology than on high power bipolar technologies.

Since rapid failure analysis is increasingly important for customers, emission microscopy remains a preferred technique for initial analysis of a leakage failure in many cases because of its ease of use and comparative speed. Liquid crystal would then be used on non-emitting failures or failures where a failure mechanism involving conductors was suspected.

While emission microscopy and liquid crystal are capable of resolving many failures separately, when used in tandem, they can significantly increase the rate of successful leakage isolation without probe and isolation. Probe and isolation should be considered a last resort for the isolation of leakage related failures due to the tedious and time-consuming nature of the technique. The different strengths and weaknesses of the two techniques make them complementary rather than competitive.

REFERENCES

1. J. Lindgren et al., International Symposium for Testing and Failure Analysis, 11 (1987).

2. D. Burgess and P. Tan, International Reliability Physics Symposium, 119 (1984).

3. N. Khurana and C. Chiang, International Reliability Physics Symposium, 72 (1987).

4. N. Khurana and C. Chiang, International Reliability Physics Symposium, 72 (1987).

5. F.J. Henley, International Reliability Physics Symposium, 69, (1984).

6. T. Taylor et al., International Symposium for Testing and Failure Analysis, 5, (1989).

PHOTOEMISSION MICROSCOPY

by Gary Shade and Kendall Scott Wills

Statement of Need

Current designs in very large scale integrated (VLSI) circuits are too complex to determine logically what is failing. Even with design for testability (DFT) in mind during the initial design stages, computer algorithms only guess at the location of the failure. In most cases the logical approach only determines the logic circuit affected. The data register, for example, may contain 64 bits at 7 transistors per bit for a total of 448 transistors. Without photoemission microscopy, a failure analyst would need to test all 448 transistors by needle probe. With photoemission microscopy, the single latch which is defective can be located. The failure is isolated to 7 transistors at most.

The nature of the semiconductor industry is to have better quality product with faster speed at a lower cost. This nature makes rapid design debug, quick failure analysis and immediate corrective action imperative. Point by point inspection of VLSI devices is no longer adequate. Even discrete components are affected by the push to bigger, faster and better. Discrete components in many ways can be more difficult to analyze than VLSI. There are fewer visual clues to help locate the problem.

The implication of what the semiconductor industry needs is simple. Failures must be located in a timely manner. All failure analysis must be accurate and detailed. Testing must be direct. As much information as possible must be gleaned from a single quick test.

Complimentary Techniques

Photoemission provides direct testing of semiconductor devices. This technique is fast and relatively low cost when compared with the cost of stroboscopic voltage contrast. Like optical microscopy, photoemission microscopy does not provide an electrical fail signature. Photoemission microscopy is limited in its ability to find defects. Shorts or any defect which can be characterized by Ohms Law cannot be found. The following techniques compliment the photoemission technique by providing more detailed information about the electrical aspects or the thermal properties of the failure.

Stroboscopic Voltage Contrast

Voltage contrast uses a scanning electron microscope (SEM) to monitor the electrical activity of a semiconductor device. Its main advantage is that the electrical signature of the failure can be determined. Like photoemission microscopy, the failure is isolated to a single circuit. The disadvantages of voltage contrast are the inability to precisely locate the fail site and its inability to force internal device nodes. It is difficult to locate the failure to a single transistor. The voltage contrast system requires a skilled operator.

Liquid Crystal

Liquid crystal is the least expensive technique to implement other than visual inspection. Two polarizers and a microscope are all the equipment that is needed. The technique is quick to set up. The device under test is placed under a microscope, liquid crystal is applied and the device is activated in such a mode that the transistors associated with the failure are turned on. The disadvantage is the limited spatial resolution and the relatively high power required at the fail site to visualize the failure.

External Electrical Testing

Device failures are caught because the device failed to perform correctly at the external pins. To do external testing there is no deprocessing of the device. An electrical signature can be developed for the failure which may limit the possible fail sites on the device. External testing in most cases cannot isolate the failure to a single transistor.

Internal Probe

Internal needle probe, like voltage contrast, can determine the electrical signature of the defect found. Unlike voltage contrast, internal needle probing can force nodes on the device. This gives the test engineer greater ability to isolate the problem to a single transistor. A problem arises with the size of structures available for needle probing. Structures as small as 0.5 micron can be probed with needles but the possibility of damage is great.

Optical Beam Induced Current (OBIC)

Optical beam induced current operates by inducing hole-electron pairs in the silicon. If there is a defect or junction, the OBIC current varies from the current of the native silicon. The current variation is converted into contrast which can be seen through the use of a video monitor. OBIC is quick and relates directly to junction damage. The primary disadvantage is the technique cannot be performed when metal covers the junction.

Electron Beam Induced Current (EBIC)

Electron beam induced current is somewhat like OBIC. Contrast is generated due to the interaction of the primary electron beam of an SEM and the defects in the semiconductor junction. The SEM monitor displays the location of the defect. SEMs have good position accuracy so the defect can be specified to within 0.5 µm. The EBIC technique only verifies a defect is present. No determination of what the defect is can be made from the EBIC signal. The major problem with EBIC is the amount of work required to ready the device for test. In most cases to perform EBIC testing, the device must be completely deprocessed.

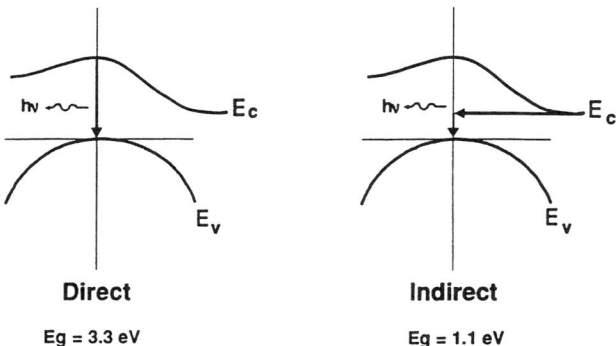

Fig. 1 Valence subband structure, and intraband transitions.

Resistive Contrast Imaging (RCI)

Resistive contrast imaging is similar to EBIC. The primary electron beam is used to excite the defect. In this case, however, the defect can be in an oxide above the substrate or be a metal to metal short. The requirement is for the SEM primary beam current to reach the substrate by some path through the defect. No current travels to the substrate through any other path. Contrast is generated by the difference in current through the defect and the current from the surrounding area. No deprocessing is required to see metal shorts. Some deprocessing may be required to see other types of defects such as poly filaments shorted to substrate. Typically this technique will not find opens.

Photoemission Theory

Photoemission microscopy is useful because of the defects it can find. A representative list of the defects which can be located by the photoemission technique are listed below:
Photoemitting defects:

- Forward biased junctions
- Avalanche junctions (reverse bias)
- Junction leakage (defects or breakdown)
- Oxide breakdown due to contamination or electrical overstress (EOS)
- Oxide leakage
- Hot electrons
- Latch-up
- Saturated transistors (bipolar, MOS)
- Electro Static Discharge (ESD) event or damage
- Crystal mechanical damage
- Bremsstrahlung radiation

Fundamentally, there are three mechanisms for photoemission that help to understand each of the above. These three types will be described first, followed by an interpretation of the first four (and most universal) defects shown in the above listing. Next a summary will be provided to aid in classifying leakage sources and their interpretation.

The three fundamental mechanisms are:

- Interband recombination
- Intraband recombination
- Bremsstrahlung

Interband Recombination

As the name implies, interband recombination is an allowable transition of the electron from the conduction band to an open energy state in the valence band. When the transition occurs, a photon is emitted. Figure 1 lists two possible energy transitions. The first is from the conduction band directly to the valence band. The example shows a direct transition energy of 3.3 eV. An indirect transition takes place when the electron moves from the conduction band at a point other than the maximum (usually at the band minimum). This type of transition requires a momentum exchange often provided by a defect, or contaminant. As most conduction electrons are at the band minimum, this transition is the most common and provides a photon at 1.1 eV.

Intraband Recombination

The energy band for holes in silicon is actually two bands representing light and heavy holes. In dynamic conditions, with hole current present, transitions can occur between these bands providing photons of low energy (<1.0 eV) (see Fig. 2).

Bremsstrahlung (Braking Radiation)

Bremsstrahlung radiation has been included here for completeness. To date no evidence has been found to prove or disprove its observation with emission microscopy. This may change with increased emphasis on quantitative techniques. For the present, it serves as a convenient means to explain the broad spectrum results often observed.

For Bremsstrahlung to occur, an electron with high energy must approach an atom, so as to affect the electron momentum and energy. The atom slows the electron down, or changes its trajectory. This change must be counteracted by a photon emission in order for energy and momentum to be conserved.

Photoemission During Junction Forward Bias (Fig. 3)

Forward biased junction current is dominated by recombination. Majority carriers (both holes and electrons respectively) are injected across the depletion region and recombine by an indirect (phonon assisted) transition. The emitted photons are centered at the bandgap energy of 1.1 eV. Although the typical detector has a low sensitivity at this energy, high forward current is sufficient to allow detection. The recombination current is proportional to the square of the intrinsic carrier concentration and exponentially to the ratio of applied voltage to temperature. Forward biased junctions occur in latch-up, incorrect device connections, bipolar designs, and in some cases junction spiking (where an undesired junction is formed).

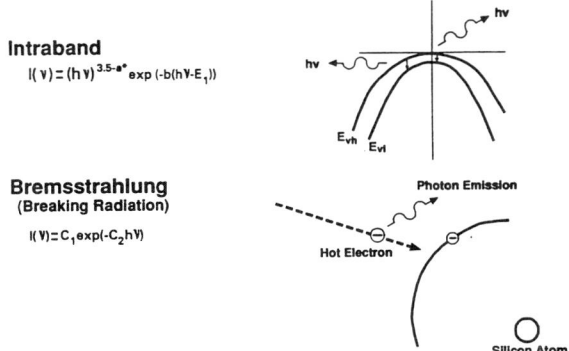

Fig. 2 Subband transitions showing emission mechanisms.

Photoemission During Junction Avalanche

During reverse bias, the thermal carriers crossing the junction may obtain sufficient energy to be out of equilibrium with the surrounding crystal lattice. The probability of collision becomes greatly increased allowing the carriers to lose some energy and return to equilibrium. The collisions can create additional electron-hole pairs thereby increasing the chance for more collisions. This process is called carrier multiplication. Once the multiplication exceeds unity, the process is considered an avalanche and junction damage is likely. As seen in Fig. 4, each carrier has a probability to recombine and produce a photon. A wide range of photon energies results starting from the bandgap energy and extending out to greater than 3.0 eV. Several authors have published representative data for reference [11, 14, 19, 29, 122]. Emission probability has been stated as approximately 1 in 10,000 or 100 ppm for electrons, which accounts for the smaller amounts of photons seen as compared to forward biased junctions. Holes have a much lower probability due to their lower mobility resulting in less energy increase as they cross the junction.

Photoemission in a Leaky Junction

Leaky junctions have two major causes: excess recombination sites in the depletion region and asperities causing non-uniform, premature breakdown. The former is more prevalent and will be discussed in greater detail. The latter is analogous to avalanche discussed in the previous paragraph.

Frequently, failures due to leaky junctions are caused by excess material (or defects) in the depletion region due to stress applied to the part. This stress may be improper processing, or incorrect bias conditions. These impurities, or defects, give rise to recombination centers which enhance the recombination current and increase the diode reverse leakage. In the extreme case, the level of impurities may reach a point that the junction is shorted as is the case for contact spiking.

Reverse diode current, for the ideal junction, is due to the drift component of current. Minority carriers within a diffusion length of the depletion region can reach the depleted zone where they are swept away by the field (i.e. drift current). Thus, the current is not limited by the field, but by the thermal generation of carriers. The mechanism that describes this behavior is

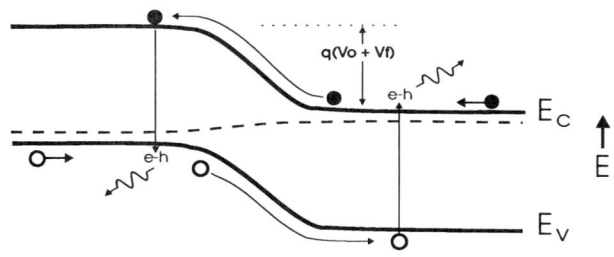

Fig. 3 Forward biased junction band diagram. The electrons cross the junction with low energies and combine with the holes in the p-type material. The photons released in the transition are in the red and near infra-red.

Shockley-Read-Hall (SRH) generation-recombination [125, 126]. However, in the case of a stressed junction, defects in the depletion region may lead to barrier lowering as described by the Frenkel-Poole mechanism [70, 83]. SRH leakage is typically undetectable because of its low level; whereas, the Frenkel-Poole leakage can reach very high levels depending on defect levels and the degree of localization as in contact spiking or melt filament formation. Frenkel-Poole leakage is linearly proportional to the field strength. Figure 5 shows the case for Frenkel-Poole barrier lowering caused by a defect within the depletion region.

Photoemission from Fowler-Nordheim Tunneling

When electrons tunnel across an oxide, they produce a photoemission spectra centered around 2.5 eV. If there were no interaction between the tunneling electrons and the oxide, the peak of the spectrum should be at 4.3 eV for n-well capacitors and 4.8 eV for p-well capacitors. However, during tunneling

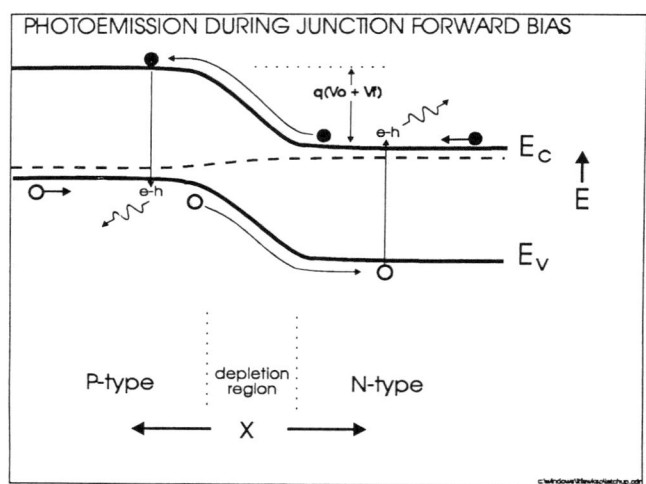

Fig. 4 Band diagram for reversed biased junctions. Peak energy levels in the visible spectrum.

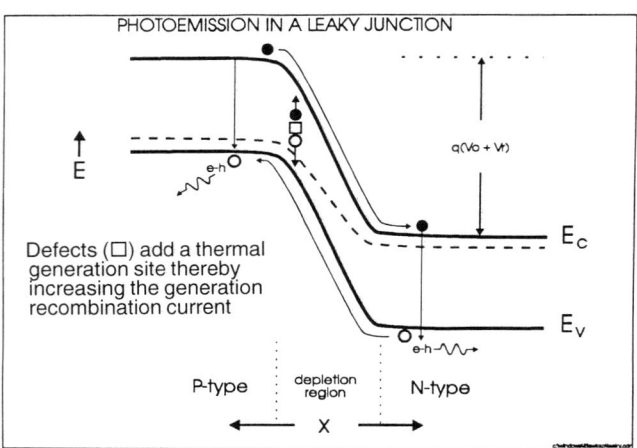

Fig. 5 Band diagram for a junction which has a defect.

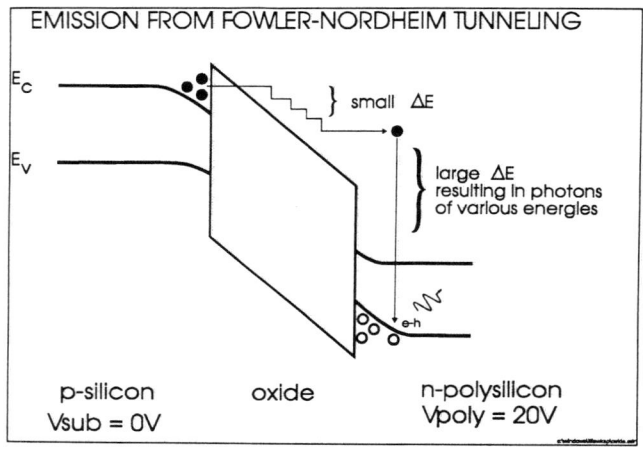

Fig. 6 Fowler-Nordheim tunneling band diagram. The photoemission peaks in the middle of the visible spectrum.

there is an interaction between the electrons and the oxide which reduces the electron energy gradually as the electron moves across the oxide. When the electron arrives in the n-type material, the peak of the recombination spectra is at 2.5 eV. The probability of electrons with higher energies decreases exponentially. The photoemission spectra above the peak should follow the trend in electron energy and fall off exponentially. The ability of the electrons to tunnel through the oxide decreases exponentially below the peak energy. Therefore, the spectra should rise in the middle and fall exponentially at both the low and high ends. The case for holes is analogous. Figure 6 shows the path of an electron as it traverses the oxide and recombines in the n-type polysilicon gate.

Considerations: Classifying Emission Concerns

A foundation for the understanding and interpreting of photon emissions has been provided in the previous discussion. What remains is to consider the causes of leakage current and how the emission microscope can be applied to detect them. A system of classifying defects is instrumental in reaching this objective.

The following list shows how the various causes of leakage current can be fit to four classes. In each case, the list makes it clear whether the defect is potentially detectable and whether or not is should be a concern.

Detectable Emissions

True concerns (Emissions that represent a weakness)

- Junction leakage
- Contact spiking
- Hot electrons (saturated transistors)
- Junction avalanche
- Latch-up
- Oxide current emission
- Polysilicon filaments
- Substrate damage

Artifacts (Emissions that are artifacts of design or test condition)

- Floating gates
- Saturated bipolar transistors
- Saturated analog MOSFETs
- Forward biased diodes

Photoemission Microscopy

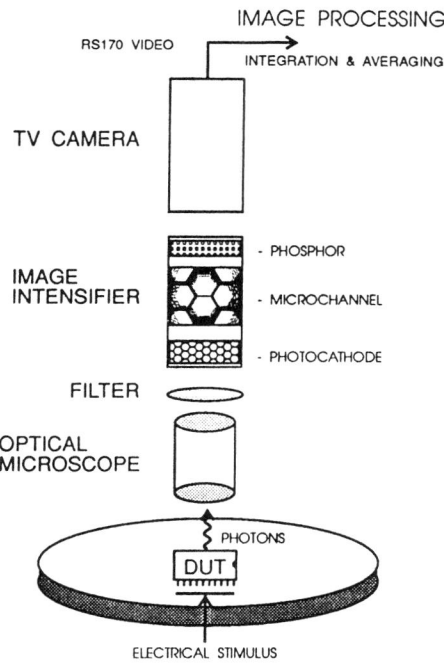

Fig. 7 Typical photoemission system.

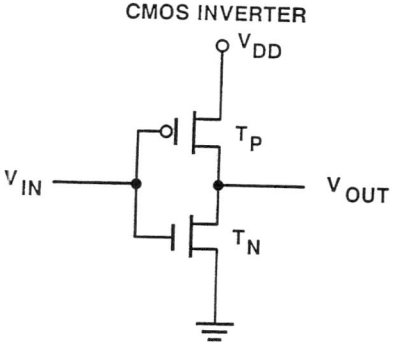

Fig. 8 (a) Circuit diagram of a typical CMOS inverter. (b) Transistor parameters for the CMOS inverter shown in Fig. 16.

Channel Length (drawn, µm)	Channel Width (drawn, µm)	Drain Structure	Gate Oxide Thickness (Å)
n-channel	1.25	30	LDD	180
	2, 3, and 4	16	non-LDD	450
p-channel	1.75	30	non-LDD	180
	2, 3, and 4	16	non-LDD	450

Non-Detectable Emissions

Non-Emitting (Leakage sites that do not have an emission mechanism)

- Ohmic shorts
- Shorted metal interconnects
- Surface inversion
- Silicon conduction paths
- Sub-threshold condition

Emission masked (Emissions that are not detectable)

- Buried junctions
- Leakage sites under metal

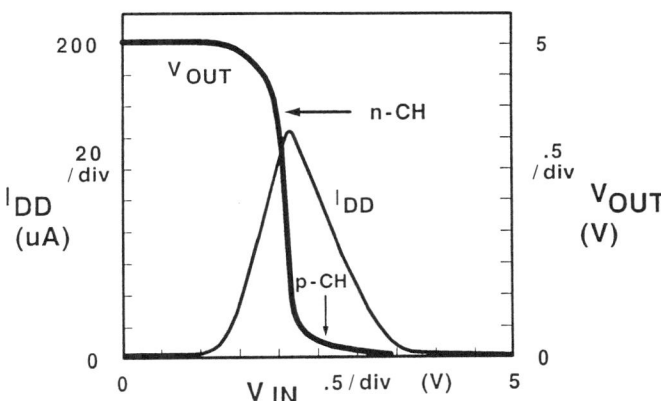

Fig. 9 Inverter transfer curve showing current and voltage.

Photoemission Equipment

The Basic System

The basic photoemission system may vary slightly, but the basic fundamentals shown in Fig. 7 will be the same. The basic system has a stage to hold the device under test (DUT). The device must be stimulated in such a manner that the defect site is active. Generally, the best results are obtained with a system which puts all the pins in a known state. Floating pins or uncontrolled pins can lead to erratic photoemission results.

An optical microscope is used to image the DUT onto a photocathode of a microchannel plate. A filter may be placed in the beam path to the photocathode to obtain the spectra described in the section on theory. A phosphor plate on the end of the image intensifier generates an image which can be seen by a video camera. Because the photon count is low, the image from the video camera is averaged over many frames and manipulated by an image capture board.

Benchmarking/Correlation of Systems

Benchmarking is a term which has been used to mean many things. In this case, benchmarking is the standardized testing of the photoemission system. Benchmarking is needed to understand when the system is not functioning correctly. Benchmarking is needed to ensure that the measurements from the system are repeatable, thus allowing process control.

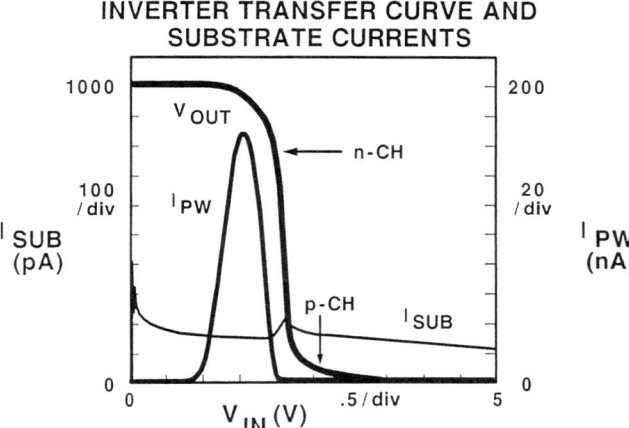

Fig. 10 Inverter transfer curves showing current and voltage as a function of the input voltage.

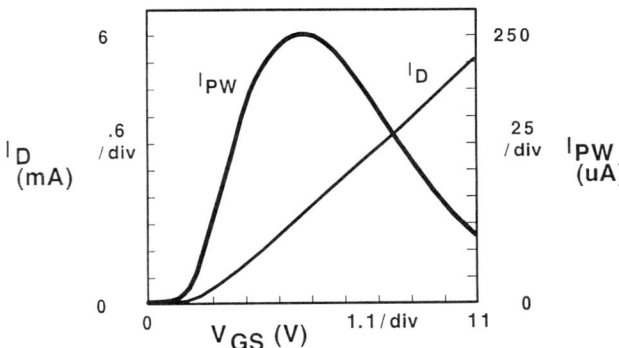

Fig. 11 Transistor currents for n-channel transistor as a function of the voltage on the gate.

Why should a photoemission system be calibrated? Who cares if the absolute value of the data is different from day to day? The answer to these questions is simple. If failure analysis labs are to be able to compare data, the data must be consistent and accurate. A failure analyst must be assured when no photoemission is seen, that the sensitivity of the equipment has not changed. The results are credible. Suppliers of photoemission equipment must be able to tell when the equipment is not performing correctly and what is wrong. Ultimately, the performance of the system must be trustworthy. So trustworthy, in fact, that unskilled operators can use the equipment.

Additional benefits can be gained as well. Benchmarking allows for a better understanding of the system, better characterization, and facilitates routine maintenance. Benchmarking increases measurement confidence and permits verifiable training. When upgrades are made to the system, benchmarking permits the verification of the upgrade status.

The desired benchmarking procedure will provide, on a routine basis, system characterization. The measurements will be traceable. The procedure would provide system diagnostics. This list of requirements can be accomplished by checking the main camera sensitivity, by checking the linearity of the main camera and determining the minimum detection limit of the main camera. A test fixture with varying spectral frequencies can determine the spectral response of the main camera and the spatial response uniformity of the main camera. The best possible benchmarking procedure would require low skill level, take a short time and be done at reasonable cost.

Technique

Analytical Procedure

The objectives of the photoemission technique are to locate potential fail sites, separate false from true failures, identify potential device weak points and to study the device physics. To meet the objectives the device must be tested according to the following procedure.

Photoemission Testing Sequence

1. Perform functional test
2. Perform IddQ testing
3. Reproduce failure on bench
4. Decapsulate or delid component
5. Visually inspect the die
6. Reconfirm bench test (preferably in the photoemission microscope)
7. Inspect component with photoemission microscope (acquire emission image)
8. Compare results to known good component
9. Isolate emission location
10. Verify leakage site by direct technique

As with any good testing technique there is some preparation to do before the device can be tested. In the case of photoemission microscopy, the most important item is the hardware. Any test sockets, cabling or test programs should be readied prior to photoemission testing. In the high pin count VLSI devices, the test sockets are difficult to wire or must be custom built. Waiting until the last minute will not produce results in a timely manner. The next most important item in photoemission testing is to compare the suspect unit to a known good unit. This verifies that the test conditions are correct and that any new emissions are abnormal. The last item is to check the test program to see if the program puts all the pins in a known state. Remember that the photoemission microscope is a tool which locates the failing circuit. Be prepared to follow-up with additional techniques to isolate root cause.

Optimizing for Success

Lack of photoemission is not necessarily an indication that the device does not have a failure. The test procedure and the

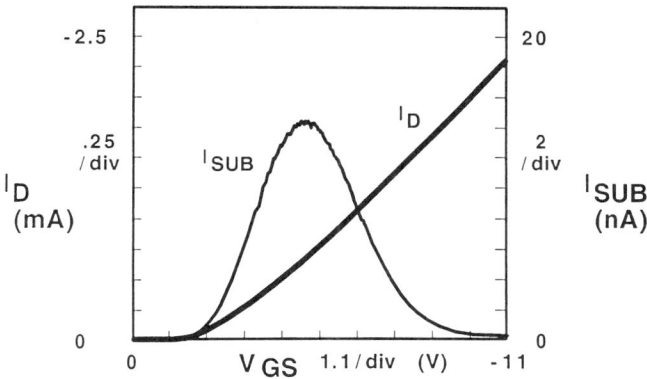

Fig. 12 Transistor currents for p-channel transistor as a function of the voltage of the gate.

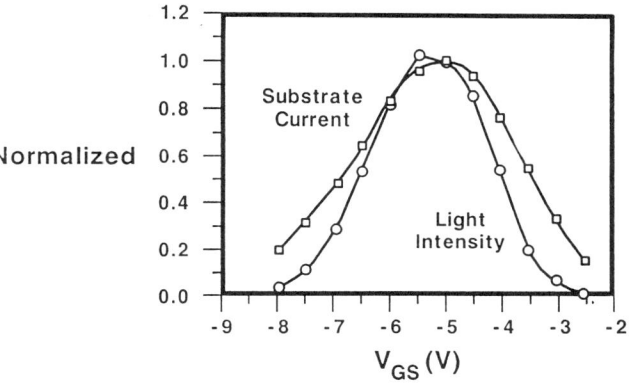

Fig. 13 Photoemission intensity as a function of applied gate voltage for a p-channel transistor. The substrate current through the transistor is superimposed upon the graph.

device must be checked before concluding that no photoemission should be seen.

What to Check When No Photoemission is Seen

1. Readjust the acquisition parameters (i.e., acquisition time) until the noise level is reached
2. Increase the device supply voltage to maximum rating
3. Increase the stress level on the failing pin
4. Increase the clock velocity
5. Vary the component temperature (i.e., for latch-up)
6. Check that component is still in the failing state
7. Try additional failures (different pin or device)
8. Review the possible leakage modes and consider the non-emitting types; are they possible?
9. Try an alternate technique
10. Alter the voltage ramp rate or the ramp rate on various pins to determine if the failure has a relationship to how the pin is driven

What to Check When There Are Too Many Photoemission Sites

When too many photoemission sites are seen, the device may have gone into latch-up or could be in an improper logic state.

1. Reduce the supply voltage and/or clock frequency
2. Reduce the acquisition time or sensitivity
3. Reduce the stress to the failing pin
4. Try to use functional vectors that stress smaller portions of the component

Success Rate

The photoemission technique has been shown to find 80% of the failures when combined with liquid crystal techniques [48]. The caveat is that the device must be tested correctly and the location of the defect must be in a region of the device which can be activated by the test. In ESD structures, for example, the defect may be in a transistor which has a gate connected directly to ground. Unless a fast high voltage spike activates the defect, it will never be detected. Defects which can be electrically activated can be located to the circuit where they reside. The photoemitting site is not necessarily at the location of the defect. This problem will be described in detail later (see "Troubleshooting"). Leakage currents below 100 nAmps per square micron are rarely seen. In some cases, the defect causing the failure will cause the entire die to photoemit. An example is when a latch-up prone device may photoemit over most of the die even though only one area of the die is the cause of the problem. The most important aspect of photoemission testing is that the defect does not need to propagate to the external pins and there is a 50% chance of finding the defect on a complimentary metal oxide semiconductor (CMOS) on power up with no additional stimulus.

Troubleshooting

With photoemission microscopy, as with any test, there are tradeoffs which must be made. For best results using photoemission microscopy, a full tester should be used. Attaching power and ground can be used on simple devices or devices where the operating conditions are not very stringent. On complex VLSI devices, a tester is a must to put all the pins at a known state.

Another problem which may be encountered with photoemission microscopy is that the operational requirements for the defect may be at a voltage level that photoemission is not likely. The best that can be done with this problem is to put the device in a failing state, then raise the power to a level where photoemission is possible. Hopefully the defect will photoemit.

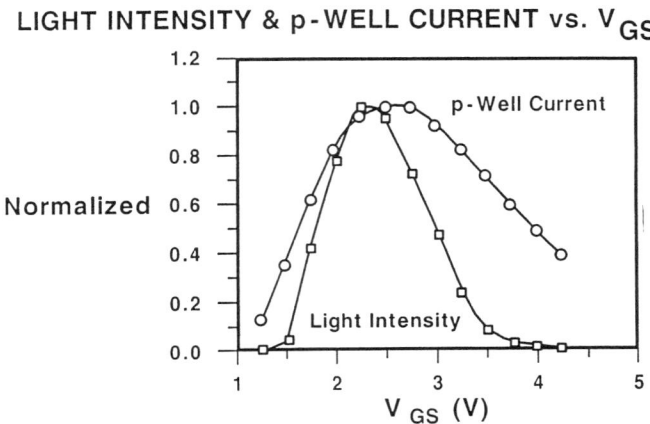

Fig. 14 Photoemission intensity as a function of applied gate voltage for an n-channel transistor. The substrate current through the transistor is superimposed upon the graph.

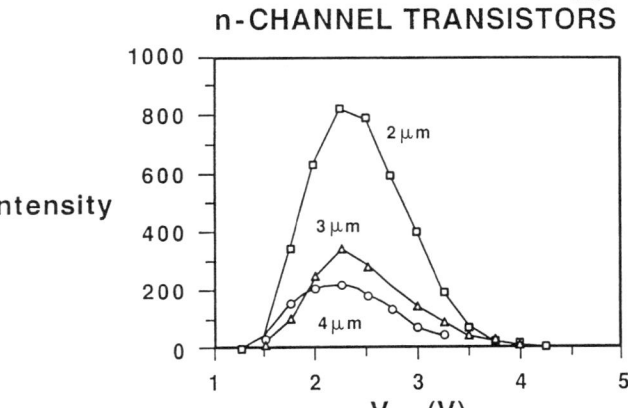

Fig. 15 Photoemission intensity vs V_{GS} for an n-channel transistor.

In some cases like the example of the ESD structure, the defect may not be able to be activated. In those cases, the only recourse is to deprocess the device looking for visual defects. Typically, in a failure of this type, the defect would not have caused a functional failure. The defect would go unnoticed until the device failed in the field.

When a defect is on a buss line pulling down the voltage, photoemission may occur at several sites in the die. In this case the fail site (photoemission site) is not the defect site. The inability of the photoemission microscope to locate the actual defect site causes a problem with analysis. Alternate techniques such as voltage contrast, internal probing and liquid crystal must be used.

Dynamic random access memories (DRAMS) can have a charge pump. Unless a reference device that is working correctly is supplied, the photoemission from the charge pump could be misconstrued as a fail site. Large clock circuits also exhibit the same problem. They can photoemit under heavy load. Only the use of a reference device will determine if the photoemission is legal.

Timing related failures, defects which hide under metal and ohmic leakage paths cannot be detected directly with photoemission technique. With the application of liquid crystal techniques, some of the hidden defects and ohmic leakage paths can be found, but timing defects cannot be found without the help of voltage contrast or internal probing, unless the timing failure causes a short which heats the device or causes photoemission. Photoemitting timing failures may occur at small dwell times with respect to the total cycle time of the sync clocks. If this is the case, a gated photoemission camera would be useful. To date little use has been made of gated photoemission techniques.

The photoemission technique cannot separate defects which are in the substrate from those which are metal shorts or oxide related. Alternate techniques are needed to resolve the conflict.

OBIC, EBIC, and RCI help to overcome the short fall of the photoemission system. The alternate techniques can help isolate the failure to the level in the device. Of course voltage contrast, internal probing and liquid crystal help as well.

Applications of the photoemission technique

Typical Photoemission Applications:

- Substrate defects (crystal defects, stacking faults, gettering damage)
- Specialized test structures
- Finished die
- Screen for "Latent" defects (finished die)
- Location of known defect
- In-process control
- Yield analysis
- Reliability studies
- Design faults
- Processing faults

To understand real world applications, a typical CMOS device will be discussed. Figure 8(a) shows a circuit diagram of a CMOS inverter. Figure 8(b) shows the transistor parameters for the inverter circuit. As the inverter changes states, the voltage on the output swings as shown in Fig. 9. There is a change in Idd during the switching which peaks when Vout is just over 3.5 V. The peak current draw occurs at the time n and p channel transistors are both slightly turned on. The substrate and well current can be monitored during switching as is shown in Fig. 10. The peak well current occurs when the n-channel transistor is most active, while the substrate current occurs while the p-channel transistor is most active.

When only the n-channel transistor is observed, the peak current into the substrate is not a linear function of the current from source to drain, Id. The relationship between well current and the voltage applied to the gate is not linear. The well current peaks at about 4.4 V gate voltage while the Id continues to climb. Figure 11 shows the relationship between well current and the voltage applied to the gate. The same is true for the p-channel transistor; the substrate current peaks at 4.4 V gate voltage while the Id value continues to climb. Figure 12 shows the relationship between the substrate change and gate voltage.

Photoemission for the p-channel and the n-channel transistors is directly tied to the substrate and well currents. Figure 13 shows the photoemission intensity for the p-channel transistor as a function of applied gate voltage. Notice the photoemission intensity tracks the substrate current. Figure 14 shows the n-channel transistor behaves in a similar manner. The photoemission intensity tracks the p-well current. In Fig. 15, the photoemission is shown to change with the size of the n-channel transistor under test.

This real world look at an inverter indicates that blindly raising the voltage to obtain more photoemission will not necessarily work. If the cause of the photoemission is related to junction breakdown or oxide breakdown, the photoemission will become stronger. If the photoemission is due to the turn-on of a transistor, the gate voltage must be made to match the peak of the photoemission curve.

Another important implication from the inverter discussion is that the total current Idd is not a good indicator of the probability of success. To get optimum conditions for photoemission, the analyst must know what he is looking for and adjust the gate voltages and device currents accordingly.

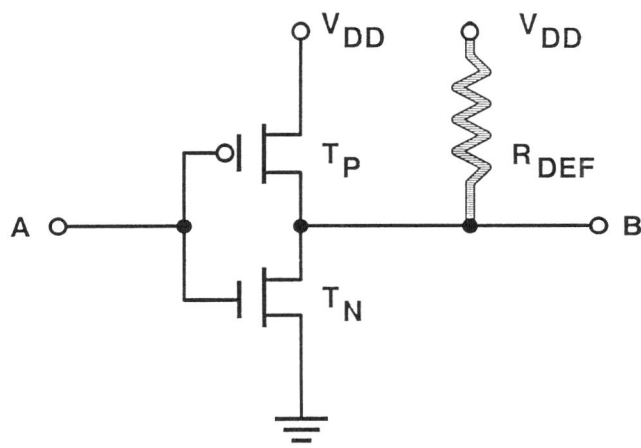

Fig. 16 Inverter with bridging defect.

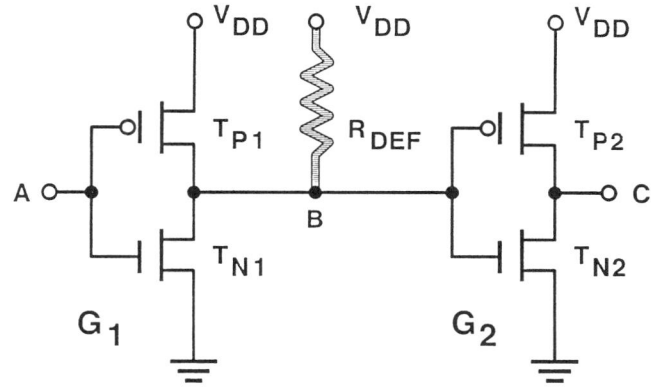

Fig. 18 Two inverters with bridging defect.

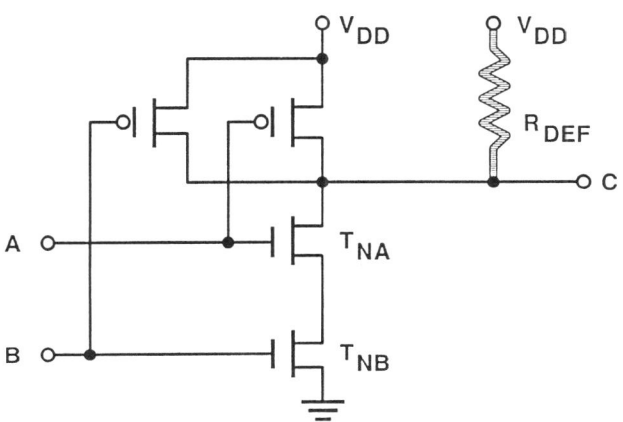

Fig. 17 Circuit diagram of a 2-NAND with a bridging defect.

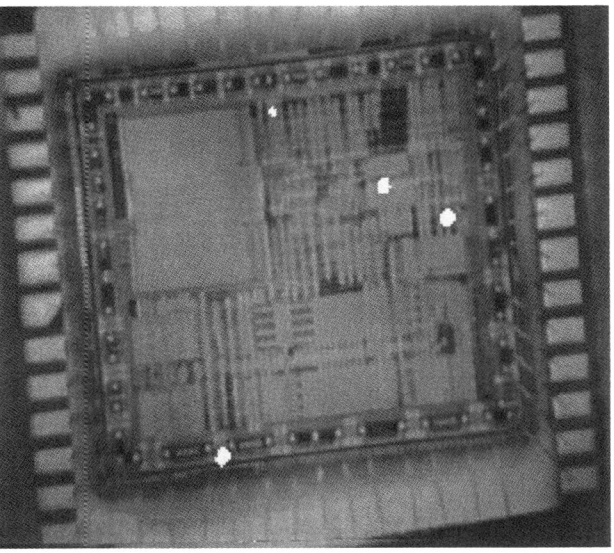

Fig. 19 Four light emitting gate shorts visible as 14K vectors are cycled repeatedly and light emission is integrated.

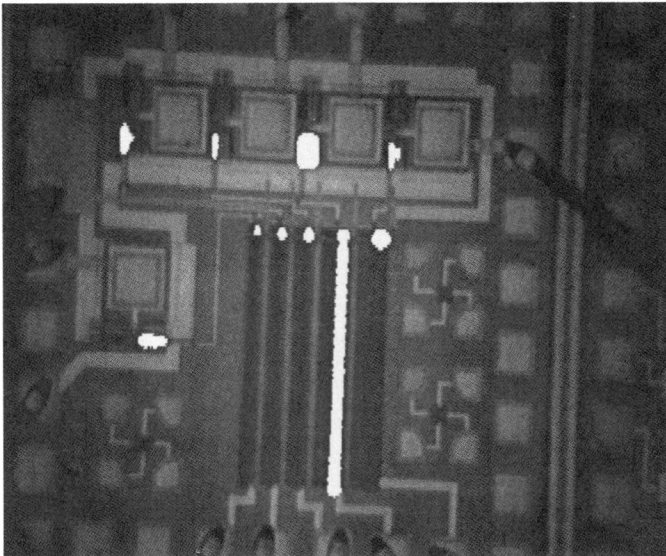

Fig. 20 Legal photoemission from a ring oscillator operating at 5V and 30 MHz.

Fig. 21 Gate oxide rupture. The flash of visible light at the instant of gate rupture. The spot in the lower image shows light emission from the gate short.

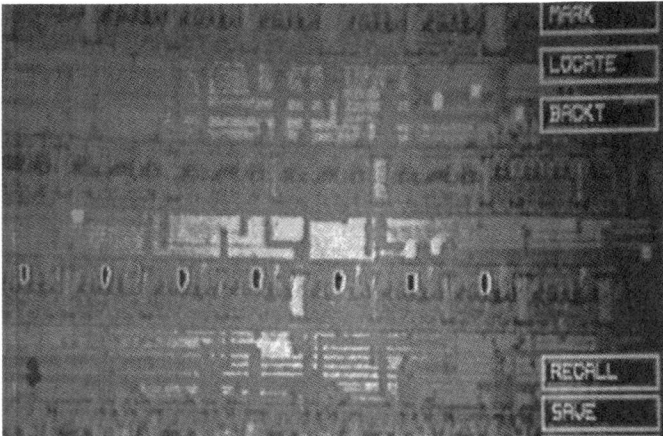

Fig. 22 Light emission from a row of flip-flops at power-up. The line of light emitting areas suddenly end on the right before the flip-flop chain does.

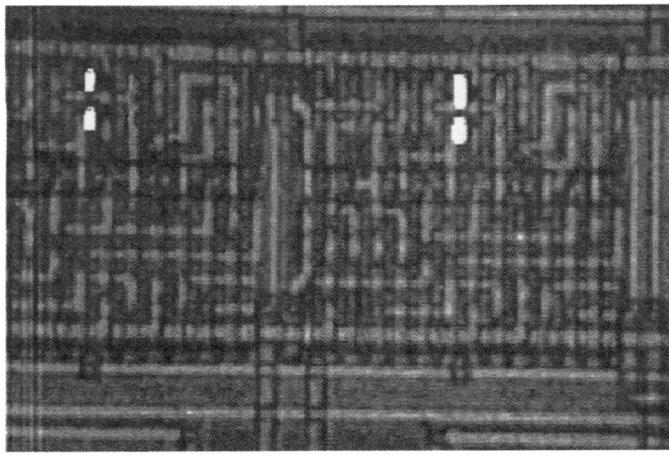

Fig. 23 Light emission from the row of flip-flops at higher magnification. The n-channel transistors are the light emission sources.

To illustrate how a defect might affect the inverter circuit, different defect configurations are shown in the following figures. It is left to the reader to consider the effects of different bias states and resistances of the defect.

In Fig. 16, the defect is connected to the output of the inverter. When the state of the inverter turns line B to ground, there is the possibility of photoemission from the n channel transistor. Depending on the nature of the defect, there is also a possibility the defect might photoemit. As long as the defect is a true resistor as shown in Fig. 16, no photoemission will occur at the defect site. Only the photoemission at the n-channel transistor will identify the problem. If the defect is connected to ground, then the photoemission will occur when B is at Vdd. Then the p-channel transistor will photoemit.

Figure 17 shows a 2-NAND with a bridging defect. When the state of the system tries to make line C go to ground, then transistor Tna will photoemit. Transistor Tnb might photoemit, but the photoemission would be less. The greater photoemission in Tna is due to the current being diverted into the substrate. Transistor Tnb will not see the same Id as Tna.

Figure 18 is more typical of a real life situation. A defect Rdef controls the voltage on line B. Photoemission will occur at different points depending upon the value of Rdef. For example, if Rdef can hold B at 3.5 V, then Tn1 will photoemit due to excessive draw to supply the current demand for the defect. Transistors Tn2 and Tp2 can photoemit depending upon their parameters. As the voltage on B is raised, n-channel transistor Tn2 would photoemit more than Tp2. If the voltage on B is lowered, the photoemission would be more prominent on Tp2. As described earlier, the photoemission intensity of a transistor is

ESD STRESS TEST

CASE #	STRESS VOLTAGE	PHOTOEMISSION YES/NO	BY	PROBLEM CURVE TRACER	PROBLEM BY SEM
1	500V	YES	GG* GH*	NONE	GATE EDGE BLOWN CONTACTS
2	500V	YES	GH	NONE	BLOWN CONTACTS GATE EDGE
3	500V	YES	GH	NONE	BLOWN CONTACT
4	1000V	NO		NONE	BLOWN CONTACTS
5	1000V	YES	GH	NONE	BLOWN CONTACT
6	1000V	YES	GH	NONE	GATE EDGE

*GG — Gate grounded
*GH — Gate high

Fig. 24 Results of photoemission testing after ESD stress using HBM.

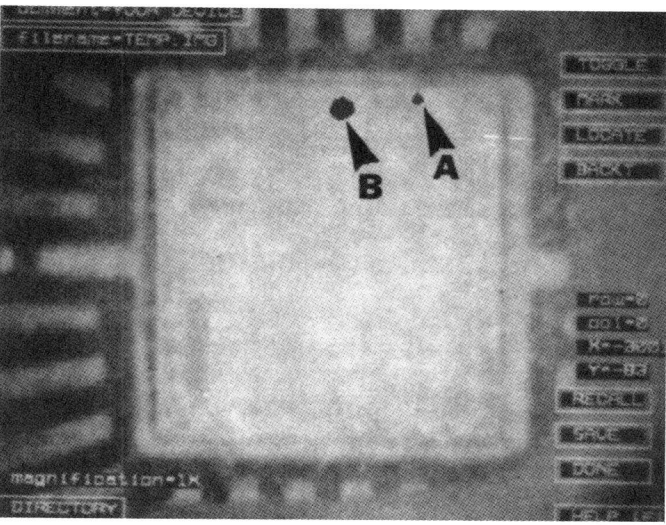

Fig. 26 Photoemission found at teh clock circuit and an internal node. The photoemission at B is not valid. The photoemission is due to latch-up.

Fig. 25 Gate oxide contamination found using the photoemission techniques.

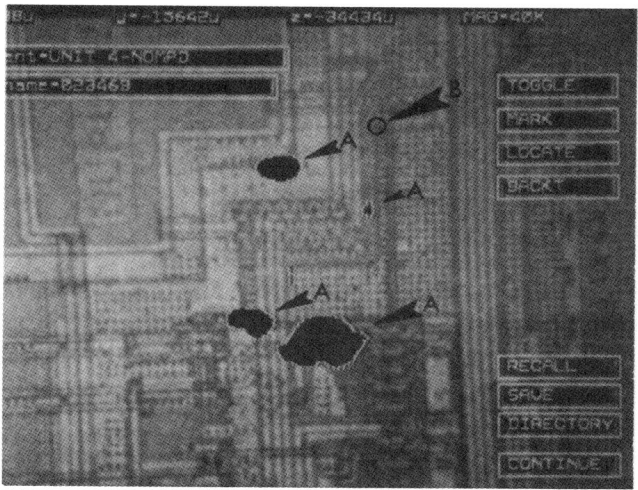

Fig. 27 Photoemission and liquid crystal testing did not locate the same defect site. In this case the defect is at the liquid crystal site.

dependent upon its parameters. Which transitor will be seen as the photoemitter transistor will depend upon the particular transistor parameters in the circuit being tested.

In Fig. 19, photoemission is seen on a microprocessor. The processor is being cycled through 14K vectors. The photoemission is being integrated over all test vectors.

The following figure, Fig. 20, shows a legal emission in a ring oscillator operating at 5 V and 30 MHz.

Gate oxide rupture can be seen in Fig. 21. The top image is the start of the ESD process. The middle image is during maximum current draw. The photoemission is due to current crowding in polysilicon. The bottom image has a white dot at the actual fail site.

The next image, Fig. 22, shows a long chain of flip-flops which photoemit. The defect is in the flip-flop to the right of the right-hand-most photoemission. Figure 22 shows how one defect can cause many photoemission sites. Figure 23 shows a closeup of the photoemitting sites. The transistors which are photoemitting are the n-channel.

ESD is a constant problem for VLSI devices. To determine the ESD sensitivity of a device, one test technique is the Human Body Model (HBM). The results of HBM testing have been varied and have been somewhat erratic. To locate the fail site, photoemission testing is very helpful. Figure 24 shows the results of some typical photoemission testing to determine the ESD fail site. The photoemission results are compared with the results of curve tracing. Notice that the curve tracer did not find any of the failing pins. Parametric testing, which is normally used as the indication of good ESD performance, did not indicate a problem. Photoemission testing did locate the fail site most of the time. In one case, the fail site could not be found parametrically or by photoemission.

Contamination in various oxides can cause photoemission. In Fig. 25, the photoemission was caused by the contamination in the gate oxide. The contamination is the black spot at the arrow.

When devices are not powered up correctly, the device may go into an unknown state. Figure 26 shows photoemission from a device which latched due to floating pins.

Complimentary techniques do not always point to the same location as the fail site. Figure 27 shows locations found by liquid crystal and photoemission. The locations do not match. In this case, the defect was at location B.

EOS and ESD can both cause photoemission. Oxide rupture, due to ESD, can cause photoemission, as shown in Fig. 28. In the image, the photoemission was caused by oxide breakdown in a gate oxide pin hole. Extreme EOS can cause massive damage. In the case of Fig. 29 and Fig. 30, the EOS event caused silicon migration as can be seen by the silicon filaments from the gate polysilicon to the substrate.

Conclusions

Photoemission microscopy is an effective tool when applied properly. When the device is powered up in a known state, the results are consistent and reliable. Care must be taken in interpretation of the cause of the photoemission. A photoemitting site can be generated by a defect which is on the opposite side of the die. The photoemission is always in the circuit which failed, not necessarily at the site of the photoemission. With proper care and complimentary techniques, the fail site can be located allowing deprocessing and root cause analysis for proper corrective actions.

Acknowledgments

Edward Isaac Cole, Jr.
Sandia National Laboratories
Failure Analysis, Division 2142
Albuquerque, New Mexico 87185

The authors would like to acknowledge Dr. Cole for contributing the figures and discussion of CMOS inverter photoemission.

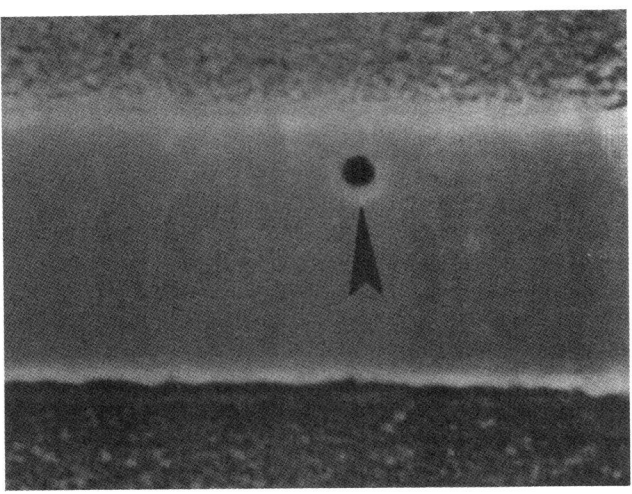

Fig. 28   Gate oxide pinhole found by photoemission microscopy.

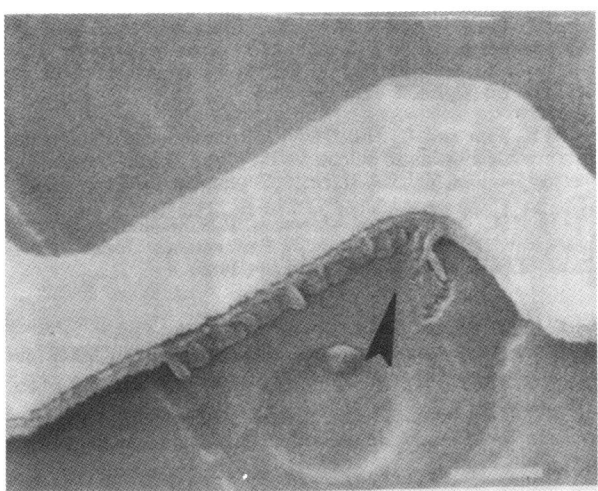

Fig. 29   Extreme EOS failure site found by photoemission.

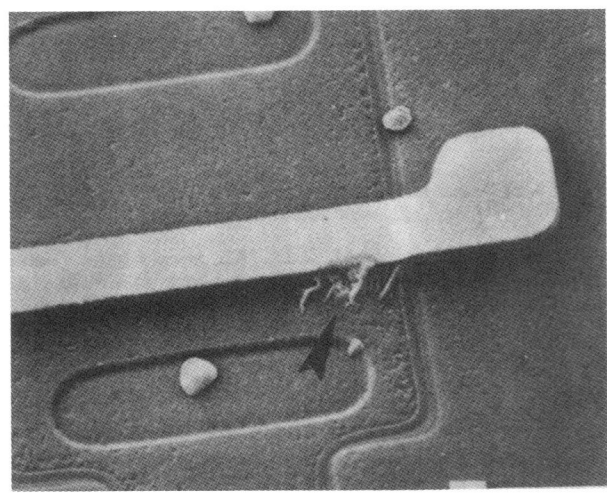

Fig. 30   Extreme EOS failure site found by photoemission.

References

1991

1. J. Sweeney, M. Phillips, P. Thomas, and V. Soorholtz, "Emission microscopic identification of nonuniform submicron transistors with hot carrier characteristics," Proc. of the Int. Symp for Test. and Fail. Analy., pp. 113-117, 1991.

2. D. Barton and M. Horry, "Case history: wafer level failure analysis of functional ICs with elevated IDDq caused by resistive gate oxide shorts," Proc. of the Int. Symp. for Test. and Fail. Analy., pp. 139-144, 1991.

3. A.N. Campbell, E.I. Cole, Jr., C.L. Henderson, and M.R. Taylor, "Case history: failure analysis of a CMOS SRAM with an intermittent open contact," Proc of the Int. Symp. for Test. and Fail. Analy., pp. 261-269, 1991.

4. K.S. Willis, D. Depaolis, and G. Billus, "Advanced photoemission technique for distinguishing latch-up from logic failures on CMOS devices," Proc. of the Int. Symp. for Test. and Fail. Analy., pp. 335-341, 1991.

5. J. Balog and M. Lin, "Failure analysis of degraded voltage regulator using light emission microscopy," Proc. of the Int. Symp. for Test. and Fail. Analy., pp. 343-352, 1991.

6. R. Mann and D. McElfresh, "Emission microscopy applied to optoelectronic emitter failure analysis," Proc. of the Int. Symp. for Test. and Fail. Analy., pp. 353-362, 1991.

7. S. Kiefer and M. Oyler, "Evaluation of the Gate defects in GaAs MESFETs by emission microscopy," Proc. of the Int. Symp. for Test. and Fail. Analy., pp. 363-368, 1991.

8. K. Symonds, M. Bahrami, and P. Skerry, "Functional failure analysis using photoemission microscopy," Proc. of the Int. Symp. for Test. and Fail. Analy., pp. 369-375, 1991.

9. K.S. Wills and E.I. Cole, Jr., "Photoemission microscopy," Microelectronic Failure Analysis Desk Reference, Supplement Two, AMS/ISTFA, Nov., 1991.

10. An Economical Approach to Correlation of Light Emission Microscopes, K.M. Baker, Proc. of the Int. Symp. for Test and Fail. Analy., 1991.

11. Characterization of Device Structure by Spectral Analysis of Photoemission, T. Wallinger, Proc. of the Int. Symp. for Test and Fail. Analy., 1991.

12. F. Magistrali, D. Sala, G. Salmini, F. Fantini, M. Giasante and M. Vanzi, Proc. IEEE Annual Int. Rel. Phys. Symposium IRPS-91, pp. 224-233, 1991.

13. F. Magistrali, D. Sala, G. Salmini, L. Zazzetti, F. Fantini, M. Giasante, and M. Vanzi, Proc. 2nd Europ. Symp. on Reliab. of Electron Devices, Failure Physics and Analysis ESREF-91.

14. K.S. Wills, et al., "Spectroscopic Photoemission Techniques to Understand Microprocessor Failures," Third International Symposium of the Physical and Failure Analysis of Integrated Circuits-Pan Pacific Singapore, 1991.

15. J. Kolzer, "Quantitative aspects of emission microscopy," Proc. CERT '91 (Components Engineering, Reliability and Test Conference), 1991, pp. 86-94.

16. Y. Uraoka, H. Yoshikawa, N. Tsutsu, and S. Akiyama, "Evaluation of Gate oxide reliability using luminescence method," Proc. IEEE Int. Conf. on Microelectronic Test Structures, Vol. 4, 1991, pp. 69-74.

17. J. Kolzer and J. Otto, "Scenario for electrical characterization of Megabit DRAMs: part II, Internal testing," IEEE Design and Test, December 1991.

18. Y. Uraoka, N. Tsutsu, Y. Nakata, and S. Akiyama, "Evaluation Technology of VLSI Reliability Using Hot Carrier Luminescence," IEEE Transactions on Semiconductor Manufacturing, Vol. 4, No. 3, Aug. 1991.

1990

19. N.C. Das and B.M. Arora, "Luminescence Spectra of an N-channel Metal-Oxide Semiconductor Field Effect Transistor and Breakdown," Applied Physics Letters, Vol. 56, No. 12, pp. 1152-1153, (19 March 1990).

20. C.F. Hawkins, J.M. Soden, E.I. Cole, and E. Snyder, "MOSFET Photon Emission for Analysis of CMOS ICs," ASM International Symposium for Testing and Failure Analysis, 1990.

21. C. Hawkins, J. Soden, E. Cole, and E. Snyder, "The use of light emission in failure analysis of CMOS ICs," Proc. of the Int. Symp. for Test. and Fail. Analy., pp. 55-67, 1990.

22. J. van der Pol and J. Koomen, "Relations between the hot carrier lifetime of transistors and CMOS SRAM products," Proc. of Int. Rel. Phys. Symp., pp. 178-185, 1990.

23. K. de Kort and P. Damink, "The spectroscopic signature of light emitted by integrated circuits," First ESREF Symp. (Italy), September, 1990.

24. J.T. May and G. Misakian, "Failure analysis of a turn-on degraded transistor using photoemission microscopy," Proc. of the Int. Symp. for Test. and Fail. Analy., pp. 69-71, 1990.

25. C. Khandekar, M. Hennis, P. Brownell, and D. Bethke, "Photoemission detection of vendor integrated circuit failures during system level functional testing," Proc. of the Int. Symp. for Test. and Fail. Analy., pp. 73-79, 1990.

26. C. Hawkins, J. Soden, E. Cole, and E. Snyder, "The Use of Light Emission in Failure Analysis of CMOS ICs," ISTFA '90, pp. 55-67.

27. N. Tsutsu, Y. Uraoka, Y. Nakata, S. Akiyama, and H. Esaki, "New Detection Method of Hot Carrier Degradation Using Photon Spectrum Analysis of Weak Luminescence on CMOS VLSI," IEEE ICMTS, vol. 3, no. 1, p. 143, 1990.

28. F. Magistrali, D. Sala, M. Vanzi, and L. Battistella, Proc. Int. Symp. on Testing and Failure Analysis, ISTFA90, Los Angeles, pp. 291-299, 1990.

29. K. de Kort and P. Damink, "The Spectroscopic Signature of Light Emitted by Integrated Circuits," Proceedings of the Workshop on Emission Microscopy, 1990, pp. 45-52.

30. N. Tsutsu et al., "New detection method of hot-carrier degradation using photon spectrum analysis of weak luminescence on CMOS/VLSI," IEEE ICMTS, Vol. 3, No. 1, p. 143, 1990.

31. B. Dosneau, R. Michel, and P. Plougonven, ESREF 90 proceeding, p. 101.

32. F.M. Roche, S.D. Bocus, B. Pistoulet, Analyse de Construction et de Defaillance, Atliere de la SEE, Cargese (1990).

33. ESREF '90, First European Symposium on Reliability of Electron Devices, "Failure Physics and Analysis," Technopolis (Bari) Italy, 2-5 Oct. 1990.

34. H. Ishizuka, M. Tanaka, H. Konishi, and H. Ishida, "Advanced method of failure analysis using photon spectrum of emission microscopy," Proc. ISTFA '90, 1990, pp. 13-19.

35. G.F. Shade, "Physical mechanisms for light emission microscopy," Proc. ISTFA '90, 1990, pp. 121-128.

36. E. Zanoni, S. Bigliardi, R. Cappelletti, P. Lugli, F. Magistrali, M. Manfredi, A. Paccagnella, N. Testa, and C. Canali, "Light emission

36. in AlGaAs/GaAs HEMT's and GaAs MESFET's induced by hot carriers," IEEE Electron Device Letters, Vol. 11, 1990, pp. 487-489.

37. M. Oyler and S. Cohen, "Gate-drain breakdown mapping in gallium arsenide power MESFET's by emission microscopy," Proc. ESREF '90, 1990, pp. 85-92.

38. A. Dallman and G. Deboy, "Characterization of Trench-Trench Punch-Through Mechanisms by Emission Microscopy," ESREF 90, 1990, p 69-76.

39. D. Weinmann, C. Boit, and J. Kolzer, "Characterization of leakage currents by emission microscopy," Proc. ESREF '90, 1990, pp. 61-68.

40. C. Boit, J. Kolzer, C. Stein, J. Otto, H. Benzinger, and M. Keritmaier, proc. CERT '90 (Component Engineering, Reliability, and Test Conference), 1990, p. 110-114.

41. C.L. Chiang, "Light emitting phenomenon in Si technology and its applications," Proc. of the VLSI Int. Education Workshop and Symposium (VIEWS), 1990, VLSI Technology Inc.

42. C. Boit, J. Kolzer, H. Benzinger, A. Dallman, M. Herzog, and J. Quincke, "Discrimination of parasitic bipolar operating modes in ICs with emission microscopy," Proc. IEEE/IRPS '90, 1990, pp. 81-85.

43. J. Lin, S. Banerjee, J. Lee, and C. Teng, "Soft Breakdown in Titanium-Silicided Shallow Source/Drain Junctions," IEEE Electron Device Letters, Vol. 11, No. 5, pp. 191-193, May 1990.

44. O. Sirich, J. Kolzer, and C. Boit, "Emission microscopy on integrated bipolar transistors," Proc. ESREF '90, 1990, pp. 93-100.

45. M. Hannelamm and A. Amerasekera, "Photon emission as a tool for ESD failure localization and as a technique for studying ESD phenomena," Proc. ESREF '90, 1990, pp. 77-84.

46. S. Wolfe, *Silicon Processing for the VLSI Era.*, Vol. 2, Sunset Beach, CA, Lattice Press, 1990.

47. Y. Hiruma and E. Inuzuka, "Applications of photon counting imaging to semiconductor failure analysis," Proc. ESREF '90, 1990, pp. 53-60.

1989

48. K.S. Willis, S. Vaughan, Charvaka Duvvury, O. Adams, and J. Bartlett, "Photoemission Testing for EOS/ESD Failures in VLSI Devices: Advantages and Limitations," ASM International Symposium for Testing and Failure Analysis, p. 183, 1989.

49. T.S. Taylor, T. Dao, T.B. Haddock, and Lawrence C. Wagner, "Leakage Detection Techniques: A Comparative Study," ASM International Symposium for Testing and Failure Analysis, pp. 5-13, 1989.

50. R. Mann and W. Chim, "Failure Characterization of Light Output Degradation in Copper Contaminated LEDs," *IPFA Proceedings*, pp. 145-157, 1989.

51. Y. Uraoka et al., "Evaluation technique of gate oxide reliability with electrical and optical measurements," in IEEE ICMTS, Vol. 2, No. 1, p. 97, 1989.

52. C. Canali, F. Corsi, M. Muschitiello, E. Zanoni, IEEE Trans. Elec. Dev., Vol. 36, No. 5, (1989).

53. T. Aoki and A. Yoshii, "Analysis of Latch-up-induced photoemission," Proc. IEEE/IEDM '89, 1989, pp. 281-284.

54. A. Toriumi, Experimental study of hot carriers in small size Si-MOSFETs," Solid-State Electronics, Vol. 32, 1989, pp. 1519-1525.

55. D.J. DiMaria, and J.W. Stasiak, "Trap creation in silicon dioxide produced by hot electrons," J. Appl. Phys., Vol. 65, 1989.

56. N. Khurana, "Second generation emission microscopy and its application," Proc. ISTFA '89, 1989, pp. 277-283.

57. R. Lemme, M. Gentsch, R. Kutzner, H. Wendt, and H. Haudek, "Defect analysis of VLSI dynamic memories," Proc. ISTFA '89, 1989, pp. 9-14.

58. C.G.C. de Kort, "Integrated circuit diagnostic tools: Underlying physics and applications," Phillips J. Res., Vol. 44, 1989, pp. 295-327.

1988

59. E. Hackbarth and D. Tang, "Inherent and Stress-Induced Leakage in Heavily Doped Silicon Junctions," IEEE Transactions on Electron Devices, Vol. 35, No. 12, pp. 2108-2118, December 1988.

60. S.C. Lim and E.G. Tan, "Detection of Junction Spiking and its Induced Latchup by Emission Microscopy," IEEE Annual Proc. of the Reliability Physics Symposium, pp. 119-125, 1988.

61. K.S. Wills, C. Duvvury, and O. Adams, "Photoemission testing for ESD failures, advantages and limitations," Proc. of Electrical Overstress/Electrostatic Discharge Symp., pp. 53-61, September, 1988.

62. S.N. Chu, S. Nakahara, M.E. Twigg, L.A. Koszi, E.J. Flynn, A.K. Chin, B.P. Segner and W.D. Johnson, Jr., J. Appl. Phys., Vol. 63, No. 3, 1 February 1988, pp. 611-623.

63. C. Canali, M. Giannini, A. Scorzoni, M. Vanzi, and E. Zanoni, IEEE Journal of Solid State Circuits, Vol. 23, No. 2, (1988).

64. K. Penner, "Electroluminescence from silicon devices—a tool for device and material characterization," Journal de Physique, Coll. c4, Suppl. 9, Tome 49, 1988, pp. 797-800.

65. M. Herzog and F. Koch, "Hot-carrier light emission from silicon metal-oxide semiconductor devices," Appl. Phys. Lett., Vol. 53, 1988, pp. 2620-2622.

1987

66. N. Khurana and C. Chiang, "Dynamic imaging of current conduction in dielectric films by emission microscopy," Proc. of Int. Rel. Phys. Symp., pp. 72-76, April 1987.

67. A. Toriumi, M. Yosimi, M. Iwase, Y. Akiyama, and K. Taniguchi, "A study of photo emission from n-channel MOSFET's," IEEE Trans. on Electron Dev., Vol. ED-34, No. 7, pp. 1501-1508, July, 1987.

68. D.L. Cook *et al.*, "Techniques of evaluating long term oxide reliability at wafer level," IEEE Int. Electron Devices Meeting, Washington, D.C., 1987, p. 444.

69. G. Auvert, Proceedings of Microsc. Semicond. Mater. Conf., pp. 563-572 (1987).

70. J.C.S. Woo, J.D. Plummer, and J.M.C. Stork, "Non-Ideal Base Current in Bipolar Transistors at Low Temperatures," IEEE Transactions on Electron Devices, Vol. ED-34, No. 1, p. 130, 1987.

1986

71. R.A. McPhee, C. Duvvury, R.N. Rountree, and H. Domingos, "Thick Oxide Device ESD Performance Under Process Variations," EOS/ESD Symp. Proceedings, pp. 173-181, 1986.

72. N. Khurana and C.L. Chiang, "Analysis of Product Hot Electron Problems by Gated Emission Microscopy," IEEE Annual Proc. of the Reliability Physics Symp., pp. 189-194, 1986.

73. C.L. Chiang and N. Khurana, "Imaging and Detection of Current Conduction in Dielectric Films by Emission Microscopy," IEEE International Electron Devices Meeting, pp. 672-675, 1986.

74. I.C. Chen, S. Holland, and C. Hu, "Oxide Breakdown Dependence on Thickness and Hole Current-Enhanced Reliability of Ultrathin Oxides," IEEE International Electron Device meeting, pp. 650-663, 1986.

75. L. Manchanda, "Hot Electron Trapping and Generic Reliability of p+ Polysilicon/SiO2/Si Structures for Fine-Line CMOS Technology," IEEE Annual Proc. of the Reliability Physics Symp., p. 193, 1986.

76. A.R. Leblanc and W.W. Abadeer, "Behavior of SiO2 Under High Electric Field/Current Stress Conditions," IEEE Annual Proc. of the Reliability Physics Symp., pp. 230-234, 1986.

77. C. Hawkins and J. Soden, "Reliability and electrical properties of gate oxide shorts in CMOS ICs," Int. Test. Conf., pp. 443-451, 1986.

78. "Live TV Pictures of transistor failure," Semiconductor Int., p. 30, March 1986.

1985

79. T. Tsuchiya and S. Nakajima, "Emission Mechanism and Bias-Dependent Emission Efficiency of Photons Induced by Drain Avalanche in Si MOFSET's," IEEE Transactions on Electron Devices, Vol. ED-32, No. 2, Feb. 1985, pp. 405-412.

80. L. Solymar and D. Walsh, Lectures on the Electrical Properties of Materials, Oxford University Press, 1985, pp. 113.

81. C. Hu, S.C. Tam, F.C. Hsu, P.K. Ko, T.Y. Chan, and K.W. Terrill, "Hot-Electron-Induced MOSFET Degradation—Model, Monitor, and Improvement," IEEE Transactions on Electron Devices, Vol. ED-32, No. 2, pp. 375-385, 1985.

82. A. Toriumi, M. Yoshimi, M. Iwase, and K. Taniguchi, "Experimental Determination of Hot Carrier Energy Distribution and Minority Carrier Generation Mechanism Due to Hot-Carrier Effects," IEEE International Electron Devices Meeting, pp. 56-59, 1985.

83. M.J.J. Theunissen and F.J. List, "Analysis of the Soft Reverse Characteristics of n+/p drain diodes," Solid-State Electronics, Vol. 28, p. 417, 1985.

84. N. Das and W. Khokle, "Visible light emission from silicon MOSFETs," Solid-State Electronics, Vol. 28, No. 10, pp. 967-977, 1985.

85. C. Hawkins and J. Soden, "Electrical characteristics and testing considerations for gate oxide shorts in CMOS ICs," Int. Test Conf., pp. 544-555, 1985.

86. K.L. Chen et al., "Reliability effects on MOS transistor due to hot-carrier injection," IEEE Trans. Electron Devices, Vol. EDL-5, p. 386, 1985.

87. N. Khurana, T. Maloney, and W. Yeh, "ESD on CHMOS devices, equivalent circuits, physical models and failure mechanisms," Proc. IEEE/IRPS '85, 1985, pp. 212-223.

88. S.D. Borson, D.J. Di Maria, M.V. Fischetti, F.L. Pesavento, P.M. Solomon, and D.W. Dong, "Direct measurement of the energy distribution of hot electrons in silicon dioxide," J. Appl. Phys. Vol. 58, 1985.

1984

89. S. Tam and C. Hu, "Hot-Electron-Induced Photon and Photocarrier Generation in Silicon MOSFET's, IEEE Transactions on Electron Devices, Vol. ED-31, No. 9, Sept. 1984, pp. 1264-1273.

90. N. Khurana, "Pulsed infra-red microscopy for debugging latch-up on CMOS products," Proc. IEEE/IRPS, 1984, pp. 122-127.

1983

91. J.M. Soden, H.D. Stewart, and R.A. Patorek, "ESD Evaluation of Radiation-Hardened High Reliability CMOS and NMOS ICs," EOS/ESD Symp. Proceedings, pp. 134-136, 1983.

92. T.C. Ong, K.W. Terrill, S. Tam, and C. Hu, "Photon Generation in Forward-Biased Silicon p-n Junctions," IEEE Electron Device Letters, Vol. EDL-4, No. 12, pp. 460-462, December, 1983.

93. T.N. Theis, J.R. Kirtley, D.J. DiMaria, and D.W. Dong, "Spectroscopic Studies of Electronic Conduction of SiO2," in *Insulating Films on Semiconductors*, J.F. Verweij and D.R. Wolters (ed.), North-Holland, pp. 134-140, 1983.

94. S. Tam, F. Hsu, P. Ko, C. Hu, and R. Muller, "Spatially resolved observation of visible-light emission from Si MOSFET's," IEEE Electron Dev. Letters, Vol. EDL-4, No. 10, pp. 386-388, October 1983.

95. P.A. Childs, R.A. Stuart, and W. Eccleston, "Evidence of optical generation of minority carriers from saturated MOS transistors," Solid-State Electronics, Vol. 26, No. 7, pp. 685-688, 1983.

96. E. Takeda et al., "An empirical model for device degradation due to hot-carrier injection," IEEE Electron Device Lett., vol. EDL-4, p. 111, 1983.

97. C. Hu et al., "Hot-electron effects in MOSFETs," IEDM Tec. Dig., p. 176, 1983.

98. F. Henley, M. Chi, and W. Oldham, IEEE 21st Proceedings of IRPS, pp. 122-129 (1983).

1982

99. J.R. David, J.E. Stictch, and M.S. Stern, "Gate-Drain Avalanche Breakdown in GaAs Power MESFETs," IEEE Trans. Electron Devices, Vol. ED-29, pp. 1548-1552, Oct. 1982.

1980

100. R.L. Sproull and W.A. Philips, Modern Physics, John Wiley and Sons, 1980, pp. 471.

1970's

101. R.S. Muller and T.I. Kamins, Device Electronics for Integrated Circuits, John Wiley and Sons, 1977, pp. 308.

102. P. Solomon and N. Klein, "Electroluminescence at High Electric Fields in Silicon Dioxide," J. Appl. Phys., Vol. 47, pp. 1023-1026, March, 1976.

103. A. Bergh and P. Dean, *Light-Emitting Diodes*, Oxford: Clarendon Press, 1976.

104. P.M. Petroff and L.C. Kimmerling, "Dislocation Climb Model in Compound Semiconductors with Zinc Blende Structure," *Applied Physics Letter*, 29, pp. 461-463, 1976.

105. J.I. Pankove, *Optical Processes in Semiconductors*, Dover Publications, Inc., 1975.

106. W. Haecker, "Infrared radiation from breakdown plasmas in Si, GaSb, and Ge: Evidence for direct free hole radiation," Phys. State Sol., Vol. 25, 1974, pp. 301-310.

107. D.V. Lang and L.C. Kimmerling, "Observations of recombination enhanced defect reactions in Semiconductors," *Physical Review Letters*, 33, pp. 489-492, 1974.

108. A. Bahraman and W.G. Oldham, "Role of Copper in Degradation of GaAs Electroluminescent Diodes," *J. Appl. Phys.*, 43, pp. 2383-2387, 1972.

109. B.G. Streetman, *Solid State Electronic Devices*, New Jersey, Prentice Hall, p. 195, 1972.

1960's

110. H. Kressel, "A review of the effect of imperfections on the electrical breakdown in silicon p-n junctions," RCA Rev., Vol. XXVIII, No. 1, p. 181, March 1967.

111. S. Wang, *Solid State Electronics*, New York, McGraw-Hill, 1966.

112. M. Cardona and F.H. Pollak, Phys. Rev., Vol. 142, p. 530, 1966.

113. D.H. Navon, "Semiconductor Micro devices and Materials," Holt, Reinhart, & Winston, 1966, pp. 142-151.

114. J. Shewchun and L.Y. Wei, "Mechanism for Reverse-biased Breakdown Radiation in p-n Junctions," Solid-State Electronics, Vol. 8, 1964, pp. 485.

115. E. Kamieniecki, "Hot Carriers in Microplasmas and their Radiation in Germanium and Silicon," Physical Status Solidi, Vol. 6, 1964, pp. 877.

116. A. Goetzberger, B. McDonald, R. Haitz, and R. Scarlett, "Avalanche effects in silicon p-n junctions II. Structurally perfect junctions," J. Appl.Phys., Vol. 34, p. 1591, June 1963

117. T. Figelski and A. Torun, "On the Origin of Light Emitted from Reverse Biased p-n Junctions," Proceedings of the International Conference in the Physics of Semiconductors, 1962, pp. 853.

118. L.W. Davies and A.R. Storm, Jr., "Recombination radiation from silicon under strong-field conditions," Phys. Rev., Vol. 121, 1961, pp. 381-387.

119. K. Kikuchi, "Visible light emission and microplasma phenomena in silicon p-n junctions," J. Phys. Soc. Japan, Vol. 15, p. 1822, October, 1960.

120. P.A. Wolff, J. Phys. Chem. Solids, 16, p. 184, 1960.

1950's

121. A. Chynoweth and G. Pearson, "Effect of dislocations on breakdown in silicon p-n junctions," J. Appl. Phys., Vol. 29, p. 1103, July 1958.

122. A.G. Chynoweth *et al.*, "Photon emission from avalanche breakdown in silicon," Phys. Rev., Vol. 102, No. 2, p. 369, 1956.

123. R. Newmann, Dash, Hall, and Burch, "Visible Light From Si p-n Junctions," Phys. Rev., Vol. 100, p. 700, 1955.

124. W. van Roosbroeck and W. Shockley, "Photon-radiative Recombination of electrons and holes in germanium," Phys. Rev., Vol, 94, 1954, pp. 1558-1560.

125. W. Shockley and W.T. Read, Phys. Rev., Vol. 87, p. 835, 1952.

126. R.N. Hall, Phys. Rev., Vol. 83, p. 228, 1951 and Vol. 87, p. 387, 1952.

Analytical Instrumentation

INSTRUMENTAL ANALYSIS TECHNIQUES

by Keenan Evans and Thomas A. Anderson

INTRODUCTION

In many instances it may be necessary to include detailed chemical characterization of a sample into the failure analysis report. The failure analyst may be required to gather this data himself or it may be necessary to send the sample to an outside laboratory for characterization by a variety of sophisticated instrumental analytical techniques. The following summary of ten common analytical methods is designed to provide both the analyst and the report recipient with a basic understanding of the various methods. The summaries are presented in outline form with a description of the Theory of Operation and Typical Applications provided for each of the techniques listed below.

A. Auger Electron Spectroscopy, AES

B. Scanning Electron Microscopy, SEM

C. Secondary Ion Mass Spectrometry, SIMS

D. Inductively Coupled Plasma-Atomic Emission Spectroscopy, ICP-AES

E. Residual Gas Analysis-Mass Spectrometry, RGA-MS

F. X-Ray Fluorescence, XRF

G. Gas Chromatography/Mass Spectrometry, GC/MS

H. Ion Chromatography, IC

I. High Pressure Liquid Chromatography, HPLC

J. Fourier Transform InfraRed Spectroscopy, FTIR

Instrumental Analytical Techniques

Auger Electron Spectroscopy, AES

Theory of Operation. Atomic core ionization of nuclides within the sample occurs as a result of interactions with the electron beam. As a result of this inner core ionization, an electron from a higher energy level (shell) within the ionized atom drops down in energy to fill the vacancy. The quantized energy difference between the two levels is then given up via either X-ray photon emission or via the radiationless Auger process yielding the *Auger electron*. (See Fig. 1.) Energy analysis of the emitted Auger electron provides elemental identification, and counting of the relative numbers of Auger electrons at the various energies provides semi-quantitative information. The technique is surface sensitive due to the limited escape depth of the Auger electrons, typically on the order of 3-30 Angstroms.

To provide information about subsurface composition and allow depth profiling capabilities, the surface is removed via in-situ ion milling (sputter etching) with an argon ion beam. Scanning or rastering of the electron probe beam also allows for elemental mapping capabilities. An ultra-high vacuum (10^{-10} Torr) in the analytical chamber is necessary in order to prevent re-adsorption of sputtered species or permanent gases onto the freshly sputtered sample surface.

Typical Applications. Auger analysis is used to characterize surface contaminants which may inhibit bondability and solderability, contribute to surface leakage or constitute visual rejects. It is also used for characterization of multi-layer structures to monitor the degree of interlayer blend or detect contaminants at various interfaces. Comparative analyses of "good" vs "bad" samples are common. The sample must be sized to a maximum of approximately 1 cm × 1 cm. Samples should not be coated. Due to charge build up at the surface from electron beam bombardment, analysis of thick insulators is limited. Auger electron spectroscopy should be considered a destructive analysis technique. Figures 2 & 3 illustrate a typical Auger survey spectrum and a depth profile.

Scanning Electron Microscopy, SEM

Theory of Operation. Under high vacuum (approximately 10^{-6} Torr), a 1 to 30 keV primary electron beam is rastered over the sample surface, creating secondary electrons. These are extracted from the sample and imaged to create a high resolution, high depth of field, secondary electron image at magnifications to 250k ×. (See Fig. 4.) Interactions of the sample atoms with the primary electron beam also result in inner core ionization of the atoms with the subsequent emission of quantized X-ray photon. Wavelength or energy analysis of the emitted X-rays provides qualitative elemental information and the relative intensity of these X-rays is used for quantitative purposes. In semiconductor materials, electron beam bombardment also creates electron-hole pairs, resulting in an electron beam induced current (EBIC), which can be imaged to correlate with

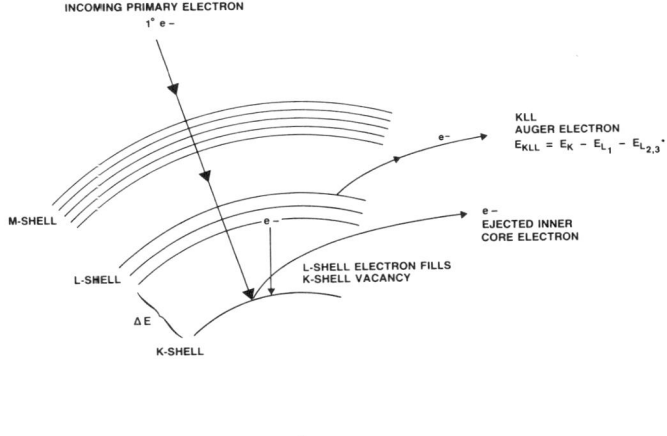

Fig. 1. Schematic representation of the Auger process.

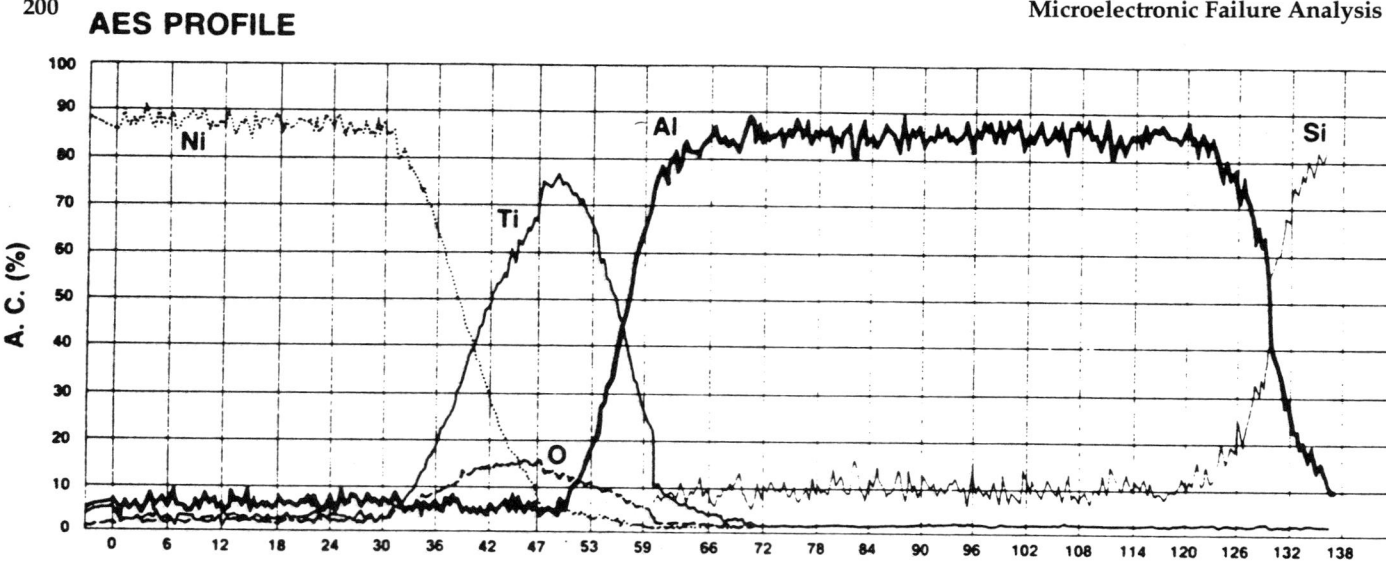

Fig. 2. Auger sputter depth profile through backmetal system of Si chip. Care should be used in the interpretation of A.C.% data since some quantification routines can result in false indication of elements at low AT%. Interface resolution can be improved by rotating the sample during ion milling. The determination of layer thickness from sputter time requires accurate sputter rates for each layer under identical experimental conditions. A surface profilometer may be used in conjunction with selective layer removal to measure true layer thickness.

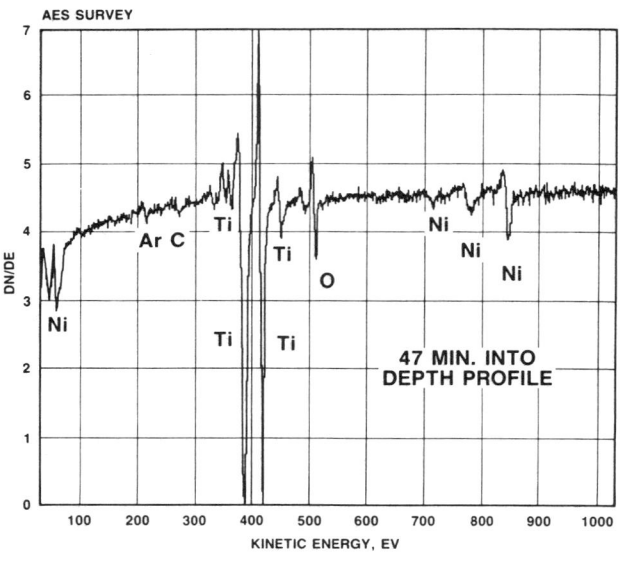

Fig. 3. Auger survey spectrum taken midway through a depth profile.

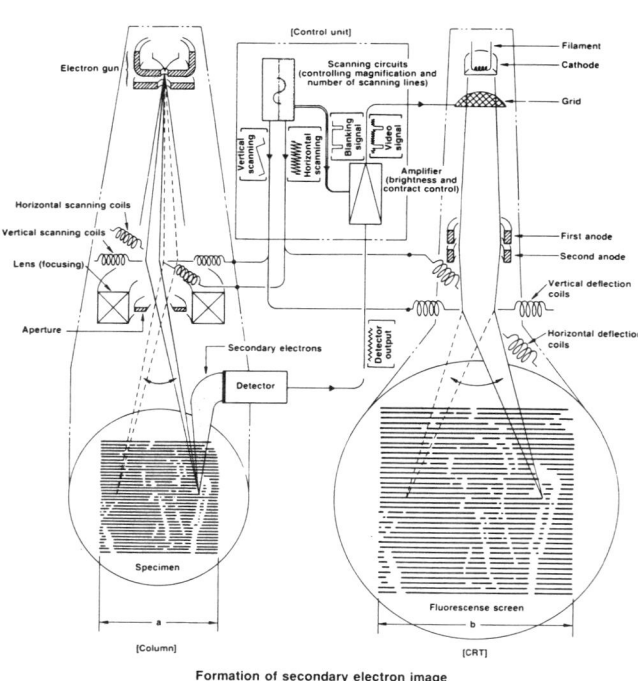

Fig. 4. Schematic of a typical scanning electron microscope.

other sample features. Various stains and sample preparation techniques are utilized to enhance contrast between regions of interest within the sample. X-ray emission from the sample and primary electron beam penetration into the sample typically range from depths on the order of 0.5-2.0 μm, varying as a function of electron beam energy and sample density. Figure 5 gives range of penetration for typical electron beam energies in Si and GaAs, and Fig. 6 illustrates depths of quanta generation.

Typical Applications. SEMs are used in the characterization of microstructural surface topography, step coverage, grain size, oxide slope, construction parameters, sample composition, contamination, structural defects, bonding defects, intermetallic formation and degree of wire bond deformation. Digital X-ray maps of up to eight elements may be

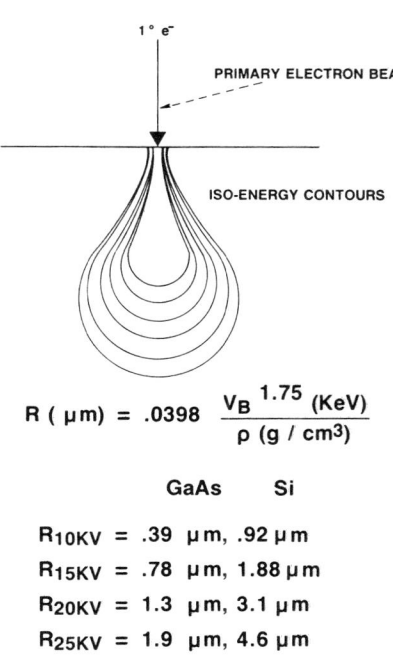

$$R\ (\mu m) = .0398\ \frac{V_B^{1.75}\ (KeV)}{\rho\ (g/cm^3)}$$

	GaAs	Si
R_{10KV} =	.39 μm,	.92 μm
R_{15KV} =	.78 μm,	1.88 μm
R_{20KV} =	1.3 μm,	3.1 μm
R_{25KV} =	1.9 μm,	4.6 μm

Fig. 5. Electron beam ranges in Si and GaAs. (Drawing adapted from Goldstein 1, 2, 3).

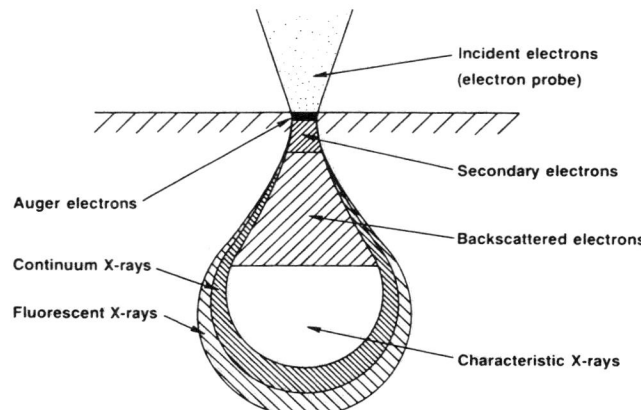

Fig. 6. Analytical depth of electron beam generated signals. (Drawing adapted from Goldstein 1, 2, 3).

generated simultaneously to document the geometrical distribution of elemental composition or contamination.

Secondary Ion Mass Spectrometry, SIMS

Theory of Operation. The basis of ion microprobe SIMS is the use of a focused primary ion beam to erode atoms from a selected region of a sample surface.

A portion of the eroded atoms undergo a charge exchange in the near surface environment, resulting in their conversion to negative or positive ions. These secondary ions are then extracted via an electrical potential and subsequently analyzed by a mass spectrometer. SIMS thus yields an analytical technique capable of elemental specificity, with detection limits in the range of ppm-ppb (atomic concentration). Figure 7 illustrates the ion beam-sample interactions and Fig. 8 displays a typical instrumental arrangement of the primary ion source and the mass analyzer. By comparison of relative ion yields in sample depth profiles to those of ion implanted reference samples in the same matrix, an approximate quantification is possible. The current density of the primary ion beam may be varied over several orders of magnitude to produce sample milling rates ranging from a few Å/min to several thousand Å/min, making the technique a powerful surface analysis, bulk analysis, thin film characterization and depth profiling tool. Minimum ion beam spot sizes are on the order of 2-5μm, thus limiting the lateral imaging resolution achievable with the primary ion sources to similar dimensions. However, SIMS is not useful for small areas. Data must be taken over much larger plastered areas in order to obtain any reasonable detection limit. Some instruments

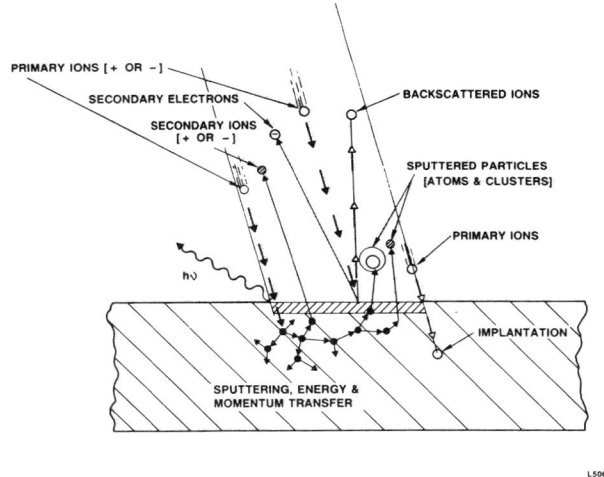

Fig. 7. Ion Beam / Sample interactions.

can analyze small areas for larger concentrations, such as in particle analysis, but dopant concentrations cannot be measured from an area of the same order of dimension as the spot site.

Typical Applications. SIMS is used to evaluate dopant profiling and trace contamination of surfaces, thin films, thick films, multilayer structures, and interfaces. It is used to measure relative levels of incorporated impurities or component elements as a function of processing parameters, correlation of trace ionic contaminant levels to electrical leakages, and mapping of impurities for correlation to device geometrical features. Samples must be compatible with the ultra-high vacuum

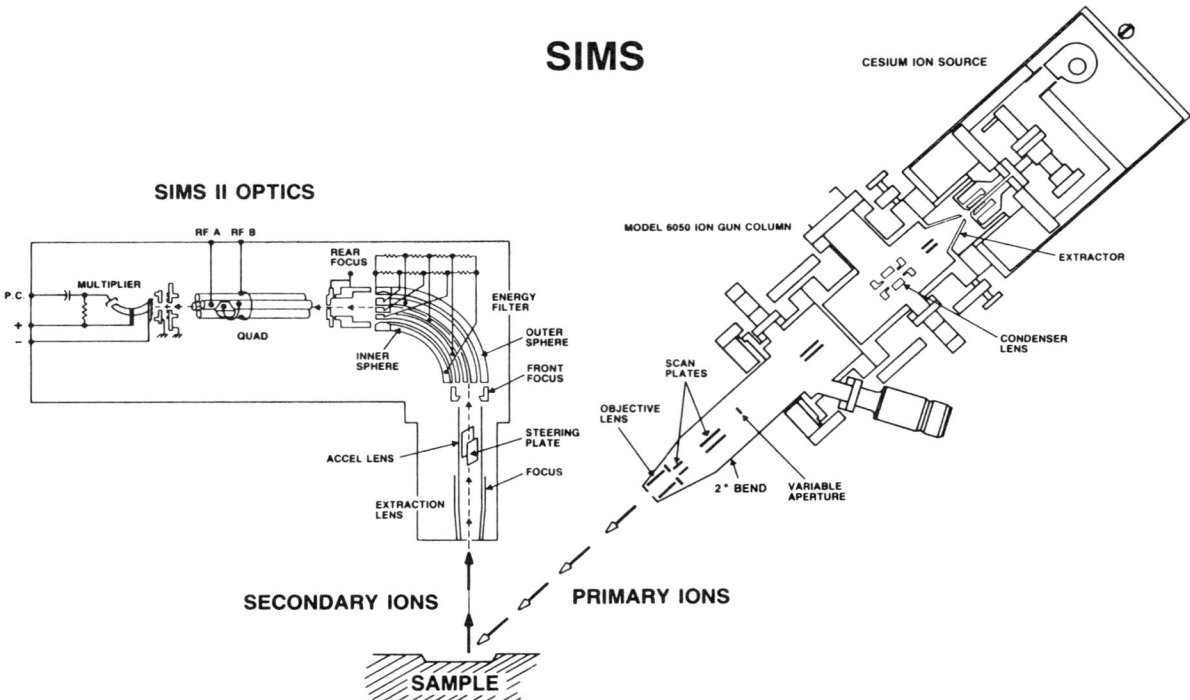

Fig. 8. Typical arrangement of a quadruple ion microprobe.

(approximately 10^{-10} Torr) of the SIMS analytical chamber that is required to prevent reabsorption of sputtered species and permanent gases onto freshly ion milled sample surfaces.

Inductively Coupled Plasma-Atomic Emission Spectroscopy, ICP-AES

Theory of Operation. Three concentric quartz tubes are centered in the water cooled coils of an RF generator. An argon gas flow is introduced into the tubes and initially made electrically conductive by Tesla sparks. The magnetic field of the RF coil is axially oriented in the quartz tubes thus inducing an annular flow of electric current in the argon plasma, as shown in Fig. 9. This produces temperatures in the range of 5,000-10,000 Kelvin as a result of ohmic heating. Proper regulation of the argon gas flows allows a stable annular plasma to be maintained above the ends of the concentric tubes. A liquid sample is aspirated into the center tube, and it flows into the plasma. The complete plasma torch configuration is shown in Fig. 10. Solvent evaporates prior to reaching the actual plasma and atomization of the elements present in the sample solution occurs in the extremely high temperature of the plasma. Thermal excitation of valence electrons to higher allowed energy states takes place. When relaxation to the ground state occurs the energy difference between the excited state and the ground state is emitted as a quantized photon. The photon wavelengths emitted are commonly visible in the ultra-violet portion of the electromagnetic spectrum (UV-VIS). By separating the various emission wavelengths with a monochromator, qualitative information is gathered. By monitoring the relative intensity at a particular wavelength and comparing this intensity to that generated by introduction of a known external standard into the plasma, quantitation is achieved. Detection limits in solution are on the order of ppb for most elements. A typical instrumental arrangement is shown in Fig. 11.

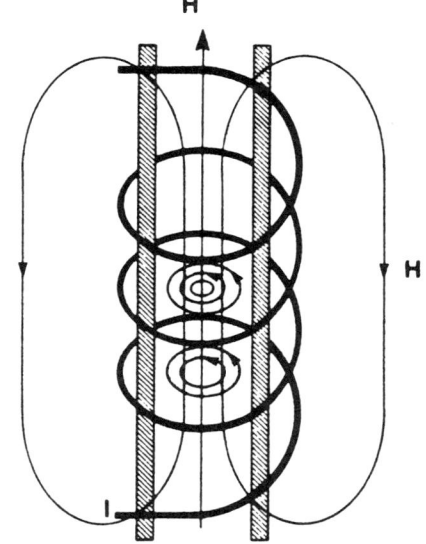

Fig. 9. Magnetic and electric field lines in a plasma torch.

Instrumental Analysis Techniques

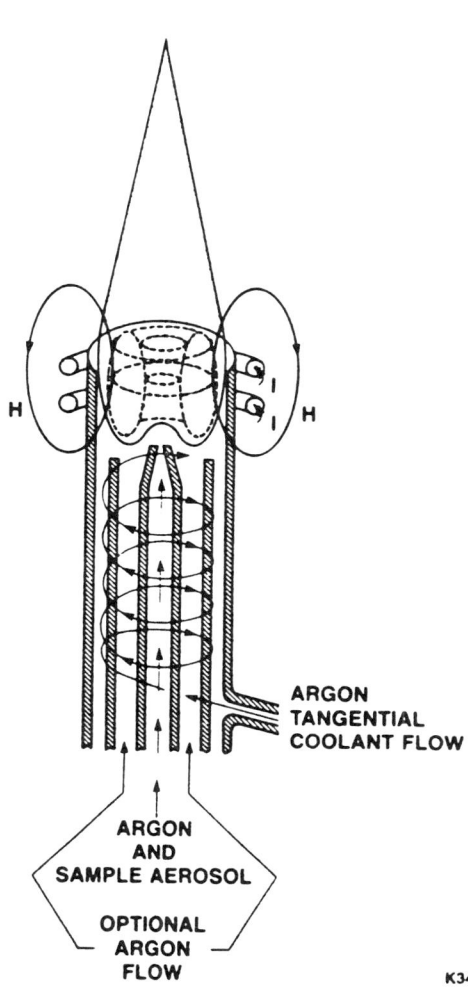

Fig. 10. Complete plasma configuration for an ICP-AES system.

Typical Applications. Trace metals in process solutions and waters, and concentration of major, minor and trace elements in alloys, thin films, solders, residues or other samples which can be digested in an appropriate solvent or mineral acid, are analyzed by this method.

Residual Gas Analysis-Mass Spectrometry, RGA-MS

Theory of Operation. Hermetically sealed packages are loaded into a sample chamber which is then evacuated to a base pressure of approximately 10^{-7} mbar. The samples are allowed to reach thermal equilibrium in the heated chamber. Then, a single sample is rotated into alignment with the package piercing assembly. A seal is made between the package lid and the vacuum system of the mass spectrometer. After an acceptable mass spectrometer background is reached, the package is pierced and the internal atmosphere of the hermetic device is sampled into the mass spectrometer for analysis. Up to 16 gas species which may have been trapped inside the package at sealing may be analyzed. Moisture levels are the primary concern. For quantitation of internal moisture levels, relative intensities of the mass

THE COMBINATION SYSTEM

A. ARGON GAS AT 12 TO 18 LITERS PER MIN.
B. SAMPLE, APPROX. 1ML/M
C. OPTIONAL TEFLON NEBULIZER SYSTEMS FOR HF SOLUTIONS AND SLURRIES
D. PLASMA 2.5 KW
E. THE FUNCTIONAL PATH PROCEEDS AS SHOWN ABOVE EITHER THROUGH THE SEQUENTIAL, THROUGH THE SIMULTANEOUS OR THROUGH BOTH SYSTEMS.

JY-86P ICP-AES
COMBINATION SYSTEM

Fig. 11. Component layout for a dual detection ICP-AES system.

18 ion current resulting from the sample are compared to those of known concentrations of moisture which have been previously generated through an NBS traceable moisture hygrometer, moisture generator and four volume calibrator.

Typical Applications. RGA is used for process monitoring of internal package atmospheres to ensure minimum internal moisture levels. It is also used for tank gas analysis. RGA may be used as a measurement device for optimization of die coat processing parameters to minimize internal package moisture, oxygen and hydrocarbon levels.

X-ray Fluorescence, XRF

Theory of Operation. In an X-ray tube, electrons are generated from a cathode and accelerated at approximately 5-50 keV to impinge on the target anode thus generating broad band X-ray emission from the deceleration of the electrons (Bremsstrahlung) in the anode target material. Characteristic fluorescent X-rays also are generated from the target material due to atomic core ionization of atoms. The X-rays from the primary source are then directed on to the sample, causing atomic core ionization of the atoms. As a result of this inner core ionization, an electron from a higher energy level (shell) within the ionized nuclide drops down in energy to fill the vacancy. The quantized energy difference between the two levels is then given up via X-ray photon emission. Energy analysis of the secondary X-ray emission (or fluorescence) yields qualitative information about the elemental composition of the sample. Relative intensity of the X-ray fluorescence is used to quantify the amounts of the various species present in the sample. Detection limits are of the order of a few hundred ppm, and all elements above atomic number 11 (Na) may be detected with typical energy dispersive systems. Some EDXRF systems may also incorporate an optional secondary target into the system to allow sample excitation by the characteristic X-rays of any of the available secondary targets, such an arrangement is schematically depicted in Fig. 12.

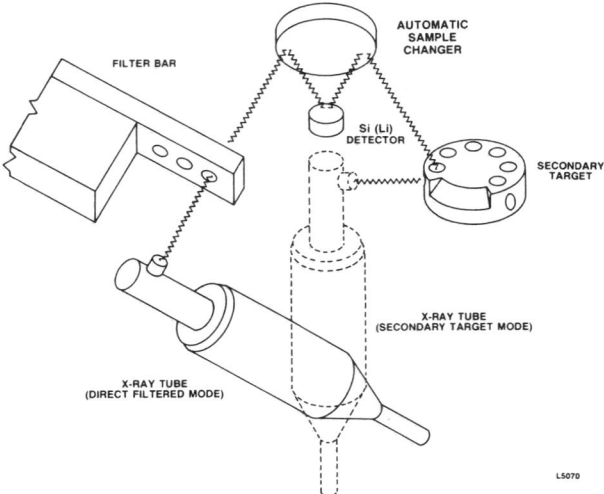

Fig. 12. Optional direct/secondary excitation in EDXRF.

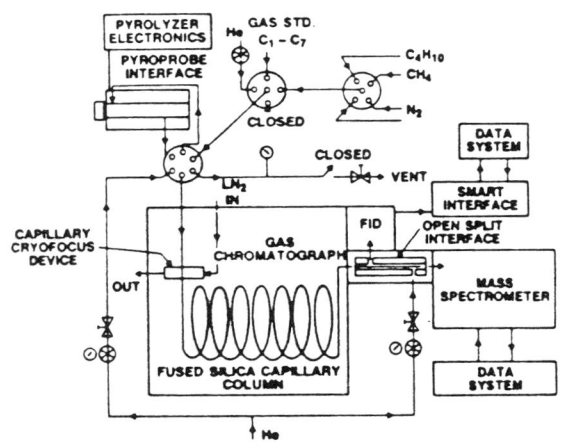

Fig. 13. Pyrolysis GC/MS Schematic. (See Reference 5)

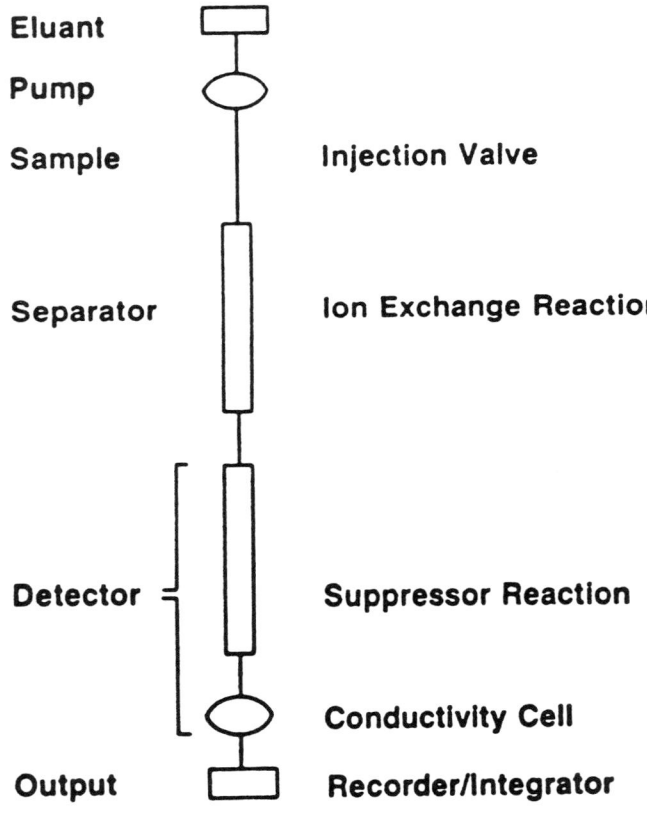

Fig. 14. Components of an ion chromatograph.

Typical Applications. XRF is employed for qualitative elemental screening of unknown samples, often for subsequent characterization by other methods. Process control analyses of % Phosphorus in P-glass, thicknesses of multilayer backmetal systems, composition ratios of binary thin films, plating thicknesses, and solder bump thicknesses can all be performed with XRF.

Gas Chromatography-Mass Spectrometry, GC/MS

Theory of Operation. A solid, liquid, or gas sample is thermally vaporized, transported in a He carrier gas, and cryofocussed onto the head of a capillary chromatographic column. A capillary column GC/MS, equipped with a capillary pyroprobe and a parallel detection system is shown in Fig. 13. The various species in the sample matrix have different relative affinities for the coating of the capillary column and the carrier gas. These properties result in a partitioning and a time separation of the various sample components as they pass through the chromatographic column, and exit to the detectors. Upon exiting the capillary column, the carrier gas containing the separated sample components is split into the flame ionization and mass spectrometric detectors. As the samples pass through the hydrogen flame ionization detector, they are ionized and collected at a cathode, resulting in a detector current. The magnitude of this detector current is monitored to be used for quantitative purposes by comparison to the currents generated by an external standard of known concentration. The portion of carrier gas that is passed through the mass spectrometer is ionized via 70 eV electrons, resulting in ion fragments that are mass analyzed by the quadrupole mass spectrometer. The fragmentation patterns are characteristic of a particular molecular species, and are compared to those of known standards for qualitative identification of the various components of the sample matrix.

Typical Applications. GC-MS can be used for the identification of molding compounds by comparing the pyrolysis-chromatography-mass spectrometry fragmentation patterns or "fingerprints" to those of known compounds. Assessment of "degree of cure" of die coat material by monitoring residual solvents and pyrolytic fragmentation patterns and comparing to "overcured" material can be performed. Optimization of curing process parameters, measurement of solvent contamination levels, polymer and co-polymer identification and relative ratios, and identification of organic contaminants on wafers and devices can be performed.

Ion Chromatography, IC

Theory of Operation. A known volume of aqueous sample solution is injected onto the head of a chromatography column packed with a pellicular ion exchange resin. Basic components of an ion chromatography system are illustrated in Fig. 14. A dilute acidic or basic eluant is pumped across the column and the various anions or cations of the sample are eluted off the column and passed through the system detectors. As the different ionic species have different relative affinities for the ion exchange resin, they spend more or less time in the mobile eluant phase depending on whether they have a lesser or greater relative affinity for the exchange resin. Thus, the ionic species exit the separation column at different times. The eluant carrying the various ionic species present in the sample then passes through a suppressor column or membrane where a second ion exchange process takes place, which results in the anions or cations of the eluant being exchanged for either OH^- or H^+, respectively. The separated ionic species of the sample are then passed through a flow-through conductivity detector in a background of de-ionized water. The relative magnitude of the resultant conductivity current is utilized for quantitative purposes and the elution time yields qualitative information. Detection limits in solution are typically in the ppm-ppb range and can be extended further with pre-concentration techniques. Ammonium, alkali, and alkaline earth cations can be determined in a single chromatographic run on the cation side, and all the common strong acid anions can be determined in a single run on the anion side.

Typical Applications. IC is better suited to process control analysis than to failure analysis. GC can measure trace ionic contamination in plant, process and rinse waters, major, minor and trace ions in plating baths and etch solutions, extractable ionic constituents of powders, encapsulation compounds, glasses, and devices via sample preparation by overnight reflux in ultra-pure de-ionized water.

High Pressure Liquid Chromatography, HPLC

Theory of Operation. A liquid sample solution or extract is injected in a known volume onto the head of the appropriate chromatographic column and transported under high pressure through the column by a liquid carrier solvent (or solvent mixture), termed the eluant. Due to various attractive forces or physical interactions with the coating or packing material of the chromatographic column, the various components of the injected sample will pass through the column at different rates, exiting at characteristic times for a particular species. The eluted species are then passed through either a refractive index detector (mg/L-µg/L) or an ultraviolet-visible photometric detector (µg/L-ng/L), or both. Qualitative and quantitative information for each of the sample components are determined through the use of external standards. In the gel permeation mode, components of the sample solution are separated on the basis of size by passing a column train of various pore sizes. The smaller (or lower molecular weight) components are able to penetrate into the smallest pores or voids of the column packing material and thus have a longer effective route through the

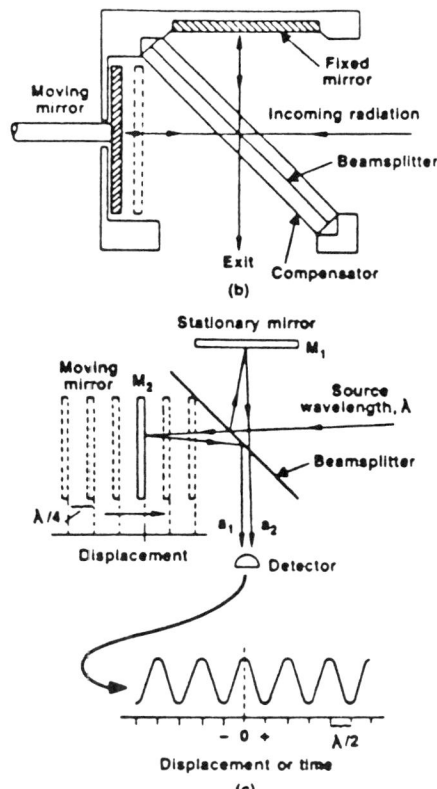

Fig. 15. Components of an FTIR system.

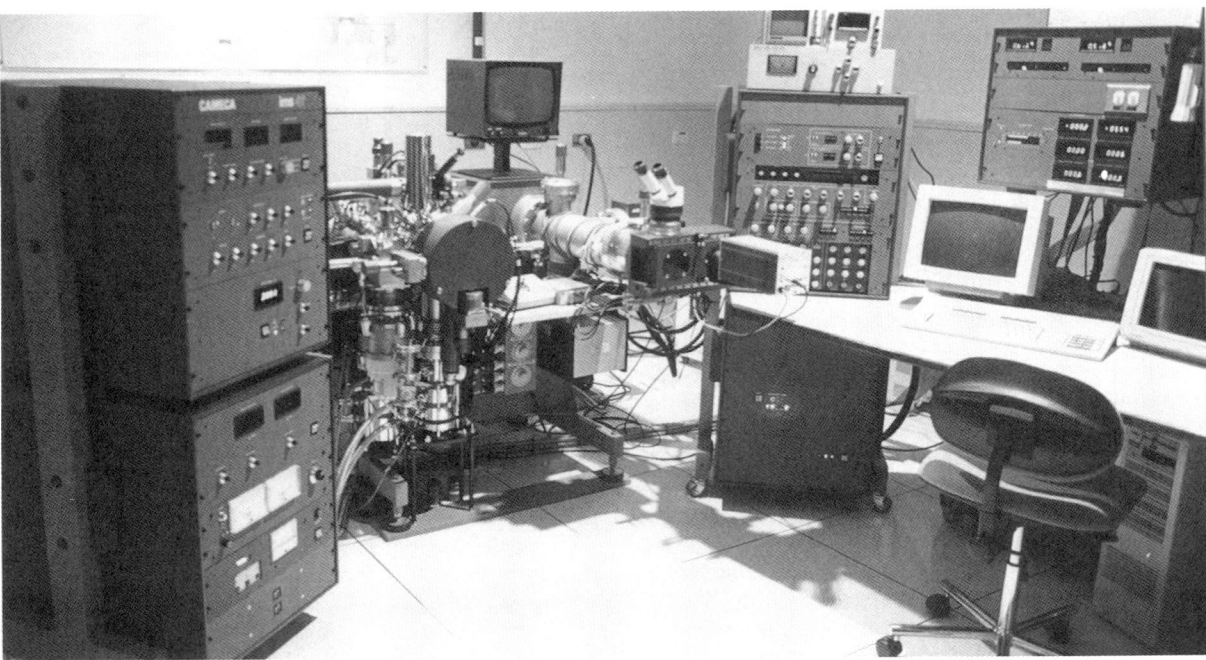

Fig. 16 Example of a SIMS system, which uses a magnetic sector to achieve a high mass resolution and a low energy spread for the identification of elements and compounds which have mass interferences, or overlaps. Examples are TiSi vs AsSi, and P vs SiH. Analysis time on this machine is expensive, and must be well justified. It costs over $1 million.

Fig. 17 Example of an ICP-MS sytsem, which uses a mass-spectrometer to analyze the mass of the atomic species from a plasma-volatilized sample. By contrast, an ICP-AES machine uses a diffraction grating to analyze the light emitted from the plasma flame containing the volatilized species. This machine is used primarily for analysis of the purity of process chemicals and deposited thin films.

column, exiting at later times than the larger (or higher molecular weight) components. Average molecular weight, and molecular weight distributions, are determined by the comparison of elution characteristics of the sample to those of a series of high purity polymer standards, usually polystyrene.

Typical Applications. HPLC is a process control tool suited for measurement of extractable organic components of encapsulation compounds and packaging materials. Component ratios of solvent mixtures and average molecular weight and molecular weight distributions of die coat materials can be evaluated.

Fourier Transform Infra-Red Spectroscopy, FTIR

Theory of Operation. The component atoms of polyatomic molecular groups are in constant motion with respect to each other, constantly changing between the molecular ground state and quantum mechanically allowed excited states due to thermal excitation. The twisting, bending, rotating and stretching motions of the atoms within a molecule occur at frequencies that are generally in the infra-red (IR) portion of the electromagnetic spectrum. (the wavelength, of this IR, lambda, is 0.7-500 µm). When infra-red light of energy coincident with the difference between the ground state and excited state of a particular molecular vibration is radiated onto a sample, the vibrations are stimulated, and the light of that particular wavelength is absorbed by the molecule (IR absorbance). By monitoring the absorbance of the various wavelengths or frequencies of an infra-red light as it is transmitted through or reflected from (IR reflectance) a sample, a characteristic infra-red spectrum can be obtained. In Fourier transform infra-red spectroscopy, the source is modulated with a Michelson interferometer. This results in a signal at the detector which has a distribution of frequencies determined by the speed of the moving mirror of the interferometer. The resulting interferogram in the amplitude-time domain is then transformed via the Fourier algorithm into the appropriate amplitude-wavelength domain, resulting in a characteristic infra-red spectrum for a particular compound. Typical instrumental components are depicted in Fig. 15.

Typical Applications. During failure analysis, FTIR can be used for the identification of extractable organic components from a device or wafer surface, and the identification of organic contaminants, including polymers, on surfaces of an appropriately reflective substrate. FTIR can identify unknown solid organic compounds, powders or residues and solvent or liquid organic sample identification.

Sample Requirements

Sample size, shape and physical state restrictions apply to all of the methods outlined above. For specific requirements of a particular technique it is best to consult with the appropriate analyst prior to sample submission. Generally, analyses performed to determine the reasons for a particular type of electrical device failure are best performed on a comparative basis. It is desirable to perform a side-by-side analysis of a "good" part and a "failed or rejected" part in order to distinguish the chemical, surface or physical sample features that differentiate them. As noted in the theory and applications sections describing the available instrumentation, many of the techniques are sensitive to elemental concentrations in the ppm to ppb range and/or to surface regimes on the order of 1-10 atomic layers. It is therefore imperative that samples to be analyzed are handled in an appropriate manner to avoid the introduction of extraneous contaminants prior to submission for analysis.

Special Note: Some of the equipment schematics in this article may be wholly, or in part, adapted from drawings contained in the operators manuals or descriptive brochures published by the manufacturers. These manufacturers include: Perkin-Elmer, Jobin-Yvon, JEOL, and Bio-Rad.

REFERENCE

1. *Metallography—A Practical Tool for Correlating the Structure and Properties of Materials*, by J.I. Goldstein, ASTM Special Technical Publication 55T, ASTM, 1974, pg 86.

2. *Scanning Electron Microscopy and X-ray Microanalysis*, by J.I. Goldstein, D.E. Newbury, P. Echlin, D.C. Joy, C. Fiori, and E. Lifshin, Plenum Press, New York, 1984.

3. *Monte Carlo Calculations on Electron Scattering in a Solid Target*, by K. Murata, T. Matsukawa, and R. Shimizu, Japan Journal of Applied Physics, 10, June, 1971, pp 678-686.

4. *Materials Characterization by Analytical Electron Microscopy*, by John E. Porter, Proceedings of ISTFA-84, pp 46-52 (see Fig. 2)

5. *Characterization of Molding Compounds Via Polymer Reconstruction Investigative Chromatopyrography*, by S.P. Rogers, K.L. Evams, and M.L. Parsons, ISTFA-84, pp 98-107 (see Fig. 2)

6. *Standard Practice for Reporting Data in Auger Electron Spectroscopy*, ASTM standard E-996-89, available from American Society for Testing and Materials, 1916 Race Street, Philadelphia, PA, 19103.

7. *Standard Practice for Reporting Sputter Depth Profile Data in Secondary Ion Mass Spectrometry (SIMS)*, ASTM Standard E-1162-87, available from American Society for Testing and Materials, 1916 Race Street, Philadelphia, PA, 19103.

8. *Problem-Solving Surface Analysis Techniques,* by Robert D. Cormia, Advanced Materials and Processes magazine, December, 1992, pp 16-23.

9. *Instrumental Methods of Analysis,* by H.H. Willard, L.L. Merritt, J.A. Dean, and F.A. Settle, D. Van Nostrand, 1981.

INDEX OF FAILURE ANALYSIS ACRONYMS

AA—atomic emission spectroscopy
ACS—American Chemical Society
AES—atomic emission spectroscopy
AES—Auger electron spectroscopy
AE—atomic emission
BC—bias contrast
CVD—chemical vapor deposition
DTA—differential thermal analysis
EDS—energy-dispersive spectrometry
EDXA—energy-dispersive x-ray analysis
EBIC—electron beam-induced current
EELS—electron energy-loss spectroscopy
EMP—electron probe microanalysis
EOS—electrical overstress
ESCA—electron spectroscopy for chemical analysis
FIB—focused ion beam
FTIR—Fourier transform infrared spectroscopy
GC—gas chromatography
GC/MS—gas chromatography—mass spectroscopy
GPC—gas permeation chromatography
HPLC—high-pressure liquid chromatography
IC—ion chromatography
IMMA—ion microprobe mass analysis
ICP/AES—inductively coupled plasma—atomic emission spectroscopy
ICP/MS—inductively-coupled plasma mass-spectroscopy
IR—infrared, IR microscopy, IR thermography
ISS—ion scattering spectroscopy
LC—liquid chromatography
LEM—laser emission microprobe
NMR—nuclear magnetic resonance

NAA—neutron activation analysis
ppb—parts per billion
ppm—parts per million
ppt—parts per trillion
PEELS—parallel electron energy-loss spectrometry
PDVC—phase-dependant voltage contrast
quad—quadrupole mass spectrometer
RGA—residual gas analysis
RIE—reactive ion etching
RBS—Rutherford backscattering spectroscopy
RBS—Robinson backscattering spectroscopy
SAM—scanning Auger spectroscopy
SAM—scanning acoustic microscopy
SEM—scanning electron microscope
SIMS—secondary ion mass spectroscopy
SLAM—scanning laser acoustic microscopy
SOM—scanning optical microscopy
SRP—spreading resistance profiling
STEM—scanning transmission electron microscopy
STM—scanning tunneling microscope
TEM—transmission electron microscopy
TDS—total dissolved solids
TGA—thermo-gravimetric analysis
TXRF—total reflection x-ray fluorescence
UV/VIS—ultraviolet-visible
VC—voltage contrast
WDXA—wavelength-dispersive x-ray analysis
XPS—x-ray photoelectron spectroscopy
XRD—x-ray diffraction
XRF—X-ray fluorescence

SCANNING ELECTRON MICROSCOPY (SEM) AND ENERGY DISPERSIVE X-RAY ANALYSIS

by John R. Devaney

Scanning electron microscopy (SEM) has been heralded as one of the most significant inspection and analytical tools to evolve since man first aided eyesight with the telescope and microscope. Its older, more familiar counterpart, the transmission electron microscope (TEM), has never achieved the widespread popular acknowledgment received by the SEM or "sem" as some call it. Rumors began to filter out of England, in the early part of the 60s, of a new, wonderful, magical, all-seeing microscope that could perform miracles! Early photos seemed to substantiate these claims. Marvelous photos of objects with great depth-of-field, good resolution, and no illumination problems appeared. In addition to these attributes, engineering reports talked of voltage mapping and displays depicting current flow in semiconductors!

Depending on one's point of view, the resulting marriage between the micro-electronics/semiconductor industry and the scanning electron microscope was either detrimental or beneficial. The new tool greatly aided the analyst-engineer in his efforts to examine and analyze devices and determine failure modes. It also made it very difficult to hide or ignore mistakes.

The SEM of today is basically third generation iteration of the laboratory models first sold in the mid 60s. These modern instruments are no longer bulky and cumbersome. They no longer require legions of mechanics and electronic technicians to keep them running. They are sleek and svelte.

Like most modern instruments, the actual operation of the SEM has been greatly simplified by complex electronics. Functions and adjustments once performed by the operator are now automatically performed by the internal electronics. Thus, what once required an operator with a full and complete working knowledge of the instrument, is now handled by an array of circuit boards and components. These advancements have made the SEM an instrument that many can operate, but few utilize to its full potential. For without the basic knowledge of the instrument and an understanding of electron beam interactions, the SEM is just an expensive camera. To the trained SEM technologist however, the instrument is much more.

When the electron beam strikes any sample, whether a semiconductor or ancient Roman coin, it produces a variety of signals. These signals exist whether or not the means are available to detect them with a particular instrument.

The basic signals created by electron beam-specimen interaction are as follows:

1. Emissive (Secondary Electron)
2. Backscattered (Primary Electron)
3. Absorbed
4. X-Rays
5. Auger Electrons
6. Beam Induced Hole-Electron Pair Generation (EBIC)
7. Voltage Field Enhanced (Voltage Contrast)

When electrons are injected into or onto a nonconductive surface the injected charge is accumulated and the specimen will "charge up". Thus under most conditions, the specimen should be somewhat conductive.

MICROSCOPE COMPARISONS

1. What are the operating parameters the SEM operator or user needs to be cognizant of to ensure maximum utilization of the instrument's potential?

2. What are the advantages of the SEM over other commonly available instruments?

3. What are the drawbacks or disadvantages of the SEM?

The typical application of the SEM in failure analysis is as "a microscope." Let's compare the SEM to the older, more common, optical and transmission electron microscopes.

Optical Microscope

The Optical Microscope, the mainstay of analysts for over 300 years, has several limiting features, especially for the microelectronic/semiconductor analyst. The useful magnification

Fig. 1 2kV, 50 degree tilt.

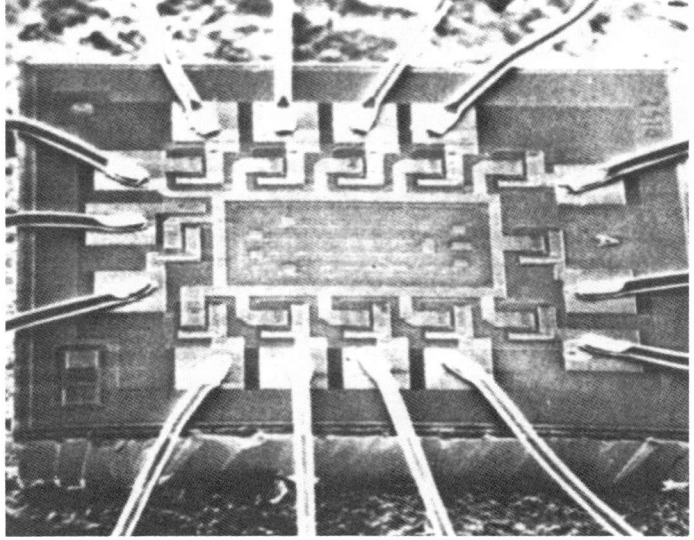

Fig. 2 3 kV.

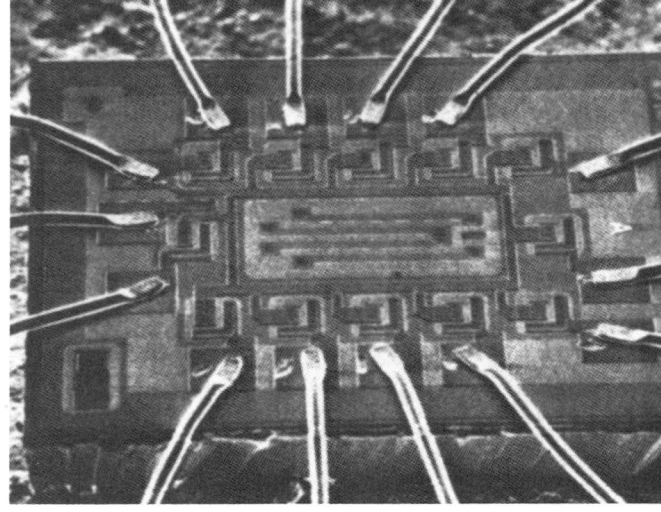

Fig. 3 5 kV.

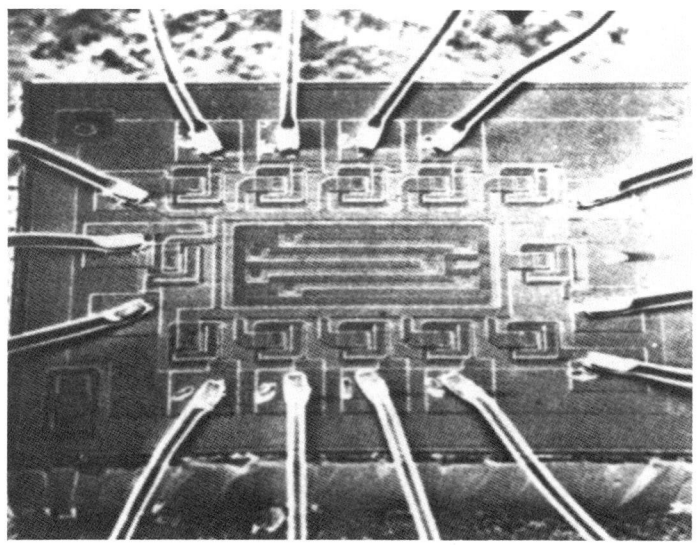

Fig. 4 10 kV.

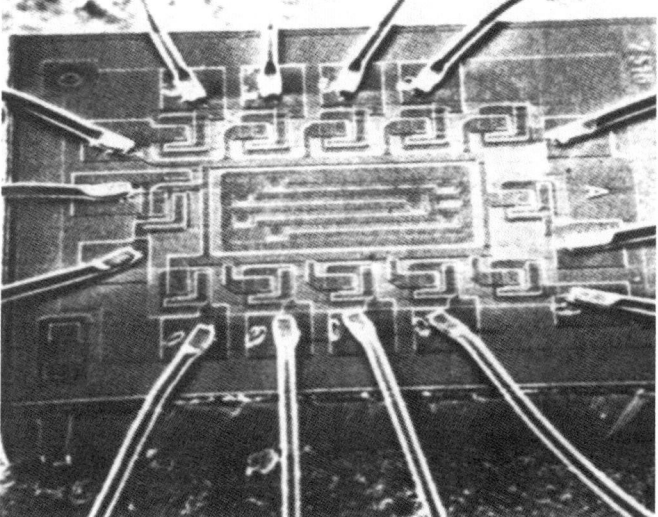

Fig. 5 20 kV.

range extends from very low, approximately 1×, to a useful high of 1,500×. If higher magnification and better resolution are needed, the microscopist can use an oil immersion lens. Even this only extends the range to 3000× and a useful resolution of 0.1 micron, or 1,000 Angstroms (Å). This has the shortcomings of contamination of the specimen with oil, a short working distance, and a sample with no vertical relief!

Transmission Electron Microscope (TEM)

It is surprising that the most important feature of the SEM is its great depth of field, or depth of focus. Although resolution (high magnification) is important, the long focal depth enables the microscopist to examine real, intact, 3-dimensional samples.

This is in sharp contrast to the TEM, which can only be applied directly to very small, thin sections of material or replicas of the sample to be examined. Sample preparation and replication is a difficult, time-consuming task if no artifacts are to be introduced.

Additionally, although the TEM has atomic level resolution (10-20 Å) and high magnification (300,000 -500,000×), it obviously has no apparent depth of field and the lower useful limit of its magnification range is 5000-10,000×.

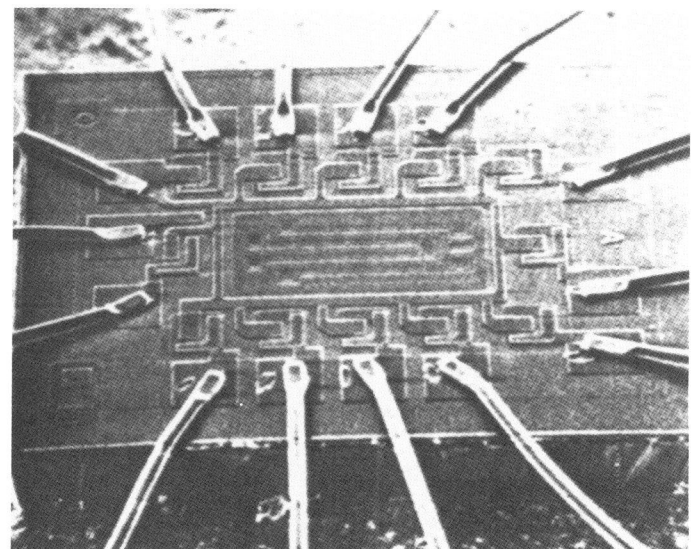

Fig. 6 30 kV

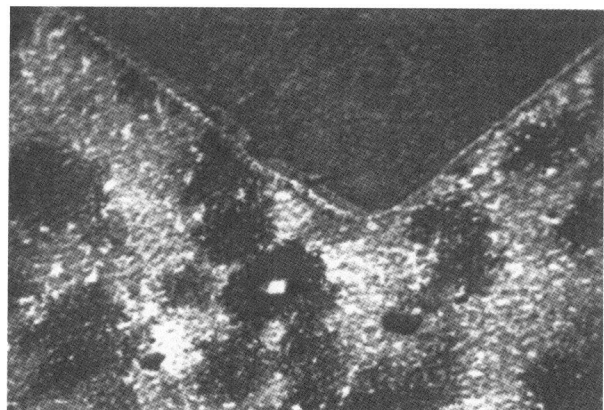

Fig. 7 2 kV, 45 degree tilt.

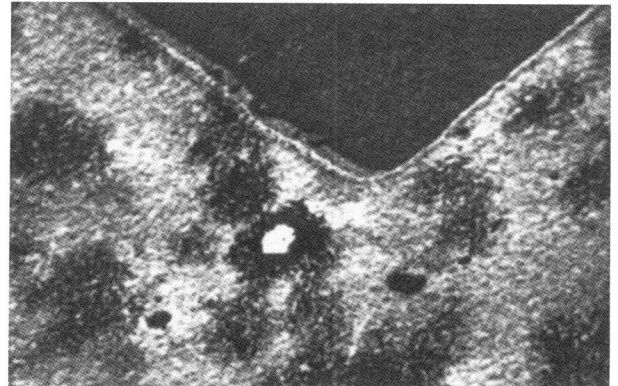

Fig. 8 3 kV, 45 degree tilt.

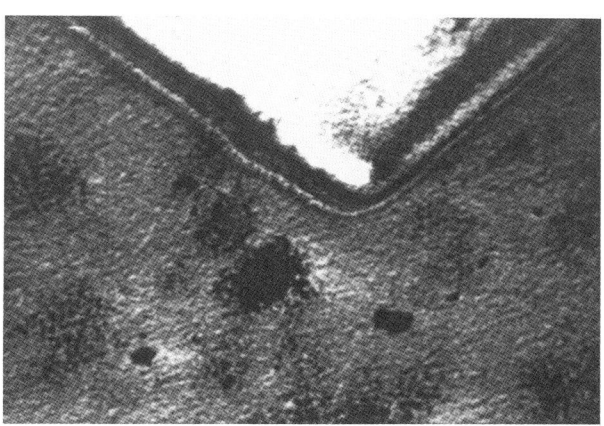

Fig. 9 5 kV, 45 degree tilt.

Scanning Electron Microscope (SEM)

The SEM, however, has a useful working range from 10× to 100,000×, a minimum attainable resolution of 20 Å, a depth of field (focus) 300 times that of an optical microscope, and good working distance. The point of focus or working distance from the final lens can be 15 to 20 mm. An obvious drawback, the specimen must withstand the rigors of vacuum!

In addition to these advantages there is one often overlooked feature which can save the working technician or engineer hours of frustrating effort.

The semiconductor or microelectronic component, for various reasons, is composed of many discrete physical elements. The typical IC, for instance, has smooth surfaced aluminum or gold wires. Gold wires have round shiny balls formed as a bond on one end. The surface of the die is quite reflective as are many portions of the package, cavity surface, bond pads, etc.

Effective illumination of such a sample for examination or photographic documentation is almost impossible. This is especially so if the analyst attempts to examine the sample in some tilted condition.

With the SEM, due to the unique properties of signal generation and collection, many of these illumination difficulties are bypassed or overcome. For these reasons, the SEM still is very widely used as a low-power camera in the range of 10-20×.

The SEM is, however, a monochromatic microscope: THERE IS NO COLOR! Color enhancement may be accomplished, however, by special computer enhancement signal processing. The effect is purely artificial. With all these comments in mind, what then are the specimen and instrument parameters of interest?

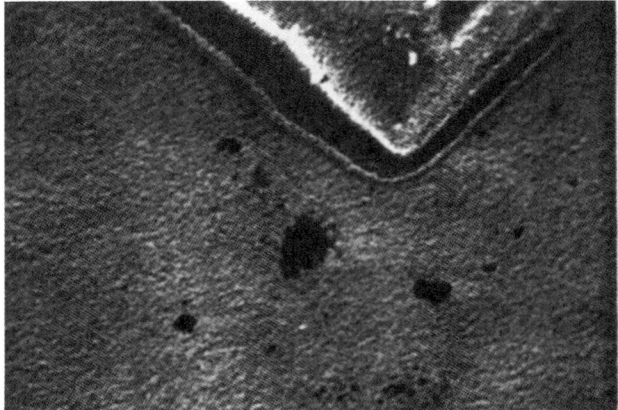

Fig. 10   10 kV, 45 degree tilt.

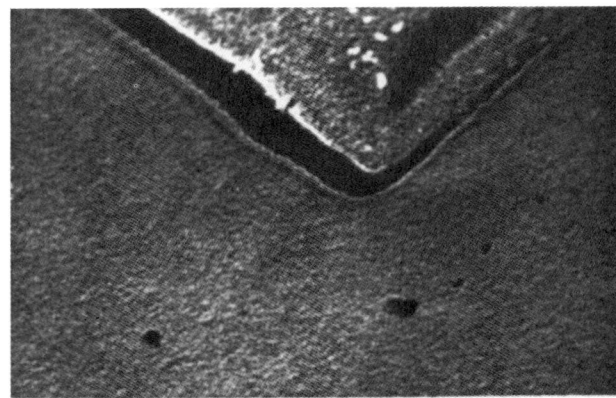

Fig. 11   20 kV, 45 degree tilt.

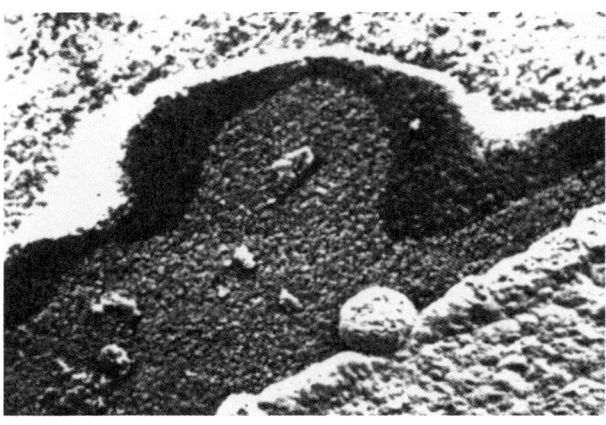

Fig. 12   20 kV, 40 degree tilt.

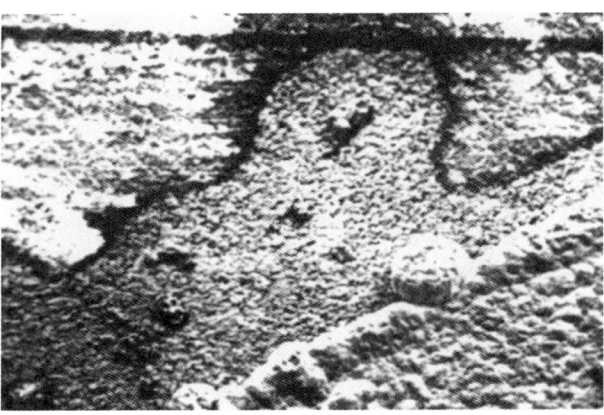

Fig. 13   10 kV, 40 degree tilt.

SPECIMEN AND INSTRUMENT PARAMETERS

The interaction of specimen and instrument parameter will be discussed in the context of the various imaging modes.

The Emissive Mode (Secondary Electron Imaging). With most modern instruments this is the standard imaging mode. The signal is comprised primarily (approximately 90%) of secondary electrons, although some backscattered electrons are collected and modify the image.

For the failure analyst, the primary consideration, in most applications, is to do nothing to alter or affect the sample. *A conducting coating is a major alteration!*

CAUTION: DO NOT COAT THE SPECIMEN, EXCEPT AS A LAST RESORT!

Examination of the sample surface directly results in significant and useful data only if properly interpreted. Surprisingly, the emissive image of an uncoated surface, especially one as complex as an IC, is parameter dependent. The parameters are instrument beam voltage and specimen tilt.

Beam Voltage. Early SEMs and even most in use today utilize a current-heated tungsten wire as a source of electrons, a "hot filament gun". Due to various electron-optic interactions, the electron beam is more sharply defined and smaller at high accelerating voltages (20-25 Kev) which results in increased point-to-point resolution. But a high accelerating voltage results in greater penetration of the sample surface by the beam. Thus, the various signals are generated from a greater volume of the sample. This penetration depth is directly proportional to the density/atomic number of the sample. For a material like aluminum, AT No. 13, or silicon, AT No. 14, this penetration depth at 25 kV can exceed 4 microns! If the beam voltage is reduced, the electron beam penetration is also reduced. This results in greatly improved contrast due to both an increase in secondary electron emission at the surface (point of interest) of the device and a decrease in background (subsurface) secondary emission. The latter is generated by backscattered electrons at considerable distances from the beam impact point. These backscattered electrons can also impinge upon other portions of the sample. If these areas are non-conducting, they can

SEM and Energy-Dispersive X-Ray Analysis

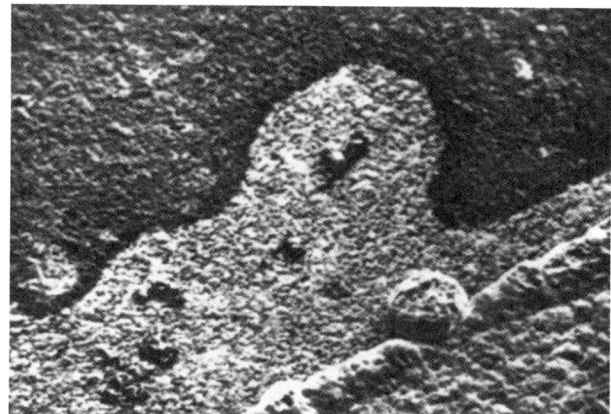

Fig. 14 5 kV, 40 degree tilt.

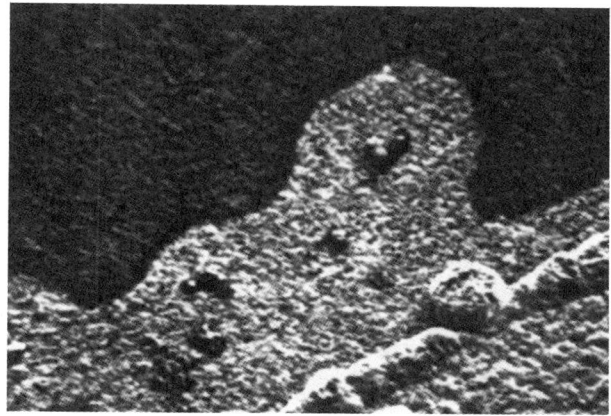

Fig. 15 3 kV, 40 degree tilt.

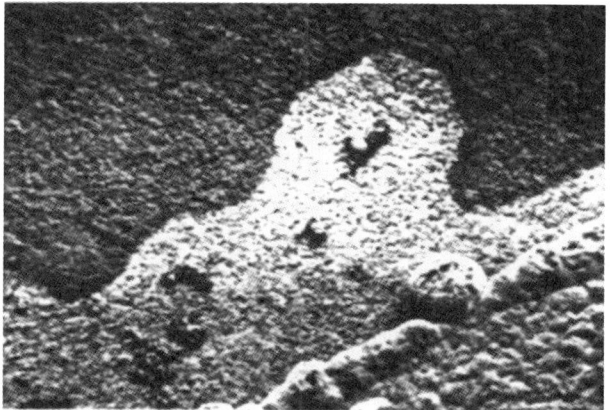

Fig. 16 2 kV, 40 degree tilt.

Fig. 17 3 kV, 50 degree tilt.

charge up and result in a drastic deterioration of the signal-to-noise ratio (in the area of observation) without the operator being cognizant of the effect. Thus, the selection of the beam acceleration voltage is dependent upon the specimen and the information desired.

Figures 1 thru 20 are examples of the difference of an image of a surface at low and high accelerating voltages. Figure 1 shows an area of a microcircuit examined at a beam voltage of 2Kev. Note the surface detail on the metal stripes. Figure 5 images the same area at a beam voltage of 20Kev. It is obvious that much of the finer surface detail is obscured. The loss of fine surface detail outweighs the apparent advantage of higher resolution. However, the use of higher beam voltages with greater penetration depths can result in information obtainable no other way. (See Fig. 19, 20).

Figure 7 shows the surface of a device covered with contamination (the image was obtained at a beam voltage of 2Kev). Since penetration and emission are shallow, only the contaminant is imaged. When the same area is examined at 20Kev, the greater depth of penetration and signal generation shown in Fig. 11 results. Imaging of the underlying surface and all traces of the contaminant are lost.

CAUTION: The analyst must constantly be aware of the fact that different beam voltages result in preferential charging and/or emission from different materials which radically alter the information content of the image (see Fig. 12 thru 16). Figures 1 thru 6 are photos of an IC surface. They illustrate the great diversity available by using different beam voltages. The specimen tilt is 50° in all six figures. The sample is not coated.

The series of micrographs in Fig. 7 thru 11 should be studied at the same time. They illustrate the advantages of using low beam voltages to examine a surface suspected of being contaminated with organic or silicone thin films. The sample is composed of thick film gold on a ceramic substrate. The dark spots in Fig. 7 are contaminant stains of varying thickness. Compare this to the apparently clean surface in Fig. 11.

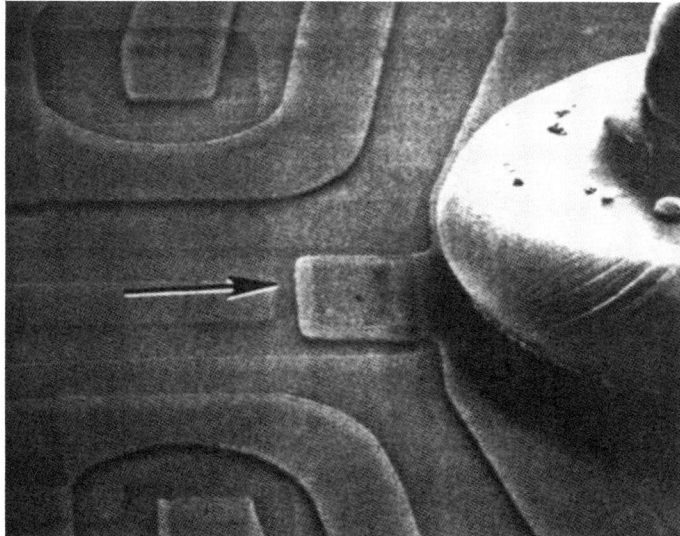

Fig. 18 30 kV, 50 degree tilt.

Fig. 19 3 kV, 60 degree tilt.

If surface contamination is suspected, always examine the sample at low beam voltages. High beam voltages will not detect surface films.

Figures 12 through 16 should be studied simultaneously. This sequence of SEM micrographs illustrates the dramatic change in photo information as the beam voltage is lowered. Most significant changes occur in the upper half of the image, which is bare ceramic. The region cutting diagonally across the lower right corner of the photo is a thick film of gold. The dark irregular shaped region in the photo center is a thin film chromium bleedout.

Figures 17 and 18 are photos which dramatically illustrate the loss of surface information on a low atomic number thin film, aluminum, at 30 kilovolts as compared to 3 kV.

Higher beam voltages do have some useful features. Figure 19 is an image of a badly contaminated IC surface. At 3 kV, the beam is charging the light element contaminant. In Fig. 20, at 20 kV, a significant portion of the beam penetrates the contaminant.

Tilt. Just as various beam voltages radically alter the emissive mode image, changes in specimen tilt also have a striking effect on the image. This is true not only for semiconductors, but many other thin film samples and samples where non-conductive films of varying thicknesses exist on the specimen surface.

The effect is due, in part, for the same reason that changes in beam voltage affect the image. As the angle between the beam and the normal to the specimen increase, the actual depth of penetration from the true surface decreases. The passage of the beam through various surface films at different angles results in varying penetration, emission, and charging rates.

Analogously, the sun illuminates and penetrates the same blanket of atmosphere (sky) at high noon and sunset, yet we see one as blue and the other as yellow or golden. The difference is angle! For specimens at very high tilt angles, more of the pri-

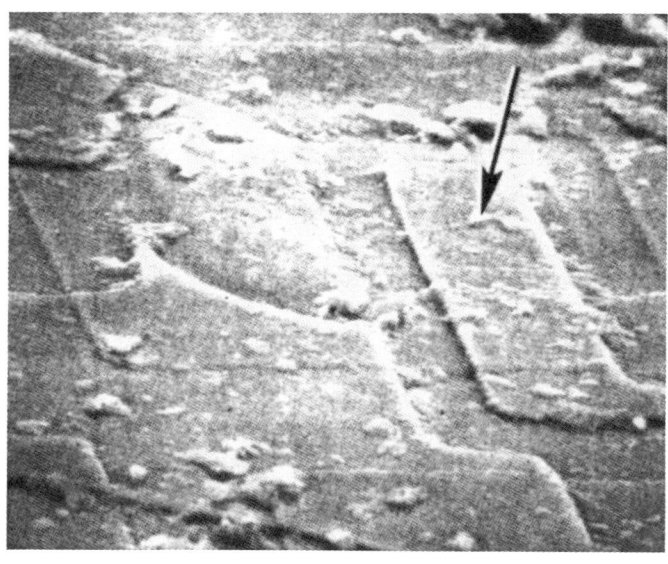

Fig. 20 20 kV, 60 degree tilt.

mary electron penetration volume is within signal escape depth of the surface. Thus more electrons are emitted. In the example of an IC, the analyst must recall that some of the enhanced contrast obtained by tilting the specimen is due to complex charging of thin oxides on the sample surface and preferential generation of the signal in the variously doped regions of the device. Figures 21 and 22 depict the effects on the image of merely tilting the specimen while keeping beam voltage fixed at 10 Kev and tilting from 0 to 75 degrees.

Another consequence of specimen tilt is illustrated in Fig. 23 and 24. Figure 23 shows a passivated microcircuit examined at

SEM and Energy-Dispersive X-Ray Analysis

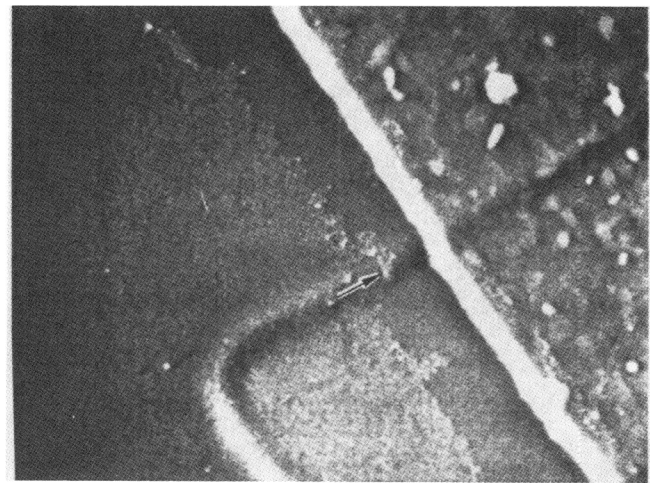

Fig. 21 10 kV, 0 degree tilt.

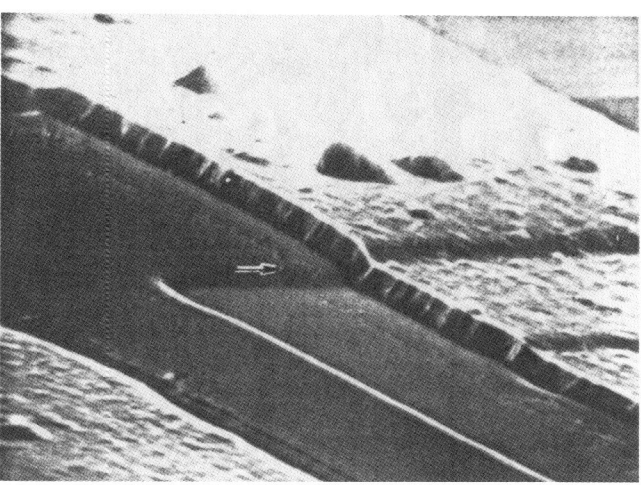

Fig. 22 Effects on image (from Fig. 21) by tilting the specimen while keeping beam voltage fixed at 10 kV, 75 degree tilt.

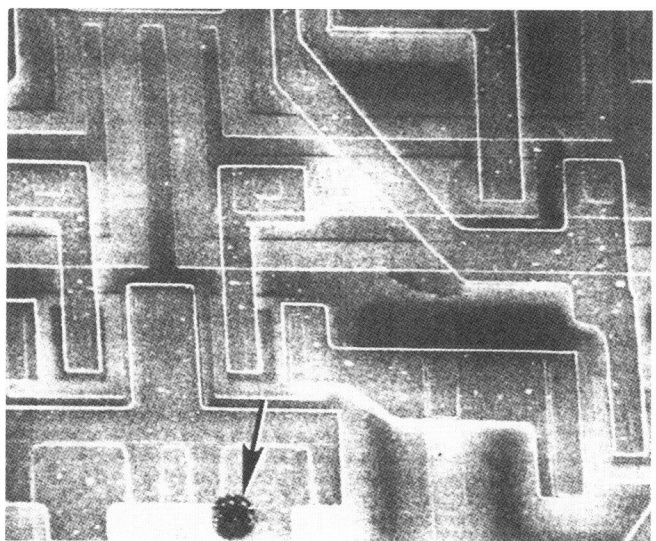

Fig. 23 5 kV, 0 degree tilt.

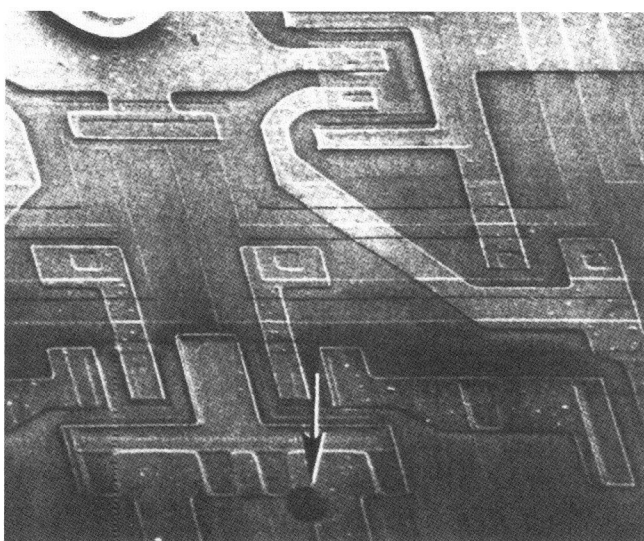

Fig. 24 5 kV, 40 degree tilt.

a zero degree tilt with no prior treatment to prevent charging. Charging typically equated with brightness (or enhanced emission) can also result in decreased emission (dark areas). Charging is thus manifested by very bright areas or spots with dark adjacent regions which results in loss of detail in both areas. Examination of the same area at the same beam voltage but a higher tilt angle (40 degrees) finds that the number of backscattered and secondary electrons has increased. This results in a reduction of residual electrons (charging) in the glass layers.

The resolution and surface detail in Fig. 24 is not spectacular, but a significant amount of information is obtained. Figures 23 and 24 are SEM micrographs of the same area at the same magnification, but two different tilt angles. Compare these two rather bland images to that of the same area at 50 degree tilt in Fig. 25.

Backscattered Mode—Primary Electrons (BSE). As with applications to samples other than semiconductors, the backscattered image is highly dependent on the constituents of the surface. More important for general applications is the fact that, unlike weak low-energy secondary electrons which can be collected out of holes and around corners, primary or backscattered electrons travel in relatively straight trajectories. Thus they are influenced very little by detector field voltages. Therefore, it is reasonable to expect them to be nearly unidirectional, and the images will have sharp shadows where the surface topography blocks the area imaged from the detector. Subtle features are often greatly enhanced by backscatter mode imaging.

Figure 26 illustrates a passivated microcircuit examined at a tilt angle of 60 degrees, in the emissive mode. This illustration

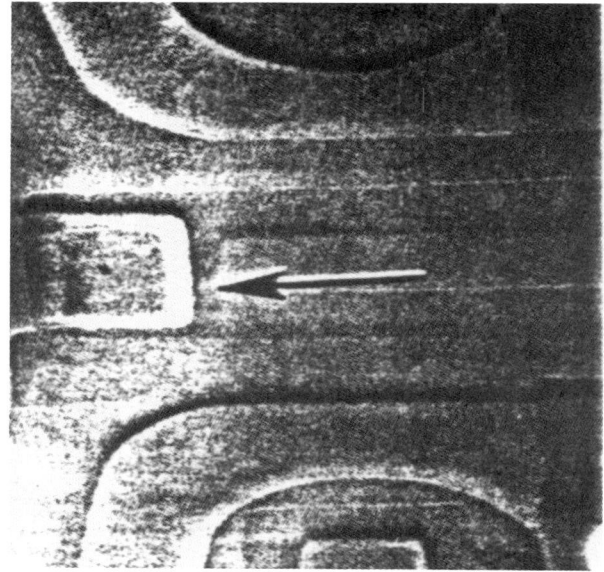

Fig. 25 Image of area shown in Fig. 23 and 24, except that angle of tilt has been increased to 50°.

Fig. 26 Passivated microcircuit examined at tilt angle of 60 degrees in the emissive mode.

Fig. 27 Passivated microcircuit examined at tilt angle of 60 degrees but imaged only with backscattered electrons.

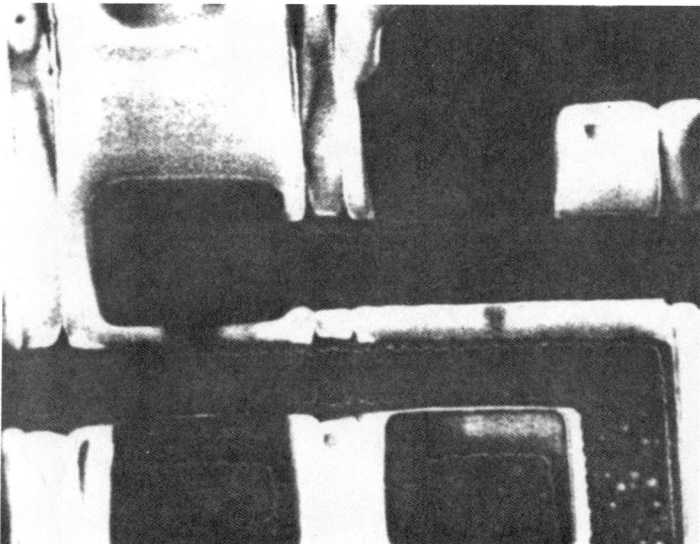

Fig. 28 Secondary emission

is followed by Fig. 27, which is a passivated microcircuit examined at a tilt angle of 60 degrees but imaged only with backscattered electrons. The two figures should be studied together.

Figures 28 and 29 show the radical differences between the same dual level passivated IC surface when imaged by secondary (Fig. 28) and backscattered electrons (Fig. 29). Preferential charging of the inter-layer glass has occurred in the secondary mode (brightness).

Voltage Field Enhanced. (See Fig. 30, 31, 32, and 33.) A special case of secondary electron imaging is "voltage contrast" imaging. This mode requires that connections be made to the device and the electrical biases be applied.

Voltage contrast is used primarily to pictorially display the electrical functions of devices. By careful adjustment of the beam accelerating voltage, aperture, beam sweep rate, and specimen electrical operation, voltage states and transitions can be detected on the device surface. This is accomplished by elec-

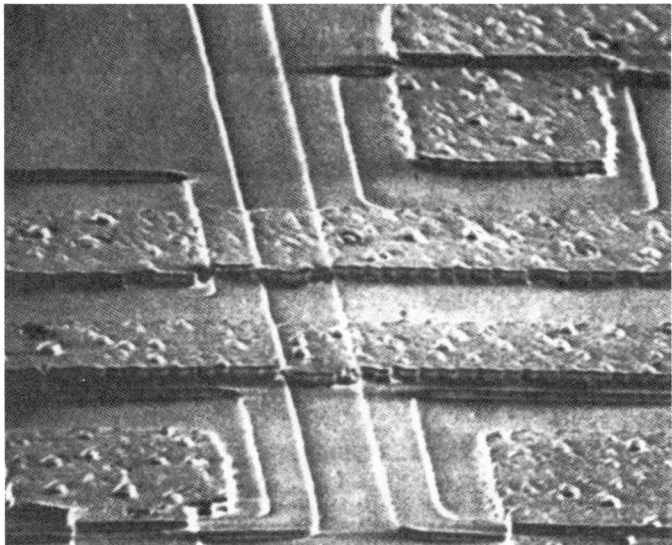

Fig. 29 Backscatter signal only, sample tilted 60 degrees.

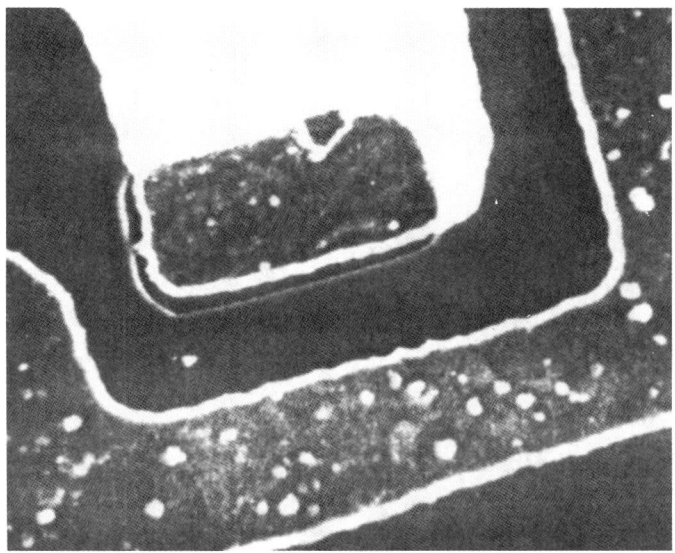

Fig. 30 Voltage contrast image of an IC surface, no glass overlay at 6 volts (black) and ground (bright metal contact). Open at step into contact window.

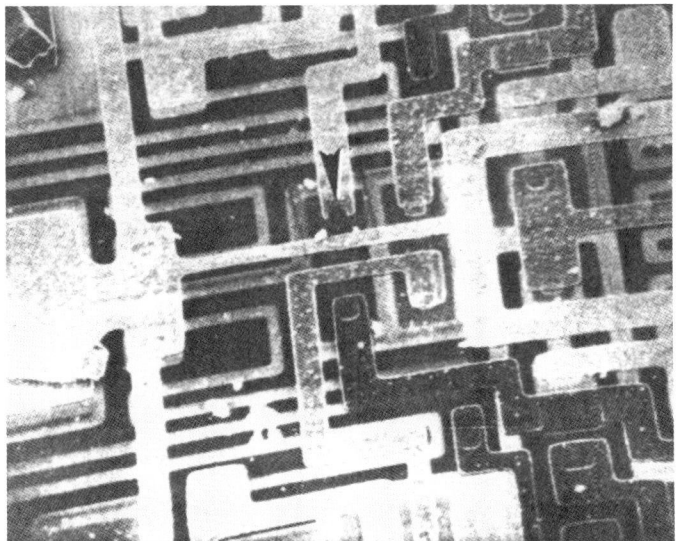

Fig. 31 Voltage contrast of a digital IC surface showing different voltage levels. Note the open at step in tunnel contact window.

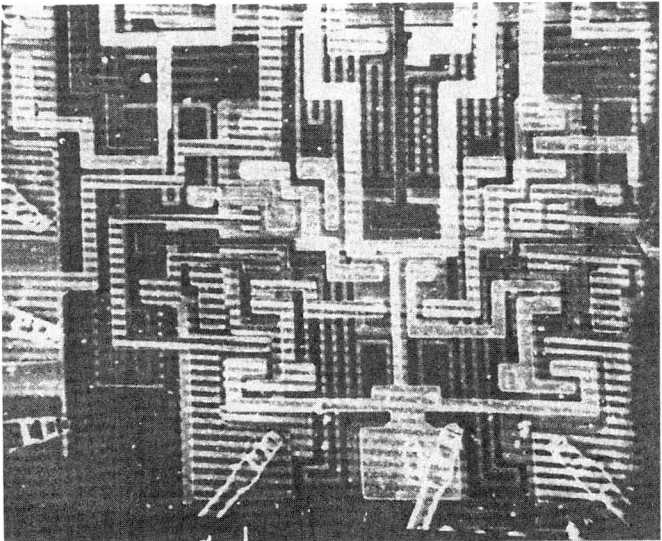

Fig. 32 Voltage contrast beat, or hetrodyne, pattern caused by pulsing the input voltages.

trically operating the device through external connections and adjusting the beam acceleration of the SEM so that secondary emission is affected by the charge states on the device. Positive areas will have a lower secondary emission efficiency than negative or ground potential areas. The more positive areas will thus appear dark in comparison to the more negative areas. By pulsing the device at a low repetition rate (approximately 10Hz), the effect is quite noticeable, and transitions from one state to another are detectable. Open circuits are very obvious since one side will be dark (positive) in comparison to the light side (ground).

Voltage contrast imaging can be performed on both single and multilayer interconnect devices with the glass passivation intact. The advantage of this mode of observation is that the glass does not need chemical treatment. The location of an open by VC, and verification of its presence, cannot be refuted because of an analytical procedure. Operation in this mode requires that the beam voltage be carefully selected for the layers of dielectric involved.

No special detector is required to achieve voltage contrast (differentials of 0.6v) on an ordinary semiconductor device. It is recommended that those unaccustomed to working with semiconductors in the voltage contrast mode experiment with well-defined simple specimens where complications due to the specimen or package can be kept to a minimum.

The writer suggests that these experiments be initially performed on a transistor or integrated circuit which has no glass passivation over the metallization pattern. In addition, the package itself should be of the TO-5 or TO-18 "can" type, where the die sits on a conductive header with little or no glass or ceramic nearby. All efforts should be made to keep the specimen as far from the final lens as possible to prevent trapping of the signal electrons. *Under no conditions should the scanned beam be allowed to strike nearby glass or ceramic insulation. Charging of these will result in the loss of observable voltage contrast.* If possible, experiments should be carried out over the entire beam voltage range starting at 2 kV up to 30 kV, to determine which condition results in the maximum contrast. Scan rate and exposure of the sample will also alter visible contrast. In addition, higher beam voltages can result in alterations in the specimen itself.

Once the operator is fully familiar with one or two "standard" specimens, he should then begin to expand his experience by working with more complex devices and less ideal packages.

In addition to these very basic techniques, the literature abounds with excellent reports detailing various modifications to the instrument and signal collector which enhance observable contrast, and in some techniques quantifies it.

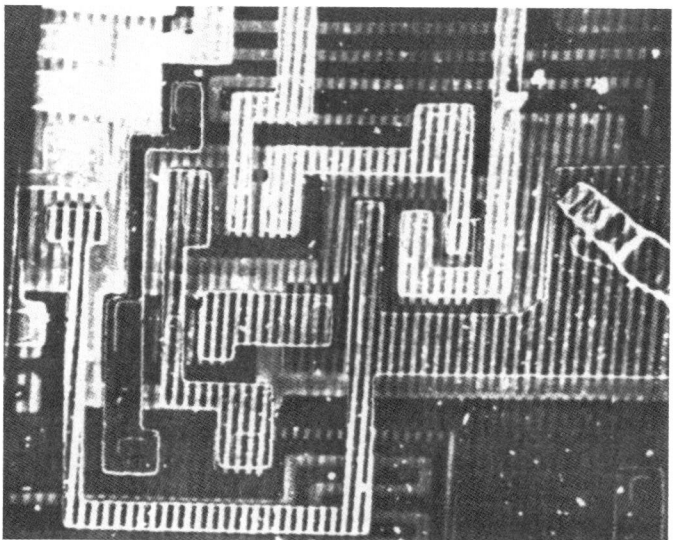

Fig. 33 Voltage contrast beat (or hetrodyne) pattern high magnification view of image in Fig. 32.

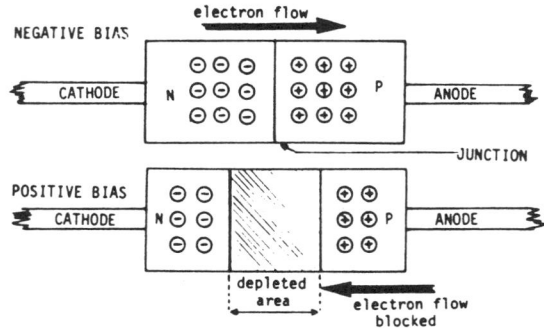

Fig. 34 Biasing a pn junction.

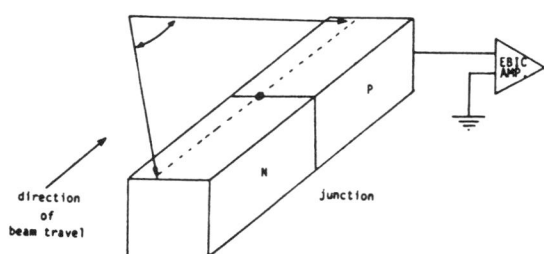

Fig. 35 Electron beam flow direction.

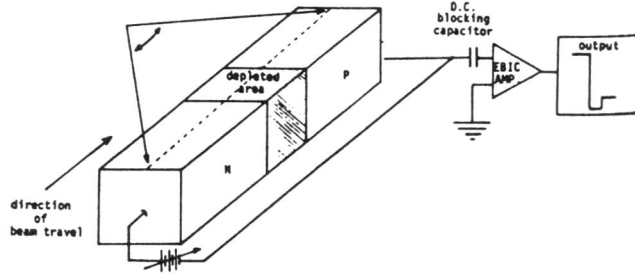

Fig. 36 Reaction of electron beam with depletion region.

Fig. 37 Micrograph of the specimen diode.

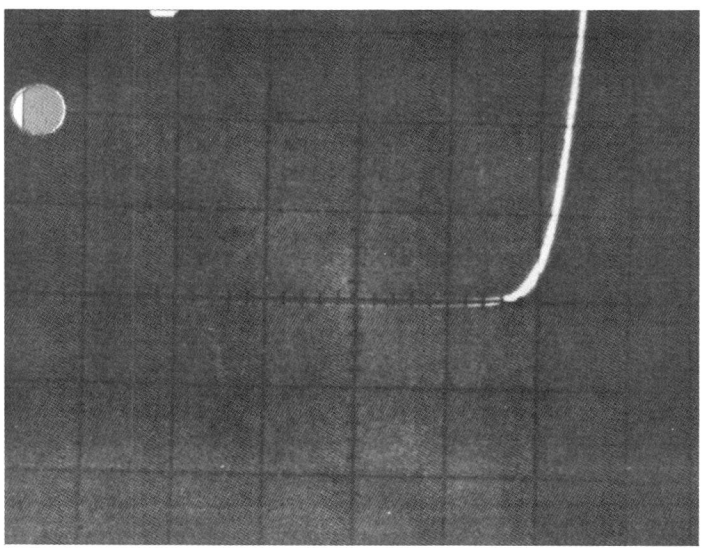

Fig. 38 Curve tracer photo of the forward breakdown characteristics

Other efforts have resulted in techniques which produce voltage contrast of a device in varying states of operation, such as alternating or pulse current mode, by strobing of the beam to the device. In fact, the creative electronics engineer is presented a basic tool with which he can experiment, and add signal processing instruments, and greatly increase the spectrum of data obtainable from the SEM.

EBIC utilizes the focused beam of an SEM to create hole-electron pairs in the examined specimens. The penetration depth of the beam is a function of the beam accelerating voltage. The induced current magnitude, on the other hand, is a function of electron beam current and the specimen.

The electrons generated flow in the specimen in a direction influenced by the depletion region and external circuitry. This induced current is supplied to an external high gain amplifier by attached wires. The amplifier and associated control circuit synchronize the signal with the x-y coordinates of the SEM's CRT display. The resultant image is manifested as light or dark regions, since the amplifier signal varies from positive to negative depending on the beam location on the specimen. (See Fig. 35.)

As a semiconductor surface is examined, the EBIC output signal is modulated by the various surface compositions. The current injection efficiency of the negatively charged beam is higher in N-doped areas than P-doped areas, and the output signal will be accordingly of higher amplitude.

The result of reverse biasing a P-N junction is that the depletion region reacts to electrical bias by acting as a barrier to current flow. The depletion region is an area depleted of free charges and thus, for all practical purposes, is incapable of supporting current flow. Theory contends that this depletion region expands with reverse bias (positive on N) and collapses with forward bias (positive on P).

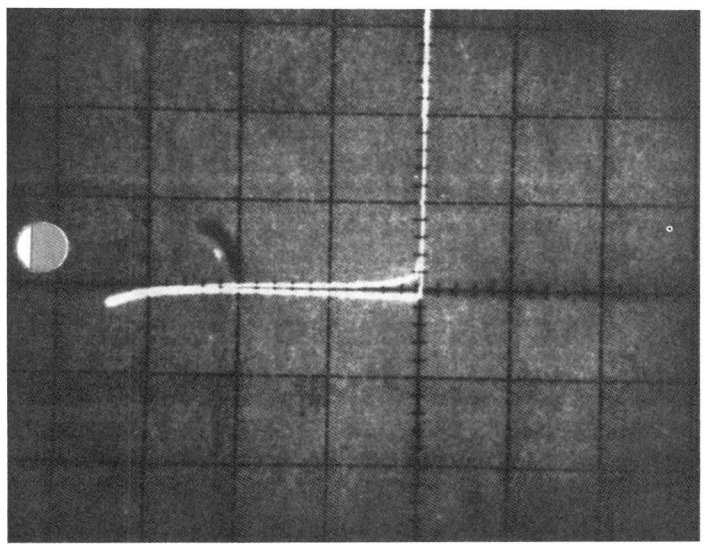

Fig. 39 Curve tracer photo of the forward and reverse breakdown characteristic of the diode. (1 mA/vert div, 200 V/hor div.)

The EBIC effect, when properly adjusted, can be utilized to examine the depletion region. The beam will react with the depletion region as though the area is a junction. Due to the last of free, available electrons, the current generation is much less in the depletion region. (See Fig. 36.)

Example

A 1N3209 power diode was selected as the experimental specimen. The choice of a power diode was made because of its rugged construction and relatively thick die. (See Fig. 37).

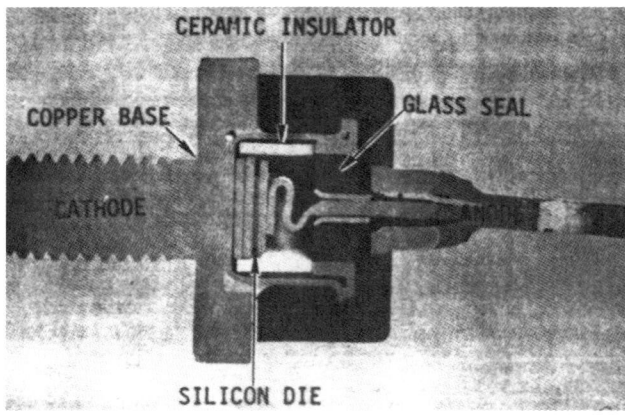

Fig. 40 Macrograph of the section of the diode.

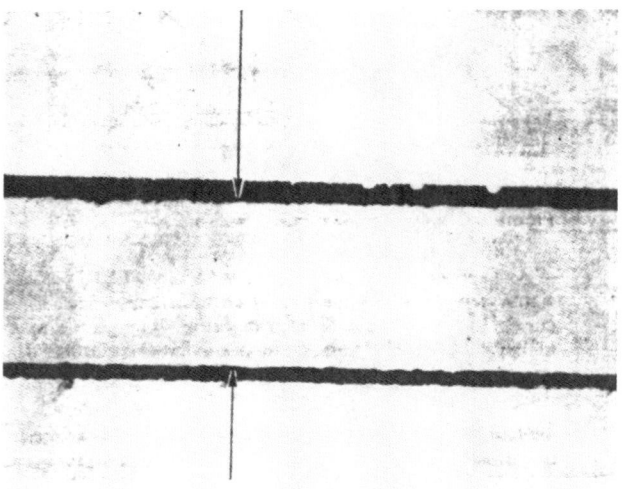

Fig. 41 Micrograph of the unetched die section.

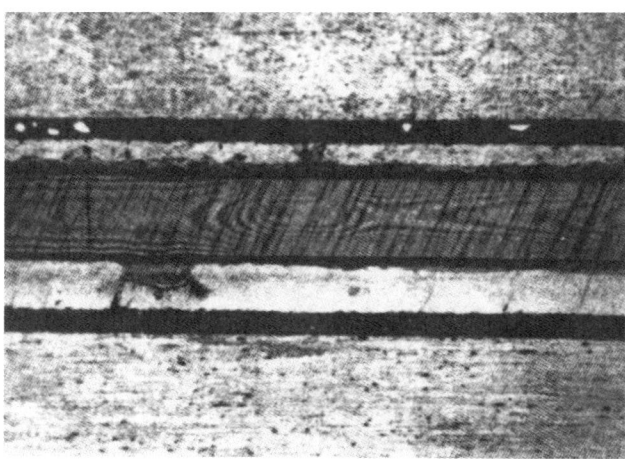

Fig. 42 Micrograph of the etched die section.

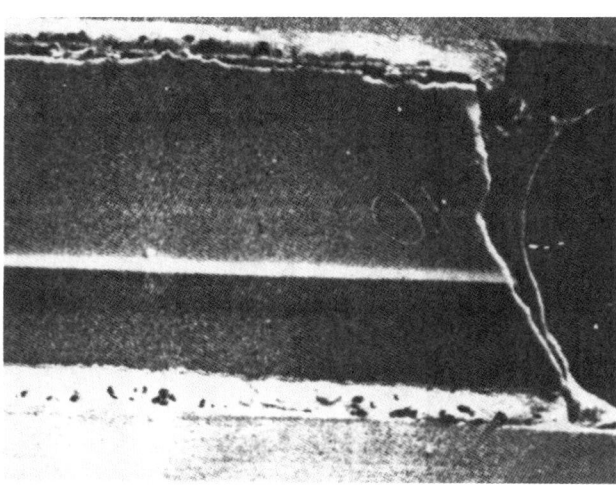

Fig. 43 SEM micrograph with superimposed EBIC image. 0 V bias.

The specimen was initially tested with a curve tracer and its operating characteristics recorded (see Fig. 38 and 39).

The device was then bisected with a low speed diamond wafering saw. (See Fig. 40 and 41). This step resulted in two experimental specimens.

The procedure continued with a fine polish of the sectioned surfaces to remove the sectioning-induced defects. Ultrasonic cleaning with acetone followed, to remove sectioning debris.

The final preparation step was a water rinse and high temperature (200 °C) bake. This allowed the polished silicon surfaces to form a protective oxide.

Curve tracer analysis revealed that both halves were still functional after the sectioning procedure. A slight degradation in reverse leakage was noted, but was not considered significant.

Specimen Examination. Microscopic examination of the sectioned die could not distinguish the junctions. One specimen half was then chemically etched to define the junction (Fig. 42). This sample was reserved for comparative purposes.

The unetched sample half was examined with the SEM. Again, without chemical etch, the junction was not detected. However, with EBIC analysis the junction area became quite visible. A definite correlation was noted between the EBIC display of the unetched sample and the optical results of the etched sample. This result demonstrated the ability of EBIC to accurately delineate junctions.

Under the influence of reverse bias, the EBIC image underwent a significant change. The junction area was replaced with the depletion region. With increasing reverse bias, the depletion region was observed to expand.

Figures 43 through 46 examine the specimen under several different reverse bias conditions.

SEM and Energy-Dispersive X-Ray Analysis

Fig. 44 SEM micrograph with superimposed EBIC image. 5 V reverse bias.

Fig. 45 SEM micrograph with superimposed EBIC image. 10 V reverse bias.

Fig. 46 SEM micrograph with superimposed EBIC image. 50 V reverse bias.

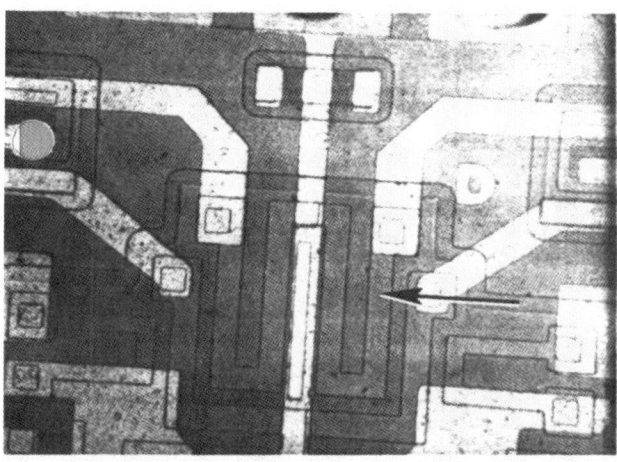

Fig. 47 Suspect resistor optical photo —device is a quad-NAND gate. Left and right sides are mirror images.

The depletion responded as anticipated, i.e., expanding with increasing reverse bias, contracting with decreasing reverse bias, and collapsing with forward bias. Once in stable conduction, the specimen would no longer react with the electron beam and no image response was observed.

The EBIC mode is relatively insensitive to charging artifacts. The package shape and materials are of minor consequence as compared to voltage contrast. A major drawback to the use of EBIC on complex devices is the proper interpretation of its data output.

The author has found the greatest success in using EBIC is in the comparison mode where EBIC maps of a good and (suspect) bad devices are obtained under identical conditions.

It should be remembered that EBIC mode imaging does not require that the device be biased with external power supplies. Any P-N junction will have an induced current flow when exposed to a scanned electron beam. The only requirements are that any signals generated be collected and fed to the high gain amplifier. Because of the extremely high gain in the amplifier, long leads are undesirable since they act as antennas, feeding spurious signals to the amplifier.

Examples of suspect devices exposed to EBIC imaging are shown in the following figures:

1. Resistor, Fig. 47 & 48
2. Shift register, Fig. 49 & 50
3. Diode, Fig. 51

Artifacts Introduced by the Beam

Before the failure analyst can use any tool or instrument successfully and with a degree of confidence, he must be com-

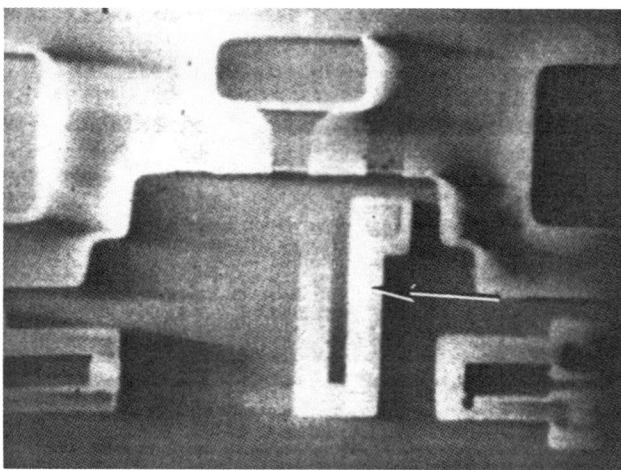

Fig. 48 Suspect resistor exposed to EBIC imaging. Identical electrical connections were made to each side. The arrow notes a resistor imaged by EBIC (bad side) not seen in the good side. Current induced resistor reached substrate thru a pinhole in the field oxide beneath that metal run.

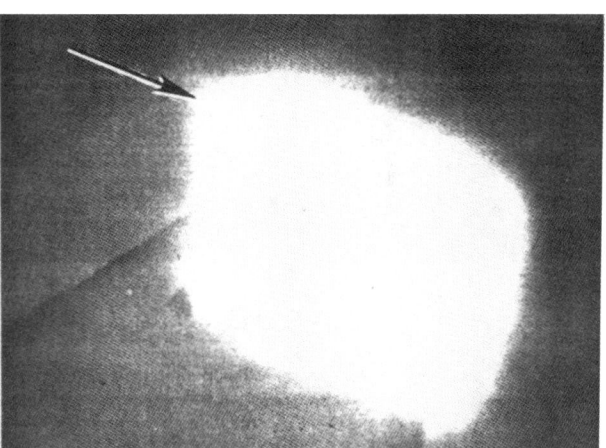

Fig. 49 EBIC image of input protection diode with anomaly on junction perimeter.

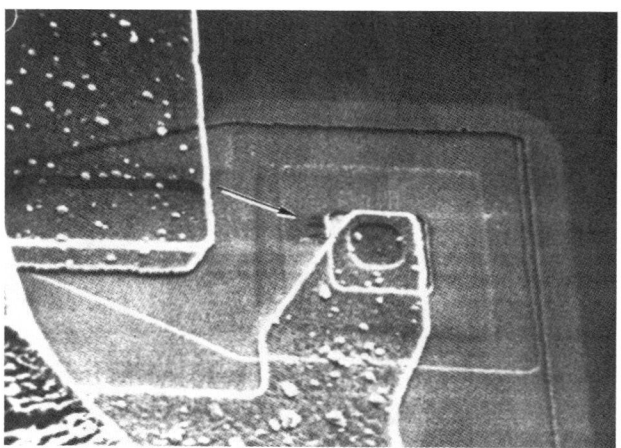

Fig. 50 Shift register EBIC image shows short across edge of protection diode diffusion, glass removed and surface etched to enhance damage.

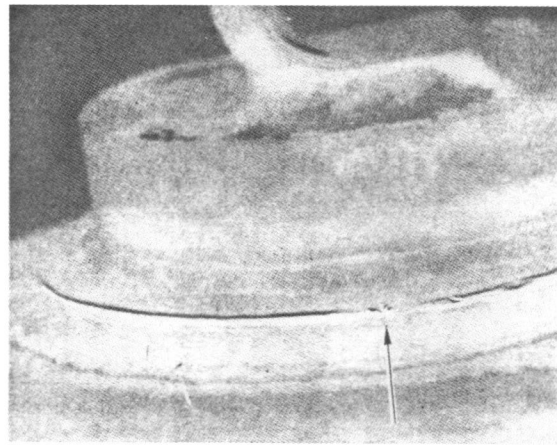

Fig. 51 Suspect diode exposed to EBIC imaging—defect across junction is the leakage site.

pletely cognizant of any and all artifacts induced by the inspection/testing tool.

Thus he must be familiar with:

1. Artifacts induced by the beam
2. Specimen preparation
   a. Surface coating
   b. Chemical processing
3. Correlation with optical microscopy

In scanning electron microscopy, three of the most common artifacts introduced by the primary electron beam are electrostatic charging, vacuum pump oil polymerization, and electron beam damage. These artifacts present special problems for examining semiconductor devices because of the materials used in their construction and the fact that active electronic components are susceptible to radiation damage.

CHARGING

Electrostatic charging is due to the presence of a negative potential on an insulation portion or point of the specimen surface. Charging is caused by the collection of electrons on the surface of the insulator. The accumulation of electrons builds up a negative space charge which can deflect the incident beam, causing severe image distortion. The presence of the surface charge will also change secondary emission greatly. A charged area can cause the darkening of the image of a nearby un-

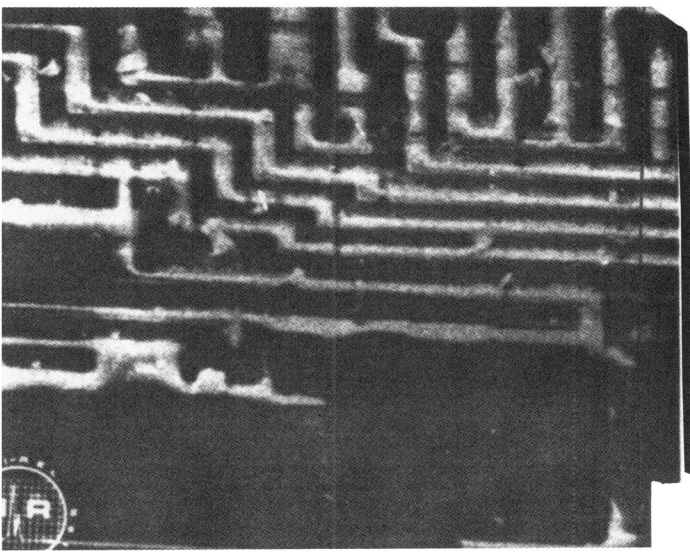

Fig. 52 Examination of passivated integrated circuit surfaces at 3 kV without a conductive overcoating results in extreme localized charging of the passivation layer.

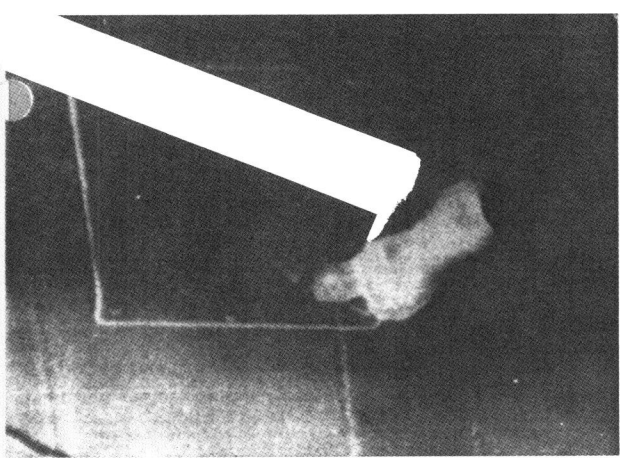

Fig. 53 Examination of an integrated circuit with an unattached nonconductive particle in the field of view results in charging of the particle and loss of information in the region around the particle.

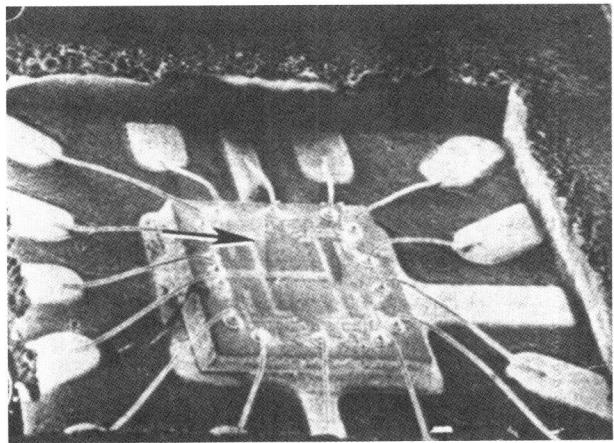

Fig. 54 Examination of a small region (dark square) of an integrated circuit die at a 10 kV beam voltage resulted in charging of a portion of the package out of the field of view by backscattered electrons (arrow).

Fig. 55 Initial image. Figures 55 and 56 show the image degradation due to the charge buildup on the package described in Fig. 54.

charged region by deflecting the emitted secondaries away from the detector. The charging effect is time-dependent, since a charged region may discharge by a breakdown to ground, and then the region can recharge.

Portions of semiconductor devices charge readily because of the insulating materials used in their fabrication. Glass, alumina, and oxide are examples of insulating materials used as substrates, packages, or passivating layers in semiconductor fabrication. These materials, along with nonconductive contaminant particles, are possible sites for charging.

In order to recognize how charging manifests itself, some examples are given. One example has already been seen in the discussion of tilt angle. Figure 52 shows glass passivation which has become charged between the metallization pattern. A single nonconducting particle whose image has become extremely bright is shown in Fig. 53. As a result of the high negative potential at its surface, the secondary electron collection has been reduced in the vicinity of the particle.

It is not necessary for an area of the specimen to be in the field of view for it to become charged! There can be a charge buildup where backscattered electrons strike a surface. Figure 54 shows a device which had been previously viewed with an accelerating voltage of 10kV at a higher magnification. The raster area during that exposure is shown by the darkened re-

Fig. 56 Degraded image.

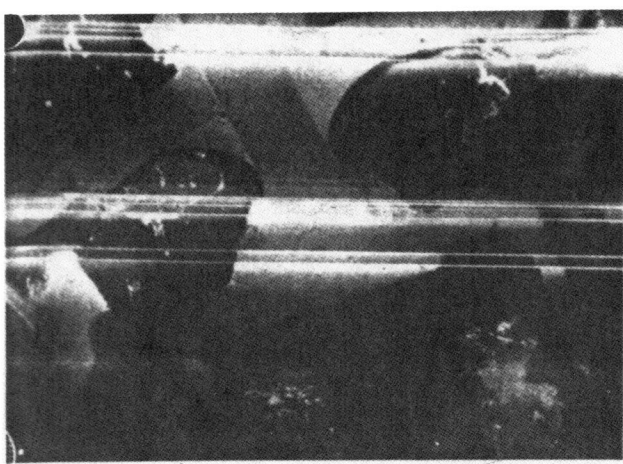

Fig. 57 Examination of a sample consisting of conducting and insulating areas without coating often results in severe image distortion due to charging and discharging in areas, which result in the bright lines in the photo.

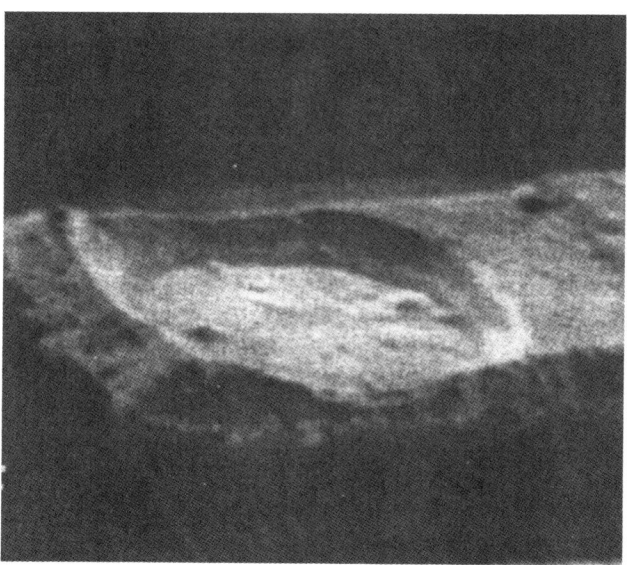

Fig. 58 Charging often is not obvious and can result in subtle degradation of the image as in this case. The resolution is quite poor. It is not difficult to see that charging can render SEM images useless. Reducing the electron beam current will reduce, but not eliminate, charging.

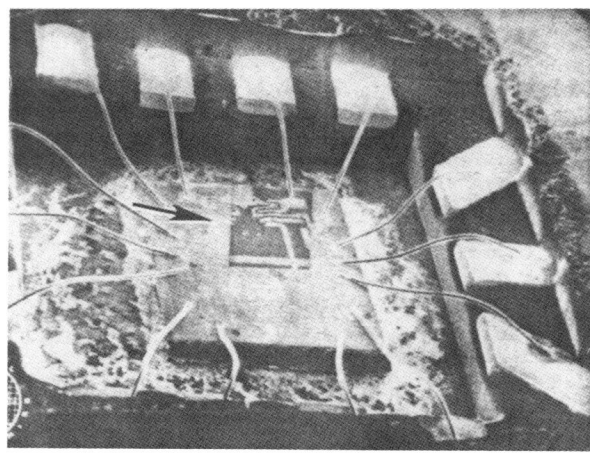

Fig. 59 Examination of the specimen from Fig. 54 at a beam voltage of 3 kV instead of 10 kV prevents charging of the package by energetic backscattered electrons.

gion on the die. The electrons scattered from this area formed a charge buildup on the ceramic package in the area that appears bright (indicated by the arrow). The darkened area around this bright region indicates the suppression of secondary electron collection in the vicinity of the charged region. The effect of this suppression can be seen more clearly in the next two figures which show two versions of the high magnification view.

Figures 55 and 56 show the image degradation due to the charge buildup on the package described in Fig. 54.

Charging is often seen as a distortion of the image in the photomicrograph. Figure 57 shows the characteristic bright areas of charging but in addition there are bright lines breaking the continuity of the image at the top and center. These are caused by unsynchronized discharges of the charged specimen surface during scanning.

Figure 58 is an example of a particularly deceptive charging symptom. There is poor point-to-point resolution, which no amount of manipulation of the SEM controls could significantly improve. In this case, there is a charge buildup in the residual glass left behind by poor etching. As a result, the primary beam cross section is distorted from its normal circular shape to

SEM and Energy-Dispersive X-Ray Analysis

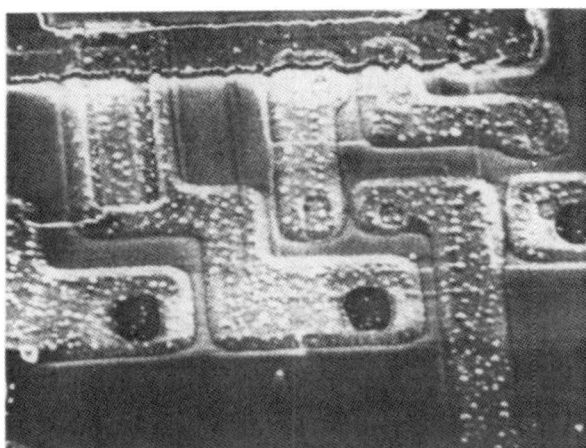

Fig. 60 Examination of a two-layer metallization system integrated circuit in an uncoated state results in preferential charging and signal collection. The dark areas are electrically isolated from the surrounding metallization, i.e., induced voltage contrast.

Fig. 61 Examination of the same sample with only backscattered signal results in the elimination of contrast due to charging effects. Considerable information has been lost, however.

that of an ellipse, which results in a loss of point-to-point resolution. This distortion is very similar to the uncorrectable instrumentation astigmatism caused by dirty apertures in the electron beam column.

It is not difficult to see that charging can render SEM images useless. Reducing the electron beam current will reduce but not eliminate charging.

A technique for reducing the effects of charging is illustrated in Fig. 59. The penetration of primary electrons into an insulating surface, and their storage on and beneath the surface, is the cause of charging. In Fig. 59 a microcircuit similar to the microcircuit in Fig. 54 is shown. However, the high magnification raster at the top of the die in Fig. 54 was irradiated using a high tilt angle and a 3 kV accelerating voltage instead of 10 kV, as in Fig. 54. The energy of the backscattered electrons is 3 keV, which is near the secondary electron crossover, unlike the 10 keV backscattered electrons which cause the charging seen in Fig. 54. In the section on specimen tilt, Fig. 7 demonstrates that specimen tilt could be used as a method to reduce electron penetration, and, therefore, charging. Using low accelerating voltage and specimen tilt is the most expedient means of preventing charging. It can be seen in Fig. 59 that another byproduct of these techniques is better contrast. The contrast in the dark square area in Fig. 59 is rich in comparison to that in Fig. 54.

Figure 60 shows a secondary electron image of a device with a charged passivation layer. Figure 61 shows a backscattered electron image of the same area. The characteristic dark areas adjacent to regions of charging is a result of the distortion in the paths of secondary electrons while higher energy backscattered electrons are not affected. Therefore, using backscattered electrons will eliminate contrast distortion due to charging. A treatment of electron beam interactions with metals and insulators and effects of specimen tilt and accelerating voltage has been given by Oatley.

Fig. 62 Examination of a clean integrated circuit surface at high magnification has resulted in polymerization of vacuum pump oil deposited on the surface in the SEM in the form of a dark square. This artifact can be removed by exposure to an O_2 plasma.

COATING OF SPECIMENS

One technique commonly used to prevent specimen charging, applying a conductive coating over the sample, has not been discussed. Because of the nature of semiconductor devices, such a coating should be a last resort and undertaken only after careful consideration of all phases of the examination to which the device will be subjected. Once coated, a device can no longer be operated, nor will use of voltage contrast and EBIC modes of SEM examination be possible. X-ray analysis can become more difficult after application of *any coating but carbon*.

Fig. 63 A microcircuit window which has been buried under large globs of aluminum from a poor evaporation process. This specimen is a total loss for examination purposes because of the excessive artifacts introduced.

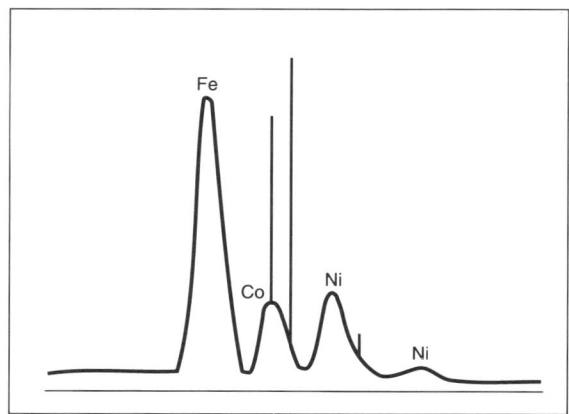

Fig. 64 Kovar, 20 kV. EDX spectrum of a Kovar particle at a beam voltage of 20 Kev. The Fe, Co, and Ni peaks are well defined with the Ni peak being higher than the low percentage Co peak.

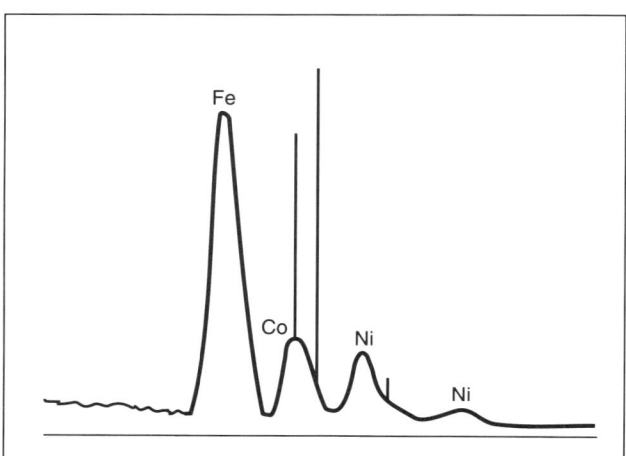

Fig. 65 Kovar, 10 kV. Same sample as 65 but at a 10 Kev beam voltage. All three elements are detected but the nickel peak is low due to poor excitation by the 10 Kev beam.

Scanning a semiconductor surface at high magnification can polymerize hydrocarbon monolayers deposited by an oil-pump vacuum system. This effect appears as a very dark square area as in Fig. 62. Such contamination is an unwanted artifact, and can be avoided by not making prolonged high magnification scans on a localized area. All focusing should be done outside, but adjacent to, the surface area to be photomicrographed. Alternatively, the hydrocarbon diffusion pump oil can be replaced with an oil that will not be polymerized by the primary beam. A good alternative is found in perfluorinated polyether oil. This oil fractures into molecules of low molecular weight which are swept away by the vacuum system. Another method is to employ a cold finger or trap to help reduce specimen contamination by capturing the oil contaminants on a cold surface. Pump oil contamination is not a problem with an SEM equipped with dry-pumped vacuum system.

The SEM primary electron beam damages the semiconductor device. Electron beam penetration into the device results in ionization processes which change the surface oxide properties and thus degrade the electrical parameters. Digital bipolar devices are susceptible to parameter changes during SEM examination; however, it is unlikely that it would cause them to cease functioning. MOS devices are readily damaged by exposure to an electron beam. Even a very short exposure can make them inoperative.

For devices for which optimum electrical performance is required after examination, the only way to eliminate electron beam damage is to not examine the device with an SEM. However, if an SEM examination must be performed, an accelerating voltage as low as possible, considering the resolution which is required, should be used. Also, the lowest practical beam current should be used and the specimen should be irradiated only as long as absolutely necessary. Appropriate alignment techniques are required so that time is not lost searching for the area of interest. If the examination does not include voltage contrast or EBIC modes, a bias should not be applied to the device junctions. It should be noted in subsequent use of the device that the electron beam could have made some irreversible changes in the device.

Semiconductor Specimen Preparation. One of the primary advantages of SEM over TEM and optical microscopy is the minimum amount of specimen preparation necessary. However, some preparation is required and if the study is a failure analysis, the microscopist must take care that the preparations do not obscure the cause of failure being studied. An examination with an optical microscope should be made of a semiconductor device immediately after delidding the package. The analyst should observe any contamination that may be present and which appears to be the result of device failure or poor fabrication. If any solder balls, fragments of broken passivation, or smeared metallization are noted, the sample surface should not be disturbed or valuable information could be lost. Photographic documentation after the optical examination is desirable.

SEM and Energy-Dispersive X-Ray Analysis

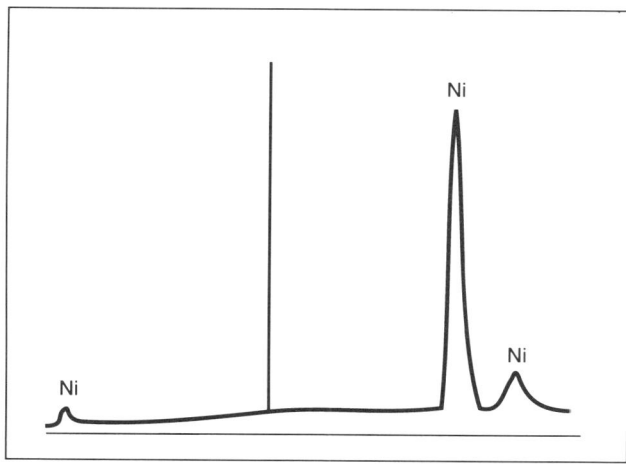

Fig. 66 Nickel, 20 kV. EDX spectrum of Nickel at a 20 Kev beam, both the low and high peaks are excited.

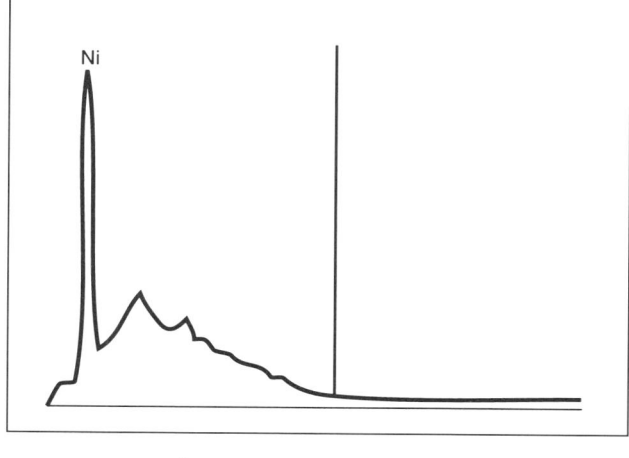

Fig. 67 Nickel, 5 kV. At a low, 5 Kev beam, only the low energy line of nickel is excited.

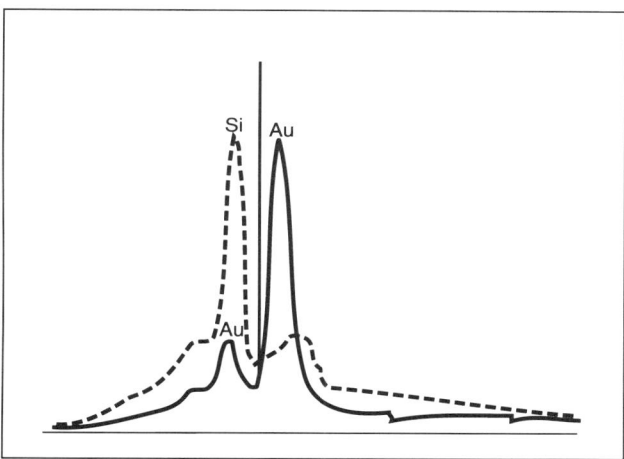

Fig. 68 Preform, 20 kV. EDX spectrum at 30 Kev of a Au-Si eutectic (bars) as compared to the same sample at 5 Kev (dots). Note the much higher silicon peak at 5 Kev, as compared to the gold peak.

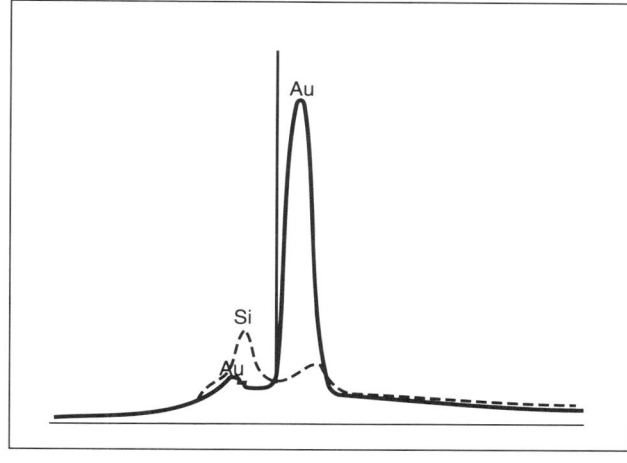

Fig. 69 Preform, 20 kV. EDX spectrum (bars) of gold at 20 Kev shows apparent silicon line at 1.65 Kev, which is actually an Au-M-Zeta line. Dot spectrum shows true silicon peak.

If the optical examination did not reveal any obvious device problems, the glass passivation layer should be removed from the semiconductor surface. Typically, the glass is removed by a wet acid or gas plasma etch. This is necessary because the glass does not replicate the underlying metallization stripes or oxide steps. Inadequate removal of the glass will leave a residue of glass on the metallization which would make topographical interpretation difficult. The residue also charges, which will normally manifest itself in an inability to focus, as illustrated in Fig. 58. The specimen should be firmly fixed to the specimen holder using an electrically conductive adhesive. Aquadag preparations and conductive silver paste can also be used. The holder should be firmly mounted in the stage and make electrical contact to it.

RECOMMENDED COATING PROCEDURES

Coating the specimen with a thin (5- to 20-nm thick) layer of carbon or metal may be used to enhance secondary emission and reduce charging. However, this deposited coating has potential drawbacks: its application is time consuming, there is die surface detail which is covered by the thin film coating, and the device is rendered inoperable. Despite these drawbacks, specimen coating is the method of suppressing charging and increasing the electrical emissivity of a specimen most often utilized. When applied with the specimen surface parallel to the rays of evaporating coating material, subtle changes in surface topography can be accentuated by shadowing.

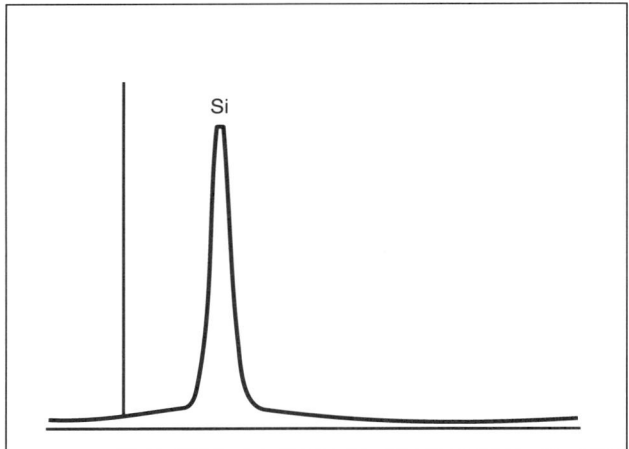

Fig. 70 Silicon, 5kV. EDX spectrum of a pure silicon sample excited with a 5 Kev beam.

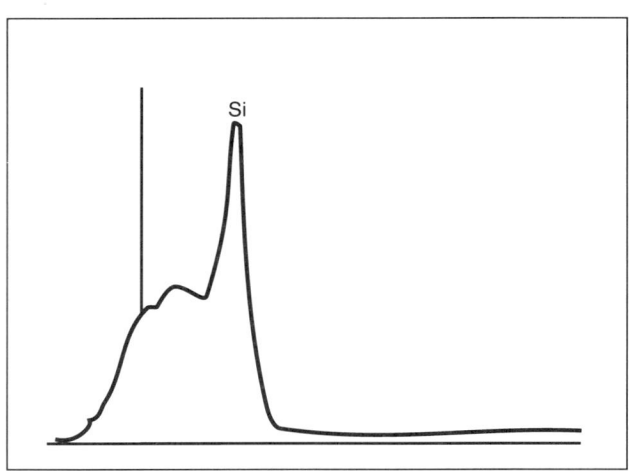

Fig. 71 Silicon, 2 kV. Same sample as Fig. 71 except at a beam voltage of 2 Kev. Silicon is still excited, but notice stray counts to its left.

Vacuum Evaporation

In order to reduce the amount of data which is obscured by specimen coating, certain guidelines should be observed. The selection of the evaporant material is the first consideration. Criteria for selection of a material include: electron emissivity, replication ability, oxidation rate, and melting point. Gold (Au) has excellent electron emissivity but, at high magnifications, can have an agglomerate (lumpy or spheroidal) appearance. Silver (Ag) has the best replicating ability, but in a few days it tarnishes and the specimen cannot be viewed. Gold-palladium (AuPd) alloy is a compromise between high electron emissivity and replicating ability, but the heat from its high melting point may alter the sample. Aluminum (Al) is a good alternative, having good replicating ability, a low melting point, and adequate electron emissivity. On a very irregular specimen surface, carbon is often used on a first layer to establish surface conductivity because its atoms deflect from surrounding surfaces and reapproach the semiconductor surface from many different angles. A second, very thin film of a noble metal is then applied to improve electron emissivity. For X-ray microanalysis, carbon can be used exclusively to establish conduction, while avoiding extraneous X-ray information.

The film should be the thinnest possible coating that will suppress charging. The best procedure is to deposit a very thin film of a noble metal (<10 nm, or 100 Å, thick) and then view the specimen in the SEM to see if the charging has been suppressed. This procedure can be repeated as often as necessary to satisfactorily suppress charging. Although very time consuming, this procedure will limit the amount of die surface data lost.

An evaporated thin film will not be uniform on a semiconductor die surface because of the irregular metallization and oxide steps. A gimbal mechanism may be used to spin the specimen through a complex movement exposing each die surface step for an equal amount of time to multiple evaporate sources, but it is still doubtful that the thin film will be truly uniform. There are two reasons for this. First, there are variations in the specimen-source distance while spinning and the thickness of an evaporated thin film depends on the inverse square law. Also, there are changes during evaporation in the nominal angle of deposition.

To prevent the evaporant source heat from altering die surface topography during evaporation, three precautions should be taken. (1) The specimen should be kept at least 3 to 4 cm from the evaporant source(s). (2) A pre-evaporation shutter should be kept between source and specimen until the coating material has begun to evaporate. (3) A coating material with a low melting point, such as aluminum, should be used.

Sputter Coating

An alternative to vacuum evaporation as a specimen coating technique is sputter coating. In the evaporation technique, the coating material is heated to a sufficiently high temperature so that it evaporates, depositing a thin film on the die surface. In the sputter coating technique atoms from the coating material are ejected when bombarded by relatively high energy particles, usually argon ions. Some of the ejected atoms will land on the die surface to be coated. The particles must be of sufficiently high energy in order to overcome the binding energy of the atoms at the surface of the target. The fast heavy particles used to erode the target are usually derived from an ionized inert gas, and are produced by three main processes: ion-beam sputtering, radio-frequency sputtering, and direct current sputtering.

The main advantage of sputter coating is that a continuous thin film is deposited, even on parts of the die surface which are complex, and those surfaces which are not directly facing the rays of sputtered atoms. This occurs due to the fact that, under the high pressures used, the coating material atoms experience many collisions and approach the die surface from all directions. This effect is achieved without using a gimbal movement. Therefore, the sputter coating technique has good surface replication ability, and there is a reduced chance of surface damage by heating. Noble metals should be used because of their emissivity. *Aluminum is not adaptable to this technique because its surface oxide prevents sputtering*. The techniques advised for specimen coating should be rigorously followed. Thin film

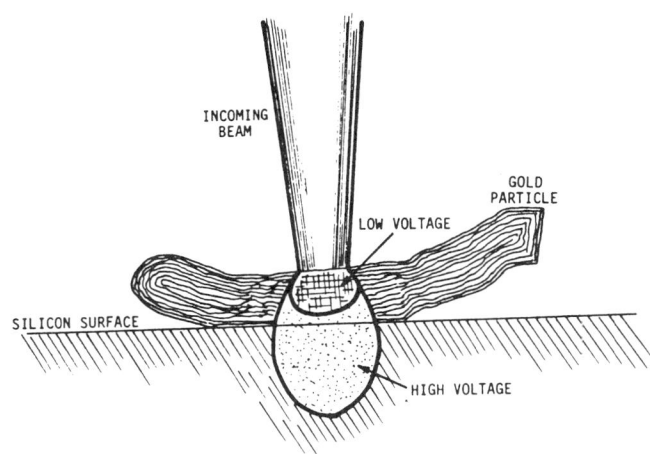

Fig. 72 Example of gold particle on silicon surface.

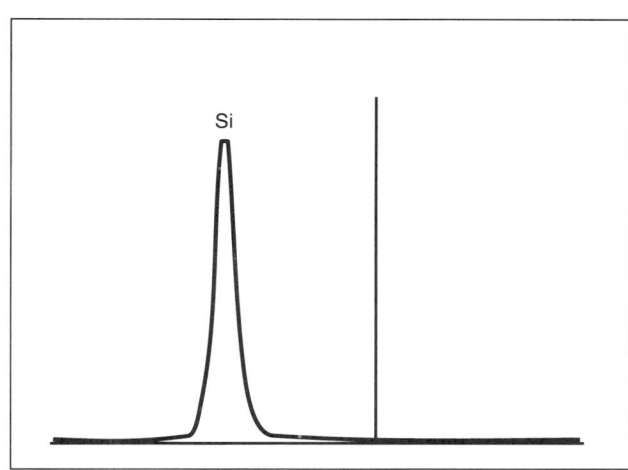

Fig. 73 20 Kev beam generated spectrum of a thin (20 nm) NiCr film on silicon dioxide. Only the silicon shows up.

evaporations which are performed improperly will ruin the possibility for SEM examination. Figure 63 pictures a microcircuit window which has been buried under large globs of aluminum from a poor evaporation process. This specimen is a total loss for examination purposes because of the excessive artifacts introduced.

SEM operation procedures must not become stereotyped for failure analysis applications if the maximum amount of information is to be obtained from the failed device. The analyst should recognize the need for SEM techniques, or imaging modes, which will yield the highest quality information. For failure analysis applications to be most productive, the SEM analyst must be fully aware of the various interactions between specimen and beam and be prepared to modify his operational modes to fit the requirements of the specimen.

In no other application does the operator need to be as informed about the subtle interactions between instrument and specimen as in semiconductor failure analysis.

Correlation with Optical Microscopy. Most often, initial phases of failure analysis include examination of the specimen by low (10×–20×), medium (100×–500×) and high power (1,000×–1,500×) optical microscopy. This is performed under both oblique and direct illumination. Unless the analyst is familiar with the SEM he can often be both misled and disappointed by the data as compared to the optical microscope. The two most common erroneous assumptions are: (1) since the SEM has better resolution it can "see" smaller defects, (2) it can image diffusion and oxide defects. Neither case is true, due to some (useful) properties of light. In the application to thin films and their defects, interference fringing and refraction often renders visible defects in the range of 3 to 8nm, which are still considerably smaller than the resolution limits of many SEMs.

Defects in the oxide layers which are obvious with optical microscopy, due to refraction or phase changes, can be quite invisible on the uncoated specimen due to oxide charging; then defects or features visible optically due to interference coloring will not be detectable in the SEM.

ENERGY DISPERSIVE X-RAY SPECTROSCOPY (EDXS)

The interaction between the energetic monochromatic electrons from an impinging electron beam and the electrons in the atoms of the specimen results in the generation of X-rays. Characteristic X-rays are always of a specific energy or wavelength and identify the elements in a specimen. The same principle is used to create the X-ray source for X-ray radiography, except the current of the electron beam is very high and the goal is the generation of a strong source of X-rays.

Based on this well-known effect, several analytical instruments have emerged to assist the failure analyst. The first, the electron microprobe, entered the field in about 1961-1962. This instrument utilized a stationary beam of electrons to excite X-rays from the sample area of interest. The X-rays were characterized by their specific wavelength with a crystal diffraction grating. The electron microprobe suffers from its selectivity and method of analysis. Various crystals are required to efficiently diffract different wavelengths of X-rays. In addition, a given crystal can diffract only one wavelength at a time. Therefore, ranging over all the wavelengths in the periodic chart is a very tedious and time consuming task. The second, much more common instrument, is an energy dispersive X-ray analyzer (EDXA) attachment to the SEM. This tool and its capabilities make use of the known binding energy in a doped wafer of silicon, which is used as a detector.

The generated X-rays impinge on the silicon detector surface. The silicon is doped with lithium. The penetration depth of the X-ray into the silicon is a direct function of the energy of the X-rays. Along the penetration track, interaction occurs between the X-rays and the silicon atoms, creating hole-electron pairs. The generation of each hole-electron pair requires a specific amount of energy. Thus a weak X-ray of a shallow penetration depth will generate a smaller current pulse than a more energetic X-ray of a longer penetration track. If the optical and physical appearances give no obvious clues to the particles' makeup, then with the energy dispersive analyzer (EDX) the analyst has to determine the particle compositions and speculate on a potential source for each.

Recommended EDX Procedure

1. The operator aligns the specimen in such a manner that a unobstructed path exists from analysis site to detector.
2. For a hybrid microcircuit, if uncoated, a low initial beam voltage should be used to locate the area(s) for analysis without charging up surrounding insulators.
3. The beam should be centered on the particle or area of interest and beam voltage increased to at least 20 or 25 Kilovolts. At 20-25 Kev, sufficient energy is available to produce X-ray emission for all elements up to lead (Pb), gold (Au), tungsten (W) and platinum (Pt) and/or bismuth (Bi). All these elements are atomic number 74 or heavier. Typically, a full spectrum count rate of 1000-1500 counts per second for 60-120 seconds will result in a spectrum of sufficient resolution to identify the major constituents of the particle.
4. If backscatter or penetration thru the particle to the underlying region is a concern, then a reduction of beam voltage may be feasible.

A beam voltage of 10 Kev is adequate to excite the chromium and iron K-α lines but not sufficiently for quantitative work. If, however, just detection and identification is required and only one of several typical elements exists of the Cr, Fe, Ni, Cu, Sn family, then the L-lines at approximately 1 Kev can be used. For confirmation of the tungsten, platinum, gold, lead group (W, Pt, Au, Pb,) then the M-series of lines between 1.5 and 2.5 Kev can be used at an incident beam voltage of 5Kev.

Caution

Due to the low energy spread at the L-series and M-series lines in the 1-3 Kev range, great difficulty is experienced in uniquely identifying the foregoing elements within a group.

For the four particles mentioned, then, the recommended conditions could be:

1. Kovar: Iron, Nickel, and Cobalt –20 Kilovolts, 1500 counts per second for 120 seconds —Spectrum must he carefully examined to spread the Kα_1 and Kα_2 lines because of the overlap between the three sets of lines.
2. Nickel: Nickel or Nickel and Phosphorus –20 Kilovolts for general spectrum (can be easily reduced to 10 Kilovolts) is also feasible since the Kα and Kβ phosphorus lines are easily excited and the Ni Lα lines are also excited.
3. Eutectic: Gold and Silicon: 20 Kilovolts will excite all the M-series gold lines as well as the first three L-series lines. If gold is confirmed then great caution must be exercised because an M series (Zeta) lines exist at 1.65 Kev, which many mistake for silicon. If the sample is fairly small, then analysis at 5 Kev is recommended; this greatly reduces penetration, and the M-Zeta gold line does not overpower the silicon line at 1.74 Kev.
4. Silicon or Silicone: (Silicone contains Silicon). If the particle does not charge up, it is probably not silicone or SiO_2. Either 3 or 5 kV is more than adequate to confirm the existence of the silicon 1.74 Kev Kα_1 and α_2 lines.

The overlapping of peaks of different elements must be well recognized, understood and expected. Although newer EDX systems have a built-in data analyzer, they are not foolproof.

Efficient X-ray generation from specific elements requires use of an electron beam voltage at least twice as energetic as the X-ray line to be generated.

Example

1. Excitation of the iron Kα line at 6.4 Kev energy requires an incoming beam voltage of a 12.8 Kev although 10 Kev is sufficient except for quantitative work.
2. Excitation of the gold Kα line at 9.63 Kev requires at least an 18.8 Kev beam so most analysts use a 20, 25 or 30 Kev beam.

Caution

The presence of gold can be easily detected with a 5 Kev beam if the analyst uses the gold M (#4l) line at 2.12 Kev. Although not as intense, it is an effective approach.

Hydrocarbon or organic residue containing chlorine is a common foreign contaminant in the electronics industry. A beam voltage of 20 Kev is more than adequate to excite the 2.71 Kev chlorine line. Once the chlorine has been detected, an analysis with a 5 Kev beam may result in a much stronger chlorine peak, indicative of the fact that the chlorine is contained in a thin surface film.

X-ray detection suffers from the same restrictions as backscattered electron detection, i.e., shadowing. X-rays, similar to backscattered electrons, travel in an absolutely straight trajectory from the emission point, and cannot be focused or collected other than by the detector's exposed surface. *Although not nearly as sensitive to sample location and tilt as the crystal spectrometer, there are usually "dead spots", that is, combinations of height and tilt which result in poor X-ray detection.* The operator must constantly monitor the X-ray spectral count rate to ensure he is not imaging and analyzing such a "dead zone".

Light elements can result in an apparent "dead zone". Elements lower than atomic number 9 are not detected by a normal EDX system with a beryllium window. Thus an oxide, carbide or hydrocarbon cannot be analyzed by energy dispersive techniques. However, inferences can be drawn about a sample. If the analyst is sure he is not in a dead zone and the total count rate drops markedly when the beam traverses from a detectable matrix to the unknown, for instance, from 1500 cps to 50 cps, then the detector is not sensing the light element X-rays. Other knowledge about the specimen can assist the operator in postulating about the unknown as to whether it is oxide or hydrocarbon.

Several other artifacts are common enough to warrant mention:

1. High intensity backscattered electrons can generate X-rays from areas other than the area of interest. X-rays can also generate X-rays in the same manner.
2. Any film used to coat the specimen will alter the observed spectrum. Carbon will alter it the least. Aluminum, gold or

SEM and Energy-Dispersive X-Ray Analysis

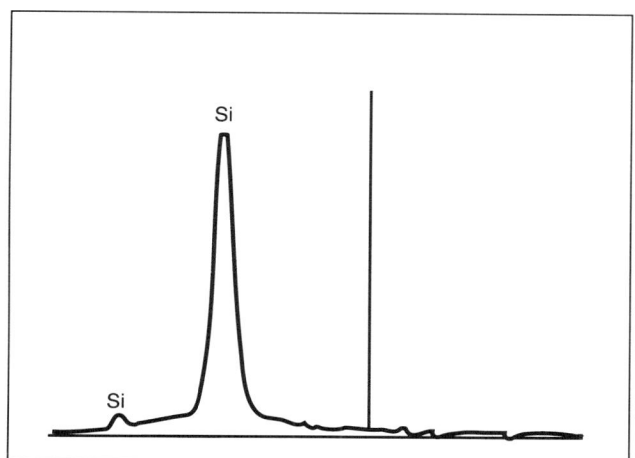

Fig. 74 10 Kev beam generated spectrum of sample from 74. Ni line at 1 Kev is starting to show up.

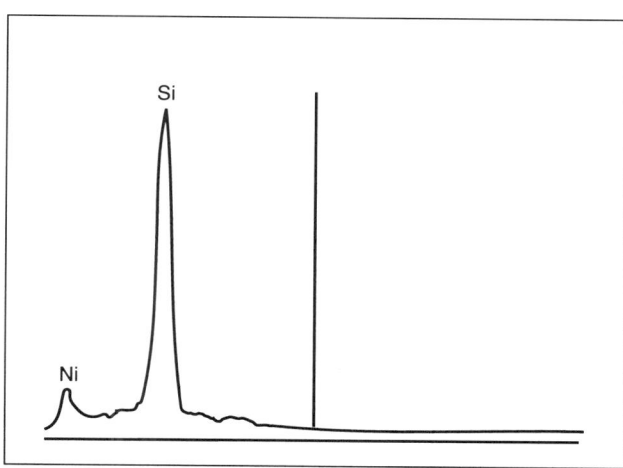

Fig. 75 At a beam voltage of 5 kEv, the low-order Ni line is becoming very obvious.

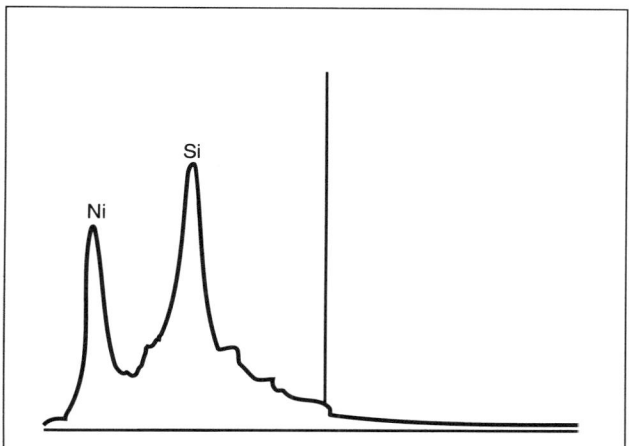

Fig. 76 At a beam voltage of 3 Kev, the Ni peak is almost as large as the silicon peak.

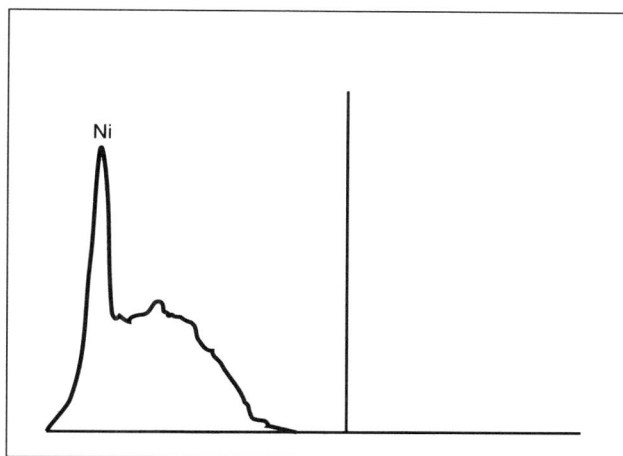

Fig. 77 At 2 Kev, the beam no longer penetrates the NiCr film to effectively excite the underlying silicon and only the Ni line is obvious.

gold-palladium should be used sparingly. Never coat with a film of the same material that is a subject of the analysis.

3. A common element in many spectra is iron. Under some conditions this is a false indication, which can be generated by the final lens pole piece.

Why, then, would the analyst want to use a low beam voltage for X-ray analysis? Penetration of the specimen is a direct function of beam voltage and sample density. Analysis of small particles or thin films is very dependent on operator experience and knowledge of the specimen. An example is illustrated in Fig. 73 thru 77 of thin film analysis.

As can be seen from Fig. 72, the gold particle on a silicon surface can easily be analyzed as containing silicon if a 20 Kev beam is used for analysis. Using the 5 Kev beam restricts penetration to the particle itself.

Because of the complexity of samples usually encountered in failure analysis, a thorough working knowledge of beam-specimen interactions is a necessity. Usually the author uses a 10 or 20 kilovolt beam to determine what elements are present. Then, the beam voltage is lowered to bracket the elements in the area of interest and so obtain an accurate spectra at the lowest possible beam voltage (Fig. 73 thru 77).

The relative intensities of spectral lines may vary significantly with changes not associated with excitation efficiency. This is especially the case with thin films consisting of relatively light elements.

As widespread and useful as the energy dispersive spectrometer (EDS) has become, it still is no panacea. Some problems, defects, films, chemicals, elements and compounds, are not readily or definitively analyzed by EDXA (EDXS).

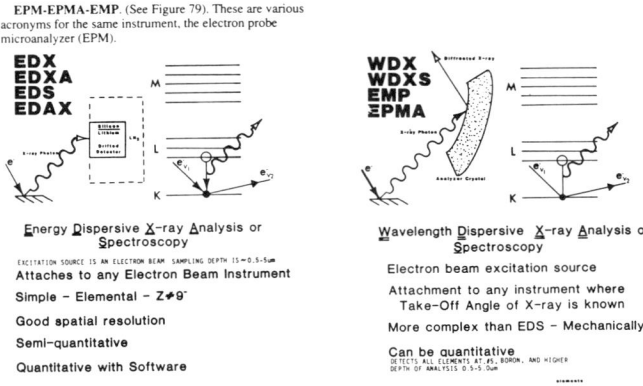

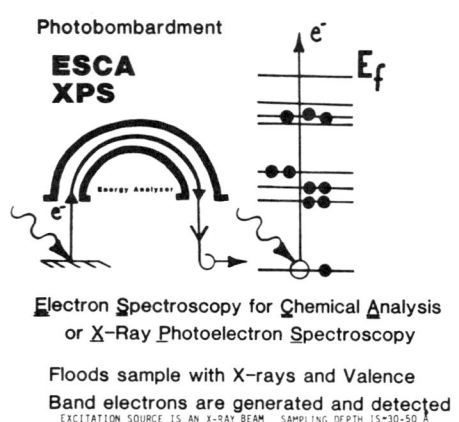

Fig. 78 (a) Energy dispersive X-ray analysis concept. (b) Wavelength-dispersive X-ray analysis concept.

Fig. 79 Electron Spectroscopy for chemical analysis concept.

Note: EDXA, EDXS, EDS, and EDX all may be used to indicate "energy-dispersive X-ray analysis." "EDAX" is a tradename for energy-dispersive X-ray detector product. Thus, instruments which only analyze thin surface films or have incredible sensitivity levels have been designed to extend our analytical capacities. The most common are:

AES - Auger (pronounced O-JAY) Electron Spectroscopy

ESCA - Electron Spectroscopy Chemical Analysis

SIMS - Secondary Ion Mass Spectroscopy

The following pages indicate the various acronyms each of these techniques is known by, and the fundamental physical mechanism by which the technique produces information about a sample. The final charts compare pertinent parameters for all the common analytical instruments.

For the majority of failure analysis, the energy dispersive X-ray spectrometer system is the basic working tool. Only after its capabilities have been exhausted, need the analyst resort to the use of more complex techniques.

EPM-EPMA-EMP. (See Fig. 78). These are various acronyms for the same instrument, the electron probe microanalyzer (EPM). The EPM is the oldest of the instrumental techniques. It was the forerunner of the SEM-EDXA system.

In review, constraints from the EPM are similar to the constraints for the SEM system. The EPM may use a light microscope or SEM image to view the specimen and may have slightly higher sensitivity for many of the elements. Sample size and shape are limited as compared to SEM.

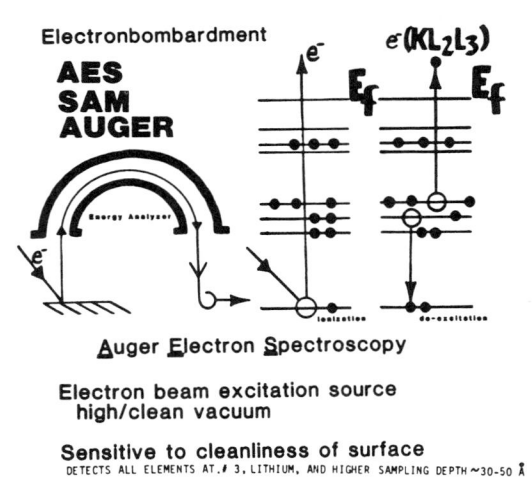

Fig. 80 Auger spectroscopy concept.

Results from EPM are often quantitative as compared to the EDXA. There is better light element detection down to boron and carbon. Determination of oxides and nitrides are a significant advantage.

ESCA. (See Fig. 79. Electron spectroscopy for chemical analysis provides valence state information of the material presentonthesurfaceofthepackage.Spatialresolutionispoor.

ESCA resolves an area as small as an integrated circuit die, no less. Usually no real visual observation is possible. Excellent resolution of the various carbon compounds are derived from ESCA. Since the excitation source is a soft X-ray beam, very shallow surface data is obtained. For this reason, ESCA is an excellent analytical tool for determining the molecular structure of polymer coatings and the identification of chemical states.

AES. Auger electron spectroscopy has an irradiation source that is a low voltage electron beam similar to that used in EDXA or EMP. The difference in the instrument techniques is in the detector. The cylindrical voltage analyzer/detector is sensitive to weak Auger electrons which are emitted from the first 15 to 30 Å of the surface.

Auger spectroscopy detects all elements from lithium upward and displays good spatial resolution. (See Fig. 81). If combined with a sputtering source the technique can provide shallow depth profiles.

SIMS. SIMS is an acronym for secondary ion mass spectrometry. The surface of a device to be analyzed is bombarded with ions which then cause secondary characteristic ions to be emitted (sputtered) from the surface. See Fig. 81, for a functional diagram. The sputtered ions are then fed into a mass spectrometer for identification.

The SIMS is the most sensitive of all techniques. In regard to minimum detectable limits, it is capable of detecting all elements and is a surface analytical tool. It is the only instrument capable of directly measuring dopant profiles in a semiconductor.

The drawbacks of SIMS are availability, expense, and poor spatial resolution.

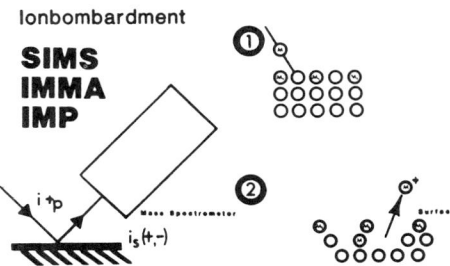

Fig. 81 Secondary ion mass spectroscopy concept.

References

Special note: Although the publication dates on some of these references are not current, they are classics in the field, and are applicable to the work discussed in this paper, as well as to current SEM tasks.

1. *Proceedings of the Symposium on Scanning Electron Microscopy,* published by the IIT Research Institute, 1967 through 1980.

2. P.R. Thornton, *Scanning Electron Microsopy*, Chapman and Hall, Ltd. 1968.

3. O.C. Wells, A. Boyde, E. Lifshin, and A. Rezanowich, *Scanning Electron Microscopy,* McGraw-Hill, 1974.

4. J.W.S. Hearle, J.T. Sparrow, and P.M. Cross, T*he Use of the Scanning Electron Microscope,* Pergamon Press, 1972.

5. K.F. Heinrich, *Electron Beam Microanalysis,* Van Nostrand-Reinhold, 1981.

6. J.J. Hren, J.I. Goldstein, and D.C. Joy, *Introduction to Analytical Electron Microscopy,* Plenum Press, 1979.

7. M.L. Meny and R. Tixier, Microanalysis, *Scanning Electron Microscopy,* Les Editions des Physique, 1978.

8. B.L. Gabriel, *SEM: A User's Manual for Material Science,* published by American Society for Metals, Materials Park, Ohio, 1985.

MICROBEAM ANALYSIS AND SEMICONDUCTOR RELIABILITY

by R. K. Lowry

1. INTRODUCTION

Microbeam analysis is an essential tool for reliability scientists and failure analysts. Electrical measurements are not always sufficient to fully understand performance of electronic devices. Indeed, every electronic malfunction has an underlying physicochemical cause. It is often necessary to supplement electrical parametric measurements with electron beam and/or ion beam imaging and analysis of the die or package under test, to fully understand causes of failure.

Microbeam analysis can be "micro" in two senses of the word. It can be micro in the z (height) dimension, in that ion and electron beams can give information about the uppermost few Angstroms of the sample. This can be true even with beams many millimeters in diameter. It can also be "micro" in the x,y dimension where finely-focused beams provide data from spatial areas fractions of a micron in diameter. Both types of "micro" beam analysis are important for electronic materials characterization. The former is important for broad areas of general film surface composition and/or cleanliness characterization. The latter is important for obtaining data from the closely-packed geometries of today's integrated circuits.

It is not the aim of this article to make surface scientists or microanalysis equipment experts out of failure analysis and reliability professionals. This article simply gives an overview of a few of the more common microanalysis techniques, and a potpourri of successful applications to reliability-oriented investigations.

The investigations discussed are chosen not just from failure analysis cases, but also from materials selection and assurance studies, studies to design and develop robust products and processes, and cooperative efforts to serve outside customers. These types of examples are included because reliability engineers

Table 1 Analytical Methods Overview

Analytical method	Incident particle, for sample excitation	Signal detected providing information about sample	Type of information (x,y) [z]	Elemental Sensitivity
Scanning electron microscopy (SEM)	electrons	secondary electrons emitted from sample atoms	visual image of topography	na
Energy dispersive and wavelength dispersive spectroscopy (EDS/WDS)	electrons	secondary and backscattered electrons and x-rays	chemical element identity, $(1\mu \times 1\mu)$ [1-5μ]	0.1-1%
Auger electron spectroscopy (AES)	electrons	Auger electrons from near surface atoms	chemical element identity $(.1\mu \times .1\mu)$ [20A]	0.1-0.5%
X-ray photoelectron spectrometer (XPS)[1]	x-rays	photoelectrons from near surface sample atoms	chemical element identity & valence $(75\mu \times 75\mu)$ [20A]	0.1-0.5%
Secondary ion mass spectrometer (SIMS)	ions	secondary ions emitted from the sample	chemical element identity $(.2\mu \times .2\mu)$ [20A]	$\geq$ppm

[1] Also known as ESCA or Electron Spectroscopy for chemical analysis

Table 2 Applications of Microanalytical Methods

	SEM	EDS	AES	XPS	SIMS	TOPIC	DIE	PKG	IMPACT
1.					X	Qualify Product Handling Material	X	X	ESD; Ionic Contamination
2.	X	X	X			Soldering		X	Connection Reliability
3.		X				Seal Glass Contamination		X	Electrical Stability
4.			X		X	Die Cleaning	X	X	Molded Part Stability
5.	X	X				Human-Sourced Contamination	X	X	Corrosion; Ionic Contamination
6.	X		X			Bond Pad Quality	X		Wire Bond, Electrical Integrity
7.		X				Contact Resistance	X		Electrical Parameters
8.				X		Laser Processing	X		Electrical Parameters
9.					X	Device Ionic Contamination	X		Electrical Parameters

are not just failure analysts any more. They play an ever-increasing proactive role, often as part of TQM/continuous improvement activities, in design and analysis for failure prevention.

In these efforts, reliability scientists must often submit samples requiring microbeam analysis work to an in-house or an outside commercial analytical laboratory. The more background knowledge they have about strengths and limitations of various methods, the more fruitful will be the working relationship between themselves and analytical personnel, and the better the understanding they will have of the completed data.

Table 1 is a general overview of the analytical methods discussed in this article. The "type of information" column indicates information provided and the extent to which the analysis method is "micro", in terms of (x,y) and [z] dimensions from which the analysis information is obtained.

Table 2 lists the topics discussed in the following pages. The examples comprehend at least one application of the listed microanalytical methods. Examples were also chosen to represent both die level and package level types of reliability problems, characterized by applying the proper analytical method. Each application is discussed briefly in terms of the reliability hazard investigated and reasons for choosing the analytical method.

Table 3 Selected Supplies for Handling Electronic Devices

Finger Cots	Sticks
Gloves	Foam
Wipers	Bags
Swabs	Staticides

Table 4 Contaminant Effects

CONTAMINANTS	EFFECTS ON DIE	EFFECT ON PACKAGES
Positive Ionics e.g., Na+, K+, etc.	Electrical Anomalies: Inversion Drift Leakage Instability	Pin-to-pin Leakage
Negative Ionics e.g., CL^-, F^-, SO_4, etc.	Metallization Corrosion Bondability	Pin-to-pin Leakage Plating Defects Corrosion
Residuals, e.g., Hydrocarbons	Film Adhesion Contact Resistance Bondability	Solderability
Particulates	Lithographic Defects Metal Shorts	PIND Test Failures

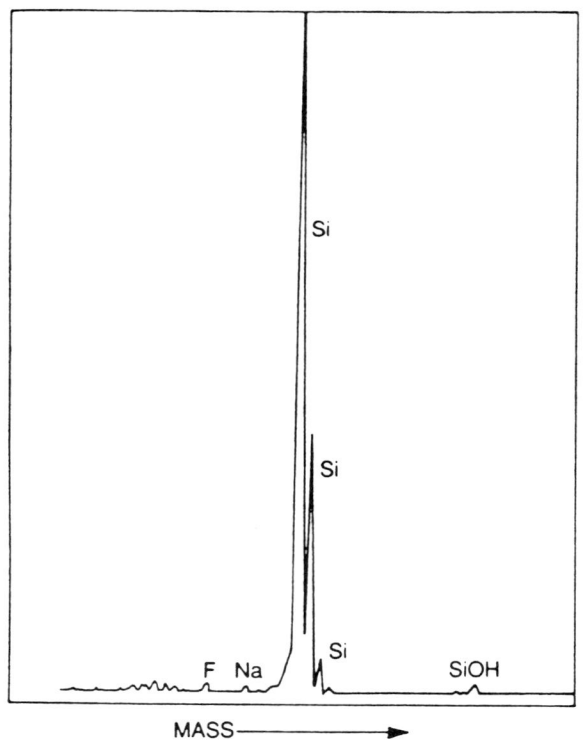
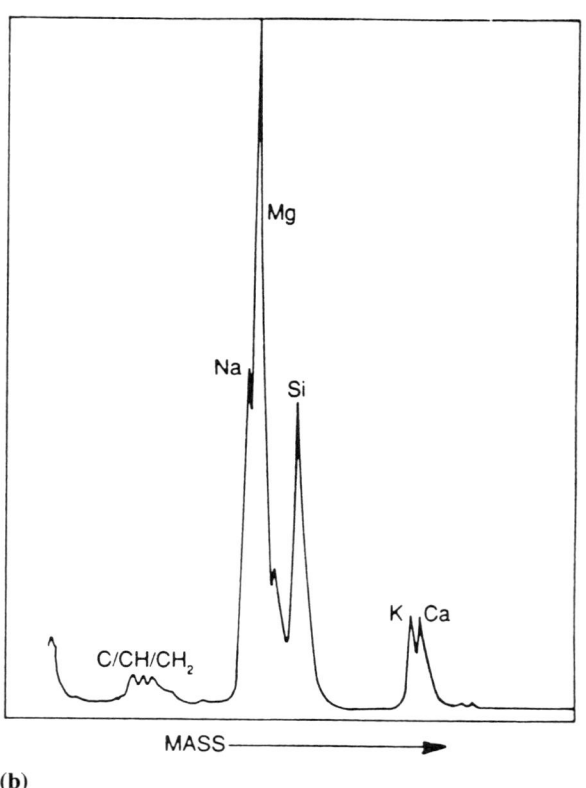

Fig. 1. (a) SIMS spectrum of untouched, deglazed silicon wafer surface. (b) Sample number 2. SIMS spectrum of deglazed silicon surface touched by finger cot.

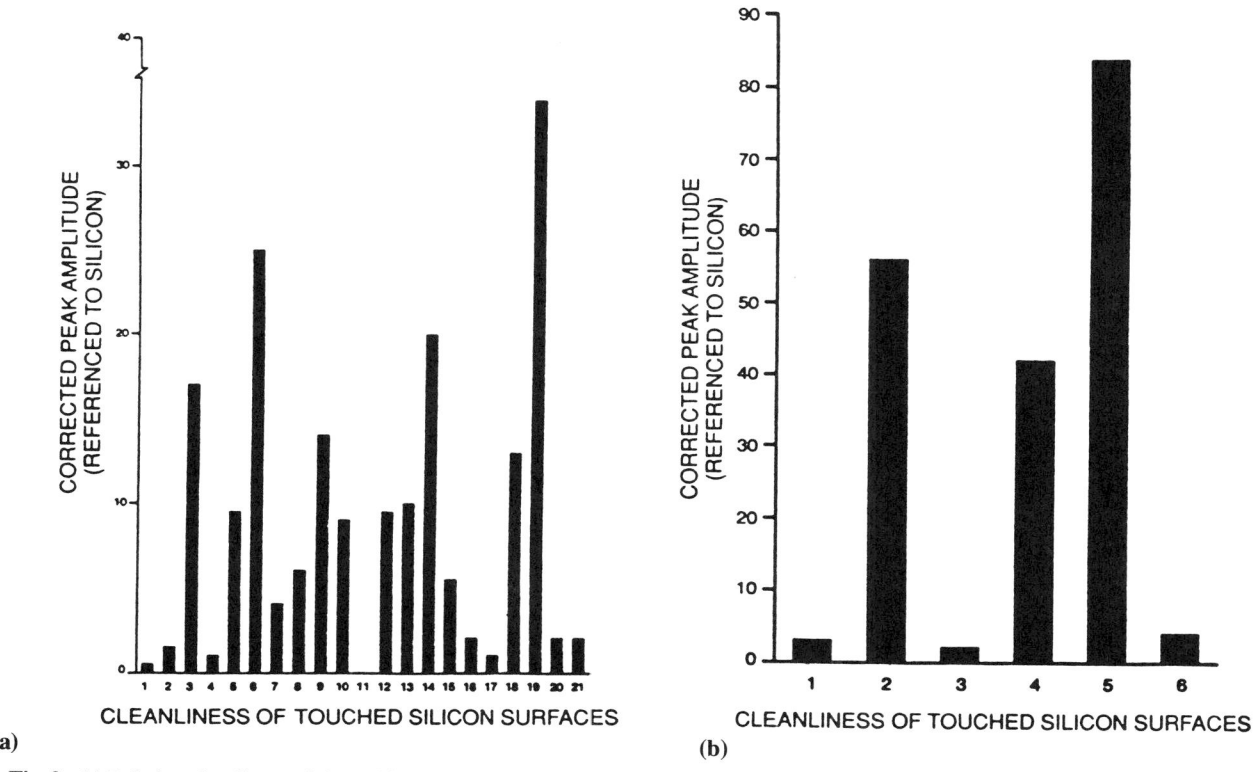

Fig. 2. (a) Relative cleanliness of clean silicon surfaces touched by finger cots. (b) Relative cleanliness of clean silicon surfaces touched by gloves.

One area of special concern regarding cleanliness of handling materials is that of controlling electrostatic discharge (ESD). Product handling materials are often treated with conductive coatings to help drain away electrical charge and keep it from building up on/in devices. Handling materials and supplies must offer the property of preventing static charge buildup which could induce ESD failure. These anti-static properties may be gained from chemical coatings. However, at the same time, they must not transfer chemical residues to surfaces which should remain clean, in order to avoid failure modes such as metal corrosion or electrical instability.

To ensure that product handling materials such as clean room gloves or sticks for shipping packaged IC's do not inadvertently contaminate products, it has been found appropriate to evaluate them for their surface contamination potential.

One of the ways to evaluate contaminating potential is to touch a clean "reference" surface with the item of interest and use surface analysis to measure the extent to which the touched surface was contaminated. This can identify types and amounts of contaminants transferred. In the materials evaluations described here, silicon wafers were used as reference surfaces. These reference surfaces were prepared by deglazing pieces of silicon wafers in 49% electronic grade hydrofluoric acid followed by ultrapure water rinse and blow dry with filtered ultrapure nitrogen. Immediately after deglazing, each clean silicon substrate was touched with a sample of the material under test using nominal fingertip pressure for 2 seconds.

Touched, and therefore possibly contaminated, silicon pieces were placed in a sample mount, along with a similarly deglazed but untouched silicon piece for reference. The silicon surfaces were analyzed by Secondary Ion Mass Spectrometry (SIMS). SIMS spectra were obtained from the touched silicon surfaces using helium excitation at 4.5×10^{-5} Torr chamber pressure with an accelerating voltage of 2 KeV. Spectra covered the 3-100 amu range. Peak intensities from the clean reference surfaces were subtracted from those obtained from the touched surfaces. The result is a measure of amount and identity of transferred contaminants, permitting a judgment of relative cleanliness of the sample materials in terms of what they transfer to clean surfaces.

Figure 1 shows an example of the information that can be obtained this way. Figure 1a is the SIMS spectrum of a deglazed silicon surface. Only silicon appears in the spectrum, with barely discernable peaks for silicon hydroxide, fluoride (from the HF deglaze solution) and sodium, which appeared universally in all the spectra. Figure 1b is the SIMS spectrum of a deglazed silicon surface touched as described above while wearing one of Supplier B's finger cots (sample number 2). Magnesium now dominates on the surface, sodium is substantially increased above background, an enhancement of organic species is apparent, and potassium and calcium appear. Clearly

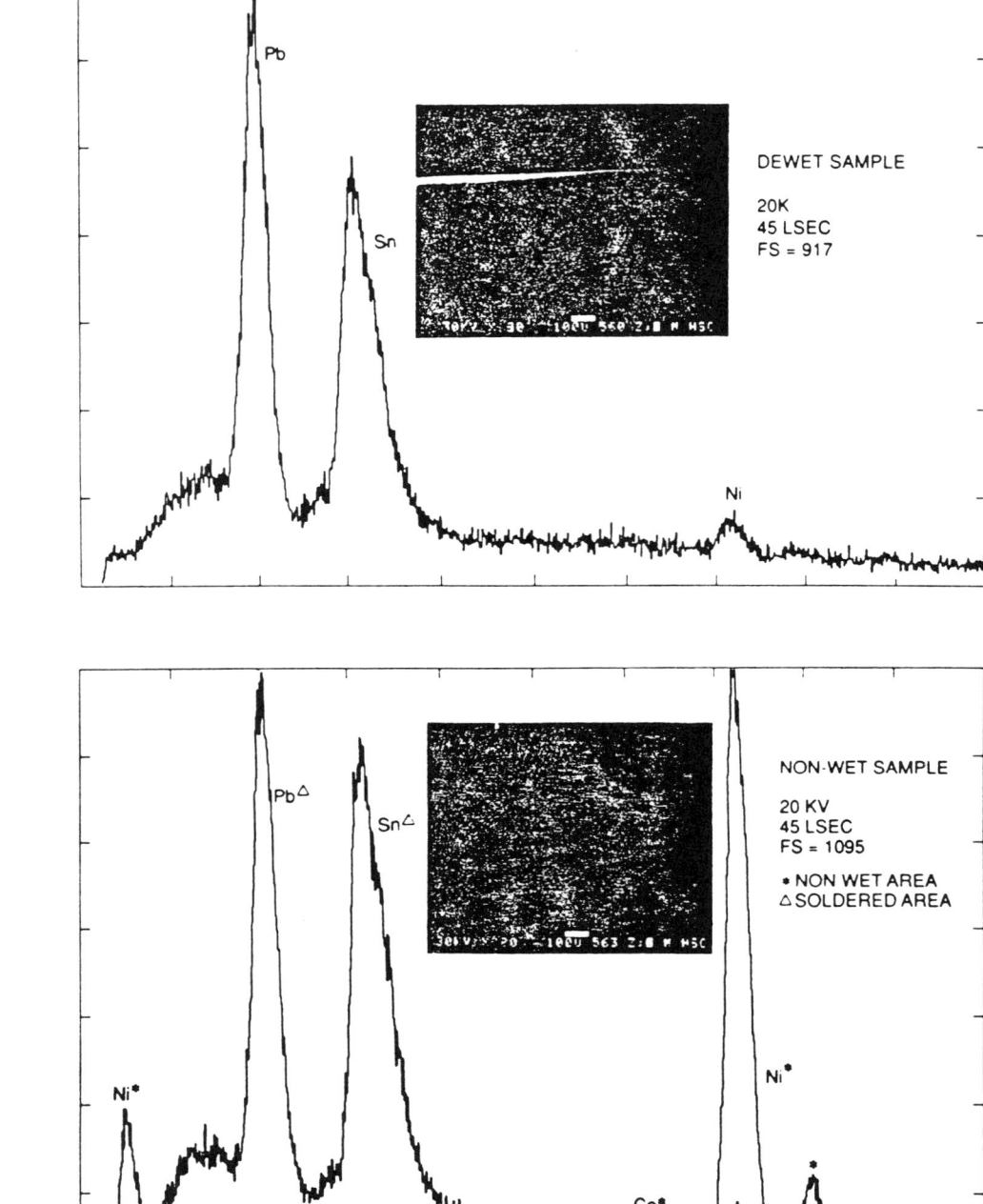

Fig. 3. (a) Photo and EDS spectrum of region of poor solder coverage on a package pin. (b) Photo and EDS spectrum of region of solder dewetting on a package pin.

Supplier B's finger cot transfers potentially dangerous ionic contaminant species to surfaces it contacts. The magnesium was likely sourced by talc, used to prevent sticking.

In controlled experimentation to select the cleanest finger cots and gloves, 6 samples of finger cots and 21 samples of gloves were subjected to these evaluations by SIMS. The peak

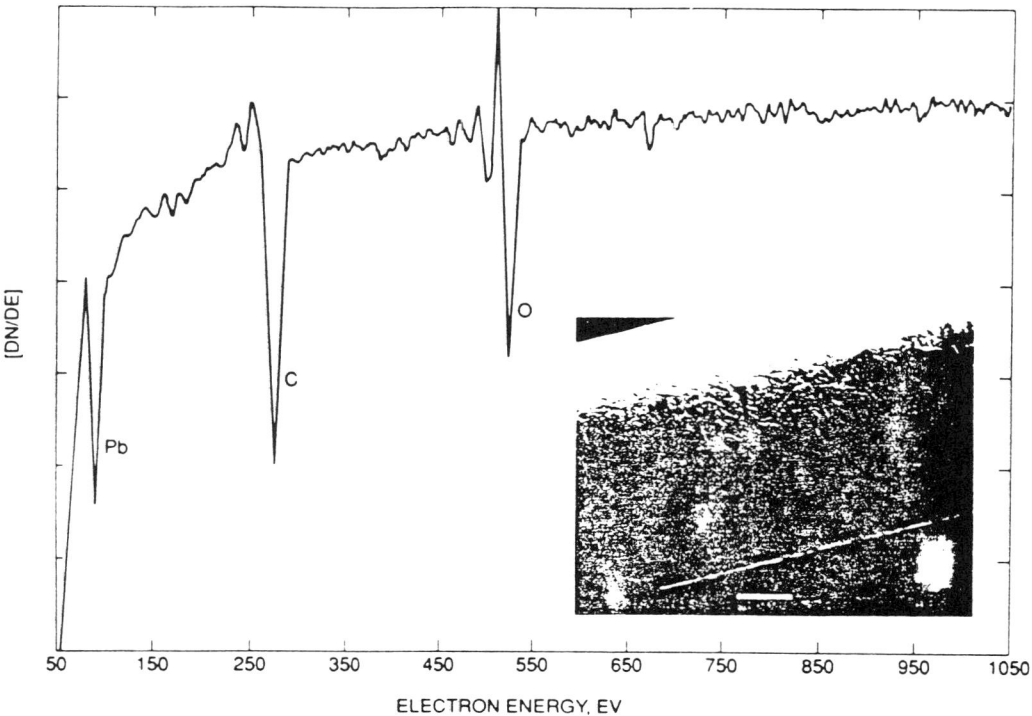

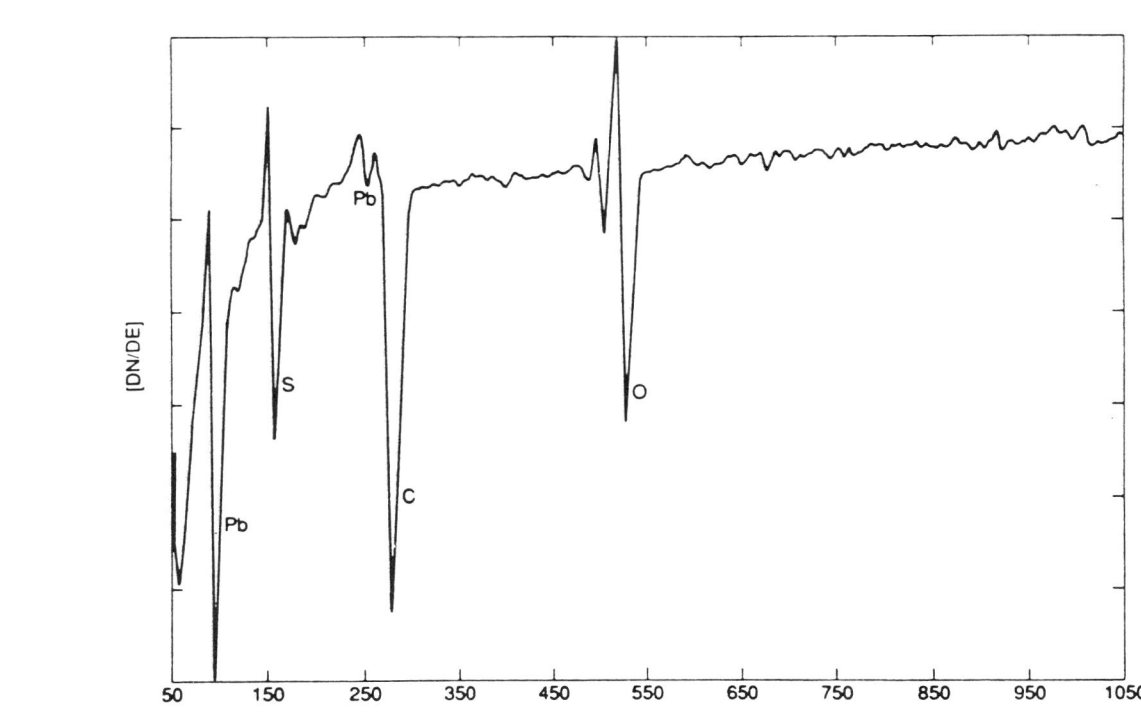

Fig. 4. (a) SEM photo of soldered region which appeared discolored under optical microscopy, and Auger survey spectrum of normal-appearing solder surface adjacent to anomalous region. (b) Auger spectrum obtained from surface of anomalous region shown in SEM photo in 4a.

amplitudes were summed and ratioed to the clean surface silicon peak intensity in order to display the relative surface cleanliness data in the bar graphs of Figure 2.

Based on these data no finger cots were qualified for use in IC assembly. Two kinds of ESD-protective gloves (Suppliers 1 & 4) were qualified for use.

While silicon was the reference surface for all transferable testing reported here, it is not necessarily the appropriate universal reference. For instance, plated metals (e.g. tin, gold, etc.) representative of package lead finishes may "receive" transferable contaminants differently than silicon. Test pieces should be chosen specifically for each situation being evaluated.

2. SOLDER WETTING

Soldering technologies for electronic connections have been a constant subject of study by reliability scientists and interconnection engineers for many years. Microbeam methods can be effectively used to evaluate soldering anomalies.

Determining the degree of solder wetting on surfaces can be done effectively using Energy Dispersive Spectroscopy (EDS) on a Scanning Electron Microscope (SEM). Photos in Fig. 3a and 3b show regions on package pins which exhibit varying degrees of poor solder coverage. Figure 3a shows a region where solder appears to be present, but with rather thin coverage. Figure 3b shows a region on a pin where there appears to be little or no solder coverage. The EDS point spectrum for the region in 3a shows a weak signal for nickel (from the pin's substrate metal). In contrast, the EDS point spectrum for the region in 3b exhibiting full solder coverage shows the same lead-tin response, while the EDS point spectrum from the center of the anomalous region shows a strong signal for nickel and no response for tin and lead. In these analyses, the relative sampling depth of EDS, which obtains signal from several microns of depth (compared to only a few tens of Angstroms for surface techniques like Auger), is used to advantage since this is really a "thick film" problem. The data characterizes very well the conditions of solder dewetting. In Fig. 3a, thin layers of solder remain, in contrast to the non-wetting condition in Fig. 3b, where no solder at all is present in the anomalous region.

Other types of anomalies include visual effects which suggest various contaminations on or in solder. Many of these, at first glance, could raise concern about the integrity of the solder coating. Figure 4 is a SEM micrograph of a local region of a soldered pin which appeared discolored under optical microscopy. The area is obviously locally roughened on the surface. Scanning Auger microprobe analysis (Auger electron spectroscopy, AES) was used to determine if the region contained significant levels of other chemical elements. The Auger spectra in Fig. 4a and 4b were taken off of, and directly on, the anomalous area respectively, after a brief 30 sec. argon ion sputter to remove surface-adsorbed species. The spectra show carbon impurity uniquely associated with the anomalous area. This carbon is present as a consequence of routine part handling and exposure to ambient conditions in the field. The roughened solder surface readily traps hydrocarbons from the surroundings; these hydrocarbons are surface-adsorbed species and are not actually an integral part of the bulk of the solder covering the pin.

The photographic and EDS data led to the conclusion that this anomalous region of the solder surface is a cosmetic aberration which contains no chemical impurities.

3. SEALING GLASS CONTAMINATION

Reliability scientists must often search long and hard for physical or chemical causes of electrical malfunctions. Parametric problems, where parts operate outside one or more of their expected electrical characteristics all or part of the time, are often the most difficult problem to associate with a physicochemical anomaly.

In this example involving sealing glass, microbeam analysis is used to identify contamination responsible for electrical leakage in operational amplifiers sealed within ceramic packages. Electrical leakage occurred on finished product sporadically. After much data accumulation, leakage was determined to be independent of all wafer processing functions. This fact, coupled with the observation that leakage seemed to be associated with particular pin-out configurations, led to an investigation of the packages. The problem was still sporadic, both between and within lots of the glass-sealed dual in-line packages. Eventually, careful Auger electron spectroscopy (AES) analyses of sealing glass between pins of samples exhibiting the problem vs glass between the same pair of pins on parts not exhibiting the problem were conducted. Figures 5a and 5b are examples of the Auger data which ultimately defined the problem. Devices with electrical leakage exhibited sulfur associated with the sealing glass; devices which were stable had no detectable sulfur on the surface of the sealing glass of their packages.

Sulfur was always found associated with the sealing glass of unstable device packages. It was never found on the sealing glass of stable devices. *The apparent leakage mechanism was the combination of chemisorbed moisture on the glass surface with the sulfur, forming a conductive chemical species which supported surface electrical conduction and therefore leakage between pins.* Through additional Auger analytical work conducted generally on the raw materials, it was found that package piece parts were received with sporadic occurrences of sulfur. All package lots had some fraction of packages whose sealing glass composition exhibited sulfur. Auger depth profiling showed that sulfur was not just a surface phenomenon. It persisted for thousands of *angstroms* into the glass. A program based on the analytical findings was initiated with the package supplier to eliminate the sulfur-contaminated sealing glasses.

This problem is an excellent example of the fact that many analytical investigations are a process and not a one-time event. The spectra in Fig. 5a and 5b could be declared definitive of the sulfur contamination problem only after months of studies on dozens of samples resulting from the ongoing, sporadic nature of the electrical leakage.

4. CHIP CLEANING

A reliability concern related to part and package qualification is the cleanliness of chips as they enter plastic molding assembly processing. Chip surfaces must be free of film contaminants to ensure intimate adhesion of mold compound. Additionally, they should be free of ionic residuals which could

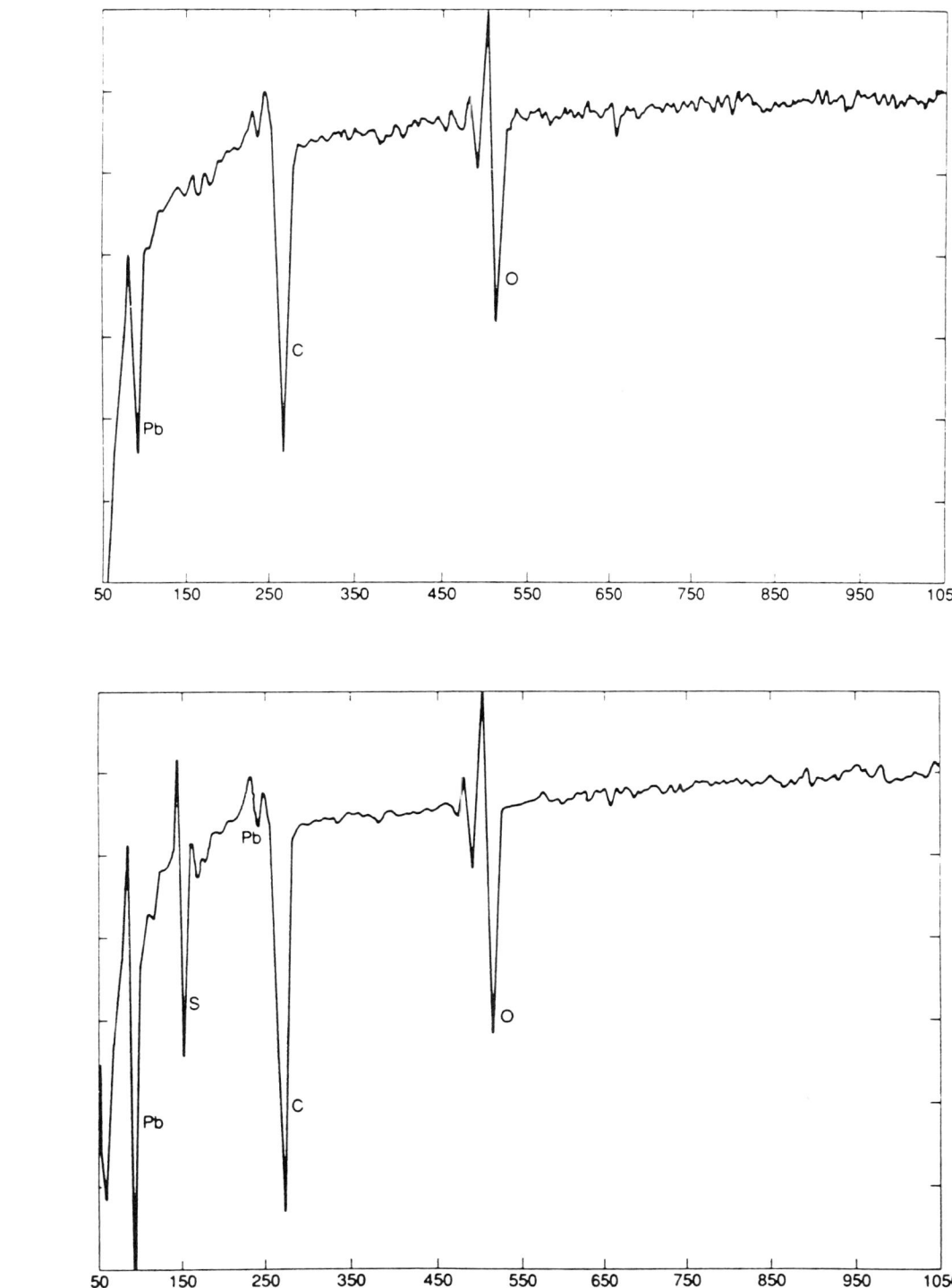

Fig. 5. (a) Auger spectrum of sealing glass between package leads of an electrically stable device. (b) Auger spectrum of sealing glass between package leads of electrically leaky device.

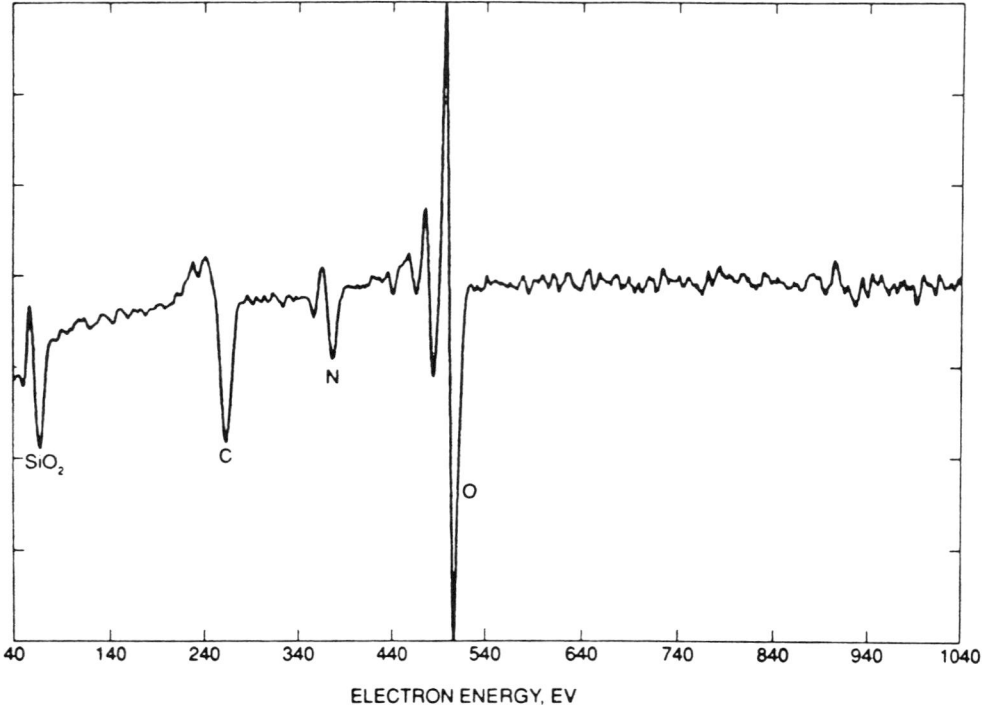

(a)

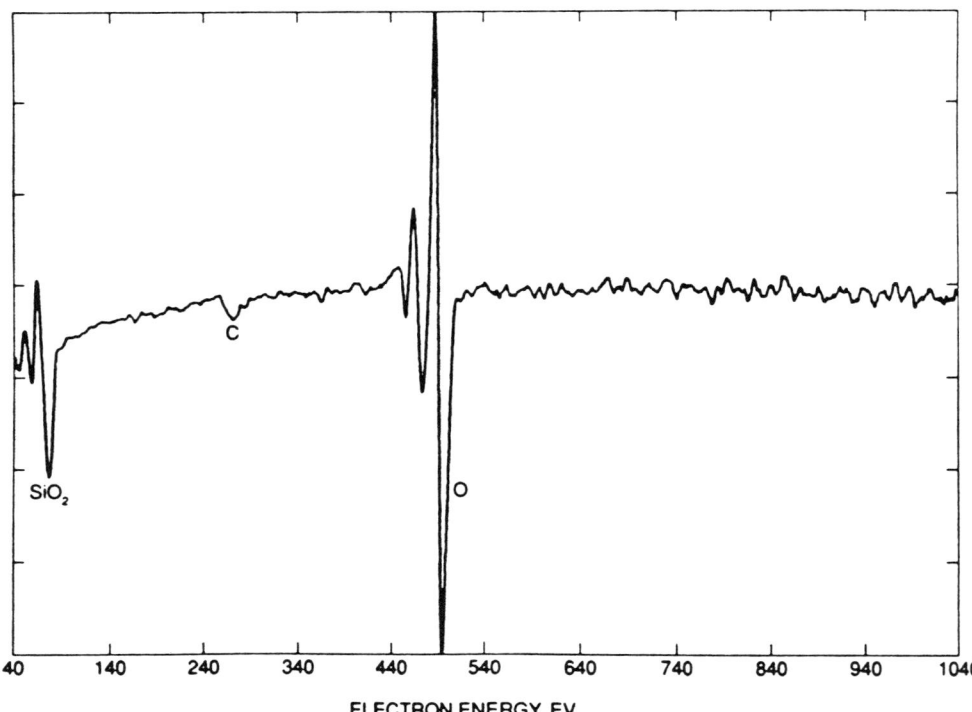

(b)

Fig. 6. (a) Auger spectrum of silicon nitride passivation subjected to UV/ozone cleaning. (b) Auger spectrum of silicon oxide passivation subjected to UV/ozone cleaning.

contribute to metal corrosion or to ionic-induced electrical parameter changes. Molding compounds by themselves are a major potential contributor to such concerns, so chip surface cleanliness is imperative to ensure reliable device operation after plastic encapsulation.

Auger and SIMS analyses of chip surface cleanliness were conducted as part of cooperative technical support for a customer purchasing large numbers of die and assembling them in plastic packages. The customer was experiencing significant assembly yield loss with the product.

The die sales mix included parts that had surface protective passivation films of both silicon nitride and silicon dioxide. Figure 6a and 6b show Auger surface analysis spectra of a nitride passivated die and an oxide passivated die respectively on the as-received passivation surfaces. The spectra are identical with respect to surface cleanliness except for the high amounts of carbon on the surface of the nitride passivation. All product had been exposed to a pre-seal UV/ozone clean. Apparently, the UV/ozone process being used was a much less effective cleaner of nitride surfaces than of oxide surfaces.

Figures 7a and 7b are SIMS surface spectra of the nitride passivated and oxide passivated die. These spectra were taken to supplement the Auger spectra because of the better sensitivity of SIMS to low levels of ions. It was confirmed that elevated hydrocarbon levels were unique to nitride surfaces. Furthermore, SIMS analysis revealed that nitride surfaces were substantially more contaminated with sodium and potassium, and even exhibited the presence of titanium.

The cause of mobile ions was traced to a pre-UV/ozone chemical clean used exclusively for nitride passivated die. Using the data from these surface analyses, the customer was able to engineer more effective and more consistent die cleans prior to plastic encapsulation, which increased assembly yield to the high 90% range.

5. HUMAN CONTAMINATION

Despite spending huge sums for building and maintaining clean room production space, IC's are often inadequately protected from one of the most ubiquitous and sinister sources of dangerous contaminants: PEOPLE! Humans are abundant sources of organic, ionic, particulate, and moisture-laden materials that pose serious threats to both yield and reliability.

IC products are susceptible to human-sourced contamination from the time they enter the fab area as wafers until, as individual die, they have been sealed into a package. While cleanroom clothing is designed to reduce the spread of operator-sourced contamination, it can be less than adequate, especially in the facial area. Furthermore, post-wafer fabrication processes, such as probe, scribing, die storage/handling, and package seal, sometimes use clean room discipline less rigorous than that in wafer fabrication.

Human-sourced contamination is chemically reactive and easily capable of producing corrosive ionic solutions. It is also a source of solid particulates that can cause physical damage in hermetically sealed packages.

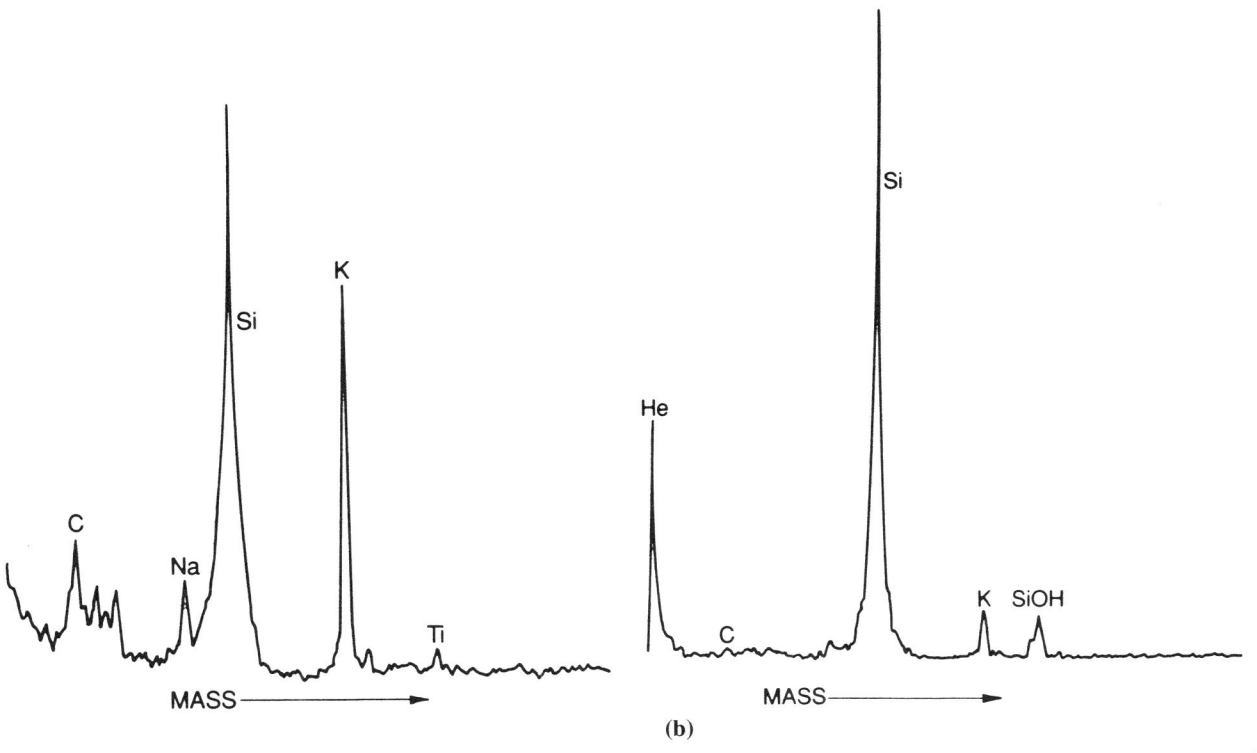

Fig. 7. (a) SIMS spectrum of silicon nitride passivation subjected to UV/ozone cleaning. (b) SIMS spectrum of silicon dioxide passivation subjected to UV/ozone cleaning.

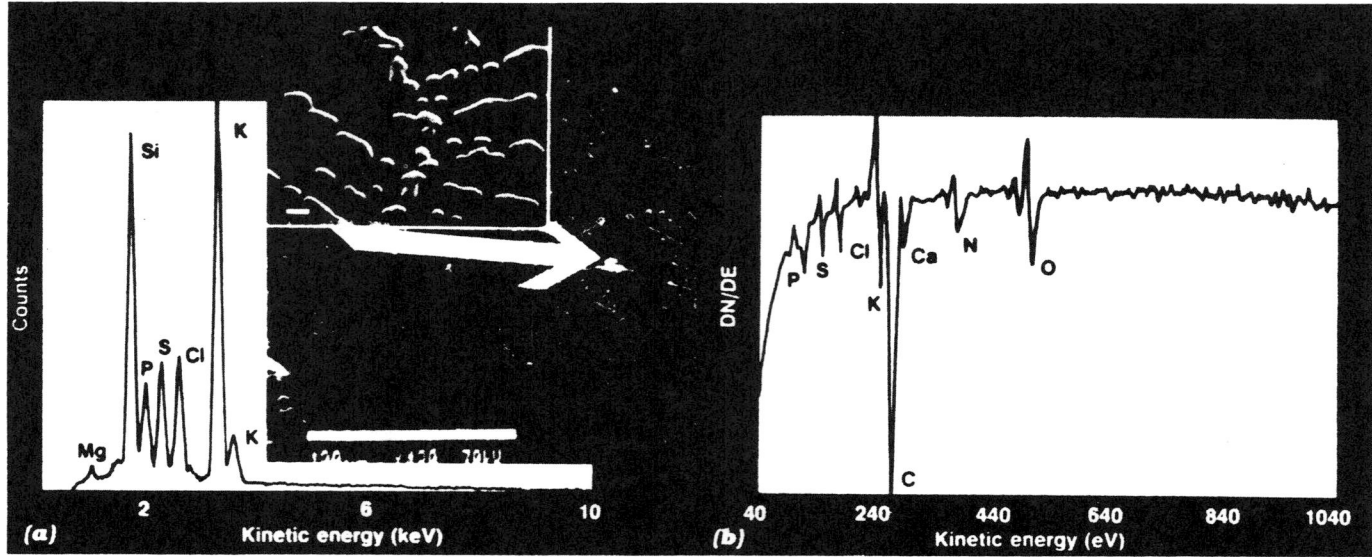

Fig. 8. Spittle contamination on a wafer, with a) the associated EDS spectrum, and b) Auger spectrum.

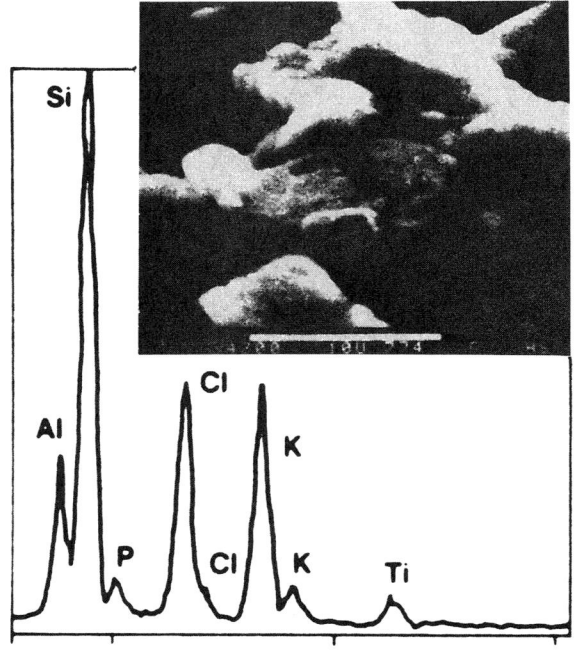

Fig. 9. Potassium chloride crystal, derived from human spittle, and its associated EDS spectrum.

On-die contamination is often first detected as "foreign material" by optical microscope inspection. When optical microscopy cannot plainly identify foreign material, it is submitted for microchemical analysis. This is usually done by **SEM/EDS**, where SEM is first used to locate and physically characterize the foreign material. EDS then provides chemical information. Characterization of three types of human-sourced contamination by SEM/EDS, spittle, perspiration, and cosmetics, is briefly described below.

Spittle

Figure 8 is a SEM photo of dried human spittle (airborne saliva droplets) on a silicon wafer. The inset in Fig. 8 is an enlargement of a small area of the patterned droplets.

The energy dispersive spectrum in Fig. 8a shows the elemental constituents to be potassium (K), chlorine (Cl), sulfur (S), phosphorus (P), magnesium (Mg), and traces of sodium (Na). The most noticeable feature is the potassium K-alpha line, indicating that a major constituent of spittle is potassium.

Complementary Auger data, shown in Fig. 8b, indicates that the surface of the dried spittle contains large amounts of carbon (C) and nitrogen (N). These elements are present in the liquid portion of the spittle. The potassium, chlorine, and sulphur are all readily visible in the Auger survey.

Another feature of spittle residue is the crystalline deposit shown by the arrow in Figure 8a. A close-up of these crystals, discovered on an actual circuit, is shown in Figure 9 along with the corresponding EDS spectrum. The spectrum shows the crystals to contain potassium (K) and chloride (Cl).

The two distinguishing features that uniquely define spittle residue are the high potassium concentration in the chemical spectrum and the presence of cubic potassium chloride crystals in the residual material.

Perspiration

Figure 10 shows a micrograph of dried perspiration on a silicon wafer. EDS analysis shows its elemental constituents to be sodium (Na), potassium (K), chlorine (Cl), and traces of sulfur (S) and aluminum (Al). Auger data in Fig. 10a and 10b support this and indicate that perspiration contains considerable carbon

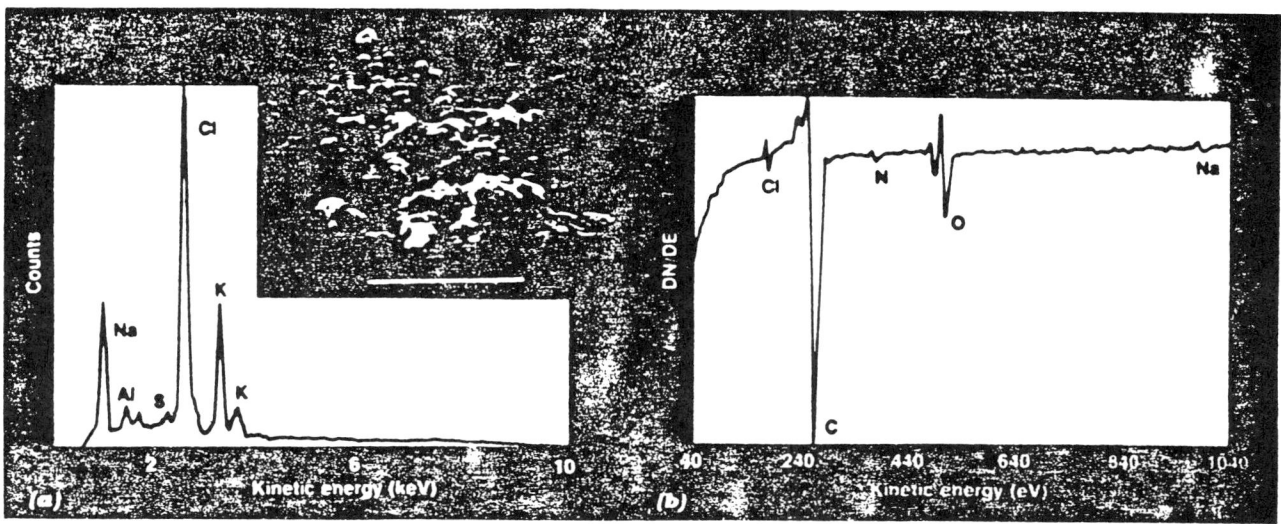

Fig. 10. Perspiration contamination on a wafer, with, a) the associated EDS spectrum, and, b) the Auger spectrum.

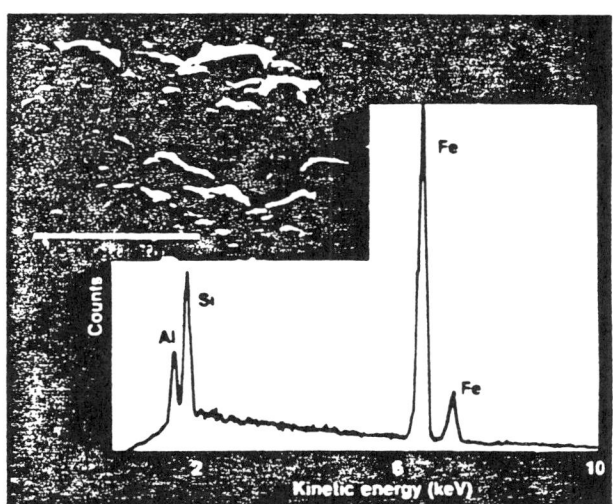

Fig. 11. Mascara deposit and associated EDS spectrum.

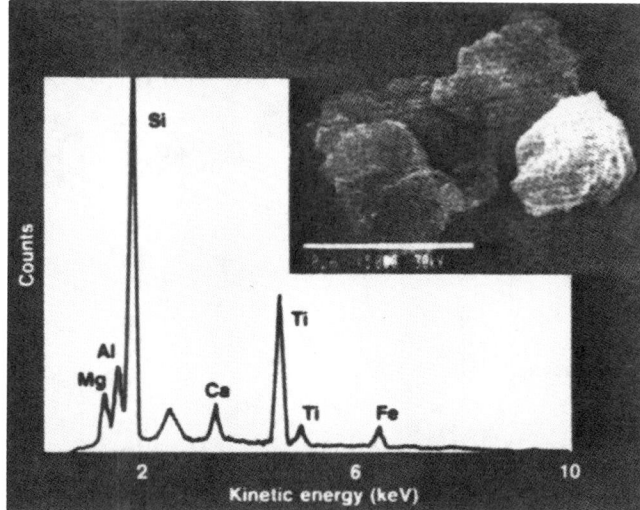

Fig. 12. Face powder deposit and associated EDS spectrum.

(C) and nitrogen (N) as well. *The main difference between perspiration and spittle is the sodium/potassium ratio. Spittle has a very low Na/K ratio when atomic sensitivity factors are considered, while perspiration has a large Na/K ratio. Perspiration also contains more nitrogen, due to release of amino acid derivatives from skin. Aluminum in perspiration is sourced by aluminum comounds used in antiperspirant formulations.*

Cosmetics

A third major type of human-sourced contamination is cosmetics. These can include mascara, facial powders, and fingernail polishes. For this study, samples of a popular mascara and a popular face powder were deposited onto a silicon wafer for analysis by SEM/EDS and Auger.

Figure 11 shows a SEM photo of dried mascara. EDS analysis of this area indicates that the mascara is composed of compounds containing the elements iron (Fe) and aluminum (Al). The prominent iron line is definitely an identifying feature. The iron is probably present as oxide, used as a pigment to give mascara its dark color. Significant quantities of carbon (C) were observed by Auger, but no other elements were identifiable. The carbon is a result of the solvent used to liquefy the mascara. Unlike the various types of body effluvia, no alkaline or halide-type ionic species such as sodium, potassium or chlorine are present in mascara studied.

Figure 12 shows a SEM photo of face powder deposited on a silicon wafer. EDS analysis shows titanium (Ti), iron (Fe), magnesium (Mg), aluminum (Al) and potassium (K). The most obvious feature of the spectrum is the large titanium K-alpha line. The titanium, as well as the iron, magnesium, and aluminum, is probably present as an oxide and is the pigment in the powder. Other researchers have found that bismuth (Bi) and barium (Ba) can be present in cosmetics, and human contamination can generally contain: Al, Ca, Cl, Fe, K, Mg, Na, P, S.[1]

Microbeam characterizations obtained from purposely contaminated substrates, and from actual on-line occurrences of human contamination on product, have been instrumental in helping reliability scientists and production engineers optimize manufacturing and product handling procedures to greatly reduce contamination from human sources.

6. BOND PAD APPEARANCE

Process engineers and reliability engineers are often dismayed by the widely varying visual appearances of wire bonding pads on IC chips. Bond pads can certainly be a source of trouble. They expose a relatively large surface area of metal, slightly recessed from the actual top surface of the chip (covered everywhere else by the passivation film), and thus are a repository for sundry contaminants. Electrical probing mechanically (if not chemically) perturbs bond pads. In a finished device, bond pads contain metallurgical wire attachments which themselves can be sources of chemical and physical anomalies, or can be compromised by various types of contamination.

Fig. 13 Photo of bond pad with mottled surface appearance.

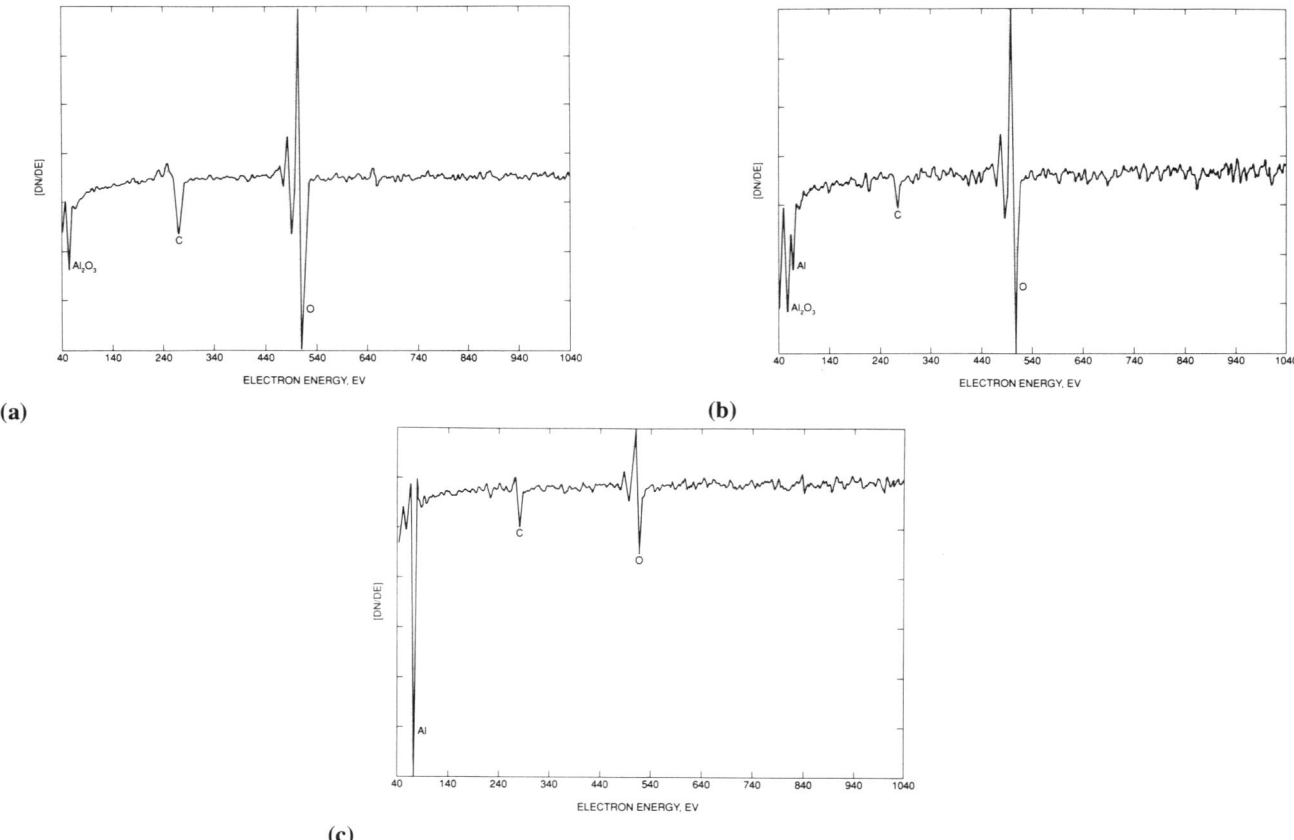

Fig. 14. (a) Auger survey spectrum of mottled bond pad surface as-received. (b) Auger survey spectrum of mottled bond pad after 0.5 minute Ar+ sputtering. (c) Auger survey spectrum of mottled bond pad after 1.0 minute Ar+ sputtering.

Microbeam Analysis and Semiconductor Reliability

The general expectation is that bond pads will present the appearance of shiny, silvery aluminum. During routine visual inspections, any appearance to the contrary raises suspicions about the reliability of the device. However, some bond pad conditions, perceived by visual inspection to be problems, are actually shown by microbeam analysis to be purely cosmetic circumstances.

As an example, Fig. 13 shows a pad whose surface appears mottled and grainy. A sequence of Auger spectra on this pad is shown in Fig. 14a, b, and c for 0, 0.5, and 1.0 min. sputter times of the pad, respectively. These spectra show what is generally expected for an aluminum pad, DESPITE its appearance. The surface is aluminum oxide with a normal amount of native

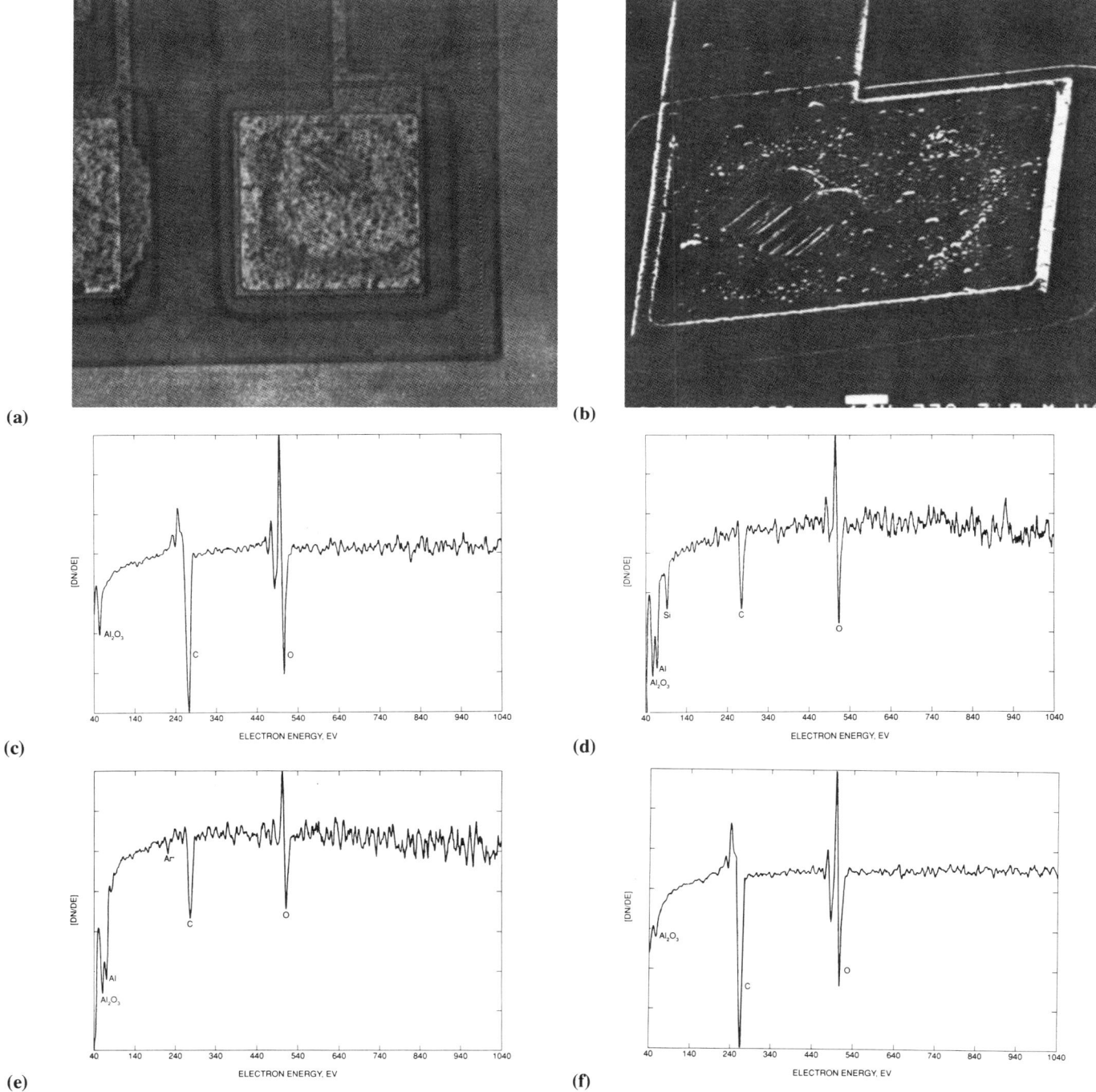

Fig. 15. (a) Optical photo of bond pad with ring anomaly. (b) SEM photo of bond pad in Figure 15a. (c) Auger survey spectrum on ring anomaly as-received. (d) Auger survey spectrum on ring anomaly after 0.5 minute Ar^+ sputtering. (e) Auger survey spectrum on ring anomaly after 1.0 minute Ar^+ sputtering. (f) Auger survey spectrum at center of ring anomaly as-received. **(continued)**

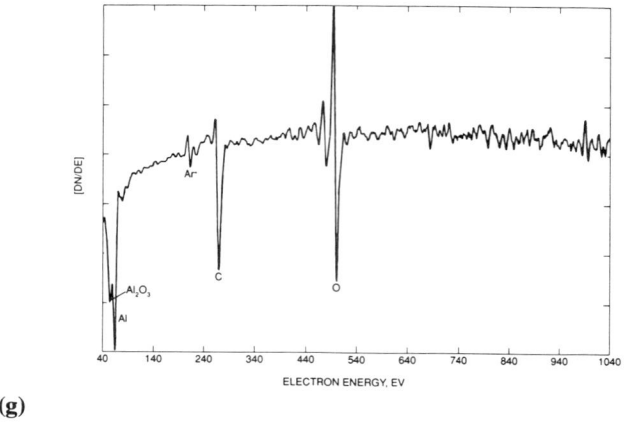 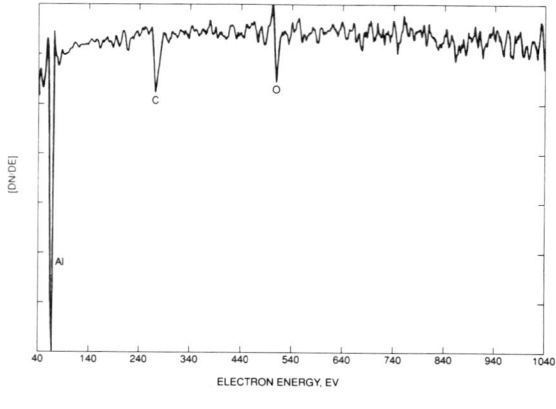

Fig. 15. (g) Auger survey spectrum at center of ring anomaly after 0.5 minute Ar$^+$ sputtering. (h) Auger survey spectrum at center of ring anomaly after 1.0 minute Ar$^+$ sputtering.

carbon (C) in Fig 14a; 0.5 min Ar$^+$ sputtering shows reduced carbon with a mixture of aluminum (Al) and aluminum oxide (Al$_2$O$_3$) in Fig. 14b; and 1 min. Ar$^+$ sputtering shows mostly elemental aluminum (Al) in Fig. 14c. This Auger study, differentiating between aluminum and its oxide as the pad surface is sputtered, takes advantage of the wide separation in Auger electron energies between elemental aluminum and its oxide, at 68eV and 51 eV, respectively.

Photos in Fig. 15a and 15b show another bond pad anomaly. This pad displays the expected shiny surface appearance, except for what appears to be a ring of material encircling the probe mark. This ring was characterized by Auger microanalysis, first at a point on the "ring", and then at a point in the center of the pad, which appears normal. The sequence of Auger profiles shows that the "ring" on the pad consists of aluminum enriched in aluminum oxide. This aluminum oxide is actually an area of pad material debris kicked up by the probe point and re-deposited around the circumference of the probed area.

The spectra in these two examples are as significant for what they *do not* show as for what they *do* show. No halide or positively charged ionic impurities are found with the pad anomalies. Thus, even though the pad surfaces appear contaminated, they are free of surface impurities which could lead to ion-induced failure events. Again, this is a common problem in IC assembly areas. It is not generally safe to assume that anomalous appearance of the metal can always be discounted. The point here is that the surface analysis can define the source of the problem and provide a scientific basis for an "accept" or "reject" decision.

7. CONTACT RESISTANCE

Reliability scientists often find that increased electrical resistance occurs where metal conduction layers make contact. When these specific contact points can be isolated, in-situ microbeam analysis, in the depth profile mode, can reveal the cause of elevated contact resistance.

Frequently, an interposed layer of carbon is found in the contact. Carbon can increase contact resistance so much that electrical parameters cannot be met. Figure 16 is a composite

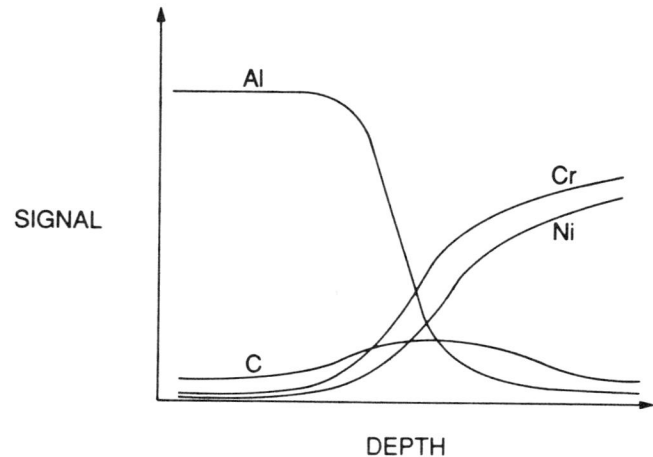

Fig. 16. Typical depth profile through a metal contact with high contact resistance.

depth profile ("averaged" from many similar types of analyses) showing the typical finding of carbon "sandwiched" between metal layers. Carbon is readily left behind by poor photoresist strips or other cleans that do not completely remove organic films. Contacts free of carbon never have high resistance.

8. THIN FILM LASER TRIMMING

Laser trimming is a process tool for in-situ fine-tuning of resistor films in devices. Achieving thin film resistors with reliable, stable electrical properties by in-situ lasing has been the subject of extensive microbeam analysis. In this example, understanding how to adjust resistance by laser alteration of the chemistry of as-deposited chromium silicide thin film was critical to obtaining devices of long term operational reliability.

Figure 17a presents X-ray Photoelectron Spectroscopy (XPS) spectra for silicon (2p) and chromium (2p) binding energies taken at the surface, 100 Å, and 200 Å deep, respectively, in a chromium silicide (CrSi) film as-deposited on thermal oxide, with no heat treatment. Figure 17b is the same set of infor-

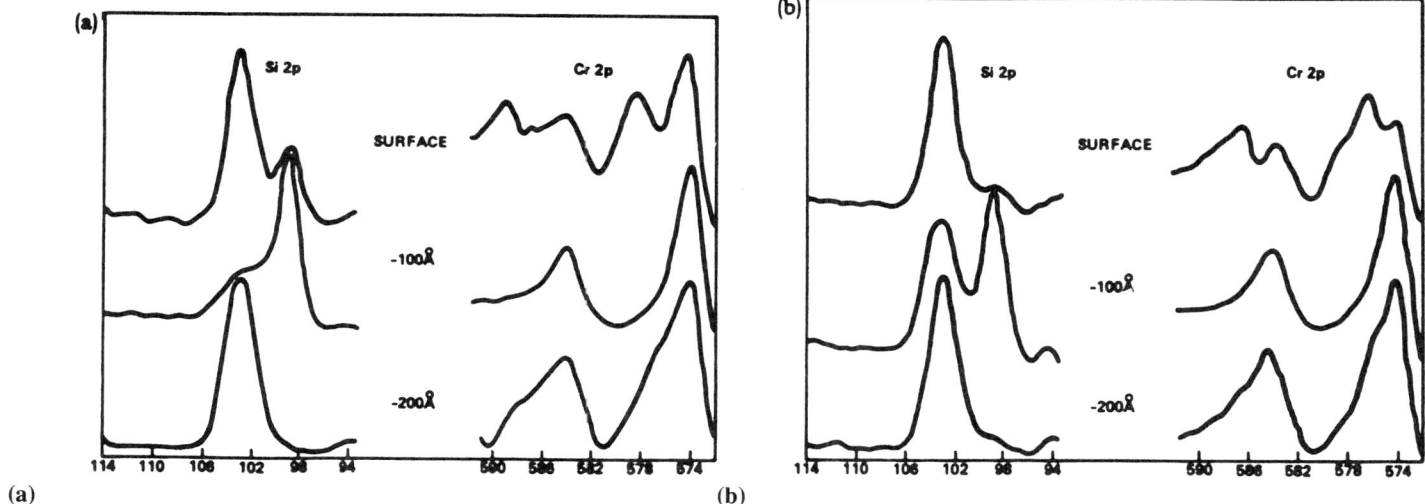

Fig. 17 (a) XPS spectra showing Si(2p) and Cr(2p) binding energies for an unbaked sample of CrSi deposited over silicon dioxide. (b) XPS spectra showing Si(2p) and Cr(2p) binding energies for a baked sample of CrSi deposited over silicon dioxide.

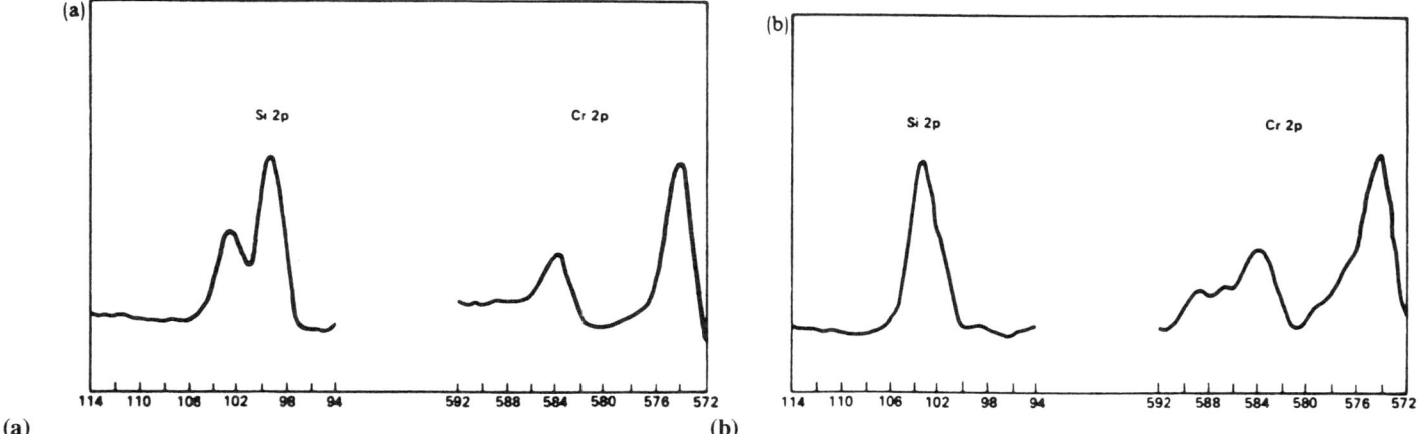

Fig. 18 (a) XPS spectra showing Si(2p) and Cr(2p) binding energies for a baked, passivated, and untrimmed sample of CrSi deposited over silicon dioxide. (b) XPS spectra showing Si(2p) and Cr(2p) binding energies for a baked, passivated, and trimmed sample of CrSi deposited over silicon dioxide.

mation for this sample after a stabilization bake. A layer composed of Cr_2O_3, $Cr(OH)_3$, and SiO_2, is detected on the surface of the unbaked sample. The baked sample shows a Cr_2O_3-SiO_2 layer free of $Cr(OH)_3$. SiO_2 is found in heavy concentrations on the surface, gradually decreasing to a minimum near the center of the film, and gradually increasing again towards the lower CrSi-SiO_2 interface.

XPS results of trimmed vs. untrimmed baked and passivated CrSi over SiO_2 are shown in Fig. 18a and 18b. These samples enabled gathering of XPS data on samples most closely resembling standard processing (i.e. baked, passivated, and trimmed). The control in this case is the baked, untrimmed sample, which represents processing up to the laser trim step. For these samples, the glass passivation was removed down to approximately 1000 Å with hydrofluoric acid. The remaining glass was removed down to the passivation-silicide interface by Auger profiling, monitoring closely the Cr and O signal to determine the actual location of the interface. The interface consisted of SiO_2, Cr_2O_3, and CrSi, as expected. The entire CrSi layer contained SiO_2, decreasing in concentration toward the center of the film, with Cr_2O_3 detected only at the upper and lower interfaces. The trimmed sample showed a diffuse layer composed of Cr in several oxidation states; Cr^0, Cr^{+3}, Cr^{+4}, Cr^{+6}, and SiO_2. These results indicate sufficient thermal energy in the laser beam to significantly alter the chemical makeup of the film. This XPS data, supporting a team effort by process and reliability groups, was an essential in a proactive process development initiative designed to build quality into the product.

9. A CONTAMINATED DEVICE STRUCTURE

Figure 19 is a photo of a test structure used as part of a yield/reliability evaluation for a process under development. It

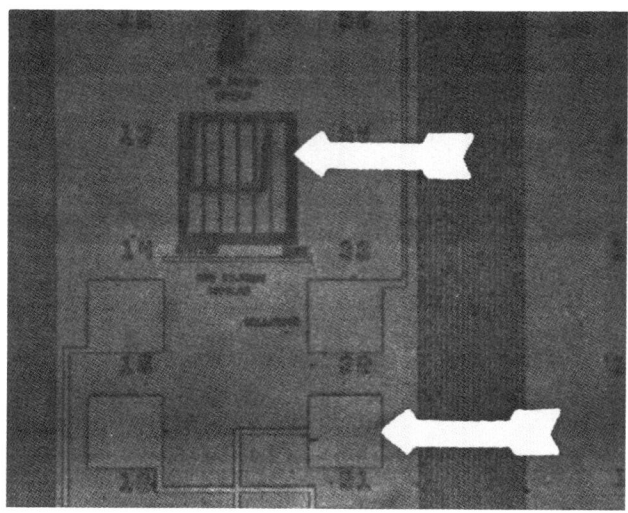

Fig. 19. Test structure. Small and large square features indicated by arrows to be depth profiled for elemental composition.

was desired to depth-profile through the two features marked by arrows in the photo. The larger feature (upper arrow) is about 166×200 while the smaller feature (lower arrow) is about 125μ on a side. These structures were analyzed by Secondary Ion Mass Spectrometry (SIMS) using O_2+ primary ions to sputter selectively through the regions of interest. The sputter crater in the upper feature is the black square favoring the upper left corner, and is $125 \times 125\mu$. Analytical information was obtained from an area within this sputter crater about 30μm on a side, providing very localized data about the structure. The lower feature was analyzed in exactly the same manner. Care was taken to ensure that the sputter crater's approximately $900\mu^2$ area, from which secondary ions were collected and analyzed, did not fall outside the boundaries of the sites of interest.

Figures 20a and b are the SIMS depth profiles for boron dopant and aluminum in these areas of interest. The upper (larger) structure is found to contain 3 to 4 times as much aluminum (as much as 10^{16} atoms of Al/cc) than the lower structure (in the N^+ layer, and near the pn junction). The presence of aluminum coincided with observed electrical anomalies. The in-situ depth profiling capability of the ion microprobe within small device features enabled the identification of impurity affecting device performance, allowing process corrections to be made. It is interesting to note, in this case, that the impurities

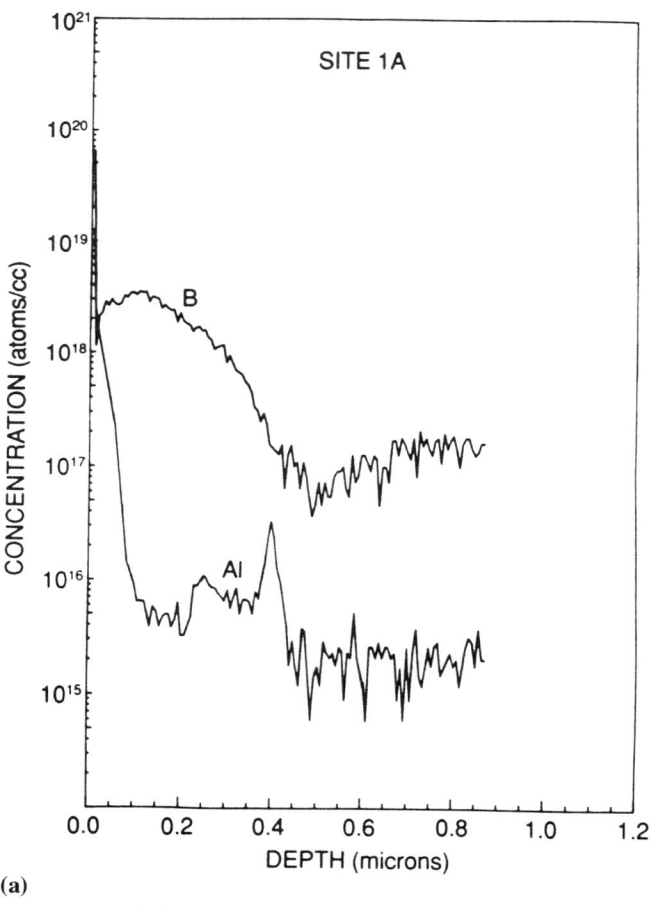

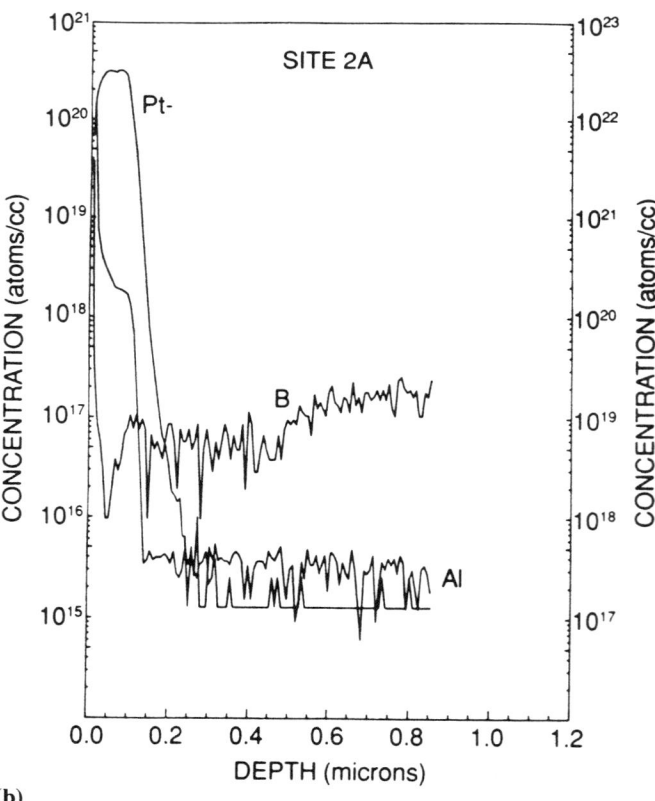

(a)   (b)

Fig. 20. (a) SIMS depth profile through small feature, (upper arrow). (b) SIMS depth profile through larger feature, (lower arrow).

responsible for the device problem was an element that is otherwise common to IC processing.

CONCLUSION

These are just a few examples of how microbeam analysis can characterize IC anomalies to help engineer high-quality processes and products as well as to understand failure mechanisms. The resourceful reliability scientist will be aware of and will use these and other microbeam techniques as supplementary tools to electrical methods in his quest for product excellence and customer satisfaction.

ACKNOWLEDGMENTS

The technical staff of the Beam Team in Harris Semiconductor's Analytical Services Laboratory contributed material for this tutorial.

REFERENCE

1. R.W. Thomas and D.W. Calabrese, *The Identification and Elimination of Human Contamination in the Manufacture of ICs*, ISTFA-85, pp 169-176.

TRANSMISSION ELECTRON MICROSCOPY: A REVIEW AND A COMPARISON WITH HIGH RESOLUTION SCANNING ELECTRON MICROSCOPY

by Jamie H. Rose, Barbara Miner, and Aldo Pelillo

INTRODUCTION

Transmission electron microscopes (TEM) have evolved over the past twenty-five years from awkward-to-operate instruments which were employed for microstructural observations to microprocessor-controlled instruments with various detectors attached, providing a range of chemical and structural information down to the atomic level. Today, metals, ceramics, semiconductors, and polymers are actively studied with TEM.

The latest generation of all-purpose analytical TEMs provide unique capabilities for materials structural and chemical analyses with exceptional spatial resolution. Microstructural imaging at magnifications up to 1,000,000 times, crystal lattice imaging below 3Å, and diffraction and chemical information from regions approaching 20Å across are nearly simultaneously possible. As device dimensions enter the submicrometer realm and with film thicknesses often much less than 1000Å thick, TEM is often required in process development and to assess the finest details of device structures. Direct observation in the TEM provides microstructural and chemical information unobtainable by other materials analysis techniques. The capabilities which distinguish TEM analysis can be summarized as follows:

1. Imaging internal structure with high resolution;
2. Observation of a class of features (crystalline defects) not readily observed by other techniques;
3. Chemical and phase analysis of features or films less than 100Å in size or thickness;
4. Atomic structure imaging.

This article reviews the instrument and types of information provided by TEM and is based in part on two previous ISTFA presentations. To accomplish this, the imaging and analytical features of the TEM will be outlined, followed by a brief description of the most common aspects of sample preparation. Next, the types of materials information provided by TEM are categorized. These considerations are then illustrated with a variety of examples demonstrating the value and place of TEM in the examination of VLSI-related thin films. Included in this last section is a comparison of TEM and high-resolution SEM for imaging device structures.

THE INSTRUMENT

Today's standard analytical TEM is a combined TEM and STEM (Scanning Transmission Electron Microscope). In the TEM mode of operation, magnetic lenses are used to image the sample in transmission in a manner analogous to a light microscope (Fig. 1). Lenses above the sample illuminate it with a defocused electron beam. Magnifying lenses below the sample focus the transmission image on a phosphor screen. Images are recorded by tilting the screen and exposing electron-sensitive film held in a chamber below.

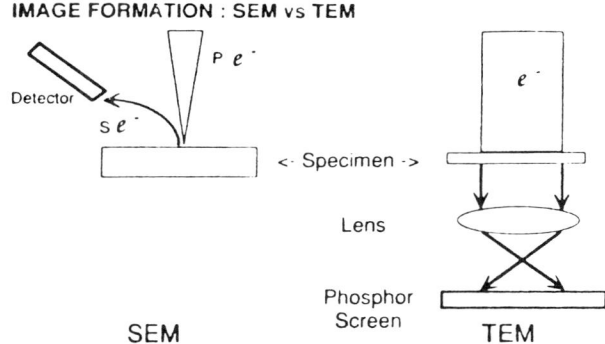

Fig. 1. A SEM forms an image by scanning a focused electron beam over a sample surface while modulating the intensity of a CRT beam in proportion with the signals of secondary or backscattered electron detectors. Imaging in the TEM is analogous to light optical microscopy with electrons replacing photons and magnetic lenses replacing glass lenses.

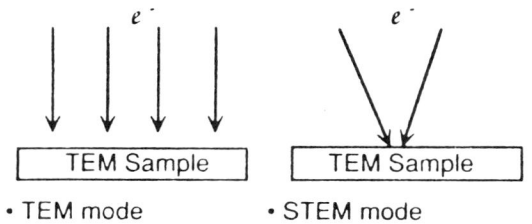

Fig. 2. Most TEMs have the ability to operate in a SEM-like mode (hence STEM mode—Scanning Transmission Electron Microscopy). In addition to the electron detectors found in a SEM, an additional detector is typically placed below the sample for transmission STEM images. Operation in STEM mode assists in probe placement and permits elemental mapping.

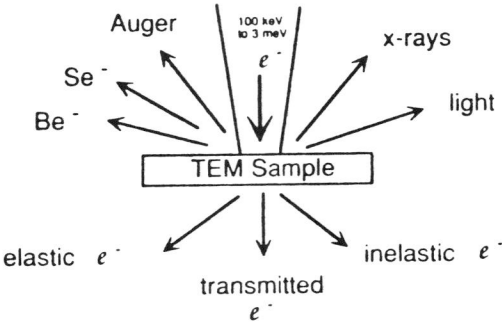

Fig. 3. While most electrons are transmitted through a typical TEM sample, various other scattered primary and ejected electrons and photons are emitted from the sample. The TEM makes use of transmitted and elastically scattered (diffracted) electrons for imaging and x-rays and inelastically scattered electrons from elemental analysis.

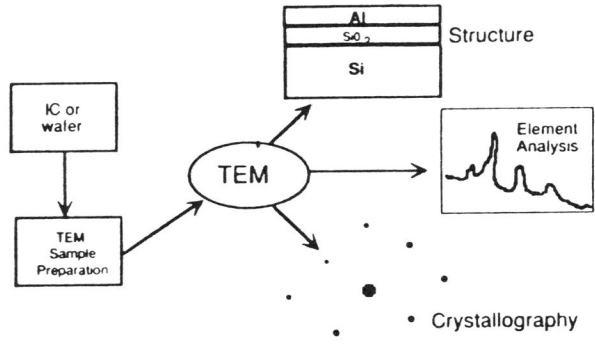

Fig. 4. After appropriate sample preparation, the TEM can quickly provide high spatial resolution microstructural, elemental, and diffraction data.

In the STEM mode of operation (Fig. 2), the microscope scans a focused electron beam over the sample, forming a video image with electrons collected by a secondary electron detector above the sample or a transmitted electron detector below the sample. Hence, it now has similarity to an SEM. This mode of operation allows placement of fine probes (about 20Å to 400Å in diameter) on regions of interest for diffraction or chemical analysis.

Chemical analyses can be performed during either mode of operation, the selected mode being a function of the size of the feature to be analyzed. STEM operation also permits production of elemental maps.

When the electron beam strikes the sample (commonly at 100 to 200 keV in TEMs), a variety of beam-specimen interactions occur which yield X-rays, light, Auger electrons, secondary and backscattered electrons, inelastically and elastically scattered electrons, and transmitted (unscattered) electrons (Fig. 3). This permits many potential applications of TEM. The following describes the types of image, diffraction, and spectroscopic data provided by TEM (Fig. 4).

Imaging

Imaging in the TEM makes use of transmitted and elastically scattered electrons, the latter being diffracted beams in the case of crystalline samples. This leads to three types of image formation, including mass thickness imaging.

Microstructural imaging (diffraction contrast imaging) employs the transmitted and at least one diffracted beam to observe the internal microstructure of a sample at magnifications typically up to a few hundred thousand times. A sample produces image contrast due to varying crystal orientation, presence of defect strain fields, or changing elemental make up. Phase morphology, grain boundaries, dislocations, and device structures may readily be observed in great detail (Fig. 5). Resolution is limited by the strain field surrounding defects, typically tens of Angstroms, though skilled microscopists employ the technique of weak beam imaging to reduce this to about ten Angstroms. This latter method revealed that disloca-

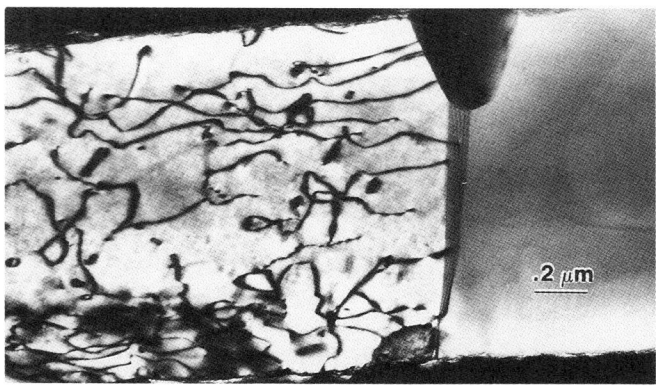

Fig. 5. Planar TEM image of aluminum—1% copper interconnect. Note the grain boundaries, dislocations, and copper rich precipitate at the grain boundary.

tions in semiconductors are typically dissociated into two partial dislocations bounding a stacking fault.

Lattice imaging (phase contrast imaging) is employed to obtain images of crystal atomic structure projected along low index directions (Fig. 6). This is typically accomplished in silicon by aligning a ⟨110⟩ crystal axis parallel to the electron beam. Interference between the transmitted and the {111} diffracted beams leads to periodicity in the image (phase contrast) corresponding to the planar spacings associated with the diffracted beams (3.13Å). Resolution below 2.5Å is available in all-purpose TEMs (below 2Å in dedicated high resolution TEMs).

This technique has been employed most heavily for study of semiconductors. Lattice imaging was first used to determine the atomic model for a crystalline defect in the late 70s (for a grain boundary in germanium) though it can routinely be employed for extremely precise internal magnification calibration, interfacial observation, and identification of defects in crystals.

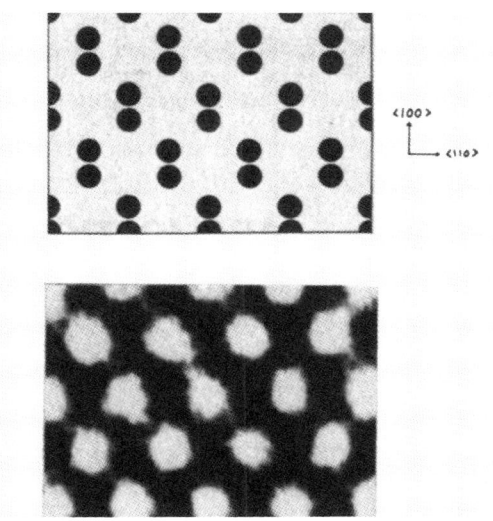

Fig. 6. Atomic model (top) and TEM lattice image (bottom) of ⟨110⟩ oriented silicon. The spacing of the "dumbbells" (1.35Å) is too small to resolve, leading to a single spot for each atomic column pair in the actual image.

Analytical Electron Microscopy

The modes of TEM operation which provide elemental, crystal structure, and phase information are collectively referred to as Analytical Electron Microscopy (AEM).[5] These techniques employ electron diffraction and spectrometers attached to the TEM. True AEMs have their designs optimized for sensitivity, small electron probe potential, and ease and flexibility of operation. For analytical studies with the best possible spatial resolution, AEMs are usually equipped for STEM operation. Here, probes down to about 20Å in diameter (5Å in dedicated STEMs) can be placed on areas of interest with the aid of the scanned image. STEM operation also permits elemental mapping and secondary and backscattered electron imaging.

Electron diffraction is readily available with a quick adjustment to the imaging lenses in the microscope (Fig. 7). This reveals crystallinity, grain orientation and texture, and phase identification. Since the electron beam can be focused to a probe well below 100Å in diameter, the technique of convergent beam diffraction, though outside the scope of this paper, provides additional crystal structure and electronic band information by analysis of details within the diffraction spots. Diffraction information can be obtained from very small regions, and requires nearly perfect crystals. This is achieved by converging the electron beam so that electrons strike the sample over a range of angles. See Ref. 6 for a review.

Elemental analyses are performed by utilizing inelastic events during which incident electrons lose discrete amounts of energy and sample elements give off characteristic X-rays. This is most commonly accomplished with Energy Dispersive X-ray Spectrometers (EDS) attached to the TEM. This technique provides sensitivity below 1 atomic %, and quantification of about

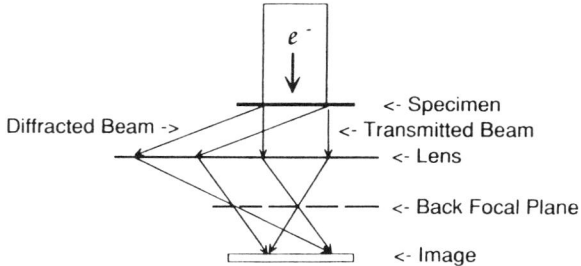

Fig. 7. By simply pushing a button on the TEM, one switches from image mode to diffraction mode. With a planar TEM sample, electron diffraction patterns can be obtained from films of any thickness. These provide phase and crystal orientation information.

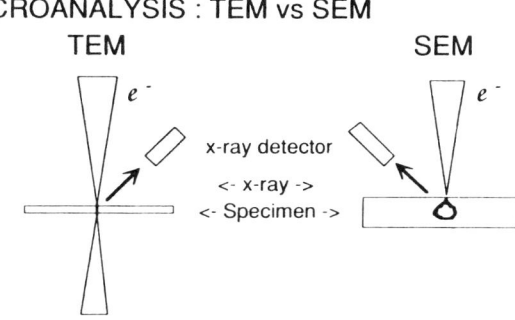

Fig. 8. While TEM x-ray data spatial resolution is limited by the probe diameter, in the SEM it is limited by the scattering and absorption of electrons in a bulk sample. Hence, TEM resolution is about 1000× finer than SEM.

10%. Elements with atomic numbers down to boron can be detected with ultra-thin window or windowless detectors.

Electron energy loss spectrometry (EELS) can be performed with sensitivities similar to EDS for all light elements. EELS also has the potential of providing bonding information, though such application requires expertise with the technique.[7] The EELS spectrometer is attached beneath the electron column. The electron energy spectrum is obtained by raising the phosphor screen and allowing an image or diffraction spot to strike the spectrometer entrance aperture. EELS is practical only for extremely thin (< 500Å) samples.

Again, the fine probe available on the TEM permits chemical analysis of regions less than 100Å across. Since such observations are performed on a sample region typically less than a few hundred angstroms thick, the beam broadening limitations of SEM work are largely avoided (Fig. 8). In addition, quantification is easier and more reliable for such thin samples since bulk correction factors are unnecessary.

TEM SAMPLE PREPARATION

As in SEM and light microscopy there are two salient sample preparation geometries planar and cross-sectional (Fig. 9).

What distinguishes TEM is its requirement for electron transparent samples—typically one hundred to a few thousand angstroms thick, depending on material, type of observations, and accelerating voltage. This is a demanding requirement, however a number of methods have been developed over the years to obtain thin sections from virtually any material. These include electropolishing for metals, chemical polishing for semiconductors, ultramicrotomy for polymers and particles, and ion milling for ceramics and multi-component samples.

VLSI studies require techniques for examining thin films deposited on silicon wafers. Planar oriented samples provide a relatively large area of view of a thin-film normal to its surface (Fig. 10a) while cross-sections (Fig. 10b) permit a detailed analysis of structure versus depth.

Cross sections are extremely useful in VLSI studies given the high spatial resolution capability of the TEM. The various layers as well as the internal microstructure of the films (e.g., grain size and shape) can be observed in detail. Wafers with uniform thin films can be prepared easily, however small device features present the problem of obtaining an electron transparent section through the object of interest. This difficulty is sometimes remedied by the production of test patterns specifically designed to aid TEM sample preparation. A test pattern has a device-like structure which mimics all process aspects of a real IC. This structure is then repeated in the X direction and is unchanging in the Y-direction. Such test patterns can be included on actual wafers for TEM evaluation of production steps. Of course, test patterns cannot be electrically tested, nor serve for failure analysis of specific device sites.

Preparation of cross sections for TEM observation, though often routine, is still under active development. However a few essential steps are typical of most approaches. Wafer sections are sandwiched with epoxy, sliced, mechanically polished, and lastly ion milled.

In addition to production of samples of high quality, reliable and rapid generation of TEM samples is required for routine support of manufacturing and process development groups. Using the basic methods outlined above, workers in various industrial labs have modified techniques to suit their particular needs. Though TEM analysis can never be as rapid as SEM analysis, it can provide routine support for problem solving and development work that requires materials analysis of high spatial resolution.

CATEGORIES OF INFORMATION PROVIDED BY TEM

As described above, TEM provides a wealth of data via images, diffraction patterns, and spectroscopic analyses. Such data gives structural, crystallographic, phase, and chemical

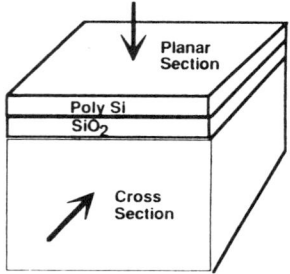

Fig. 9. For wafer or device observation, TEM samples are typically prepared for planar or cross sectional view of the surface films.

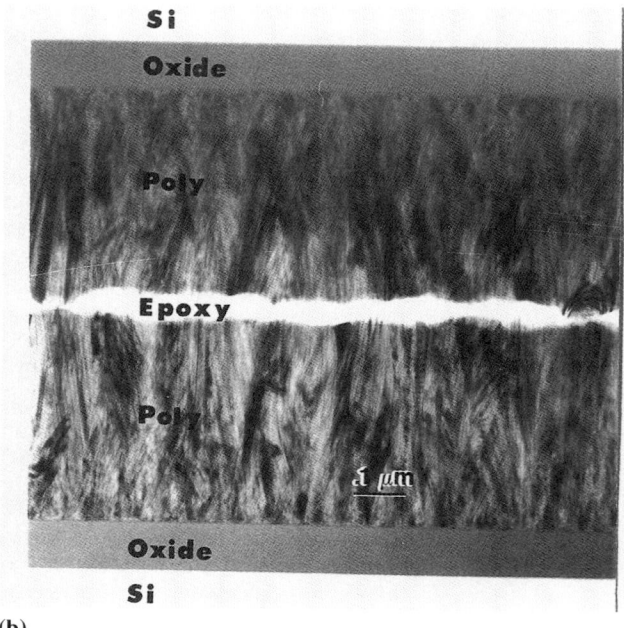

(a) (b)

Fig. 10. TEM micrographs of a CVD polycrystalline silicon film deposited on oxide. a) planar section and b) cross section. Two wafer sections are face-to-face in the cross section.

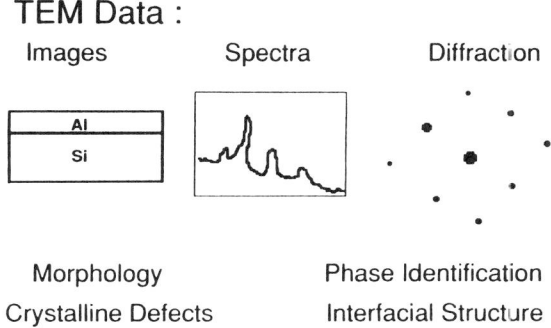

Fig. 11. Through the analysis of TEM data (images, spectra, and diffraction patterns) one obtains information about device morphology, crystalline defects, phase makeup, and the nature of thin film interfaces.

composition information (all with high spatial resolution). TEM provides materials analysis of such variety and detail as a consequence of its many modes of operation and its sample requirements.

For analyses of silicon-wafer-based thin-films, the salient types of materials information provided by TEM can be categorized into four areas, discussed in the following. This can be used for comparison to other materials analysis techniques and in determining when it is appropriate to seek TEM analysis.

Configuration. TEM imaging provides feature observation with the highest possible resolution. Device structures can be viewed in detail, film thicknesses can be measured with high precision, and grain structure can be evaluated. This area of application of the TEM is the most easy to interpret, but its use requires considerable expertise in crystallography. However, in device cross-section imaging applications, there exists much overlap in the information provided by TEM and SEM. Comparisons of the two methods for this application are instructive—one should be careful to select the most appropriate method for the problem at hand.

The new ultra-high-resolution SEMs, which use field emission filaments and in-lens imaging, combine the traditional strengths of SEMs with the higher resolution that has previously been attainable only with TEMs. Resolution of these new SEMs is ~ 1-10 Å, which is more than sufficient for many imaging tasks. They can be run at low kV, and excellent images can be obtained even from uncoated samples. There are several advantages of using the ultra-high-resolution SEM over using TEM; sample preparation time is much lower, field of view is much larger, depth of focus is large (i.e., device surface imaging), and recorded images are available quickly. On the other hand, TEM provides highest resolution of interfaces and, unlike SEM, does not require surface topography to distinguish different material regions at high resolution. However, one should not apply TEM to device imaging applications where SEM will suffice.

To understand which jobs must be performed on the TEM and which can be satisfactorily completed on the SEM, or the ultra-high-resolution SEM, the analyst must understand the different contrast mechanisms under which images are formed

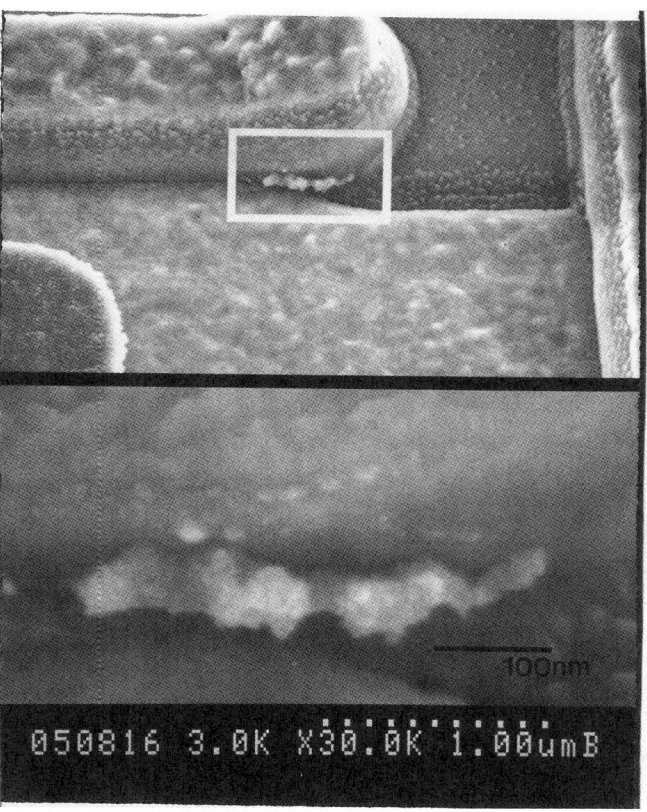

Fig. 12. SEM image of an isolated TiN stringer from a section of patterned wafer. The sample was uncoated and unstained. The image was recorded with a beam energy of 3.0 kV. The upper view of the stringer gives it relationship to the surrounding structures at a magnification of 30 kX, while the higher magnification view gives details of the stringers at a magnification of 150 kX.

and know what types of information are important. As described earlier, a TEM is NOT a SEM with higher magnification capabilities. Much information in a TEM image is fundamentally different from that in a SEM image. Some analytical problems require information that is uniquely available from SEM images, some problems require information that is uniquely available from TEM images, and some questions can be resolved with either technique; in the final case the SEM, with its quicker sample preparation time, should be used. The examples section of this article compares the use of TEM and SEM in this area.

Phase Identification. In addition to the configuration of the material components in a device, the identity of these components is often required. Thin film reactions and second phase precipitation are common. With chemical and diffraction analysis, the phase identity of a feature less than 100Å across can sometimes be determined. Such work on VLSI device cross-sections is impossible with an SEM.

Crystalline Defects. All crystalline materials typically possess point, line, and plane defects (e.g. grain boundaries, dislocations, and stacking faults). Such defects often have significant influence on device processing and performance.

 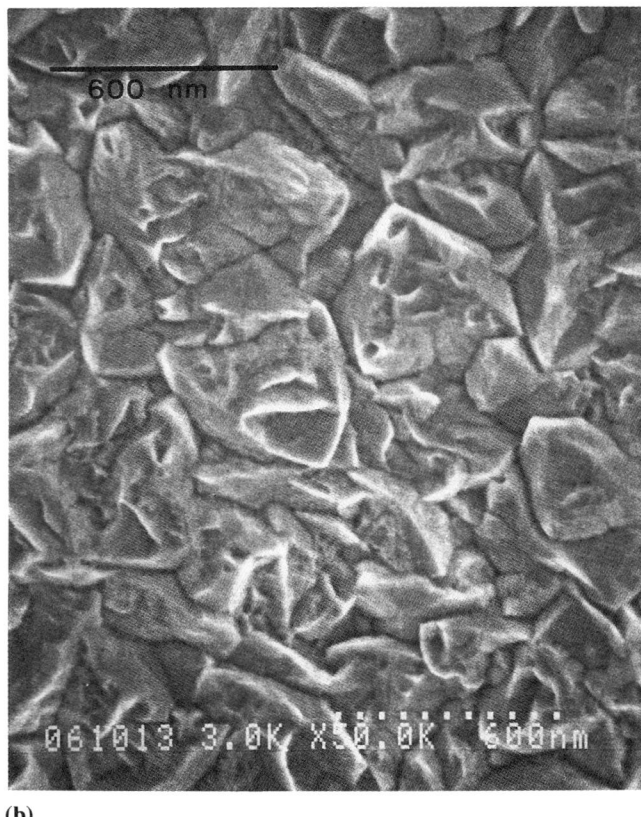

Fig. 13. (a) Plan-view TEM image of tungsten grains taken at 300 kV; magnification of 52 kX. (b) Plan-view SEM image of tungsten grains taken at 50 kX using a beam energy of 3.0 kV to optimize information from the surface.

For example, anomalous diffusion profiles can result from the presence of stacking faults while dislocations can serve to short a pn junction. Diagnosis of such problems is possible with TEM by direct imaging of these features. However, interpretation of the images of such features often requires the assistance of an experienced TEM scientist. This image analysis can be provided by the microscopist performing the TEM work.

Interfacial Structure. In this area, TEM can produce images with information not available by any other means. With lattice imaging performed on cross sections, the atomic level configuration of a semiconductor interface can be viewed and its influence on neighboring thin-film defects can be evaluated. Such information may prove vital in device analysis; device feature dimensions are now small enough that interfaces have a significant effect on electrical performance. Features will eventually be so small that interfacial effects will dominate.

ILLUSTRATIVE TEM ANALYSES

In illustrating the above categories, multiple use will be made of some case studies; usually, more than one type of information is required to evaluate a problem or provide process support. Most discussion will concentrate on configurational observations due to the value of imaging in device related studies. The focus in the following discussion will be on the data and its analysis rather than the technological significance of the structures.

Configuration. (Note: All ultra-high-resolution SEM images were recorded on the cold-cathode field emission in-lens Hitachi S900 SEM. TEM images were recorded on a JEOL 2000FX and a Philips CM30.) Use of TEM and SEM micrographs for device feature observation is common. (Ref. 8 catalogues many excellent TEM micrographs of device cross sections.) The following compares and contrasts TEM and SEM device imaging applications.

An SEM sample, even for in-lens, ultra-high-resolution SEMs, is larger than a TEM sample; more area is available for analysis. SEM analysis has always been used to search for abnormalities, defects, and residues. The higher resolution now available allows smaller defects to be imaged. Figure 12 is an image of a titanium nitride (TiN) stringer located in a particular position only on specific die. Several dozen RAM cells were viewed before any stringers were found. TEM is not suited to searches for such isolated defects. Part of the strength of SEM in device surface observations is due to its great depth of focus. The SEM is particularly well qualified for analyzing several layers of structure and the relationship of the layers to one another. In Fig. 12, the location of the stringer is also important.

The TEM is a transmission imaging instrument—a TEM image is inherently a two-dimensional projection of the sample structure. For planar orientation samples (see the tungsten film in Fig. 13a), there is no topographical or surface information in the image. Grains of tungsten appear as various shades of gray

Transmission Electron Microscopy

depending on their crystallographic orientation with respect to the electron beam direction. The black-appearing grains are oriented such that some zone axis is nearly parallel to the electron beam. If a particular orientation of grains is important, an-

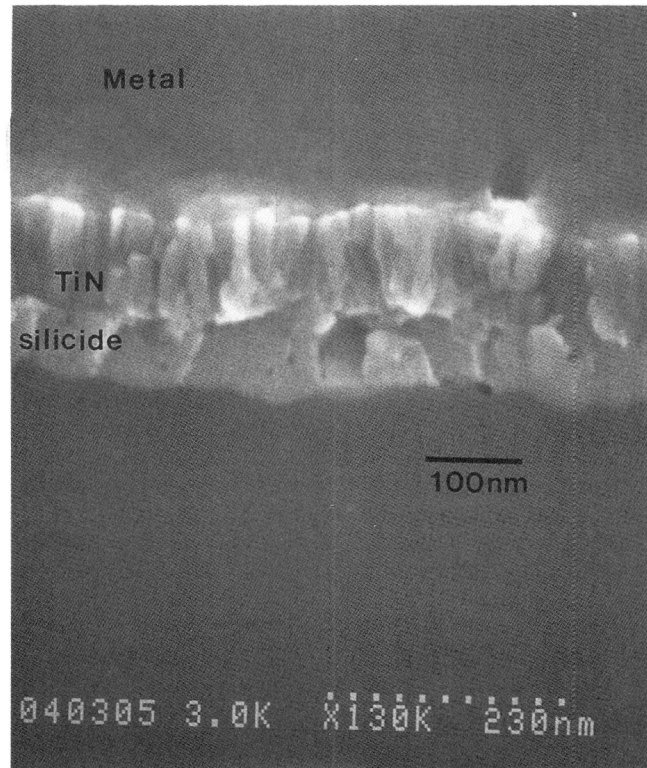

(a)

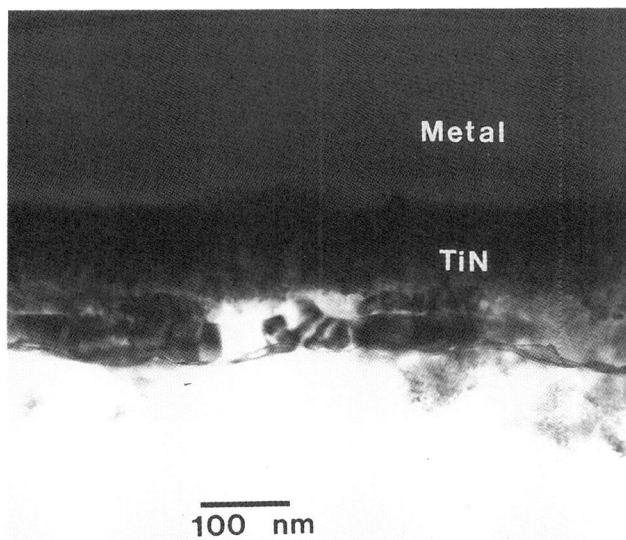

(b)

Fig. 14. (a) Cross-section SEM image of the metal layers. The sample was unstained and uncoated. Beam energy was 3.0 kV; magnification is 130. (b) Cross-section TEM image of the same metal layers. Magnification of 120 kX.

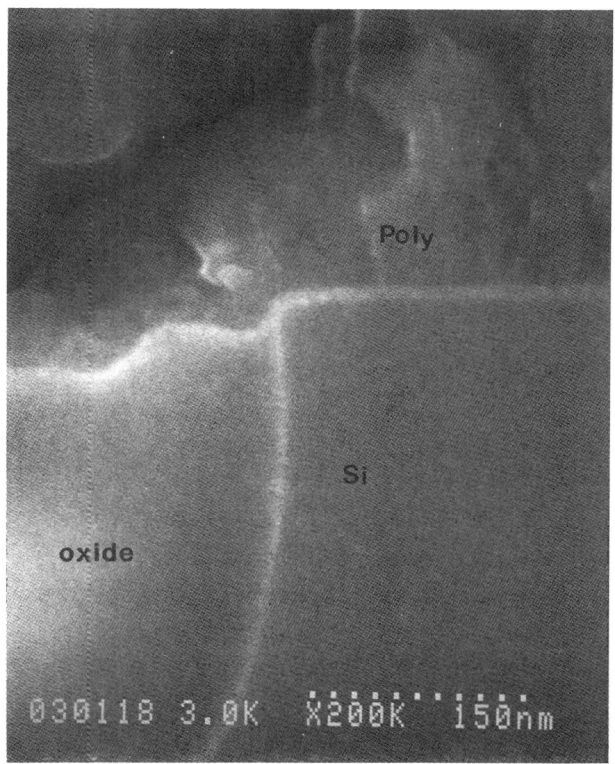

(a)

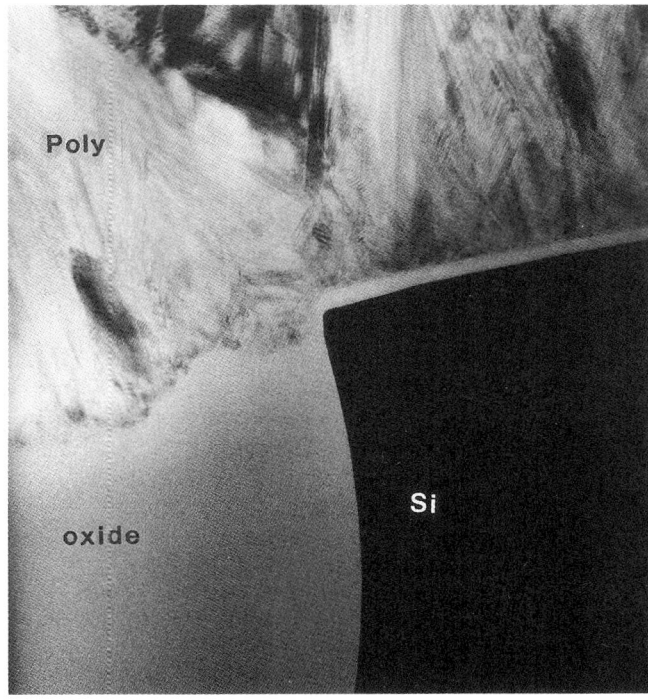

(b)

Fig. 15. (a) Cross-section SEM image of the Si corners at the edge of an oxide-filled trench. The image was recorded at 3.0 kV. The sample was uncoated. The oxide was highlighted using a 5 second dip into diluteHF. (b) TEM image of the same corners.

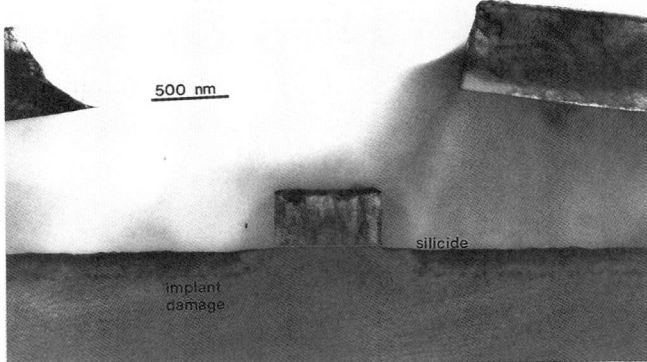

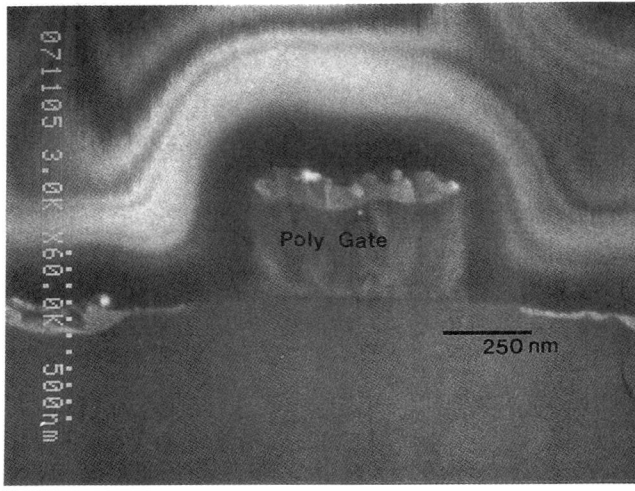

Fig. 16. (a) Cross-section TEM image of the gate and active areas of a transistor from a set of electrical test structures. Silicide covers the active area; end-of-range implant damage is visible beneath the silicide. The spacer is barely differentiated from the surrounding field oxide. The original magnification of this image was 10 kX. (b) Cross-section SEM image of a similar area. The sample was not capped, coated, or stained. Beam energy was 3.0 kV.

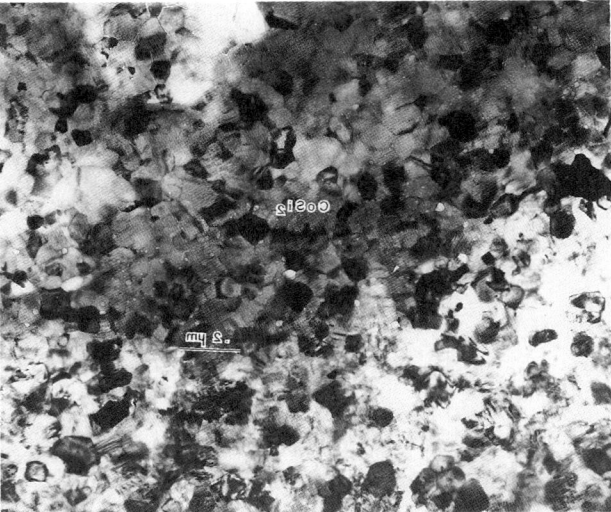

Fig. 17. Planar sample of $CoSi_2$ film.

nular dark field TEM images, which show all the grains at a particular orientation, could be collected. In the SEM, the image formed by secondary electrons that are emitted from the surface of the sample has a three-dimensional appearance, as shown in Fig. 13b. The highest sample regions reflect more electrons into the detector (that is located above the sample) and appear brighter in the SEM image. This image of tungsten is from the same sample as the TEM image in Fig. 13a. Grain boundaries are also visible in this ultra-high-resolution SEM image of uncoated tungsten. However, there is no information available about the orientation of the tungsten grains and one can not be sure all grain boundaries are observed. If the surface roughness is important, the analysis is a SEM job. Grain size analysis could be done from either image, though TEM is more accurate (when properly done) and is required for fine grain sizes. Analysis of the relative orientation of grains is a TEM job as is any detailed study of grain boundaries.

The ultra-high-resolution SEM image shown in Fig. 14a clearly shows the location of the silicon, titanium silicide (silicide), TiN, and metal layers. The roughness of the silicon/silicide interface can be measured. The contrast within the silicide layer relates to how the sample was cleaved, which can have some dependence on the crystallography within the layer. The entire field of view contains useful information. In the TEM image shown in Fig. l4b, the TiN layer is too thick to observe its internal structure. The TEM image is from the same sample and contains additional information. The silicide layer is composed of two separate layers of crystals; the contrast between the layers results from the differing orientation of the grains and the grain boundaries between the layers. The surface roughness of the silicon/silicide interface can be measured from either micrograph.

One of the advantages for TEM analysis of ICs is that they are formed on a large, perfect single crystal. Whether the substrate is GaAs or Si, the substrate is always present in cross-sections as a reference for calibrating magnification (if the TEM can resolve the lattice fringes), camera length for diffraction patterns, and for precise tilting of the sample so that the interface is observed exactly perpendicular to the electron beam. This orienting of the interface is completely independent of the orientation of the original cleave of the sample or the angle of the grinding or ion milling.

Figures 15a and b show the corners of an oxide trench. Figure 15a is the ultra-high-resolution SEM image, recorded at 200 kX. The sample is uncoated and recorded at 3 kV to minimize noise from secondary electrons created in the bulk of the sample. The shape of the trench corner is obvious; the gate oxide is clearly resolved. Figure 15b is a TEM image from the same sample recorded at the same magnification. The shape of the trench corner and the continuity of the oxide around the corner can clearly be seen in either image. Both TEM and ultra-high-resolution SEM provide information about the shape of the corner and the continuity of the oxide around the corner. If

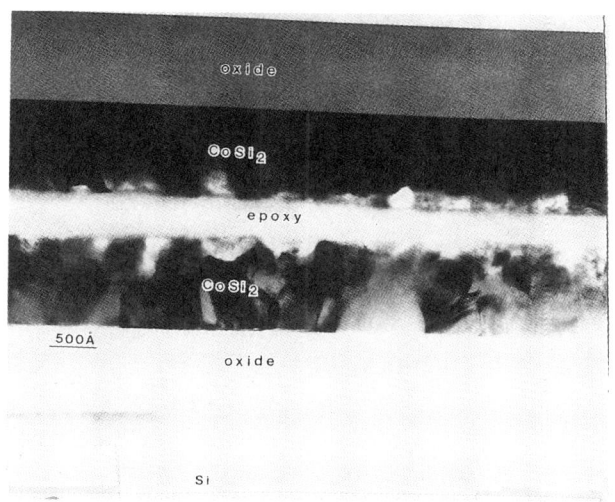

Fig. 18. Cross section of CoSi$_2$ sample in Fig. 17. Note voids near the surface of the film.

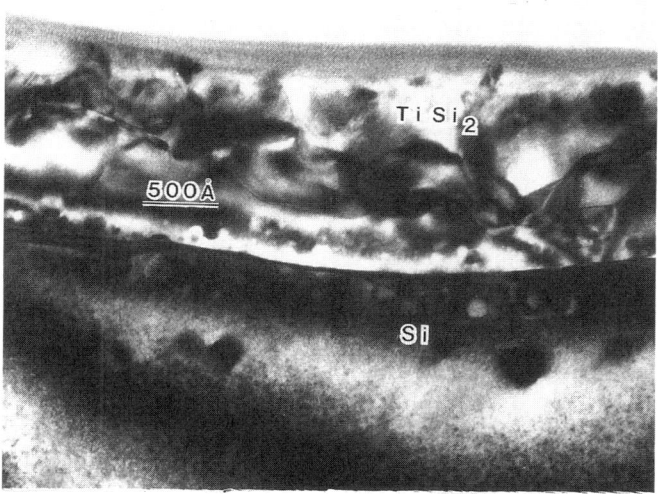

Fig. 19. Cross section of an annealed TiSi$_2$ film.

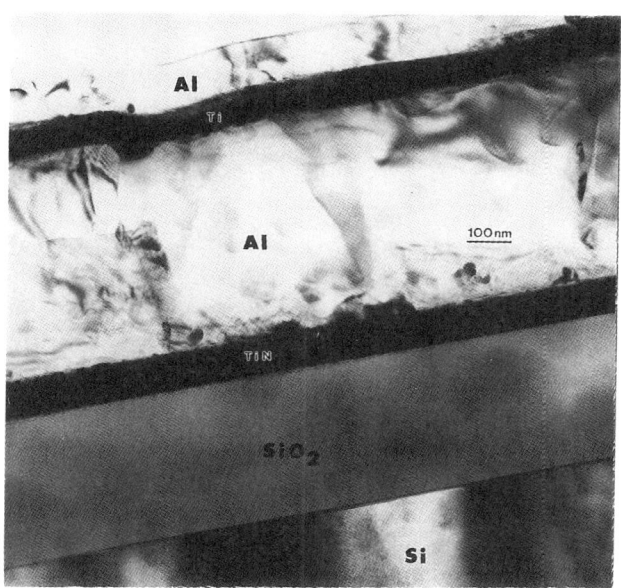

Fig. 20. Bright field image of a cross section of an as-deposited wafer. The detailed structure of the thin films is revealed by such samples.

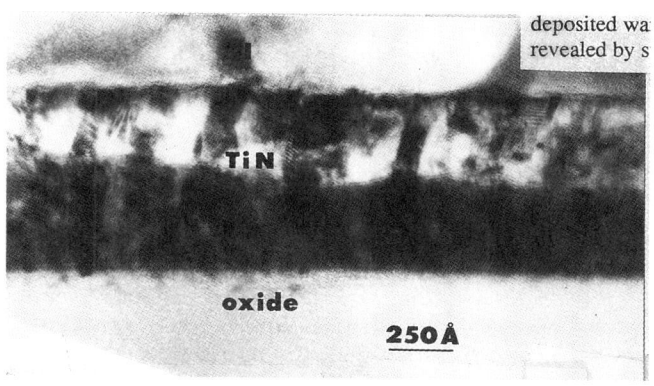

Fig. 21. High magnification view of a TiN layer showing columnar grain structure.

preparation. A higher magnification view of a similar area is shown in Fig. 16b from the ultra-high-resolution SEM. This sample is uncoated, and so a halo effect from charging surrounds the poly gate. The silicide over the active area is obvious; the space is delineated from the field oxide. This sample was unstained and uncoated. If the appropriate stain were used, a qualitative dopant profile would be visible in the active area and differences between types of oxides would be more dramatic.

Other device features which are unexpected and invisible with other techniques can be found with TEM. In a planar sample prepared from a cobalt silicide thin-film, light-colored objects about 200Å across are found scattered through the film (Fig. 17). In a cross section prepared from the same wafer, these features are revealed to lie near the surface of the film (Fig. 18). These objects are voids, the identification being confirmed by observing the sample from different angles. Voids have been observed in a variety of silicides. The behavior of such voids

variability of the corners is important, than the ultra-high-resolution SEM should be used. If an exact measurement of the gate oxide at particular points is important, the TEM must be used to ensure that the oxide is not shadowed, the interface is exactly parallel to the electron beam, and the silicon lattice plane can be used as an internal magnification calibration.

In a final comparison, Fig. 16a gives a low-magnification TEM image of a transistor. The poly gate is capped with silicide. There is also silicide over the active areas. The oxide spacer on either side of the gate is barely differentiated from the field oxide. The aluminum lines were broken during sample

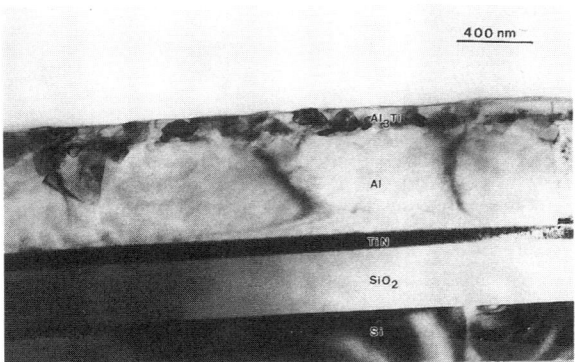

Fig. 22. Bright field image of a cross section of an annealed wafer. Aluminum has reacted with the titanium layer to form Al₃Ti.

must be monitored to make certain they do not trap chemicals on the wafer surface during processing steps.

Composition and Phase Identification. At times, the chemical identity of a device feature is unknown or must be confirmed. Conventional TEM imaging alone does not provide this information. Chemical composition can be obtained by use of the spectrometers described earlier.

As useful as chemical composition data is, at times it leaves ambiguities as to the phase identification of a given structural feature. Often, data is too inaccurate to distinguish between phases of related composition while a material of fixed composition can have a variety of stable or metastable crystal structures. This problem is solved with TEM by obtaining selected area or microdiffraction patterns. The crystal planar spacings extracted from the patterns (analogous to X-ray diffraction analysis) are compared to known spacings for the suspected phases. As an example, note the cross section of an annealed $TiSi_2$ film in Fig. 19. Prior to annealing, the film composition was that required for $TiSi_2$. However, diffraction analysis revealed a change in planar spacings subsequent to the annealing. The as-deposited film had the metastable C49 structure which transformed during heat treatment to the equilibrium C54 phase.[14] This transformation is significant since the C54 phase (formed above 800°C) has electrical properties superior to C49.

Figure 20 presents a multilayered interconnect scheme in as-deposited condition—a silicon wafer with oxide film followed by TiN, Al, Ti, and Al. The TiN film is smooth, continuous, and columnar grained (Fig. 21), while the Ti layer is continuous, although wavy. The Al has grown with an undulating surface, while the Ti has a uniform thickness and follows the undulations of the Al. Such undulations confuse analyses by thin-film techniques which average over relatively large sample areas, e.g., Auger Electron Spectroscopy (AES) or Rutherford Backscattering Spectroscopy (RBS). The non-flat interfaces blur the apparent location of the interface in depth profiles from such techniques. This could falsely be interpreted as a gradual, rather than abrupt, interface.

After annealing, the Ti layer reacts with neighboring Al (Figs. 22 and 23) to form a layer with a distinctly new structure.

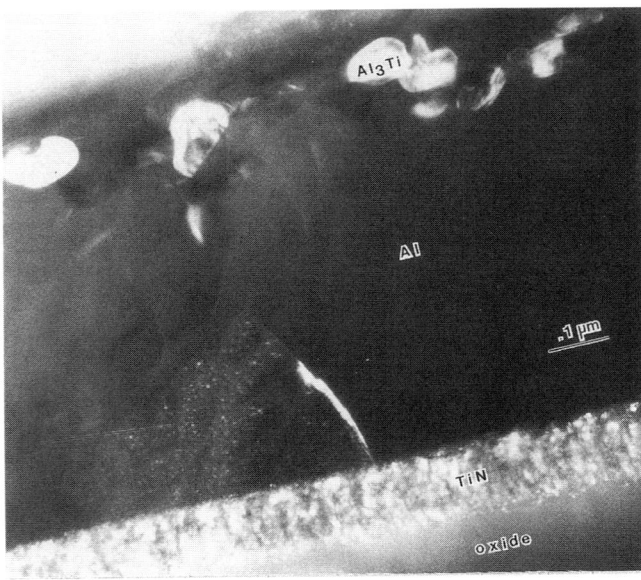

Fig. 23. Dark field image revealing grains of Al₃Ti.

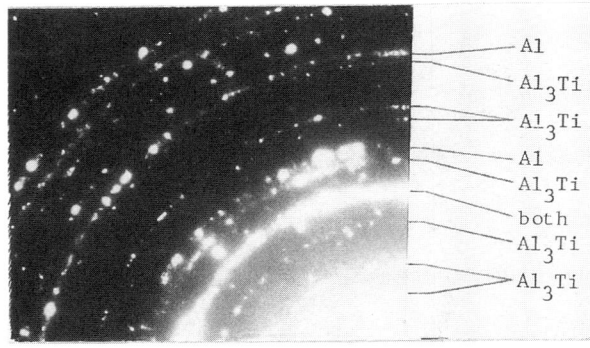

Fig. 24. Selected area diffraction pattern from annealed sample exhibits polycrystalline rings for Al and Al₃Ti.

The columnar grained Ti has been replaced by grains of irregular shape (most obvious in the dark field image, Fig. 22). This film appears to be discontinuous, though the shape and orientation of its grains make this uncertain. Further analysis of the composition of this new film requires observations beyond imaging. A number of phases are possible. However, since excess Al is available, Al₃Ti is expected if the reaction is allowed to attain an equilibrium state. This phase is desired since it is very stable against electromigration damage. The presence of this phase was confirmed by diffraction patterns taken from planar samples (Fig. 24). A ring pattern is obtained since many grains contribute to the diffraction pattern. Some Al film overlaps the Al₃Ti film so that rings are present for each material. Electron diffraction is very useful for planar spacing analysis in thin films due to the strong interaction of electrons with matter. Such detailed data is difficult to produce with X-ray diffraction techniques for very thin films.

The multilayered interconnect design discussed above is an attempt to prevent circuit opens caused by electromigration. This phenomenon causes voids to appear at grain boundary triple junctions and surface intersections. With two Al layers separated by an electromigration resistant Al_3Ti layer, current can bypass a void in one Al layer by utilizing the other layer for conduction. In the unlikely case of neighboring voids in both Al layers, the Al_3Ti serves as a current path for a short distance. TEM confirms the production of the desired structure and in the event of future failures would show why the failure occurred. Though TEM is more commonly applied to research and process development studies, failure analysis can be performed on completed devices. Figure 25 shows micrographs revealing the failure mechanism in a PROM device. Such devices are programmed by sending current through a fusible link, leading to the formation of insulating oxides in the formerly metallic link. In this case the link was composed of nichrome, NiCr. In this study, some fusible links still conducted after programming. SEM observations of the exterior of the device showed no differences between good and bad links. However, TEM comparisons of these fusible links revealed that metallic filaments were reforming across the fuse gap in bad samples, shorting the link.

Crystalline Defects. Defects typically encountered in device thin films include precipitates, dislocations, stacking faults, twin boundaries, and grain boundaries. These can all have direct or indirect effects on device performance by influencing the processing of the device or its electrical properties. Although grain size can be thought of as a structural feature, grain boundaries are in fact crystalline defects—the boundary between misoriented grains. These defects are invisible to the

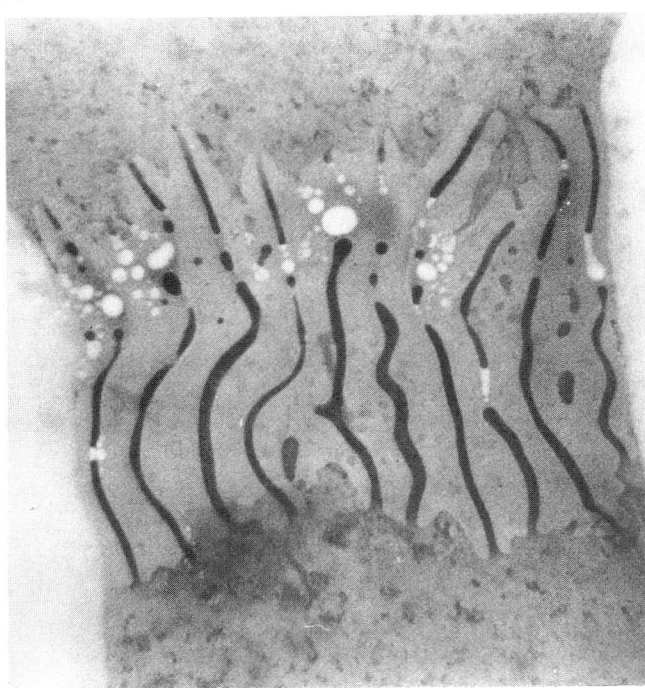

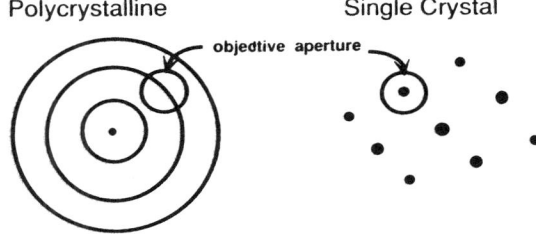

Fig. 26. By use of an aperture in the TEM, dark-field images can be formed by utilizing only a portion of the diffracted electrons. For single crystals, images are formed from a single diffracted beam while for polycrystalline samples a portion of a diffracted ring is used.

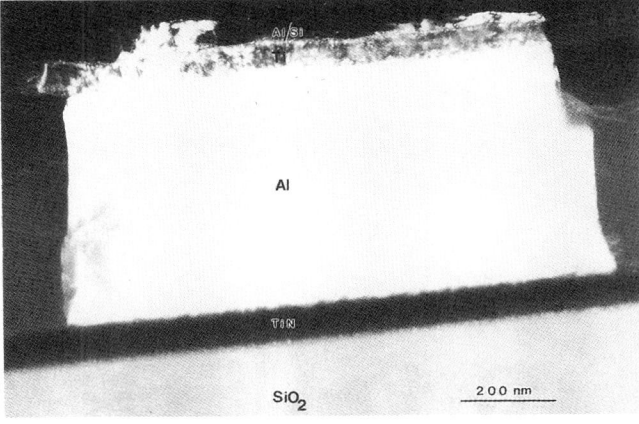

Fig. 25. Good (left) and bad (right) PROM NiCr fusible links. The bad link is shorted by metallic filaments, as determined by EDS analysis.

Fig. 27. Dark field image of the sample in Fig.20. The aluminum is very large grained.

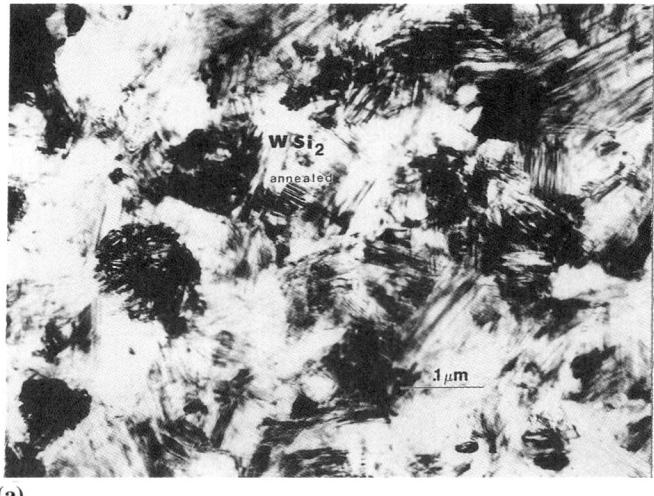 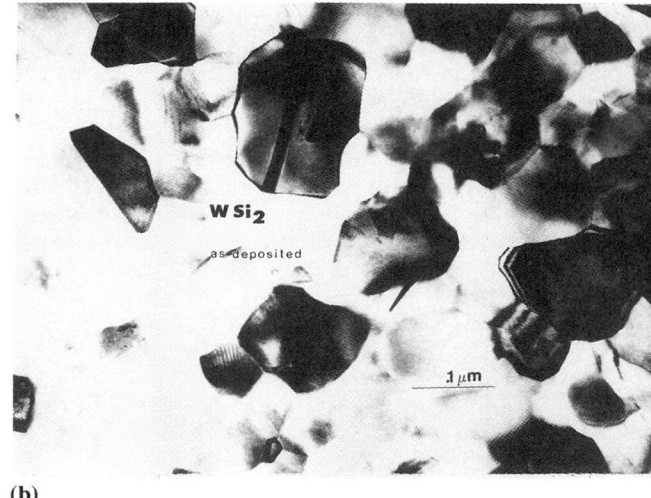

Fig. 28. (a) Planar sample of sputter deposited WSi₂ film. (b) WSi₂ film after annealing.

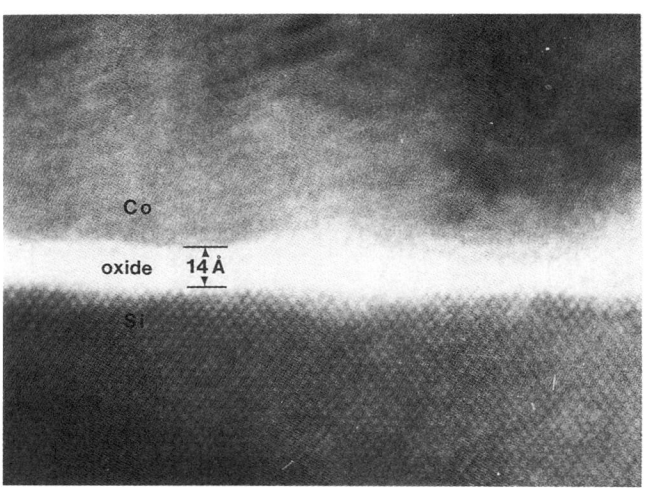

Fig. 29. Lattice image of silicon substrate with native oxide and Co film.

SEM, unless they intersect the sample surface AND have some surface topography at this intersection.

Dark-field images have many uses. In one application, by forming an image with diffracted electrons, high contrast is produced between grains of differing orientation or phase (Fig. 26). In Fig. 27, a large Al grain is seen. The Al grains span the full thickness of the film and are wider than they are thick. The boundaries between Al grains are perpendicular to the wafer surface. Also, see Fig. 23, discussed above.

Like grain boundaries, dislocations provide an enhanced diffusion path. An aluminum interconnect typically has a high dislocation density due to the large mechanical stresses encountered during the thermal cycles of IC processing (Fig. 5). Such dislocations may have a role in electromigration in narrow interconnects.[16] Returning to the transistor of Fig.16, note in the TEM image that underneath the silicide, there is a row of dislocation loops that define the end-of-range of the original implant (not the junction depth). The depth of these crystalline defects is dependent on the dose, type of implant, and implant energy.

Sputtered films often contain a high fault density. Figure 28 shows micrographs of planar TEM samples of a tungsten silicide (WSi_2) film, as-deposited, and after annealing. A high density of stacking faults and twin boundaries (the striations in the grains of the as-deposited sample) are largely removed by annealing. The behavior of these defects is readily observed in the TEM. Note the same types of defects in the polycrystalline silicon film of Fig. 10.

Interfacial Structure. Though the atomic structure of an interface may appear to be an esoteric topic, dimensions employed in devices are approaching a size where scattering of electrons at interfaces will have a significant influence on the electrical performance of thin-films. Figure 29 is a lattice image of a cross sectioned silicon wafer with a deposited (Co) film. The native oxide layer had not been etched prior to deposition of the metal. The silicon is oriented for observation of a $\langle 110 \rangle$ projection; the spots in the image correspond to columns of atoms in the silicon. The cobalt film planar spacings were beyond the resolution of the TEM used for this micrograph while the oxide is amorphous, hence no periodic structure is seen in these layers. The boundary between the Co and oxide is observed because these materials are of different atomic number. This permits a thickness measurement of the oxide layer to an accuracy within a few angstroms.

The lattice imaging capability of the TEM is gradually extending from basic materials studies to become a routine problem-solving tool in device fabrication and failure analysis. Even with "conventional" TEM imaging, the ability to study the many critical device interfaces with great detail has proven of great use in device development and analysis.

Transmission Electron Microscopy

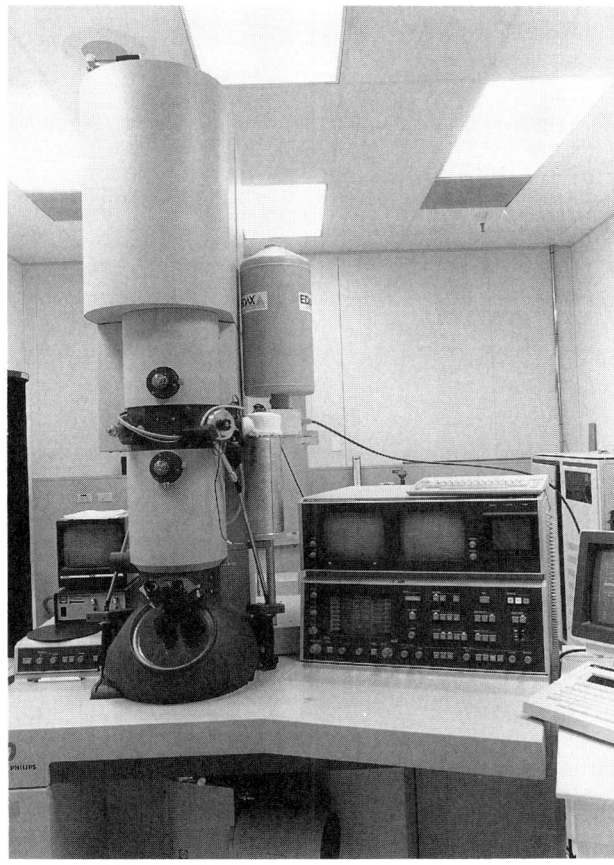

Fig. 30. View of a typical scanning transmission electron microscope system, consisting of major subsystem components from 3 manufacturers. The main TEM is in the left foreground, with the large 300 kV power supply in a pressurized tank filled with SF_6 gas immediately behind. The EDXA detector and Dewar are in the center of the photo, and the PEELS detector is in the kneehole of the desk. Notice that the operator looks through a thick leadglass window directly at the phosphor screen. With its 23 Å resolution, this large and expensive machine is a necessity in manufacturing sub-micron ICs, and in the failure analysis of these products.

SUMMARY

The TEM has a unique ability to observe internal structure and chemistry with resolution to a few Angstroms. Due to the extensive use of SEMs in device analysis, comparisons have been given for there instructive value. While the SEM is essentially a surface imaging instrument, the TEM observes internal microstructure by imaging with electrons which have traversed the sample. Internal structure not observable with SEM can be examined in great detail with TEM. Minimal beam spreading in thin samples permits chemical analyses from regions approaching the beam diameter—less than 100Å. Diffraction data can be obtained from crystalline films. Structural observations in the latest generation of field emission SEMs are approaching the resolution of TEMs. The purpose here has not been to determine which technique is better; the two are complimentary and appropriate realms of application must be delineated. For problems where either technique is capable of providing the necessary information, SEM is preferred because of timeliness and lower costs.

The role of SEM is well established in the minds of the process developer, fabrication engineer, and failure analyst. Why is this not so for TEM? This is in part due to the specialized nature and less common availability of the technique, making it less familiar than the SEM to many engineers. TEM is also perceived as being so time consuming that it is impractical in manufacturing and failure analysis. In addition, full application and interpretation of TEM requires experience in materials science. However, these reasons are invalid for ignoring the technique, because many problems can only be solved through TEM study. Additionally, deeper materials understanding of many device related phenomena often only comes through studies which include TEM observations. The portion of problems in VLSI manufacturing which require TEM for solution are continually increasing. Manufacturers who do not pay sufficient concern to microstructural details will find themselves at a disadvantage to competitors with well developed TEM capabilities. Failure analysis applications of the TEM hinge on the ability to obtain thin sections from pre-specified device areas. Since the present smallest device dimension of interest is about 0.5μm, one requires the ability to cross section with a precision of about 500Å. This requirement applies to SEM cross sections as well! [17] A very few laboratories are presently performing such work. The recent introduction of Focused Ion Beam (FIB) methods for final thinning of TEM samples may prove a useful aid in such precise cross sectioning. The demands of failure analysis on a submicrometer scale will continue to be the primary driving force for advances in sample preparation methods.

In summary, the following are questions the engineer might ask himself/herself. Has the desired structure been produced? Have desirable or undesirable thin film reactions occurred? What defects have formed? Are there potentials for failure? What was the failure mechanism? When a device process development or manufacturing problem is encountered which would benefit from information obtained at high spatial resolution, TEM should be investigated as a potential problem solving tool. Armed with TEM analyses, the engineer can adjust design, process, and materials selection for improved performance and reliability.

REFERENCES

1. J. H. Rose, N. Riel, and R. Flutie, Proceedings of ISTFA 1987, pg. 95.
2. J. H. Rose, J. Kowalik, and N. Riel, Proceedings of ISTFA 1988, pg. 145.
3. J. W. Edington, "Practical Electron Microscopy in Materials Science", TechBooks, Herdon, VA (1976).
4. O. L. Krivanek, Chem. Scr. 14, 78-83, (1979).
5. D. B. Williams, Mat. Res. Soc. Symp. Proc., Vol 31, pgs. 11-22 (1983).
6. J. W. Steeds in "Introduction to Analytical Electron Microscopy", pgs. 387-422, J.J. Hren, J. L. Goldstein, and D. C. Joy (Eds.), Plenum Press, New York (1979).
7. David C. Joy in ref. 2, pgs. 223-244.

8. R. B. Marcus and T. T. Sheng, "Transmission Electron Microscopy of Silicon VLSI Circuits and Structures", John Wiley and Sons, Inc., New York (1983).

9. John C. Bravman, Ron M. Anderson, and Michael L. McDonald (Eds.), "Specimen Preparation for Transmission Electron Microscopy of Materials", Mat. Res. Soc. Symp. Proc., Vol 115 (1988).

10. Ron M. Anderson (Ed.), "Specimen Preparation for Transmission Electron Microscopy of Materials II", Mat. Res. Soc. Symp. Proc., Vol 199 (1990).

11. Richard Flutie, Mat. Res. Soc. Symp. Proc., 62, 105 (1986).

12. R. Pinizotto, F. Y. Clark, and M. L. Jarvis, Mat. Res. Soc. Symp. Proc., 62, 9 (1986).

13. A. E. Morgan, E. K. Broadbent, M. Delfino, B. Coulman, and D. K. Sadana, J. Electrochem. Soc., 134, 925 (1987).

14. Robert Byers and Robert Sinclair, J. Appl. Phys., 57, 5240 (1985).

15. D. Gardner, T. Michalka, K. Saraswat, T. Barbee, P. McVittie, and J. Meindl, IEEE J. of Solid State Circuits, 20, 94 (1985).

16. J. H. Rose, J. Lloyd, A. Shepela, and N. Riel, Proc. 49th Ann. Meeting EMSA, 820 (1991).

17. R. Anderson, Proc. 49th Ann. Meeting EMSA, 828 (1991).

18. S.J. Kirch, R. Anderson, S.J. Klepeis, Proc. 49th Ann. Meeting EMSA, 1108 (1991).

APPLICATIONS OF FTIR SPECTROSCOPY TO SEMICONDUCTOR FAILURE ANALYSIS

by Christopher D. Gondran and Ronald A. Carpio

Fourier Transform Infrared Spectroscopy (FTIR) is utilized extensively on a macroscopic as well as microscopic scale for both qualitative and quantitative analysis of inorganic and organic molecular species. The infrared spectrum is not only a unique fingerprint of the material being analyzed but it also provides information on the structural units present.

Common uses in the semiconductor industry include determination of the interstitial oxygen and substitutional carbon in processing wafers, epitaxial film thickness measurements, monitoring the phosphorus in phosphosilicate thin films and both phosphorus and boron in borophosphosilicate thin films, measurement of the hydrogen content in oxide and nitride thin films, and the identification of particles and residues. FTIR is utilized for incoming materials characterization, trace gas analysis, carrier concentration measurements, and the characterization of silicon surfaces after cleaning operations to list only a few of the lesser known applications. It is probable that the results of such FTIR measurements will be used at various times by the FA Engineer to help trace the source of device problems. So it is advisable to become at least acquainted with all these applications.

However, the most common use of FTIR in the FA Lab will for the identification of particles and residue contamination found on the partially or fully processed device, for the purpose of determining the origin of the contamination or to judge the probable effects of the contamination on device performance.

Thus, emphasis in this paper will be focused on the capabilities and limitations of FTIR for this particular application.

INTRODUCTION TO FTIR SPECTROSCOPY

Fourier transform infrared spectroscopy is a branch of vibrational spectroscopy. The sample is illustrated with infrared from a source such as a Glowbar® or Nernst Glower. Molecules will absorb radiation at characteristic frequencies in the infrared region of the electromagnetic spectrum whose wavelength extends from 2.5 microns to 50 microns or whose wavenumber (number of waves per centimeter) extends from 4000 cm^{-1} to 200^{-1}. *The wavenumber (cm^{-1}), which is used extensively in infrared spectroscopy instead of wavelength or frequency, is simply the inverse of the wavelength.*

$$\text{Wavenumber} = \frac{1}{\text{Wavelength}} \qquad \text{(Eq. 1)}$$

There are 3N-6 degrees of freedom for a non-linear molecule, where N is the number of atoms in the molecule. Those vibrations which occur with a change in dipole moment will absorb radiation at characteristic frequencies in the infrared region of the spectrum and will be observed in the infrared spectrum. Those that occur with a change in polarizability will be observed in the Raman Spectrum, which is thus complementary to infrared spectroscopy. In both cases, a spectrum is produced which is as unique as a fingerprint for the material being studied.

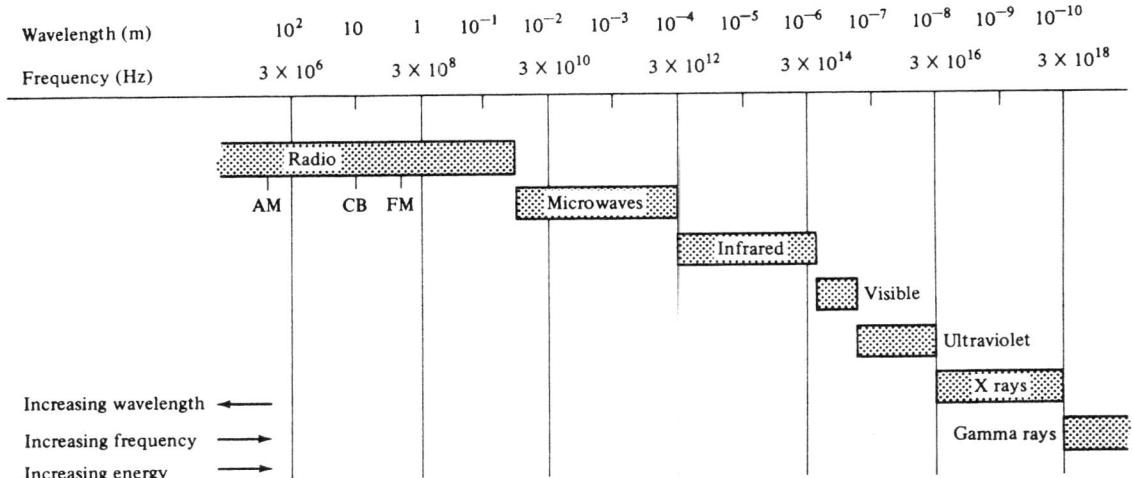

Fig. 1 The electromagnetic spectrum, showing the relative position of infrared between microwaves and visible light. (Adapted from Fundamentals of Organic Chemistry, by H. Richey, Prentice-Hall, 1983.)

Table 1 Commonly encountered bonds and their absorption regions.

Bond	Characteristic absorption
p=0	1316 cm^{-1}
Si–0	1080 cm^{-1}
B–0	1370 cm^{-1}

In general, the FA Engineer can consult a reference text which tabulate the more common absorption frequencies for major vibrations of interest. In order to identify unknowns, most modern FTIR Systems can be equipped with reference libraries and a computer algorithm will allow the library to be rapidly searched and the best full spectral match or matches will be displayed. It should be noted, however, that no library currently on the market covers all possibilities which will be encountered in the course of semiconductor failure analysis work. This is especially true of plasma residues, which are unique. Moreover, in the case of a mixture of components, complexities are expected in interpretation. Nevertheless, it will always be possible to explain the more prominent vibrational bands which will provide clues as to the identity of the residue.

Table 1 provides an example of some commonly encountered boron, silicon, and phosphorus bonds and the absorption regions where the vibrational modes are expected. It should also be noted that frequency shifts of vibrational modes can be expected and are diagnostic of interactions as are changes in intensity and band shape. Moreover, in the solid state one must also consider phonons, or collective lattice motions, which can give rise to vibrational modes. In the case of polymer analyses, the infrared spectrum has bands which arise from absorption of energy by relatively localized parts of the chemical molecules in the sample. These parts are the functional groups contained in the polymer chain.

EXPERIMENTAL CONSIDERATIONS

Tremendous advancements have been made in the area of infrared spectroscopy in the past two decades due to the advent of (1) Fourier Transform Infrared Spectrometers, (2) computerized data processing procedures, (3) modern sampling techniques, and (4) the development of microscopic capabilities. These advancements are largely responsible for the widespread use of infrared spectroscopy in the semiconductor industry. The limitations of infrared spectroscopy can largely be attributed to the use of a black-body source and thermal detectors.

Almost all modern infrared spectrometers used in a semiconductor facility today are of the Fourier Transform Type. This type of instrument makes use in most cases of a Michelson Interferometer, and they appeared in the market in the 1960s. The FTIR concept was adopted largely to overcome the instrumental limitations of dispersive spectrometers based on gratings, and prior to the 1950s on prisms. These limitations included poor resolution, stray light, wavelength inaccuracies, and lack of sensitivity and speed.

To most effectively utilize FTIR, the FA Engineer should be acquainted with not only the specific instrument hardware and software but also sampling techniques used for obtaining the

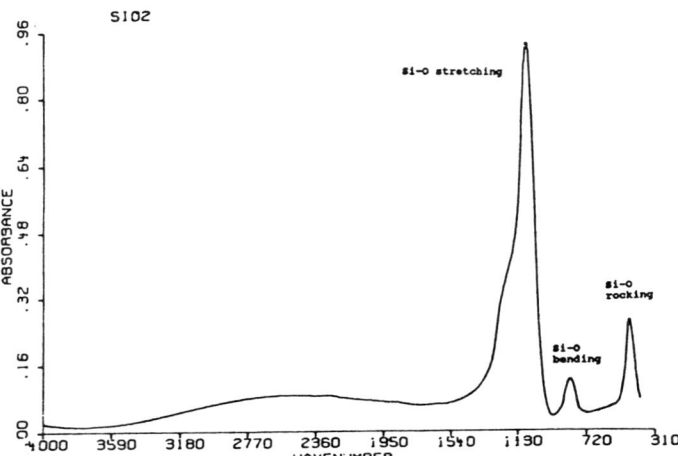

Fig. 2. Infrared spectrum of a silicon dioxide (SiO$_2$) thin film.

infrared spectrum. There are six techniques commonly used, which are transmission, external reflection, internal reflection, diffuse reflection, emission, and photoacoustic spectroscopy. Some of these modes are shown diagrammatically in Fig. 1. The transmission mode is the most popular while the emission mode is still in its infancy.

Transmission measurements should be utilized whenever possible, since the ease of obtaining and interpretation of the spectral information is simplified. Moreover, quantitative measurements are readily made from transmittance spectra by the application of Beer's Law, which is stated in the following equation:

$$A = eLC \qquad (Eq. 2)$$

where A= -log %T where %T= I/I$_o$ and in turn I$_o$ is the incident intensity and I is the transmitted intensity, e = molar extinction coefficient, L = the path length, and C = the concentration of the absorbing species

Generally, the incident beam is perpendicular to the sample surface, but this need not be the case. The key consideration in the use of the transmission mode of sampling is to ensure that the sample is thin enough, or diluted with a non-absorbing medium, to prevent total absorption of the beam. There are a variety of approaches for thinning or diluting the sample. One convenient approach is to dissolve the sample in a suitable volatile solvent, such as acetone, and to place the resulting solution on a silver chloride (AgCl) plate. The material of interest will be deposited on the support plate when the solvent evaporation is complete. Another potential complication that should be considered in the spectrum is the superposition of interference fringes upon the absorption spectrum. The separation between the fringes in terms of frequency f$_2$-f$_1$ is related to the thickness of the film or substrate having parallel faces, t, by the equation

$$f_2 - f_1 = N/(2nt) \qquad (Eq. 3)$$

where N is the number of interference fringes between the two frequencies, and n is the refractive index of the medium. These fringes are due to reflections at the interfaces where changes in refractive index are experienced. Such a fringe pattern will be observed for example if epitaxially grown silicon is present on top of bulk (Czorchralski-grown) silicon, and the doping concentrations are sufficiently different that there is a difference in refractive index. The equation above is also useful in determining the cell thickness.

Reflection techniques must often be used for studies of solids and thin films where it is inconvenient or not possible to prepare a sample for transmission work. This is often the case for FA work. Specular reflection is utilized when the surface is highly reflective, such as for metal surfaces, where the incident beam is reflected at the substrate interface. No beam penetration into the sample occurs, and the angle of reflection is equal to the angle of incidence. The bands will actually be derivative in shape. In cases where there is a thin absorbing layer on the surface of a more reflecting substrate, one must deal with the technique known as reflection absorption spectroscopy (1). When the angle of incidence is greater than 60 degrees, we are dealing with a special case of reflection absorption spectroscopy which is known as grazing angle reflection spectroscopy. The incident radiation is polarized parallel to the plane of incidence, and the angle of incidence is slightly less than 80 degrees. Extremely thin films on highly reflective surfaces can be studied using the grazing angle attachment, since the effective path length is much larger than the actual thickness. For example, an organic layer <10 nm thick can readily be detected on a metallic substrate using a grazing angle attachment. Diffuse reflectance is utilized for powders and rough surfaces, where the scattered radiation occurs in all directions.

Internal reflection, also known as attenuated total reflection or ATR, makes use of a crystal in contact with the sample. (2) It is desirable to select the crystal to ensure that there is some beam penetration into the sample at the crystal/sample interface. Intensification of the spectra will occur by multiple reflections. ATR is often useful for studies of liquids as well as solids. The contact between the sample and the internal reflection element is a critical factor in ATR. Thus, ATR is most useful for solid samples which are very flat and have a low degree of surface roughness. One of the benefits of ATR is that it makes use of multiple reflections which serve to maximize the band intensity, whereas in the case of external reflection a single reflection may be required to produce a usable spectrum. It is often possible to vary the angle of incidence between the IR beam and the ATR crystal and obtain depth profiling information. ATR spectra appear similar to transmission spectra, but the band intensities change significantly from one end of the spectrum to the other, which is due to the varying depth of penetration into the sample by the IR beam. In fact, the most often cited problem with reflection spectra is that they are generally distorted and hard to interpret. Reflection spectra of multi-film structure can be quite complex. Some of the techniques used in spectroscopic ellipsometry for simulation of spectra can be applied to infrared reflection spectra analysis.

There are a variety of accessories on the market which are utilized for determining spectra in the transmission or reflection modes. A current listing of suppliers can be found in the *Analytical Chemistry Lab Guide* and the *American Laboratory Buyers' Guide*. These accessories fit into the sample compart-

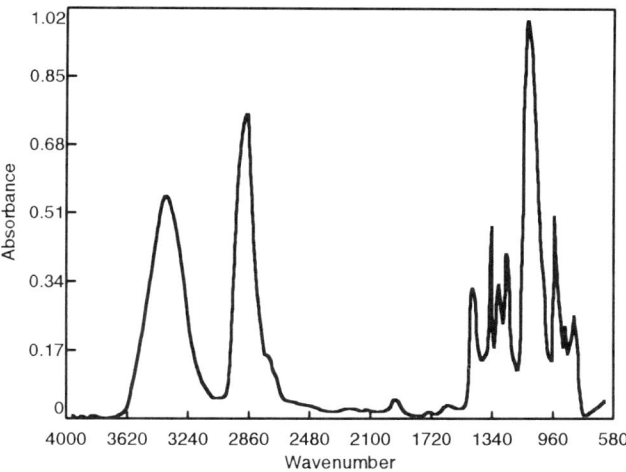

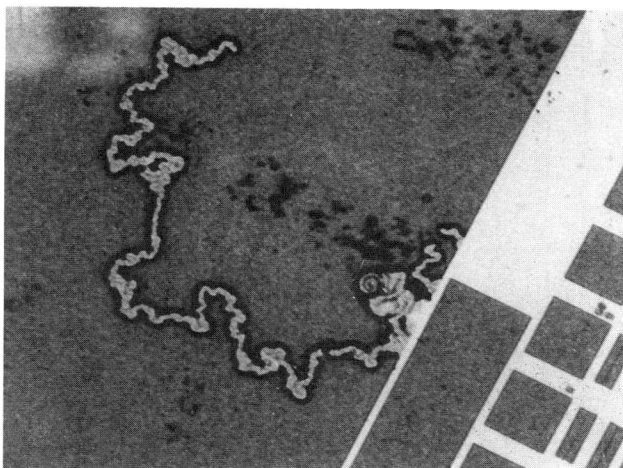

Fig. 3. Polymerized ethylene glycol on an integrated circuit, identified by infrared microanalysis.

ment of the spectrometer and generally require some alignment unless they are made specifically for the instrument that is being utilized. As a general rule, these accessories can be utilized if the sample is 200 microns or greater and at least 1 micron in thickness in size. Samples between approximately 200 microns and 20 microns, which is the diffraction limit, are handled with a special microscope attachment designed for transmissive and reflective modes of spectral acquisition. Special sampling accessories can be used in conjunction with an infrared microscope, and these accessories fit on the microscope stage. More information on microsampling is provided below.

Other general considerations which should be made in determining spectra is the selection of an appropriate reference. In some cases, the spectra of the sample is referenced against air. In other cases, such as the determination of a transmission spectrum borophosphosilicate glass (BPSG) thin films, it is desirable to use the wafer upon which the film was deposited or an identical wafer from the same lot as the reference. The spectrum of the reference will be mathematically subtracted from

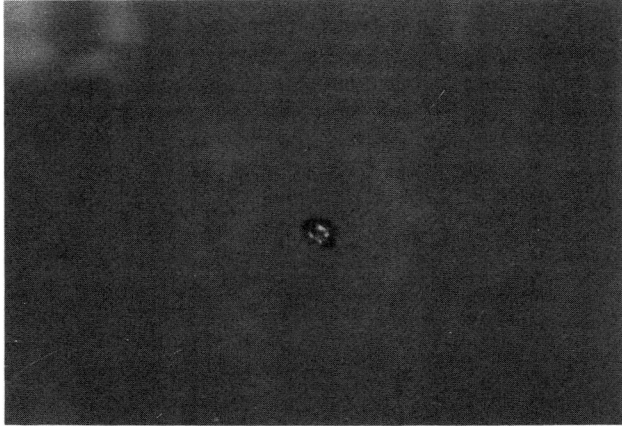

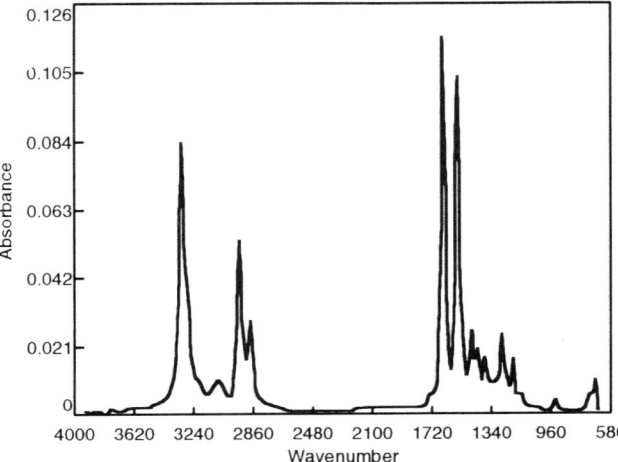

Fig. 4 Contaminant in photoresist identified as a polyamide (Nylon®) particle.

the composite spectrum leaving the spectrum of the BPSG film only. This will aid in interpretation and often in quantification. In the case of specular reflection being used for residue analysis on a bond pad, the reference chosen might be an area of the metal surface having no residue film.

Optimization of the signal-to-noise ratio can usually be achieved by increasing the number of scans. The number of scans generally must be changed by a power of 2 in FTIR, such as 526 or 1028. Of course, the time for the analysis will increase as the number of spectral scans is increased. The signal-to-noise ratio improvement generally goes up as $N^{0.5}$ where N is the number of scans. Other parameters which often must be optimized include sample purging, sample alignment, and selection of the interferometer data acquisition and Fourier transformation process itself.

The two most widely used **detectors** in the semiconductor industry are the liquid nitrogen cooled mercury-cadmium-telluride (mer-cad telluride, or MCT) photoconductivity detector and the triglycine sulfate (TGS) proelectric bolometer detector, which can be operated at ambient temperature. The MCT detector can be of the wide band or the narrow band type. Microsampling generally requires a narrow band MCT detector, which makes it impossible to go to wavenumbers much below 600 cm^{-1}.

Modern data analysis systems afford the opportunity for spectral searches and comparisons as has already been noted. Software generally also exists for smoothing, spectral subtraction which is very useful for observing small differences, baseline correction, and other more advanced spectral analyses such as establishing derivatives or performing spectral deconvolution and bandshape analysis.

For more complete details on FTIR theory and instrumentation, the reader is referred to the more common references in this field. (3,4,5,6)

SPECIFICS OF MICROANALYSIS

In general, infrared microscopy has been widely used in a variety of fields besides semiconductor failure analysis, such as forensics and polymer science. It identifies or characterizes particles, fibers, and inclusions or imperfections which are embedded in a larger matrix and to perform spectral mapping. This technique allows spectra in many cases to be determined with less than a nanogram (10^{-9} gm) of material. Although the main discussion here is focused on the application of FTIR microscopy to particle analysis in the semiconductor industry. The reader who is interested in semiconductor applications of infrared microscopy to spectral mapping such parameters as epitaxial thickness, interstitial oxygen concentration, and boron and phosphorus oxide measurements is referred to the article by Krishnan.(7)

Particle Analysis

Proper isolation of the particles for study is key to the success of microanalysis. This should be accomplished in a manner which prevents the introduction of external contamination, which can lead to erroneous conclusions. This often means working within an area having laminar flow hoods. In cases where it is not practical to remove the particle or residue from the matrix or substrate, it will be necessary to use the reflection mode unless the substrate has a suitable transmittance. Typical examples of infrared microanalysis are demonstrated in Fig. 2 and 3.

The particles, if removed, must be placed on a support and then placed on the X-Y stage of the microscope. This support could be a silicon wafer itself or it might be a transmissive substrate such as an AgCl crystal or a potassium bromide (KBr) "salt plate". Another approach is the use of the Diamond Anvil Cell (available from High Pressure Diamond Optics, Arizona) in which the particles, fibers, residue, etc. is compressed under high pressure to a thin layer. In the use of this accessory, it is important to check for window parallelism before the application of pressure; good parellism is indicated by the appearance of a barely noticeable interference pattern. The nature of the sample will dictate not only the support but also whether the sample is studied in the transmissive or reflective mode.

The next step is to position the sample at a point where the IR beam will traverse during the analysis. To make this possible an IR microscope is designed with both an optical and a microspectroscopy system. Using familiar optical microscopic techniques, the desired particle or sample area is positioned and

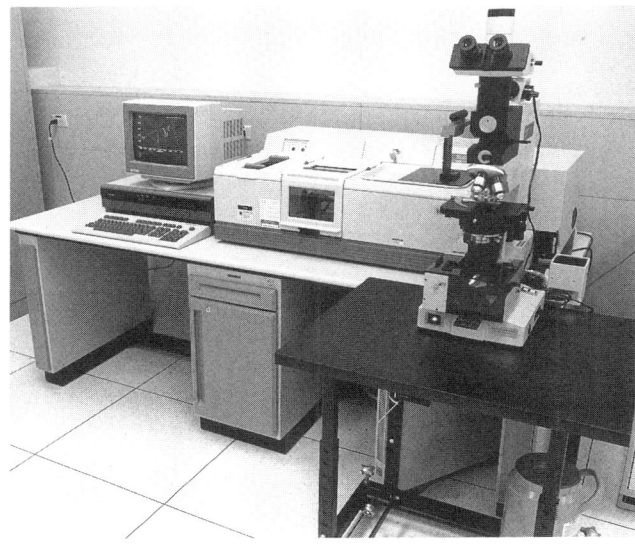

Fig. 5 An example FTIR system, with the main analyzer at center. The microscope in the foreground is equipped with both glass and Cassegrain optics. The system has a library of over 50,000 compounds, including organics and inorganics from commercial chemical suppliers, steroids, coatings, solvents, and flavors and fragrances.

brought into optical focus by using stage and focus adjustments. Once this is accomplished, the image is masked with a fixed or adjustable aperture. Most IR microscopes are equipped with variable apertures which can mask a sample of any shape. As the size of the aperture is decreased, the spectral noise level will increase.

The IR microscope and the optical system share a common intersection point at this remote image, so then it becomes a simple matter to simply covert the microscope from the optical viewing mode, which can either be transmissive or reflective, to the corresponding IR microspectroscopy mode. This conversion is most often achieved by rotating a knob or pulling a lever. To obtain a satisfactory spectrum, one must then optimize the acquisition parameters in a manner which is similar to that used for obtaining a quality spectrum for a macro-size sample.

In selecting an infrared microscope, the key parameters that must be considered are the numerical aperture, magnification, and the working distance. Since the numerical aperture is a measure of the collection efficiency, it is directly related to the sensitivity of the measurements. Cassegrain optical systems which are based on reflecting optics are utilized in such microscopes since ordinary lens material used as objectives in optical microscopes do not have acceptable transmittance properties for IR radiation.

COMPARISON OF FTIR WITH OTHER ANALYTICAL TECHNIQUES

It is very important for the FA Engineer to be able to select the appropriate analytical tool for the problem being investigated. This can best be accomplished by acquiring information on the sensitivity and capabilities of the wide array of analytical tools which are now available. In the case of particulate contamination problems, the role of FTIR must be compared to a number of other readily available instrumental techniques. Chemical composition can often be established by an energy dispersive X-ray system (EDS) attached to a SEM. X-Ray fluorescence (XRF) is also useful if the contamination area is larger than 100 microns. Auger spectroscopy (AES) or scanning Auger spectrometry (SAM) can provide elemental identification with more sensitivity than an EDX system. Secondary Ion Mass Spectroscopy (SIMS) is even more sensitive than Auger Spectroscopy, but lacks the spatial resolution of Auger, especially if the Auger has field emission capability. Molecular level information can be sought with FTIR if the contaminants are greater than 20 microns in size. When the residual contaminants or particles approach or exceed 150 microns in size, electron spectroscopy for chemical analysis (ESCA) can also be utilized. Often useful data can be determined with particles as small as 10 microns with the FTIR.

If the particles are between 10 microns and 0.5 microns, the Raman Microscope is the clear choice. Moreover, the Raman spectral information will provide a valuable supplement to the infrared data for molecules of the particles which are larger than 10 microns. Besides improved size resolution, the Raman Microprobe has certain other advantages over FTIR which should be considered. The preferred method for determining Raman spectra is in the reflectance, or more appropriately the backscattering, mode in contrast to FTIR. No interpretative difficulties are encountered with Raman spectra which have been determined in the backscattering mode as is the case for FTIR. Less sample preparation is generally required for determining Raman spectra. Aqueous solutions present no problems for Raman spectroscopy but are difficult to analyze by for FTIR.

In general, the Raman spectra of ionic inorganic materials are sharp while the FTIR spectra are very broad. On the other hand, there are presently a number of advantages of the FTIR technique over the Raman technique. Raman Microprobes are not as readily available as FTIR Systems. The Raman instrumentation is also more expensive, and the need for high intensity laser excitation is accompanied by safety considerations. The Raman spectral data bases are less extensive than the FTIR data bases. There are also fewer individuals trained in the use of Raman spectroscopy. In the case of organic analysis, fluorescence problems can be encountered with the Raman technique. No comparable problems must be overcome in organic analysis using FTIR.

There are fewer instrument manufacturers which specialize in Raman spectroscopy in comparison to FTIR systems, and most of the Raman instruments are not domestic companies. In addition, it is generally necessary to construct a Raman System from components. Thus, at the present time and for sometime into the future the FTIR approach will be the method utilized for determining vibrational information which can be used for the identification of contamination sources in the semiconductor FA Laboratory.

SELECTED REFERENCES

1. Carter, R.O., III, C.A. Gierczak and R. A. Dickie, Applied Spectroscopy, 40, 649, 1986.

2. N. J. Harrick, Internal Reflection Spectroscopy Wiley Interscience, New York, 1967.

3. P.R. Griffiths, Chemical Infrared Fourier Transform Spectroscopy, Wiley, New York, 1975.

4. R. J. Bell, Introductory Fourier Transform Spectroscopy, Wiley, New York, 1975.

5. J.R. Ferraro and L. J. Basile (eds.), Fourier Transform Infrared Spectroscopy, Vol1, Academic Press, New York, 1982.

6. J. R. Ferraro and K. Krishnan, Practical Fourier Transform Infrared Spectroscopy: Industrial and Chemical Analysis, Academic Press, Inc., 1990.

7. K. Krishnan Characterization of Semiconductor Silicon Using the FT-IR Microsampling Techniques, in Infrared Microspectroscopy Theory and Applications, R.G. Messerschmidt and M.A. Harthcock, eds., Marcel Dekker, Inc., New York, New York, 1988.

Failure Modes and Mechanisms

FAILURE MECHANISMS IN INTEGRATED CIRCUITS

by Dan Corum, Swee Yong Khim and Kendall Scott Wills

CHIP RELATED FAILURE MECHANISMS

Contamination

Contamination failures are by far the most common IC failure mechanisms. The root cause for the majority of defects can be traced to particles or etch residue. Contamination related defects can be categorized into three groups: foreign contamination or particles, process etch residue and ionic contamination (non-visual).

Mobile ionic contamination, such as sodium (Na), is difficult to isolate. This type of contamination causes transistor Vt shifts and junction leakage that will usually recover after a high temperature bake. The mobile ions that collect at the PN junctions and gate oxide will disperse after baking but will return under biased operating conditions.

The origin of foreign particles is usually traced to process equipment and personnel. The example shown in Fig. 1 is a pre-passivation particle that caused a leakage path between adjacent aluminum metal traces in a peripheral circuit which was isolated by photoemission microscopy. The EDX spectrum of this particle revealed the presence of iron and nickel, which is indicative of a stainless steel composition, most likely from some process equipment after metal patterning.

Process etch residue is usually associated with particles but the contamination that blocks the etching process is removed during additional processing. The example shown in Fig. 2 shows residual nitride and oxide under the polysilicon layer and across the diffused interconnect. This residue caused a resistive path as a result of blocked silicide and implant in the diffusion area. This residue was caused by contamination during the field plate patterning.

OXIDE PINHOLES

Oxide pinholes and gate oxide breakdown are also very common failure mechanisms. Oxide pinholes are typically associated

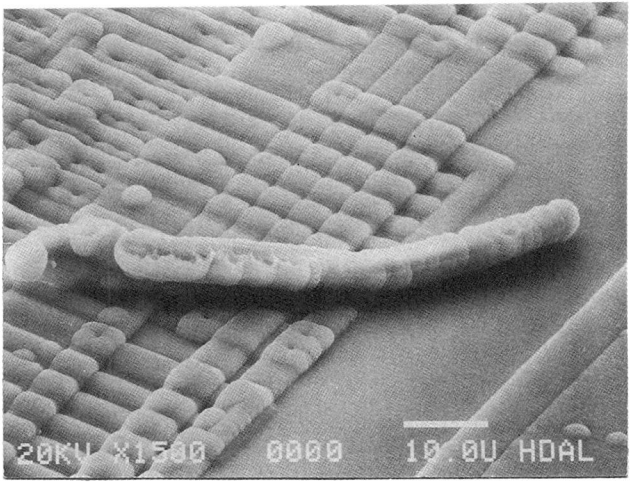

Fig. 1 Pre-passivation particle

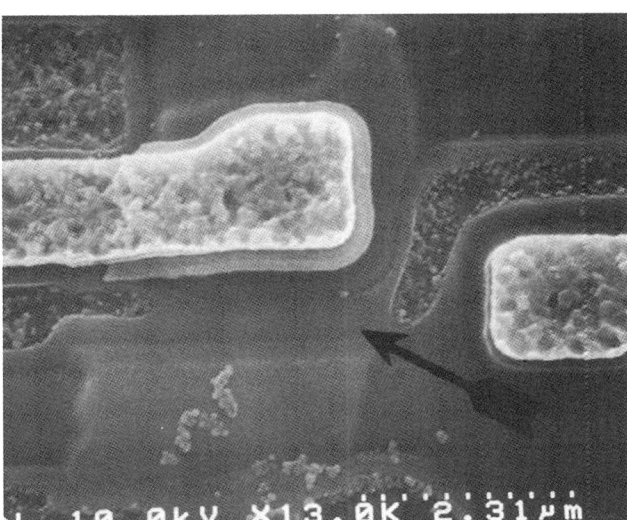

Fig. 2 Residue contamination

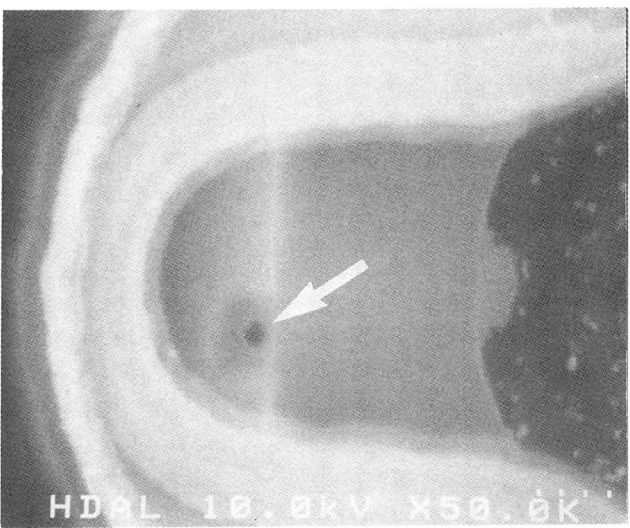

Fig. 3 Gate oxide pinhole

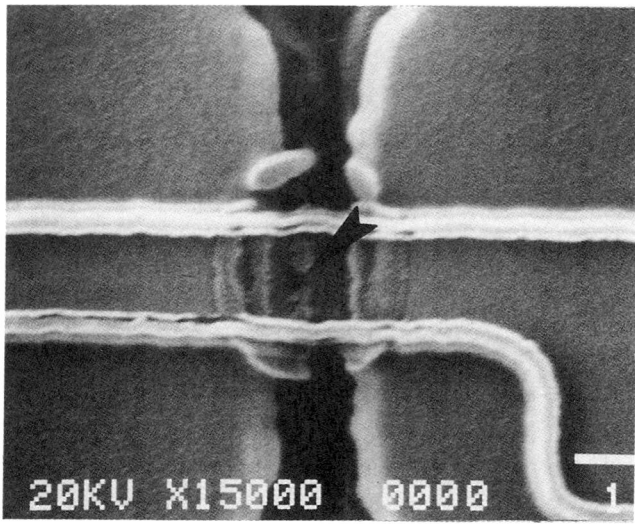

Fig. 4 Buried N⁺ oxide pinhole

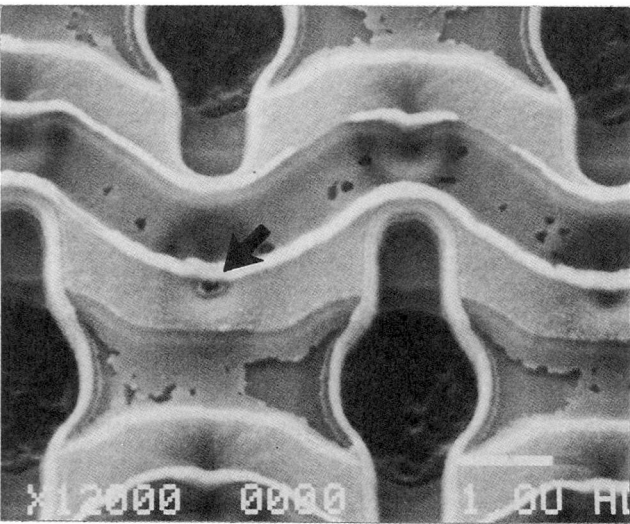

Fig. 5 Interlevel oxide pinhole

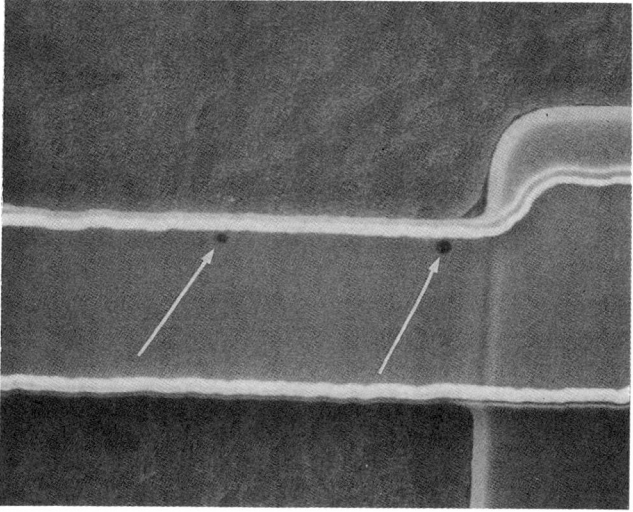

Fig. 6 Oxide breakdown (EOS/ESD)

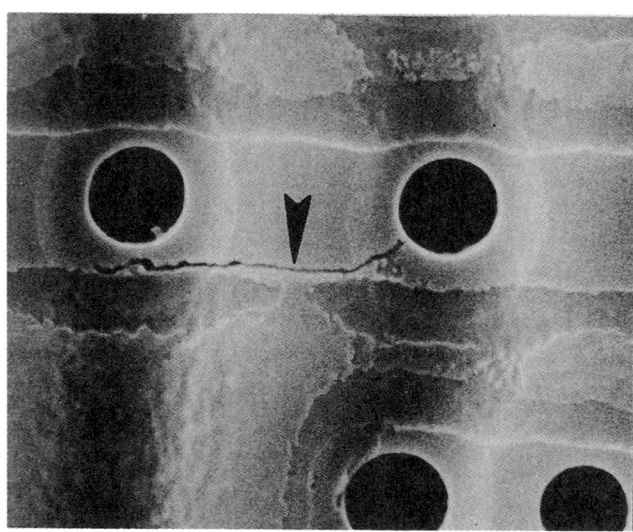

Fig. 7 Multilevel oxide crack (BPSG)

with particles, silicon dislocations or other process related problems.

A classic gate oxide pinhole is shown in Fig. 3. In this case, there is no evidence of contamination although a tiny particle during gate oxidation was suspected. The shadow around the oxide hole is the decorated bulk silicon that was created during polysilicon removal. This oxide pinhole created a leakage path from the poly wordline to the storage cell implant (logic 0 fail).

In Fig. 4, a buried N⁺ oxide pinhole is shown, that was associated with contamination. The pinhole created a leakage path from the poly interconnect to the underlying VCC guard ring. Auger analysis of the contamination revealed the presence of carbon and nitrogen. The presence of these two elements suggests residual photoresist and nitride were present before BN+ oxidation.

An interlevel oxide (ILO) pinhole is shown in Fig. 5. This pinhole is associated with a dielectric void over the trench opening, most probably resulting from a trench fill oxide problem. The ILO pinhole created a leakage path from the poly wordline to the underlying poly 1 field plate.

Gate oxide pinholes are sometimes associated with electrical overstress (EOS) or electrostatic discharge (ESD). The oxide pinholes shown in Fig. 6 were induced after ESD stress testing. Pinholes of this type are attributed to gate oxide breakdown and usually occur at the edge of the gate electrode.

OXIDE CRACKS

Oxide cracks are often caused by mechanical damage to the chip surface. The crack presented in Fig. 7 occurred at the mul-

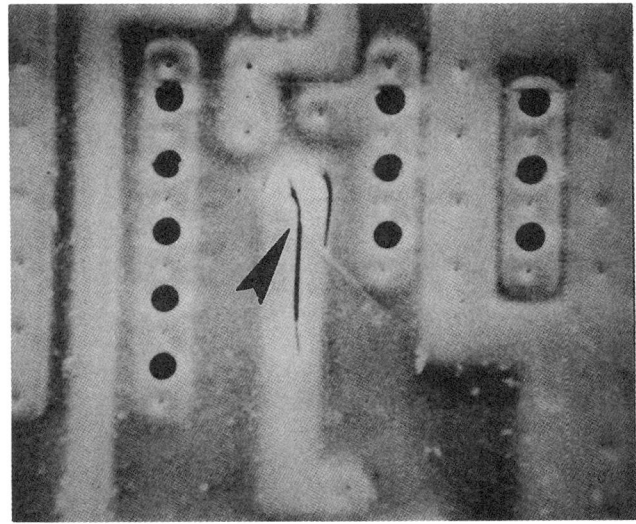

Fig. 8 MILO crack

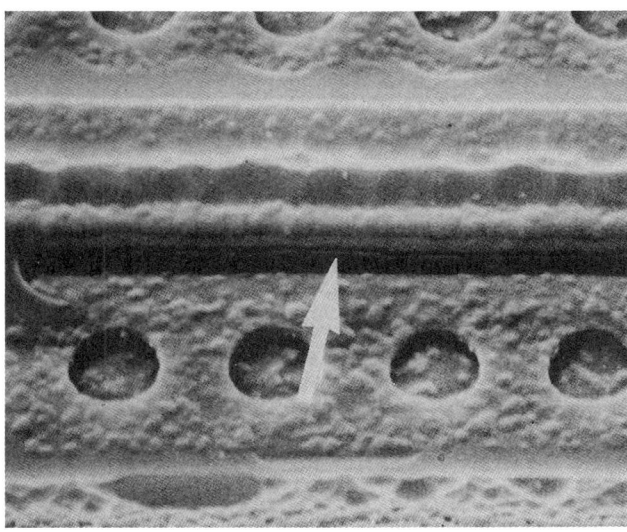

Fig. 9 Side wall oxide crack

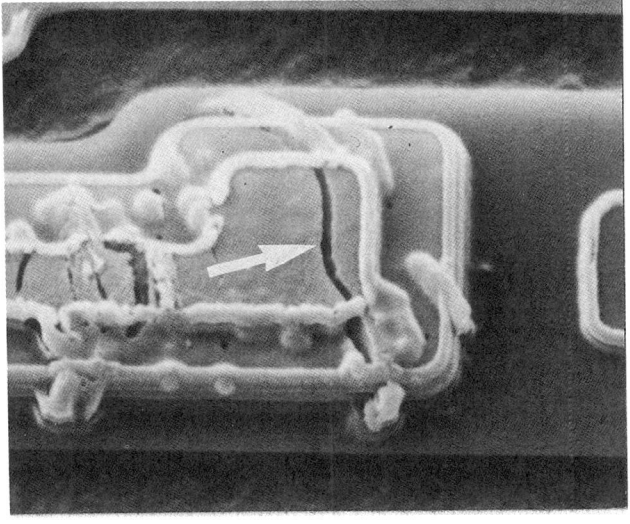

Fig. 10 ILO crack (contamination)

Fig. 11 Gate oxide crack

tilevel oxide (MLO) and runs parallel to the metal line. The cracks created a leakage path from Vcc metal line to the underlying polysilicon interconnect. Additional analysis of these types of failures revealed severe PO "bread loafing" around the metallization. The poor PO coverage made it susceptible to cracking and the cracks would propagate through the MLO layer.

The crack in Fig. 8 occurred at the metal interlevel oxide (MILO). The crack could not be isolated with emission microscopy due to the wide metal 2 trace that covered the area. Liquid crystal analysis was successful in isolating the metal 2 to metal 1 leakage which was caused by this crack. The MILO layer shown in this example is comprised of three layers: CVD oxide 1, spin on glass (SOG) and CVD oxide 2. The cause for these cracks was determined to be film stress.

Figure 9 shows a crack in the side wall oxide at an N-channel transistor. The cracked sidewall caused a Vt shift in the transistor as well as source to drain leakage. The crack was caused by stress associated with polymer contamination after polysilicon etch prior to side wall oxide deposition.

A crack involving the poly interlevel oxide is shown in Fig. 10. The crack occurred as a result of film stress associated with contamination.

A gate oxide crack in a large capacitor is shown in Fig. 11. This crack is possibly an etch decoration that resulted after polysilicon etch. The cause was attributed to severe oxide thinning at the birds beak area, which broke during deprocessing. The oxide breakdown caused polysilicon to diffusion leakage which was isolated by emission microscopy.

OPEN METAL

Open metal defects are typically caused by lithography-related problems. When AlSi alloy metallization is used, there is a potential for electromigration and stress induced metal voiding.

The metal void shown in Fig. 12 was taken from an electromigration test structure. The open in this case was attributed to electromigration. The open shown in Fig. 13 is an example of a stress induced metal void. The open in this case was caused by excessive compressive stress in the passivation layer. The stress induced voids often look similar to electromigration and are difficult to distinguish. Stress induced voids will occur over a wide range of the die and will be found on low current and unbiased metal lines. Electromigration voids will occur only in high current circuitries.

OPEN CONTACTS

Open contact failures have various failure mechanisms including contamination. Some examples of contact failures are shown in Fig. 14-17.

Figure 14 shows a typical contact lithography failure where the contact pattern was severely undersized or missing. The arrow points to the distorted contact that became resistive after reliability testing. The logic failure caused by this contact defect was very

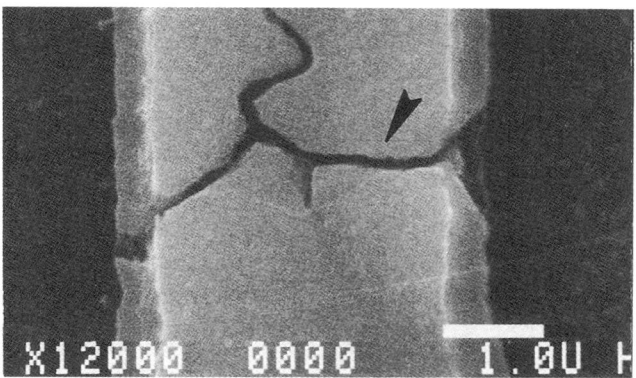

Fig. 12 Metal void (electromigration)

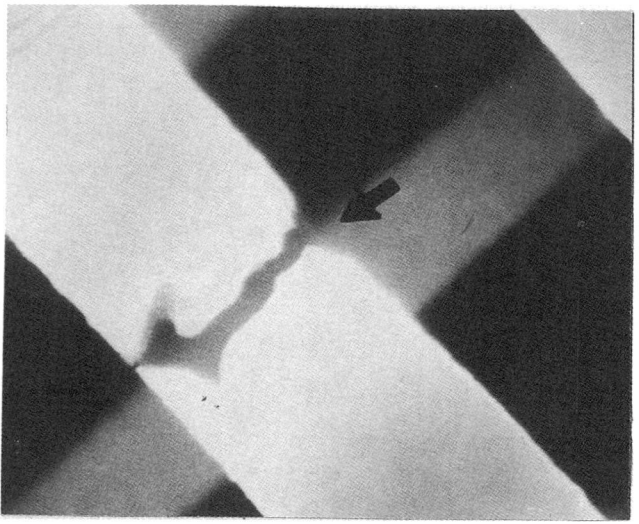

Fig. 13 Metal void (stress induced)

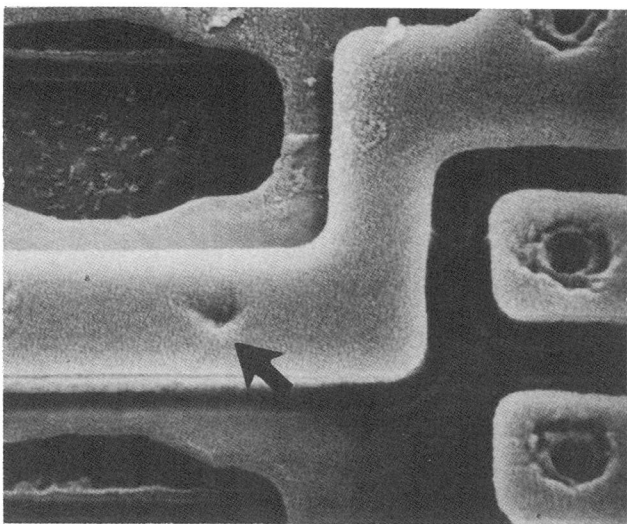

Fig. 14 Contact lithography defect

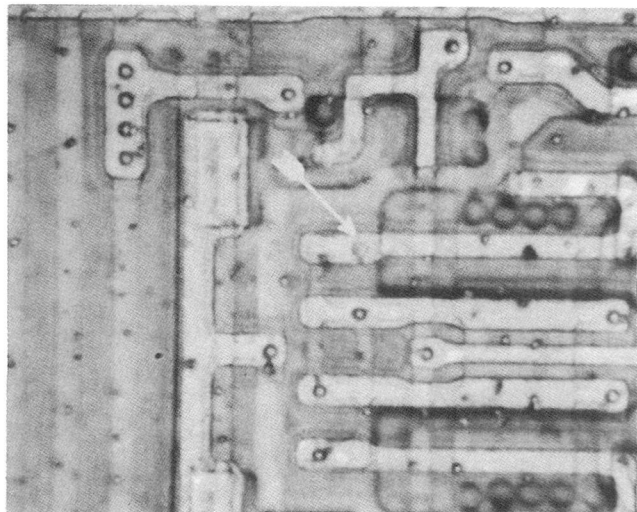

Fig. 15 Incomplete contact etch

Failure Mechanisms in Integrated Circuits

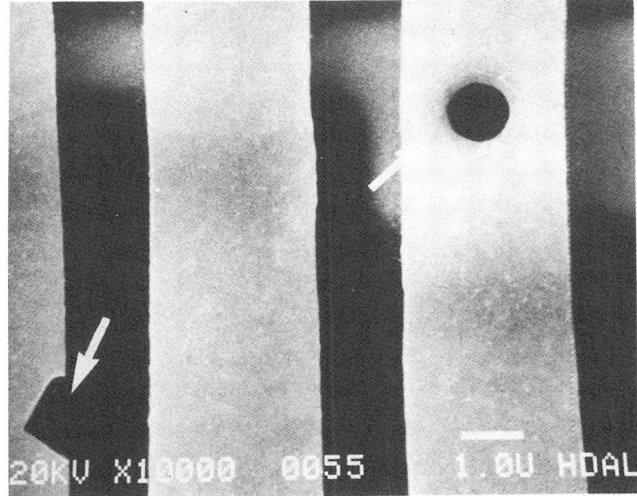

Fig. 16 Stress induced contact void

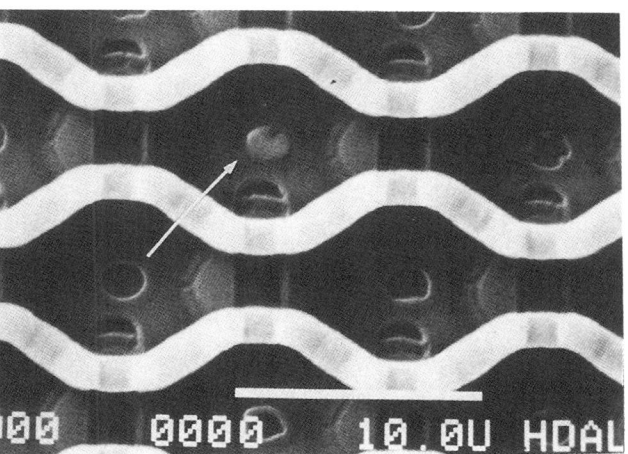

Fig. 17 Excessive Si epitaxial growth

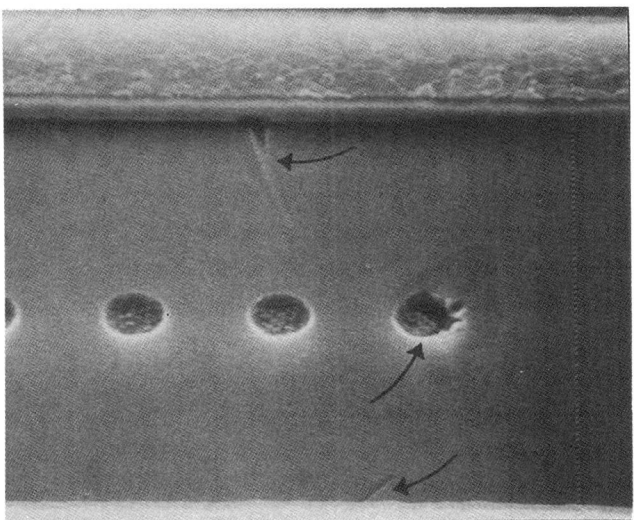

Fig. 18 Blown contact (EOS/ESD)

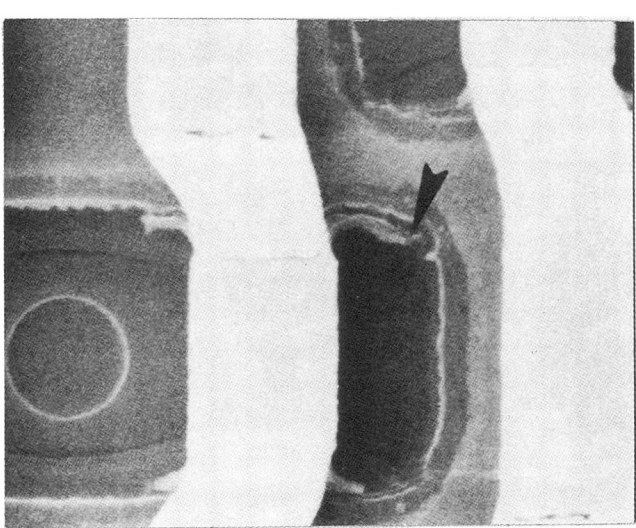

Fig. 19 Polysilicon etch filament

temperature and voltage sensitive (cold temperature and low voltage were the worst case).

Figure 15 shows an optical photo of a discolored contact caused by incomplete oxide removal. These types of contact failures are similar to the lithography failures. The incomplete oxide removal failures are difficult to document because they appear normal under SEM inspection. The best method to show the discoloration is with color optical micrographs, which indicate film thickness variations with color shifts.

Open contacts can also be caused by excessive PO stress, as was mentioned in the stress induced metal voiding case. An example is shown in Fig. 16. The dark contact at the failing contact, when compared to surrounding contacts, is indicative of contact voiding. Typically, when stress induced contact voiding occurs, there is also metal trace voiding present.

Resistive contacts in AlSi (aluminum-silicon) metallization have also been attributed to excessive Si epitaxial growth as shown in Fig. 17. The silicon in the aluminum-silicon alloy is relatively resistive. During contact sintering, this silicon combines with the chip silicon in the contact area. If the large amounts of the alloy silicon consume the contact area, then the contact becomes resistive. The contact failure In this example caused four adjacent EPROM bits to fail programming.

Contact Shorts

Shorted contacts are usually associated with electrical overstress (EOS) or electrostatic discharge damage (ESD). The example shown in Fig. 18 is a blown contact after ESD testing. This example also shows gate edge breakdown.

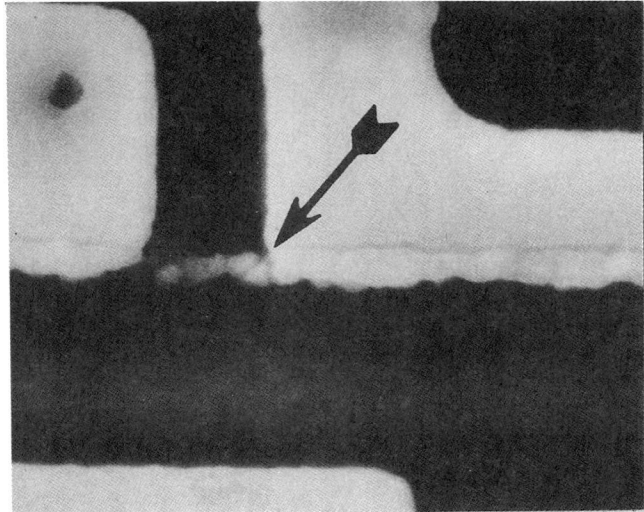

Fig. 20 Metal etch filament

Fig. 21 Poly melt filament (EOS/ESD)

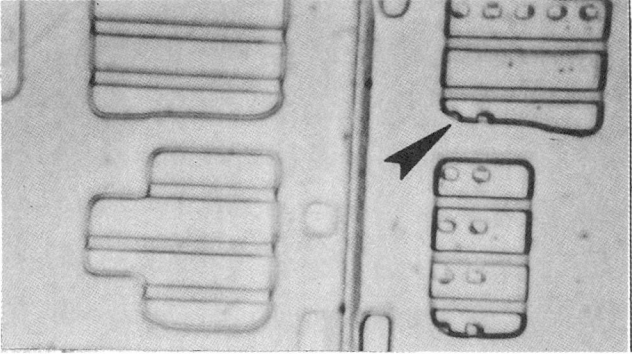

Fig. 22 Diffusion lithography defect

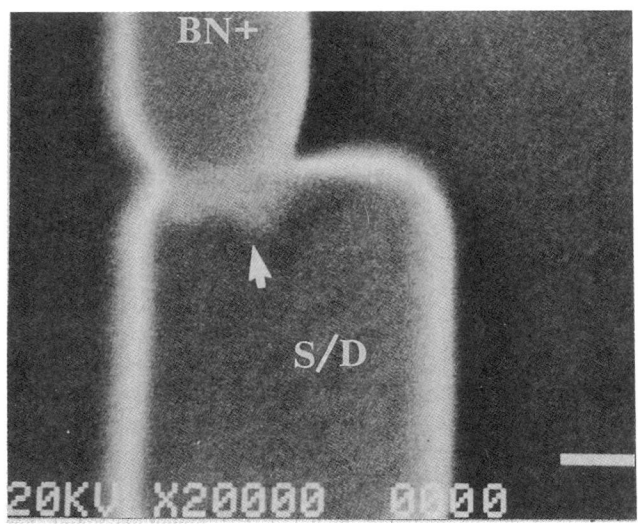

Fig. 23 Blocked implant

Another example of a contact short is known as contact spiking. Contact spiking can occur in AlSi metal systems when the silicon content is low or when the sintering temperature is too high. *Contact spikes can occur during deprocessing if hot chucks are used above 450 °C.* Spiked contacts are similar to blown contacts but generally are characterized by tiny pits in the contact.

FILAMENTS

Filament is a term typically associated with etch residue or incomplete etching. The term filament has also been used to describe poly melt filaments which are caused by EOS/ESD damage to the gate edge.

A polysilicon filament is shown in Fig. 19. The filament in this case caused a leakage path from the poly wordline to the cell S/D diffusion. Typically, the root cause for etch filaments is topology and etch related. A more isotrophic etch during poly patterning would allow removal of the poly in the crevices created by the topology of the lower levels. The poly filaments fail after the gate oxide underneath breaks down. Since this device uses a self aligned gate process, the small filament is exposed to the implant. The implant will pass through the thin filament and create damage in the underlying gate oxide which will then be prone to failure.

A metal filament is shown in Fig. 20. This example was isolated by emission microscopy, and failed after reliability testing, which suggests it was not a direct short. The filament residue between the metal lines was determined to be titanium-tungsten alloy particles. At time zero, these Ti-W particles were isolated by the oxide passivation. Over time, this extremely thin passivation dielectric broke down and a leakage path was created.

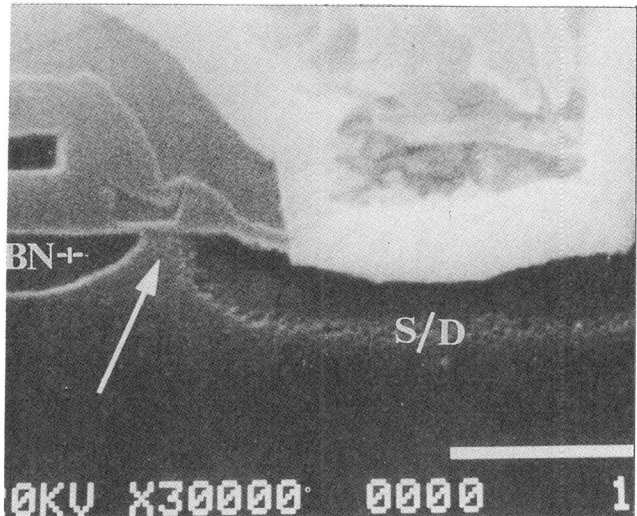

Fig. 24 Blocked implant

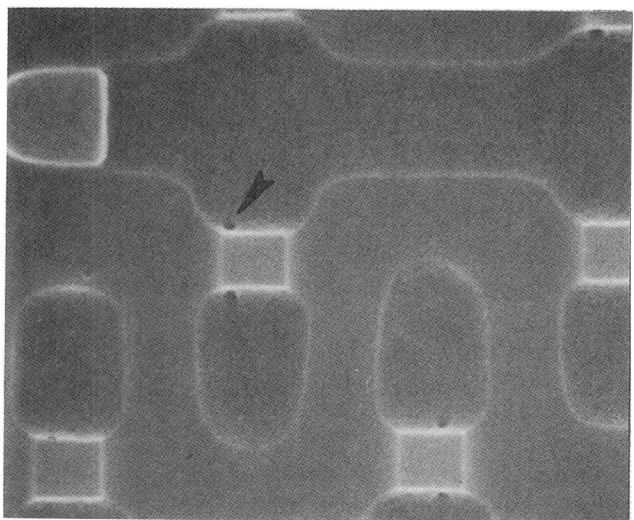

Fig. 25 Silicon dislocations

A polysilicon "melt" filament is shown in Fig. 21. This example is from an ESD failure. The poly melt filaments occur at the gate edge, and resemble tiny tree roots. The material transfer is typically from diffusion into the polysilicon, which is noted to be the evacuation of silicon from the diffused area. Gate oxide breakdown is usually (but not always) associated with poly melt filaments.

DIFFUSION DEFECTS

There are several diffusion related defects that are non-visual, for instance, abnormal diffusion depths, channel lengths, diffusion profiles, etc. These types of problems depend on very detailed electrical characterization and sophisticated analytical tools. Discussion of some of these defects is considered beyond the scope (and length) of this paper. Therefore, only three generic diffusion defects are presented. Figure 22 shows an example of a lithography defect documented at substrate level. The arrow points to a set of contacts that were isolated using liquid crystals. The distorted diffusion pattern caused the contact to leak to the substrate. The cause of the patterning defect was traced to a worn mask that exhibited chrome peeling.

Figures 23 and 24 are examples of implants blocked as a result of poly residue in the field plate opening. Figure 23 shows a substrate level micrograph after staining with a dislocation stain. This highlighted the blocked source/drain implant, which was intended to overlap into the buried N$^+$ diffusion. This blockage created a resistive path between the source/drain diffusion and the BN$^+$ interconnect. Figure 24 shows a similar blocked diffusion example, as in Fig. 23, after a very light junction stain.

SUBSTRATE DEFECTS

Substrate defects can be caused by oxygen precipitation during crystal growth, metal impurities in the silicon, or implant damage or stress induced during wafer processing. A dislocation can be present in the substrate without causing a failure.

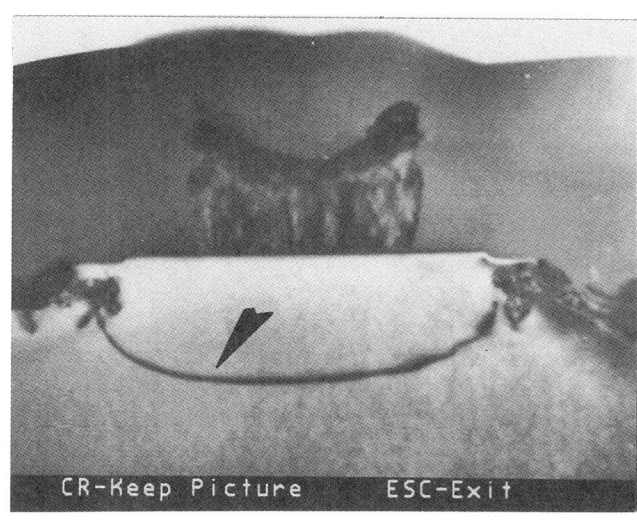

Fig. 26 Silicon dislocations

The dislocation must be electrically active to induce a failure. Most of the silicon dislocations which occur naturally are rendered inactive by H_2 sintering.

The example in Fig. 25 shows implant related dislocations at the gate region of P-channel transistors. Figure 26 is a TEM cross section of a similar failure, showing a loop dislocation in the silicon from source to drain under the channel region.

TRENCH DEFECTS

Defects in DRAM trench capacitors can be observed by etching the bulk silicon from the chip backside and viewing the exposed trenches in the SEM. Figure 27 shows a distorted

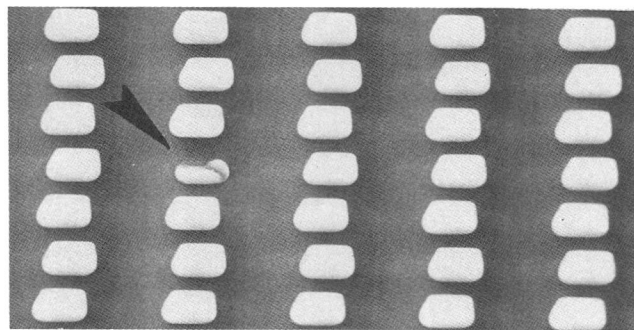

Fig. 27 Distorted trench (particle related)

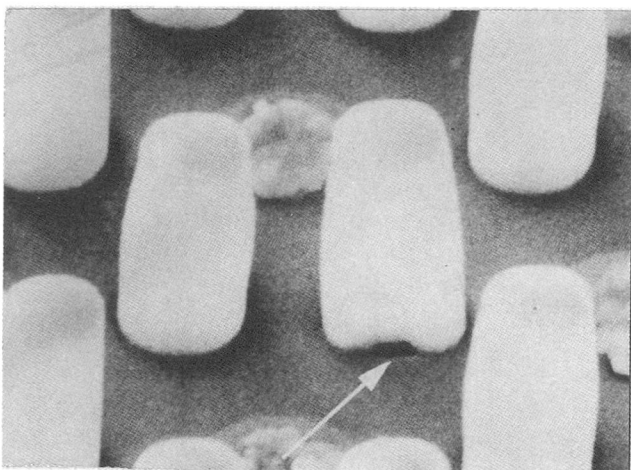

Fig. 28 Trench oxide defect

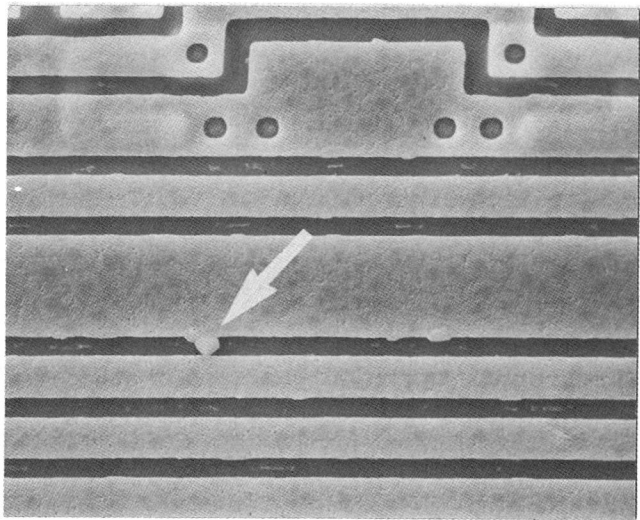

Fig. 29 Metal side hillock

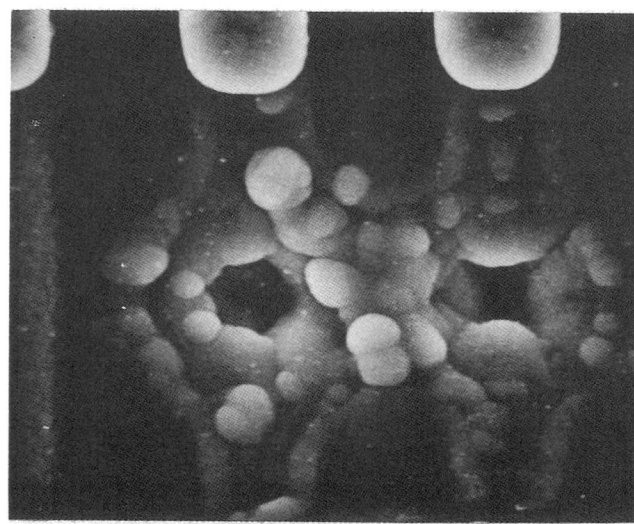

Fig. 30 Laser repair defect (overpowered)

trench capacitor that caused a single bit failure. The distorted trench was caused by particulate contamination prior to trench etching.

A trench oxide defect is shown in Fig. 28. The silicon etchant decorated the oxide pinhole at the base (top edge trench) of the failing capacitor.

METAL HILLOCKS

Metal hillocks are common on AlSi metallizations. When these hillocks occur on the sides of wide metal lines they have the potential to cause leakage paths to adjacent metal lines.

The micrograph in Fig. 29 shows a metal side hillock causing leakage between a large metal line and an adjacent metal line. These hillocks are caused by high temperature processing after metal patterning, such as contact sintering and passivation deposition. These hillocks can be controlled by applying a thin layer of CVD oxide to suppress hillock growth during additional processing.

LASER REPAIR DEFECTS

Laser repair related defects have become more common as more devices use redundancy in memory arrays. The laser fuses are usually polysilicon lines that are fused open with a pulse, to set a specific memory address to be replaced by a redundant address.

The laser repair defect shown in Fig. 30 is an example of a grossly overpowered laser blow. The round particles present are laser "slag" particles blown out of the substrate. The particles are separated by a thin passivation dielectric since the re-

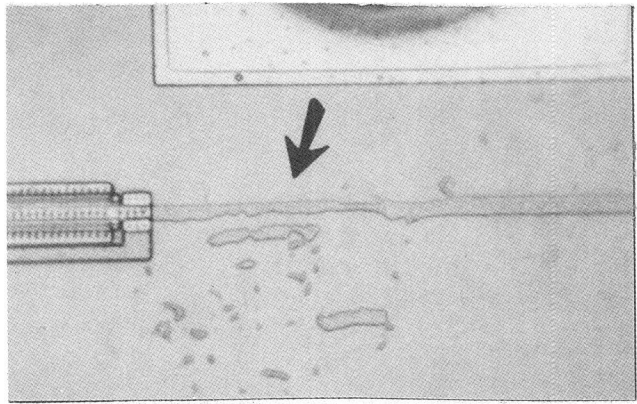

Fig. 31 Wafer fab scratch

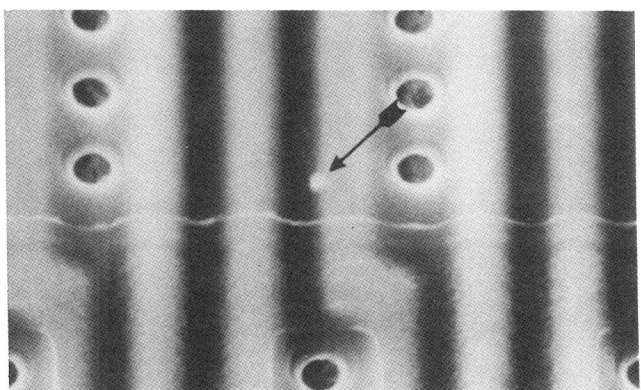

Fig. 32 Silicide extrusion

Bonding Issues

Bond Issues	Causes	Distinguishing feature
Displaced (or Misplaced wire)	Wire feed	Check stitch
	Handling	On long leads
	Bond	Hi loop or long wire
	Molding	Meets mold front at 90°
Broken at Stitch, Neck and Mid Span	Tight wire/handling/ Crack package/ Trim and form	
	Mold stress/voids	Grain growth, PO damage
	Mold stress	Grain growth, EOS-hi I_{DD}
Intermittent contact	PO contamination	Spotty intermetallic growth
	Edge chip of wafer	Irregular PO coverage on IC
	Lead contamination	Discoloration on leads
	Coplanarity	Bent lead
	Trim	Short lead
Wire short	Mold release pin	Specific wire, sharp edge
	Handling jig	One corner or end, many wires
Misplaced ball	Bonder electrical noise	Misplaced ball
Club foot	Displaced wire/ball	Club foot
	Bonder electrical noise	Smashed ball bond
	Bonder	Small ball
	Bonder	Large ball
Low bond strength	Wire feed	Low loop height, nicks in wire
	Delamination	Chipout into Silicon, corner
	Over bond	Flat ball, cracking in underlay
	Low mash force	Round ball, low intermetalic
	Silicon pptn	Large silicon nodules
High bond strength	Handling	High loop height

Fig. 33 Bonding issues: their cause and their distinguishing physical features

pair in this example was performed prior to PO deposition. This thin dielectric is prone to breakdown, and leaky laser fuses can result. Other laser repair related defects include misaligned laser blows and under-powered laser blows.

Wafer Fab Scratches

Scratches to the chip surface can occur at any level during wafer processing due to handling or equipment malfunction. An example of a water tab scratch is shown in Fig. 31. The scratch in this example caused an input to substrate leakage failure, which was isolated by emission microscopy. The scratch was determined to have occurred sometime after diffusion patterning and before moat silicidation. The scratched area indicated by the arrow appears to be particles, but were determined to be "pieces" of diffused and silicided areas.

Silicide Extrusions

Silicided polysilicon lines have become very popular in reducing the poly interconnect resistance. One problem associated with this technology involves titanium silicide ($TiSi_2$) and BPSG MLO.

The micrograph in Fig. 32 shows a $TiSi_2$ extrusion through the MLO layer, causing a short from the poly layer to the above metal line. The silicide extrusion was generated during the BPSG high temperature reflow and densification process. Although the defect appears to be silicide related, the root cause was primarily the MLO processing.

ASSEMBLY FAILURE MECHANISMS

General Comments

The major process steps in IC assembly are Wafer Backgrind, Saw, Chip Mount, Wire Bonding, Molding, Lead Finishing and Trim, and Form and Symbolization. Typical overall Assembly yield is about 98-99%. Despite this high yield, the assembly process can still have many failure mechanisms under accelerated reliability testing. The accelerated testing is done by the application of high temperature, high humidity, high voltage or a combination of these. In some cases, mechanical stresses or thermomechanical stresses are applied to screen for any structural flaws or design weaknesses.

Typical failures are bond wire continuity, package cracks, chip cracks, corrosion, passivation cracks, interface delamination, leakage and parametric failures. We will examine some of

Chip and Package Cracks

Chip and Package Crack Issues	Causes	Distinguishing feature
Crack die and Package cracks	backgrind damage	rough backside, high Rt
	saw	incomplete cut, chip
	wafer mount-poker pin	Pin mark, crack thru mark
	trim and form	scuff marks, imprint delamination and int pkg crack
	test handling	scuff marks
	surface mount	package cracks at die edge
	mold voids	crack passes through void
	molding and mold voids	distortion of mount pad
	mount epoxy voids	epoxy voids
	mount epoxy coverage	incomplete coverage, misalignment
	COL LOC and Conventional	
		tape adhesion
		tape size
		chip alignment
	design/lay out weakness	po at scribe
	interaction with saw	metal at scribe

Fig. 34 Chip and Package Crack issues: their cause and their distinguishing physical features

PO Damage

PO damage issues	Causes	Distinguishing feature
Localize PO defect Point defect	Mold Compound Filler and Resin	compressive fracture cracking at brittle levels
Linear Fragment	Design/layout	metal over poly, long metal runs
Point defect	Mount	displaced segment
Linear Scratch	Mold	scratch lines along mold flow
Collapse and localize cracking	Stress	PO slide, fracture at corners
Point defect	Backgrind	H or X pattern damage at topologically high areas
	Bond-wire	touches PO during bond
Corrosion	PO stress/voids	x-section xtics sharp geometries, corners

Fig. 35 PO damage issues: their cause and their distinguishing physical features

(a)

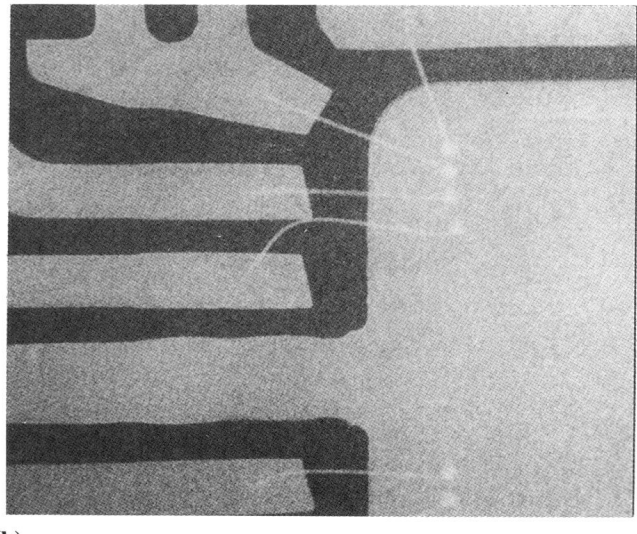

(b)

Fig. 36 (a) Example of good vs bad stitch orientation. (b) X-ray of the wire

them sufficiently to quickly determine where the failure originated in the assembly process.

The electrical signature of bond-related failures is typically continuity problems, followed by shorts or leakage. Some of the devices will show "input hi" or "input lo" level failure. Generally, these will be detectable if an X-ray is done to check for wire or bond related issues. In the case of intermittent contact, performing a continuous test with a temperature ramp will bring out the failure. Bond strength problems due to low and high bond strengths are detectable by X-ray when the screen is done with a side view shot. They are indicators of potential yield and reliability problems.

The electrical test signature of a cracked die can be continuity fails, high current consumption, or leakage fails. Usually, if a leakage failure occurs, it will occur on more than one pin. If the breakdown is measured, it will be much lower than normal and will show a strong time dependence, decreasing with the application of bias. If the cracks are on the back side, and did not

Failure Mechanisms in Integrated Circuits

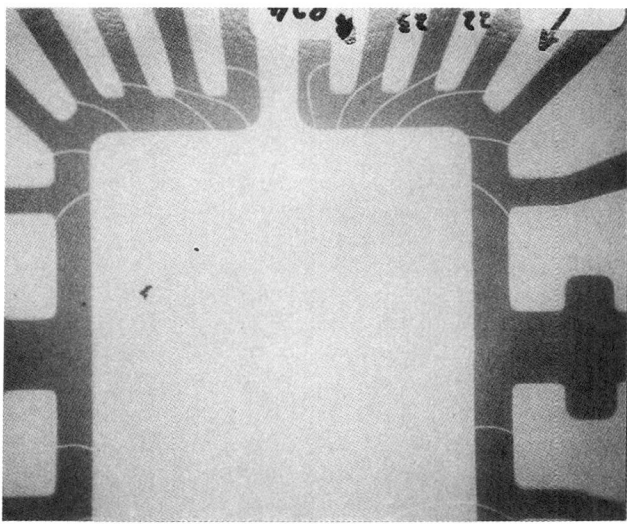

Fig. 37 Example of swept wire due to mold flow (from the top)

propagate to the surface, it will show degradation in its functional characteristics, for example, a collapse in the voltage box of the part.

Scanning Acoustic Microscopy (SAM) is non-destructive, and is the least invasive means of examining a package or chip suspected of having cracks. X-ray inspection can also reveal cracks, but this needs to be implemented with many views, as its success greatly depends on the crack surface orientation. Fluorescent dye penetration, followed by recovery of the fracture surfaces, is also a good method of analysis. The recovery of the fracture surfaces and examination of the fracture pattern is usually done through the use of microsectioning techniques.

The electrical signature of failures due to PO damage for localized point defects in memory devices ranges from single bit cell failures to row and column failures.

Recovery of the imprint of the mold is key if identification of filler as the cause is required. Locating PO defects can be done by the use of a metal etch or a Na ion drop test. Partial etching of the PO followed by inspection will also reveal crack patterns that are not obvious, or regions of stress through the crack patterns. X-sections using metallographic techniques can be used to examine PO profiles.

SUMMARY

Examination of the failure and failure region and understanding the failure scenario allows us to distinguish possible or likely causes from a host of possible causes. In the case of the 3 issues highlighted, analysis of the fracture patterns and fracture surfaces is a key to help distinguish many different causes and sources of failure. Sometimes associated material failure or anomalies can clue us to the root cause. An example is the presence of PO slide for continuity failures due to broken wire. Examination of the failure should also probe deeper past the original report or first reported fail label or failure mode detected at test. For example EOS failures can show up as bond intermetallic issues due to the high temperature generated by the EOS failure.

(a)

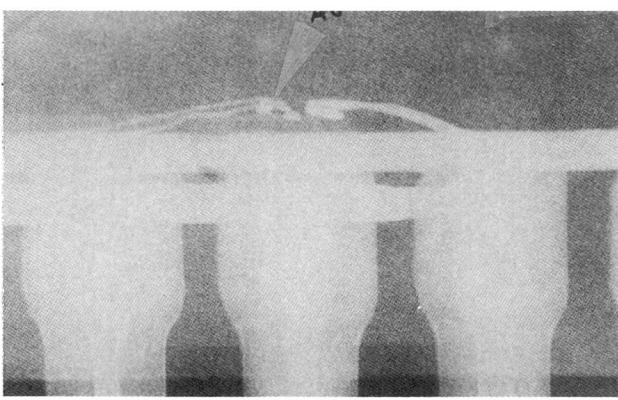

(b)

Fig. 38 (a) Example of displaced wire due to handling (left side). (b) Example of displaced wire due to handling (right side)

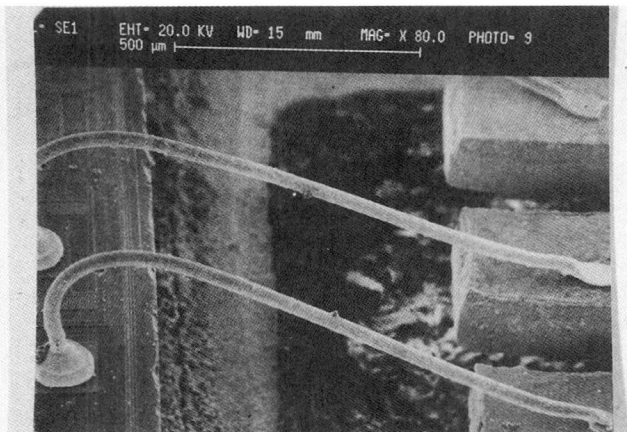

Fig. 39 Example of broken stitch

EXAMPLES AND CASES

The bond wire is shorted to an adjacent wire or pin, and X-ray shows that the wire is swept. Examination of the stitch orientation will show if this is due to the wire feed tensioning

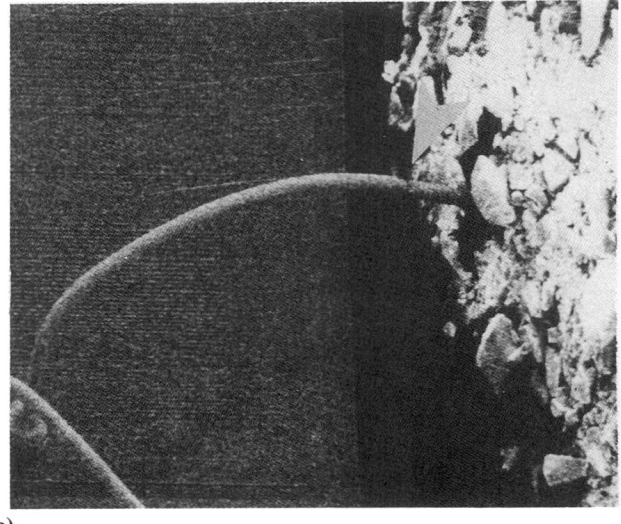 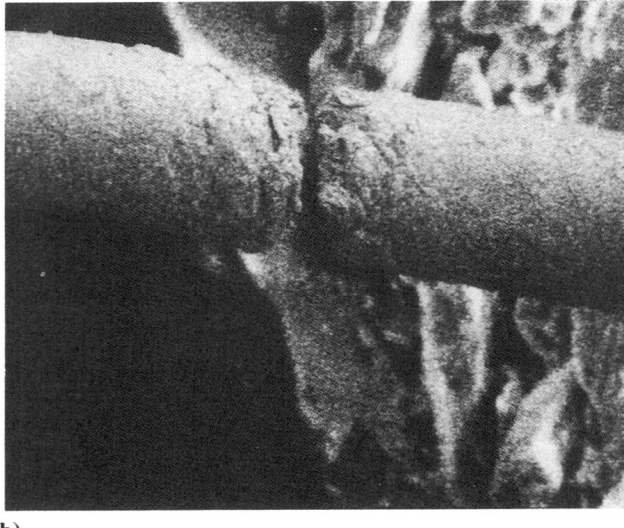

Fig. 40 (a) Clean fracture on Gold wire. (b) View of the fracture at high magnification

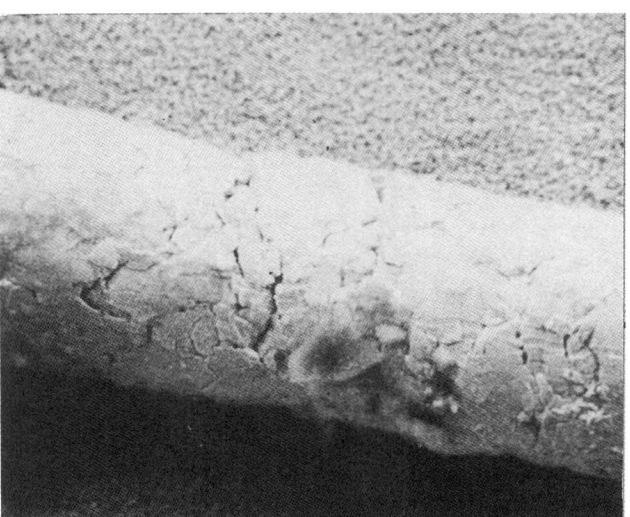 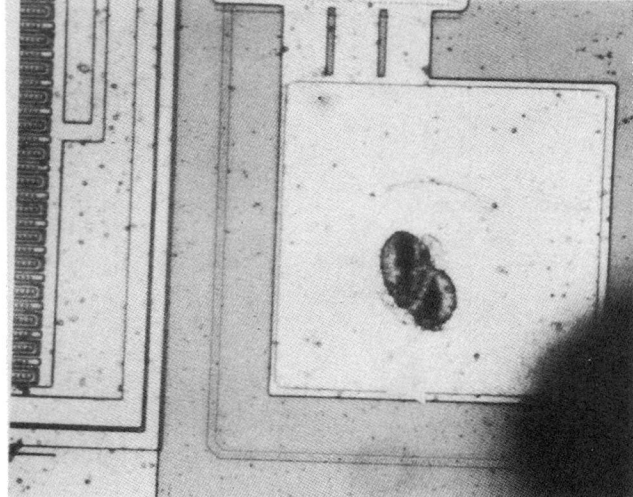

Fig. 41 Example of grain growth on gold wire

Fig. 42 Example of spotty intermetallic formation

mechanism. During bonding, the wire fed out during the bond sequence is longer then the total final bonded wire length, and this excess is retracted by a tensioning mechanism. If the wire stitch orientation is good, then check if the adjacent wire loop height is higher then normal. This is a general indicator of handling.

If the wire deformation is due to mold sweep, then generally the wires most badly affected are at the gate end of package, and the wire orientation will meet the mold flow front in a sidewards or "broadside" orientation. For minimum interaction, the wire orientation should be perpendicular to the mold flow front. Estimation of the mold flow front should then be done through mold flow simulations, or by rule of thumb, from the package X-section information. Alternatively, a short shot can be done;

that is, stopping the ram and freezing the mold flow pattern during the transfer molding process, to examine the flow pattern.

Wire deformation can also occur due to the distortion of the leadframe or vibration of the leadframe during the loading and unloading from jigs. Generally, they will show a group of wires with a deformed profile. This can be confined to one corner or one end of the device. High loops can also be found on some of the wires and sometimes some of the wires will also touch the edge of the chip.

The stitch can fail due to package cracks propagating into the leadframe area. Delamination of the lead finger from the mold compound is another common cause of the stitch failure. This loss of adhesion can come about from the trim and form process, or during a lead rework process. Use of a scanning

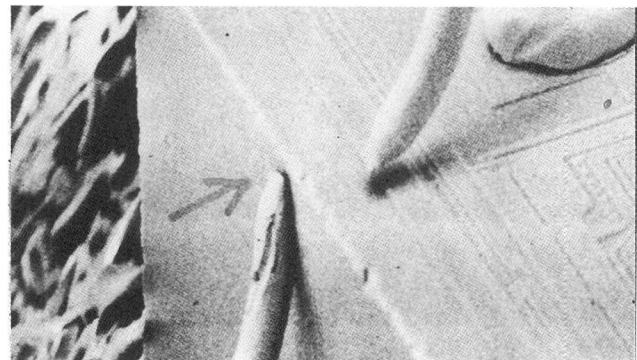

Fig. 43 Continuity fail due to Mechanical damage on Wire

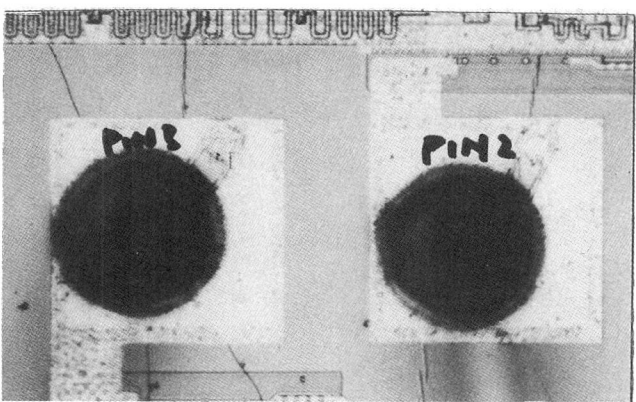

Fig. 44 Silicon Damage due to High Bond Mash Force and Probe Damage

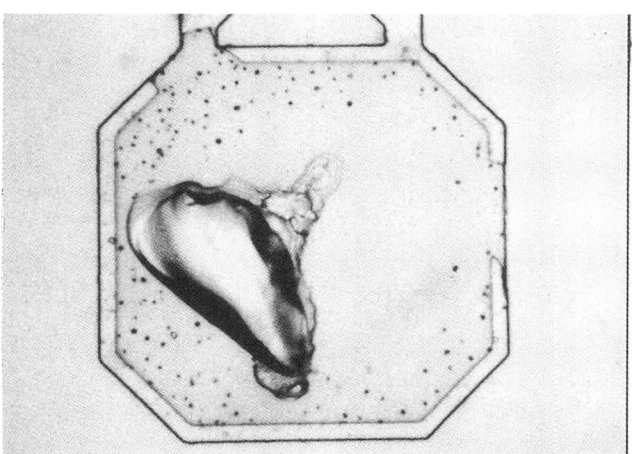

Fig. 45 Chipout into silicon

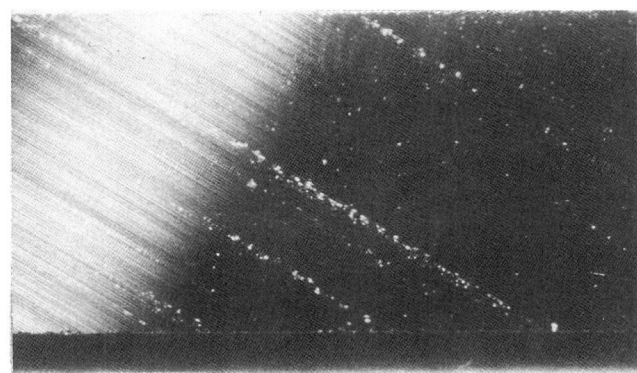

(a)

(b)

Fig. 46 (a) Backgrind damage in silicon. (b) High magnification view of the damage

acoustic microscope will allow you to non-destructively examine the interface area for evidence of the delamination.

Here the wire breaks cleanly in midspan. This clean break is unusual for a ductile material like gold, and is an indication that the fracture occurred in the package, and cut across the gold wire. Typically the crack tip of a fracture is under very high stress and the relative crack movement is small. Generally at this stage of the analysis, the package has already been decapped, and evidence of the crack may not be available. If a slot decap was done, the crack area normally etches much faster, providing supportive evidence that a fracture surface already existed there. If the break in the wire was due to a void at that point, evidence of slight elongation before fracture should be observed.

The grain growth is indication of hi-temperature exposure or the presence of high-stress. Normally this type of failure occurs under accelerated reliability testing. Sometimes, checks need to be made for evidence of EOS as a cause. This can be done by further testing of the device for current consumption and leakage. If any exists, it is likely that EOS resulted in high power dissipation and heating, causing grain growth and weakness at the neck area. Examination of the intermetallic interface between the gold ball and the aluminum bond pad should also show thick intermetallic growth. If the cause of the grain

Fig. 47 Chipped Silicon at Die Edge

growth is high stress from the mold compound, then evidence of PO slide should be found in the corners of the chip.

Incomplete PO removal during the pad opening process, or insufficient cleaning after the backgrind process, can leave behind contamination at the bond pad. This can prevent the formation of a reliable ball bond. Typically if the layer formed is greater than 75 µm, reliability problems will be encountered. Generally examination of the bond pad surface under the SEM at appropriate accelerating voltage to reveal the surface details, should show whether a layer is present. Energy Dispersive X-ray Analysis (EDXA) can also show if additional material is present, as can Auger analysis. If the ball bond is pushed away we can usually observe spotty intermetallic formation. These types of continuity failures can show recovery if the bias is set high during testing or curve tracer examination. Discoloration of the bond pad is an obvious sign that there is extraneous material present. However, this must be distinguished from discoloration due to low level corrosion, which sometimes occurs. Sometimes, chips on the edge of a slice can be picked up. These tend to show large variation in the PO removal at the bond pad. Examination of the PO in general will show if this is true. The PO is generally rough, and looks dark under an optical microscope.

The presence of a sharp discontinuity along the wire at the point where it breaks is a sign that it was caused by a machine part, and in this case (refer to Fig. 40), it is due to a mold ejector pin. The cause can be due to timing error or a pin that is much longer than usual. The fail pin is above the mold ejector pin location.

During the bond process, a high mash force is required to establish good contact, the transmission of the ultrasonic energy to the bond pad, and the gold ball for the formation of the intermetallic. The maximum shear forces are not acting at the interface, but much deeper inside the ball. For example, if there is an absence of bond pad metallization, then it is possible for high stress to cause fracture and resulting cracks into the underlying silicon.

(a)

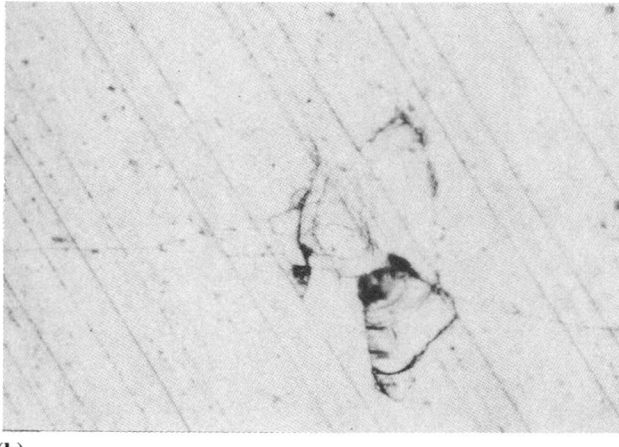

(b)

Fig. 48 (a) Damage and Crack Features in Silicon (40×). (b) Hi-magnification view of the Fracture (400×)

When the ball is golf club-shaped is it due to bonder-cap movement and wire tip in cap knocking into other bonded wires. This can happen because the actual head movement is different from the programmed head movement. Alternatively, it can be well formed at flame off time, but wire is not centered when it is bonded. When the ball is very flat, usually the mash force is very high, or gold ball is soft during touch down. In this case the contact area is large, so there is no concentration of forces and a poorer ball bond results. Ball bond size is not well correlated to damage at the bond pad.

The silicon in the aluminum which precipitates out into nodules are sources of stress concentration, and usually are the initiation sites for chipouts. The presence of a precipitate-free metal like Ti:W, for example, can remove all sources of localized stress concentration by acting as a buffer, eliminating this failure mechanism.

Checking the leakage current value can be tied back to the failure mechanism. If the leakage is very gross, then most likely a chipout into silicon occurred. If the leakage is in the 6µA range at 5V bias, then the failure is probably due to cracking in the interlevel dielectric with no chip out into the silicon. If the leakage is lower, and can recover, and is worse with temperature, then it is probably due to trace level moisture. In this case physical analysis will not show any anomaly.

Failure Mechanisms in Integrated Circuits

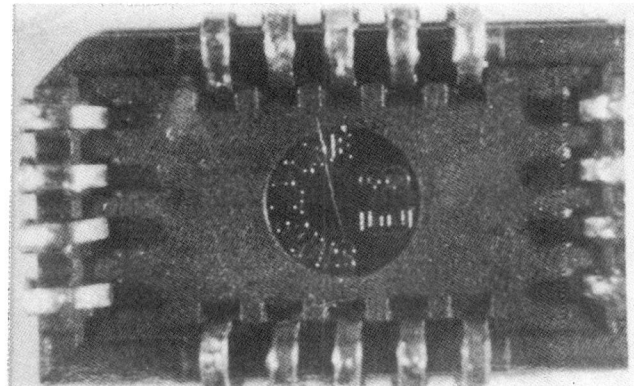

Fig. 50 Typical Package Crack Pattern

Fig. 51 Package Cracks due to Mold Voids

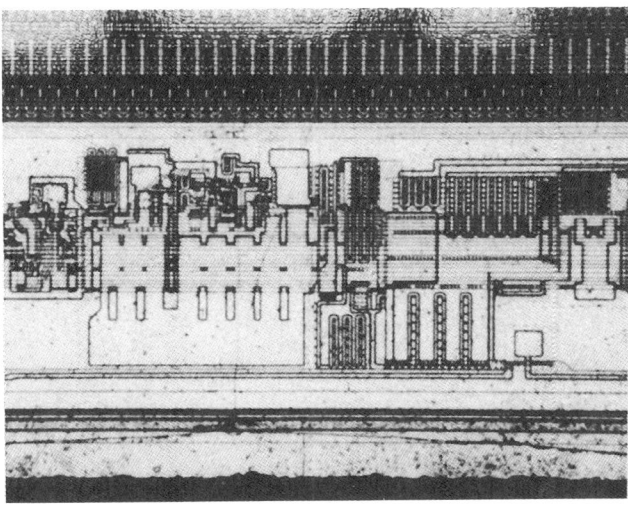

Fig. 49 (a) Scuff Marks/Package Cracks and Die Crack .(b) Crack Die after Package Removal

Fracture in silicon usually originates at the site of a flaw. This flaw can come from the backgrind operation, which leaves behind deep grooves. It can also come from the saw operation due to incomplete cuts, or from a roller and breaking operation. During the chip mounting operation using a tape process, the chip is displaced upwards for a pickup mechanism by a pin on its bottom side. This can cause damage at that location if excessive force is applied. If fracture occurs due to backgrind damage, typically this will correlate with Rt, the maximum surface roughness. The crack line will pass through a region where this roughness is obvious. The chip will normally break into 2 parts with a fracture line running across the whole chip, lengthwise or breadthwise, and parallel to other striations, if observable. If the fracture occurred due to incomplete cut or roll break, then the fracture usually occurs across the chip, splitting it into a top half and bottom half.

If the chip crack was from the saw operation, the fracture origin can be pinpointed normal to the top edge of the die and sometimes to damage on the side of the chip. Sometimes there is interaction between the sawing operation, and the layout and material present at the top surface of the chip. For example, the

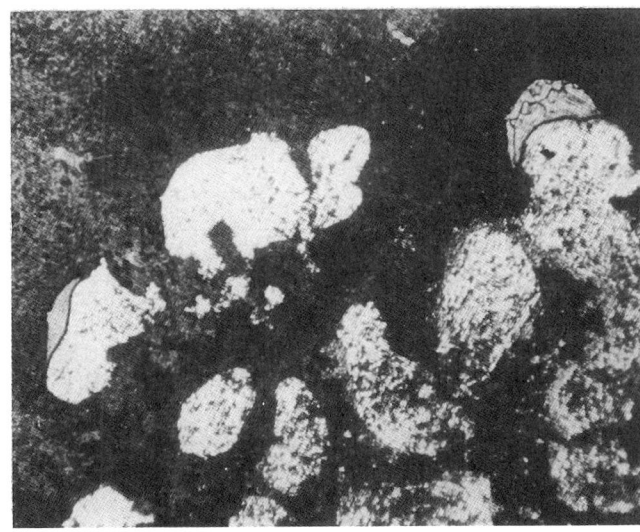

Fig. 52 Example of Resin Bleed on the underside of the Chip

presence of PO over the scribe area or certain metallization structures are detrimental to the saw quality and may cause flaws at the edge of the die.

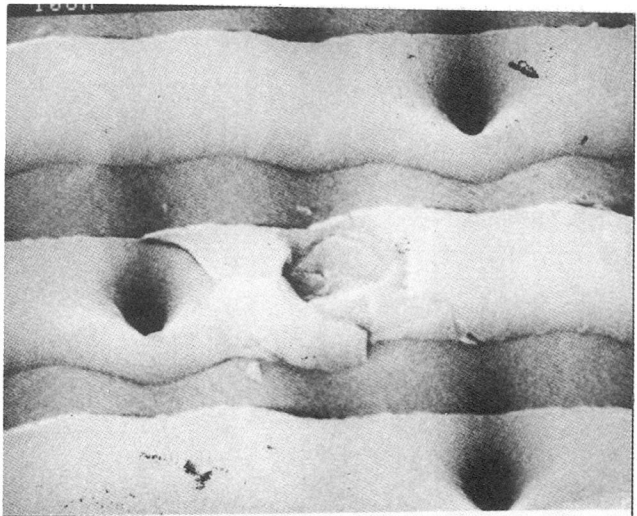

Fig. 53 PO damage

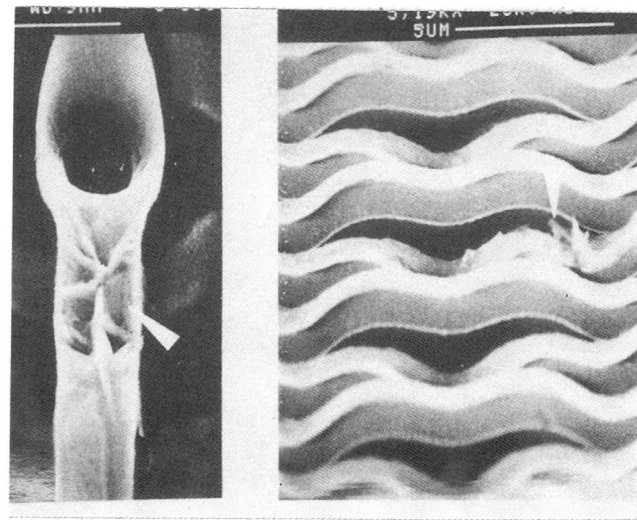

Fig. 54 Filler Induced Damage at the metallization level

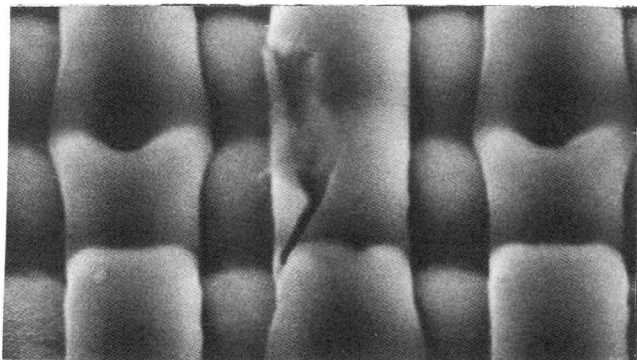

(a)

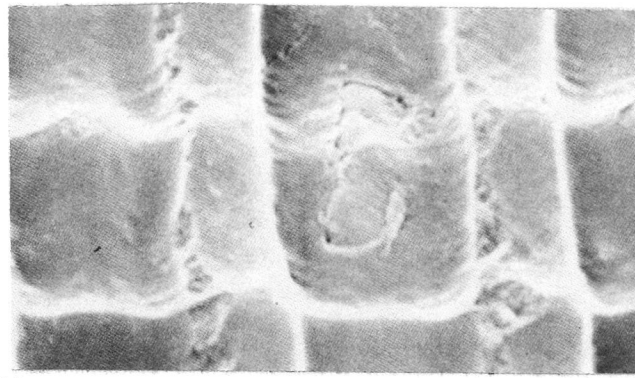

(b)

Fig. 55 (a) PO Damage induced by filler (b) Imprint of the chip Surface

Crack patterns in a cross form are characteristic of damage from the poker pin. One of the crack surfaces will pass through the poker pin mark, which is generally located in the center of the chip on the underside. Sometimes the chip will fracture only after surface mount or exposure to some thermomechanical stress.

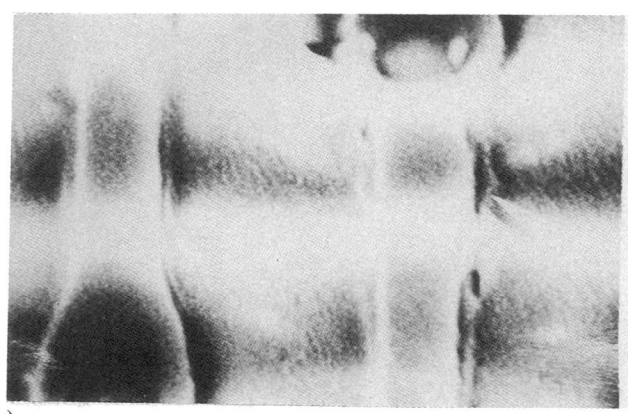

(a)

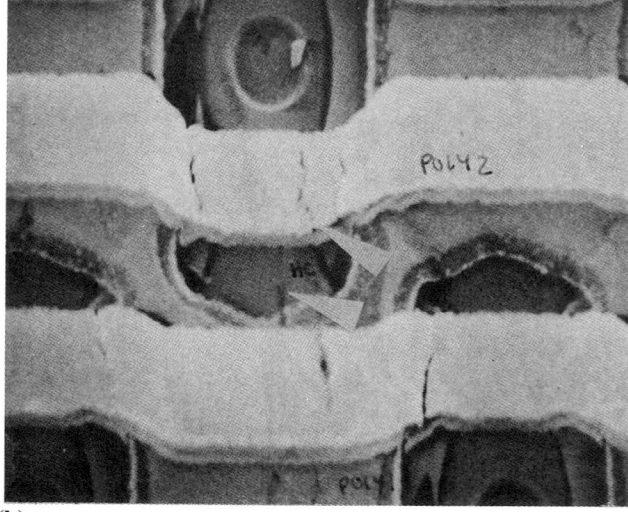

(b)

Fig. 56 (a) Damage at lower Interlevel Dielectric. (b) Damage at the Polysilicon level

Failure Mechanisms in Integrated Circuits

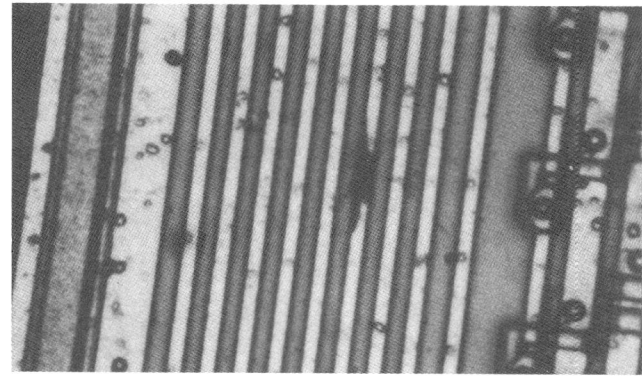

Fig. 57 Metal Shift (by filler)

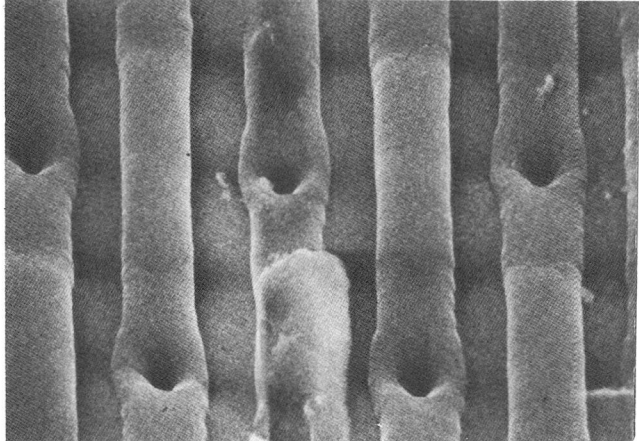

Fig. 59 PO Damage due to Chip Mounting

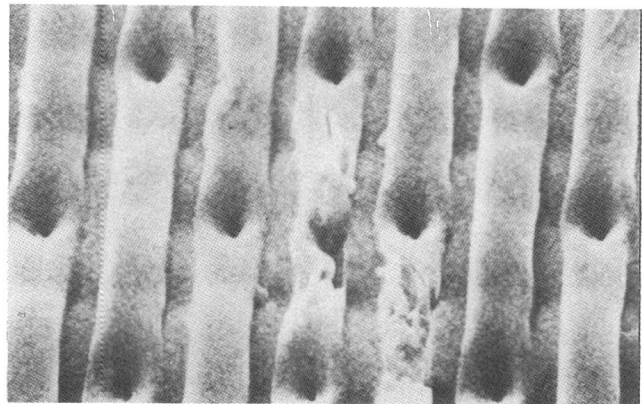

Fig. 58 PO Damage due to Silicon Particles at Mold

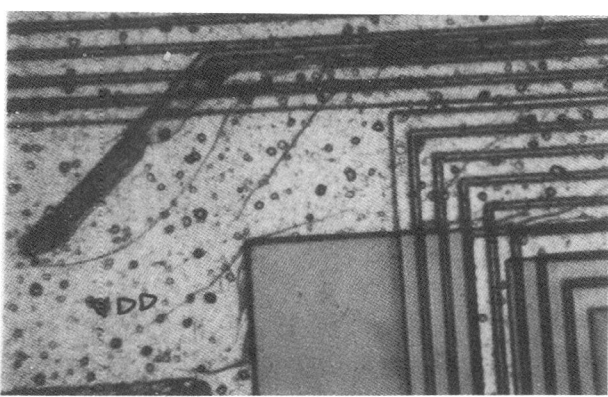

(a)

(b)

Fig. 60 (a) PO Damage due to Aluminum Slide at Chip Corners (optical) (b) PO Damage due to Aluminum Slide at Chip Corners (SEM)

It is possible, under certain loading conditions, for the chip to fracture with the package intact. In manufacturing, this can happen during the time the leads are trimmed and formed. Sometimes, evidence of the damage can be seen by carefully examining the package exterior for tool imprints or scuff marks. The crack pattern is a simple line which is fairly straight. If package cracks occur, then the crack pattern in the chip is usually more complex, but will have a crack line that follows the external package crack line.

Package cracks (Fig. 49 to 51) due to a surface mount operation are due to the expansion of the moisture absorbed by the package, and present at the interfaces. Depending on which is the weak interface, and where the moisture is present, different crack patterns result. These cracks will generally not cause malfunctions in the chip, and the IC may still be fully functional. Sometimes, however, the high stress generated causes failure in the IC due to particulate contamination. This is probably due to the high localized stress and the presence of the contamination or flaw which lowers the structural strength of the material. Scanning acoustic microscopy will easily reveal the presence of these internal package cracks.

Crack patterns on the chip surface (Fig. 53 to 60) are usually curved crack lines, and do not totally cut across the surface of the chip. The crack origin is usually traceable to the back side of the chip and near a void in the die attach material. If this happens on the edge of the chip, there will be evidence of resin bleed (Fig. 52) into the underside of the chip.

The electrical test signature for this type of failure is a row and column pattern with a background of scattered bits. Because it has such a distinct signature, we are easily able to identify these failure types, and quantify their effect, or failure rate, from variation in process parameters. Identification of the root cause requires the recovery of the plastic imprint and obtaining the EDX spectrum of the particle. This may prove that the filler in the mold compound was causing the brittle fracture of the PO and the underlying interlevel dielectric (ILD) and the polysilicon. The underlying gate oxide is also shorted to the polysilicon. The distinct signature is the brittle fracture pattern at the PO level and the deformation of the metallization. In some instances, the PO and metal may be intact, but the polysilicon will show only the brittle fracture. In some cases, only a very tiny puncture mark is seen at the PO level.

These failure types typically occur at autoclave testing. The cause is the relative movement of the mold compound with respect to the chip surface. The filler material, with its relatively hard points, pushes and displaces a piece of the PO, and the underlying metallization, at a topologically high point. This occurs due to the relative movement of the two surfaces. The mold compound expands significantly when it absorbs moisture. Quite often, corrosion will take place due to the damage to the PO at the site of the damage.

The PO surface looks damaged but this time, there is no brittle fracture. Examination of the region surrounding the damage surface will show tiny scratch lines aligned in the direction of the mold flow. This is due to the particle, usually silica or mold dust, being pushed over the chip surface at the time of molding. The damage extends only over a short distance, as the particle is most likely absorbed into the epoxy resin during the molding process. Topologically high areas usually show the scratch marks.

Again, the PO surface looks damaged. Sometimes a whole piece of the PO is missing, and the metallization may also be missing. In the case of barrier metallization, sometimes the chip is still functional, as the lower level metallization is still intact. Generally, a piece of the PO is missing and scratches are absent. The affected area is normally on the topologically high side.

The type of PO damage shown in Fig. 57 is accompanied by deformation in the aluminum and damage to the underlying dielectric layers. Extensive damage is normally found in the corners of the die and decreases in severity and extent as one moves away from the corners. Normally, accompanying damage to the ball bonds can be found, present due to the high shear forces that act on them. These will cause fracture into the silicon. Scanning acoustic microscope examination will show delamination in the corners. This type of damage is highly accelerated by temperature cycling.

As shown in Fig. 60(a) and 60(b), in some cases, the damage in the PO is characterized by cracks in the corner of metallization runs. This crack is due to the expansion and contraction of the aluminum on the PO. They typically occur on the corners of metallization runs, and will occur even on chips in ceramic packaging. This shows the cracks will occur even with the additional stress of the plastic packaging. Once the PO cracks, corrosion takes place in the case of plastic packaging. Thickness of the PO or the metallization thickness is a strong factor.

On EPROM devices, PO defects will show up as data loss failures, since the EPROM cell is very sensitive to low level contamination. In Fig. 61, these PO defects show up as characteristic H- or X-type cracks, and on multilayer PO systems they will sometimes only show up on the lower level PO. They can be decorated out by a partial etch and inspection. These crack patterns are characteristic of damage due to the wafer backgrind operation.

DATA CAPTURE, STORAGE, AND RETRIEVAL

Data Capture

Information comes in many forms. It can be graphic, textual, and/or audio. The first problem of the information engineer is to capture the information in such a manner that the data contained within can be retrieved easily.

Capture of graphic information requires sophistication in software and hardware. Graphic information is received as large printed drawings, photographs, analog signals from test equipment, digital data from test equipment, or computer graphic files.

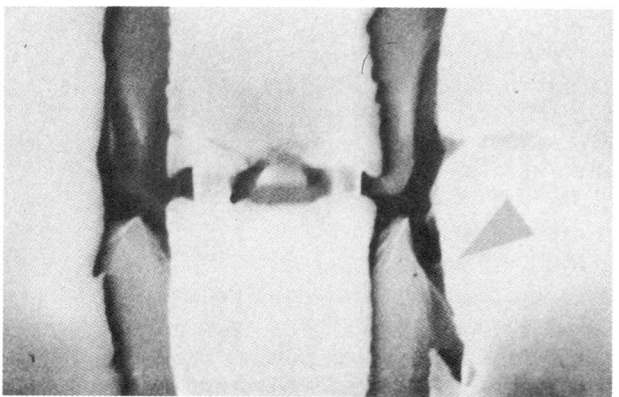

Fig. 61 PO Damage with X- or H-shaped Cracks

Receipt of a large drawing, which is to be stored on a computer, requires the drawing to be converted into a form recognized by the computer. Scanners are used for this purpose. The output of the scanner is a computer file, which is in Tagged Image File Format, or "TIFF".

Photographs are converted into a computer-recognized form by scanners using a process similar to that by which large drawings are converted. Photographs can also be converted into digital format by image capture boards which reside in the computer. Image capture boards accept an input from a video source, such as a video camera. A video camera is connected to the video capture board. The camera is focused on the photograph. Software associated with the image capture board is instructed to capture the video image and store it in a computer file.

As yet, there is no one image file format which has clear dominance in the personal computer (PC) market. Graphic images are stored in PCX, PIC, TIFF, GIF or BMP formats. The file extension indicates which image file format is being used for that particular file. In the UNIX world TIFF appears to be the clear winner. For other applications such as FAX and storage of large digitized drawings, TIFF appears to be the format of choice. A phone survey of graphics users through the SEMATECH Competitive Analysis Group found that all the above file formats are being used, but where communications between companies is required, TIFF is the preferred format. One company is using the TIFF format as the central graphics format for incorporation in an integrated factory automation system.

Automated test equipment like scanning electron microscopes (SEMs) produce an analog signal which can be intercepted by an image capture board, just as the video signal from a video camera is captured. The resultant digital image is stored as with video capture.

Some of the more advanced test equipment contain their own techniques of image capture or generation. The output from such equipment can be transferred to the central computer system for storage, but may require some translation before the file is useful. Photoemission equipment and automated curve tracers are examples of equipment where the data is available in digitized format.

Programs such as AUTOCAD®, FREELANCE®, or HARVARD GRAPHICS® generate graphic files which may require translation before they can be restored for use in other applications.

Textual information is the easiest form of data to capture, yet even textual data provides challenges. Capture of textual data requires input by an author into some digital text medium. Word processors such as WORDSTAR®, MICROSOFT WORD®, or WORDPERFECT® are used for this purpose. The output from the word processor is stored in a computer file in ASCII format, or a word processor specific format. Translation of the word processor file may be required before the data contained in the file can be used by another application. A phone survey of word processor users through the SEMATECH Competitive Analysis Group found that WORDPERFECT was the preferred word processor. However, judging from the competition in the Windows® word processor market, a word processor/file format which all can agree on, is still a long way off.

If the textual information is already printed, the information engineer is presented with a more difficult challenge. They must convert the printed textual matter into a format which can be stored by the computer. Scanners convert the textual matter into digital information which the computer can understand. Optical character recognition software is used to convert the digital scanned information into data which can be recognized by a word processor as text. The data is stored in the same manner as it would have been, if it had been generated in the word processor format.

Where content searches of the scanned information from a printed page will not be necessary, the scanned document is left in the scanned digital format. The scanned document is then stored in the same manner as a graphical image. The preferred file format appears to be TIFF.

Audio information is the most difficult medium to capture in such a way the user can retrieve useful data. Audio information has been captured by records, magnetic tape, digital audio tape (DAT) and optical disk. Each of the recording techniques records the data as heard, with no attempt to translate the information into data which can be searched. In the near future, voice recognition software will be able to translate the spoken word into text for storage in a word processor format.

DATA STORAGE

The discussion of information capture focuses on techniques to acquire information and convert it into a file format which can be stored by the computer system. The actual storage techniques and storage medium become very important to the success of electronic information management. On the computer where the data is originally captured, a large hard drive (HD) should be used because graphic files are typically large (450K bytes or more for a single TIFF image). Large HDs are required to handle even 1 days' worth of data. A 120 megabyte HD or larger is required for the local system.

Audio information has similar file size problems as graphic information. If both graphic and audio data are to be on the same computer system, the HD would need to be larger than that specified previously. Consider doubling the drive size.

Information is not useful unless it can be transferred to other individuals. One way to handle the situation is to locate an extremely large HD on a server which is part of a local area network (LAN). For storage of approximately two weeks of data, one production site uses a 2.04 gigabyte HD on a server. Backup for the HD is done with a 2.2 gigabyte tape.

Information backup, where loss or damage is not permissible, requires extreme measures. Write-once-read-many (WORM) drives provide a useful solution. WORM drives come in many sizes and styles. For permanent data on a local computer, one or two 800 megabyte WORM drives can be installed. One company uses the WORM drives on a local computer to act as a defect dictionary. The defect dictionary is used as a training tool. For servers and mainframe applications, terabyte optical drives are available. Companies use the terabyte drives to handle the massive amounts of data required; examples of other applications requiring such massive storage would be the airline industry, consumer credit, and law enforcement.

The choice of system configuration for information storage is dependent upon the application for which the data is to be used. If the data is to be used only on the local computer, a 120 megabyte HD with a WORM drive is adequate. Information which must be shared by many computers will require large HD's on each computer with one 2.0 gigabyte HD on a server. Computer systems, which allow access to each of the other computers on the network, can reduce the size of the server HD, but must increase the local HD capacity. The biggest problem seen on most computer systems is undersizing the HD. When judging the needs of a computer system, be aggressive when sizing the HD. Expect the worst possible in maximum data volume to be stored, then multiply by at least 4.

DATA RETRIEVAL

Assume all the necessary data has been collected. The data must be presented to others in a format which is informative. Graphic data with the use of proper file format translators can be viewed on a computer screen. Graphics resolutions of 512×512 pixels or better, is enough to make the image appear to be photo-like if the number of levels of color/gray scale is great enough. Today's video graphics adapter (VGA) cards offer at least 16 colors. Advanced or super VGA cards offer up to 32,000 colors and 1024×768 pixel resolution. More advanced video cards used for computer animation, image capture, and image enhancement can present millions of colors with up to $4,000 \times 4,000$ pixel resolution. The choice of the graphics display is determined by the user's job requirements.

Viewing graphics, whether on the local computer or remote computer, is important for rapid communication of ideas. Individuals with the proper graphics display adapter for the graphics being viewed, can pull the image up on their own computer. The graphics display adapter does not necessarily need to be the same as that used to capture the image. The graphics display adapter must be able to display the image from the file format used for graphics storage through some translator.

Computers which are stand-alone must use magnetic media such as floppy disks, or removable HD to transfer the graphics image to the user. If the computer is on a LAN, transmission of the graphics image is no different than displaying the image from the local HD on the user's PC. If the user has more sophistication and is connected to a wide area network (WAN), then the image can be transferred to the user's computer and displayed from the local HD. The power of the WAN or LAN should not be overlooked. Data transmission is fast and accurate.

If the graphics image is required to be rendered into hard copy, video printers are available in the market place, which can receive input from the graphics board of the computer. The video printer, upon instruction from the operator, prints a quality graphic image of what is being seen on the computer monitor. The size of the printed video image is from approximately $4" \times 5"$ to $10" \times 10"$. Text can be superimposed on the graphic image as desired to comment the image.

Textual data can be retrieved on any computer system as long as the proper means of translating the word processor file is available. The individual who stores the original data should be responsible for ensuring compatibility of the storage format with users needs. If the user requires an ASCII format, then the individual who stored the data must translate from what ever word processor file format they have to the users format. The best hard copy output today for general textual data is with a laser printer. Laser printers can also output graphic data such as charts, graphs and photos. The output is typically at 300×300 dots per inch (DPI), a resolution which is of mediocre quality when used for photo reproduction, in most cases.

With current printing technology, combining text with graphics compromises the quality of the graphics and the look of the printed document. Until both information formats can be printed with equal quality using the same print device, text should be printed with a laser printer and graphics should be printed with a graphics printer such as a video printer.

Audio output is still in its infancy. In the market place there are cards which provide the capability to capture audio, and later play back the captured audio signal. The capability of the technology stops there. Currently, there is no working voice recognition technique which is universal enough to allow for information input directly from audio to textual data. Software exists to convert textual data to audio output, but the quality is still somewhat lacking. Other audio output, such as music, can be generated through the audio boards. In computer audio, this is perhaps the most advanced and remarkable aspect. The capability to emulate musical instruments or abstract sounds is outstanding even for a low cost audio generator. The failure analysis report in the future will contain voice and/or sound to help the reader understand the content of the report.

The combination of graphic display, textual output, and audio forms the basis for the emerging technology called Multimedia. This new technology promises to allow information to be output in such a manner that the author will seem to be in the room. ISTFA will some day be highly automated with true multimedia displays of the sessions, instead of the lectures as given today.

**Typical Standardized Data
2 Examples**

Database item	Data 1	Data 2
What was observed	Thick layer/structure	Crystalline structure
Long dimension:	3.14	0.2
Short dimension:	0.2	0.2
Units:	Micron	Micron
Responsible materials:	Oxide	B
Where seen (level):	ILO	MLO
Level affected:	Poly2	MLO
How material affected:	Short	unknown
Phenomenon:	n/a	n/a
Common name:		
Design location:	Around trench	Random in MLO
Discussion:	Typically in a cluster.	Typically in a cluster.

Fig. 62 Standardized data for two example defects

DATA TRANSMISSION

Communication of information seems to be a simple task. Computers are made to transmit digital information from computer to computer. This can be done by JAN, FAX, modem (telephone lines), or WAN. There are transmission protocols such as XMODEM, KERMIT E-Mail and X400 which attempt to address one portion, or another, of the communication question. Users can determine which transmission protocol is suitable for their application by discussion with the local information management expert.

More important to the transmission of data is the ability to ensure security of intellectual data. Local security experts and company policy need to be followed when communicating failure analysis information. Organizations such as SEMATECH, where many companies are using common data, yet do not want their own proprietary data mixed with the shared data, are trying to understand the overall security issue.

Be aware that transmission of data to other countries is controlled by Customs, just as a piece of equipment is controlled. Before transmitting of data to another country, know the requirements of that country. The requirements change rapidly, following the world political situation, and penalties for noncompliance can be severe.

A WORKABLE SYSTEM

One workable system (other systems are in operation; see references 1, 2, 3, and 4) is being used in Texas Instruments. The basis of the system is an image capture computer connected to a LAN. The server has a 2.0 Gauge Byte HD. All computers on the server are connected to the server by Ethernet. Images are captured on the image capture system, then downloaded to the server for inspection by any engineer requiring the data. Equipment such as the SEM and optical microscopes have similar image capture equipment, which is also connected to the server.

For communication to the outside world, the LAN is connected to three WAN configurations. The first is a modem network which uses standard PC dialup file transfer techniques to transfer compressed files to other sites. The second network is an X400 protocol system where E-Mail and other WAN communications techniques are possible. The third WAN is the internal company file transfer system. In this process the file is loaded into a company internal file handling system and transmitted to another site.

On the receiving end, all computers use SVGA to view the graphic images and MICROSOFT WORD to view the text. Laser printers are used for output of the text and video printers with a 3×5 format are used to print graphics which need photographic quality resolution.

This system has increased the ability to communicate and reduced the cycle time for communication. Reports can be generated at one location and transmitted to a second location. The transmission time varies from a few seconds to several minutes, but is no longer days, even when transmitted to locations in other countries.

DATA STANDARDIZATION

Information is only as good as the conclusions which can be drawn from its use. If the information is structured so the data is difficult to retrieve or the data is inaccurate, the database must be restructured. One possible structure of the data is given below, with two examples of what data might be input.

Notice the first item is "What was observed". In the data column the contents say for example "Crystalline Structure". A common problem with normal databases, as well as with everyday speech, is the amount of data reduction. Normally "Crystalline Structure" would be called "Contamination", although the comment that contamination is seen is true. The term contamination is too broad. A more descriptive term is required. Void, missing geometry, crystalline structure or particulate, provides a more useful description of what was observed. A database search letter can be more selective in choice of contamination, or define synonyms to combine types of observations. Ultimately, conclusions are more accurate because the data is more descriptive.

Descriptions of the defect are of use, but many times the size of the defect will provide information as to the source of the defect. In other cases the size of the defect determines process capability. For this reason the size of the defect is included in the data.

For lack of a better term "responsible material" is used to describe the material in the defect. For example, if the defect contained Ca, Na, P and C, the defect would come from a human source. If the defect is found to be a particulate with only Si and O present, the defect would be SiO_2. The knowledge that the defect site is an oxide tells the process engineer that a process needs to be cleaner.

Failure analysts are interested in finding the root cause of the defect. To find the root cause, it is required that there be an understanding of the position of the defect in the various layers of the semiconductor structure. The database item "where seen" attempts to quantify the location of the defect in the Z direction.

If the defect lies only in one level, no other levels are affected. A missing poly silicon geometry is an example of a single layer failure. In most instances, the defect would not be in a single level, but would affect multiple layers within the semiconductor device. In this case, the database item would contain multiple values. Poly 2, Multilevel oxide (MLO), Metal 1 or Moat would be examples of levels which could be affected. All levels affected should be listed in the database.

When discussing defects, knowledge that a defect is present is not sufficient to prove the defect caused a failure. The defect must electrically affect the device. The database item "How material affected" shows the analyst has an understanding of the electrical aspects of the device, and that the defect actually caused the failure. In a semiconductor device there are only a limited number of ways a defect can have an effect. They are a short, open, or leakage. The failure can be intermittent, or can recover. The precise way the defect affected the device, should be determined during fail site isolation.

In some cases, the way the defect was generated is a well known phenomenon. The database item "phenomenon" refers

Electrical Failmode Information

Device

Family	Subgroup	Device Type
Part Type	Shrink	Revision

Failpoint in Qra Process

Wafer Fab	Assembly	Burn In
Life Test	Reliability	Monitor
Customer	Qualification	

Continuity (per Pin)

Input Leakage Hi (per Pin)

Input Leakage Low (per Pin)

Tristate Leakage (per Pin)

Fig. 63 Required electrical failmode information

Perform Tests as Needed to Understand Failmode

Algorithm	6Bit M6AP Data/ Fail Signature	Logic State "0", "1" or Both	Temperature	Voltage	Freq

Standardization in terminology is required for the algorithm, bit map and fail signature
For Microprocessors and Microcontrollers a minimum set of device algorithms is listed below.

Fig. 64 Drop through test form-standardized for ease of computer entry

Minimum Set of Device Algorithms

Init	Logic
Array	
SRAM	EPROM
EEPROM	ROM
Analog	
D/A A/D	

Fig. 65 Microcontroller algorithms for understanding both electrical "fail signature" and failsite isolation

Failsite Location Techniques

Describe the results of each test.
   Visual inspection — refer to the standard defect nomenclature for the method of describing what was seen in the visual inspection.
Package
Die (by level)
Acoustic microscope
SEM (by level)
Photoemission
TEM
Bond pull
Bond shear
Type of decap method
Die bomb results
Hermeticity results
Laser isolation

Fig. 66 Possible techniques for failsite location

to methods of defect formation such as electromigration or Fowler-Nordheim tunneling.

Scientific classification of defects, as with plants, leave a cold impression of the defect. No matter how many database items might be included in a database, the defect can never be described completely. To help reconcile the problem, a database item "common name" is used to give the visual description missing in the rest of the data. This field is mainly for analyst training and improved visual perception of the defect.

If the analyst needs to comment further on the defect, a "discussion" item is in the database structure. In this item more detail can be given on the defect. This item is not intended to be used for data sorting.

Perform tests as needed to understand fail-location

Equipment: Photoemission Microscope					
Algorithm	Bit Map Data/ Fail Signature	Logic State "0", "1" or Both	Temperature	Voltage	Freq
Equipment: Needle Probe					
Algorithm	Bit Map Data/ Fail Signature	Logic State "0", "1" or Both	Temperature	Voltage	Freq
Equipment: Liquid Crystal					
Algorithm	Bit Map Data/ Fail Signature	Logic State "0", "1" or Both	Temperature	Voltage	Freq

Standardization in terminology is required for the algorithm, bit map and fail signature.
For Microprocessors and Microcontrollers a minimum set of device algorithms is listed below.

Fig. 67 Drop through test form-standardized for ease of computer entry

To improve current device quality and prevent failures in the future, knowledge of where failures occur and how to prevent them, is important. The item "design location" provides a way of indicating where in the device the fail point occurred. By careful data analysis, potential design/process limitations can be found. In the above example, the location is given as "around trench". This is not a good description of the design location. The preferred description would be TNX300, a node location on a design database. Then include a more descriptive term such as "around trench". The standardization of defect description in a form easily searched is the first phase of developing an expert system[5]. For failure analysis, the techniques used to isolate the defect site are as important in training as the actual defect description. Presented in the following section, is an example of how to standardize on a description of the electrical signature and the techniques used to identify the failsite. The terminology is specific to the MOS division of Texas Instruments and should be changed to match the terminology of the company involved. The JEDEC 14.6 committee is trying to standardize reports, procedures, and eventually, terminology.

Before an analysis can start on any device, preliminary information is required. The information is rudimentary and is intended to provide the analyst with the data necessary to properly start the failsite isolation process.

More sophisticated testing is required for functional failures than for parametric failures. The data given so far will allow for failsite location of parametric failures, but will not work for functional failures. To supplement the parametric fail information given thus far, a series of functional tests must be done.

Production programs only supply a general idea of the type of functional failure being analyzed. Time during production testing prohibits more detailed data acquisition. A product engineer must: test the device more rigorously than a production engineer. When doing more detailed testing, characterization can not stop at the first fail but must "drop through" to the next set of tests. Recording the results of drop through test helps the failure analyst determine the location of the failsite.

The results of drop through tests recorded in Fig 63. are from past history, failures are known to change with bake, exposure to light, plasma etch, and decap. To account for the possibility the device characteristic has changed since the previous tests, an additional data column is necessary for the table. The column defines at what point in the analysis the device is being tested. Is the device being tested before or after bake? Did the device heal at this step? Testing should be datalogged in the same way after each of the reliability tests, life tests, temperature cycle tests or autoclave tests.

For the purposes of training and development of an expert system, the techniques used to locate the failsite after the electrical fail signature is understood must be recorded. The following is a partial list of the possible failsite location techniques and how the data might be recorded:

When working with techniques which require electrical stimulation of the device such as the photoemission microscope, liquid crystal, voltage contrast and needle probe, to say the test passed or failed, is not sufficient. A full understanding of the electrical test which was performed is required. A table such as the one shown in Fig. 67 is required. In the table, the result is described in terms of the design location as discussed in the standard defect nomenclature. Affects of bake, exposure to light and life test must also be considered when recording the data for each of the techniques.

With the data structured as given in the previous discussion, expert systems can be developed, information can be retrieved for accurate analysis, and the data can be used in training new analysts.

ACKNOWLEDGMENTS

Many thanks to the members of the Houston Device Analysis Laboratory (HDAL) and to the members of the Singapore Failure Analysis for their contribution to this paper.

Referenced within this text are copyrighted names. The names belong solely to their respective owners.

REFERENCES

1. *The Advent of the Paperless Failure Analysis Information System*, by Steve Morris, et.al., ", Proceedings of the International Symposium for Testing and Failure Analysis, p 81, 1990.

2. *Practical Electronic Photography In Failure Analysis*, by Dominic E. Bottaro, , Proceedings of the International Symposium for Testing and Failure Analysis, p 97-110, 1990.

3. *Desktop Publishing for Failure Analysis*, by David Wilson, Proceedings of the International Symposium for Testing and Failure Analysis, p 107-111, 1990.

4. *VAXcamera for Paperless Laboratories*, by Lloyd G. Powell, et.al., Proceedings of the International Symposium for Testing and Failure Analysis, p 113-116, 1990.

5. *Computers in Failure Analysis*, Laurel M. Bellay, et.al., Proceedings of the International Symposium for Testing and Failure Analysis, p 89-95, 1990.

SEMICONDUCTOR FAILURE MODES AND MECHANISMS

by Thomas W. Lee

Corrosion can result in opens, leakage, and functional failure modes. The activation energy is 0.3 to 0.6 eV. Various electrochemical effects may occur over varying lengths of time, dependent upon field strength, concentration, temperature, and presence of moisture. Corrosion can be caused by moisture in packages where phosphorus-gettered passivation is used to protect the device from ionic contamination, especially in concentrations exceeding 10%. The presence of moisture in the packages can be a result of insufficient hermeticity, inadequate process cleaning, or moisture which has been entrapped during the packaging process.

Crystal defects can result in leakage and functional failure modes. In this mechanism, field dependent degradation takes place faster in and along defects, such as stacking faults, swirl defects, etc. Carbon contamination, from graphite heaters, oxygen clusters, quartz crucibles, or from the Czorchralski furnaces growing the original silicon crystals, can cause these defects.

Electrical overstress (EOS) can result in opens, shorts, leakage, functional, power SOA, parameter shift, and instability failure modes. This mechanism results when gross overstress damages or destroys the device at the moment of exposure to excessive energy. Also, marginal overstress can result in gradual degradation which leads to catastrophic failure at a later time. The physical evidence of EOS is usually melt-throughs (craters), fused wires, and fused metallization runs. EOS can be caused by misapplication of the device, resulting in excessive voltage, current, peak energy, conducted charge, or operating temperature. It is possible that inflated specification claims can be a factor. EOS can be aggravated by the presence of defects which reduce device capability, but such defects are usually destroyed at the time of catastrophic failure, and evidence is lost.

Electromigration can result in open and functional failure modes. In electromigration of the first type, metal atoms move, or are pushed, **in the direction** of the "electron wind" by momentum exchange with the charge carriers, and voids move in the opposite direction. Voids may collect at steps in the metal, and create opens. Electromigration of the first type takes place in aluminum. It can be caused by faulty design rules, thin metallization, poor step coverage, or marginal overpowering, all of which allow the current density to be too high in the conductor runs. Aluminum can be doped with copper to increase its current carrying capacity.

Electromigration can also result in shorts. Metal atoms, most notably silver, may move in the direction of the electric field, but the resultant dendrites may propagate by accumulation of metal atoms at their tips, lengthening themselves **in the opposite direction** of the electric field. The silver migration mechanism may happen quickly, in as little as a few seconds, and also make take place in structures in which silver is present only as a layer, or an alloy. The cause of the mechanism is the use of silver in packages which are likely to be exposed to moisture in sufficient quantity to aggravate this mechanism.

Electrostatic discharge (ESD) can result in opens, shorts, leakage, functional, or parameter shift failure modes. This mechanism is in the form of damage to the device, when an impulse of energy is inadvertently applied at a sensitive pin. The energy results in a short due to oxide puncture or junction melt, degradation due to hot electron injection, or parameter shift due to poly resistor damage. This mechanism can be caused by device handling methods which are not in conformance with ESD procedures. An example of nonconformance could be the presence of improper materials where devices are handled. Devices could be used where a field site has high ESD generation capability.

Hot electrons can result in shorts, functional, parameter shift, and instability failure modes. In this failure mechanism, electrons originating in the conducting channel of a MOS transistor are accelerated by the D-S electric field at normal Vdss, and multiply current by impact ionization. Some of the electrons in the channel may be energetic enough to cross the silicon-oxide barrier, and become trapped in the oxide. This trapped charge may result in threshold shifts and instability. Time-dependent dielectric breakdown (TDDB) may occur. This mechanism is caused by scaling of device dimensions without a corresponding reduction in the circuit supply voltage, consideration in DRAMS, SRAMS, CPU's or other devices utilizing sub-micron geometris.

Inversion can result in functional failure modes. In this mechanism, charge spreading in the field oxide inverts the insulating region. The activating energy is 1.0 eV. This failure mechanism can be caused by inadequate clean-ups, and the presence of an excessively thin field oxide layer.

Metallic particles can result in short, functional fail, and intermittent failure modes. The mechanism responsible is eventual dielectric breakdown, which may be due to the random location of a conductive particle in a high-field region of the device. The high-field region may also result in the gettering of particles. This mechanism is typically caused by rough handling during assembly, and poor assembly process control.

Moisture and contaminants can result in failure shorts, leakage, and functional failure modes. When present in sufficient concentration, moisture can result in surface leakage paths. The presence of moisture can be the result of inadequate monitoring of moisture in the sealing atmosphere (RGA). Additionally, remnant process solutions on piece parts, or poor package hermeticity in the presence of temperature cycling and available moisture from humidity, can be causes.

Oxide defects can result in shorts, leakage, and functional failure modes. Pinholes are the most common oxide defects, and they may lead to oxide breakdown. Activation energy is 0.3 to 0.5 eV. Pinhole oxide defects may be caused by mask scratches, which may result in incomplete diffusion masking. Poor photoresist development, particles on masks and in photoresist, and bubbles in etch stations, can all cause oxide defects and pinholes.

Radiation damage can result in leakage and functional failure modes. Energetic particles and rays may leave ionized tracks in silicon, which become leakage generation sites, and upset device switching. The damage causes bulk leakage in the same manner as with material defects. The particles and rays may also discharge storage circuit elements, causing soft errors. Improperly annealed ion implantation damage can also result in this failure mechanism. The rays and particles can originate from isotope decay in silicon or packaging material, or exposure of the device to radiation flux, such as in Kr^{85} leak testing.

Slow trapping can result in functional, parameter shift, instability, and increased MOS threshold voltage failure modes. In this mechanism, carriers tunnel into traps near the silicon-oxide surface and cause MOS threshold voltages to increase. The activation energy is typically 1.0 to 1.3 eV. This mechanism can be caused by improper sintering of the aluminum.

Solid-solid diffusion can result in opens, shorts, functional, and power SOA failures. In this mechanism, metal atoms move according to the concentration gradients in the semiconductor structure, in the presence of temperature. Gold-aluminum (Au-Al) "purple plague" intermetallics and aluminum-silicon (Al-Si) junction "spiking" are two examples. This mechanism is caused by the absence of diffusion barrier materials in the device structure, or excessive thermal conditions. Aluminum can be doped with silicon to eliminate the concentration gradients. Barrier materials, such as titanium nitride, TiN, may be used.

Surface conduction can result in functional, and power/SOA failure modes. The gradual accumulation of charge on exposed gate oxide can create inversion and leakage. The activation energy for this mechanism is 0.5 to 1.0 eV. The cause of this mechanism is misalignment of the gate metal, which exposes the oxide.

REFERENCES

1. *Reliability and Degradation*, McGraw-Hill.

2. *Integrated Circuit Quality and Reliability*, Eugene R. Hnatek, Viking Labs, Marcel Dekker, 1987.

FAILURE MODES AND MECHANISMS OF NON-SEMICONDUCTOR ELECTRONIC COMPONENTS

by Martin J. Johnson and Scott E. Smith

Non-semiconductor electronic components, such as resistors, capacitors, inductors, and crystal oscillators are often given decreased emphasis in failure analysis texts in favor of extensive discussions of semiconductor devices. However, it is widely known that these components are used in virtually every circuit to provide the resistance, capacitance, inductance, and frequency signals necessary for semiconductor devices to function. Surprisingly, the level of technology used to produce these components may rival that used in semiconductor device fabrication. Additionally, many labs report that the volume of work performed on capacitors, for example, far outweighs that of ICs.

The perspective presented in this section is that of the communications and automotive electronics industry. In these highly competitive markets, high reliability under adverse environmental conditions is crucial, but product cost must also be controlled. The components discussed fit these criteria, but most of the failure modes and mechanisms described here are also applicable to the more specialized components used in other industries. We have divided the *failure modes* for each component type into three categories. Two are catastrophic, and cause the component to be completely non-functional; "open" (high resistance) and "short" (low resistance). The third category is "parametric," which describes a significant, but non-catastrophic, change in a critical electrical parameter of the device. Open and shorted components are frequently cases of the same *failure mechanisms* which cause parametric failure.

RESISTORS

General Failure Modes and Mechanisms of Leaded Resistors

Most resistors used today are two leaded axial devices with the leads protruding from a cylindrical body. The devices may have their leads bent, or formed, prior to hand or machine insertion into the circuit board. These operations frequently cause mechanical damage to the device leads, body, or the interface between them, resulting in failure. External visual inspection of a failing device as well as familiarity with the board assembly processes will help in identifying any failures of this type.

Carbon Composition Resistors

Construction. The body, or element, of a carbon composition resistor is produced by pressing carbon powder to form a rod, which is then sheared to the proper length. The resistance value and power dissipation rating of the resistor depends on the density of the carbon powder and the rod diameter. Copper wires with anvil-shaped ends are inserted into holes in the ends of the rods to form the resistor's external leads. The wire, including the embedded portion, is tin or tin/lead plated to make it solderable and to prevent oxidation of the bare metal. The lead-rod assembly is then further compressed to form a bond between the pressed carbon and the device's external leads, then coated with a protective layer, typically Bakelite® or epoxy resin. This coating serves as a dielectric barrier and also provides additional mechanical strength for the resistor during handling. Bakelite® coating is commonly used in less expensive devices, where it may leave the ends of the carbon element exposed at the point of lead insertion. In comparison, molded epoxy coatings completely encase the device, sealing off the ends of the element from the environment.

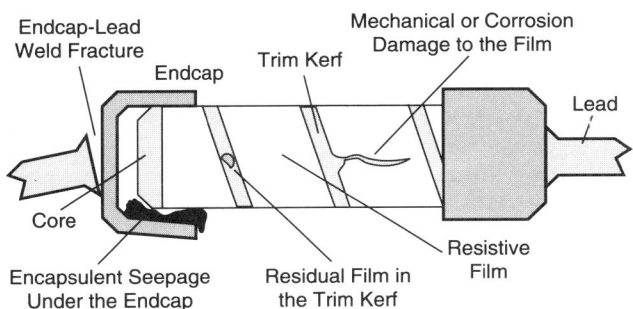

Fig. 1 Common failure mechanisms of film resistors

Fig. 2 Photo of actual film resistor, as made, coating removed

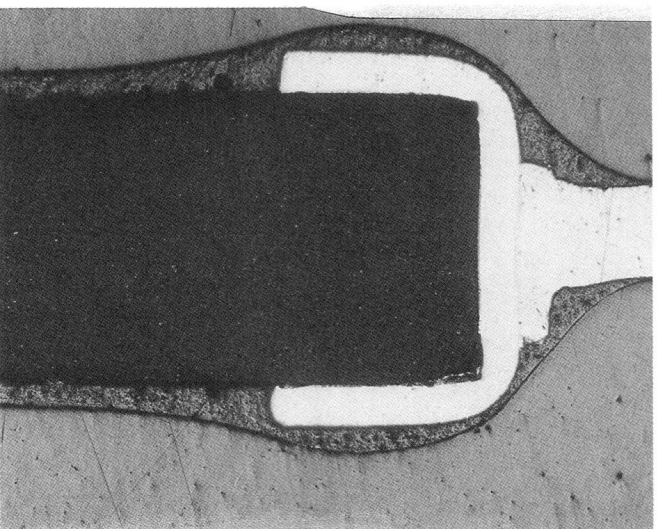

Fig. 3 Cross section of film resistor

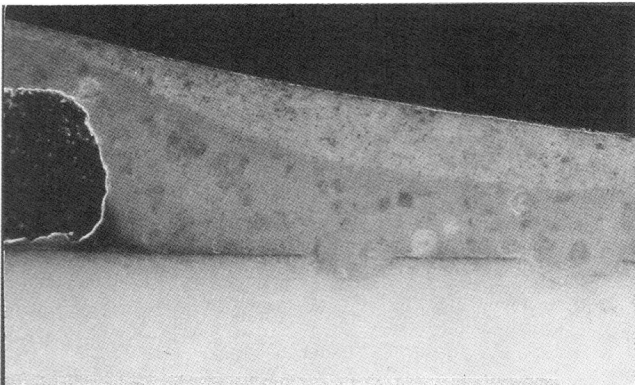

Fig. 4 Cross section of film resistor showing laser kerf, film and 2 layers of coating

Opens. Because of their simple construction, open carbon composition devices usually occur due to mechanical damage or abuse during leadforming or board insertion. Separation of the mechanical contact between the lead and carbon element is visually indicated by a fracture where the lead enters the element. A fracture in this area may exhibit itself as an open, intermittent, or a high value resistance. A high resistance could result from the loss of intimate contact with the element and/or oxidation of the lead end in the contact area.

Shorts. Shorted carbon composition resistors technically cannot occur. Voltage overstress can cause carbon arcing within the element, creating a very low resistance path within the element, while current electrical overstress or excessive power dissipation will heat the device to the point of fully sintering the pressed powder carbon element. Both these phenomena result in a sharp reduction in resistance, and are commonly referred to as "shorts."

Parametric Failures. Thermal, chemical, and mechanical processes can result in changes in the resistance value of a carbon composition resistor. Permanent resistance decreases of up to 7% can be caused by thermal exposure during soldering. The exposure has the effect of additionally sintering the carbon element, reducing its resistance. If the carbon element of a resistor is not fully sealed by its encapsulant, the device may absorb assembly process solvents or atmospheric moisture and exhibit either an increase or a decrease in resistance. This is because of two competing mechanisms: water absorption into the carbon element decreases the resistance while moisture-induced oxidation or corrosion affecting the carbon element-to-lead bond increases the resistance. The effect of water absorption may disappear over time or with baking as the device literally "dries out." *Carbon composition resistors are becoming obsolete in the industry, with the exception of some special applications, due to these inherent failure mechanisms.*

Film Resistors (Carbon, Metal, and Metal Oxide)

Construction. Film resistors consist of a round ceramic rod as a core or mandrel, onto which a resistive film of conductive material is sputtered or evaporated. Steel endcaps which are pressed on or crimped over the ends of the core, and copper or copper-clad steel leads are butt-welded to the endcaps. The film used may be carbon, a pure metal or alloy, or a metallic oxide, depending on the application's cost and performance requirements. The component body is coated with a high dielectric enamel or epoxy. This coating should cover the endcaps and weld area of the leads as well as the body of the component. The external leads are generally tin or tin/lead coated, usually by a dip process.

The gross resistance of the device is determined by the resistivity of the film while its specific value is "trimmed" or "scribed" by a laser or abrasive disk technique. Laser trimming is performed by a pulsed laser beam hitting the film surface while the device is rotated and moved laterally, producing a spiral trim pattern. Continuous resistance measurements are made as this trimming takes place, allowing resistors with repeatable, tight tolerance values to be produced. Abrasive disk trimming utilizes a disk, spinning at high speed, which is brought into contact with the film on the core. The core is rotated and moved laterally, similar to a device being laser scribed. Abrasive wheel trimming is less precise and must be carefully monitored to produce comparable results to laser trimming. Of particular concern in both processes is the presence of residual film remaining in the trim area (the "kerf"), from insufficient laser power or a worn abrasive wheel.

Opens. Construction defects and mechanical and electrical abuse are common causes of opens in film resistors. Separation of the endcap to lead welds because of a poorly controlled weld process or from mechanical abuse during board insertion or leadforming may result in an intermittent or permanently open condition. Endcap contact to the resistive film can be compromised by the coating enamel seeping under the endcap if its viscosity is poorly controlled. The difference in thermal coefficients of expansion (TCE) between the enamel and metal endcap during thermal cycling can then cause the endcap to lose contact with the resistive film. A poorly formed endcap

Failure Modes and Mechanisms

will have the same result. Gross electrical overstress may also vaporize the resistive film or cause it to blister or separate.

Chemical corrosion can produce open metal film resistors and is generally unique to metal film resistors. Carbon film and metal oxide film resistors are much less susceptible to chemicals. The enamel or epoxy coating, while not hermetic, provides a good barrier to atmospheric moisture. Moisture, a metallic film, and a voltage across the device can combine to cause galvanic corrosion of the film. In comparison, metal oxide and carbon film resistors are generally not susceptible to corrosion. Metal oxide resistors are often used as replacements for wirewound or carbon composition resistors because they can easily withstand up to four times their rated power in surge applications. *By contrast, carbon composition resistors can withstand up to ten times rated power under surge conditions.* Under surge conditions, the ceramic core acts as a heat sink for the power dissipated by the film. However, delaminating of the film from the core due to properly deposited or contaminated film will result in separation of the film and failure under surge conditions.

Shorts. Film resistors commonly short from construction and application defects and damage. Missing or voided encapsulant coating covering the endcaps can allow shorting of the exposed endcap to an underlying circuit board trace or produce a path for arcing in a high voltage application. Internal arcing across the trim could produce a short or decrease in resistance, particularly if residual film material is left in the trim kerf. This is most likely to occur in applications where electrostatic discharge (ESD) or other high voltage transients occur. The spiral pattern of the trimmed resistive film may also make the device *act like an inductor* during fast rising transients, contributing to circuit instability.

Parametric Failures. Resistance changes in film resistors are often due to mechanical damage to the film during the trimming operation or from poor electrical contact between the film and the endcap. In carbon film resistors, trimming damage can cause flaking of the carbon off the core due to a damaged or worn abrasive disk or poor laser power control. This damage is characterized by a missing area of carbon along the trim kerf. Corrosion in metal film resistors may result in parametric value changes as well. (See Fig. 1 to 4.)

Wirewound Resistors

Construction. Wirewound resisters consist of a cylindrical core of mandrel of fiberglass, ceramic (commonly alumina, Al_2O_3), or plastic. The resistance wire used is a Nichrome alloy, with dopants added to adjust the resistance, temperature characteristic, and application. The wire is wound around the core and secured to the metal endcaps by welding it to the endcap, or by crimping it between the core and endcap. The leads of the device are then welded or crimped to the endcaps, and the component is encapsulated in a resin such as epoxy, a molding compound, or a glass frit-filled ceramic ("sand bar" resistor).

Opens. The most frequent cause of open wirewound resistors is mechanical damage. Lead forming or board insertion damage may result in fracturing of the core, separation of the lead from the endcap, separation of the resistance wire along its length, or separation of the wire from the endcap.

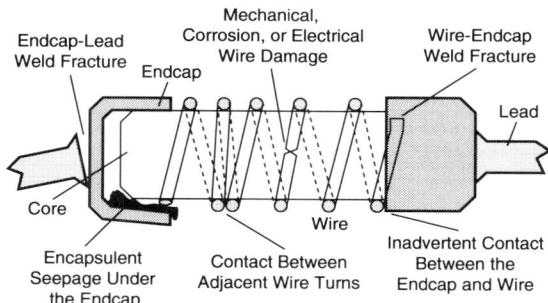

Fig. 5 Common failure mechanisms of wirewound resistors

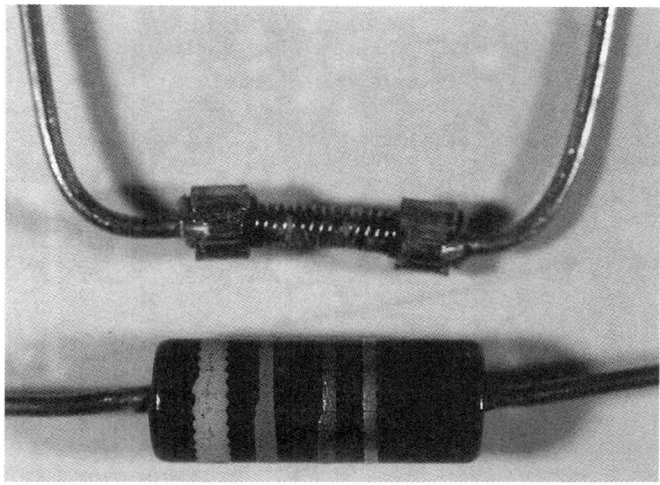

Fig. 6 Common wire wound resistor packages, as made, decapsulated

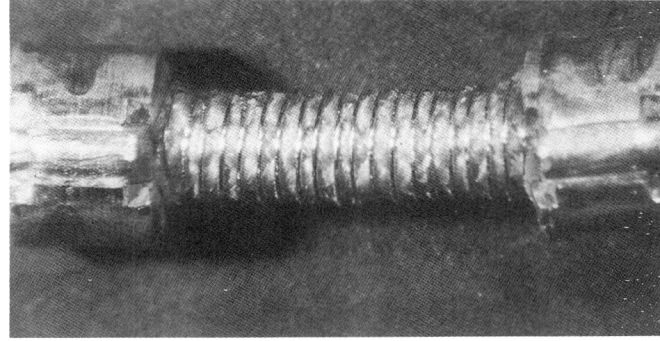

Fig. 7 Crimp ring end termination

The interface between the end termination and resistance wire is a common failure location. On wirewound resistors with welded resistance wires, a poorly controlled weld process will result in fracture of the weld (from insufficient weld power) or of the wire itself (from excess weld power). Other wirewound resistors rely on the pressure contact of the endcap with the re-

Fig. 8 Crimp endcap end termination

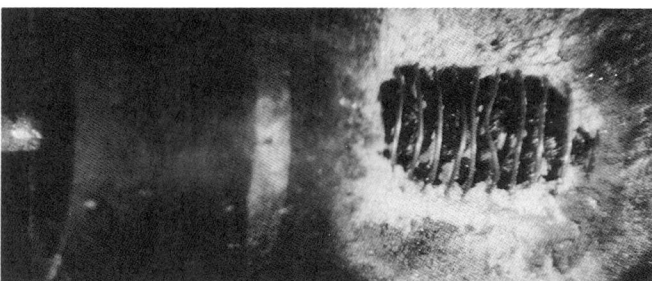

Fig. 10 Failed aprt showing winding deformation

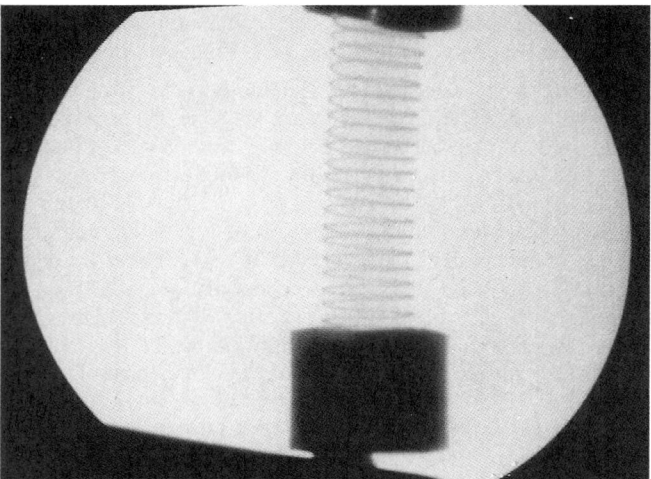

Fig. 9 X-ray of wire wound resistor

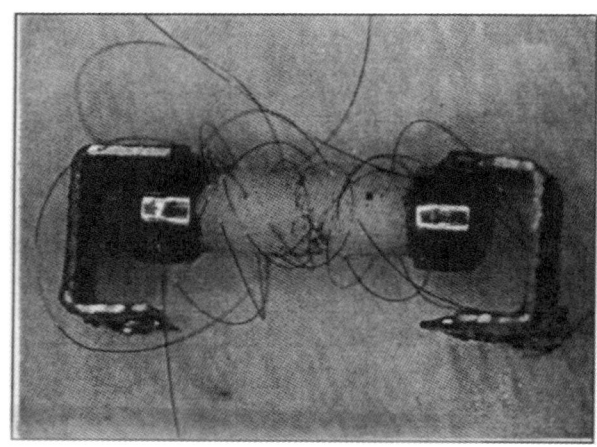

Fig. 11 Example of tight windings (after decap)

sistance wire by mechanical crimping. If the crimping process, or viscosity of the encapsulant is not well controlled, encapsulation materials may seep beneath the endcap and separate the endcap from the wire during thermal cycle stressing.

Separation of the wire within the winding also causes opens in wirewound resistors. Gross electrical overstress (EOS) damage may result in the fusing of the wire. This type of damage may be the result of excessive current for the size of the resistance wire used or result from high voltage arcing between irregularly spaced turns of the resistance wire. A kink in the wire incurred during winding may cause a tensile break in the wire at the spot due to thermal expansion of the core. Corrosion of the wire by ionic contaminants may also occur, especially since most encapsulates are not truly hermetic. Excessive tension during wire winding, especially common in high-value resistors, may cause separation of the resistance weld or separation of the wire itself.

Shorts. Shorted wirewound resistors are uncommon and may actually be large changes in resistance value rather than a true dead short.

Parametric Failures. If the spot weld of the wire to the endcap is located near the lead terminal, intermittent contact between wire and the endcap edge may cause resistance increases or decreases of a few percent of the resistor's value. Poor endcap crimping of the wire may allow intermittent contact between the last few wire turns and the endcap, also causing intermittent resistance shifts. Partial wire corrosion will cause the resistance to increase as the effective diameter of the wire decreases at the corrosion site. Likewise, intermittent contact between irregularly spaced wire turns will result in a decrease in the wire's effective length and thus a decrease in resistance. Contamination of the encapsulation material or package exterior may cause current leakage, especially after exposure to moisture and voltage bias conditions, but this often disappears after a high temperature bake operation. (See Fig. 5 to 11.)

Flat Thick and Thin Film Resistors: Discrete Chips and Networks

Construction. Thick and thin film resistors are widely used in two forms: discrete chip resistors and multiple resistor networks. Thick film components are more common than their thin film analogs because of their lower cost and simpler process requirements. *Thick film* components begin with a ceramic substrate (typically alumina) which provides the component's

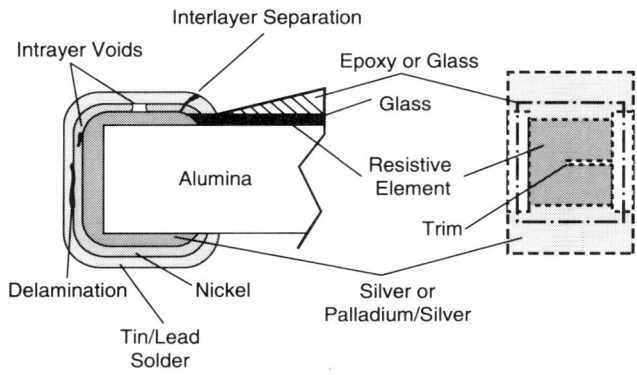

Fig. 12 Common failure mechanisms of surface mount chip resistors.

structural strength. A conductive metallization, commonly platinum/palladium/silver (Pt/Pd/Ag), palladium/silver, or silver, is silk-screen printed on the surface to provide contact between the resistive material and the external circuit board. The resistive material, primarily ruthenium oxide (RuO_2), is then screen printed onto the substrate and trimmed to the correct resistance value using an abrasive wheel or laser beam. A glass and/or epoxy encapsulant is used to cover the resistive material to prevent mechanical damage, and to provide a clean smooth surface for component marking. The substrate is then fractured along mechanically or laser produced scribe lines to separate it into individual devices. The solderable end terminations on chip resistors are produced by printing or dipping the end surfaces with palladium/silver or via a sputtering process. This layer is then plated with a nickel (Ni) diffusion barrier layer and then a tin/lead (Pn/Sn) solderable layer. Network components have terminal leads attached to the substrate and are frequently epoxy or enamel encapsulated by dipping or via a fluidized bed process.

Thin film components are used where their greater precision or high temperature stability is needed. The construction of thin film components differs from that of comparable thick film devices in that the conductors are first sputtered or evaporated onto the substrate, instead of screen printed. The construction of thin-film resistors (and capacitors) is precisely the same as in many integrated circuits, such as I.C. The conductors are photolithographed and etched into the desired configuration, then a resistive material such as nichrome is deposited, a pattern is photolithographed, and then etched into resistor patterns. The resistors are typically laser trimmed to the correct resistance value, yielding a high degree of precision in these devices.

Opens. Open components may result from mechanical, thermal, metallurgical, chemical, or electrical mechanisms. Fracturing of the substrate may result during component placement or board insertion. Fracturing may also occur during thermal cycling if there is a poor match between the thermal coefficient of expansion (TCE) of the component and the circuit board it is mounted on. A conformal coating may also exert undue stress on the thick or thin film during thermal cycling and result in an open.

Fig. 13 Cross section showing metal layers, resist ink, and passivation layers

Fig. 14 Chip resistor, EOS failure

In thick film devices, the silver metallization used to connect the resistive layer to the end termination is susceptible to leaching by the tin/lead solderable layer if the nickel barrier layer is incomplete or fractured during assembly or soldering. This can ultimately result in separation of the end termination from the ceramic body or an open device with no obvious external cause.

The resistive element itself can be separated, due to several causes, and cause an open. Electrical overstress (EOS) damage from excessive current may cause fracturing, initiating near the end of the trim due to current-crowding at that point and propagating across the width of the device. Thermal cycling may similarly cause propagation of microcracks present in the trim area due to poor control of the trim process. Contamination of the substrate or firing problems may cause the resistive material to delaminate from the substrate and fracture, especially under high thermal or electrical stresses.

Shorts. Shorted thick/thin film resistors are uncommon and usually represent large decreases in resistance value, rather than a true dead short. A decrease in resistance generally indicates the presence of an ionic contaminant or alternative conduction path, such as that due to metal migration or silver (or other metal) dendrite formation. Dendrite formation or electromigration may also indicate the presence of an ionic contaminant.

Parametric Failures. In their less severe form, many of the above mechanisms can cause parametric resistance increases in thick and thin film resistors. These include TCE mismatch between the component and circuit board, EOS damage, propagation of trim-induced microcracking, and resistive element delamination. Contamination in the encapsulation material or on the package exterior can cause parametric failures by decreasing resistance. Both think and thick film resistors can be damaged by electrostatic discharge (ESD) (Fig. 12 to 15).

Variable Resistors and Potentiometers

Construction. Variable resistors and potentiometers consist of a resistive medium on a substrate which has one fixed contact and one which is allowed to move along the length of the resistive medium and make contact via a wiper. The resistive medium can either be a wound length of resistance wire on a ceramic or fiberglass cylindrical core, a thick film resistive ink on a planar ceramic substrate, a cermet element, or a conductive polymer film deposited on a plastic or phenolic base. Motion or rotation of the wiper changes the overall resistance value of the component. The resistive media of these components are similar to wirewound or thick film resistors, thus are susceptible to the same failure modes and mechanisms. In addition to these, the quality of the electrical contact between the wiper and resistive media affects component performance.

Opens. Open devices may occur from damage to the resistive element via mechanical or electrical overstress damage, wire corrosion, voids or bubbles in the thick film element, or excess wear of the element by the wiper. The wiper may fail to make proper contact due to mechanical deformation, contamination of the element surface by flux or other material, as well as the formation of an insulating film on the wiper or resistive medium contact surface by degradation or aging of lubricants commonly applied. *Some lubricants are known to be chemically changed or completely stripped from the device by process cleaning solvents.* The removal of lubricants paves the way for mechanical wear failures.

Shorts. Shorts or low resistance may be the result of corrosion or contamination of the contact between the external lead or termination and the wiper. Other sources of contaminants are degradation of the housing materials during high temperature processing or operation and the entrapment of process solvents or fluxes.

Parametric Failures. The formation of an insulating film on the wiper contact or corrosion of the wire resistive element may cause increases in resistance. Intermittent operation is far more common than a parametric value defect and may result from wiper contact surface contamination, uneven or irregular element height, or mechanical deformation or damage to either the wiper or element.

CAPACITORS

General Failure Modes and Mechanisms

Capacitors are manufactured in a wide range of body styles and technologies, from surface mount ceramic capacitors to leaded metal can electrolytic capacitors. Leaded capacitors are susceptible to handling and insertion damage, similar to that outlined for leaded resistors. Surface mount capacitors are relatively fragile and can be damaged by handling, placement, or rapid temperature changes as applied in soldering processes. Film and some types of electrolytic capacitors are especially susceptible to excessive temperatures and can melt or dry out if not processed or applied correctly.

Ceramic Disk Capacitors

Construction. The dielectric material used in ceramic disk capacitors is commonly barium titanate. Barium titanate can be grown from a melt, and pulled by the Czorchralski technique, like silicon and other materials, then sliced into wafers. In another process, an aqueous slurry of the ceramic powder is formed into a sheet, air dried, and separated into the individual disks. The disks are then fired to a ceramic at a high temperature to remove the organic binder material and to solidify the ceramic. Electrodes, commonly silver, nickel, or copper can be plasma or flame-sprayed, or screened on and the assembly is fired to sinter the metal layer. The external leads are then soldered on, or attached with a conductive silver-filled epoxy. The component is then encapsulated in a epoxy or phenolic protective coating. Disk capacitors are often impregnated with a high temperature wax as a moisture barrier.

Opens. The most common cause of opens in ceramic disk capacitors is separation of the external lead from the electrode.

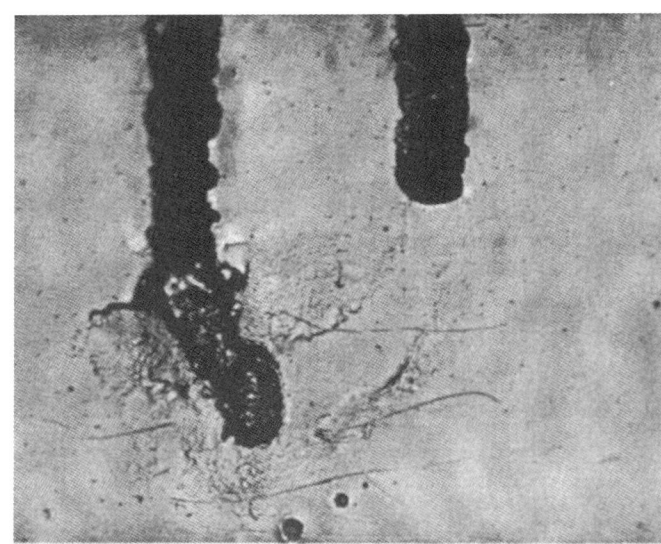

Fig. 15   Chip resistor, EOS at trim defect

This may occur due to mechanical abuse of the device during handling or board insertion or from thermal shock fracturing of the joint caused by differential expansion due to the large difference in the thermal coefficient of expansion (TCE) of the metal wire and the ceramic disk. It is not uncommon for soldering heat, conducted through the leads, to cause desoldering of the leads. This most often occurs in physically small devices soldered on assemblies with a large thermal mass.

Shorts. As is the case with most capacitors, shorting of ceramic disk capacitors is frequently caused by mechanical or electrical damage to the dielectric. Dielectric breakdown or punch through is the result of exceeding the dielectric strength of the material during high voltage exposure. Dielectric breakdown normally occurs at voltages in excess of two and one half times the rated voltage. Fracturing of the dielectric from handling or thermal shock may result in shorting by allowing migration of the electrode or lead attach material under the capacitor's electric field, especially in high humidity conditions. Disk capacitors can be internally fractured by dropping them on the floor. Pre-existing dielectric fractures may result in punch through damage at voltages far below the rated dielectric breakdown voltage.

Parametric Failures. Low capacitance value is frequently caused by fracturing of the dielectric material. Similarly, dissipation factor may be changed by fractured dielectric or moisture entrapped in the encapsulation.

Leakage may occur due to dendrites or whiskers of the electrode or lead attach material migrating across the edge of the dielectric disk to the other electrode. Most encapsulants are not hermetic, and moisture penetration in the presence of sufficient voltage but limited current (to avoid fusing the dendrite open) must be present for this to occur. (See Fig. 16 and 17.)

Multilayer Ceramic Capacitors

Construction. Like ceramic disk capacitors, the dielectric material in multilayer ceramic capacitors is predominantly barium titanate, with rare earth oxides added to achieve the desired dielectric properties. Other base ceramics are also used for specialized applications, such as capacitors with high Q, intended for RF circuits. Multiple ceramic layers are separated by silver or palladium-silver metallization layers ("plates") to form the rectangular body of the component. The device end termination layers are silver or silver/nickel/solder depending on the application environment. The end termination layers are formed by screen printing, dipping, or electroplating. If the device is leaded, metal leads will be soldered to the device and it will be encapsulated in a resin coating, typically an epoxy.

Opens. Open multilayer ceramic capacitors frequently result from damage to the external leads or end terminations or by poor control of construction processes. Mechanical abuse of leaded components may fracture the solder joint attaching the external lead to the end termination. Exposure of the device to excessive heat may result in fracture of the solder joint or a loss of contact with the metallization layers within the ceramic. The presence of porosities or discontinuities in the nickel barrier layer may allow the solder to "leach" the silver layer away from the ceramic, which may also resulting in solder joint fracture or a loss of internal contact. The alignment of the metallization layers to the ends of the body while the capacitor layers are being built-up is crucial: poor alignment may allow marginal or intermittent contact between the metal layers and end termination layers.

Shorts. Shorts or low resistance failures of multilayer ceramic capacitors frequently result from mechanical damage to the dielectric layers and manufacturing defects. The ceramic material itself is somewhat brittle and is susceptible to fracturing from excessive thermal gradients during soldering. Most capacitor manufacturers recommend that components not be subjected to thermal gradients in excess of 3 °C/second, or temperature shock changes of more than 90 °C during board assembly processes. Similarly, fracturing of the ceramic during handling, insertion, or automatic placement may occur. The fracturing may cause immediate failure by allowing adjacent

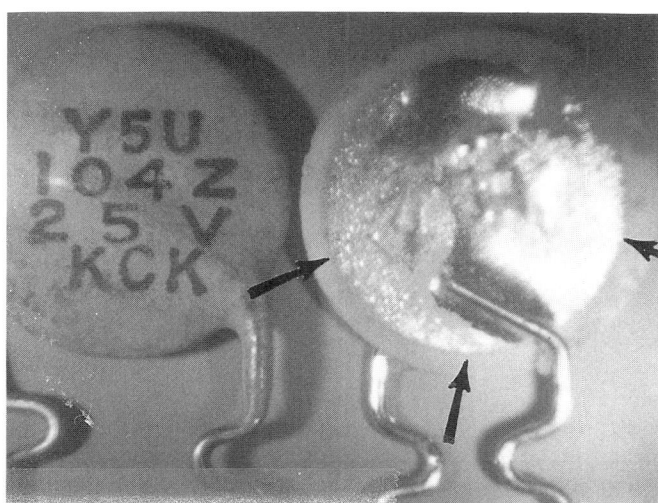

Fig. 16 Disk capacitor with edge margin

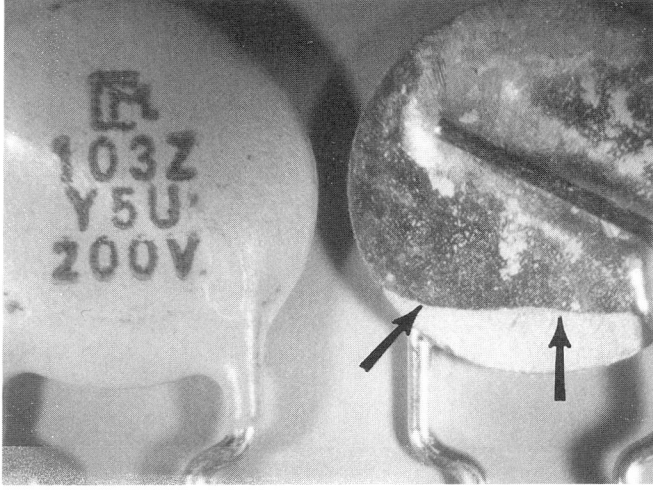

Fig. 17 Disk capacitor with no edge margin

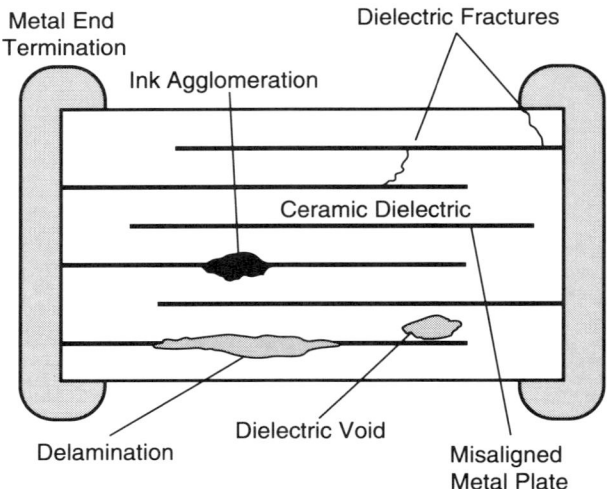

Fig. 18 Common failure mechanisms of surface mount monolithic ceramic chip capacitors.

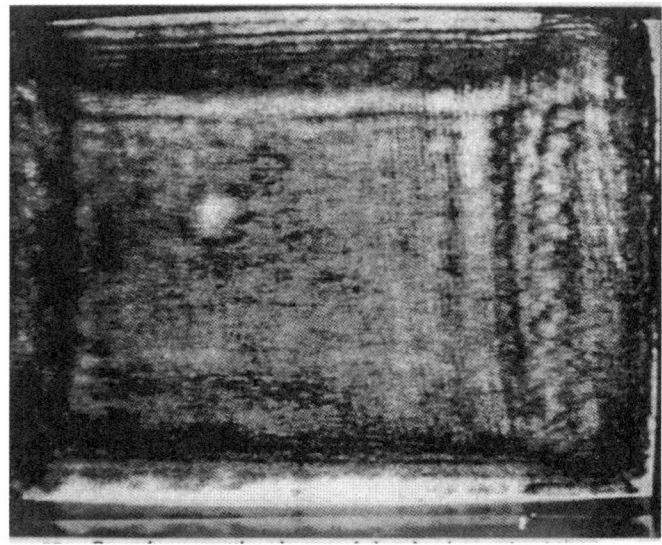

Fig. 19 Scanning acoustic micrograph showing internal void

Fig. 20 Cross section of chip capacitor, verifying void

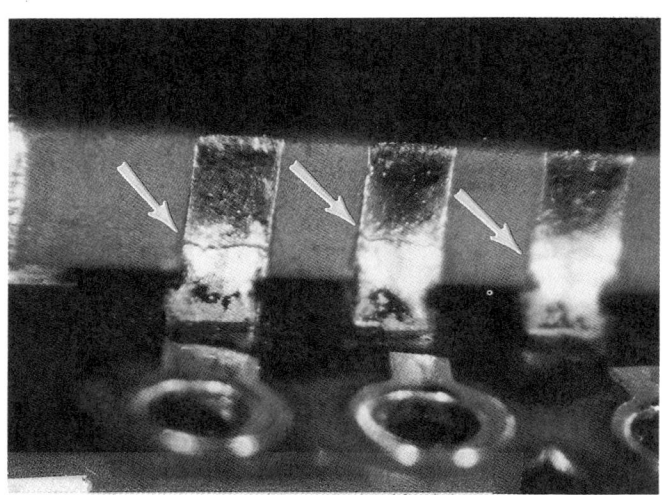

Fig. 21 Capacitor array with cracked solder joints

layers to touch or may produce latent damage, which can result in field failures. Another form of intra-layer misalignment may allow a metal layer to bridge the length of the body and short the two end terminations.

Multilayer ceramic capacitors are frequently used as suppressors to prevent high voltage transients from damaging semiconductor components, or as filters to prevent the transmission of noise or RF energy to the circuit. However, exceeding the dielectric breakdown voltage of the ceramic may produce internal fracturing and metallization migration that will cause immediate shorting or latent leakage paths, depending on the energy of the transient. Damage due to high energy electrical overstress will cause gross melting or fracturing of the ceramic material that is readily visible during component cross-sectioning.

Parametric Failures. The most common parameter to be affected by defects is the capacitance value. Changes in capacitance without changes in other parameters are commonly caused by poor end termination contact with internal metal layers due to layer misalignment or from a change in an inherent material property referred to as "aging." Aging is due to the long term exposure of a dielectric material to electrical bias and results in a decrease in component capacitance over time. The higher the dielectric constant of the ceramic, the more pronounced the aging effect. "De-aging" can be accomplished by heating the device above its Curie temperature or exposing the component to its maximum DC rated voltage.

Failure Modes and Mechanisms

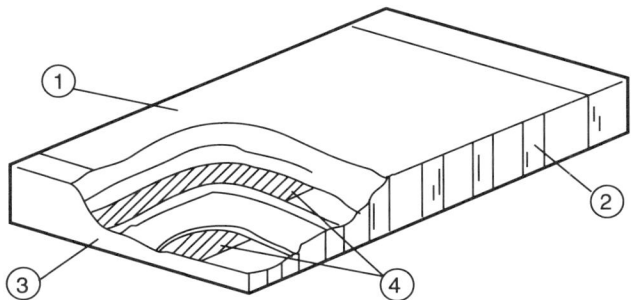

Fig. 22 Schematic of capacitor array devices

Thermal or mechanical fractures between adjacent metallization layers provide an excellent leakage current pathway, especially in the presence of moisture, allowing the migration of the electrode metallization to occur and produce dendrites. The ceramic layers themselves are susceptible to defects during device manufacture, including inter-layer delamination, pockets of excess metallization ("ink agglomerations"), voids, or excessive porosity of the dielectric, all of which can accelerate dendritic growth. Delaminations and dielectric fractures and voids are detectable by acoustic microscopy, while cross sectioning provides a positive identification and verification of the failure mechanism.

Other parameters which may change or degrade are leakage current, dissipation factor (DF), and equivalent series resistance (ESR). Failures may be caused by the dielectric failure mechanisms described above as well as external contamination or moisture entrapment in the encapsulation. Electrostatic discharge at high levels has been shown to cause reduced capacitance by causing discontinuities in electrodes. The high inrush current associated with this type of event can cause metal migration. Many suppliers recommend limiting inrush current to less than 50 milliamps (see Fig. 18 to 24).

Film and Foil Capacitors

Construction. Film and foil capacitors have similar exterior appearances but differ in the materials and processes used to manufacture them. Both types may be of a wound or stacked construction. Foil capacitors are produced by interleaving two sets of a thin metal foil and a dielectric film while film capacitors are produced using a dielectric film which has had a metal film applied by evaporation. Because of this, the metal thickness is typically much less in film capacitors than in the foil type.

Wound capacitors are individually formed by winding the foil and dielectric film layers around a small diameter mandrel which is removed after the operation. The windings are then heat sealed. By comparison, stacked devices are bulk-formed by winding the dielectric and foil pairs on a very large diameter drum. This sandwich is then cut into individual segments which approximate a flat stack of "plates." For both processes, the metal layers are offset so that alternating layers are exposed at the ends to form electrodes onto which metal may be plasma or flame-sprayed to form an electrical contact for the end termination. External leads are then soldered or cold welded to the met-

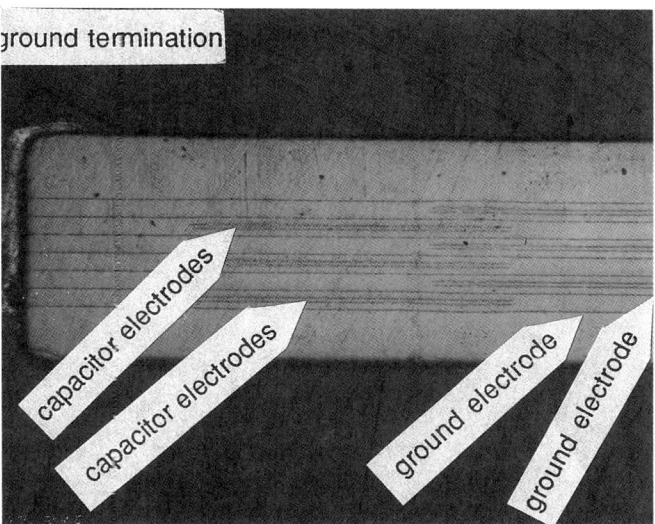

Fig. 23 Cross section of array showing electrodes and internal construction

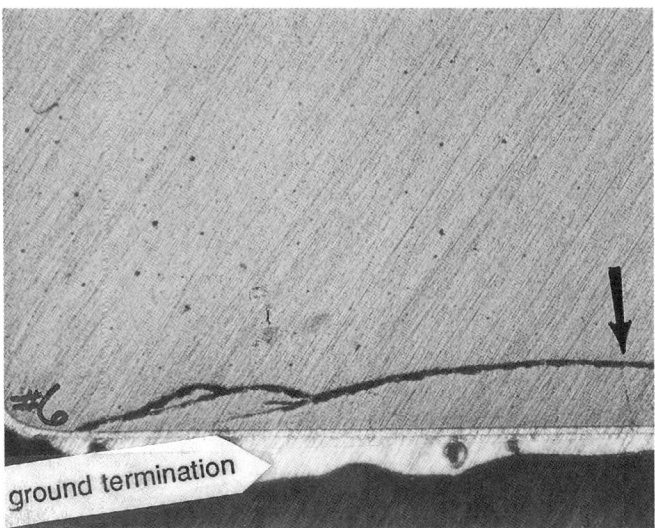

Fig. 24 End termination cracks due to part and substrate CTE mismatch

al end termination. The completed component may be epoxy dipped or sealed in a plastic box for protection from environmental stresses.

Opens. End termination and lead attachment defects are the most common causes of open devices. Leads which are poorly attached to the end metallization can manifest themselves as open or intermittent devices. The lead attachment may easily be damaged by mishandling during lead forming or automatic insertion. Detection of open or intermittent connections due to lead attach problems almost always occurs after insertion and soldering, making this problem nearly impossible to screen out at the component level. A general assessment of the robustness of the lead attach can be performed at the component level by a destructive physical analysis (DPA). The electrical resistance of a poor attachment can add to the real portion of the complex impedance of the device, resulting in a high dissipation factor (DF) value, but this is a subjective and inconsistent method of assessing lead attach. Corrosion of the electrode can result from a broken epoxy seal, and an open will eventually result. Atmospheric moisture, particularly from coastal areas, can be aggressive enough to cause the corrosion.

Shorts. Shorts are rare in film devices due to the self-healing qualities of the metallized film. Film capacitors may exhibit shorts if the available current is not sufficient to vaporize the metal film around the damage or defect, and "heal" the short. Shorted devices, or ones with high leakage are commonly seen in high impedance circuits or coupling circuits. Foil devices can become shorted due to dielectric breakdown caused by overvoltage punchthroughs, or defects in the dielectric film. Miscutting of the device during manufacture may allow adjacent electrodes to be connected to the same end termination. The end metallization is applied in a process called "shooping" where a hot metal is sprayed onto the end of the device. Improperly controlled temperature or spray force may cause the sprayed metal to change the electrode spacing resulting in a leakage path and eventual shorting. External causes of high leakage or shorting, such as surface contaminants or the presence of an ionic material on the outer body, should be evaluated before destructive analysis methods are employed.

Parametric Failures. Changes in device capacitance values are often caused by the loss of metal in the electrode area. Metal loss can be due to corrosion, electrical overstress, or manufacturing defects. Electrical overstress damage can result from high levels of AC, or ripple, current in filter applications, which gradually vaporizes the metallization.

The dielectric film thickness, and thus capacitance, may be changed by various mechanisms. Moisture absorption may swell some types of dielectrics, effectively increasing the electrode spacing and thus decreasing capacitance. Excessive heat from internal power dissipation, soldering, cleaning, or the application environment may also melt or deform many dielectrics resulting in capacitance shifts and/or increased leakage current.

Dissipation factor (DF) value and leakage current are affected by moisture absorption by the dielectric. Moisture intrusion in film devices is more damaging than in foil devices owing to the thinness of the metal film relative to the foil in a foil device. Moisture attacks aluminum films readily. Film and foil capacitors may have leakage failures from voltage punch through of the dielectric. The damage will appear in foil capacitors initially as high leakage which will continue to degrade in the presence of a DC bias voltage via metal migration, finally resulting in a short. Similarly damaged components in high impedance AC applications will continue to appear as high leakage failures without degrading further. Film capacitors are also susceptible to voltage punches, however, the thinner metal of the film devices will easily vaporize in the vicinity of the short,

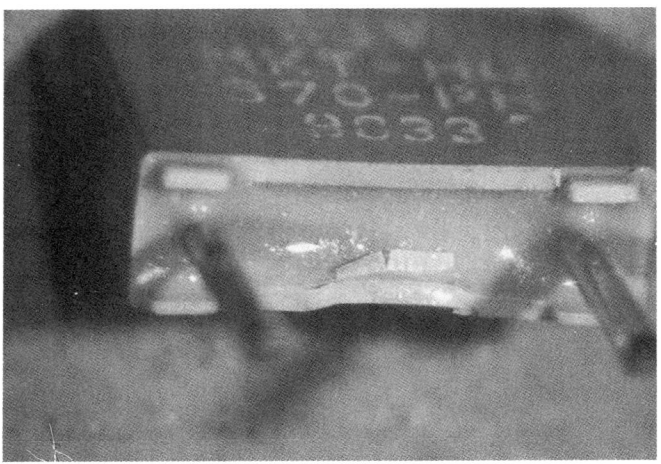

Fig. 25 Wound film capacitor encapsulated in a "box" package

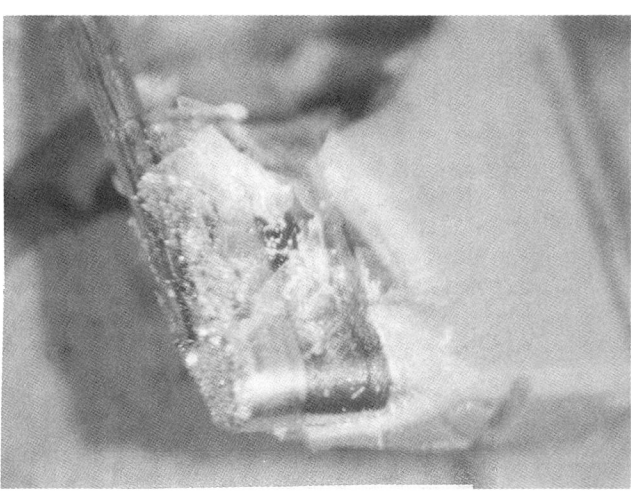

Fig. 26 Decapsulation technique for box packages

Fig. 27 Internal film unwinding

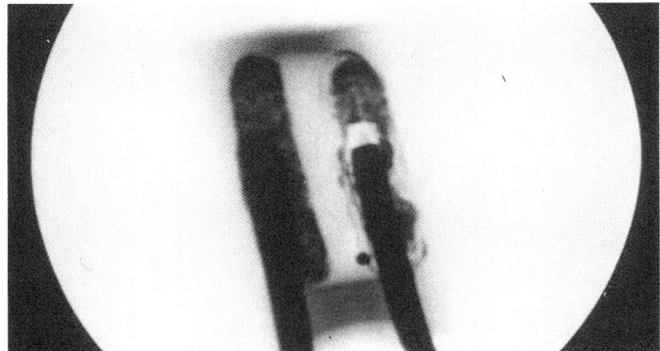

Fig. 28 X-ray showing internal lead detachment after auto insertion

causing the film device to exhibit a self-healing quality. Film devices which have been exposed to multiple voltage punches will eventually exhibit permanent loss of capacitance which can be identified by areas of the metallized film which have been vaporized (see Fig. 25 to 28).

Aluminum Electrolytic Capacitors

Construction. Aluminum electrolytic capacitors are manufactured in two general types: axial, with one lead protruding from each end; and radial, with both leads protruding from one end. Many electrolytic capacitors have "vented" cans: an "X" or "V" is scored in the top surface of the can. This vent will provide a deformation or rupture point in the can to prevent a potentially dangerous gas explosion in the event of a thermal or electrical overstress condition.

The capacitor is formed by winding two sheets of etched aluminum foil, typically 1 to 3 mils (1 mil = 10^{-3} inches) thick. The foils are chemically etched to increase their surface area. This produces a component with a smaller foil and package for a given capacitance. A porous paper separator maintains constant spacing between the foil electrodes. The paper is saturated with a liquid electrolyte to form the basis of charge conduction between the plates. An aluminum oxide (Al_2O_3) layer grown on the surface of the foils acts as the capacitor's dielectric. The two foils and two strips of the separator paper are rolled together into a tight coil. Several windings of plastic tape are attached to terminate the windings. The distance between the plates (foils), the oxide thickness, and the surface area of the plates determine the capacitance of the device while the oxide thickness alone determines the operating and surge voltage rating. The component's other parameters are largely characteristics of the electrolyte, the tightness of the winding, the actual geometry of the foils as wound, and the location of lead attachment.

Axial devices are made with one lead (cathode or negative) internally welded to the end of the aluminum can and the other lead (anode or positive) protruding through a rubber end seal connected internally to the anode foil. The cathode connection is formed by a narrow aluminum ribbon 5 to 10 mils thick mechanically connected to the cathode foil and then welded to the inside of the end of the can. In radial devices, both anode and cathode leads are typically copper clad steel wire with a solder coating, welded to aluminum stems internally. The wires protrude through the silicone or neoprene rubber end seal. The end seal is roll-crimped into the can to seal the liquid electrolyte inside; in both styles. The aluminum stems extend into the windings and are mechanically attached to the foils at approximately one-half to one-third of their length. The aluminum stem is necessary to maintain an internal material compatibility in the presence of the electrolyte and prevent a galvanic interaction.

Opens. Opens may be caused by chemical corrosion of the foils or electrodes. The contamination may be introduced with the electrolyte during the manufacture of the device. The introduction of atmospheric moisture into the can may initiate a catalytic process involving the electrolyte and aluminum, with the water acting as an catalyst. Another source of contamination is that of chemical intrusion through the rubber end seal. These contaminants are usually board cleaning agents such as methylene chloride, Freons, 1,1,1-trichlorethane, and other chlorofluorocarbons.

Shorts. True shorting in aluminum electrolytic capacitors is uncommon, except as the result of mechanically shorted foils. Shorts can be induced by extreme overvoltage which will breakdown the dielectric oxide and arc between the foils. Reverse polarity exposure can have the same effect as extreme overvoltage and is probably more common as a failure mechanism. Most shorts are really extreme manifestations of high leakage. This is prevalent in older devices or in applications where the electrolyte is likely to have dried out. In these cases, the electrolyte is no longer present to rebuild the oxide layer and the device continues to deteriorate.

Parametric Failures. All electrolytic capacitors will degrade parametrically over the life of the device. Typical parameter shifts result in decreased capacitance (>20%), increased DF (>200%), and increased leakage. These shifts are not abnormal over the life of the device and while they may cause an assembly to fail, these shifts are the result of "wearout" mechanisms, not substandard or defective devices. An assignable cause can usually be attributed to premature occurrence of these wearout mechanisms. Equivalent series resistance (ESR) also increases. It is a measure of the impedance and

Fig. 29 Aluminum 'lytic decapsulation—radial leaded part

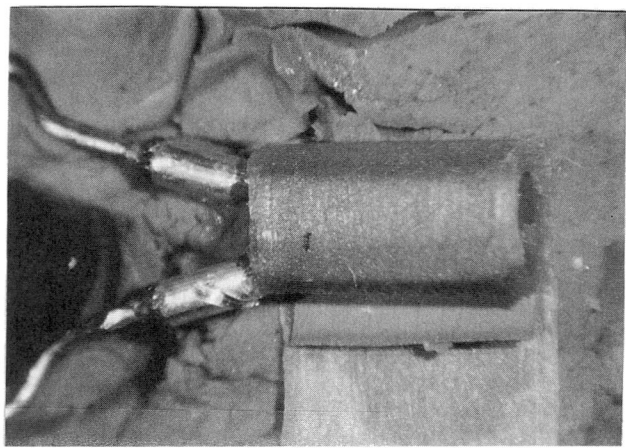

Fig. 30 Aluminum 'lytic foil unwinding—radial

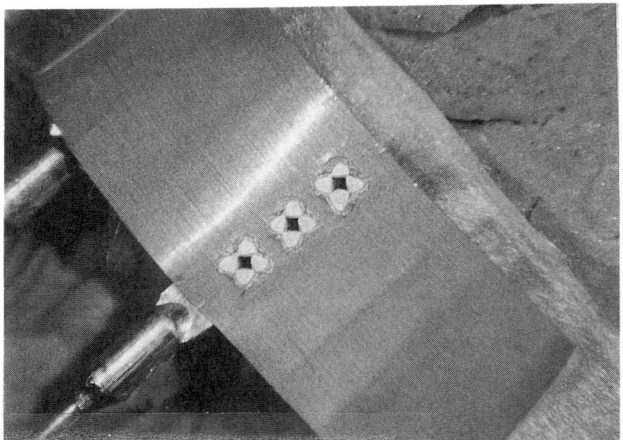

Fig. 31 Aluminum 'lytic lead/foil attachment

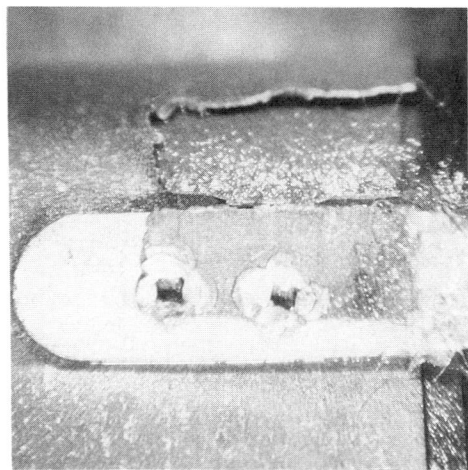

Fig. 32 Aluminum 'lytic foil breakage—supplier assy defect

reactance of the device mathematically derived from the capacitance and the dissipation factor and is a commonly given capacitor parameter.

Several mechanisms can affect capacitance alone or in combination with other parameters. Exposure to a current-limited voltage significantly higher (>40%) than the rated voltage will cause the device to be "reformed" at the higher voltage. The reforming process thickens the oxide on the foils, effectively increasing the distance between plates and reducing the capacitance. This failure mechanism is usually accompanied with slight degradation in DF and a decrease in leakage at the rated voltage. Loosening or stretching of the plastic termination tape will cause the windings to loosen and increase the distance between plates, decreasing the capacitance. The loss of electrolyte through extended operation at high temperature, a leak in the can, heat generated internally by high ripple currents, or by extended service of the component will also cause severe decreases in capacitance.

Increases in DF normally indicate a change has occurred in the amount or chemistry of the electrolyte. The electrolyte is the charge transfer medium whereby the plates are coupled. Any decrease in the charge transfer efficiency, such as exposure to low temperature, will significantly affect the performance of the device. Loss of electrolyte due to aging or exposure to high heat is the most common cause of DF failures. Chemical contamination of the electrolyte by water or halogenated solvents will cause loss of capacitance, and almost always result in a catastrophic failure.

Leakage failures can be of three types. Recoverable leakage failures are those which result from long periods of disuse where the device has had no bias voltage applied. A process called reforming is recommended by the suppliers which rebuilds the aged dielectric. The leakage in this case recovers to initial or better than the initial value and does not diminish the expected operational life of the part. Decreases in leakage are normally not considered failures but may indicate that the part

Failure Modes and Mechanisms

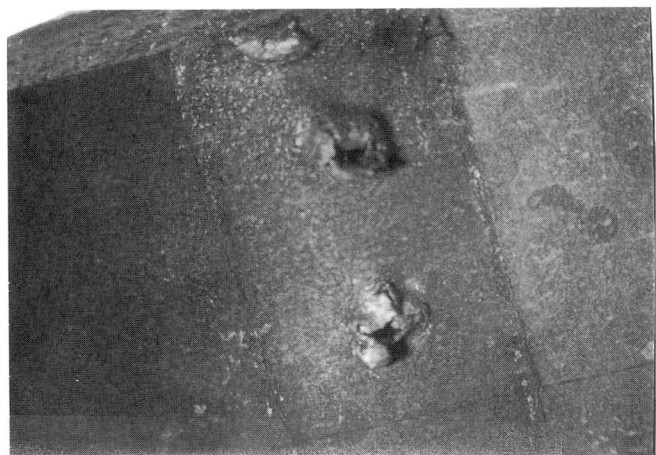

Fig. 33 Aluminum 'lytic poor lead attach—supplier assy defect

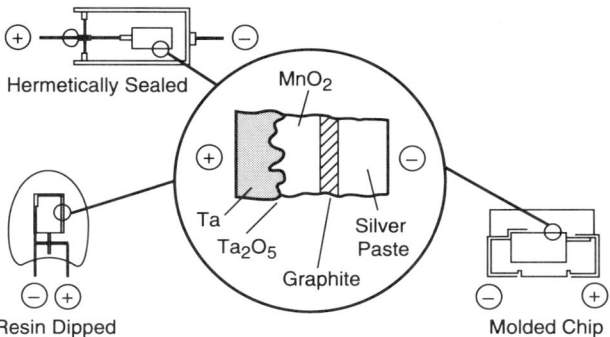

Fig. 34 Structure of solid tantalum capacitors.

Fig. 35 Chip tantalum capacitor with poor cathode lead attachment

was biased at a voltage exceeding the rated voltage. Increases in leakage which do not decrease after forming are due to breakdown of the dielectric oxide. The causes of this are exposure to reverse polarity causing massive oxide damage (either DC or AC, such as in a signal path), or the lack of electrolyte with which to form new oxide (due to extended use or high temperature exposure) (see Fig. 29 to 33).

Tantalum Electrolytic Capacitors

Construction. Tantalum slug capacitors are made by a complex process of sintering and chemical treatments to produce a component with a relatively high value of capacitance per unit volume. Powdered tantalum is pressed into a pellet into which a tantalum wire is embedded. This pellet is then sintered to achieve a robust but very porous slug. Tantalum pentoxide (Ta_2O_5) dielectric is chemically formed on the porous surface. A manganese dioxide (MnO_2) cathode layer then is chemically formed over the slug, producing the cathode layer. The cathode is carbon coated, silver coated, attached to a metal can with a soft solder or conductive epoxy, and encapsulated by a transfer molding process. The anode lead (the tantalum wire) is welded to a copper clad steel wire or copper lead which then protrudes from the package, since tantalum wire is not solderable.

Opens. Because of the robust construction of the pellets, opens can only occur in the internal attachment of the electrodes to the leads. The weld of the tantalum anode wire can break, usually the result of mishandling or poor weld, or the delamination or fatigue of one or more of the cathode layers will result in an open device.

Shorts. Device shorts are more common than opens. Shorts within the pellet are due to a breakdown of the dielectric. The same mechanisms that cause parametric leakage failure can also cause catastrophic leakage failures; or shorts. Common causes of shorts the application of reverse polarity voltage, either by incorrect bias or by application of an AC signal which reverse biases the device by one volt or more. Reverse bias on a device normally produces immediate catastrophic failure. In these cases, the device can reach very high temperatures and ignite, melting its solder joints and burning the components around it. Shorts can also be introduced by packaging. The lead wire welded to the tantalum anode wire sometimes has a "tail" which can short to the cathode. This failure is commonly seen after soldering or thermal cycling, and is readily diagnosed via x-ray.

Parametric Failures. Due to the electrode formation processes, parametric changes in capacitance, DF, and leakage are not common. Parametric shifts are indicative of major internal damage and may be the precursor of catastrophic failure. Internal damage to the oxide after power cycling can occur due to the large differences in thermal coefficients of expansion (TCE) between the sintered tantalum powder, the tantalum pentoxide dielectric, and the manganese dioxide cathode. The differences in TCE of the basic materials of the slug are the basis for supplier cautions regarding bottom side wave soldering or thermal shock cycling. Small cracks in the oxide will also cause increased leakage. This damage is permanent and not self healing as in aluminum electrolytic capacitors. Increased leakage can lead to localized heating where the oxide cracks occur and may lead to eventual catastrophic failure. Increases in DF are usually caused by degradation of the cathode attachment. Delamination of the carbon layer from the MnO_2 or other defects in the cathode electrode system will cause in the real portion of the capacitor impedance to increase. Insufficient solder attachment of the pellet to the can has also been known to cause dissipation factor (DF) failure, especially after thermal cycling. Leakage increases are frequently due to defects in the oxide layer. These defects may be induced by mechanical deformation, excessive temperature, or reverse polarity. As alluded to earlier, leakage increases are indications of internal damage and may lead to eventual catastrophic failure (see Fig. 34 and 35).

CRYSTALS

Quartz crystal oscillators, or simply "crystals," produce the stable frequency signals needed for the correct functioning of clocked circuits, such as microprocessors. While integrated components containing the crystal oscillator as well as other components, functioning as drive circuits, are available, the discussion here will be confined to the crystal oscillator itself.

Construction. The piezoelectric effect forms the basis of operation of the quartz crystal oscillator. Piezoelectricity is the tendency of many substances, especially crystalline materials, to mechanically distort when subjected to an electric field. Conversely, these materials also produce a voltage across their surfaces when mechanically distorted. When placed in an amplifier feedback circuit which is arranged (tuned) to have no gain except when the crystal produces a phase shift, the crystal will mechanically oscillate, and the circuit will produce an extremely stable AC signal. The shape, size, and other physical properties of the crystal material determine which mechanical oscillation modes are the most efficient, and require the least input signal. These are the resonant frequencies of the crystal oscillator. For example, a circular drum surface's "fundamental" frequency occurs when the middle area of the surface oscillates as a whole in and out of the drum. The "first overtone" or "first harmonic," a weaker resonant frequency, occurs when half of the drum's surface moves in while half moves out. Similarly, a crystal oscillator will have a fundamental frequency, as well as many overtones which occur at higher frequencies. It is important to note that small changes in the mass or dimensions of the crystal element will cause a significant change in the frequency of its oscillation.

Although many materials are piezoelectric, the oscillating element in modern circuits is commonly synthetic or natural quartz, or crystalline silicon dioxide (SiO_2). The crystal element's dimensions, shape, and the amount of surface metallization added during manufacture determine its frequency. Both round disk and rectangular strip element shapes are common, while other oscillator designs utilize a crystal element in the shape of tuning fork. The element metallization functions as electrodes, and is sputtered or evaporated onto the crystal. The metallization is typically silver, though chromium, aluminum, or gold (for high stability applications) can also be used. The element is supported at each side by a post (commonly nickel plated steel), and is frequently cemented to the post using a silver-filled epoxy or polyamide. The supporting posts may extend through the base of the device to act as the external leads or they may be welded internally to the leads. The leads protrude through a Kovar base, isolated by glass seal. Kovar is chosen because its thermal coefficient of expansion (TCE) is close to that of glass. A steel shield, or can, is welded or soldered to the base after filling the compartment with an inert gas (commonly 90% nitrogen/10% helium) or a vacuum. The external leads are plated with tin/lead solder, tin, or silver for solderability.

Opens. By far the most common cause of quartz crystal oscillator opens is fracturing of the quartz element. The brittle quartz may be fractured during component handling or board insertion or damaged during the element manufacturing process. Component handling damage frequently originates at the point furthest from the supports, where the element may impact the inside of the shield. Assembly damage commonly originates where the element and post meet. The edge of the quartz element, like a semiconductor die, must be free of nicks or other damage or it will be susceptible to damage at levels of acceleration or shock stresses which would not affect a good device. The orientation of the element within the shield must be controlled carefully to prevent contact between the two, which will result in fracturing or damping of the crystal's oscillation.

Both construction defects and mechanical abuse to other structures in the package can result in open components. Exposure to excessive vibration or shock, insufficient or uncured cement, or poor post design may allow the element to separate from the post or cause the post itself to fracture. Similarly, excessive vibration, poor post-lead weld, or excessive lead forming or insertion force can separate the post to lead bond.

Several mechanisms can separate the electrode metallization from the element, resulting in an open device. Contamination of the crystal element or metal source can result in poor adhesion and delamination of the metallization, as can operation of the crystal at drive power levels in excess of its specification. Scratching of the metallization during the frequency adjustment or device assembly operations will have the same effect.

Shorts. Mechanical damage can result from internal contact between the crystal support post, lead, and shield and result in shorting or failure to oscillate. This damage may be the result of exposure to excessive force or shock during handling or use, poor assembly practices, or internal post to lead weld failure.

Parametric Failures. Crystal frequency and starting resistance are the two parameters most often affected by damage or defects. The starting resistance represents the device's resistance to beginning to oscillate at low levels of drive power, normally only milliwatts. Many of mechanical damage mechanisms described above cause changes in frequency or starting resistance in addition to shorts or opens. These include fracturing of the element during crystal assembly or handling damage and element contact with the interior of the shield.

The mass of the crystal element also affects its oscillation frequency. Contamination on the element, post, base, or shield or from contaminated sealing gas can result in the deposition of films or particles on the element while will change its frequency. Similarly, a poor shield to base weld or fracturing of the lead seal glass by component mishandling can admit moisture or contaminants which will also change the frequency. Operation of the crystal in excess of its rated maximum temperature can cause outgassing of the cement which has the same effect as the introduction of external contamination. Drive power levels far in excess of the crystal's recommended levels, typically 0.5 milliwatt maximum, can cause partial or complete delamination of the metallization from the element. The starting resistance, which relies on coupling of the metal electrodes through the piezoelectric quartz, will increase with poor alignment of the electrode areas relative to each other.

Temperature stability of the crystal's electrical parameters across its operating temperature range is often crucial. At some narrow temperature ranges, however, a crystal's resistance may increase significantly with a small change in temperature. Such behavior is referred to as an activity dip, and is the result of the crystal design, assembly processes, or contamination. Fracturing of the element, excessive compressive or tensile stress on the quartz element from the mounting posts, and moisture or other contamination of the sealing gas can cause activity dips. Rectangular strip-type crystals are especially susceptible to activity dips depending on the length-to-width-to-thickness ratios of the element. Electrical testing of crystals across the full application temperature range at 1 °C increments is needed to identify activity dips (see Fig. 36 to 39).

INDUCTIVE COMPONENTS

Construction. Inductive components are commonly constructed of soft annealed copper wire wound around a magnetic or non-magnetic core. Some applications such as RF tuned circuits utilize "air core" coils. These may also be wound on plastic bobbins into which a ferrite slug is threaded for tuning purposes. Magnetic cores are normally of a sintered iron powder or iron ceramic; a ferrite. The wire usually has a thin polyurethane or polyester enamel insulation layer and is soldered, welded, or crimped to the external terminations of the component. Depending on the application, this assembly may then be coated by a silicone or epoxy encapsulant.

Recently, some manufacturers have introduced chip-type inductors and transformers, utilizing thick-film screening technology to place alternating layers of ferrite and conductor paste on an insulating ceramic substrate. These devices have failure modes and mechanisms similar to chip resistors and capacitors, and will therefore not be discussed separately.

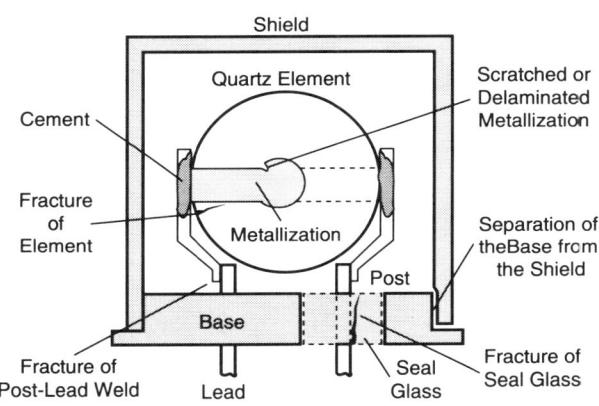

Fig. 36 Common failure mechanisms of quartz crystal oscillators

Fig. 37 Crystal metallization defects

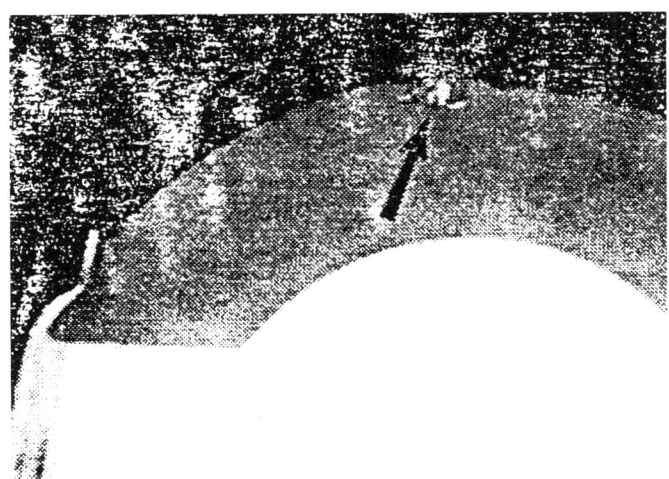

Fig. 38 Crystal edge finish defect—chip

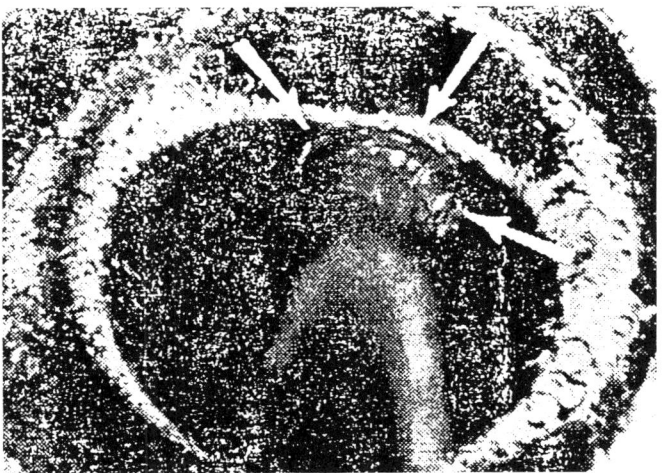

Fig. 39 Crystal glass feed through defect

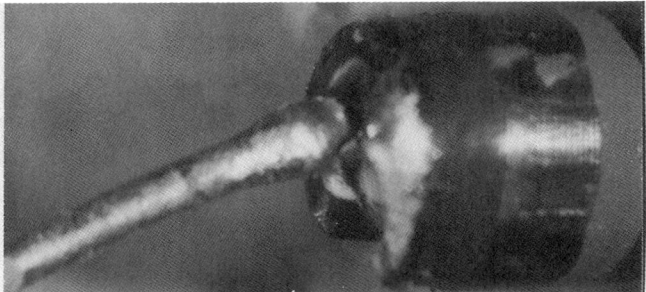

Fig. 40 Ferrite core inductor with lead breakout

Opens. The main mechanisms causing open inductive components are damage to the copper winding and its connections to the component terminations. The copper wire is itself susceptible to corrosion, mechanical damage, or in extreme cases, electrical overstress damage.

The wire insulation serves to protect the copper wire from environmental effects as well as insulate between adjacent wire turns. Failure of the insulation to keep out sulfur and chlorine-containing compounds will result in corrosion of the copper wire, forming compounds that can reduce the wire's effective diameter and thus increase its resistance. Sulfur and chlorine-containing materials are commonly used in solder fluxes and board cleaning agents, and must be thoroughly removed from the component to prevent corrosion failures.

Device assembly processes can also contribute to insulation failures. The technology used to wind the wire around the cores is frequently unsophisticated and not under statistical process controls (SPC), or preventative maintenance programs. Wear or damage to winding mandrels may result in nicking of the insulation, or of the wire itself. Similarly, poor process control during application of the insulation material may result in voids or thin areas. Visual inspection of the wire after unwinding will often result in the discovery of the defect source.

The solder joint, weld, or crimp which joins the wire to the external termination can be separated by mechanical abuse or poor control of component assembly processes. Poorly controlled solder processes can result in the fracture of the wire-to-termination joint due to a "cold solder joint" if the termination is not heated sufficiently to truly wet with solder. Similarly, insufficient or excessive weld energy may result in weld fractures or partial or complete separation of the wire itself at the weld site. Mechanical crimp contacts are susceptible to the separation for the same mechanisms. All these connection techniques may also be adversely affected by mechanical damage during component encapsulation, board assembly, or field use.

Inductive components such as transformers frequently have the highest mass and size of all components on a circuit board. This can result in solder process defects due to the large mass which must be heated, flexing damage to the underlying substrate due to the large mass which must be supported, as well as stresses to the terminations and board from vibrations and accelerations experienced by the application. Additional mechanical supports or attachments often address these issues successfully.

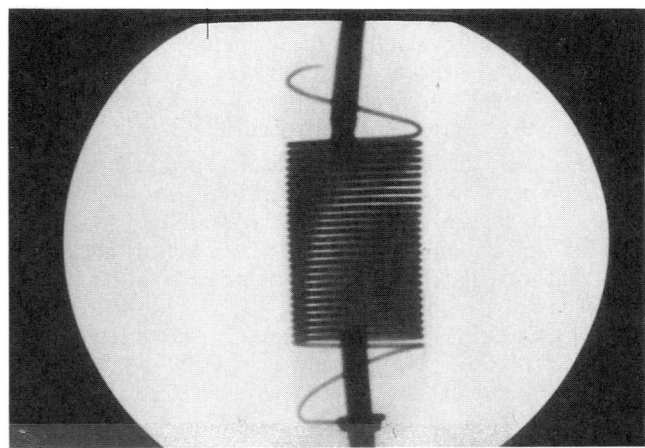

Fig. 41 X-ray showing a winding termination defect

Fig. 42 Ceramic core surface mount chip inductor

Shorts. Scratched or worn insulation can allow contact between adjacent turns or the wire and its metal shield, resulting in changes in the component's inductance and resistance. The leakage path resulting from these sources may cause localized heating of the component, resulting in catastrophic short or open failures during field service. Worn insulation can result from loose windings which abrade either from mechanical vibration or due to physical movement from electric field fluctuations.

Parametric Failures. The component encapsulation application process may cause failures if it disturbs the wire windings or the connections to the terminations. This may result in decreases in the resistance and inductance values. This mechanism is analogous to "bondwire sweep" in plastic encapsulated integrated circuit components. Failure mechanisms in large or power handling inductive components are similar to those of the signal or RF variety with the addition of several core failure mechanisms. Separation in the laminations of steel cores will cause the device to run excessively hot due to the degraded efficiency of the magnetic coupling, and can also produce noise and hum from lamination vibrations. Likewise, any effect which causes the laminations (magnetic path) to deteriorate will have the same effect. This includes formation of rust between the laminations, as the formation of iron oxide forces the laminations apart. This can stretch the windings and cause wire breakage. Laminated steel cores are varnish coated for two purposes, to help maintain the lamination integrity and to prohibit the intrusion of moisture between the windings. Ferrite cores can crack causing loss of efficiency and greatly affect the efficiency of the device. Cracked cores can be the result of thermal stress, mechanical damage, or windings wound too tightly. The expansion of large diameter wires with high temperatures over a fragile or poorly made core can cause core fracture (see Fig. 40 to 42).

FAILURE ANALYSIS PROCEDURES AND TECHNIQUES

The failure analysis of non-semiconductor electronic components depends strongly on an understanding of the context and construction of the failing components. We have chosen to include a discussion of one possible failure analysis methodology that will help ensure that a careful analysis will be completed.

Background of the Failure

A thorough understanding of the background conditions of a failure are frequently crucial to accurately and efficiently conducting the failure analysis of a non-semiconductor component. Some guidelines are given below.

Test the part (determining what failed). Remember that the electrical conditions and component symptoms should suggest the nature of the component defect or damage.

Clarify the reported failure symptoms. How did the component act in the circuit, and how was the determination made that this was the defective component? Was the failure confirmed "in-circuit" or after removal? If the failing component was placed in a functioning circuit, did the failure follow the part?

Identify which parameters to test. Those parameters relating to the circuit performance should be closely scrutinized. Could marginal parametric behavior of other circuit elements have adversely affected the performance of the failed part?

Determine which test limits should be used. Test limits given by the supplier are gross or maximum/minimum values at a set of conditions reflecting the maximum ratings of the device. These normally do not have a direct correlation to the conditions seen by the device in the application. Compare the parameters of the failed device to a known "good" device at test conditions approximating levels observed in the application.

Relate the defect to the failure. What component parameters could account for the circuit symptoms? The parameter found defective should account for the symptom seen in the circuit. If it does not, you may have not found the cause of the failure. Defects seen must pass the "common sense" test for relationship to the failure symptom. The cause of a defect which is not related to the symptom may be faulty analysis technique or the part may have been damaged when removed from the circuit.

Understand the construction of the part. Document the construction and materials used in the fabrication of the device. Gain an understanding of the methods and practices used to construct the part, and the ways in which the part can fail. Determine construction practices which may relate to the failure.

Determine the conditions at the time of failure. Were there abnormal line conditions (e.g. a large inductive line load starting or stopping, power line surges) which might coincide with the time of failure? Was an electrical storm in the area at the time (i.e. lightning, line surges, or spikes), especially for systems connected to telephone lines? What was the temperature, humidity, and atmospheric pressure? Is the operating atmosphere corrosive, dusty, salty, industrial (i.e. air pollution)? Was this part thermal cycled, part of a life test, or an electrical stress test?

Identify any other defective components. Could damage or parametric variation of other components in the circuit cause the failure of the device in question? Could a unique circuit condition or failure of other components in the circuit cause the component in question to become bad?

Determine the assembly process history of the application. What kind of flux was used, how much, and how was it applied? How hot and how long was the preheat/solder profile? Are there other processing steps in which heat or solvents are used (e.g. chip component glue cure)? Was it automatically or manually inserted and what were the conditions at the insertion station? Is there a lead forming operation associated with the part, and if so, how well was the machine controlled? Is this a hand assembled or soldered part? Where are the test points in the process and what level of test is performed at each?

Identify how the component is used in the circuit. What is it that the part is being expected to do? Is the component capable of doing what's expected of it? Does this appear to be the right component for the application?

Failure Analysis Techniques

Electrical test. Tools of particular usefulness in analyzing non-semiconductor electronic components include digital volt/ohmmeters for gross defect determination, curve tracers for resistance and capacitance characteristics and intermittent conditions, LCR meters or other appropriate test equipment for parametric evaluation, and network analyzers for RF components.

Investigation/destructive analysis. Some helpful techniques used in the failure analysis of non-semiconductor components include radiography, visual inspection via low and high power optical microscope, hand tools for mechanical decapsulation, metallographic sample preparation equipment for grinding and polishing, a low speed saw, and chemical decapsulation equipment.

Failure Analysis Tips

If possible, determine if the part is sensitive to finger pressure, temperature, or mechanical flexure. The behavior of the part in the circuit under these conditions may guide you to a probable failure mechanism. Intermittent behavior or temperature sensitivity indicates possible lead or element connection defects.

Intermittent behavior can be caused by fractured solder joints, cold solder, or a fracture in the printed circuit board plating at the barrel. All of these failures will be "fixed" when the part is changed. Handling that can aggravate the condition or cause internal damage will lead to a faulty root cause conclusion.

It is sometimes preferable to have the complete application assembly submitted for analysis so that the device may be studied before removal from the application substrate. *If the part is carelessly removed from the circuit, it may not be possible to distinguish between damage to the part caused by removal and the true cause of failure.* Avoid exposing the component to excess heat during desoldering. Some indications of unacceptable removal practices include:

- Mechanical damage to the leads or detached leads.
- Mechanically deformed or scarred part bodies.
- Missing leads from components, unless accompanied by information indicating the lead attachment was weak.
- Leads which have the barrels of plated-through holes still attached indicate that excess force was used to remove the part.

To verify that the part is indeed defective, it should be inserted into a functioning circuit. If the part is defective, the failure should follow the part.

If the device has a cavity, it may be advisable to fill the cavity with epoxy or potting material to hold all internal components in their relative positions for sectioning or decapsulation.

Analysis and Documentation

Investigation. Observe and record or photograph all anomalies such as discolored areas, deformed members, burn or arc marks, etc. Your root cause determination must account for all observed deviations from the norm. If several scenarios will account for the failure artifacts observed, the least complex, most straight-forward explanation is the most likely cause.

Reporting your findings. The report should provide a brief historical account of what happened. Describe how the failure manifested itself and whether this was an isolated event. Summarize your findings based on your observations and any inputs you may have had from suppliers or technical resources such as reference literature. Draw conclusions, if appropriate, based on your findings. The conclusions must account for all artifacts found in your investigation. Your conclusion may be that no conclusion could be reached based on the evidence. Make recommendations to alleviate future failures. Include reliability assessment information if warranted for units which may have escaped to the field.

The level of technical language and concepts should be geared for the intended audience. If necessary, write the report body in technical terms and attach an "Executive Summary" to the front summarizing the "bottom line" results in less technical language.

REFERENCES

Though not as common as semiconductor failure analysis guides, there are several good component and failure analysis references which discuss the failure modes and mechanisms non-semiconductor components. These include:

1. *Failure Analysis, Mechanisms, Techniques, and Photo Atlas* by J. R. Devaney, G. L. Hill, and R.G. Seippel (Failure Recognition Services, Inc., 1983)
2. *Electronic Materials Handbook, Volume 1: Packaging* (ASM International, 1989.

Another good source of information about non-semiconductor components are the applications guides and bulletins published by the component suppliers themselves. These guides are especially helpful when working with surface mount components, which are frequently subjected to stresses far in excess of those experienced by comparable through-hole devices.

LIMITING PHENOMENA IN POWER TRANSISTORS AND THE INTERPRETATION OF EOS DAMAGE

by J. Thomas May

INTRODUCTION

This article presents power transistor failure theory from the perspective of the failure analyst. Sufficient theory is covered to explain why different categories of EOS result in disparate damage to the die. The phenomena that limit the capabilities of transistors are thoroughly discussed. These phenomena include: debiasing due to the resistance of the emitter finger metallization and the resistance of the base region; thermal biasing; the spread of the conduction process associated with turn-on and turn-off; and base widening. These phenomena, acting alone or in combination, result in hot spot formation when the transistor is operated beyond its rated capability. It is explained that, depending on the overstress category, the location of the failure site may be at particular locations within the emitter diffusion or at the outside edge of the die. The significance of the melt-thru being under the emitter bond or at the end of an emitter finger is noted. Melt-thru size and bond wire damage are also discussed.

Power transistors are used in most electronic systems. While excellent reliability is the norm, it is not uncommon for a rash of failures to occur due to misapplication, especially during the first production run of a new design. A common cause of such failures is electrical overstress (EOS). In order to determine the fundamental cause of the overstress, circuit designers need to know the exact failure mode and mechanism. Because of the wide variation of transistor designs from different vendors, and because of the limited information that has been published on the subject of interpretation of EOS damage, the failure analysis often leads to ambiguous conclusions regarding the precise nature of the overstress. The purpose of this paper is to explain the physics of EOS failure in power transistors as well as how to interpret the damage on the die to determine the EOS category. Failure resulting from high energy transients, such as lightning and other gross abuse, are not discussed in detail.

LIMITING PHENOMENA IN POWER TRANSISTORS

The high current and high voltage operation of power transistors result in several limiting phenomena that are unimportant in the operation of small-signal transistors [Ref 1,2,3]. These phenomena include (1) resistive debiasing due to both the lateral resistance of the emitter fingers and the lateral resistance of the base region, (2) thermal biasing, (3) the spread of the conduction process associated with turn-on and turn-off, and (4) base widening. These phenomena, acting alone or in combination, cause the formation of hot spots on the die when the transistor is operated beyond its rated capability.

Resistive Debiasing. Resistive debiasing occurs during the normal operation of the transistor and is the result of small IR drops in the semiconductor and metallization, thereby reducing the forward-bias across the base-emitter junction. A voltage drop of $\ln(2)*k*T/q$ (approximately 18 mV) will result in a 50% reduction in current density. Some of this voltage drop results from the emitter current flowing through the emitter metallization, some results from the base current flowing through the base fingers, and some results from the base current flowing through the active base layer. The effect of resistive debiasing is graphically depicted in Fig. 1. Figure 1(a) shows the effect of the lateral resistance in the emitter fingers and Fig. 1(b) shows the effect of the lateral resistance in the active base region (the base spreading resistance). In this figure, as well as in all illustrations in this paper, the npn transistor structure is shown. The resistance of the active base layer is strongly effected by the magnitude of the reverse voltage bias across base-collector junction. Increasing this reverse-bias causes the base-collector depletion region to expand into the active base. This effect pinches and thins the base and increases the effect of resistive debiasing. Figure 2 illustrates base pinching and shows why excessive current density occurs along the edge of the emitter diffusion. Power is thus dissipated primarily at the edges of the emitter diffusion.

The consequences of resistive debiasing can be summarized as follows: The lateral flow of emitter current through the emitter fingers has the effect of debiasing the ends of the emitter fingers. This effect is negligible for small emitter currents but more significant for large currents. The lateral flow of base current through the base fingers has the effect of debiasing the sections of the base-emitter junction located at the ends of the base

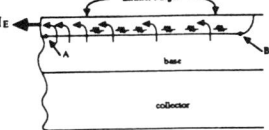

a) Effects of Large Currents in the Emitter Fingers. The base-emitter junction is more forward-biased and the current density is greater at point A (near the emitter bond pad) than at point B (the end of the emitter finger).

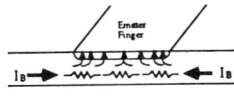

b) Effects of Large Currents in the Active Base Layer. The current density is greater along the edges of the emitter fingers because the base-emitter junction is more positively biased.

Fig. 1 Cross-sectional drawings showing the mechanism of resistive debiasing.

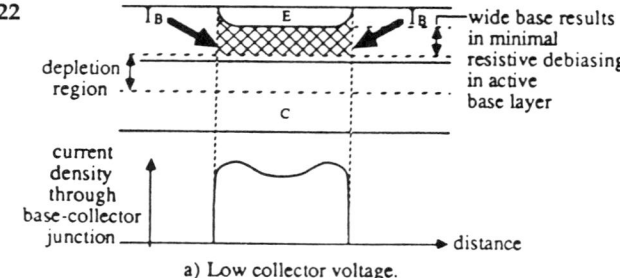

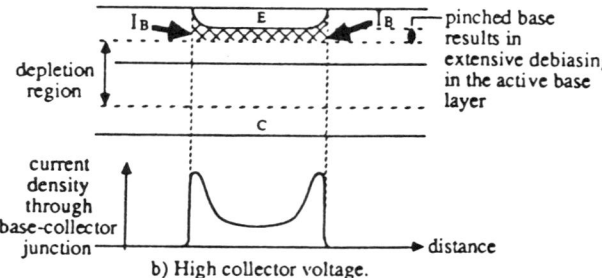

Fig. 2 Cross-sectional drawings of transistor, showing effects of base pinching.

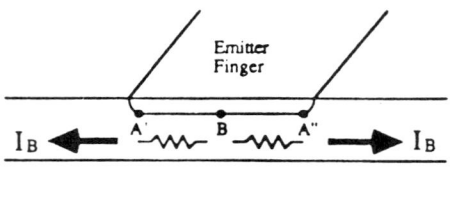

Point B is more forward biased than points A' and A".

Fig. 3 Resistive over-biasing during turn-off.

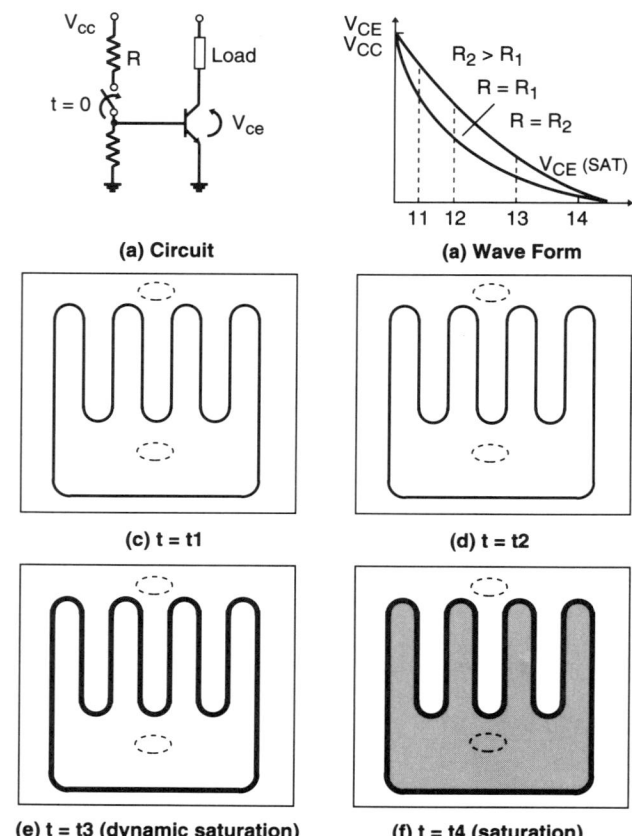

Fig. 4 Drawings of a representative transistor die, indicating conditions during various stages of turn-off.

fingers and farthest from the base bond. Current flowing through the active base layer debiases the middle of the emitter geometry. The degree of debiasing increases with the lateral distance from the edge of the base-emitter junction. The debiasing effect is less for small base currents and low collector-emitter voltages than for large base currents and high collector-emitter voltages.

Resistive Over-biasing. During turn-off, the base current flows in the reverse direction until the stored base charge is removed. The IR drop in the active base layer caused by this reverse base current causes over-biasing in the middle of the emitter. This results in less efficient removal of base charge in the middle of the emitter than at the edges [Ref 4]. Hence, the most remote interior part of the base-emitter junction turns off last. This phenomenon is shown in Fig. 3.

Thermal Biasing. Thermal biasing arises because a hot area of a forward-biased pn junction conducts more current than a cooler area. For a constant forward voltage bias across a typical pn junction at room temperature, the current will approximately double for a 9 °C increase in temperature [Ref 5]. The increased conduction causes the hot spot to become even hotter. This induces even more conduction, and thermal runaway will occur in the absence of other stabilizing effects. Note that the temperature coefficient of the resistivity of silicon has a stabilizing effect for temperatures less than the intrinsic temperature and has a destabilizing effect for temperatures greater than the intrinsic temperature. Refer to the article "Thermomechanical Effects of Overstress" for explanatory graphs. In the former case, the degree of resistive debiasing increases with temperature and in the latter case, the degree of resistive debiasing decreases with temperature. The effect of thermal biasing can be minimized by a good collector heat sink and a good emitter heat coupler (such as a thick solder layer with a heavy metal clip) to force the temperature of all areas of the emitter to be equal.

Thermal biasing, if not counterbalanced by other effects, will destroy the transistor. The consequences of thermal biasing are most pronounced when the transistor is operating near or beyond its maximum rated power rating as the heat sink and heat coupler become unable to maintain sufficiently small lateral temperature gradients.

Turn-on. The turn-off process is illustrated in Fig. 5. When a transistor is switched on (its base-emitter junction switched from zero or reverse-bias to forward-bias), the entire base-emitter junction does not turn on instantaneously and uni-

Limiting Phenomena in Power Transistors

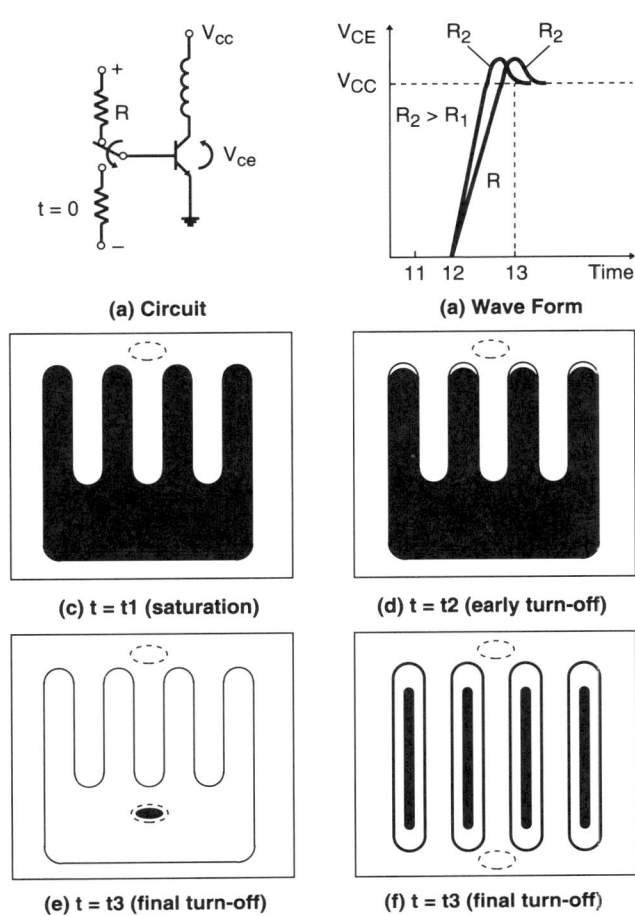

Fig. 5 Drawings of representative transistor die, showing conditions during various stages of turn-off.

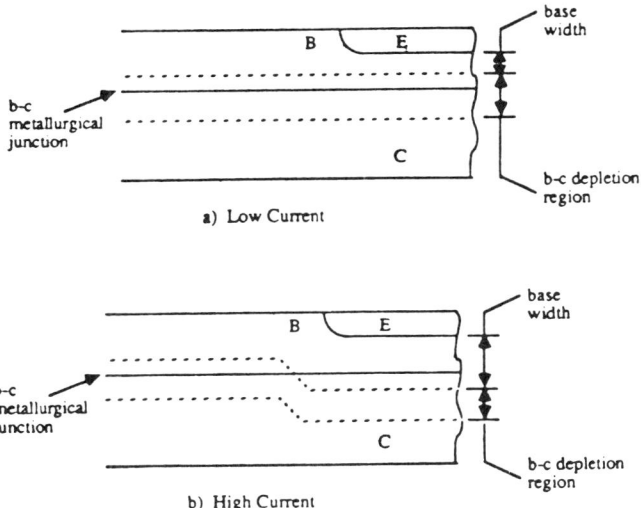

Fig. 6 Base widening.

formly, refer to Fig. 4a and 4b.. This is due to both resistive debiasing in the active base layer and time delays caused by the combination of the distributed base-emitter capacitance, and the base layer resistivity.

The first parts of the transistor to turn on, and the parts most vulnerable to failure, are the ends of the emitter fingers that are physically nearest to the base bond pad. Next, the edges of the emitter turn on (Fig. 4d). The conduction process then spreads toward the middle of the emitter fingers (Fig. 4e). If the transistor is constructed so that the emitter diffusion lies under the bond pad, then the area under the bond pad will be the last to turn on. The turn-on process is illustrated in Fig. 4. Note that in this figure, as well as in all illustrations in this paper, the base and emitter bond areas are shown as dashed circles in the sketches of the transistor die. Non-instantaneous turn-on can result in excessive power density at the ends of the emitter fingers. This effect is especially troublesome when the transistor is switching a high voltage, capacitive load. Under these conditions, the active base layer is pinched and base spreading resistance is a maximum. This larger resistance results not only in a longer turn-on delay but also in more resistive debiasing which prevents the middle emitter geometry from conducting. Hence,

almost all the power can be dissipated in the tips of the emitter fingers. This effect is aggravated by inadequate base drive during turn-on.

IMPORTANT

Notice that in 5e and 5f, two different die geometries are used; 5e has all fingers connected together in a "hand" pattern, while 5f has five separate emitter diffusions connected together dy the metallization. These drawings represent the two major device types. Notice that the drawings in Fig. 9 are paired together in the same manner.

Turn-off. When a transistor is switched off (Fig 5a and 5b) (its base-emitter junction switched from forward-bias to zero or reverse-bias), the entire base-emitter junction does not turn-off instantaneously. This phenomenon is caused by time delays resulting from the distributed base-emitter capacitance and the base layer resistivity and by the resistive over-biasing. The reverse current through the base metallization and the base layers debiases the base-emitter junction nearest the base bond pad and over-biases the junction farthest from the bond base bond pad. Hence the first parts of the transistor to turn off are the ends of the emitter fingers that are physically nearest to the base bond pad. The turn-off process then spreads to the edges of the emitter fingers and from there to the middle of the emitter geometry. Hence, the region of the emitter most remote from the base-emitter junction and the base bond pad is the last to turn off and most susceptible to excessive power dissipation. These limitations are most troublesome when an inductive load is being turned off when the base has been over-driven prior to switching.

Base Widening. Base widening is illustrated in Fig. 6. Base widening is a high current effect which is usually insignificant in power transistors. The base width grows wider as the collector-emitter current increases because the base-collector depletion region moves towards the collector [1,6]. This movement of the base collector depletion region is a consequence of space charge neutrality: The minority carriers in the base are the same charge as the ionized impurity atoms in the base (i.e., negative

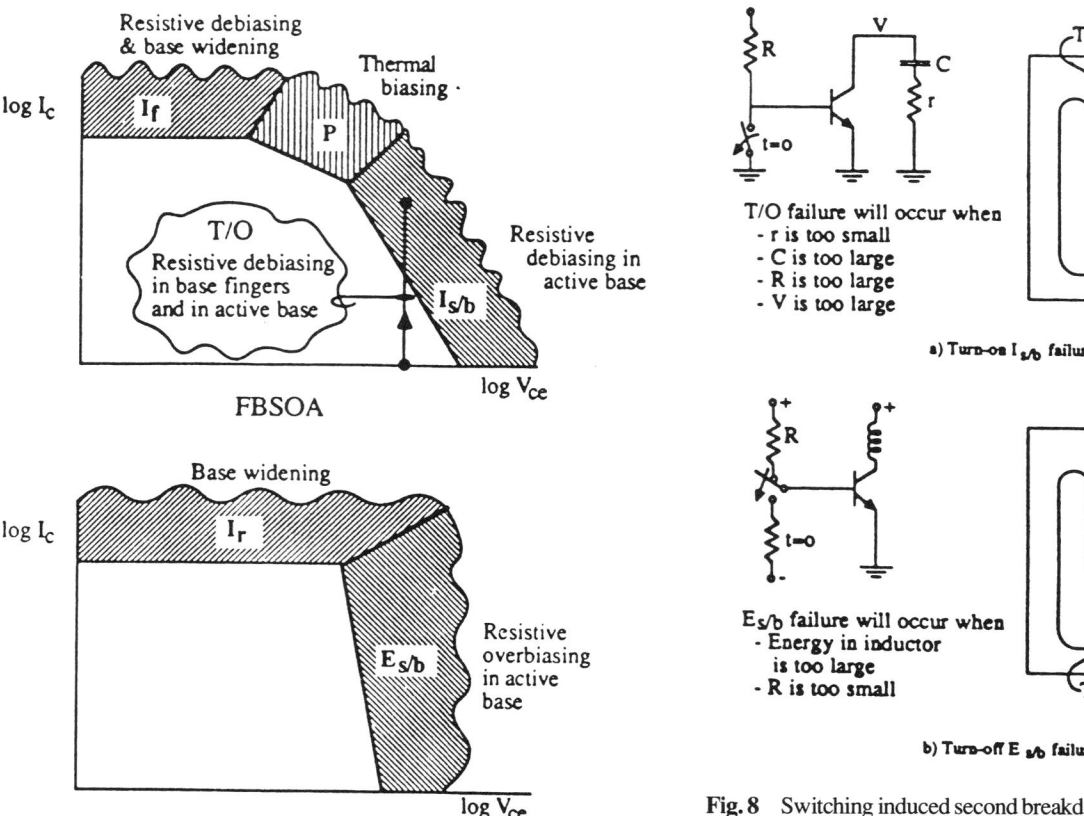

Fig. 7 EOS categories defined by SOA curves.

Fig. 8 Switching induced second breakdown.

for a npn transistor and positive for a pnp transistor). When the density of minority carriers approaches the density of impurity atoms, the base-collector depletion region does not need to reach as far into the base to uncover the required amount of charge. Hence, part of the base-collector depletion region at low current becomes the active base at high currents.

If base widening occurs, the transistor's beta decreases and the stored charge increases. In switching applications, this causes an increase in turn-on time, which increases power dissipation. Base widening thereby compounds the effect of excessive current density and limits the capabilities of the transistor.

EOS CATEGORIES

Transistor EOS can be categorized in terms of the Safe Operating Area (SOA) curves. The familiar FBSOA is shown in Fig. 7(a). Transistor manufacturer's specification sheets often show multiple FBSOA curves. The d-c SOA curve is always shown, and curves for various combinations of pulse width and duty cycle are usually shown. Safe operation can be ensured only if the transistor is operated inside this curve. The region denoted "I_f" represents the area beyond the maximum current limit. Similarly, "P" denotes to the maximum power limit and "$I_{s/b}$" denotes the second breakdown limit. Turn-on failures are a special subcategory of "$I_{s/b}$" failure and are described in Fig. 7b and are denoted by "I_r" and "$E_{s/b}$". $E_{s/b}$ failure results from switching an inductive load as shown in Fig. 8b. Such failures will be denoted by "T/O" and represent a sudden increase in collector current at a high collector voltage as indicated by path 1-2 in Fig. 7(a). Similarly, the failure regions for the RBSOA are shown in Fig. 8(b). ("$E_{s/b}$" refers to the parameter which specifies the maximum inductive energy that the transistor can safely switch.) These six EOS categories will be used in the following discussions. Note that the terms "forward-bias" and "reverse-bias" refer to the bias of the base-emitter junction. Two additional EOS categories are V_{ces} and V_{ceo} failures which occur when excessive voltage (possibly a voltage spike) is applied to the emitter and the base terminal opened, respectively.

The effects of the limiting phenomena for the various EOS categories are summarized in Table 1.

HOT SPOT FORMATION

EOS failures usually involve a portion of the die reaching a critical temperature at which irreversible damage occurs. This critical temperature might be the melting point of silicon or the silicon/metallization eutectic temperature. The high temperature region of the die is usually quite localized with damage typically confined to a single, small melt-thru on the die. In other situations, the damage area may be larger but confined to the area around the emitter bond or to a single emitter finger. An explanation for this localized damage will now be explained.

Localized damage is caused, in part, by the fact that the nor-

Table 1 EOS Category I/S Limiting Phenomenon Chart

EOS category/limiting phenomena	Resistive debiasing due to lateral IR drop in emitter finger	Resistive debiasing due to lateral IR drop in active base	Resistive debiasing due to lateral IR drop in base fingers	Thermal biasing	Non-instantaneous turn-on	Non-instantaneous turn-off and resistive over-biasing due to IR drop in base	Base widening	Avalanche breakdown of weakest point of base-collector junction
I_f	3	2	3	1	1	1	3	1
P	2	2	2	3	2	1	2	1
$I_{s/b}$	1	3	1	1	3	1	1	1
T/O	1	3	3	1	3	1	1	1
I_r	2	2	2	1	1	2	3	1
$E_{s/b}$	1	1	1	1	1	3	2	1
V_{ces}	1	3	1	1	1	1	1	3
V_{ceo}	3	3	1	1	2	1	1	1

Note: 1= Insignificant effect; 2 = Moderate effect, and; Signficant effect.

mally positive temperature coefficient of silicon resistivity becomes negative at the so-called "intrinsic" temperature. The intrinsic temperature is defined as the temperature where the thermally generated carrier concentration equals the doping concentration. When the temperature of a hot spot increases to the point where the intrinsic concentration equals the semiconductor doping concentration, the silicon resistivity develops a negative temperature coefficient which causes the hot spot current density to further increase. Phenomena that can initiate hot spot formation include non-instantaneous turn-on and turn-off, localized base widening [Ref 7], and thermal and resistive debiasing. Thus, the total current will be diverted from cooler parts of the die to the hottest spot on the die. Further aggravating the situation, the thermal conductivity of silicon decreases with increasing temperature, which further increases the temperature of the hot spot. This phenomenon is sometimes called "current hogging," or filamentary conduction. The hot spot that is formed is sometimes called a mesoplasma. (High temperature regions in semiconductors that are flooded with holes and electrons are called plasmas.) Unless counterbalanced by some stabilizing effect (such as intentional resistive debiasing from emitter ballast resistors), thermal runaway will occur and a melt-thru will form. The melt-thru will usually contain significant concentrations of metal and short the emitter to the collector with an effective resistance of one ohm or less. The mechanical stresses induced by the melt-thru will often cause the die to crack and micro-fractures can often be observed radiating from the failure site. If the current is limited by the external circuitry, it will flow through the low-resistance short circuit and very little power will be dissipated. The part will cool and no further damage will occur. The failure analyst will find a small, well-defined melt-thru. If, on the other hand, the current is not limited by the external circuit, then a large current can flow from the emitter bond to the melt-thru and cause more extensive damage.

A fairly common type of damage found on failed power transistors is evidence of melted silicon extending from the emitter bond to some point on an emitter finger, perhaps extending to the end of the latter. Such damage has been experimentally observed to start as a point in the emitter finger and progress to the emitter bond [8]. Large craters in the emitter can ensue. Such damage can be explained by the following sequence: (1) A small melt-thru occurs at some point in an emitter finger, perhaps at the end of the finger. (2) The emitter metallization over the melt-thru fuses so that current cannot be maintained through the original melt-thru. (3) The hot spot moves toward the emitter bond as the emitter metallization melts and fuses. After the die has cooled, the emitter finger exhibits signs of having been melted, and craters in the emitter are often observed.

LOCATION OF THE FAILURE SITE ON THE DIE

When transistor failure occurs as a result of being operated marginally beyond its safe operating area curve, the likely location of the initial failure site is highly dependent on the type of the overstress. Figure 9 shows the areas of likely failure for eight different categories of EOS (I_f, P, $I_{s/b}$, T/O, I_r, $E_{s/b}$, V_{ces}, and V_{ceo}) for a pair of familiar emitter geometries. The multiple discrete geometry is to the left of the pair and the interdigitated is to the right. The reasons for the preferred locations will now be explained for the eight EOS categories.

"I_f"

The dominant limiting phenomena for this maximum current region are resistive debiasing, base widening and the metallurgical properties of bonds and bond leads. Base widening tends to lower beta and hence increase the effect of the IR drop in base fingers. Base widening also causes higher temperature operation due to the increased power dissipation associated with switching transitions. This higher temperature, in turn, increases the resistance of the emitter and base regions, which further increases the effect of resistive debiasing. As shown in Fig. 9(a), the location of the melt-thru can be expected to be near the emitter bond wire, where the effect of emitter resistive debiasing is minimal. While fused bond leads would not be expected for operation marginally into the "I_f" region, other types of bond failure such as mechanical fractures from wire twitching (resulting from low frequency current pulses) [Ref 9] are possible.

"P"

Operation in this maximum power region results in a high die temperature as a consequence of the excessive power dissipation. Because of the high temperature operation, thermal biasing is more likely to be troublesome than in the other regions. Also, the high temperature of the die will reduce the effective SOA so that normal transistor operation can possibly be outside of the reduced SOA. Hence, other types of failure such as "$I_{s/b}$," "$E_{s/b}$," etc. can conceivably be induced. Failure in the "P" region usually results in discolored emitter metallization over most of the die. As shown in Fig. 9(b), the failure site is usually near the emitter bond wire and in the middle of the die due to the

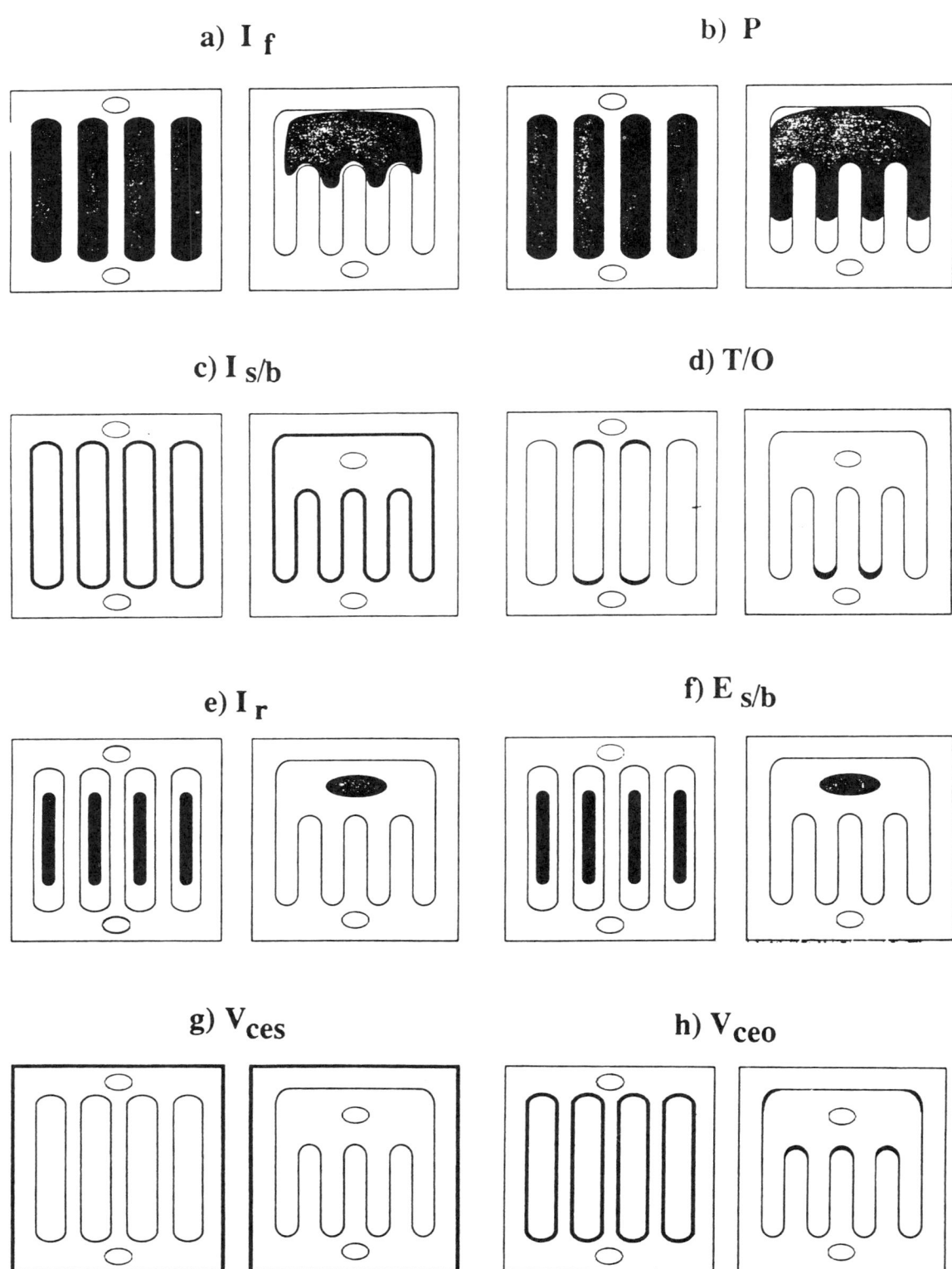

Fig. 9　Likely location of initial failure site.

effects of resistive and thermal debiasing. A large crater near the emitter bond may result. A common cause of excessive power dissipation is a distorted base drive caused by a malfunction elsewhere in the circuit.

"$I_{s/b}$"

The dominant limiting phenomena in this high voltage, second breakdown region are resistive debiasing due to the lateral IR drop in the active base layer and the non-instantaneous turn-on of the base-emitter junction. The failure site will be at the edge of the emitter as shown in Fig. 9(c).

"T/O"

The dominant limiting phenomena for high voltage, turn-on, second breakdown is non-instantaneous turn-on of the base-emitter junction and resistive debiasing caused by the IR drop between the base bond and the emitter finger tips. As shown in Fig. 4 and 9(d), the failure site is typically on the tip of an emitter finger, which is physically near the base bond wire where the conduction process is initiated.

"I_r"

The dominant limiting phenomena for this maximum current RBSOA region is base widening and metallurgical properties of the bond and bond leads. Base widening causes excessive charge to be stored in the base, which, in turn, causes the device to turn off slowly and hence dissipate excessive power. As shown in Fig. 9(e), the failure site will be that part of the die that turns off last. Note that failure does not necessarily occur when the transistor is operating in the RBSOA "I_r" region but can occur during the turn-off process. Base overdrive (see Fig. 5b) can also cause this type of failure because a longer turn-off time caused by increased saturation results in excessive power dissipation.

"$E_{s/b}$"

The dominant limiting phenomena for high voltage, turn-off, second breakdown is the non-infinite turn-off time and resistive over-biasing. As shown in Fig. 5 and 9(f), the last area to turn off and the area most vulnerable to failure is the middle of the emitter.

V_{ces}

The current conduction will be due to the reverse breakdown of the base-collector junction which, as shown in Fig. 9(g), is normally weakest at the outside edge of the diode.

V_{ceo}

The base-emitter junction will become forward-biased as it conducts the Iceo current. Since the source of the Iceo can be any part or all of the base-collector junction, the edges of the base-emitter junction receive more Iceo current than the interior parts. This, combined with the effects of base pinching, results in current crowding around the edges of the emitter diffusion. Any effects of resistive debiasing in the emitter fingers would tend to make the failure occur near the emitter bond [Ref 10]. See Fig. 9(h).

Size of Melt-Thru

The hot spot current in amperes can be crudely estimated as the melt-thru radius in mils. For example, a melt-thru radius of 3 mils indicates a hot spot current of approximately 3 amperes [Ref 11].

BOND WIRE FAILURES

Vaporized Bond Wires

A bond wire will "vaporize" if subjected to a large, short duration, current pulse. Probably a more common cause of vaporized bond wires is high voltage arcing. When a bond lifts or opens from the die due to excessive temperature, the high voltage will cause an arc to be established to complete the circuit. The voltage across the arc may be ten volts or more. Heat generated in the arc will cause the wire to vaporize. The vaporizing process starts at the original open circuit and progresses towards the ends of the wire.

Fused Bond Wires

If the external circuit does not limit the short circuit current to less than that which fuses the bond wire, then the bond wire normally fuses after a melt-thru has occurred. If gold wires melt, they usually leave balls at the edges of the melt. Aluminum wires sometimes oxidize and exhibit a shrivelled appearance. The end results in both cases is an open, or very high resistance, circuit. Often, it is useful to know the critical fusing current for a bond wire. An approximate formula [Ref 12] for the fusing current, I(amperes) for an infinitely long aluminum 1% silicon wire of diameter D(mils) is $I = 0.4*D$. Hence, an infinitely long, 1 mil Al wire will fuse at 0.4 amperes and a 10 mil wire will fuse at 4.0 amperes. Note that for long wires, the fusing current varies more nearly as D than as D squared, primarily because the generated heat is mostly conducted through the surface area by radiation (which varies as D) and not through the cross sectional area of the wire (which would vary as D squared). For a short wire of length L(mils), an approximate relationship $I = (60*D**2)/L$. Hence a 50 mil length of 1 mil Al wire would fuse at 1.2 amperes and the same length of 4 mil wire would fuse at 19.2 amperes. Note that for short wires the fusing current varies approximately as the diameter squared. This is because the heat is conducted primarily through the cross-sectional area of the wire (which varies as D squared) to the ends of the wire. Bond wires can be assumed to be infinitely long if their length-to-diameter ratio is in the order of 100 or greater. For most cases in failure analysis, the formula for the infinitely long wire can be used as an approximate value. Similar relationships exist for gold wires [Ref 13]. Also, the analyst should be aware that for short pulses, wires can conduct much larger currents than the d-c fusing current [Ref 14]. When the pulse duration is small enough that the heat transferred from the wire is negligible compared to the heat used to increase the temperature of the wire, the wire can conduct very large current for short time periods. In such adiabatic case, total energy from the pulse (not the instantaneous $R*I**2$ power) is the critical quantity that fuses the wire.

To determine the significance of a fused emitter bond wire, the failure analyst must analyze the circuit to determine the maximum d-c current that would flow if the transistor were replaced with a short circuit. If this short circuit current is less than the bond wires' fusing current, then the wire probably failed due to an excessive d-c current resulting from an overvoltage condition.

THERMAL FAILURES

Excessive die temperature will result if a transistor is operated at too high an ambient temperature or if the cooling system malfunctions. The higher die temperature results in a reduced effective SOA. The normal operating point may therefore by outside the reduced SOA so that failure would occur. For this reason, thermal problems must be eliminated before concluding failure due to EOS.

SUMMARY

The theory of power transistor failure has been presented from the perspective of the failure analyst and a method of interpreting EOS damage has been explained.

REFERENCES

1. *RCA Solid-State Devices Manual*, SC-16, 1975, 87-103.
2. Sorab K. Ghandhi, *Semiconductor Power Devices*, John Wiley & Sons
3. Warren Schultz, "Power Transistor Safe Operating Area—Special Considerations for Switching Power Supplies," Motorola Application Note AN-875, 1982.
4. Ghandhi, pp. 173-4.
5. Ghandhi, p. 138.
6. Ghandhi, p. 162-7.
7. P.L. Hower, D.L. Blackburn, F.F. Oettinger, and S. Rubin, "Stable Hot Spots and Second Breakdown in Power Transistors," *PESC '76 Record*, 1976 Power Electronics Specialists Conference, pp. 234-246, June 1976.
8. Thomas W. Lee of Motorola, Phoenix, AZ, private communications.
9. John Gaffney, "Internal Lead Fatigue Through Thermal Expansion in Semiconductor Devices," IEEE Trans. Electron. Devices, Vol., ED-15, p. 617, August 1968.
10. Thomas W. Lee, "The Interpretation of EOS Damage in Power Transistors," in *Proc. ISTFA*, 1983, p. 171, Photo 21.
11. Ghandhi, p. 27
12. The Secon Wire Company data sheet, circa 1972.
13. *Semiconductor Measurement Technology*, National Bureau of Standards Special Publication #400-1, July 1—Sept. 1973.
14. J.S. Smith, "Electrical Overstress Failure Analysis in Microcircuits," in Proc. 1978 IEEE Reliability Physics Symposium, 1978, pp. 44-45.

ESD DAMAGE SIMULATION AND FAILURE MECHANISMS

by Thomas W. Lee

BACKGROUND

Pareto diagrams of failure analysis data typically show that the most frequently occurring failure mechanisms in failure analysis (FA) are electrical overstress (EOS) and electrostatic discharge (ESD). Through questions posed by report recipients, it is known that the significance of the appearance, and in particular, the SIZE, of a meltthrough site in the determination of EOS versus ESD damage is generally not well understood. However, this understanding, through efforts shown here, is improving over the years. A common question asked of analysts who present their report recipients with evidence of electrical damage, is, "Was it EOS, or ESD?" Additionally, the frequency with which electrical damage is a conclusion in FA reports may cause all reports to be viewed with scepticism.

SEM photos of electrical damage sites are most helpful to customers. However, customers usually seek more specific information to help them in their search for the causes and solutions to ESD problems, once the problem is isolated by FA. Values such as approximate ESD voltage and polarity of the discharge, can help them more accurately identify the cause of the problem. Unfortunately, estimating these variables from the physical evidence found in damaged devices is inexact. Further, speculation, particularly when not identified as such, can lead to eventual loss of credibility; "Everything is electrical overstress". This is especially true when multiple or varied interpretations are possible, and the cause determination breaks down into contests of informed opinion. The following procedures can help alleviate this problem.

THE VALUE OF ESD SIMULATION IN F/A REPORTS

Experience has shown that customer confidence in reports is a strong contributor to the success of the problem-solving process. When accurate determination of the cause of electrical damage is important, the relatively small investment of time and effort necessary to simulate damage on a good part is rewarded by vastly more accurate interpretation. Report recipients will likely feel that investigative efforts on the problem are more likely to be targeted and productive.

A report in which simulations are used to form conclusions, should include side-by-side photos of real and simulated damage areas. Confidence in conclusions about electrical damage, when they are based on obvious similarity between the physical appearance of damage simulated on good parts, and damage found on returned parts, will be improved.

Over time, the collection of electrical damage simulation data in the FA laboratory will isolate device types having high sensitivity with respect to others. Externally generated or published ESD sensitivity data can be used as comparative references in reports. Over the past few years, an enormous amount of ESD sensitivity work has been performed and is available from the EOS/ESD Association, and Reliability Analysis Center (RAC), in particular, the VZAP 2 publication (Ref 1). Using these additional resources, the device being analyzed can be confidently placed on a relative sensitivity scale with respect to others in the industry. Figure 2 shows a typical example of this sensitivity scale.

Careful management of the magnitude of this effort is necessary to ensure that the demarcation between an FA lab which performs simulations as a part of the analytical process, and an electrical test lab, is clear. The two best indicators are level of effort and equipment set. The analyst time dedicated to the simulation of failures should be apportioned to only those reports where the extra work is justified by customer needs or criteria. The equipment set present in the FA lab should be limited to that specifically devoted to ESD or overstress testing, and should be carefully chosen so it does not duplicate equipment available in electrical test, electrical characterization, or reliability test facilities.

Objections are quite likely to be raised about the cost of sacrificing a good device for an ESD simulation, especially in complex ICs such as the example shown in Fig. 1. However, the confidence resulting from this work may be a good return on investment. Additionally, the device manufacturer has access to devices that are functional, and testable, but otherwise unsale-

Fig. 1 ESD threshold testing of complex ICs can be expensive. Thresholds can be low; from 4 kV in the R4000, 1 kV in drams and gate arrays, to as low as 500 V in some flash memories.

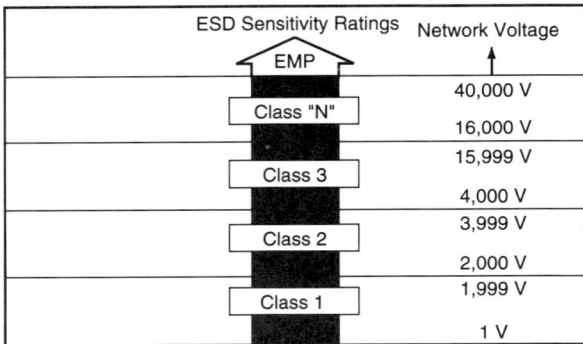

Fig. 2 ESD sensitivity rating scale, which applies to all device types.

able. Examples are devices which have undergone various reliability tests and devices with cosmetic defects.

This article concentrates on the simulation of ESD damage in particular because ESD damage constitutes a significant portion of electrical damage found in FA labs. The equipment necessary to perform ESD simulations, when compared to production ATE, is relatively unobtrusive, inexpensive, and unique.

ESD TEST METHODS

Manual ESD testers may be applied to discrete device ESD measurements. The device can be monitored with the curve tracer and pulsed until it fails. The step-stress, or "threshold" method is recognized as valid by both MIL-STD-883A and MIL-STD-750C, and is well suited to this failure analysis task. This method finds the true failure voltage in a single device, rather than bracketing it by successive approximation with a large sample of devices. An example test circuit is shown in Fig. 4.

There are other important differences between the threshold and categorization methods. With the categorization method, the devices must be carried between the ESD tester and the production ATE. With the threshold method, the device junction reverse characteristics are continuously monitored for changes as pulsing is performed, at steadily increasing voltage levels. The most sensitive or vulnerable structure is exposed to testing. Rather than using ATE to simultaneously look for input pin degradation along with functional testing, threshold testing continuously monitors the device for *any significant degradation* in the characteristics, regardless of whether or not the value has moved out of specifications. At this subtle point, ESD degradation has already begun.

No attempt is made to pulse the device until it becomes a dead short, because the term "short" is not well defined. So decreased breakdown voltage, increased leakage, and truncated or soft knees, are interpreted as an indication of device failure. Figure 3 shows typical characteristics which can be monitored.

Devices are becoming more complex. Microprocessors,

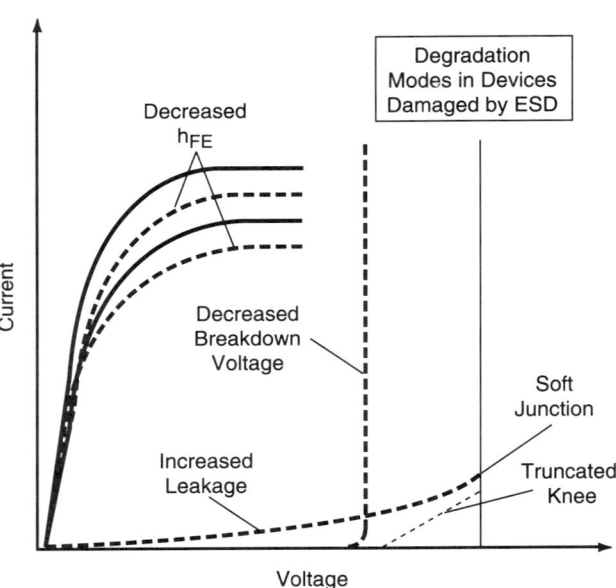

Fig. 3 Composite illustration showing some possible degradation characteristics in device junctions.

ASIC, memories, gate arrays, and other types of complex logic circuitry may need to be evaluated by a lab. Hybrids, and other modules such as SIMMs may need evaluation. The ESD test must be performed at each pin, and there may be 10's to 100's of pins. Additionally, pin *combinations* need to be evaluated. This requires a larger and more complex ESD test system than for discretes. The minimum ESD tester size and cost for a lab is then a function of the complexity of the product analyzed. When the ESD damage site usually occurs within the input protection network (i.e., the network design is effective), degradation can again be monitored with the simple circuit and criteria shown in Fig. 4.

However, the ESD energy may bypass the protection network in some circuits and cause damage within the array itself before the network degrades, causing a parametric or functional failure. In this case, the detection of an ESD failure during simulation work may require return to the categorization method, with accompanying footwork for carrying the devices between the ESD tester and an ATE system. Which method to use is a choice which must be made by collaboration between the FA engineer and the customer. Certainly, the assumption that the input network design is effective in the prevention of transmission of destructive energies to the circuit interior is reasonable.

DISCHARGE NETWORKS

Pulsing is performed with a network specified by the requestor. The most common network is the human-body model (HBM), because it was standardized first, and is used in all government documents. The circuit diagram of this model appears in Fig. 4. However, as many as 40 other models are in use, and the component values differ. Most ESD testers have provision for plug-in interchangeability of

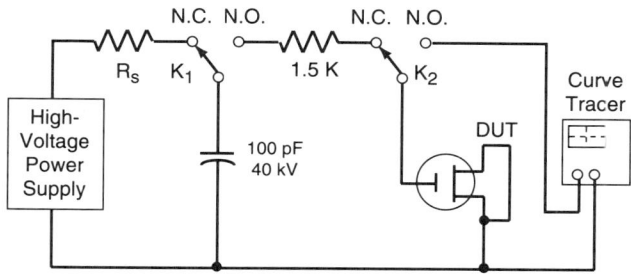

Fig. 4 Circuit diagram of an ESD step-stress, or threshold, test circuit using a curve tracer for failure indication.

model networks. Below are listed example models which have been used:

Model Name	Cap, pF	Res, Ω	Notes
Charged device	various	various	
Energy quintupler	500	300	
ECMA	150	20	pushcart simulator
ECMA	150	1000	human-body
Field-induced	various	various	
Floating device	various	various	
IEC	150	150	
Machine 1	50	0	
Machine 2	200	0	Japanese model
MIL-STD-750;883	100	1500	human-body model
NEMA	100	1500	
Opt hum body	200	1500	
SAE	300	5000	human body
UL	250	1500	human body

In the data shown in this article, only the HBM was used. A drawing of the idealized output waveform of the HBM network is shown in the applicable military specifications (Ref 2). The most important parameter of this wave form is that it is *critically damped*.

With the threshold method, the voltage is manually adjusted in increasing increments until any change in the reverse characteristics of the junction or structure under test is indicated on the curve tracer. Again, this includes increased leakage, decreased breakdown voltage, soft knees, and truncated knees. Figure 3 is a composite drawing of the characteristics of a junction, showing some of the failure modes which degradation due to ESD exposure can produce. Notice that opens and shorts are also failure modes in discrete devices tested by the threshold method. This procedure is more stringent and accurate than simply detecting out-of-spec leakage values, but it cannot detect some parametric changes, such as hFE in discrete transistors, and functional failures in complex ICs.

It is clear that the voltage increment between tester pulses, or step size, is an important contributor to the accuracy of the tests. The resolution of the tester in this configuration is limited by the number of pulses applied at a given voltage increment. Using a very small voltage step would decrease error, but greatly increase test time, especially for devices with large threshold

E^+B^-	E^-B^+	C^+B^-	C^-B^+	C^+E^-	C^-E^+
220 μJ					
	3200 μJ				
		200 μJ			
			8450 μJ		
				2450 μJ	
					2450 μJ

Fig. 5 Table showing ESD energy necessary to damage 2N918 transistor junctions in all possible configurations.

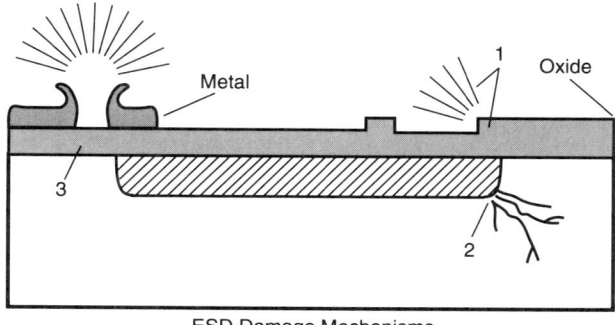

ESD Damage Mechanisms
–Bipolar–
1 Metal Transport Across Junction
2 Silicon Melting in Bulk
3 Fusing of Metal or Poly

Fig. 6 Cross-section drawing of ESD damage mechanisms in bipolar structures.

voltages. There is a direct tradeoff between test time and data accuracy. Furthermore, stress hardening, or even recovery of parameters, has been observed.

Previous work has well established that diode and transistor junctions are approximately an order of magnitude more sensitive to ESD energy when reverse-biased than when forward-biased. Figure 5 shows example data for the 2N918 bipolar transistor. If an evaluation on a sample of the device to be tested showed that the emitter-base junction was the most sensitive configuration, only the E-B, reverse-biased, case was pulsed and evaluated. The same consideration is true for the structures in IC ESD protection networks. Most of these networks consist of diodes or field-plate zeners to the rails, along with series poly resistors for current limiting. For a thorough review of input protection network schemes, refer to the book by Antinone et al. (Ref 4) This verifies the test procedure set forth in –883 and –1686, 50.1.2.

ESD SENSITIVITY CLASSIFICATION

As shown in Fig. 2, the ESD sensitivity of all devices, both active and passive, is classified into the following four categories:

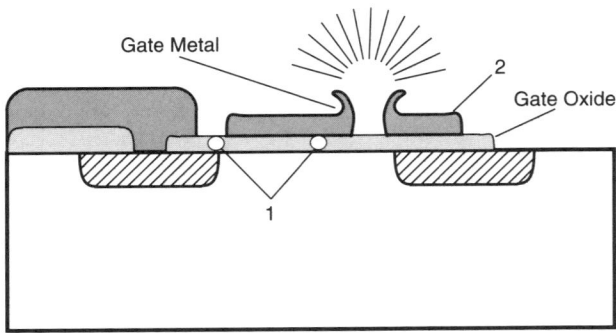

Fig. 7 Cross-section drawing showing some ESD damage mechanisms in MOS structures.

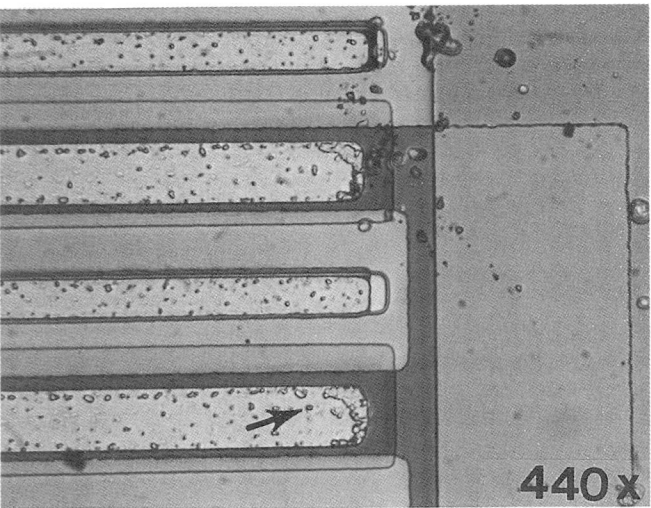

Fig. 8 Optical micrograph of bipolar structure with ESD melt-through in extremity of emitter finger. (440×)

- Class 1—0 to 1,999 V
- Class 2—2,000 to 3,999 V
- Class 3—4,000 to 15,999 V
- Class N—16,000 V and above

These categories are well suited to indicate how carefully a device must be protected, and handled. The threshold test can quickly and accurately place a device into one of these categories. The actual threshold voltage found should also be reported to your customer. Depending on the needs of your customer, stating that a device "has an ESD damage threshold of 3.14 kV" may be more useful than stating that the device "is a class-2 part". The position of the device's ESD threshold in this classification system may be utilized as an indicator of which the ESD generation site in the factory was responsible for the damage (Ref 3). Determination of the satisfactory performance of the test circuit and threshold method can be made by comparison of the FA lab measurements with published data.

Through the performance of analyses and evaluations which required ESD data and damage simulations, a database containing about 400 discrete device types was assembled. The domain graph in Fig. 11 shows the results of ESD testing in this lab over a two-year period of time. It shows the relationships between device die size, device technology and ESD sensitivity. The motivation for showing this graph is to dispel the common perception that physically large devices have high resistance to ESD. Experience tells us that this is not true with large, leading-edge ICs. It is also not true of zener diodes (small die, high ESD damage threshold), and MOS power transistors (large die, low ESD damage threshold). Notice their relative positions on the graph. Examination of DOD-STD-1686 will show tables which indicate which of the discrete technologies are ESD sensitive in the range from 100 volts to 15 kV, and agreement is apparent between Fig. 11 and this publication. Examples of sensitive device technologies are: most MOS IC technologies, hot-carrier diodes, Schottky-barrier diodes, and small signal JFET, MOSFET transistors, and GaAs devices. The following discussions partially explain the position of the various technologies on the domain graph. The testing and graph was originally extended to 40 kV because of military and automotive customer requests for specialized testing (Ref 5,6,7,8,9).

FAILURE MECHANISMS

The two drawings in Fig. 6 and 7 show idealized cross-sections of bipolar and MOS structures. Typical failure mechanisms are superimposed on these illustrations. In bipolar devices, a very common failure mechanism is the formation of a melt-through hole. The mechanism for the formation of this hole is explained in detail in the Thermomechanical Overstress section of this section, and in Dr. May's article.

Figures 8 and 9 show a typical ESD meltthrough hole. Note its small size with respect to EOS damage, which is frequently visible to the unaided eye. The energy of the pulse may be sufficient to transport metal across the junction and cause a short. This phenomenon has been termed the "silvery track", and it appears in area 1 of the bipolar drawing. The junction breakdown most likely occurs where there are small radii of curvature and electric field strength is most intense, such as at sharp edges. This is shown in area 2 of Fig. 6. Finally, any energy remaining in the external charged static source capacitance will cause joule heating and possible melting of metal runs, shown in area 3 of Fig. 6.

In MOS devices, puncture of the thin gate oxide is a very common mechanism. The gate oxide field intensity at normal operating voltages is already 10^4 to 10^6 V/cm, and overvoltages due to ESD rupture the film by exposing it to even greater field strengths (10^7 V/cm). This is shown in area 1 of the MOS drawing (Fig. 7). As in the bipolar case above, any energy remaining in the static source after forming the meltthrough hold may then damage the poly or metal runs, causing them to change resistance or melt and fuse open.

ESD Simulation and Failure Mechanisms

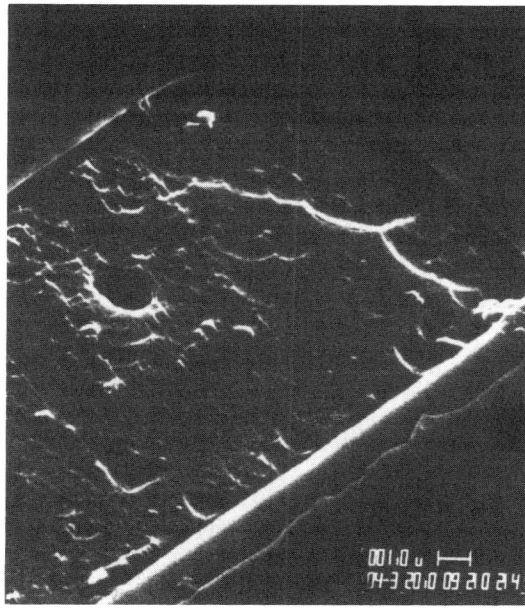

Fig. 9   SEM view of emitter shown in Fig. 8. (4,400×)

Figure 10 shows an example of ESD damage to a poly resistor in the protection network of a smart power IC. This damage was induced during device threshold testing, and the passivation has been chemically etched away. Notice that the ESD energy vaporized a trail of polysilicon from the centerline of the resistor, resulting in only a small increase in resistance. *This small increase was detectable during ESD threshold testing in the FA lab, but would have likely passed parametric testing.* Nevertheless, it has been proven that damage begins to occur at this level. This is endorsement for the FA threshold ESD test method.

CONCLUSIONS

1. For best ESD problem solving, show side-by-side field failure and simulated failure SEM photos.
2. The primary difference between ESD and EOS damage is the size of the damage feature. In ESD damage, a SEM must be used to view the damage area, and delineating etches must frequently be used. With EOS damage, the area can frequently be observed with the unaided eye, or in a low-power optical microscope. The amount of energy involved in the formation of ESD damage is also orders of magnitude smaller.
3. The factors which seem to influence ESD sensitivity are; small die size, small active area, high switching speed, low input capacitance, thin oxides, small feature size, and light doping.
4. It is difficult to detemine if an EOS failure has been initiated by ESD, because the ensuing EOS energy will very likely be concentrated in the location of the ESD damage, and destroy it. You must look in the sample for ESD (lightly) damaged devices which for some reason, were not exposed to the EOS energy.

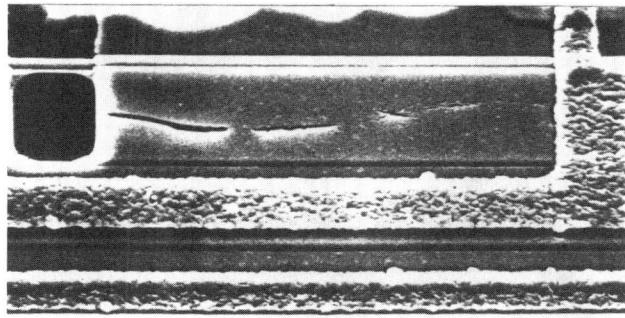

Fig. 10   SEM view of ESD damage to poly resistor. (902.8×)

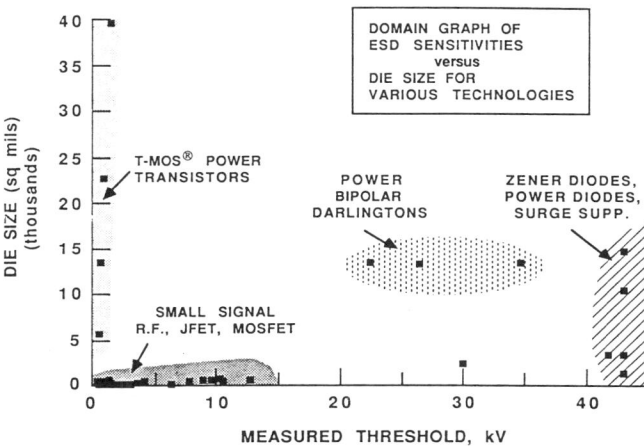

Fig. 11   Domain graph showing ESD sensitivities for various technologies vs die size.

Damage due to ESD can definitely result in EOS in some circuits. An example is the shorting of the gate and source of an MOS power transistor, The short may result in a voltage divider bias network between drain and source, internal to the device. If the signal source is not "stiff", that is low impedance, and the drain supply is not current-limited, the first turn-on of the circuit will place a device in a steady "on" state directly across the power supply.

CORRECTIVE ACTIONS FOR ESD

Customers alerted to the presence of an ESD problem through failure analysis can investigate their manufacturing and handling areas for sources of charge generation. Examples of high voltage ESD sources may include: CRT displays, plastic machine rails, carpeted offices, and certain types of clothing. The use of a hand-held static meter can locate voltage generation sites.

Corrective actions are widely available, and include proper conductive or dissipative material choices, humidity control, grounding, worker educational programs, detector networks, and charge neutralizing ion grids and blowers.

FRINGE BENEFITS

The sensitivity of devices needs to be known because ESD sensitivity could likely become a specification. Customers could begin to make purchasing decisions which are partially based on device ESD hardness, and the failure analysis ESD simulation activity could help in the evaluation process.

FA REPORTS

Use of simulations with the techniques outlined in this article can help you improve report effectiveness, particularly where suspected ESD damage is a major (expensive) problem, and conclusion of ESD damage is under question.

REFERENCES

1. *Electrostatic Discharge (ESD) Susceptibility of Electronic Devices (VZAP-2)*, by William K. Denson, Reliability Analysis Center, RADC/RAC, Griffiss AFB, New York 13441. (Presents tabulated data for I.C.s and discretes, along with data calculated with the Wunch-Bell model, and a preface discussing failure mechanisms and test methods).
2. *MIL-STD-883A*, method 3015.
3. *DOD-HDBK-1686*.
4. *Electrical Overstress Protection for Electronic Devices*, by Robert J. Antinone et al, Noyes Publications, 1986, pp 17-26, 75-80, 140-377.
5. *Economical ESD Testers for The Identification of Discrete Device Damage Thresholds*, by T. W. Lee, Evaluation Engineering Magazine, pp 80-99. (Details of the construction and application of testers for measuring ESD damage thresholds to 15 kV, and includes a nomograph for conversion of ESD test voltage to stored energy on page 98).
6. *Determination of the Threshold Failure Levels of Semiconductors Due To Pulse Voltages*, by D.C. Wunch and R.R. Bell, I.E.E.E. Transactions on Nuclear Science, Vol. NS-15, No. 6, 1968, pp 244-250. (A large study of the failure levels of semiconductors, presenting a power per unit area model for the prediction of thresholds)
7. *Electromagnetic Detection of Nuclear Explosions*, by Robert W. Cotterman, IEEE Transactions On Nuclear Science, pp 99-103. (Shows the general waveform of an NEMP event).
8. *Response of a System*, by Dr. Norman Rudie, Defense Science and Electronics Magazine, June, 1986, pg.31. (Includes a table showing the failure energy threshold for various discrete devices, including semiconductors).
9. *Electromagnetic Pulses: Potential Crippler*, by Eric J. Lerner, IEEE Spectrum, May, 1981, pp 41-46.
10. *Electrostatic Discharge (ESD) Protection Test Handbook*, published by the Keytek Instrument Corporation, Wilmington, Massachusetts, 01887, 1983, 1986.
11. *ESD Testing for ICs*, Second Edition, 1990, Keyter Instruments, Inc.
12. *Analysis of Electrical Overstress Failures*, by Jack S. Smith, in proceedings of 11th Annual Reliability Physics Symposium (IRPS), 1973, 1973, pp 105-107.
13. *Modeling of Electrical Overstress in Silicon Devices*, by N. Kusnezov and J.S. Smith, in proceedings of EOS/ESD Society, 1979, pp 133-139.

THERMOMECHANICAL EFFECTS OF EOS

by Thomas W. Lee

INTRODUCTION

Electrical overstress, or EOS, in a semiconductor device of any type usually takes the form of a hole or crater in the silicon die. Metallization is usually fused and alloyed with the underlying silicon. The device junctions may be totally shorted. Usually, failure analysis stops with this observation of catastrophic, and obvious damage. However, through research, it has been learned that the conditions responsible for this damage can be deciphered, to a degree, from the interpretation of the physical damage present. Its extent, its location, and the portions of the device which are involved all provide important clues. The article by Dr. Tom May preceding this one describes this method of interpretation by physical location thoroughly.

There is a second type of EOS, where the effects are not as obvious. The entire sample group can be studied for example devices in which damage is not total, as described in the above paragraph. In these devices, subtle damage or indications of degradation may reveal information about the conditions which resulted in the failure of the group. The analyst inspecting the surface of the device may find that the metallization has become darkened, and rough in texture. If the die was glassivated, a crack in the glass may allow the aluminum to extrude through the crack like toothpaste. X-ray shadowgraphs of the die bond solder alloy coverage show that the coverage is perfect. However, closer examination of the solder may show it to have enlarged grains, and it may also be extruded, oozing from beneath the die to some degree. This effect may be found in R.F. and microwave devices, power integrated circuits, and power transistors.

This phenomena is a life-limiting physical process in all devices made from silicon, silicon dioxide, aluminum, and soft solder, regardless of the company or country of origin. It is interchangeably identified as "thermal fatigue," "wearout," "reconstruction," or "restructuring." It can act alone to physically degrade a device, or it can be present as an indicator of abnormal conditions just prior to a failure. It is important to recognize and understand this wearout or failure mechanism.

The interpretation of the conditions responsible for this damage are not easily deciphered, but simulations can be somewhat more informative than in the case of catastrophic EOS. Accurate interpretation of the conditions resulting in such physical damage to semiconductors, and power transistors in particular, requires a brief review of their construction and principles of operation.

SEMICONDUCTOR DEVICE CONSTRUCTION

In our discussion, we proceed in a vertical section from the top of the die to the header. Figure 1 is a photograph of a typical power transistor, with its gross features identified. Figure 2 is a cross-sectional drawing of a hypothetical device, in which many typical metals and layers in common use have been su-

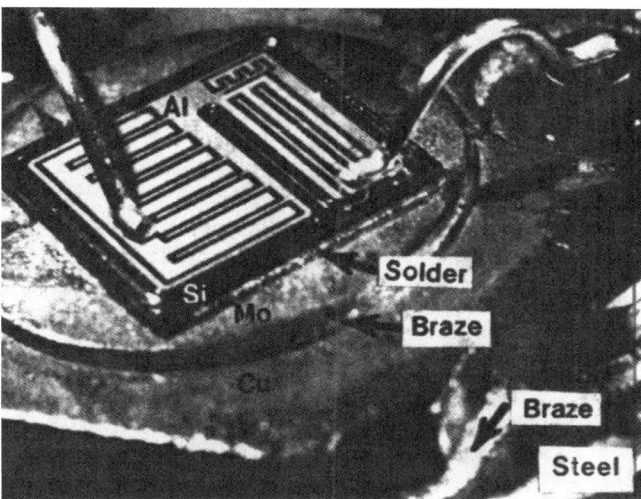

Fig. 1 Optical micrograph of the interior of a typical power transistor, with major structural features identified.

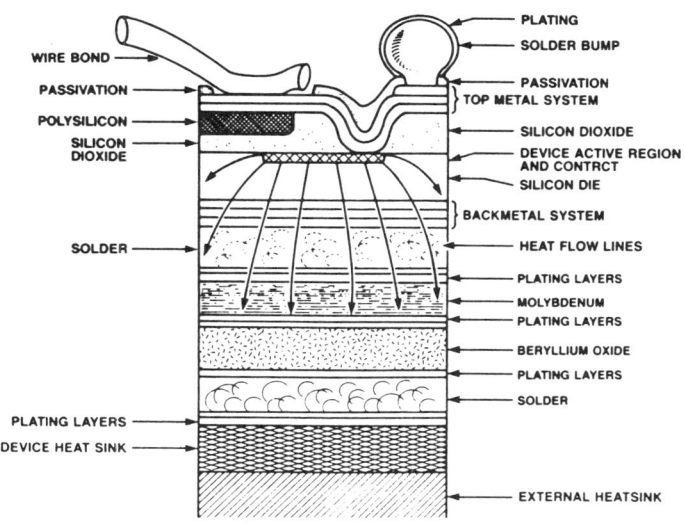

Fig. 2 Hypothetcial power device cross-section showing possible material layers (vertical dimensions not to scale).

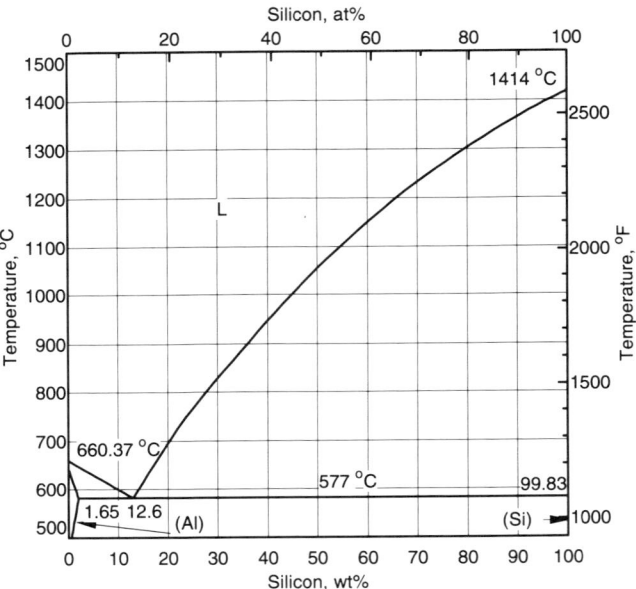

Fig. 3 Phase diagram for the aluminum silicon system. The area where the sintering of contacts takes place is marked by the arrow.

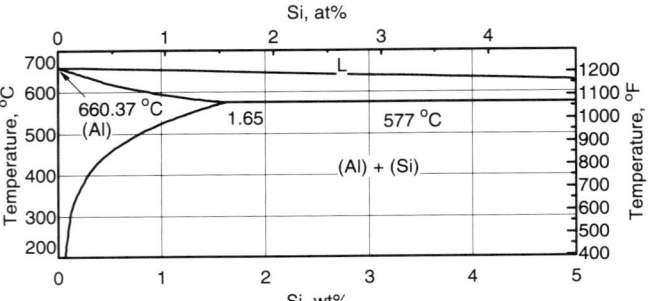

Fig. 4 Expanded area of the Al-Si diagram. Notice that alloying can take place below the eutectic temperature.

Silicide	Sintering temperature (°C)	Resultant resistivity (μΩ-cm)
Pd₂Si	400	30-35
CrSi₂	700	~600
PtSi	600-800	28-35
TiSi₂	900	13-18
VSi₂	900	13.3-66
CoSi₂	900	18-20
NiSi₂	900	50-118

Fig. 5 Resistivities and typical temperature of formation of some of the silicides.

perimposed into one device to illustrate the thermomechanical complexity of typical semiconductor devices.

Most devices have a thin deposited glass layer, or glassivation, over the surface to physically protect (passivate) it from damage, and to act as a shield against the penetration of contaminant ions. This deposited glass may be doped with phosphorus to neutralize the passage of contaminant ions to the device surface. It can be deposited by evaporation, RF or magnetron sputtering, or by chemical vapor deposition (CVD), in which a reaction which results in the formation of glass takes place in the vacuum chamber. Silicon nitride (Si_3N_4) can also be used.

The top surface of the die is occupied by the conductor runs, most commonly composed of aluminum (Al), which may also contain small percentages of silicon and/or copper. Al is a relatively easy metal to deposit. It can be evaporated or sputtered. Other metal schemes are also used, especially when a solderable surface is the end result. For example, some manufacturers evaporate, sputter, or plate the Al with a metal sequence ending with copper (Cu) or gold (Au) at the surface. A barrier metal, such as nickel (Ni), is usually placed between Cu and the Si.

Aluminum readily forms an ohmic contact with P-type silicon. All silicon (Si) surfaces quickly passivate themselves with a thin oxide skin only a few tens of Angstroms thick, which is an insulator. Aluminum on N-type silicon forms a Schottky barrier. Contact to the aluminum is made by "sintering" the Si and Al at a temperature well below the alloying temperature, 577 °C. This process is also called "densification" of the Al. The term "sintering" originally indicated the heating of a metal powder to a temperature just below its eutectic point, to bind the grains. It is an inaccurate description of the process of forming contacts in semiconductors because the melting point of Al metal is not closely approached. Instead, alloying of a small percentage of aluminum with the silicon occurs at temperatures well below the eutectic line. The Si-Al phase diagram in Fig. 3 and 4 shows this area at the extreme left-hand end of the eutectic line. The thin oxide skin is removed, and the thin Al-Si alloyed region makes an ohmic contact. If heated to greater than 577 °C, Al would alloy with the Si to form a P-type semiconductor, and short the device.

In power devices, the thin films of Al are typically 10 to 20 kÅ thick. The emitter and base diffusions in bipolar devices are each about 0.1 mil thick. Because of their thinness, these films have significant resistances. The entire die, however, is much larger than these dimensions because a minimum thickness, usually 8 to 13 mils, is required so that the wafers can be handled without breaking. Additionally, a thick lightly-doped collector is required for high breakdown voltages. Figure 2 is a cross-sectional drawing of a hypothetical die, where the layer thicknesses are exaggerated. *It is important to remember that the die is generally 200 or more times thicker than the thin films on top that are the site of activity.* Aluminum wires, from 1 to 20 mils in diameter, are used for connection between the thin film conductors on the die surface die and the header pins, or posts. The aluminum on the die can be doped with copper to increase its current-carrying ability. The aluminum in the wires can be doped with Si, Mg, or Pt to improve its characteristics. Small gold wires are also used, particularly in small-signal transistors,

RF transistors, and ICs. The bonds can be thermocompression ball bonds, and stitch bonds, or ultrasonic wedge bonds. Other metal systems can be used over the aluminum, and may have a solderable top metal. One or more layers of barrier and adhesion metals may be used between the aluminum and the solderable top metal. Contact to the die can then be made with a clip, soldered wires, or in flip-chips as solder balls, also illustrated in Fig. 2. Wires may also function as inductors, in thin- and thick-film structures that are a part of the internal tuning networks in R.F. power devices and R.F. power modules.

Again refer to the drawing in Fig. 2. The silicon die must be fastened to a header for electrical contact to the backside of the die, and to remove heat. Epoxy may be used where ambient temperatures and power dissipation is low, but metal fastening systems are necessary for power-handling devices. The metals can be plated, sputtered, or evaporated onto the back of the die. It is helpful to think of Si as basically little different from glass in its physical properties. The metals in most solders do not adhere well to silicon at low temperatures. The backside of the die must be coated with a metal, or metal system, which is simultaneously solderable, and adherent to silicon. A single metal layer often cannot be made to satisfy both conditions. Backmetals vary from a single layer of a solderable metal such as nickel, to elaborate systems consisting of more than one metal layer and containing metal silicides, adhesion layers, and interphased layers. Figure 5 shows the properties of some of the metal silicides that could be present. At the formation temperature of many silicides, the dopants may "push", or continue to diffuse, depending on the dopant species, and alter or ruin the device. As a result, backmetals usually adhere by mechanical, not chemical, processes. Figure 6 is a chart showing the mechanical properties of some of the metals and materials which compose semiconductor devices. A variety of metals can be chosen for backing because of various desirable properties with respect to the aforementioned functions. For example, aluminum (Al) is used because its contact resistance is very low. Chromium (Cr) is used because it is tough and unreactive. Titanium (Ti) is used because it has a unit cell size close to that of silicon, making a strong bond. The resistivity of the metals themselves, the contacts, and the silicides are all important to this process.

A few metals will alloy with silicon at higher temperatures to form alloys. Gold, gold-tin, and gold-germanium eutectic (hard) solders were formerly used for many part types. They melt at temperatures above soft solders (about 300 °C), but far below temperatures which will move diffusions (about 950 °C). Their cost became prohibitive in the 1970s and 1980s, when gold prices rose to several hundred dollars per ounce. Now, these expensive solders are usually limited to R.F. and premium products. The lead-based soft solders were explored by almost every manufacturer. These lead alloy solders also contain, but are not limited to, antimony (Sb), bismuth (Bi), indium (In), silver (Ag), and tin (Sn). Antimony and Bi harden solder alloys. Silver prevents the scavenging of Ag from thin layers in die backmetal schemes when it is used. Indium, one of two metals known which wet glass, is used to promote adhesion and improve temperature cycling capability. Tin lowers the melting point. Some attention must be paid to the solder alloys which are available from the manufacturers. Understandably, lead-tin alloys are by far the most popular solders. Figure 6 also includes the properties of some of the metals used to make solder alloys.

Reexamining Fig. 2, a semiconductor device may be viewed as a virtual sandwich of materials and metals, and the particular materials found are chosen because of specific physical and chemical properties.

THERMOMECHANICS OF DEVICE OPERATION

The operation of all semiconductor devices generates heat. Various resistances present in the backing metal system, the silicon bulk, the epitaxial or doped layers, the contacts, thin surface layers, and metal connecting runs all generate IR drops when current is flowing. As feature sizes decrease, these problems become increasingly important.

With the exception of Bi, a temperature increase causes all the materials to expand slightly; cooling does the opposite. Any differential in expansion rate, in this case the difference in coefficient of expansion between the silicon and the aluminum, will result in the generation of stress, and strain within the material with the lowest yield strength. It will be the material to be deformed. The forces are enormous, a fact not well appreciated because of the smallness of the features and devices being discussed. A metal expands with a force equal to that required to compress it the same amount. This value is Young's Modulus, and it easily translates the tiny movements into large forces; tons/square inch (see Fig. 29). In Al, the force with which heat sinks are extruded is comparable to those within the aluminum top metal under some conditions.

Consider the case of aluminum deposited over oxide. Examining Fig. 6 shows that the aluminum has the highest coefficient of expansion of the two materials. This differential will generate large forces. The metallization will absorb the stress inter-

Thermomechanical Properties of Semiconductor Structural Materials		
Material	Coefficient of Linear Expansion (PPM °C^{-1})	Thermal Conductivity (WATTS CM^{-2} °C^{-1})
Metals		
Aluminum (Al)	25	2.37
Chromium (Cr)	6.0	0.91
Copper (Cu)	16.6-18	3.98
Germanium (Ge)		1.5
Gold (Au)	14.2	3.15
Kovar	4.5-5.9	0.17-0.34
Lead (Pb)	29	0.346
Molybdenum (Mo)	5	1.4
Nickel (Ni)	13-14	0.899
Silicon (Si)	3	0.835
Silver (Ag)	17-19	4.27
Tin (Sn)	8.5	0.2
Tungsten (W)	4.5	1.78

Fig. 6 Table showing some of the physical properties of materials used in semiconductor packages.

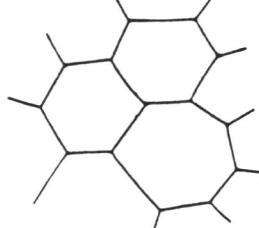

Fig. 7 Drawing of the crystal grains in an Al film, with triple points at intersections.

nally, and deform because there are differences in coefficients of expansion within the film itself. This occurs at its smallest structural discontinuities, the crystal grain boundaries. Figure 7 shows the hexagonal crystal grains in a metal film. Their size can vary tremendously, depending on the film composition and conditions of deposition. These individual crystals in the film expand at different rates along their three crystallographic axes. They are more or less evenly distributed, and randomly oriented. As they expand, the crystal grains in the film make collisions with each other where expansion rates have the highest vector sum in adjoining grains. These collisions raise the aluminum away from the device surface. The material pile-up is called a triple-point, or a hillock. Figure 8 is a SEM view of a large central hillock amid many smaller ones.

In glass passivated devices, this movement tends to be confined, but the metal may flow, and ooze, from beneath the edges, attesting to the tremendous forces involved. The glass may also crack, and the aluminum can penetrate, being extruded like toothpaste. These phenomena graphically verify the large forces developed. Comparatively, aluminum heat sinks are forcefully extruded through dies at 800 to 900 °F (425 to 480 °C), and pressures of 250 to 5500 tons/in^2.

DIE BONDS

Solder alloys, as established earlier, are composed of a mixture of grains or crystals of different metals and alloy phases, each with their own particular coefficient of expansion. Bismuth and Ga, which on occasion may occur in solder alloys, are the exceptions to this general rule; they actually expand upon cooling. Unfortunately, all the metals in alloys all expand at slightly different rates when heated; refer again to Fig. 6. As in the aluminum film, the material with the highest coefficient of expansion will move more than that with a low coefficient. The relative movement of the metals is tiny, usually a few ppm of length per degree. Thermal cycling causes the expansion and contraction of these grains, and the alloy slowly weakens with age.

At this point, with all these considerations, one may wonder how a semiconductor device works at all! In reality, the discussion of these conditions are no more consequential than, say, a tire manufacturer stating the tread life of a tire. We are all aware of the myriad mechanisms by which tires wear out. No one pushes a tire beyond its limits by, say, operating it at twice its rated load. The effects would be easily guessed. Similarly, in

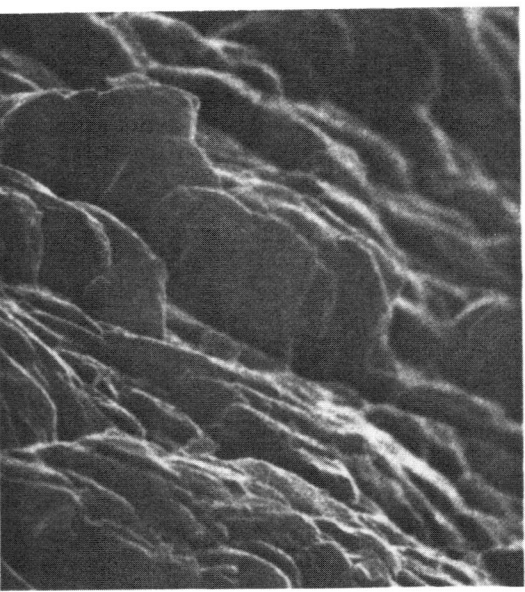

Fig. 8 SEM photo showing a group of pronounced hillocks in an Al film after stressing. (800×)

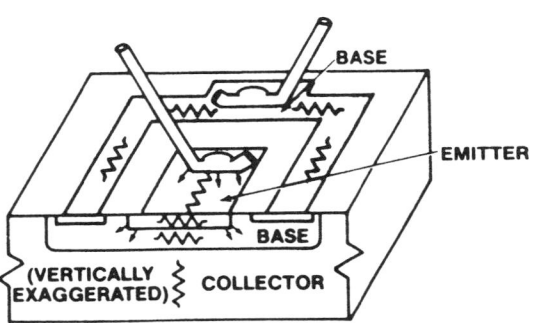

Fig. 9 Cross-sectional drawing of a transistor, with representative lumped elements representing layer resistances for equivalent circuits (Ebers-MOLL, etc.)

transistors, reliable operation is dependant upon conservative device design and rating. Examples of such parameters are maximum soldering temperature, RBSOA and FBSOA, Es/b, thermal cycling, etc. The specification sheets guide the circuit designer and ensure that the metal films are not stressed to failure within a period of time over which the device can be reasonably expected to be operated. Its use is obsoleted by technologically superior new products, not by catastrophic wear-out failure.

THERMOMECHANICAL OVERSTRESS

Refer to the drawing in Fig. 9. This drawing shows a hypothetical cross section of a device where the inherent resistances are lumped for illustration. Suppose the aluminum is 20 kÅ thick, and a finger is 20 mils wide and carries one amp, then the current density is 10^4 A/cm^2. The emitter finger can easily be 5

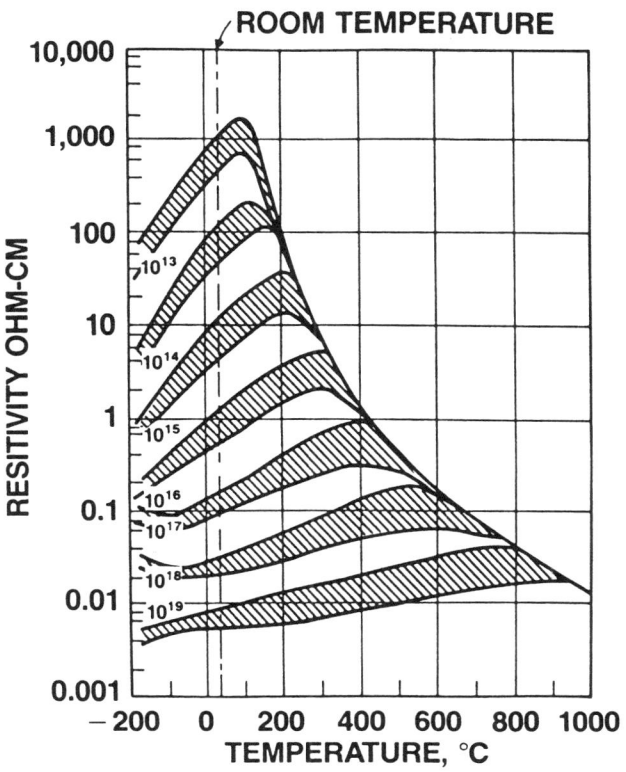

Fig. 10 Runyan's curves showing the very rapid drop in resistivity of doped Si with increasing temperature.

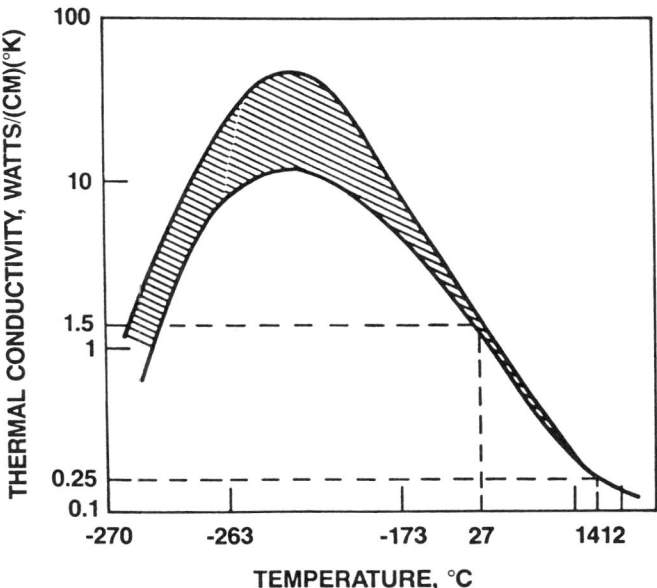

Fig. 11 Runyan's curves showing decreases in thermal conductivity of silicon with increasing temperature.

times as long as it is wide, or 5 squares long. At 1.4×10^{-2} ohms/square, this is 70 milliohms. In a five amp device, assuming current is distributed equally, there would be a 70 millivolt drop along the length of this finger. This is enough voltage to make a significant difference in conduction between the finger origin, and finger tip.

Similar considerations also apply to the emitter and base diffusions below the metal. A typical emitter doping level is about 10^{19} atoms/cc, and this will result in a resistivity of 10^{-2} ohm-cm in N or P-type material, comparing favorably with the value for the aluminum film. The more lightly-doped base will have drops and conductivity modulation, and these effects in both the emitter and base combine to make current distributions in the emitter nonuniform when the current is high, particularly when the device is operated at, or slightly beyond, its maximum ratings. These drops lead to resistive debiasing, one of the most important effects which can cause imbalance in the conduction of the device, and affects the sharing of current among its various structures. Conduction at the origin of the finger will be slightly greater than at its tip.

This process can be regenerative as the power applied to the device increases. Increasing temperature has two important effects in semiconductor devices: it decreases resistivity and increases thermal resistance of the silicon. Figure 10 shows the effect of temperature on the resistivity of silicon doped to a wide range of resistivities over the temperature range of –200 °C to 1000 °C. The resistivity decreases sharply in all cases. These are Runyan's curves. The carriers normally present are supplemented by thermally generated carriers, and the increased number of carriers will cause the transistor hFE to increase in the area that is hot.

The second effect is somewhat less pronounced. Figure 11 shows the second effect of increasing temperature in a semiconductor. The thermal conductivity drops. The plot is fuzzy because the results found by a number of researchers have been combined into one graph and the ranges are shaded-in. Note that silicon has a negative coefficient of thermal conductivity, with an increase in temperature, of only about one order at reasonable temperatures, while the change in resistivity can easily cover 2 orders. Combining the effects of increasing temperature in semiconductors, shown in the two figures, some conclusions involving thermomechanical behavior may be reached. Because of these two physical properties of silicon, an area of a transistor, usually close to the emitter wire bond, will tend to become hotter than its surroundings, and thermally isolated.

Heat travels in semiconductors by the phonon conduction process, a type of lattice vibration. During pulsed operation, the relatively low speed allows the heat to become confined in a small region of the device when power pulses are narrow.* A hot spot developed because heat is generated faster than it can be dissipated. At a critical point, carriers are thermally generated, and, the base injection no longer controls the collector current. This is thermal runaway. Depending on circuit conditions, the hot spot may also quickly develop temperatures at which the aluminum and silicon form an alloy, and the device is quickly shorted and destroyed, or "cratered."

---

*Note: the speed of sound in air is approximately 330 meters/sec. The speed of sound in silicon is 3,000 to 6,000 meters/second.

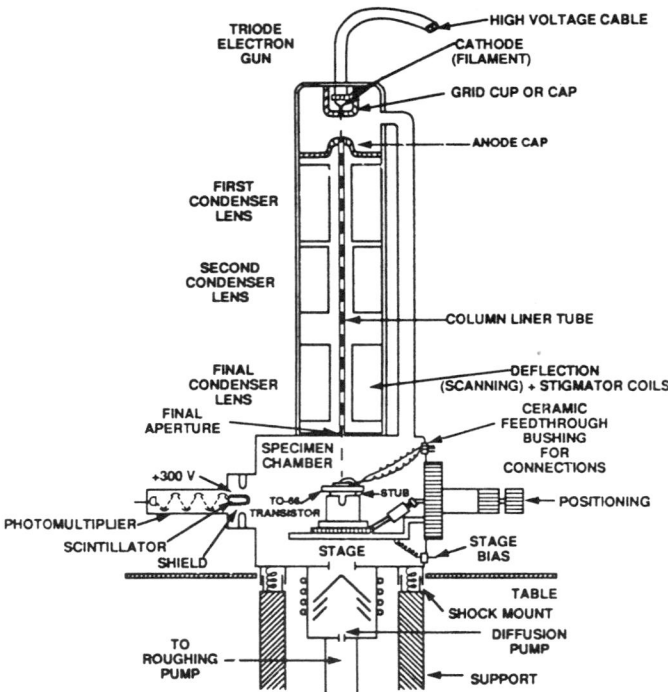

Fig. 12 Simplified cross-sectional view of typical scanning electron microscope. Note potential applied to shield in front of scintillator to attract electrons. This potential is variable in some SEMs.

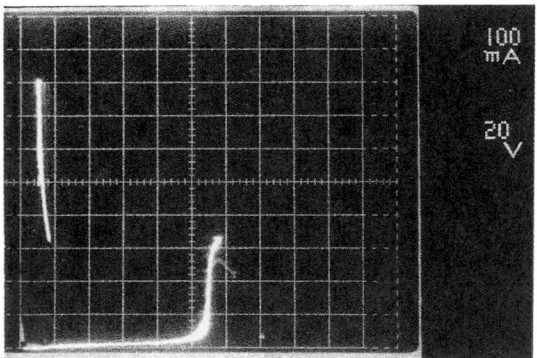

Fig. 13 Curve-tracer of display of V-I characteristic of a bipolar power transistor in second breakdown.

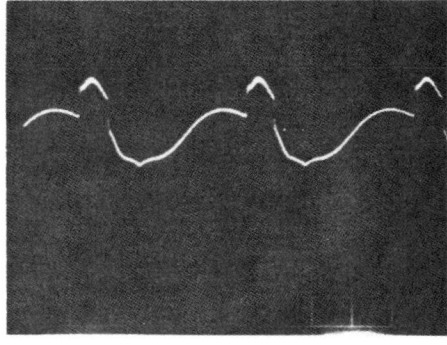

Fig. 14 Current waveform in device in Fig. 13. Pedestals are generated when device switches to low-voltage state.

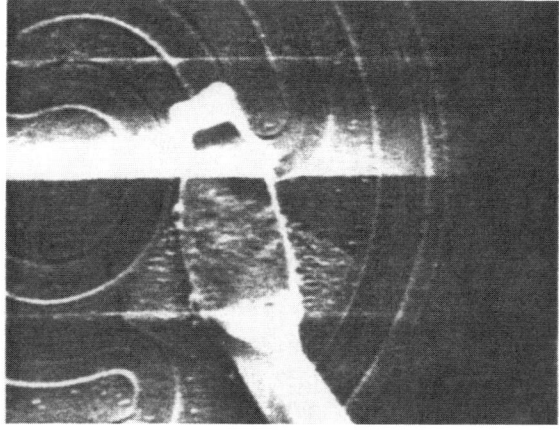

Fig. 15 SEM view of device when power is applied from curve tracer.

It is this area of operation where device damage varying from slow degradation to instantaneous meltthrough can occur, and where the aging effects in the metal films and solder alloys are tremendously accelerated. This results in shortening of the device life, degraded characteristics, or catastropic failure. The aging of the metal film is caused by the same fundamental thermally driven mechanism described above, and the metal surface in the second breakdown spot will be microscopically identical in appearance to that found in a thermal or power-cycling failure.

Optical microscopy of the thermal cycling or power cycling failures generally shows that the metallization appears dark over some portion of the die surface, usually including some or all of the emitter region. Scanning electron microscopy (SEM) of the metallization at a magnification of a few hundred, and at a low angle, will show roughening of the metal surface. Inspection of the metallization at higher magnification will show the presence of numerous hillocks. If testing is continued to device destruction, the resistivity of the metal film will increase, and its mechanical integrity will be weakened until an open develops.

SIMULATIONS

In the past, experimental accelerated life test data was generated with the standard tests and fixtures, or with a special fixture which heated devices with their own power dissipation, then cooled them rapidly with a Freon spray. Repeat of these experiments with the immersion of devices in a Freon bath, accompanied by deliberate overstressing on a curve tracer, suggested that the aging mechanisms described before could be greatly accelerated in time. The only difference between these two tests and standard power cycling, in terms of physical changes in the metallization, was the area of metallization invvolved This observation was responsible for the following experiments.

It was reasoned that the superior depth of focus of the SEM could enable a study of the effects of device degradation taking place under conditions of EOS in a very short period of time. Further, the entire event could be videotaped, and editing with time compression could be performed to adjust the real-time event length to be comfortable. Figure 12 is a cross-sectional view of a SEM for convenience in the following discussions.

Various small power devices were mounted to a specially-machined specimen mount with a large thermal mass. The leads ran through the chamber door connector. The collectors were operated at ground potential, and the emitters, in the case of NPN devices, were operated at high negative voltages with respect to ground. This was to ensure that the large collector flange was at ground potential. The devices were forced into second breakdown (Es/b) operation, as shown by the curve tracer Ic vs Vce plot in Fig. 13, and the Ic waveform shown in Fig. 14. They were not allowed to draw enough current to form a destructive meltthrough, and were manually maintained in this marginal overload condition throughout the experiment.

The initial SEM image was unusable, as it had bright lines through it, as shown in Fig. 15. It was reasoned that the beam was being deflected by the fields caused by the high emitter voltage. The backscattered electron image was examined. The energy of backscattered electrons is higher than that of secondary electrons, as shown in Fig. 16, and they are less easily deflected by stray fields. However, the image was still of unacceptable quality.

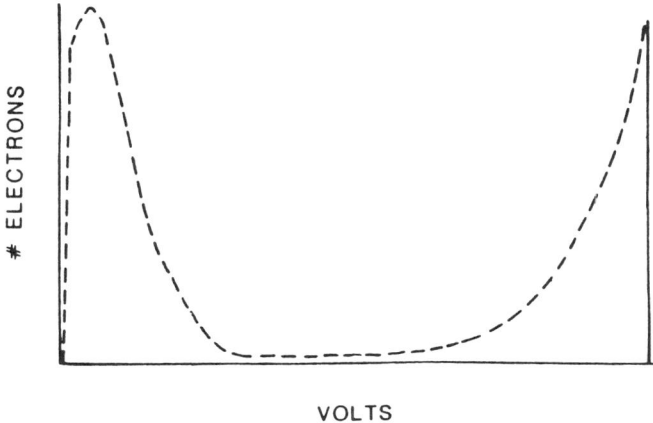

Fig. 16 Distribution of electron energies yielded from an incident beam in a SEM.

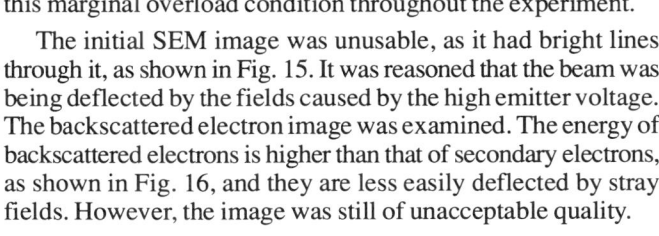

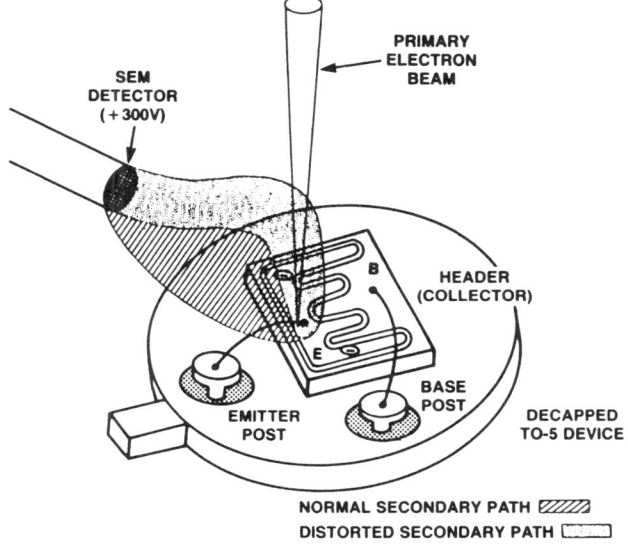

Fig. 17 Illustration of one effect of device bias on returning secondary and backscattered electrons in the SEM.

Fig. 18 View of device with Faraday shield in place.

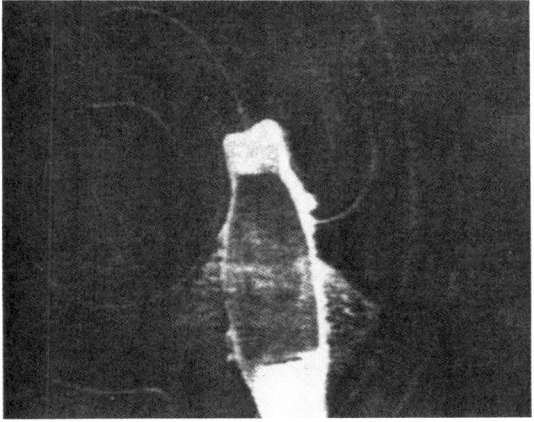

Fig. 19 SEM view of same device as Fig. 15, but with Faraday shield in place.

The interference problem was solved by constructing a Faraday Ring over the top of the device, whose operation is illustrated schematically in Fig. 17. The ring shielded the secondary electrons from the strong fringing fields from the device header. A photo of the ring appears in Fig. 18. The image was then usable with secondary emission, as shown in Fig. 19.

Operation of devices in the SEM with this improved image showed that the metallization could be quickly degraded by the mechanisms described earlier. Figure 20 shows a typical example. The operation of more samples of various device types, shown in the before-and-after Fig. 21-24, resulted in the appearance of the degradation effect in all cases. The SEM images of the degradation progress in all devices was videotaped, and the tape edited.

The metallization was stripped from the surface of a sample stressed by this method, and the surface studied. Figure 25 shows the darkened surface due to the nonreflectivity of the deformed metallization. Figure 26 shows the emitter contact region with the metallization removed, and the sintering pattern was found to be exaggerated in size and extent in the area where metallization reconstruction was most advanced, as shown in

Fig. 20 Detailed SEM view of metallization in emitter pad window of device after stressing. (500×)

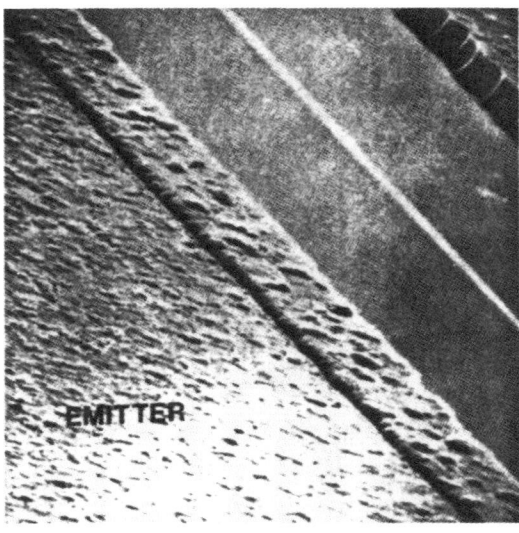

Fig. 21 Surface of power device at start of power cycling (800×)

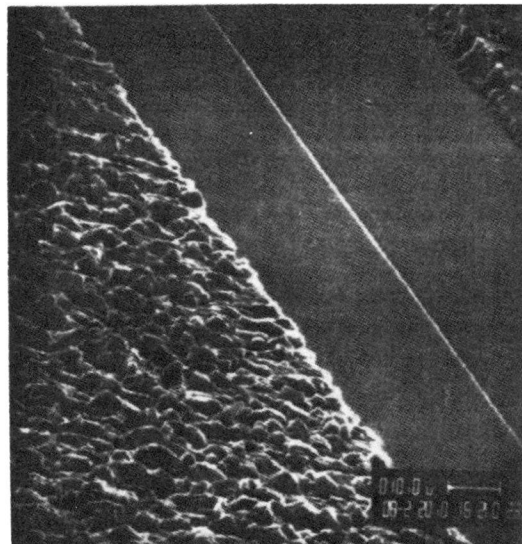

Fig. 22 Surface of power device after exposure to cycling. (800×) Compare with Fig. 21.

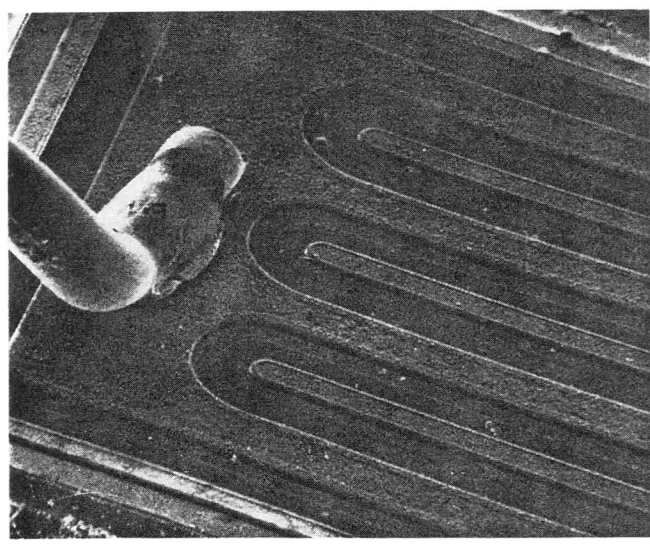

Fig. 23 Surface of device before powering in SEM. (2N5190) (75×)

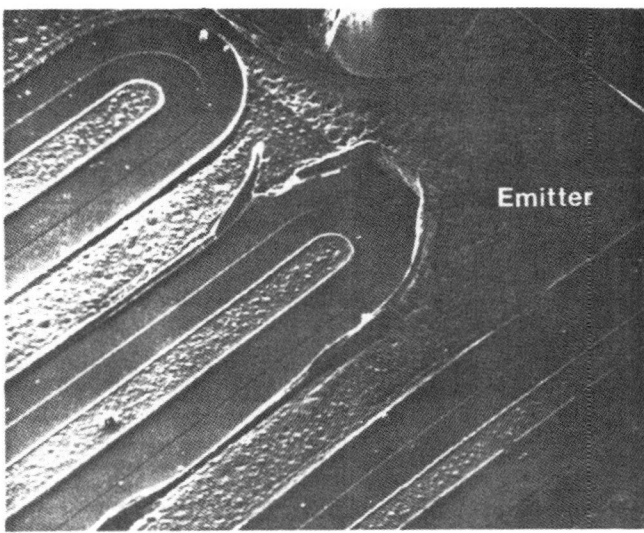

Fig. 24 Surface of power device after powering in SEM. (125×)

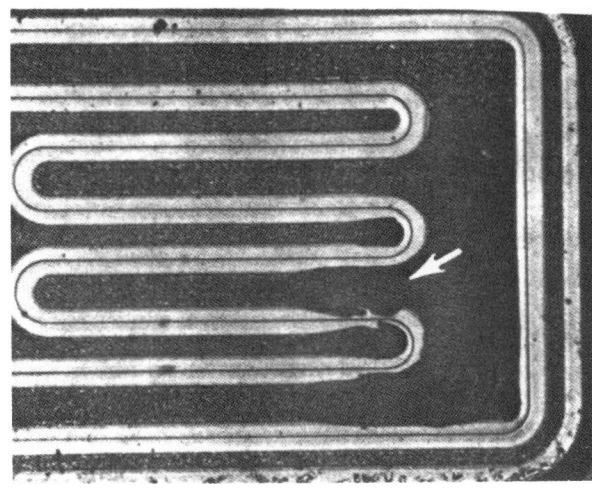

Fig. 25 Optical micrograph of die surface following overstress in SEM. (40×)

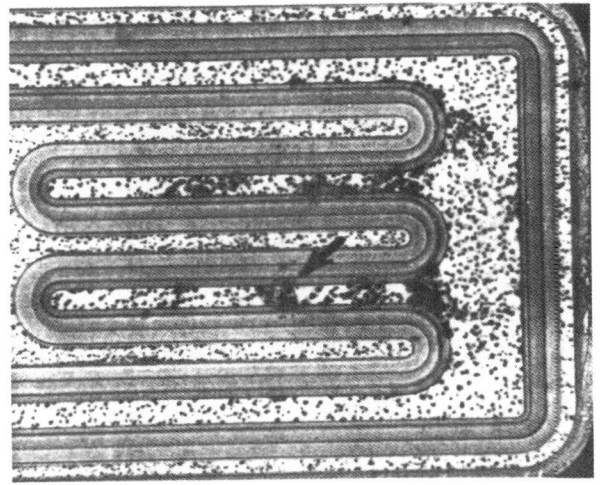

Fig. 26 Die in Fig. 25 following stripping of metallization (175×)

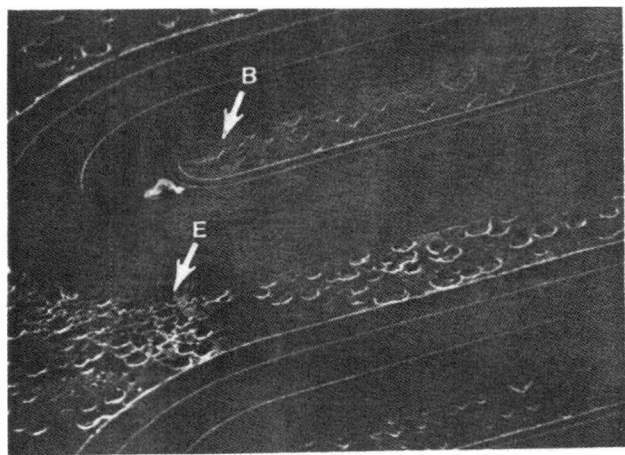

Fig. 27 SEM view of emitter region beneath most advanced metal reconstruction. (245×)

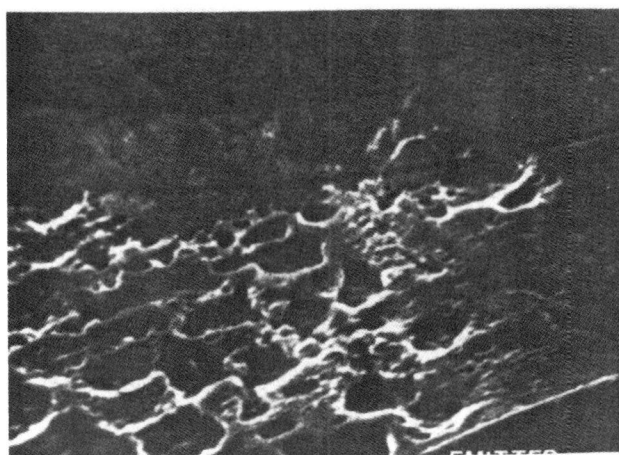

Fig. 28 SEM detail of area shown in view in Fig. 27, showing concentration and enlargement. (1050×)

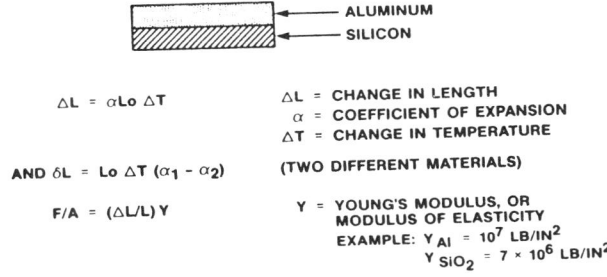

Fig. 29 Magnitude of forces in thin films.

Fig. 27 and 28. It was evident that the shallow alloying mechanism by which contacts are formed was continuing to operate in this test device, and the silicon was slowly being consumed dissolving into the aluminum emitter metallization Notice how quicky the effect took place, in only minutes.

The implied local instantaneous temperature was between 200° and 500 °C.(Fig. 4) Althouth the measurement of the temperature of microscopic site with IR thermography is well refined, the measurement of the peak temperature at the silicon-aluminum interface (referring again to Fig. 27 and 28) is virtually impossible. Thus, the enlargement of the sintering pattern is very important to this investigation. Keep in mind that simply maintaining a high temperature is not sufficient. It must be cycled with each pulse.

RESULTS

It was shown that the degradation mechanisms operating in a power transistor under conditions of marginal EOS are as predicted by theory. A hot spot forms, and all activity is concentrated in that area until a destructive meltthrough forms. Evidence of hot-spotting in a device can be positively identified by examining the metallization for areas in which hillocks are unusually numerous or large. Finally, the effects of aging of a semiconductor device due to this rapid thermal cycling could take place in a vacuum, because it could be done in the chamber of a SEM. The primary failure mechanism is an electrically powered thermomechanical effect. The effect of temperature cycling, whether a mechanical transfer of the parts from a high to low temperature environment; or from the alternate application and removal of current, is cumulative and indistinguishable. Power cycling will cause failure somewhat faster than thermal cycling because the devices carry current through the weakening metal film, and cycling can proceed at a much higher rate because physical transfer of the devices is not required.

The magnitude of stress generated is determined by the combined effects of maximum temperature, temperature change, rate of change, and coefficient of expansion of the metals involved. The die surfaces may have differing appearances, with the aluminum over the entire die roughening with exposure to thermal cycling, while power cycling generally confines the effect to the emitter region. As the power pulse narrows, the affected area on the emitter decreases in size. Again, heat travels in semiconductors by phonons, and the relatively low speed, approximately the speed of sound, allows the heat to become confined in a small region of the device when power pulses are narrow.

The effects of this wearout are not necessarily confined to the surface of the die. The heat travels through the silicon and is removed through the metal header. The metal grains in the solder alloy reconstruct, but the yield strength of the lead-based alloy and its components is less than that of the silicon die.Consequently, the solder alloy may expand and ooze from beneath the die. Figure 29 shows an example of this type of wearout mechanism. In some cases, this expansion may crack the die.

Refer again to Fig. 29. The difference in the CTE between two materials in contact with each other generates the force which drives the gradual tearing apart of the material with the lower yield strength. For example, the CTE of the alead in the die bond is about 29 PPM/°C, while the CTE of silicon is only about 3 PPM/°C. Thus, when heated, lead, tin, etc. expand more than the silicon, and the softer lead is stressed and deformed. Once again, this is a purely thermomechanical phenomenon. No moisture, air, solvents, or chemicals are necessary as participants. Early immersion boil experiments showed specifically that freon, methylen chloride, acetone, isopropyl alcohol, and methyle do not have any effect on this mechanism.

DEVICE CAPABILITY

The total number of thermal cycles that any semiconductor device can withstand is a complicated function of the following parameters:

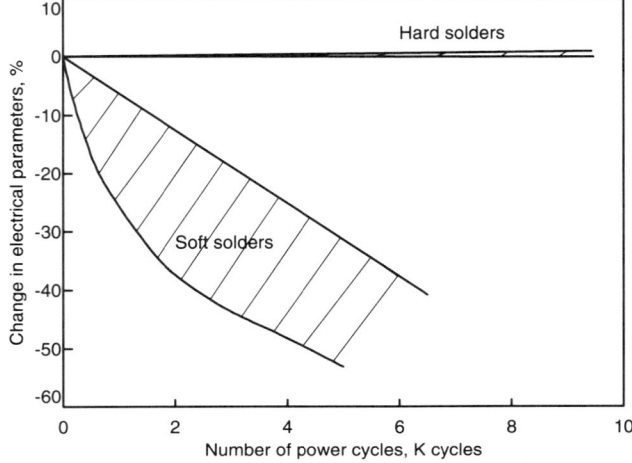

Fig. 30 Olsen and Berg's graph of the effect of power cycling on transistor parameters, excerpted from the IEEE Tran CHMT, June 1979, by permission.

Fig. 31 Die bond degraded as a result of excessive power cycling (80×).

Thermomechanical Effects of EOS

- percent of maximum rated power being dissipated,
- pulse width,
- ambient temperature,
- materials used in device construction,
- delta T

Several published papers deal with the capability of devices under temperature cycling. With the foregoing discussions, we have established that that the effects of the thermal expansion mismatch, when combined with a large number of thermal cycles or large temperature excursions, or both, tear the device apart internally, both at the surface and at the die bond. All devices, regardless of company or country of origin, can be made to fail in this manner. These mechanical wearout phenomena are so universally prevalent in electronic devices that they also apply to, for example, electron tubes, capacitors, and transformers which are operated under pulsed conditions.

The solder deterioration effect has been studied in detail. In their paper on die bond alloys, Olsen and Berg show that tem-

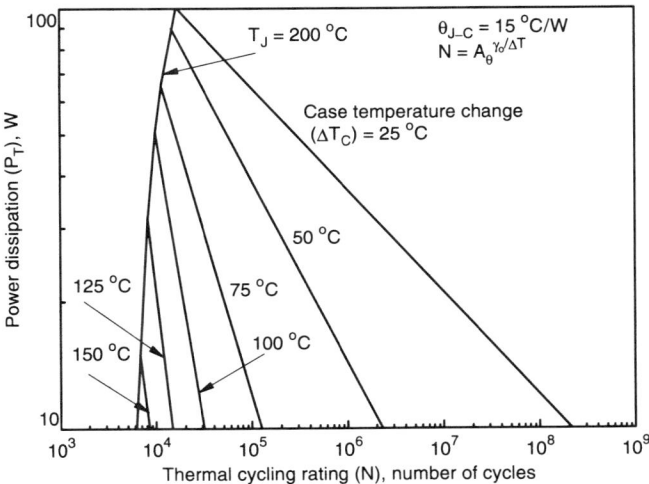

Fig. 32 Chart showing the effect of power cycling of transistor lifetime. Excerpted from the RCA High Reliability Devices Databook, SSD-207B, of 1974, Fig. 2-21, by permission.

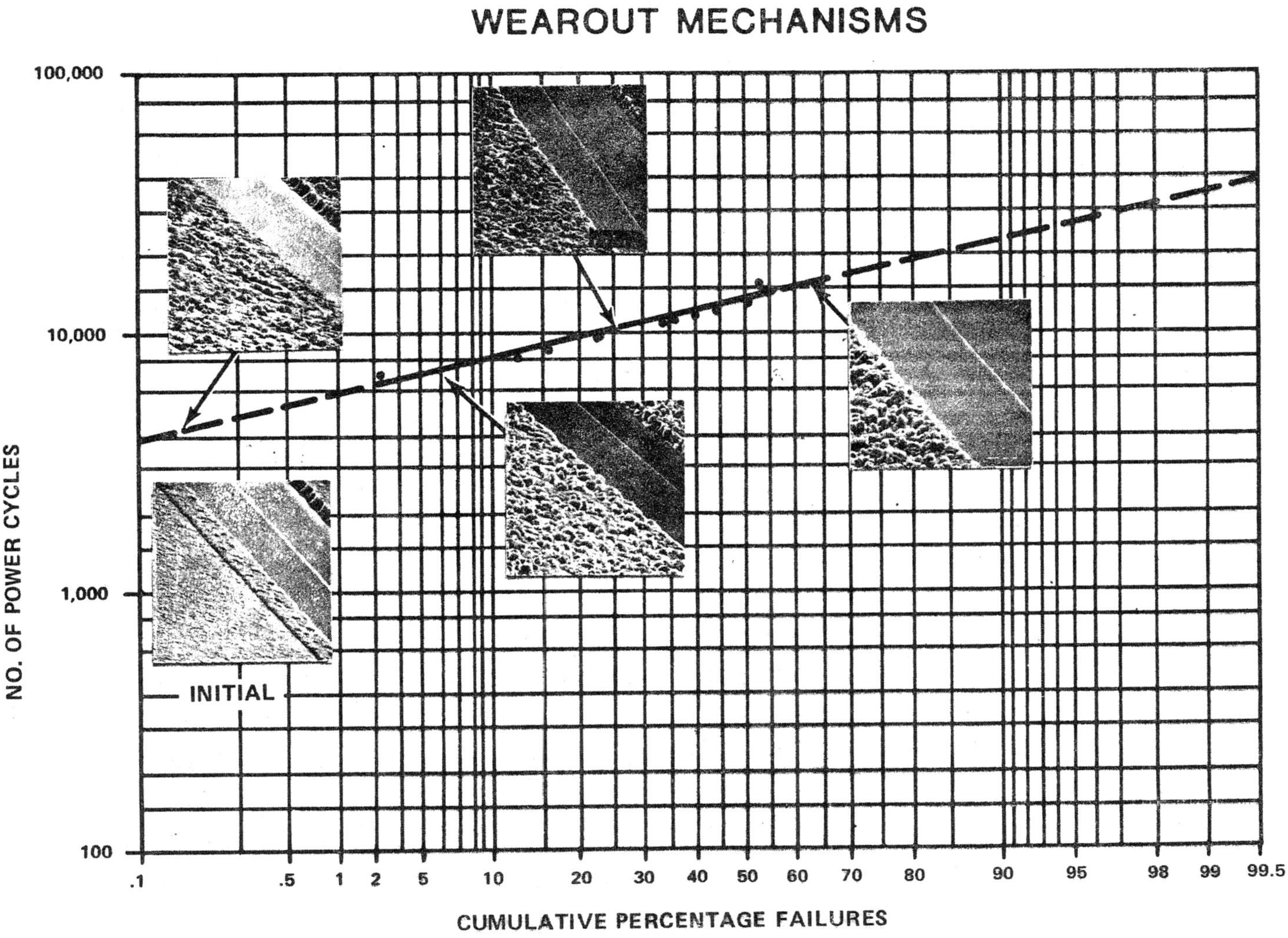

WEAROUT MECHANISMS

Fig. 33 Weibull chart of cycled devices with SEM photos of metallization. (600×)

perature cycling quickly reduces the capability of devices with soft solder die bonds, summarized in the adaptation of the graph shown in Fig. 30. The relative change in electrical parameters for transistors having hard and soft solder die bonds is dramatic

However, the lifetime of transistors with hard solder die bonds is not infinite, as Fig. 30 might imply. The stress generated by the silicon die-copper header mismatch, the driving force, moves to the die itself in hard-solder system. Because silicon is brittle and "notch-sensitive", electrically inactive mechanical imperfections can be propagated into cracks in the die. For this reason, a layer of metal having a CTE midway between silicon and copper, such as the molydbenum button shown in Fig. 1 and Fig. 2, is added. The stress is reduced, and resistance to cycling degradation improves. This increase in device lifetime is obtained at a slightly increased device piece-parts cost, but is amply rewarded by better performance.

Circuit Application Design for Reliability

The graph in Fig. 32 is excerpted from the RCA High Reliability Databook, SSD-207B of 1974. It shows the result of an extensive study with power devices in an actual circuit situation, a series-regulator type power supply. This plot clearly shows the degradation of electrical parameters in soft-soldered devices with increasing power dissipation and increasing number of cycles. Such data is precious and difficult to acquire. What it shows is that derating of components can help to extend their life expectancy beyond the product life cycle, eliminating the device as a reliability concern.

A Weibull plot is presented in Fig. 33, which shows cumulative percentage of failures of power transistors as a function of number of overstress power cycles, performed in a more conventional manner; a typical life-test. SEM photos of the metallization are superimposed on the graph. The gradual degradation of the metallization on the devices is evident from the increasingly roughened aluminum surface. These devices are being intentionally overstressed to failure, to evaluate their capability. They are not defective; they have been thermomechanically aggravated such that their end-of-life degradation mechanisms have been activated. If the test had been continued, all devices would have eventually failed.

There is no fundamental difference in the mechanisms operating in the good and failed devices when removed from any stage of this test. They are mechanically and electrically identical, with the exception of certain high-current or high-power tests, such as $V_{(SAT)}$, until the actual moment of catastrophic failure. The damage to the metallization in failures removed from test varied only in degree; the mechanism was always the same.

Catastrophic failures are due to the weakening of the metallization beneath the emitter wire bonds, which progresses until the wire bonds lift and create an open. Unless prevented by careful and elaborate experimental control, this may be followed by arcing and a destructive meltthrough. As discussed earlier and shown in Fig. 31, the solder alloy may have expanded and weakened until the die power dissipation is lowered, and the device goes into destructive thermal runaway. A typical example of device which failed under these conditions in the field is shown in Fig. 34; notice the thermal involvement of the small base fingers.

Corrective Action

In all cases, the lifetime of any device, generically speaking, is extended by generous, but not extravagant, derating. Careful attention to data such as the foregoing, from reliability and product engineering departments, combined with attention to the published device thermal derating charts and close association with the customer and his application, can avoid overstress

Fig. 34 Typical field failure caused by excessive power and thermal cycling. (100×)

Guide to the Interpretation of Physical Damage in Electrically-Damaged Devices

Physical Evidence	Most Probable Cause	Conclusion (Reported As)
Reconstructed metallization over entire die	FBSOA exceeded, device running too hot or excessive cycling	EOS./, FBSOA
Reconstructed metallization in characteristic area of emitter	RBSOA exceeded with any type of base drive, current-limited breakdown BVceo(sus) operation	EOS/RBSOA
Meltthrough hole in characteristic area of emitter	Es/b operation or repetitive high-voltage transient	EOS, second breakdown
Meltthrough hole in unusual area of emitter or base with edge arcovers	Non-repetitive transient in application	EOS, high-voltage transient
Bonding wire fused, die usually electrically OK	Forward-biased junction(s)	EOS, current pulse, reverse
Bonding wire fused, meltthrough hole in die, metal alloyed	Massive overstress	EOS, cause unknown
Tiny hole in oxide or silvery track visible in oxide between junctions	High-voltage, low energy	ESD event, test transient
Shorted junction or gate—no other physical damage is obvious	ESD event with damage located in silicon bulk	ESD event

Fig. 35 EOS Interpretation Guide

Thermomechanical Effects of EOS

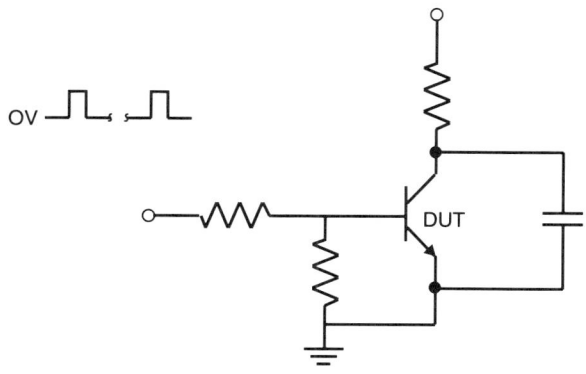

Fig. 36 Circuit showing capacitor discharge test.

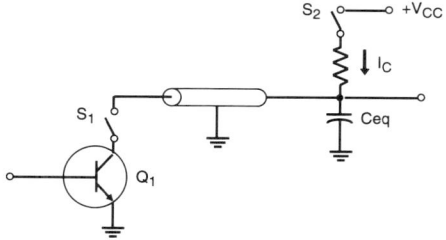

Fig. 37 Circuit showing mechanism resulting in test failures at test.

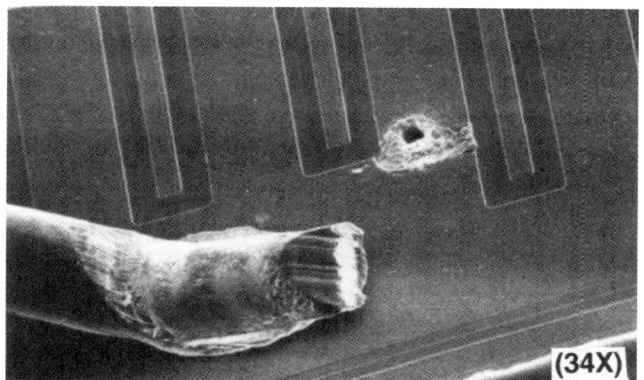

Fig. 38 Result of high-voltage transient on die. (34×)

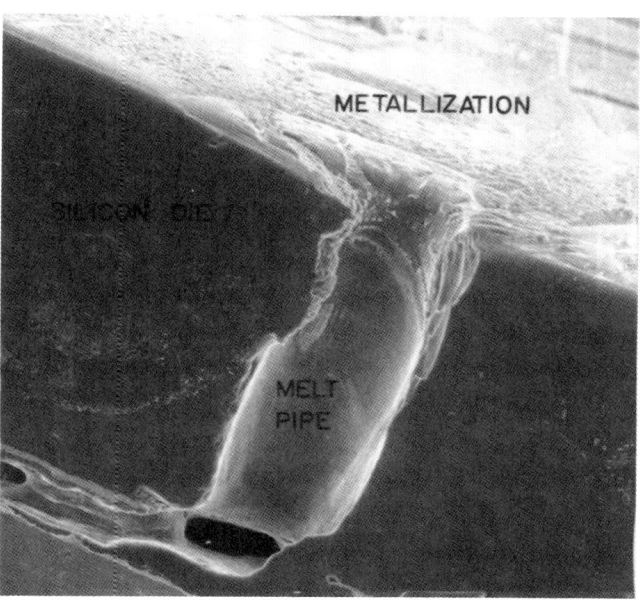

Fig. 39 Cross-section of a typical meltthrough in a die. (100×)

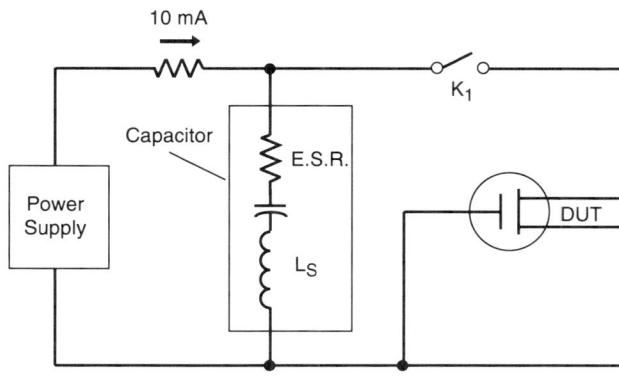

Note: K_1 Should be a Large Mercury Contactor.

Fig. 40 Circuit for forward biased impulse testing.

Fig. 41 Typical forward bias impulse failure. (130×)

conditions and assure a reliable product application of the semiconductor power device.

INTERPRETATION OF CATASTROPIC OVERSTRESS PHYSICAL DAMAGE IN FA REPORTS

The deciphering of the conditions responsible for the failure of devices under these conditions is usually somewhat complicated. More than one of the forms of overstress presented in these discussions are usually present simultaneously. Careful study of the circuit, device, and system in which failures occur can often produce better FA results than examining the device

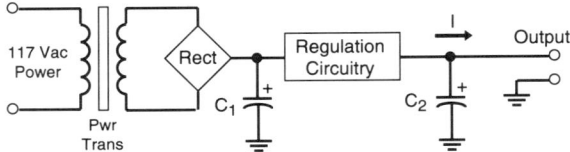

Fig. 42 Circuit arrangement which can result in impulse failures.

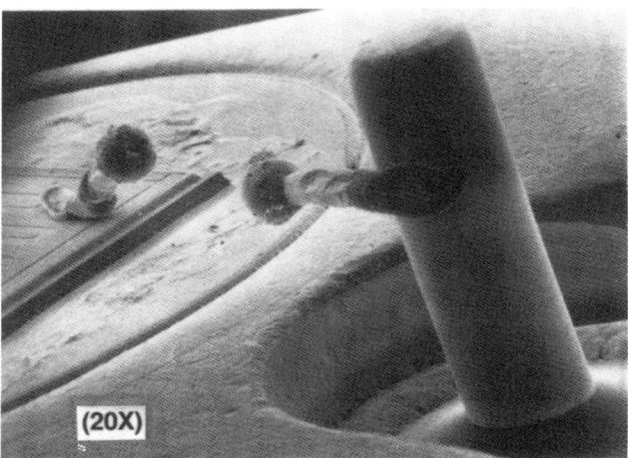

Fig. 43 Bonding wire melted due to forward bias impulse. (20×)

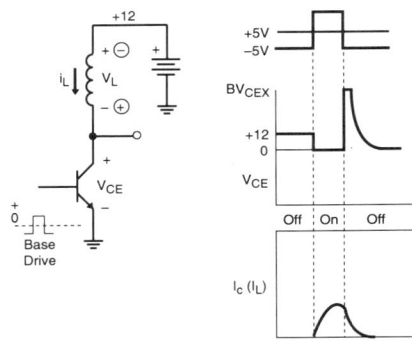

Fig. 44 Waveforms resulting from unsuppressed inductive tran-

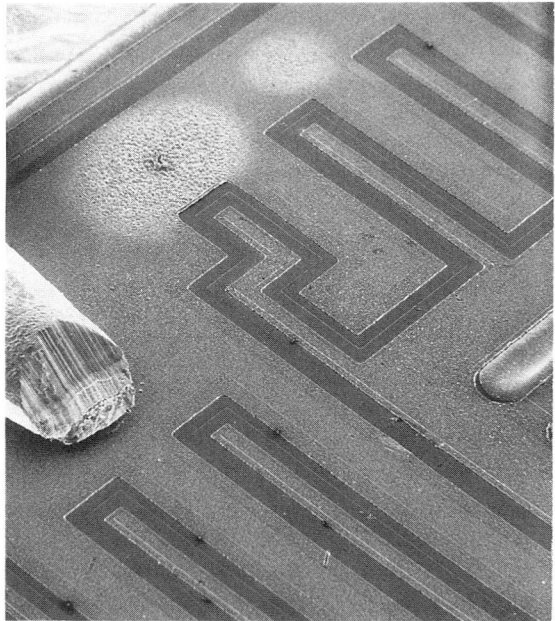

Fig. 45 Damage to due resulting from condition in Fig. 44, when energy is below instantaneously destructive level. This damage was deliberately induced with a curve tracer. This is a simulation. Compare to Fig. 47.

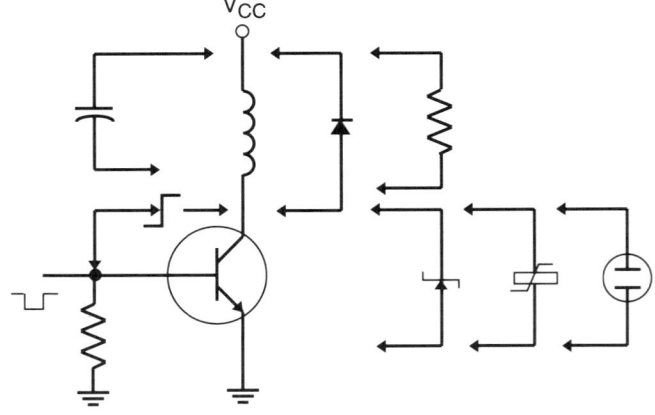

Fig. 46 Transient suppression possibilities for inductive switching.

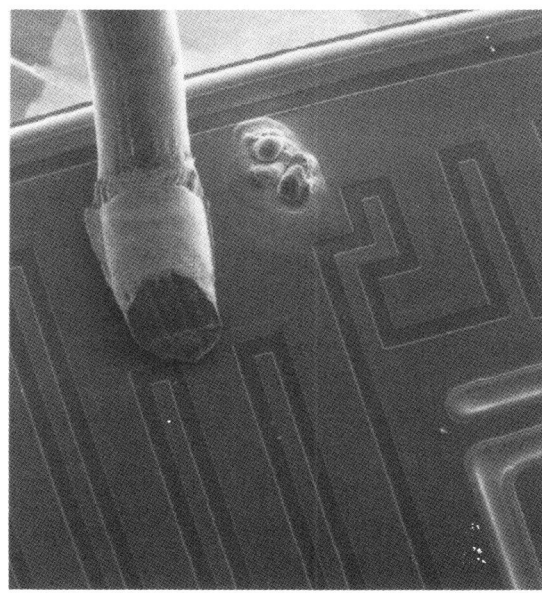

Fig. 47 Actual field failure of a device from a drive for a 3-phase brushless motor in a loomfeeder. (20×) (compare to Fig. 45).

by itself. Simulations can also lead to data which can help pinpoint the problems.

The chart in Fig. 35 is intended to show the types of conditions under which some failures may be generated, and the following pages address some of the circuit and electrical conditions which can generate failures.

Figures 36 through 39 address the case where a capacitance is allowed to suddenly discharge into a device when the voltage is in excess of the C-E breakdown voltage, but the energy in the capacitor is small; of the order of 10s of millijoules. The result is usually a small meltthrough hole in the device, causing a short, with no other associated damage. If the external circuit impedance is low, and shunt current is available, the die can experience extensive damage (a crater), and the small meltthrough hole evidence may be lost.

Figures 40 through 43 show the result of the discharge of a capacitor in circuitry where the device junctions are instantaneously biased forward (in the case of the C-B junction), or reverse (in the case of the E-B junction) (reverse of normal). The result is usually a fused bond wire, and probing of the die probably shows it to be good to specifications. Devices plugged in backwards during test, and devices subjected to reversed battery or power supply faults, can exhibit this failure mechanism.

Figures 44 through 46 show a much more usual case, where the unsupressed (or unsnubbed) kickback from an inductive load is allowed to hammer the device into avalanche breakdown, but not sufficiently to cause second breakdown with its crowbarring of the power supply, and instantaneous catastrophic failure. Gradual degradation with **delayed** catastrophic failure will usually result. Examination of the unfailed, but aged, devices from the application will occasionally show the beginning signs of degradation in the metallization; **exactly** the same mechanism exercised in the tests producing the damage shown in Fig. 24. The "characteristic area" will vary in diameter with the pulse width, die design, and many other factors.

Figure 46 shows some of the many device protection schemes that may be used with a power switching device. The accidental omission of the protection device, oscillation in a snubber circuit, insufficient switching speed of the protection device, or a defective protection device can all result in failure in a protected application.

WAS IT CURRENT, OR VOLTAGE?

Probably one of the most often-asked questions following the pronouncement of EOS as the failure mechanism in a device, this question should be more easily answerable following the reading of this article. Figure 48 is an SOA chart with **typical**, not absolute, examples of transistor dice of various designs which have likely failed under the conditions indicated on the chart. This summary overview should help to decipher which of the device maximum ratings; current, voltage, power dissipation, temperature cycling, etc was likely exceeded when EOS failures cannot be explained by any device defect.

WAS IT EOS OR WAS IT ESD?

The most effective demarcation between EOS and ESD is the amount of energy which can be delivered to the device. The human body, and the human body model test, can only deliver microjoules to a few millijoules of energy. A power circuit, however, can deliver orders of magnitude more energy. Since it is the energy which causes the melting and vaporization of a volume of silicon to create a catastropic failure, ESD damage sites will be far smaller than EOS damage sites. Consequently, ESD damage must be found with a SEM, and EOS damage can be found with an optical microscope.

REFERENCES

1. *Protecting Computer Systems Against Power Transients,* by Francois Martzloff, IEEE Spectrum, April, 1990, pp 27-40.

2. *Surface Reconstruction of Aluminum Metallization - A New Potential Wearout Mechanism,* by E. Philofsky, K. Ravi, E. Hall, and J. Black, IEEE Proceedings of the Reliability Physics Symposium, 1971, pp 120-128.

3. *Thermal Fatigue in Power Transistors,* by G.A. Lang, B.J. Fehder, and W.D. Williams, IEEE Transactions on Electron Devices, Sept, 1970.

4. *Solder Fatigue Problems in Power Packages,* by James F. Burgess et.al, in the IEEE Transactions on Components, Hybrids, and Manufacturing Technology, Vol CHMT-7, No 7., December, 1984, pg 405.

5. *Properties of Die Bond Alloys Relating to Thermal Fatigue,* by Dennis R.Olsen and Howard M. Berg, IEEE Transactions on Components, Hybrids, and Manufacturing Technology, Vol CHMT-2, No. 2, June 1979, pp 257-262.

6. *Quantitative Measurement of Thermal-Cycling Capability of Silicon Power Transistors,* by L. J. Gallace, Power Transistor Application Note AN-6163, RCA Solid State Division, Somerville, New Jersey.

7. *Protecting Computer Systems Against Power Transients,* by Francois Martzloff, IEEE Spectrum, April, 1990, pp 27-40.

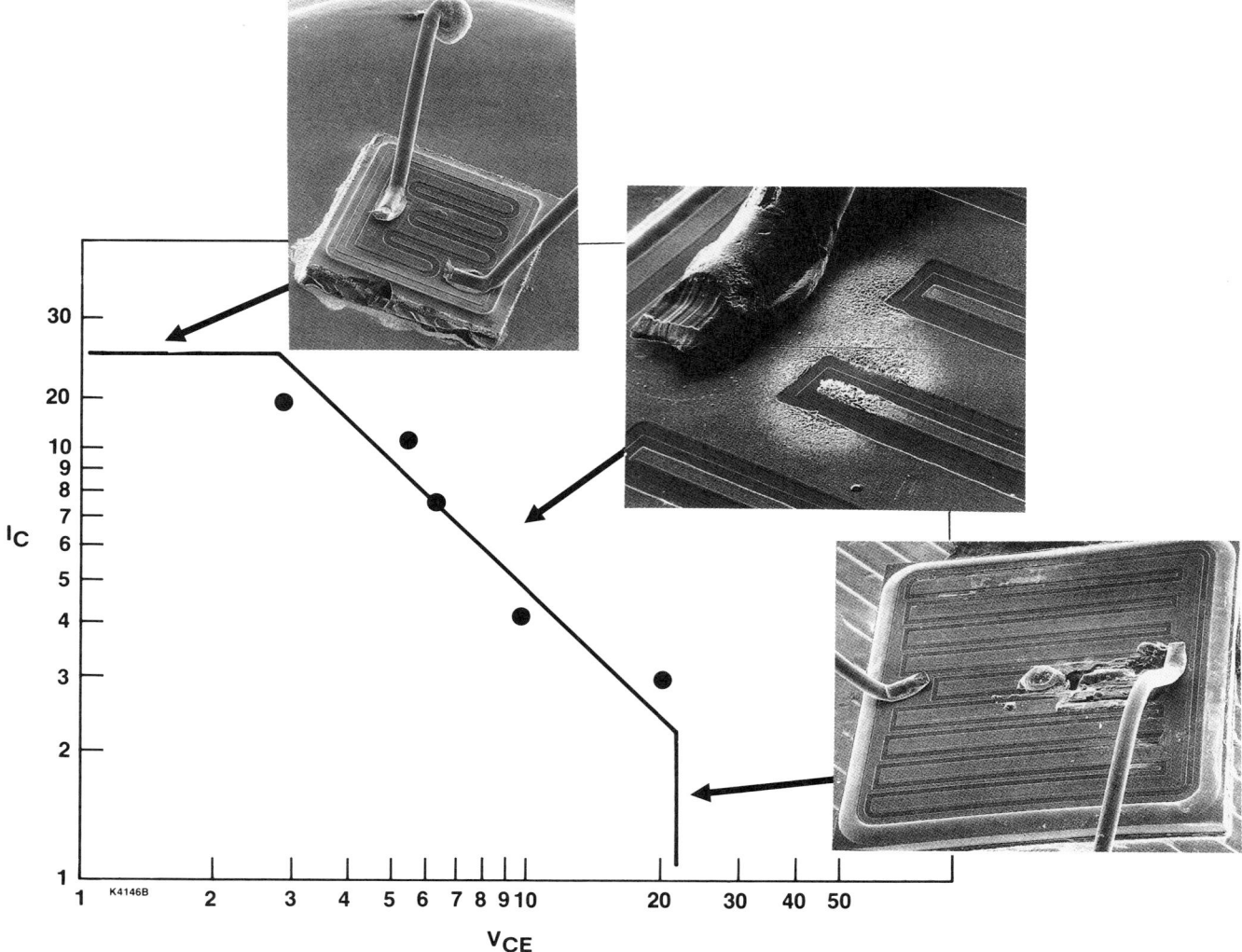

Fig. 48 SOA chart with typical power transistor damage. Appearance superimposed.

8. *Thermal Fatigue in Power Transistors*, by G.A. Lang, B.J. Fehder, and W.D. Williams, IEEE Transactions on Electron Devices, Sept, 1970.

9. *SEM Study of the Dynamics of Electrical Overstress (EOS) in Power Transistors*, by T.W. Lee, T.T. Guthrie, and G.P. Thome, Proceedings of ISTFA 1984, pp 221-226.

10. *Power Transistor Technology, and Safe Operating Area,* Chapters 1 and 2 in the Fairchild Power Data Book, 1976, pp 1-3 to 2-12. (Details the differences between double-diffused, single-diffused and epi-base transistor structures, design tradeoffs, and SOA in terms of I/sb and Es/b.)

11. *Limiting Phenomena in Power Transistors*, RCA Solid-State Devices Manual, (SC-16), 1975, pp 86-102.(Discusses base-width modulation, thermal debiasing, thermal cycle life, design tradeoffs, and variables included in the Es/b rating system.)

12. *Focus on Power Transistors and Thyristors,* by Louis Grossman, Electronic Design Magazine, #23, 8 November 1977, pp 52-60. (This article details the physical, electrical, and processing tradeoffs used by power device engineers in a non-technical manner and cautions against the misinterpretation of published specifications, including Es/b.)

13. *Reverse-Bias Second Breakdown in Power Transistors,* by D. L. Blackburn and D. W. Berning, in Electrical Overstress / Electrostatic Discharge Symposium Proceedings, (EOS-I), RADC Reliability Analysis Center, 1979, pp 116-121. (Describes a test set which uses a high-speed clamp consisting of 16 sweep tubes connected in parallel to divert current from the device under test when it switches into second breakdown.)

14. *Transient Damage In Power Transistors* by T. W. Lee, Proceedings of ATFA-79, pp 130-135 (capacitor discharge simulation methods used to simulate field failure damage.)

15. *Transient Protection Devices*, by Chin-Lin Chen, in I.E.E.E. International Symposium on Electromagnetic Compatibility, 1975, pg 248.

16. *Electrical Overstress Failure Analysis in Microcircuits*, by J. S. Smith, International Reliability Physics Symposium, 16th Annual Proceedings, 1978, pgs 41-46.

17. *Handbook of Thermophysical Properties of Solid Materials*, Pergamon Press, 1961.

18. *Handbook of Material Science*, by Charles T. Lynch, CRC Press, 1974.

19. *Semiconductor Power Devices*, by Sorab K. Ghandi, Wiley, 1977. (Contains general discussions on the source, type, and effect of various defects on the charactristics of power devices in Chapter 6. This text is "required reading" for anyone associated with the manufacture of power devices.)

20. *Focus on Power Transistors and Thyristors*, by Morris Grossman, in Electronic Design Magazine, No. 23, Nov 8, 1977, pp 52-60.

21. *The Influence of Circuit and Device Parameters on the Switching Performance of Power Transistors*, by P.L. Hoover and K.S. Tarneja, Power Conversion International, pgs 10, 12, 14, 16, and 21.

22. *Basic Design Considerations for Power Transistors*, chapter 6 in the RCA Solid State Power Circuits Designer's Handbook, 1971, pp 80-112.

23. *Physical Basis for Power Transistor Ratings*, chapter 7 in the RCA Solid State Power Circuits Designer's Handbook, 1971, pp 113-149.

24. *Limiting Phenomena in Power Transistors and the Interpretation of EOS Damage*, by J. Thomas May (in this volume).

25. *The Application of Molybdenum Contacts for Improved Second Breakdown Performance*, by E.B. Hakim, Proceedings of IEEE, Vol 54, Jun 1966, pg 880.

26. *Second Breakdown in Power Transistors Due to Avalanche Injection*, by B.A. Beatty, Surinder Krishna, and Michael S. Adler, IEEE Transactions on Electron Devices, Vol ED-23, No 8, Aug 1976, pp 851-857.

27. *MIL-HDBK-217B*, U.S. Government Printing Office.

28. *On US Mil-Hdbk-217 and Reliability Prediction*, by Charles T. Leonard, IEEE Transactions on Reliability, Vol 37, No.5, December, 1988, pp 450-452.

29. *Reliability Prediction—Use It Wisely*, by Anthony J. Feduccia, IEEE Transactions on Reliability, Vol 37, No.5, December, 1988, pg 457.

30. *Determination of the Threshold Failure Levels of Semiconductor Diodes and Transistors Due to Pulse Voltages*, by D. C. Wunsch and R.R. Bell, in the I.E.E.E. Transactions on Nuclear Science, Vol. NS-l5, No. 6, pgs 244-247.

31. *Basic Integrated Circuit Engineering*, by Hamilton & Howard, McGraw-Hill, 1975, p. 245 and 246.

32. *Pulse Power Failure Modes in Semiconductors*, by D. M. Tasca, I.E.E.E. Transactions on Nuclear Science, Vol. NS-l7, No. 6, December, 1970.

33. *A Thermal Damage Model for Bipolar Semiconductors*, by G. A. Hjellen and T. J. Lange, I.E.E.E. Symposium on Electromagnetic Compatibility, 1977, p. 444.

34. *Second Breakdown in Power Transistors due to Avalanche Injection*, by Beatty, et.al., in the I.E.E.E. Transactions on Electron Devices, Vol. ED-23, No. 8, August, 1976, p. 852.

35. *Mesoplasma Breakdown in Silicon Junctions*, by A.C. English, Proceedings of IEEE, Mar 1963, pp 500, 501.

36. *Microplasmas in Silicon,* by D. J. Rose, 7. The Physical Review, Vol 105, No. 2, 15 Jan 1957, pp 413-147. (Reverse breakdown in silicon junctions was shown to consist of a microplasma of 5-600 Ångstroms dia, in which current was conducted in discrete pulses.)

37. *Thermal Instabilities and Hot Spots in Junction Transistors*, by R. M. Scarlett, W. Schockley, and R. H. Haitz, Physics of Failure in Electronics, Vol 1, 1963, pp 194-203. (A dated, but complete review of the physics of second breakdown. The experimental work used temperature-sensitive paints to find hot spot temperatures of 300 - 400 °C.)

38. *Avalanche Effects In Silicon P-N Junctions. I. Localized Photomultiplication Studies on Microplasmas*, by R. H. Haltz, A. Goetzberger, R. M. Scarlett, and W. Shockley, Journal of Applied Physics, Vol 34, No 6, June 1963, pp 1581 - 1590. (Using a tiny light spot, carrier multiplication in microplasmas was shown to be as high as 106, microplasma diameter was shown to be a function of current, and microplasmas were found to have a dark core.)

39. *Second Breakdown and Crystallographic Defects in Transistors*, by H. A. Schafft, G. H. Schwuttke, and R. L. Ruggles, J. IEEE Transactions on Electron Devices, Vol ED-13, No 11, Nov 1966, pp 738-742. (The susceptibility to BVCEO second breakdown in 1500 transistors containing various defects was studied by means of x-ray diffraction microscopy) (The entire volume ED-13 is interesting reading.)

40. *Physical Investigation of the Mesoplasma in Silicon*, by A. C. English, IEEE Transactions on Electron Devices, Aug/Sep 1966, pp 662-667. (The properties of mesoplasmas were studied by observing melts in Zener diodes, and the mesoplasma properties are investigated with a computer program.)

41. *Second Breakdown—A Comprehensive Review,* by Harry A. Schafft, Proceedings of the IEEE, Vol 55, No 8, Aug 1967, pp 1272-1288. (A monumental summary of research on the subject from 1946 to 1967. Contains 134 references.)

42. *Stable Hot Spots and Second Breakdown In Power Transistors*, by P. L. Hower, D. L. Blackburn, F. F. Oettinger, and S. Rubin, PESC Record, 1967, pp 234-246. (Presents temperature contours of devices obtained with an infrared microadiometer.)

43. *Electrical Breakdown in Solids*, by N. Klein, a chapter in volume 26 of Advances in Electronics and Electron Physics, edited by L. Marton, Academic Press, 1968, pp 309-424. (A complete work covering the mechanisms of breakdown in semiconductors and insulators. Has 183 references.)

44. *Analysis of Requirements In Reliability Physics*, Alfred L. Tamburrino, Physics of Failure in Electronics (PFE) Vol 2, RADC, New York, 1964, pp 13, 16, 18.

45. *Silicon Semiconductor Technology,* by W.R. Runyan, McGraw-Hill, 1965. (This text book covers all phases of extraction of silicon and growth of crystals, doping diffusion, and optical, electrical and other physical properties.)

46. *High Current Transient Induced Shorts,* J. S. Smith, 9th Annual Proceedings Reliability Physics, 1971, pp 163-171.

47. *Basic Design Considerations for Power Transistors,* and *Physical Basis for Power Transistor Ratings*, chapters in RCA Power Circuits Designer's Handbook, Technical Series SP-52, 1971, pp 95-149. (Covers all concepts of power transistor design, multiple epitaxial (π-u) and multiple-emitter techniques. The second chapter discusses avalanche breakdown algebraically, RBSOA, FBSOA, and introduces forward bias capacitance-discharge as a useful test because it "approximates actual circuit conditions".)

48. *Non-Destructive Screening for Thermal Second Breakdown*, by Dante M. Tasca, et. al., IEEE Transactions on Nuclear Science, Vol NS-19, No 6, Dec, 1972, pp 57-67. (Among a large matrix of tests to screen switching diodes for second breakdown vulnerability, capacitors are placed in parallel with devices taken into second breakdown.)

49. *Reliability and Degradation*, edited by M. J. Howes and D. V. Morgan, Wiley, 1981. (An excellent study of degradation mechanisms in metal films; interfacial layers, interdiffusion, analytical methods, III-V devices, et cetera. Chapter 1, on metal films, has 457 references.)

50. *Safe Operating Area Information*, a section of the Motorola Power Device Handbook. Motorola Technical Information Center, 1982.

51. *Understanding Power Transistor Dynamic Behavior...and Power Transistor Safe Operating Area...*, by Warren Schultz, Motorola Application Notes AN-873 and -875 respectively, June and Sept 1982. (Discusses power transistor limitations in terms of dv/dt and Cob, concluding that snubbing networks are a cure, and shows that conditions resulting in power transistor failure may be subtle effects of the application conditions.)

52. *Atmospheric Electricity*, by B. F. J. Schonland, Meuthen & Co., London, 1953.

53. *Surge Voltages in Residential and Industrial Power Circuits*, by F. D. Martzloff and G. I. Hahn, in the I.E.E.E. Transactions on Power Apparatus & Systems, Vol. PAS-89 No. 6, July/August 1970, pg 65.

54. *Modeling of Electrical Overstress in Silicon Devices*, by N. Kusenov and J.S. Smith, in the Proceedings of the EOS/ESD Symposium, 1979, pp 133-139.

55. *Improved Cost Effectiveness and Product Reliability Through Solder Alloy Development*, by Dennis R. Olsen and Keith G. Spanger, Solid State Technology, Sept 1991, pp 121-126.

56. *RCA High Reliability Devices Databook,* SSD-207B, RCA Corporation, Sommerville, New Jersey, 1974.

57. *Thermal Cycling Rating System for Power Transistors, by Wally D. Williams,* RCA Application Note 4612, available from Harris Semiconductor Communications Department, Melbourne, Florida, 32902.

Electrical Overstress in Integrated Circuits

by J. Thomas May

The formulas used in this discussion are available in any college textbook, and can be found in many high school texts as well. They are not complicated at all. Anyone involved in interpreting electrical overstress damage is encouraged to study these heat transfer equations until gaining an intuitive familiarity with them. The values that the formulas predict should vary from actual results. There will be some errors because the temperature effects of the resistance, and the effect of the glassivation on the metal has intentionally not been considered. This discussion is intended to give an intuitive understanding.

Figure 1 is an oblique combination cross-section and surface drawing, showing a metal wire, with a bond connecting to a stripe or run of aluminum metallization, sitting on top of a piece of silicon (see following page). The PN junction at the end indicates that the end of the aluminum stripe is sitting effectively at ground potential. In the cross section at the side, "W" is the width of the aluminum, "D" is the thickness of aluminum, and "X" is thickness of the SiO_2. The purpose of this article is to explain why the fusing current current of a wire in series with metallization is a complicated function of the pulse width.

Let's start by considering region "A" of the graph, which is the DC fusing current of the wire. The formula for the fusing current of a typical short bonding wire is $60\, D^2/L$. For a 1 mil diameter wire that is 60 mils long, this works out to be 1 amp. If you look at the graph, you will see that the right part of the curve, labeled "A," is a constant 1 amp. If you no longer have DC, and you apply a millisecond pulse, you will find the wire will take considerably more than 1 ampere, which leads us to region "B".

Region "B" is the adiabatic heating of the wire. Adiabatic means that the wire is heated so fast that the heat does not have time to flow out of the wire to the cooler ends, which function as a heat sink. All of the energy, the I^2Rt, gets absorbed in the wire, and causes the temperature of the wire to rise. In the first equation energy absorbed is the change in temperature times "C", the specific heat, times the mass. In equation 2, the energy dissipated, is the power I^2t multiplied by $\rho'L/A$. When you equate the energy absorbed to the energy dissipated (set equation 1 equal to equation 2), you can see that the L's cancel out. It is important to see that the fusing current varies as the inverse square root of the pulse width. The equation in the rectangle is the equation that's plotted on the If vs t curve.

Next, consider region "C" of the curve, the non-adiabatic heating of the metallization. This is analogous to DC fusing current on the wire. In that case, the heat flows through the wire to the heat-sunk ends. For the region "C" metallization, the heat flows from the metallization, through the dielectric layer that is separating the metal from the silicon, and into the die. We will equate the power dissipation in the metallization to the heat conducted out. The power in is I^2R (equation 1). The K factor, Kx, is thermal conductivity, which is likened to the heat conductivity of the wall of a house. The SiO_2 is analogous to the insulation in your house. When the power dissipation is equated to the heat conducted out, the L's cancel out again, and you can solve for the current. You can see that it is a constant, not varying with the pulse width. Here, we have worked out that for 1 micron thick aluminum, you get 1.8×10^3 amps/cm^2. For a 2-mil wide stripe (0.002 in. or 50 μm), perhaps an output pad on a power device, the metal would handle 9.2 amperes before it fuses. This is a lot of current. This is because SiO_2 is a really good thermal conductor, moving heat away from the metal quite readily.

Region "D" of the curve represents adiabatic heating of the metallization. The math involved is almost exactly the same as that in the case of the wire. The energy absorbed is equated to the energy dissipated. The current is inversely proportional to the square root of the pulse width. This is again worked out for 50 micron wide metallization stripe, the equation which is plotted on the graph. Notice that the current required to fuse the metal in this region is very high, greater than 10 amperes. For a practical situation in an integrated circuit, another structure would burn out before the metallization. Suppose you were to put 10 amperes through a reverse-biased junction. The power density would be so high that the heat would cause the junction to burn out before the metallization. Region "D" is more of a theoretical curve; and normally, something else burns out before the metal.

Refer to the graph of log current vs log time, once again. There are three critical times separating the major areas; A-B, B-C, and C-D. You can derive a formula for these critical times. You can write an equation for the "A" side, and one for the "B" side, and equate them. As you can see, if you solve for "t", it turns out to be 14.4 mSec for the dimensions assumed. Similarly, the critical time separating regions B-C, and C-D, are 170 μsec, and 1.7 μsec, respectively. The 170 μSec critical time is Equation 4 of Jack Smith's famous 1978 IRPS paper.

So, in the case of this particular wire and stripe, if the pulse is faster than 170 μsec, the metallization will fuse. If it is slower than 170 μsec, the bonding wire will fuse. This is probably the most useful of the critical numbers. The C-D number is not practical for typical integrated circuits.

Next consider the temperature profile graphs which show temperature as a function of location (Fig. 2). In all cases, the sketch has the metallization on the left, and the bond wire on the right. The temperature is sketched as a function of the location. The top one is the DC case. The very hottest spot is in the center of the bond wire, and the ends of the bond wire are relatively cool.

The middle sketch is labeled 170 μsec and 9.2 amperes.

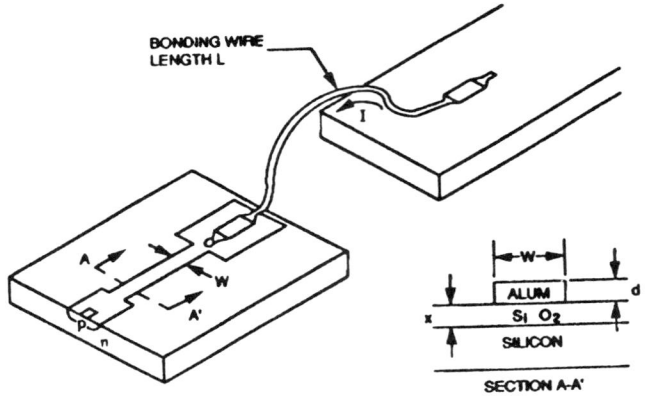

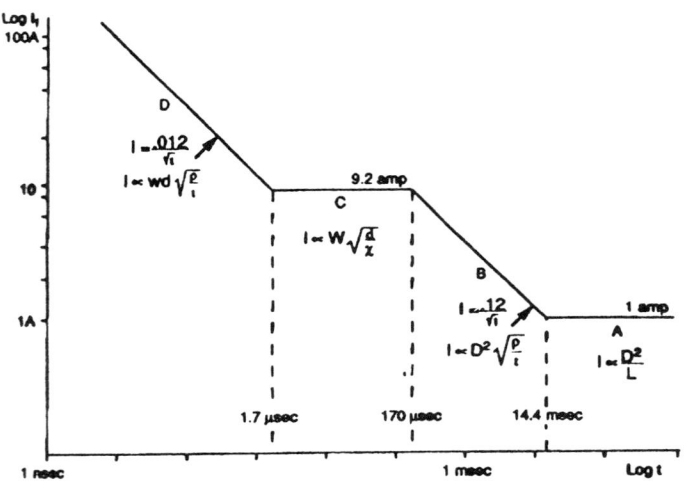

Fusing current as a function of pulse duration of a 1 mil x 60 mil alum wire in series with a 1 µ x 50 µ run of alum metallization on a 1 µ $Si O_2$ layer

Region A: DC Fusing current of wire.
(Non adiabatic)

For
$L < 100D$ $\quad I_f \doteq 60 \dfrac{D^2}{L} \dfrac{amp}{mil} = 2.36 \, E4 \dfrac{D^2}{L} \dfrac{amp}{cm}$

For $\quad D = 1$ mil dia
$\quad L = 60$ mil long
$\quad I_f = 1$ amp

Region B: Adiabatic heating of wire.
Energy involved in increasing temperature by ΔT

① Energy absorbed $= \Delta TC$ mass $= \Delta TC\rho$ vol
$\quad = \Delta TC\rho AL$ where $A = \dfrac{\pi}{4} D^2$

② Energy dissipated $= Pt = I^2 R \; t = I^2 t \rho' \dfrac{L}{A}$

① = ② $\quad \Delta TC\rho AL = I^2 t\rho' \dfrac{L}{A}$

$\dfrac{\Delta TC\rho A^2}{t\rho'} = I^2$

$I_f = \dfrac{\pi D^2}{4} \sqrt{\dfrac{\Delta TC\rho}{t\rho'}}$

for 1 mil Alum wire at melting temp.

$\boxed{I_f = \dfrac{1.2E\text{-}1}{\sqrt{t}} \sec^{\frac{1}{2}} amp}$

Region C: Non adiabatic heating of metallization.
Power involved in increasing temperature by ΔT

① Power in $= I^2 R = I^2 \rho' \dfrac{L}{dw}$

② Heat out $= \dfrac{K_x LW \Delta T}{\chi}$

① = ②

$I^2 \rho' \dfrac{L}{dw} = \dfrac{K_x LW \Delta T}{\chi}$

$\dfrac{I^2}{W^2} = \dfrac{d K_x \Delta T}{\chi \rho'}$

$I = W \sqrt{\dfrac{K_x d \Delta T}{\chi \rho'}}$

for 1µ Alum on 1µ oxide at melting temperature

$\dfrac{I_f}{W} = 1.8E3 \dfrac{amp}{cm}$

for 50µ stripe $\boxed{I_f = 9.2 \text{ amp}}$

Region D: Adiabatic heating of metallization:
Energy involved in increasing temperature by ΔT:

① Energy absorbed $= \Delta TC$ mass $= \Delta TC \rho \, dwL$
② Energy dissipated $= Pt = I^2 Rt = I^2 t \rho' \dfrac{L}{wd}$

① = ②

$\Delta TC\rho dwL = I^2 t\rho' \dfrac{L}{wd}$

$\dfrac{\Delta TC\rho dwdw}{t\rho'} = I^2$

$I = wd \sqrt{\dfrac{\Delta T \rho C}{t\rho'}}$

for 1 µ Alum 50µ wide at melting temp.

$I_f = \dfrac{1.2E\text{-}2}{\sqrt{t}} \text{ amp sec}^{\frac{1}{2}}$

Fig. 1 Oblique combination cross-section and surface drawing.

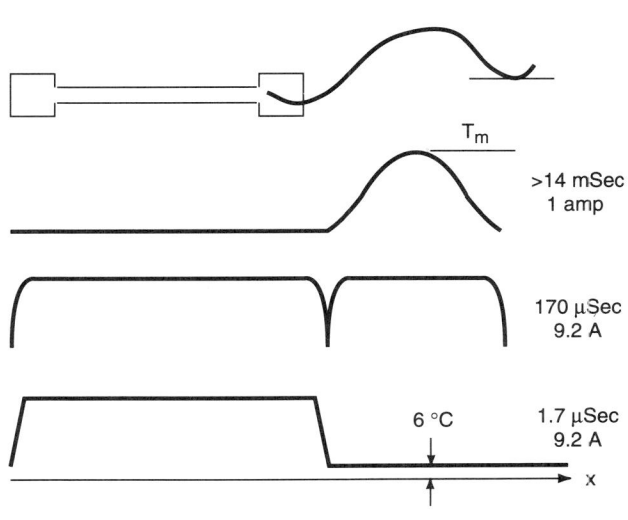

Fig. 2 Temperature as a function of location for metallization stripe

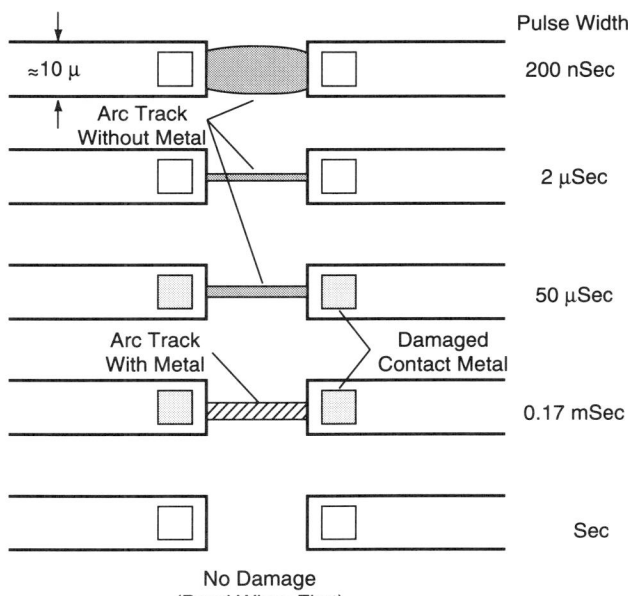

Fig. 3 Arc tracks

Table 1 Properties of Materials

	Units	Alum	Gold	Si	SiO$_2$
Specific Heat, C	$\frac{J}{°KG}$	0.856	0.128	0.7	1.0
Density, ρ	$\frac{g}{cm^3}$	2.70	19.3	2.33	2.27
Elec. Resistivity, ρ'	OHM-CM	2.63×10^6	2.1×10^6	N/A	N/A
Melting Temp. Above 25 °C, ΔT	°K	635	1020	1390	1675
Thermal Cond., K	$\frac{W}{cm \, °K}$	2.37	3.1	1.5	0.014
Thermal Diffusivity, κ	$\frac{cm^2}{sec}$	0.86	1.18	0.9	0.006
Elec Resistivity at Melting, ρ'_m	OHM-CM	11.1×10^6	13.7×10^6		

Imagine that you slowly increase the current of 170 μsec pulses until something fuses. Just before the aluminum melts, the temperature profile will be approximately as sketched. The aluminum will be right at the melting point in both the stripe and the wire. Next, imagine that you use a 1.7 μsec pulse width, and increase the current to produce a failure. Just before failure almost all of the metal along the stripe would be at the melting point. The bond wire would be much cooler, due to its greater thermal inertia.

The last drawing, which shows the arc tracks (Fig. 3), is condensed from Jack Smith's paper for your convenience. Up at the top, there is a very wide arc track, which would correspond to 200 μsec and faster pulses. This would correspond to an ESD pulse. When you make the pulse longer, the track gets thinner, then wider. At 2 μsec, you will notice discoloration at contacts. If you exceed 50 microseconds, you get silvery arc tracks, one of the most common effects of EOS that analysts observe. You also observe discoloration caused by overheating, where the metal contacts the silicon. If there is no damage whatever, it may indicate that the bond wire was fused first.

Finally, the table lists the various properties of the materials. This is useful is you want to substitute values into the equations that correspond to your particular IC. You are advised to use caution, because the values listed are for room temperature, and are therefore good only for first estimate, as explained in the opening paragraph.

REFERENCES

1. *Electrical Overstress Failure Analysis in Microcircuits*, by Jack Smith, Proceedings of the International Reliability Physics Symposium (IRPS), 1978, p 41-46
2. *University Physics*, by F.W. Sears and M.W. Zemansky, 1965
3. *The D.C. Fusing Current and Safe Operating Current of Microelectronic Bonding Wires*, by J. Thomas May et. al., ISTFA-89, p 121-131

RELIABILITY FOR THE FAILURE ANALYST

by David Burgess

PURPOSE

Failure analysts are key people in achieving the reliability the electronics world has enjoyed for the last few years. They have done this by providing the understanding required to eliminate failure causes. The work of the failure analyst is involved with the fixing of problems encountered when attempting to manufacture a product properly. The analyst is also involved when failures are encountered in the field. Figure 1 shows the typical "bathtub" curve describing product failure rate with time.

Despite this intimate role we have played, many analysts have little time to develop fluency in the mathematics of the quality and reliability they have helped create.

The purpose of this discussion will be to define and give an understanding of the terms of quality and reliability with respect to failure analysis.

At the expense of being rigidly correct, the discussion here will simplify quality and reliability statistics as they are commonly used. Just as a feel for the statistics of cards helps a poker player, a feel for reliability can help a failure analyst make wiser choices.

Quality is freedom from initial defects when the product is delivered. Reliability is how long the defect-free status is maintained.

TYPICAL TERMS

- AOQL
- AQL
- Confidence interval
- LPTD
- MTTF
- IFR
- Fit
- Normal
- Log normal
- Chi-squared
- Poisson
- Weibull
- MIL-STD-883
- MIL-STD-105
- Infant mortality
- Wearout failures
- Device-hours
- Acceleration factor

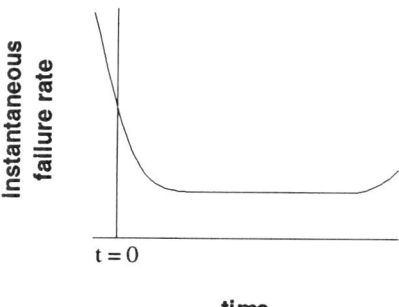

Fig. 1 Example of a typical "bathtub" failure rate curve. At the extreme left, infant mortality and early life failures constitute a decreasing failure rate. In the center, random or useful rate failures are constant and of low magnitude. At the right side of the curve, wearout failures have begun to occur. The product design should be obsolete long before wearout failures begin to occur.

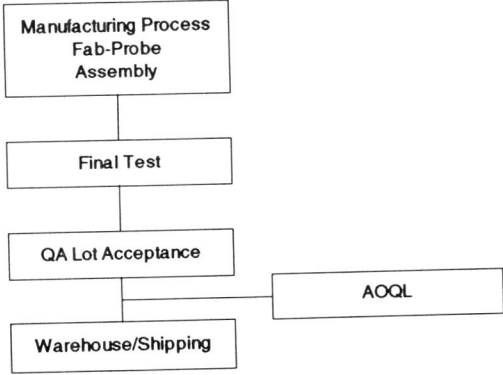

Fig. 2 Process flow block diagram, showing position of AOQL measurement.

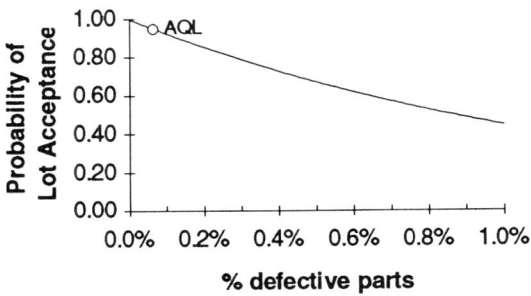

Fig. 3 Illustration of AQL.

QUALITY

At the manufacturer, measurement is done by QA sample randomly selected final test. Sample sizes depend on a sampling plan. An established and commonly used sampling plan is MIL STD 105, which specifies sample size as a function of desired AQL level and the lot size.

HOW IS AOQL MEASURED?

AOQL is a demonstrable fact, while AQL and LTPD are artifacts of a particular sampling plan which define only the user's and supplier's risks (a total of 100%) that product containing a given proportion defective will be accepted for delivery.

QA sampling is done at 0.25 AQL level. Sample sizes are determined by long established tables (MIL STD 105) according to lot size.

How can 0.25% AQL demonstrate 100 ppm? (0.25% = 0.25 / 100 = 2500 ppm)

WHAT DOES IT MEAN THAT THIS LOT PASSED 0.25 AQL SAMPLE?

If an accept on zero plan is used, we know n acceptable parts were chosen at random, what is the defect percent? If the probability that a part is defective is p, the probability of the lot being accepted is:

Prob. passing $= (1 - p)^n$
$0.1 = (1 - p_{90})^n$
$0.4 = (1 - p_{60})^n$

From the plot, only 10% of the samples will come from lots with defects $p > p_{90}$. Solve directly or estimate values by the approximation $(1 - x/n)^n \cong e^{-x}$. Since $e^{-2.3} \cong 0.1$, the 90% confidence estimate is $p_{90} \cong 2.3/n$. [p_{90} has been named the LTPD (lot tolerant percent defective.)] Similarly, $e^{-0.92} \cong 0.4$ implies that $p_{60} \cong 0.92/n$. The actual 60% or 90% confidence estimate is less than p_{60} or p_{90}.

Estimative AOQ requires data from many lots—say over a month's time.

L = # lots sampled

LR = # lots rejected

LR/L = % lots rejected

S = Total # samples per lot

N = Total # samples

D = Total # defective devices

for $D = 0$

$LR/L = 0$

60% confidence AOQL $\leq$ AOQL$_{60}$ = 0.92/N

For $D > 0$

This method uses a multiplier derived from the χ^2 distribution. The arithmetic is simple.

$$AOQL = \frac{D \cdot CF}{N}$$

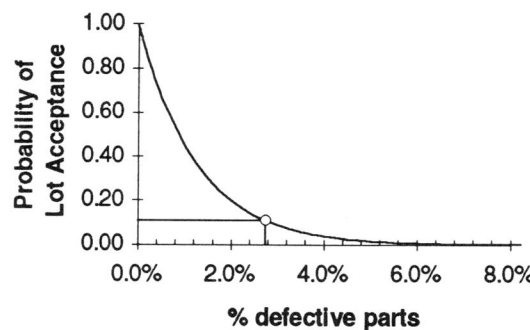

Fig. 4 Probability of lot acceptance as a function of percent defective parts.

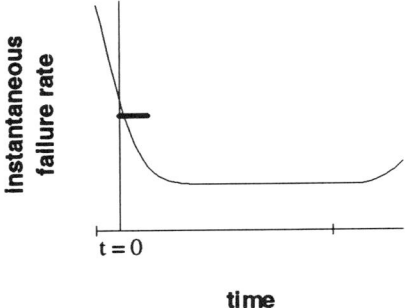

Fig. 5 Probability of instantaneous failure as a function of time.

where D is the number of defects, N is the total sample size, and CF is the confidence factor multiplier from the table below:

D	CF	
	60% confidence	90% confidence
1	2.02	3.89
2	1.55	2.66
3	1.39	2.23
4	1.31	2.00
5	1.26	1.85

EARLY FAILURE RATE (EFR) (ACCELERATED)

Accelerated-stress testing is legitimately used to estimate, and in some aspect and cases to control, both early and long-term reliability, not only the early. It is the only way to increase confidence in future reliability before the product is delivered and used.

Because of small number of failures, failure rate is assumed constant during a specified, but somewhat arbitrary, time. Typical are (a) 48 hours, 125 °C, 7V and (b) 168 hours, 125 °C, 7V.

EFR may be given as ppm value for the specified stress-time. In this case all calculations are exactly the same as for quality measurement. We are simply defining a test that extends over time.

For example, suppose 1000 devices per month are selected from inventory with no failures detected, $D = 0$:

$$EFR\ (60\%) = \frac{0.92}{1000} \times 10^6 = 9200\ \text{ppm}$$

The calculation from a year of tests at 1000 devices/month for $D = 0$ and $D = 1$:

$$D = 0 \qquad EFR\ (60\%) = \frac{0.92}{12{,}000} \times 10^6 = 720\ \text{ppm}$$

$$D = 1 \qquad EFR\ (90\%) = \frac{1 \times 2.02}{12{,}000} \times 10^6 = 1667\ \text{ppm}$$

EFR expressed as failure rate at normal use conditions (55 °C, 5V)

$$EFR = \frac{EFR \cdot 10^{-6}}{(HRS)(AF_{\text{TEMP}})(AF_{\text{VOLT}})}\ \text{in failures/hour}$$

Acceleration factors AF_{TEMP}, AF_{VOLT} will be discussed later.

Use of cumulative lot-test data to estimate quality and/or reliability is acceptable for a stable design and manufacturing process; it is not for varying ones. Cumulative data do not influence the estimate of quality/reliability for a given lot of product.

INTRINSIC FAILURE RATE (IFR)

Again, because of small number of failures, failure rate is assumed constant during a specified time. Typical is 1000 hours, 125 °C, 7V. IFR is expected as a failure rate at normal use conditions.

$$IFR = \frac{CF \cdot (\#\ \text{Fails})}{(HRS)(AF_{\text{TEMP}})(AF_{\text{VOLT}})} \times 10^9$$

DISCUSSION POINTS

1. Failure rate is shown for times $t < 0$, because the device is defined during the manufacturing process. Oxides are grown, contacts are made, etc. Defectives are removed at burn-in screen, other environment or electrical screen or at final test. These defects are yield loss.
2. Ratios EFR/IFR greater than 1 suggest the need for more/better production screening.
3. Failures occurring at final test, and during EFR and IFR are probably defect-generated. Reduction of value depends on decreasing defects.
4. If a semiconductor company's data shows AOQL < 100 ppm, observation of several failures by a customer is an unlikely event unless reject criteria, test conditions, or unique handling and processing conditions are involved. The failure analyst should not overlook lot history.
5. One "fit" is one failure in 10^9 device-hours (or unit-hours). If semiconductor company reliability monitor data show EFR $\cong$ IFR < 100 FITS, the observation of several fails by the customer suggest more than just statistical variation. Check lot history, special handling, testing, and fail criteria.
6. If field failure rate >> estimated failure rate, many reasons are more likely than statistical variation:
   a. AF factors. Using one accelerator has shown above, can be accurate only when the predominant failure mechanism is known and is properly modeled. The resulting number is useful if all conditions are met.
   b. Process variation has made other mechanisms significant.
   c. Field conditions include temperature cycles, humidity, and hot carriers not included in mode.
   d. Just as acceleration factor must be matched to failure mechanism, production screens are also mechanism specific. Screen will not be effective if defect types change.

REFERENCES

1. *Toshiba Semiconductor Reliability Handbook*
2. D.L. Burgess and O.D. Trapp, *Failure and Yield Analysis Handbook*, Technology Associates, 51 Hillbrook Drive, Portola Valley, California 94205. 415-941-8272.
3. J.M Juran and F.M. Gryna, *Juran's Quality Handbook*, McGraw-Hill, 1988.
4. T.P. Omdahl, *Reliability, Availability, and Maintainability (RAM) Dictionary*, ASCQ Press, 1988.
5. G.E.P. Box, W.G. Hunter, and S. Hunter, *Statistics for Experimenters*, Wiley, 1978.
6. D.J. Klinger, Y. Nakada, and M.A. Menendez, *AT & T Reliability Manual*, Van Nostrand-Reinhold, 1990.

BASIC SILICON FRACTOGRAPHY

by Dan Pote

When a silicon die is cracked, a significant amount of information regarding the nature of the damage remains on the resulting fracture surfaces (Fig. 1). Fractography deals with the interpretation of these surface features. The various markings commonly observed on silicon fracture surfaces are documented along with brief explanations regarding their formation.

RIB MARKS

Rib marks (Fig. 3 through 6), often called clamshell markings, are generally considered to show the instantaneous location of the crack front [5]. They form perpendicular to direction of crack propagation, and are usually concave toward the fracture origin. Wallner lines are a form of rib mark that are produced when the running crack front intersects singularities along the edge of the die. Intersecting rib marks of different curvatures result when the crack reaches a significant fraction of the local speed of sound [6] (see Fig. 18). Arrest lines, or Hesitation lines—which are more prominent rib marks—form when the crack temporarily stops and starts, and are difficult to differentiate from common rib marks.

RIVER LINES

River lines appear as thin lines merging in the direction of crack gravel (see Fig. 8 through 11). They are actually formed through the converging of parallel fracture planes, and are probably the single most useful form of evidence used for determining the crack origin in fractured silicon. The markings themselves vary in size and shape and can appear anywhere on the fracture surface. Although river marks in LiF and other crystals have been confirmed to initiate at screw dislocations, no such correlation has been found in silicon [7].

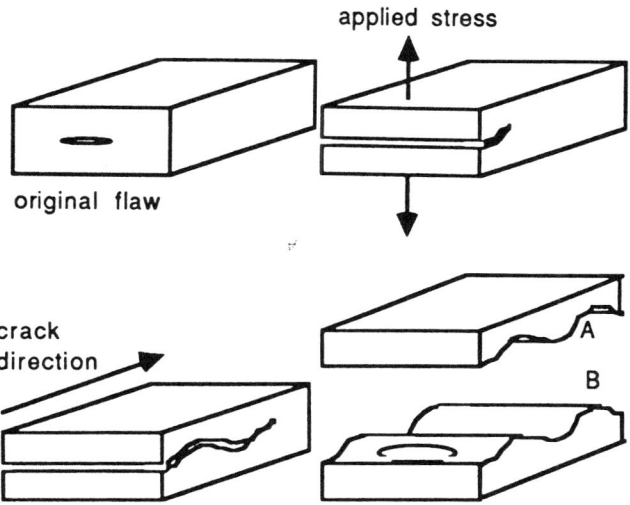

Fig. 1 Crack propagation and formation of fracture surfaces A and B.

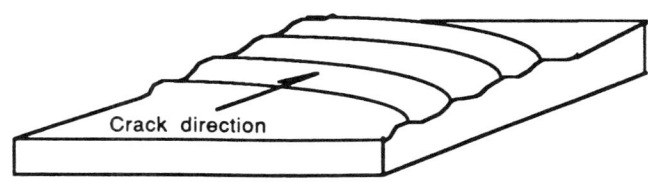

Fig. 2 Rib marks.

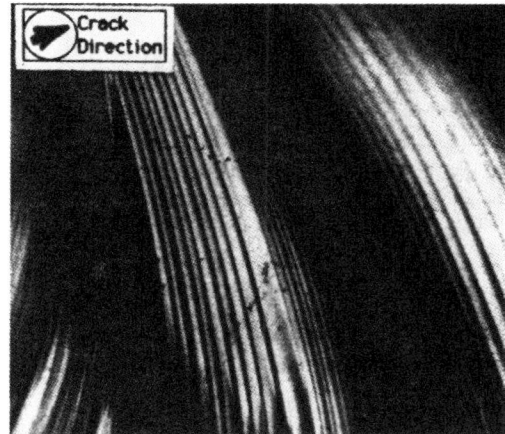

Fig. 3 Rib marks. (200×)

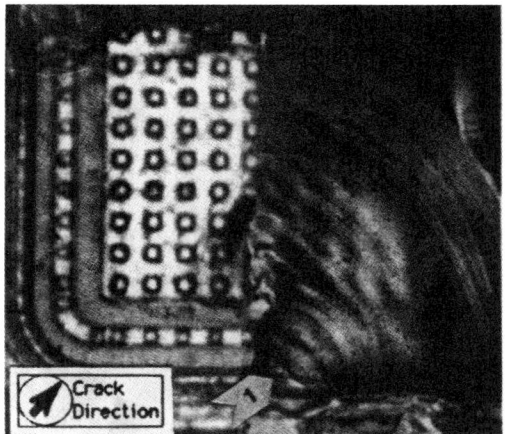

Fig. 4 Rib marks surrounding fracture origin (1). (200×)

Fig. 5 Rib marks from four separate fracture origins merging as crack propagates. (80×)

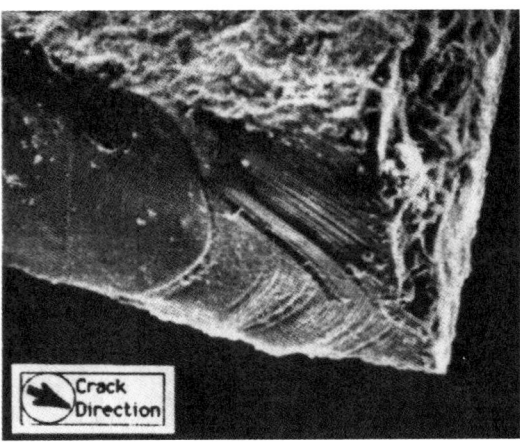

Fig. 6 SEM photo of rib marks at edge of vertically cracked die.

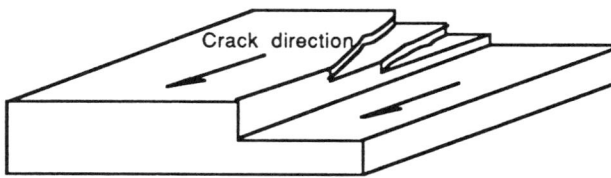

Fig. 7 River lines.

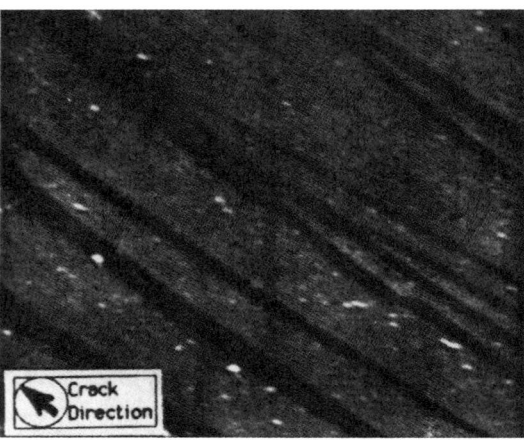

Fig. 8 SEM photo of typical river lines showing convergence of parallel fracture planes. (2800×)

TEAR MARKS

Tear marks form when the crack front splits onto multiple planes, as shown in Fig. 12. These cleavage steps appear as thin lines perpendicular to the local crack direction, and are generally considered to form as a result of the crack front reaching a critical velocity, when lattice disturbances ahead of the crack front cause the crack to split onto different levels [13]. Tear lines form radially near the crack origin. Tear marks that terminate with a bayonet-like discontinuity are referred to as lance marks, and are formed when the principle force changes direction during tear mark formation (Fig. 15 and 16).

DETERMINING THE FRACTURE ORIGIN

Locating the fracture origin, or origins, of a cracked silicon die simply involves working backwards from any directional information observed on the fracture surface. A high-power microscope is sufficient for fracture tracing, although in-depth study of the actual fracture origin usually requires an electron microscope.

River lines and rib marks are the most common markings observed on laterally cracked dice, and alone are enough to determine the origin of fracture. In vertically cracked dice, rib marks are often the only form of directional evidence. If river lines and rib marks appear to contradict each other, the river lines should be trusted; tear marks can often curve like rib marks, and rib marks can be sufficiently linear to imitate tear marks. Decapsulation and handling of a device can cause further crack propagation and chip-

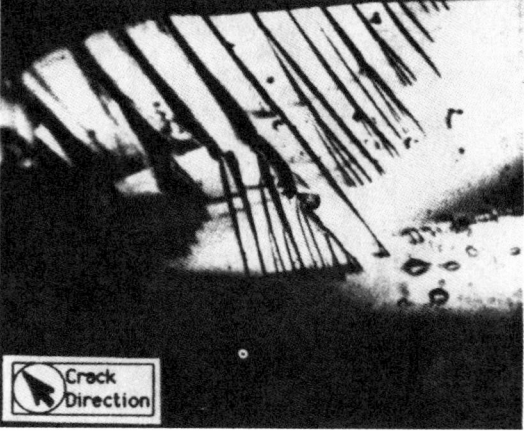

Fig. 9 River lines on edge of cracked die.

Basic Silicon Fractography

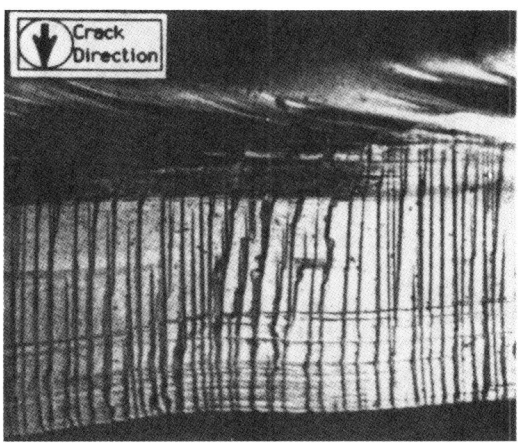

Fig. 10 Rib marks at top of photo leading into a group of parallel river lines. (200×)

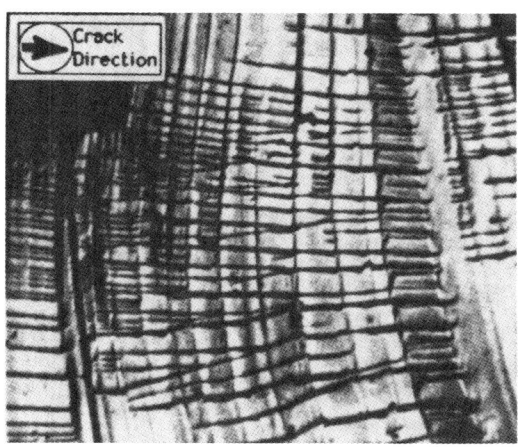

Fig. 11 Coincident formation of both river and rib marks. Note rib marks in this case are not bowed away from crack origin.

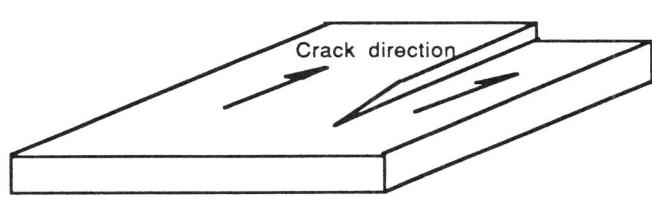

Fig. 12 Tear mark resulting from splitting of crack front onto two levels. (After Haneman & Pugh.)

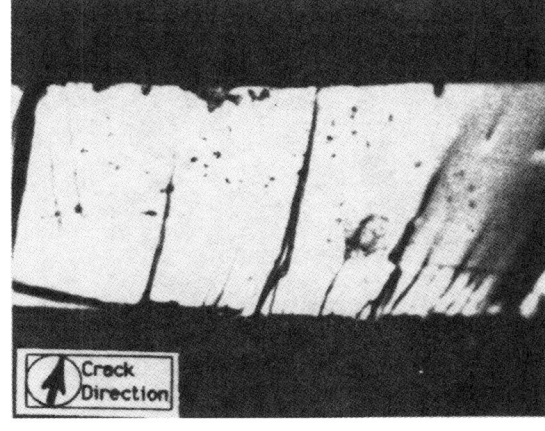

Fig. 13 Tear marks running parallel to crack direction. (120×)

Fig. 14 Tear marks extending radially from fracture origin. (80×)

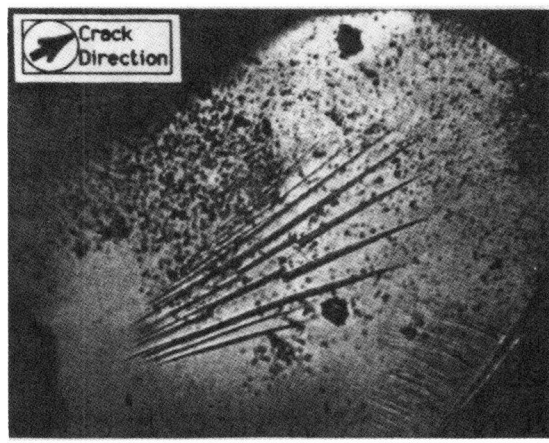

Fig. 15 Lance marks. Bayonet-like formations parallel to crack direction. (200×)

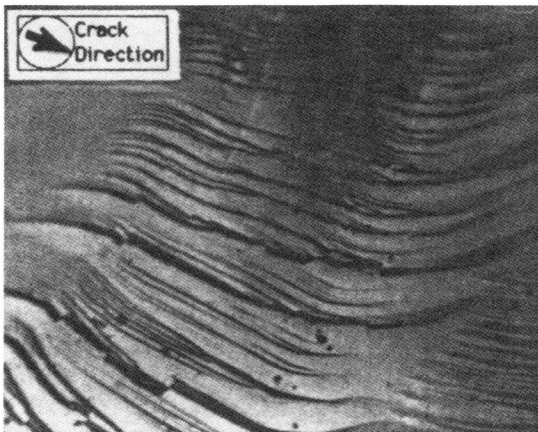

Fig. 16 Crack moves from left to right, progressing from a smooth region, then forming tear marks, then eiver lines terminated with lance marks. (300×)

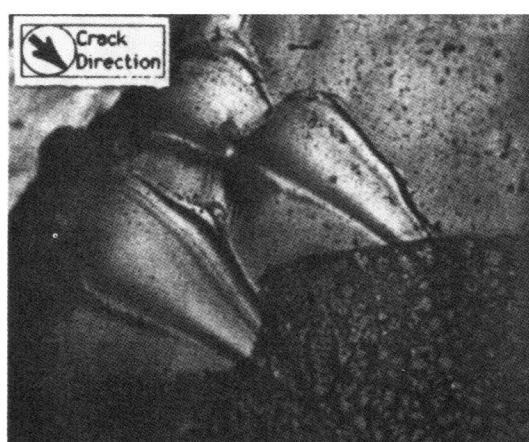

Fig. 17 Conch-like markings pointing in direction of crack travel. (200×)

Fig. 18 Smooth mirror region surrounding fracture origin (1), bordered by Wallner lines. [110] fracture surface. Vertically cracked die.

outs especially around the edges of the fracture surface, leading to erroneous local direction evidence. Fracture tracing examples are presented in Fig. 19 through 23.

FRACTURE ORIGINS

Silicon, as a brittle solid, tends to break in tension at a pre-existing flaw or micro-crack on its tensile surface. Where fractography deals with using fracture surface topology to make inferences about crack origin and velocity, fracture mechanics deals with more quantitative aspects of crack propagation. Although even a brief discussion of fracture mechanics is beyond the scope of this discussion, one useful equation relating crack length and critical fracture stress is presented in Fig. 24. The equation deals with semi-elliptical cracks; however, an edge crack—one that traverses the entire specimen length—is equivalent to an elliptical crack with an a/c ratio of 0.

The parameter K_{IC} is a material property called the Mode I fracture toughness. Appendix B lists the K_{IC} values for major crystal planes in silicon. The preference for (111) fracture is due

Fig. 19 Chip-out at heel of base bond wire. Rib marks (1) surround fracture origin (2). River lines (3) appear at bottom of chip-out. Damage in upper corner appears to be secondary.

to the lower fracture toughness of this plane.

The strength of silicon depends on both the stress distribution and inherent flaw distribution on the tensile surface. Hawkins et al. [4] have investigated die strength as a function of back-processing method. They found that the critical fracture stress at the back of the die varied from 50 to 300 MPa, depending on etch time and whether the dice were lapped or ground. Etching tended to blunt large flaws and remove small flaws, effectively strengthening the die. Lateral fractures often initiate at flaws introduced by the sawing process. Figure 25 shows the edge texture of a somewhat rougher-than-usual saw die. The equations presented in Fig. 24 are used to compute the approximate local tensile stress necessary to cause fracture originating at a large flaw shown in the photo. In this case, as is often done,

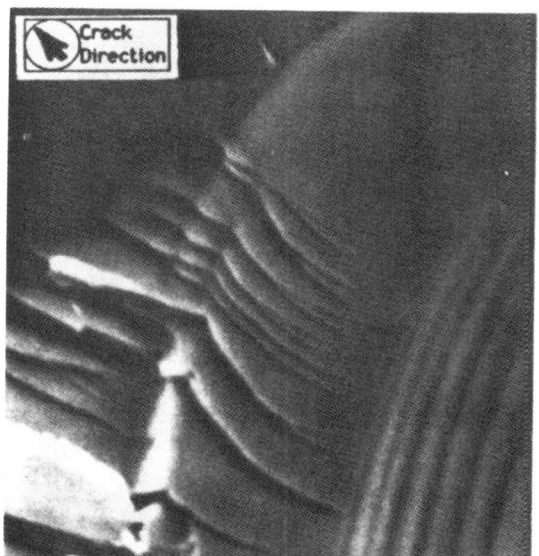

Fig. 20 Close-up of river marks shown in Fig. 19.

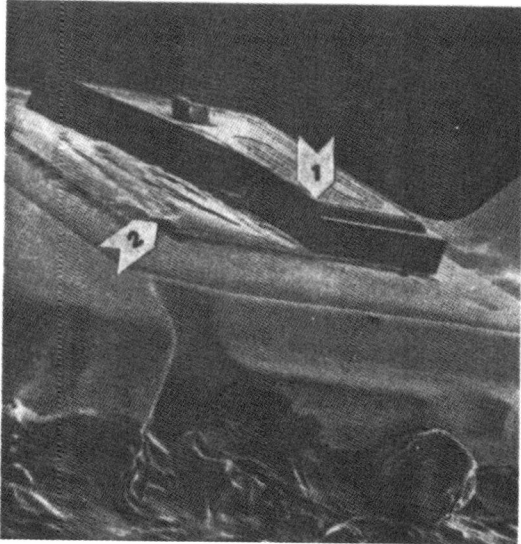

Fig. 21 Fractured die. Arrow #1 indicates location enlarged in Fig. 22 (below). Arrow #2 shows actual fracture origin. Crack began as a lateral fracture while being soldered.

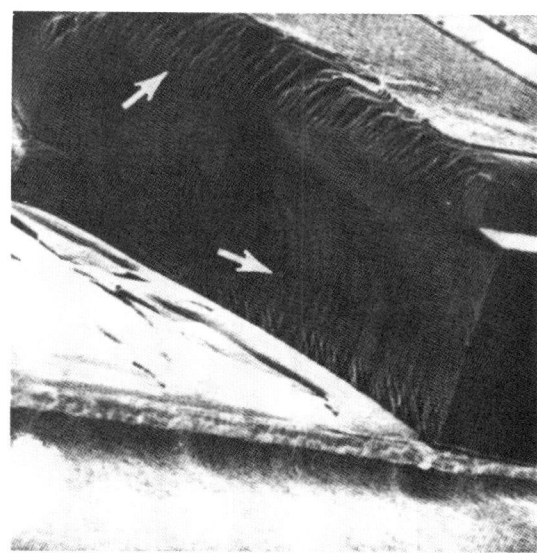

Fig. 22 Close-up of die edge shown at left. Arrows indicate local crack direction suggested by rib marks at bottom right and river lines near top surface of die.

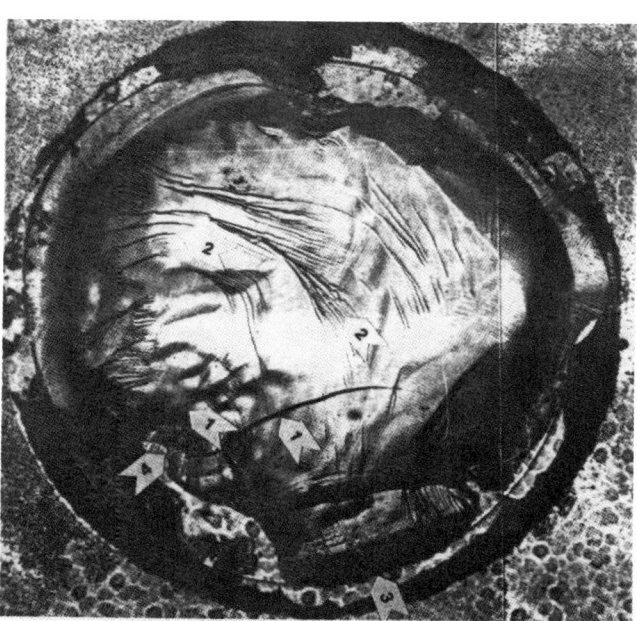

Fig. 23 Silicon "divot" remaining after center region (circular contact area) has been torn away. Study of fracture surface shows: (1) Rib marks, perpendicular to crack growth, (2) River lines, converging in direction of crack growth, and (3) Fracture origin. Damage induced most likely by tensile force (out of page) combined with a torque around point (4).

the crack shape is assumed to be semi-elliptical. The theory predicts that a local tensile stress of 500 MPa, as a result of a contact force or thermal bending or other loading conditions, will result in catastrophic lateral fracture of the die.

FRACTURE MIRROR

The classic brittle fracture origin is shown in Fig. 26: a smooth, usually semi-circular region surrounding the original flaw, bordered by tear marks (hackle). The mirror radius and fracture stress are related by the equation [1]:

$$\sigma_f \sqrt{R_m} = B$$

where R_m is the mirror radius, σ_f is the mode I fracture stress at fail-

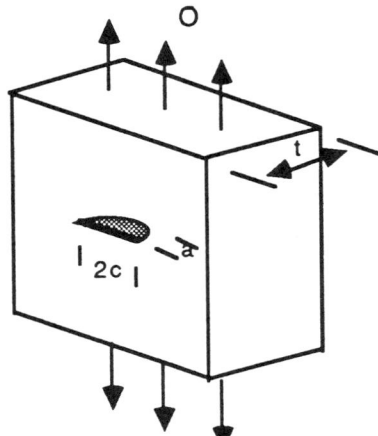

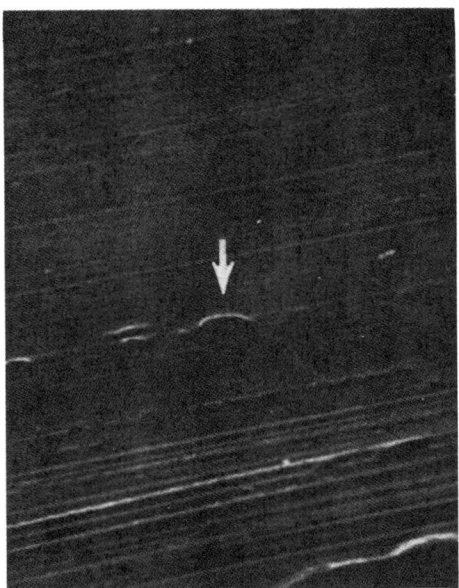

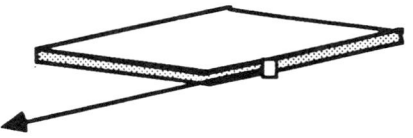

Estimate of critical fracture stress

$$\sigma_c = \frac{\Phi \, K_{IC}}{1.12\sqrt{\pi a}}$$

$K_{IC} = 0.82$ MN/m$^{3/2}$
(lateral cracking on {111} plane)

Estimating flaw shape from SEM photo:
a ≈ 1 μm, 2c ≈ 4 μm, Φ = 1.211

σ_c ≈ 500 MPa

Fig. 24 Formulas relating critical stress, fracture toughness, and crack geometry for semi-elliptical or "thumb-nail" crack. Assuming: linear elastic behavior, mode I fracture, plane strain conditions, t >> a, and small scale yielding. [1, 12].

Fig. 25 Left: SEM photo of edge of power transistor die showing typical saw damage. Right: Estimation of local, mode I tensile stress required to cause fracture at flaw indicated by arrow.

ure, and B is the constant for a given material.

Also, the ratio of mirror radius to original flaw size is a constant for a given material and flaw shape (approximately 13.0) [6]. Although the fractures of actual silicon devices rarely form a distinct, semi-circular fracture mirror from which meaningful quantitative information can be extracted, References 3 and 6 are good exceptions.

Although silicon has a preference for fracture on the [111], [110], and [100] planes, in that order (due to their respective fracture toughness values), in most real fracture situations in semiconductor devices, especially lateral fracture, the crack moves along a seemingly unsystematic series of directions. Figure 27 shows the three general modes of crack propagation. In brittle materials the crack tends to move in a direction that maximizes mode I (opening) and minimizes mode II (shear or sliding) propagation, as shown in Fig. 28.

Ideally, then, the crack initiates at the largest inherent flaw in the region of highest tensile stress, sometimes forming a smooth mirror region. The crack then propagates perpendicular to the applied force or along a major crystal plane roughly perpendicular to the applied force, whichever is easiest. As the crack velocity increases or the applied force changes direction, the crack splits onto multiple planes or causes secondary cracking. The crack halts when it exits the die or when the crack tip enters a region of local compression, as would be expected in cracks induced by contact damage or localized heating.

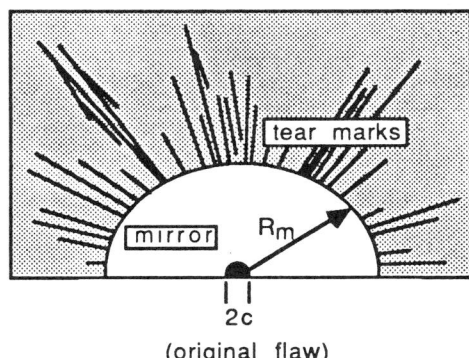

Fig. 26 Ideal fracture origin.

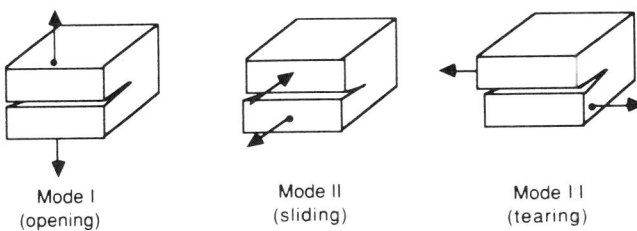

Fig. 27 Modes of crack growth.

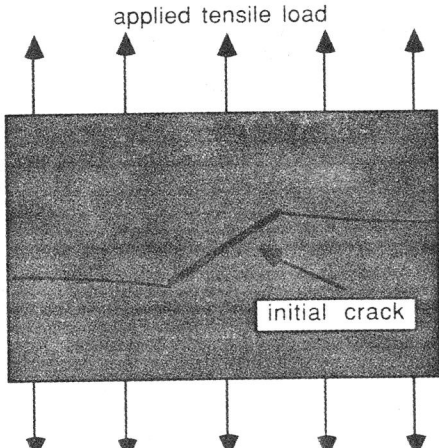

Fig. 28 Crack path in brittle plate during tensile loading of pre-existing crack (bold line). Crack changes direction in preference for mode I fracture. (After Erdogan, F. & Sih, G.C.)

REFERENCES

1. *Fractography of Ceramic and Metal Failures*, ASTM STP 827, J.J. Mecholsky, Jr. and S.R. Powell, Jr., 1984
2. *Fracture of Brittle Solids*, B.R. Lawn and T.R. Wilshaw, Cambridge University Press, Cambridge, U.K., 1975
3. C.P. Chen and M.H. Leipold, "The Application of Fracture Mechanics to Failure Analysis of Photovoltaic Solar Modules."
4. G. Hawkins, H. Berg, M. Mahalingam, G. Lewis, and L. Lofgran, "Measurement of Silicon Strength As Affected By Wafer Back Processing," IEEE/IRPS, 1987, p 216.
5. L.D. Dyer, "Fracture Tracing in Semiconductor Wafers," Semiconductor Processing, ASTM STP 850, 1984.
6. C.G.M. van Kessel and S.A. Gee, "The Use of Fractography in the Failure Analysis of Die Cracking," ISTFA 85.
7. "Creation of Cleavage Steps By Dislocation," Gilman, Transactions of the Metallurgical Society of AIME, June 1958, p 310
8. C.G.M. van Kessel, S.A. Gee, and J.J. Murphy, "The Quality of Die Attachment and Its Relationship to Stress and Vertical Die Cracking," IEEE Transactions on Components, Hybrids, and Manufacturing Technology, Vol. CHMT-6, No. 4, Dec. 1983.
9. K. Miyake, H. Suzuki, and S. Yamamoto, "Heat Transfer & Thermal Stress Analysis of Plastic Encapsulated ICs," IEEE Transactions on Reliability, Vol. R-34, No. 1, Jan 1973
10. W.A. Brantley, "Calculated Elastic Constants for Stress Problems Associated With Semiconductor Devices,", *J. Appl. Phys.*, Vol 44 (No. 1), Jan 1973.
11. C.P. Chen and M.H. Leipold, "Fracture Toughness of Silicon," American Ceramic Society, Ceramic Bulletin, , Vol 59 (No. 4), 1980, p 469-472.
12. "Fatigue and Fracture Mechanics Tutorial," ISTFA, 1988, R.O. Ritchie.
13. "Tear Marks on Cleaved Germanium Surfaces," Haneman, Pugh, *Journal of Applied Physics*, Vol 34 (No. 8), 1963.

APPENDIX A: SILICON CRYSTAL

Crystal Plane Index System

1. Find intercepts of plane on x, y, and z axes.
2. Reduce the reciprocal of these three numbers to the smallest three integers having the same ratio. Intercepts at infinity designated by zero.

Example: Plane having intercepts 1, 1, 2, = (1/1, 1/1, 1/2) = (2, 2, 1) plane

- Parentheses designate a single plane or set of parallel planes: (100)
- When plane cuts axis on negative side of origin, dash is placed over index: ($\bar{1}$00).
- Symmetrical planes are denoted by curly brackets: {100} for set of cube faces.
- Square brackets indicate a vector direction: [100] is the x axis.
- In cubic crystals, the [hkl] vector is perpendicular to the plane (hkl).

[Source: "Introduction to Solid State Physics," C. Kittel, John Wiley & Sons, 1976.]

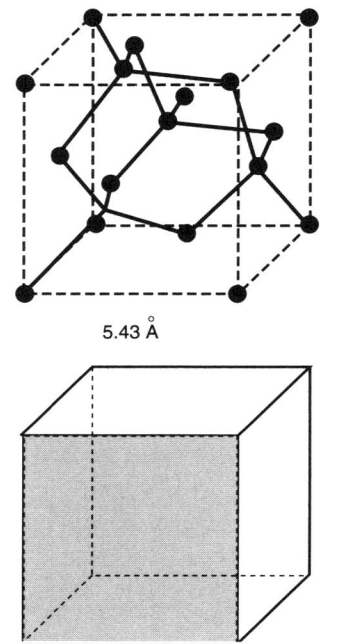

Fig. 29 Crystal structure of silicon and important crystal planes.

APPENDIX B: MECHANICAL PROPERTIES OF SILICON

Fracture Toughness
(a) Young's Modulus(b) and Poisson's Ratio (b)

	Crystal plane		
	{100}	{110}	{111}
K_{IC} (MN/m$^{3/2}$)	0.95	0.90	0.82
E(c) (10^5 MPa)	1.302 <001>	1.302 <001>	1.213
	1.689 <011>	1.875 <111>	
v(d)	L = <010>	L = <1$\bar{1}$1>	
	M = <$\bar{1}$10>	M = <$\bar{1}\bar{1}$2>	
	0.279	0.182	0.262

(a) C.P. Chen and M.H. Leipold, American Ceramic Society, Ceramic Bulletin, Vol. 59 (No. 4). (b) W.E. Brantley, J. Appl. Phys., Vol. 44 (No. 1), Jan. 1973. (c) Minimum and maximum values given with respective directions within the plane. (d) Longitudinal stress in direction L and transverse longitudinal strain along M.

Thermal Conductivity and Specific Heat

T (°K)	100	200	300	400	600	800	1000	1200	1500
k (W/m K)	884	264	148	98.9	61.9	42.2	31.2	25.7	22.7
c_p (J/kg K)	259	556	712	790	867	913	946	967	992

[Source: "Fundamentals of Heat Transfer,, Incropera and Dewitt, John Wiley & Sons, 1981.]

Linear Coefficient of Thermal Expansion

T (°K)	100	200	300	400	500	1000	1200	1500
a (10^{-6}/°K)	–0.339	1.406	2.616	3.253	3.614	4.258	4.384	4.556

[Source: "Properties of Silicon," INSPEC, Institute of Electrical Engineers, 1988]

Failure Analysis Reporting

TOPICS IN KNOWLEDGE-BASED FAILURE ANALYSIS

by Christopher L. Henderson

Failure analysis is a knowledge-intensive discipline. Historically, failure analysts have relied on mentors (experts) to teach the discipline of failure analysis. Although printed material and training courses have helped to increase the availability of expert knowledge, there is still a considerable need for wider dissemination of failure analysis knowledge. This problem grows more critical as integrated circuit technology continues to advance. Knowledge-based systems can provide a unique ability to disseminate failure analysis information to the failure analyst.

A HISTORICAL PERSPECTIVE ON EXPERT SYSTEMS

Artificial Intelligence

Expert system research is a branch of the discipline of artificial intelligence. Artificial intelligence (A.I.) in turn is a branch of computer science. Although you may not normally think that the study of artificial intelligence has produced much over its 30 years as an active field, it has produced several important products used extensively today. The windowing systems, made popular by MacIntosh, then implemented by SUN, Microsoft, and the like, are the result of artificial intelligence work done by Alan Kay and his associates at Xerox, Palo Alto Research Center (Xerox PARC). Other important tools to emerge from the study of artificial intelligence are the LISP and PROLOG computer languages, used not only for artificial intelligence (AI) applications, but also for drafting, graph search, and other engineering tasks.

The LISP language was proposed by John McCarthy c. 1960 and developed by several universities and companies in the 1960s and 1970s. LISP is an acronym for LISt Processing. PROLOG is a computer language conceived by J. A. Robinson about 1965. The majority of the development occurred between 1975 and 1979 at the University of Edinburgh. PROLOG is an acronym for PROgramming in LOGic.

Expert System Development

Expert systems began to appear in the 1960s, as researchers attempted to describe human reasoning in terms of computer program code. The earliest expert systems were developed in traditional programming languages such as FORTRAN. Traditional programming languages did not work well for expert systems, because expert systems are largely symbolic, that is, word or character based. One of the earliest successful expert systems to be developed was MYCIN [1]. MYCIN was a system designed to diagnose and prescribe treatment for spinal meningitis and other bacterial infections of the blood. This system took some 20 man-years of effort to build. One of the earliest expert systems in the field of electrical engineering was a system called EL [2]. EL helped tutor students with the task of determining voltage levels and current values in an electrical circuit. One of the more famous expert systems in electrical engineering is XCON (also referred to as RI). XCON, developed jointly by Carnegie-Mellon University and Digital Equipment Corporation, configures VAX computer orders [3]. The majority of XCON was developed between 1980 and 1985. XCON is one of the most extensively used systems in industry today; consequently it has produced a wealth of information regarding the development, maintenance, and fielding of a large expert system.

XCON is a relatively large system, containing some 5000 rules, or pieces of information. It took approximately 10 man-years of effort to develop the system, and it takes approximately a man-year per year to maintain the software (add and test new rules as new computer hardware and software options become available for the VAX family of computers). XCON was written in a LISP-like language called OPS. OPS (Original Production System language) is a high-level language written in LISP to facilitate the development of expert systems. XCON did not put configuration order entry clerks out of business; rather it provided a means for the entry clerk to quickly check standard orders (XCON could configure a VAX computer in 90 seconds, whereas an order entry clerk took approximately 20 minutes). The clerk was then free to concentrate on more complex order configurations. XCON can successfully configure 95% of all VAX computer orders. Although XCON has its drawbacks (large amount of maintenance and upkeep), overall, the system has been quite successful [4]. It stands as a target for which to aim in today's expert system development.

Knowledge-Based Systems for Failure Analysis

In the past several years, several companies have begun development of knowledge-based systems for failure analysis. The first such systems to appear were systems associated with the testing of ICs. [5,6]. Additionally, knowledge-based system development has occurred in conjunction with intelligent design validation tools. Companies such as IMAG/TIM3 Labs and Schlumberger Technologies have developed products that employ heuristics when diagnosing circuits [6,7,8]. A number of related knowledge-based system have also been developed. These include reliability expert systems [9], optical defect inspection [10], and interactive fault correction systems [11]. More interesting to the overall subject of failure analysis, several systems have been developed to assist in the failure analysis of integrated circuits [12,13,14,15]. These four knowledge-based systems will be discussed in more detail later.

THE USES FOR A KNOWLEDGE-BASED FAILURE ANALYSIS SYSTEM

Capture Failure Analysis Expertise

The primary motivation in creating a knowledge-based system of any kind is to capture and retain expertise. Any manager who has had the misfortune of losing his or her most knowledgeable staff member knows the value of retaining expertise. Failure analysis is a discipline that has a long learning curve; *it typically takes five years or more for an individual to become proficient in performing failure analysis on ICs. Much of this is due to the fact that the failure analyst must know something about a number of disciplines, such as chemistry, physics, electrical engineering, and mechanical engineering.* To become proficient in device recognition alone can take many years. By and large, failure analysis has been a skill taught by companies, not by universities. These factors, coupled with a scarcity of talented failure analysts, make the problem of retaining knowledge of failure analysis an important problem to address. If the expertise of the experienced analysts could be captured and retained effectively, the loss of an experienced analyst could be less catastrophic.

Train New Failure Analysts

In conjunction with retaining failure analysis expertise, there is also a problem associated with the training of new failure analysts. If the turnover rate is high in a failure analysis laboratory, then considerable time must be invested in training new failure analysts. This training can be costly in terms of courses and trainer time. Furthermore, the experienced analyst must coach the inexperienced analyst to ensure that he or she is doing the work correctly. This reduces the effectiveness of the more experienced analysts; they are not able to focus on the more difficult analyses. As a result, either the backlog of failure analysis work grows, or the quality of the work suffers. If a knowledge-based failure analysis system could serve as a training tool, then the pressure on the experienced analysts could be lightened, freeing them to concentrate on the more advanced and difficult failure analysis work. This work is much less likely to receive benefit from a knowledge-based system.

Centralizing Failure Analysis Information

A third use of a knowledge-based failure analysis system is to centralize failure analysis information. Historically, this has been accomplished through the use of books. Books such as Rome Lab's publication on Failure Analysis Techniques [16], HiRel Lab's Failure Analysis, and this book [17] seek to centralize failure analysis knowledge and make it more accessible to the inexperienced analyst. A knowledge-based system can provide two main advantages that books on failure analysis cannot provide.

First, the knowledge can be made available as part of each analysis if the system is interactive. Knowledge of a particular technique or procedure is useless if the analyst cannot recall or locate the information. Even experienced analysts cannot always remember particular techniques or recall where they read about them. A knowledge-based system can bring that information to the screen, making it available to the analyst.

Second, the analyst may not know how to apply the information to the particular analysis being performed. A knowledge-based system can be constructed to describe the applicability of the technique to a particular analysis.

HOW FAILURE ANALYSIS KNOWLEDGE IS CODED INTO A COMPUTER

Embedded Heuristics

As an example, most of the diagnostic software that is coupled with electron beam probing contains embedded heuristics. Embedded heuristics are "rules of thumb" that are coded into software to increase the effectiveness of the diagnostic software. Most of the time these heuristics are coded in the same language as the diagnostic software. The heuristics employed in the Schlumberger tool IDA (written in C) [7]. More specifically, in electron beam testing, it is preferable to examine top level metal whenever possible to measure waveforms. This heuristic can be coded into the diagnostic software, providing more accurate waveform measurement, hence a better diagnosis.

Structured Query Language (SQL)

Structured Query Language (SQL) is a method developed for accessing data in a relational database. This method forms the backbone of most management information systems in use today. The SQL method is sufficiently powerful to provide "smart" access to a database. SGS-Thomson has developed a computer-aided reliability analysis system using this approach [9]. Texas Instruments successfully employs this technique to scan a large failure analysis database to retrieve similar failure analysis cases [14]. This allows the analyst to initially compare symptoms with already completed analyses to provide information on how to proceed with an analysis. An example SQL query from the Texas Instruments system that searches the database for all 1989 gold wire bond failures is shown below.

- Select failmech3, failmech1, count(*)
- From job,unit,fallmech
- Where job.jobnum=unit.jobnum
- And failmech3=failcode
- And status='comp'
- And datecomp between '1989-01-01' and '1989-12-31'
- And failmech2 like 'gold bond fall.%'
- Group by failmech3,failmech
- Order by 3 desc

Expert System Shells

In recent years, a number of expert system development tools have become available on the market. The advantage of an expert system development tool (or shell) is that the developer does not need to know how to program in a low level artificial intelligence language such as LISP or PROLOG. A developer can quickly begin to write rules and not worry about writing an inference engine (a computer code to process the rules). Today's expert system shells are powerful hybrid shells that allow a combination of objects and rules to create applications. The disadvantage of these shells is that flexibility and speed of operation is sacrificed for high level development. While speed is not a major concern for failure analysis applications, flexibility might be. A shell has to be carefully chosen to ensure that the desired control structure can be implemented. At least three companies have developed knowledge-based failure analysis systems using an expert system shell: Allied-Signal (Kappa-PC), Univ. of Arizona (CESM), Texas Instruments (Aion Development Software), and Sandia Labs (Level 5 Object).

Single Knowledge Base

Several knowledge-based failure analysis systems use a single knowledge base. Because the amount of knowledge required to perform expert-level failure analysis is very large, a single-knowledge-base system must have structure. The Texas Instruments system uses a single, highly structured, knowledge base (an Aion interface to a relational database). A highly structured knowledge base is necessary in order to provide a uniform method for retrieving information.

A second method for using a single knowledge base is to scope the problem narrowly. Several University of Arizona graduate students developed an expert system to perform automatic visual defect inspection on wafers [10]. The classification system uses 64 rules to distinguish six types of visual defects: voids, probe marks, scratches, bridging, cracks, and particles. This type of system has the advantage of incorporating highly specialized detail, but addresses only a portion of the overall subject of failure analysis.

Multiple Knowledge Bases/Blackboard Techniques

The University of Milan/SGS-Thomson failure analysis assistant for linear integrated circuits [13] and the Sandia Labs Integrated Circuit Failure Analysis Expert System (ICFAX) [15,19] use a blackboard architecture with multiple knowledge bases. The blackboard architecture is a method for allowing independent knowledge bases to process information and post results asynchronously into a central global database known as a *blackboard* [18]. This allows failure analysis knowledge to be structured around different analytical techniques, providing modularity and structure. A blackboard system also reduces the "combinatorial explosion" problem associated with the number of possible paths that an analysis can take. The following code is an example of a rule used in a blackboard architecture:

RULE for e beam no netlist good die dfi fail scan die dbl step

- IF (CCVC Results = "no anomalies" OR CCVC Results = "not known")
- AND (Light Emission Simple Setup Results = "no light emission
- detected" OR Light Emission Simple Setup Results = "not known")
- AND e beam prober OF Voltage Contrast
- AND ate connection to e beam prober OF Voltage Contrast
- AND vector set available OF Voltage Contrast
- AND NOT access to netlist OF Voltage Contrast
- AND access to good die OF Voltage Contrast
- AND NOT dfi located failure OF Voltage Contrast
- AND scan die OF Voltage Contrast
- AND type of anomaly found OF Voltage Contrast IS open at a double step
- THEN Dynamic Voltage Contrast Results:= "open at a double step"
- AND CHAIN "icfax"

The "Dynamic Voltage Contrast Results" variable on the blackboard is modified as a result of rules firing in separate modules, in this case a rule from a module on voltage contrast testing.

Artificial Intelligence Languages

Artificial intelligence (AI) languages have greater flexibility, in terms of control, and an increased speed of execution, as compared to expert system shells. Unfortunately, the development cycle is much longer because the inference engine (rule processor) must be written. Languages such as LISP and PROLOG are the most popular languages with which to perform development. Several failure analysis systems have been developed using PROLOG. The first such system is an expert system used to diagnose VLSI memories by IBM France [5]. The developers choose PROLOG because of its flexibility and ability to interface to C language programs. A second system that uses PROLOG is PESTICIDE [6]. PESTICIDE is an expert system that interfaces with e-beam probing equipment to provide additional knowledge to aid in debugging complex VLSI circuits. The following piece of code is an example for several rules written in PROLOG:

- rule((fails_continuity(continuity_test):-shorted),80).
- rule((fails_continuity(continuity_test):-open),90).
- rule((fails_functional(functional_test):-fails_memory_test),90).
- rule((fails_functional(functional_test):-fails_logic_test),90).
- rule((fails_timing(timing_test):-fails_propagation_delay),70).
- rule((fails_timing(timing_test):-fails_max_frequency),80).
- rule((fails_timing(timing_test):-fails_rise_fall_time),90).
- rule((fails_parametrics(parametric_test):-fails_input_high_level_leakage),90).
- rule((fails_parametrics(parametric_test):-fails_input_low_level_leakage),90).
- rule((fails_parametrics(parametric_test):-fails_output_high_level_current),40).
- rule((fails_parametncs(parametric_test):-fails_

- rule((fails_iddq(iddq_test):-fails_iddq_output_low_level_current),40).
- rule((fails_iddq(iddq_test):-fails_iddq_single_vector),90).
- rule((fails_iddq(iddq_test):-fails_iddq_multiple_vectors),100).

Probably the most comprehensive knowledge-based failure analysis system developed from an artificial intelligence language is a system called IDA (Intelligent Diagnostic Assistant) [12]. IDA was developed by Cape Systems Inc. on a LISP-based machine (TI Explorer), to assist in the diagnosis of faulty hybrids. Although the system is quite flexible, it requires considerable rule entry from the purchaser.

A REVIEW OF KNOWLEDGE-BASED SYSTEMS FOR FAILURE ANALYSIS

IBM France System

- Developers: T. Viacroze, G. Fourquet, and M. Lequex
- Computer Platform: IBM PC or compatible.
- Linked via RS232 to ATE
- Software: Developed in PROLOG
- Approx. Rule Count: 120
- Function: Software interfaces to ATE testers to analyze test data and diagnose the type of failure occurring on the memory device.

Cape Systems Inc. Intelligent Diagnostic Assistant

- Developers: Michael Kagan (Sipex Corporation), Joseph Kudish, and Alon Zelsion (Cape Systems Inc.)
- Computer Platform: Texas Instruments Explorer. Linked to test equipment and a wafer prober via RS232, IEEE-488, and Ethenet.
- Software: Developed in LISP
- Approx. Rule Count: Delivered as an empty shell (user inputs rules). Application system at Sipex contained approximately 150 rules.
- Function: Software interfaces to test equipment and a prober to guide internal failure analysis of hybrid microcircuits.

IMAG/TIM3 Labs PESTICIDE

- Developers: M. Marzoukl, and B. Courtois
- Computer Platform: SUN 3/160 workstation interfaced into E-beam equipment (2nd generation system developed on a BULL DPX-5000 workstation)
- Software: Developed in PROLOG
- Approx. Rule Count: N/A. No rules in the system. Knowledge maintained as specifications on device structure, behavior, and function.
- Function: Software interfaces to E-beam probing system and CAD database to provide intelligent internal VLSI troubleshooting.

University of Milan/SGS-Thomson Failure Analysis Assistant for Linear Integrated Circuits

- Developers: G. Boella, P. Mussio, M. Piccoli (University of Milan), and P. Mauri (SGS-Thomson)
- Computer Platform: Workstation
- Software: Still in development (not finalized as of last published work in 5/90).
- Approx. Rule Count: Still in development
- Function: Software provides comprehensive assistance for failure analysis of linear integrated circuits; interfaces with SPICE (analog simulator), behavioral simulators, and layout validation tools

Texas Instruments' FDAL

- Developers: Laurel Bellay, Jimmy Smith, Steve Vaughn, Susan Powell, Thomas Taylor and Sanjiv Lakhanpal
- Computer Platform: IBM PC and compatible computers linked to a mainframe computer
- Software: Job submission program written in Turbo PASCAL, Database on mainframe in SQL format, AION Development System functions as the expert system software.
- Approx. Rules Count: N/A. Access is via SQL commands. Database contains in excess of 10,000 records with 52 fields each relating device fabrication, assembly, etc.
- Function: Software allows pareto analysis of database and suggests best course of analysis to pursue.

Sandia Laboratories' ICFAX

- Developers: Chris Henderson and Jerry Soden
- Computer Platform: IBM PC compatible computers
- Software: Level 5 Object Expert System Shell, Taner Research (Layout Plots), CAD-CAM Group (Schematic Plots), and Microsoft Windows Help Compiler (Help Files).
- Approx Rule Count: 8900
- Function: Software provides comprehensive guidance through the failure analysis process for CMOS ICs (3 micron channel length, 1 level metal, 1 level polysilicon).

Allied Signal's FACES (Failure Analysis Component Expert System)

- Developers: Tom Tetlow and Howard Dicken
- Computer Platform: IBM PC and compatible computers
- Software: Kappa-PC Expert System Shell
- Approx Rule Count: 300 in prototype version
- Function: Software guides analyst through a "safe sequence" of analytical procedures for discrete bipolar transistors.

THE FUTURE OF KNOWLEDGE-BASED

SYSTEMS

Knowledge is Power

Frances Bacon coined the phrase *"Nam et ipsa scienta potestas est"*, which translated says, "Knowledge is power." This motivation, probably more than any other, will drive the future of knowledge-based systems. In the future, the companies that can acquire and utilize knowledge efficiently and correctly will be the companies that survive. This statement applies to the discipline of failure analysis as well. If we are to help our companies survive and even prosper, we must be able to acquire and utilize failure analysis knowledge efficiently and correctly. The world of semiconductor technology is becoming too complex to ignore the issue of knowledge retention. *There are simply too many technologies to analyze and too many procedures to remember; the analyst cannot usually retain all this information in his head.* Knowledge-based systems for failure analysis provide a promising solution to this growing problem. With those thoughts in mind, the future is likely to bring interesting developments in the area of knowledge-based systems.

More Comprehensive Systems

Most of the knowledge-based systems previously mentioned deal with specialized sets of problems within the discipline of failure analysis (ATE test, optical inspection, E-beam probing, etc.). While systems like these will continue to proliferate, scientists will develop more comprehensive systems dealing with the entire failure analysis process, an enormous job.

There are still serious hurdles to overcome, however. The acquisition of failure analysis knowledge is a slow and tedious task. It may take many years to properly elucidate some aspects of failure analysis knowledge (turn "human" knowledge into computer code). The validation of complex, knowledge-based systems is extremely difficult. Once the knowledge is in computer code format, the code must operate correctly. The verification and testing of this code is not a trivial task.

Automatic Report Generation

One area where knowledge-based systems can aid the failure analyst is the area of report generation. Once the analyst has finished his or her work, he or she faces the mundane task of generating the failure analysis report. If a comprehensive knowledge-based system is used during the analysis, the significant steps and procedures are contained within the log of the session. This information can be used to automatically generate a failure analysis report. Such a report has the potential of being more accurate than the report generated from the analyst's memory.

Integrated Diagnostic Environments

The Integrated Diagnostic Environment, a term coined by Schlumberger, is a concept for electronically linking the equipment in a failure analysis laboratory through a network to increase the productivity of failure analysts. Both Schlumberger Technologies and Knights Technology are developing hardware and software to make this concept a reality. This concept meshes well with knowledge-based failure analysis systems. The knowledge-based failure analysis system could function as an interface to the integrated diagnostic environment, providing expert knowledge of techniques and procedures directly to the console of each piece of failure analysis equipment. While the concept is an exciting one, there are still barriers to overcome.

First, not all failure analysis equipment interfaces to computers, let alone the same *type* of computer. Second, failure analysis equipment which is made to interface to an integrated diagnostic environment is expensive. The cost may be prohibitive for many failure analysis facilities. Third, some failure anallysts may be reluctant to use such software and hardware.

CONCLUSION

The history of artificial intelligence and expert system development sets the stage for the next generation of knowledge-based systems. The motivation and rationale for creating failure analysis knowledge-based systems is substantial. Several companies and universities have already demonstrated knowledge-based systems that are directly applicable to failure analysis. Although these systems are still quite simple, the stage is set for future development. As failure analysis knowledge increases, failure analysts must find new ways of capturing that knowledge and preserving it. Knowledge-based systems may provide the key for doing so.

REFERENCES

1. B. G. Buchanan and E. H. Shortliffe, eds., *Rule-Based Expert Systems: The MYCIN Expert Proc. 29 the Stanford Heuristic Programming Project*, Addison Wesley, Reading, MA., 1984.
2. G. J. Sussman, *Electrical Design: A Problem for Artificial Intelligence Research*, Proc. of the 5th Int Joint Conf on Artifical Intelligence, pp 894-900, 1977.
3. J. McDermott, *RI: A Rule Based Configurer of Computer Systems*, Technical Report, Carnegie-Mellon University, Department of Computer Science, 1980.
4. J. Bachant and J. McDermott, *RI Revisited: Four Years in the Trenches,"* The AI Magazine, Fall 1984, pp 21-32.
5. T. Viacroze, G. Fourquet, and M. Lequex, *An Expert System for Help to Fault Diagnosis on VLSI Memories*, Int. Symp. for Testing and Failure Analysis (ISTFA), Nov. 1988, pp. 153-160, (IBM France, test diagnosis).
6. M. Marzouki and B. Courtois, *Debugging Integrated Circuits: A.I. Can Help*, Proc. 1st Test Conference, pp. 184-191, April 1989. (IMAG/TIM3 Lab, E-beam coupled with AI to diagnose IC failures, named PESTICIDE).
7. A. C. Noble, *A Diagnostic Assistant for Integrated Circuit Diagnosis*, Proc. 3rd European Conference on Electron and Optical Beam Testing, pp. 78-85, Sept. 1991. (Schlumberger, Integrated Diagnostic Assistant).
8. M. Marzouki, and F. L. Vargas, *Using a Knowledge-Based System for Automatic Debugging Study and Performance Analysis*, Proc. 3rd European Conference on Electron and Optical Beam Testing, pp. 110-117, Sept. 1991. (IMAG/TIM3 Labs, reference to PESTICIDE).
9. P. Mauri, *Computer-Aided Analysis of Integrated Circuit Reliability*, Proc. NATO Advanced Research Workshop—Semiconductor Device Reliability, June 1989, pp. 127-136 (SGS-Thomson Integrated Circuit Reliability Expert System).
10. S. D. Chi, B. P. Zeigler, and T. G. Kim, *Using the CESM Shell to*

Classify Wafer Defects from Visual Data, Proc. SPIE—The International Society for Optical Engineering, Nov. 1989, pp 66-77, (U. of Arizona, Optical defect inspection).

11. J. Krol, *ClRCOR—An Expert System for Fault Correction of Digital NMOS Circuits*, European Conference on Circuit Theory and Design, Sept 1989, pp. 674-676. (University of Gdansk, Poland; design validation expert system called CIRCOR).

12. M. Kagan, J. Kudish, and A. Zelzion, *A New Method to Diagnose Chip-and-Wire Hybrids*, Hybrid Circuit Technology, Vol 6, No 2, pp 15-20, Feb 1989, (Cape Systems, Inc., Hybrid diagnostic experiment systems).

13. G. Boella, P. Mussio, P. Mauri, and M. Picolli, *Design of an Automatic Assistant for Failure Analysis of Linear Integrated Circuits*, Proc. of the Tenth International Workshop—Expert Systems and Their Applications, pp 137-151, May 1990. (SGS-Thomaon/University of Milan, Linear integrated circuit failure analysis expert system).

14. L. M. Bellay, P.B. Ghate, and L.C. Wagner, *Computers in Failure Analysis*, Proc. Int. Symp. for Testing and Failure Analysis, Nov, 1990, pp 89-95. (Texas Instruments, Intelligent database system called FDAL).

15. C. L. Henderson and J. M. Soden, *ICFAX, An Integrated Circuit Failure Analysis Expert System*, Proc. 29th International Reliability Physics Symposium, pp 142-151, Apr, 1991 (Sandia National Laboratories, Integrated circuit failure analysis expert system)

16. E. Doyle, Jr and B. Morris, eds, *Microelectronics Failure Analysis Techniques—A Procedural Guide*, IIT Research Institute, 1980.

17. J.R. Devaney, G. L. Hill and R.G. Seippel, *Failure Analysis Mechanisms, Techniques, and Photo Atlas*, Failure Recognition and Training Services, Inc., Monrovia, Calif, 1986.

18. G.F. Luger and W.A. Stubblefield, *Artificial Intelligence and the Design of Expert Systems*, Benjamin/Cummings Publ. Co. Inc., pp 651-563.

19. "Sandia Developes Failure Analysis Expert System, Semiconductor International, August 1991, p17.

SUGGESTIONS FOR A WELL-WRITTEN FAILURE ANALYSIS REPORT

by Thomas W. Lee

HISTORY

Should be as complete as possible:
- Implications of problem
- History of problem
- Seriousness or level of problem
- Sources of parts
- Circuit diagram showing use of part
- Occurrence of associated failures
- Occurrence of known overstress conditions
- Phone numbers of analyst, lab manager, and primary contacts who are information providers and recipients.

PROCEDURE

- Detail proceeds from coarse to fine; from general to specific.
- Analysis is complete and logical. There are no obvious "holes" in the sequence or reasoning due to missing information.
- There is sufficient information about the analytical procedure present, such that the recipient could reproduce the analysis based on the photos, data, and procedure descriptions.

PHOTOS, PLOTS AND IMAGES

- Photos and text follow each other in logical sequence.
- Photo quantity is sufficient to document course of analysis from beginning to end.
- Photos proceed in logical order so that reader is "zeroed-in" to the defect or area of interest with photos of increasing magnification.
- Photos are well-exposed and contrasty, but not too dark, so that reproduction on a properly working copy machine will produce informative copies, not square, dark blotches.
- All photos present are referred to by number in text discussions.
- Photo captions are sufficiently descriptive such that report may be followed by reading just the captions, in case the photos and the text are separated.
- A magnification is included on all photos.
- Don't load up a report with excessive or irrelevant SEM photos to imply authoratativeness.
- Don't take SEM photos of features which look peculiar or interesting. Relate the photo to the failure mechanism, and prove it.
- Images on SEM photos are "top lighted", and "straight with respect to the world", not shown at peculiar or impossible angles which could nullify or confuse the meaning of the photo and the report.
- Do not include color images in a report which will be copied on a black-and-white machine in a manner such that the data will be lost. An example is a color spectral plot, where one element is plotted in yellow, and the other in red. Both will reproduce as black, and the information will be lost to the reader.
- All graphs are labeled and their significance to the report explained.
- All graph axes are labeled, and they preferably start at zero. Whenever possible, axes should be labeled in relatively comprehensible units, not peculiar units that were manipulated in order to plot a straight line.

PROCEDURES

- Chemical reagents and concentrations are named. If a procedure is proprietary, say so. Avoid the use of questionable conductions.
- References for critical procedures or methods are provided. Use industry standard procedures whenever possible, and name the standard and where to get it in the report.
- Length and temperature of bakeouts are given.
- Electrical characterization is photodocumented; hand-drawn characteristics imply a poorly-equipped lab.
- Data resulting from advanced analytical procedures is accompanied by succinct explanations of the method itself, the exact area surveyed, and the relevance of the data in the form in which it is presented and approximate accuracy.
- Good-bad comparisons; data, photos, plots, etc. are used to prove significance of findings.
- Critical terms are defined—for example, "short" does not mean just an offscale reading of the test instrument set to the specified scale.
- The analyst asks himself, "why", and documents the answers until they are no longer reasonably answerable or pertinent to the analysis.
- Attempt to repair or restore the device during the analysis

TEXT

- Text is in complete sentences, and intelligent English.
- Don't depend too heavily on the word processor and a spell-checker. With few exceptions, it cannot compose correct English. Spell-checkers will dutifully check the spelling of a totally wrong word.
- Text contains ample verbiage to describe analytical procedures used, but does not ramble or pontificate.
- All automated test data printouts are supplied as supporting documentation. Curve tracer measurements are referenced to this data.
- Complex failure mechanisms named in the report are succinctly explained for readers familiar with electrical engineering, but not necessarily fluent in device or reliability physics, or advanced analytical procedures.
- The failure is explained in terms of position in the circuit.
- Corrective action addresses containment and disposition of affected parts on hand.
- Corrective action addresses long-term effects or risks of mechanism identified.

EOS AND ESD

Where the importance of the report justifies the time and effort, simulations are used in the case of EOS, ESD, mechanical shock, or thermal damage. Simulations result in V, I, E data, as well as SEM photos of the failed device and a simulated failure. The "voltage or current" question is answered, and the answer is explained, possibly with references. The "EOS or ESD?" question is answered and proven.

REFERENCES

1. MIL-STD-883-Method 5003, *Failure Analysis Procedures for Microcircuits*
2. MIL-STD-1546B.
3. MIL-STD-1580.
4. MIL-M-38510G, Appendix A.

Appendix

ISTFA SUBJECT INDEX

by Thomas W. Lee

How to Use This Index:

Below is an alphabetical list of key words or headings under which related articles are listed. Locate the subject of interest in this list initially. Many cases of multiple uses of terms have been resolved in this listing.

Following the title, is the year of the volume in which the article appears and after the colon, the number of the first page of the article. Some additional references from other sources are occasionally included for your convenience.

- accelerated testing
- acoustic microscopy
- alloying defects
- alpha particles
- analog circuits
- analytical electron microscopy (AEM)
- analytical pyrolysis
- analysis, surface
- atomic force microscope (AFM)
- Auger energy spectroscopy ((AES
- automated testing
- backmetal
- backside etching
- batteries, Ni-Cd
- beam leads
- bond pads
- bonding wire, aluminum
- burn-in testing
- capacitance testing
- capacitors
- capacitors, ceramic
- capacitors, metallized film
- capacitors, tantalum
- cathode ray tubes (CRT)
- cathodoluminescence
- ceramic resonators
- charge-coupled devices (CCDs)

- chemicals, purity of
- CMOS (complimentary metal oxide-silicon)
- coax cable
- cordwood
- cold-start effects
- construction analysis
- contacts
- contamination
- corrosion
- cross-sectioning (see microsectioning)
- crystal oscillators
- data retention
- decapping (delidding)
- decapsulation techniques
- defects, crystallographic
- defects, layout-dependant
- defects, polysilicon
- defects, surface, oxide
- delamination, metal
- delamination, plastics
- delineation (etching, decoration, staining)
- die bond
- die coating
- die cracking
- dielectric breakdown
- DPA (destructive physical analysis)
- DRAM (dynamic random access memory)
- drift
- EBIC electron beam induced current
- EDXA, EDS energy dispersive spectroscopy
- EDXRF energy dispersive-x-ray fluorescence
- EGA (evolved gas analysis)
- electrical characterization
- electromigration
- electron multipliers

- emission microscopy
- encapsulant material
- E-beam testing
- EOS (electrical overstress)
- epoxy
- EPROM (erasable programmable read only memory)
- equipping the FA lab
- ESCA (electron spectroscopy for chemical analysis)
- ESD (electrostatic discharge)
- electroplating, nickel
- electromechanical devices
- etching (of junctions) see "delineation"
- excimer lasers
- expert systems
- failure analysis methodology
- failure mechanisms
- fault detection, location, isolation
- fault detection theory
- ferroelectric liquid crystals
- field crystallization
- field failures
- filler particles
- fits
- flip-chip bonds
- fluorescent die testing
- focused ion beam (FIB)
- FTIR (Fourier transform infrared spectroscopy)
- fuses
- gallium arsenide (GaAs)
- gas analysis
- gate arrays
- gate oxide
- glass deposition
- glass voiding
- gold metallization
- HAST (highly accelerated stress testing)
- heat pumps

- HEMTs
- hillocks (Al)
- holography
- hybrid circuits
- ICP-AES (inductively coupled plasma-atomic emission spectroscopy)
- image analyzers
- image processing
- IMMA—ion microprobe mass analyzer
- IMPATT diodes
- inclusions
- indium phosphide (InP)
- ion beams
- ion migration
- ionic contamination
- inductance testing
- infrared microscopy
- infrared thermal imaging
- interconnects
- IR detectors
- isolation
- isolation, photoresist
- isolation, probe
- ISS (ion scattering spectroscopy)
- lasers
- laser diodes
- laser microscopy
- latch-up
- lead crystals
- leads (component leadwires)
- leak testing
- leakage current
- LED (light-emitting diode)
- LIMS (laser ionization mass spectrometry)
- liquid crystal techniques
- LMMA, LAMMA (laser microprobe mass analysis)
- low temperature testing
- LSM (laser scanning microscope; confocal LSM, etc)
- magnetic decoration
- magnetic media
- marking
- MESFET (metal semiconductor field effect transistor)
- memories
- metallization
- metallography (also cross-sectioning, microsectioning)
- microanalysis—overviews
- microsectioning
- microwave devices
- MLCC (multilayer ceramic capacitors)
- MMIC (monolithic microwave integrated circuit)
- microprocessors
- moisture resistance
- moisture sensors
- MOS capacitor
- MOSFET (metal oxide semiconductor field effect transistor)
- nodules, in silicon
- noise
- optical beam-induced current (OBIC)
- optical microscopy
- optoelectronics
- oxide defects
- oxide trapped charge
- packages, IC, ceramic
- particles
- passivation cracking
- pattern recognition
- Peltier devices
- phos-glass (phosphosilicate glass)
- photoacoustic spectroscopy
- photoemission
- photoresist
- photoresist; masking of specimens
- PIND (particle impact noise detection)
- pinholes
- planar cell capacitor
- plasma-enhanced chemical vapor deposition (PECVD)
- plasma etching and reactive ion etching (RIE)
- plating
- PLCC (plastic leaded chip carrier)
- polysilicon
- power transistors
- probe and isolation
- Procedural Guide, RADC
- process improvement
- profiling, depth
- printed wiring (circuit) boards—PWB or PCB
- PROM (programmable read only memory)
- radiation damage/effects
- RAM (read-only memory)
- RAMAN microprobe
- RBS (rutherford backscattering spectrometry)
- radiation effects
- relays and switches
- reliability evaluation
- resistors, thick-film
- resistors, thin-film
- RF transistors
- RGA (residual gas analysis)
- RIE (reactive ion etching)
- ROM (read only memory)
- sample preparation
- SAM (scanning acoustic microscopy)
- Schottky diodes
- SCR (silicon controlled rectifier)
- scratch test
- screening and testing
- SEM (scanning electron microscope)
- sensors, in situ
- shear test (die)
- shear test (wire bond)
- shift register
- sichrome (SiCr)
- silicon
- SIDP (sputter ion depth profiling)
- silicides
- silicon-germanium
- silicon nitride
- SIMS (secondary ion mass)
- single bit failures
- single-event upset (upset)
- silver migration

ISTFA Subject Index

- SLAM (scanning laser acoustic microscopy)
- SOS (silicon-on-sapphire)
- specimen preparation
- spectrometry
- solar cells
- solder, indium
- solder failure (non die-bond)
- solder heat testing
- solder joints
- solderability
- space-level parts
- spin-on glass (SOG)
- SRAM (static random access memory)
- staining (of junctions) see "delineation"
- step coverage
- STM (scanning tunneling microscope)
- strain
- stress induced voiding
- surface mount devices (SOT, SOIC)
- TAB (tape automated bonding)
- tantalum capacitors
- tunneling current microscopy
- TEM (transmission electron microscopy)
- temperature sensors
- TEOS
- testing
- testing, high temperature
- test chips
- TGA (thermogravimetric analysis)
- thermal balancer
- thermal wave imaging
- thermistor bolometers—see bolometers
- thermoelectric devices
- thick films
- thin film resistors
- threshold shift
- tin whiskers
- titanium-tungsten (TiW)
- traveling-wave tubes (TWTs)
- tunneling current microscopy
- ultrasonic inspection
- vapor bubble test
- vias—see contacts
- vibration effects
- voltage contrast
- WDXA (wavelength-dispersive x-ray analysis)
- welds
- whiskers
- wire bond
- wire fracture
- wire pull
- wire fusing current
- x-ray diffraction
- x-ray radiography
- XPS (x-ray photoelectron spectroscopy)
- Zener diodes

accelerated testing

A New Accelerated Humidity Test for Plastic Encapsulated ICs **82:65**

Correlation of Accelerated Test Failure Modes with Solder Joint Integrity and Performance **81:84**

Failure Mode Analysis and Life Time Estimation of Traveling Wave Tubes **80:196**

Failure Mechanisms in Microwave Solid State Devices **80:193**

Screening, Testing, and Reliability of Charge Coupled Devices **80:185**

Accelerated Retentivity Tests of 16K EPROMS **80:180**

Semiconductors Stress Testing at High Temperature **80:172**

Accelerated Testing of Commercial Grade Microcircuits in Plastic and CERDIP Packages **80:158**

acoustic microscopy

Evaluation of Methods for Delamination Detection by Acoustic Microscopy in Plastic-Packaged Integrated Circuits **92:425**

PGA Failure Analysis Using Acoustic Microscopy and a Novel Sample Preparation Technique **92:321**

C-AM Analysis of Plastic Packages to Resolve Bonding Failure Mode Miscorrelations **92:315**

Evaluation of Aluminum Wirebonding in Discrete Transistors Using Scanning Acoustic Microscopy (C-SAM) **92:307**

Acoustic Microscopy (A.M.) Applications **92:301**

Scanning Acoustic Microscopy: Reliability in Processing High Power Semiconductor Devices **91:119**

Reliable Delamination Detection By Polarity Analysis of Reflected Acoustic Pulses **91:49**

Non-Destructive Failure Analysis of IC's Using Scanning Acoustic Tomography (SCAT) and High Resolution X-ray Microscopy (HRXM) **89:69**

Identification of Package Defects in Plastic-Packaged Surface Mount IC's by Scanning Acoustic Microscopy **89:61**

Nondestructive Evaluation of Thermally Shocked Plastic Integrated Circuit Packages Using Acoustic Microscopy **88:211**

Non-Destructive Inspection of Voids in Silver Hard Solder Layer of Power Transistor **88:47**

Characterization of Die Attach Integrity Using Destructive and Nondestructive Techniques **88:77**

Advanced Failure Analysis Techniques for Multilayer Ceramic Chip Capacitors **87:25**

Acoustical Microscopy as a Tool for Nondestructive Testing of Finished Devices **87:21**

A Novel Method of Evaluating Moisture Resistance of Soldered Plastic Encapsulated LSI by a New Ultrasonic Inspection System **86:173**

Scanning Laser Acoustic Microscopic (SLAM) Study of Die Attach Integrity **85:202**

Nondestructive Inspection of Encapsulated Capacitors: Flaw Growth Detection of Parts Submitted to 85% Humidity, 85 °C **85:193**

Practical Applications of Acoustic Microscopy in Failure Analysis **85:187**

Non-Destructive Evaluation by Acoustic Microscopy as Applied to the Failure Analysis of Microelectronics **85:181**

Acoustic Microscopy: A Nondestructive Tool for Bond Evaluation on Tab Interconnections and Die Attach **84:243**

Acoustic Imaging—A Non-Destructive Test for Module Inspection **83:199**

Acoustic Visualization of Stressed Ceramic Chip Capacitors **82:215**

alloying defects

Degradation of Zener Diodes Caused by Alloying Irregularities **86:145**

alpha particles

Alpha Flux Measuring and Soft Error Evaluating Techniques **80:30**

analog circuits

Automated Test Design for Analog Circuits **77:17**

analytical electron microscopy (AEM)

Materials Characterization by Analytical Electron Microscopy **84:46**

analytical pyrolysis

Analytical Pyrolysis and Evolved Gas Analysis of Electronic Polymers **86:11**

Characterization of Molding Compounds via Polymer Reconstruction Investigative Chromatopyrography (PRI_) **84:98**

analysis, surface

Surface Sensitive Analytical Techniques for Failure Analysis **77:236**

atomic force microscope (AFM)

Advanced Methods for Imaging Gate Oxide Defects With the Atomic Force Microscope **92:267**

Auger energy spectroscopy (AES)

Case Histories of Microelectronic Device Analysis Using Auger Electron Spectroscopy **89:373**

Identification of Modules in Dual-Layer Aluminum-Copper Metallization on a Programmable Read Only Memory (PROM) **87:251**

Solderability Problem with Solder Coated Package Leads **87:201**

Surface Analysis of Contamination in Thin Film Coatings **87:155**

Characterization of Multi-Layer Backmetal Systems via Multi-Technique Instrumental Analysis **87:137**
Analysis of On-Line Organic Microcontaminants in Semiconductor Assembly Plants **86:1**
The Failure Analysis of Reed Relays by Scanning Auger Microscopy (SAM) **86:29**
Analysis of VLSI Metallization Microstructure by High Resolution Mechanical Cross Sectioning and Auger Analysis **85:98**
Crystalline Growth on RF Tuning Screws **85:10**
Environmental Corrosion Studies of Cobalt Thin Metal Film for Magnetic Recording Media **85:6**
Auger Analysis of Refractory Metal Silicides **84:12**
Digital Processing of Auger Data **84:13**
Voltage Measurements on Semiconductor Devices Using Auger Spectroscopy **83:57**
Failure Analysis of Electroplated Nickel on Ceramic IC Packages Using Electron Beam Techniques **83:182**
Microcircuit Failure Analysis Using a Scanning Auger Multiprobe **83:53**
Sputter Ion Depth Profiling Limitations of Microelectronic Materials **82:8**
Scanning Auger Analysis of IV Semiconductor Heteroepitaxial Interfaces **78:165**
Development of Failure Analysis Techniques Including Auger Spectroscopy for CMOS/SOS LSI Devices **78:162**
Analysis of Bonds and Interfaces with Auger Electron Spectroscopy **77:246**
Surface Analysis by Auger Spectroscopy and ESCA **76:9**

automated testing
A Structured Approach for Failure Analysis of a 256k BiCMOS SRAM **89:167**
GRAPHATT/GSI—A Paperless Approach to VLSI RAM Failure Analysis with Automatic Wafer-Stager Control from Graphic Bit-Fail Map Displays **89:37**
Automating Failure Analysis Pin-to-Pin Leakage Measurements **88:227**
An Automatic LCD Defect Inspection System Using an Area Image Sensor **88:189**
Die Attach Screen on A.T.E. **84:147**
Automated Electro-Optical Testing for High Resolution and High Performance Analog VLSI Imagers **83:257**
Software Model for Parallel Test Systems **83:250**
Testing of a Microprocessor with Pseudo-Random Vectors **83:247**
Problems Encountered during Automated Testing of Microelectronic Devices **82:140**
An Automated Device Characterizer Adds the Edge to Successful Failure Analysis **82:162**

backmetal
Characterization of Multi-Layer Backmetal Systems via Multi-Technique Instrumental Analysis **87:137**
Factors Resulting in Bond Delamination of Au Backmetal Si Die Eutectically Bonded to Ni/Au **83:263**

backside etching
Back-Etch: An Effective Tool for Characterization and Failure Analysis of MESFET Devices **88:235**
Characterization of Polyimide Defects Utilizing a Backside Etch/Delamination Technique **86:223**
Backside Etching Techniques **85:147**

batteries, Ni-Cd
Failure Analysis of Ni-Cd Battery Cell for Space Application **80:148**

beam leads
Process Controls to Increase the Effectiveness of Beam Lead Devices in High Rel Applications **77:251**

birefringence
Use of Birefringent Patterns to Determine the Strain induced in Silicon Chips by Bonding **77:36**

bolometers
Reliability of Thermistor Bolometers **77:295**

bond pads
Characterizing Integrated Circuit Bond Pads **92:165**

bonding wire, aluminum
Failure Analysis of Bonding Wires in Power Transistor Modules **91:237**
Ultrasonic Aluminum Wire Bonding and Lead-Indium Soldering to Gold Alloy Thick Film Conductors—Performance and Failure Mechanism **77:227**
Behavior of 1 Mil Aluminum Wire at High Temperature for Hybrid Microcircuits **77:94**

burn-in testing
VLSI Design Considerations for Burn-In **91:377**
'FITS'—An Analytical Method of Choosing Burn-in Processes **82:53**

capacitance testing
The Use of Infrared Imaging and Measurements of Inductance and Capacitance for Location of Faults in Multi-Layer Ceramic Circuit Boards **92:201**

capacitors
Capacitor End-Cap Dissolution Testing by Solder Reflow **83:222**
Capacitors, Thermal Rating/Derating (AC-DC Operation) **80:96**

capacitors, ceramic
Acoustic Visualization of Stressed Ceramic Chip Capacitors **82:215**
MLCS Outperform Metallized-Film Capacitors Under Actual Operating Conditions **82:209**
Identification of Ceramic Capacitor Shorts by Voltage Contrast in Scanning Electron Microscope **82:203**
Extended Electrical Characterization of Ceramic Capacitors Failing Under Low-voltage Conditions **82:194**
New Tools in the Failure Analysis of Multilayer Ceramic Capacitors **81:269**
Solder Coating of Ceramic Capacitors; Wettabililty Problems **81:111**
Low Voltage Failures of Monolithic Ceramic Capacitors and Their Screening Method **81:105**
Low-Voltage Failure Mechanisms for Ceramic Capacitors **81:101**
Innovative Screening for Ceramic Capacitors to Remove Failure Mechanisms **80:237**
A Comparison of Screening Techniques for Ceramic Capacitors **80:230**
A Low Voltage Screening of Ceramic Capacitors from Leakage Failures **80:225**

capacitors, metallized film
Failure Causes and Reliability Improvements in Metallized Polycarbonate Capacitors **79:77**

capacitors, tantalum
Reliability Evaluation and Failure Diagnosis of Tantalum Capacitors Using Computer-Aided Frequential Analysis and Design of Experiments **92:189**
Life Estimation for Slug Tantalum Capacitor **86:213**
Reliability Problem of the Wet-Slug Tn Capacitor (CLR-79) **81:117**
Life Evaluation of Solid Tantalum Capacitors Under On-Off Cycler Conditions **80:87**
Unshackle the Solid Tantalum Capacitor **79:89**

capacitors, trench
Failure Analysis of Trench Capacitor CMOS DRAMs: The Backside Approach **91:225**

cathode ray tubes (CRT)
Failure Analysis of Shorts in Color TV Tubes **86:163**

cathodoluminescence
Electron Beam Analysis of Laser Diodes and LEDs Using EBIC and Cathodoluminescence **89:87**

ceramic resonators
Failure Analysis of a Ceramic Resonator Exhibiting Excessive Epoxy on the Element and/or a Fractured Element **92:197**

charge-coupled devices (CCDs)
Reliability of a CCD Video line Store **82:182**
Screening, Testing, and Reliability of Charge Coupled Devices **80:185**
Reliability Evaluation of imaging CCD's **78:70**

chemicals, purity of
Contaminants in Process Chemicals **77:42**

CMOS (complimentary metal oxide-silicon)
The Analysis of Design Related Subtle and Intermittent Electrical Failure Modes in CMOS Devices **92:247**
1.0 µm CMOS Process improvement Case Study **92:101**
Case History: Failure Analysis of a CMOS DRAM with an Intermittent Open Contact **91:281**
Reliability Evaluation of CMOS RAMs **82:133**
Analysis of a Unique Failure Mode in a CMOS PROM **81:234**
Failure Analysis of 1K × 4 CMOS RAM **81:228**
Simulation of Stuck-Open Faults in CMOS Integrated Circuits **81:53**
Problems Associated with Testing CMOS RAMS Using ATE **80:266**
Parallel Testing for CMOS Microcircuits **80:247**
Failure Modes and Analysis Techniques for CMOS Microcircuits **77:75**

coax cable
Vacuum Effect on Coax Cable Outer Ribbon Conductor **83:301**

cold-start effects
Investigation of Bipolar Microcircuit Cold Start Problem and Its Impact on Reliability **84:85**

construction analysis
DPA Rejects—Are They Really Failures? **77:213**
A Refined Part Dissection Analysis System Continues to Assure Part Integrity **77:209**
Construction Analysis of Electronic Parts **77:198**

contacts (vias)

Case History: Failure Analysis of a CMOS DRAM with an Intermittent Open Contact **91:281**

An Investigation of Open-Via Failures in ASIC Devices **91:129**

Design and Process Issues for Elimination of Device Failures Due to "Drooping" vias **91:97**

Study of Failure Mechanisms on Electrical Contact Surfaces **80:15**

contamination

Model for Ionic Contamination Induced Failure of a Semiconductor **92:237**

Case History of Epoxy Contaminated Wire Bond Failures on Space Shuttle Hybrids **92:61**

The Failure Analysis of a Bimetallic Thermostat **89:369**

Failure Analysis of Vendor Produced CMOS Modules **89:99**

Surface Analysis of Contamination in Thin Film Coatings **87:155**

Analysis of On-Line Organic Microcontaminants in Semiconductor Assembly Plants **86:1**

Analysis on a Cause of Rare Contact Failure in Sealed Contacts **86:153**

The Failure Analysis of Reed Relays by Scanning Auger Microscopy (SAM) **86:29**

The Identification and Elimination of Human Contamination in the Manufacture of ICs **85:169**

Determination of Particulate Contamination in Gate Oxide **85:48**

Facet Degradation in High Power (AlGa)As Laser Diodes from −20 °C and 50 °C Life Tests **85:31**

Hybrid Microcircuit Failures Caused by Nickel Plating **85:1**

Fourier Transform Infrared Spectroscopy (FTIR): Its Application to Failure Analysis **84:53**

Failure Analysis of Laser Optical Components **83:178**

Application of the Raman Microprobe to Quality Control in IC Manufacture and to Crystalline Analysis of Structural Ceramics **82:1**

Direct Lateral and in-Depth Distributional Analysis for Ionic Contaminants in Semiconductor Devices Using Secondary Ion Mass Spectrometry **79:148**

cordwood

Failure Analysis Techniques for Epoxy Encapsulated Multilayer Board, Cordwood Constructed Modules **77:107**

corrosion

Investigation of Corrosion Failures for Computers by Using an Original Method **92:259**

Failure Analysis of a Resistance Temperature Detector (RTD) **89:385**

Degradation Mechanisms of GaAs MESFET Devices in High Humidity Conditions **89:141**

Failure of an Embedded Bond Wire in a CERDIP: Reliability Implications **89:133**

Pin-Hole Failure Analysis of Plastic-Based Magneto-Optical Disk **89:223**

Corrosion Mechanism Analysis of Salt Spray Test and SO_2 Test on Gold Plated Connector Contact **88:205**

Highly Accelerated Stress Test (HAST) on VLSI Plastic Components **88:167**

Evaluation of NiCd Cell Terminals' Durability for Alkaline **87:233**

Deterioration of ZnO/SiO_2 Diode Packages in High Humidity **87:207**

Moisture-Resistance Test Using Unsaturated Pressure Cooker Equipment **86:189**

Testing and Analysis of Photovoltaic Modules for Electrochemical Corrosion **86:39**

Corrosion Mechanism of Thin Film Disk Magnetic Recording Structures **86:35**

Analysis of Fractures in Half-Size Crystal Can Relay Moveable Contact Arms **84:196**

A New Accelerated Test for Moisture Resistance on IC's **84:108**

Fourier Transform Infrared Spectroscopy (FTIR): Its Application to Failure Analysis **84:53**

Microanalytical Characterization of Semiconductor Device Contamination Using the Laser Microprobe Mass Analysis (LAMMA) Technique **83:63**

Stress Corrosion Cracking of Dual In Line Package Leads **82:346**

Search for a Test Simulating Indoor Corrosion of Electrical Contacts **82:115**

crystal oscillators

Test Techniques Useful in Identifying Sources of Anomalous Behavior in Quartz Crystal Filters and Oscillators **77:178**

data retention failure

Filler Particle induced Data Retention Failures in Plastic Encapsulated EPROM's **92:391**

decapping (delidding)

Opening Techniques for IC Ceramic Packages **83:100**

decapsulation techniques

Semiconductor Device Decapsulation Using a CNC Milling machine **91:77**

Plastic Mold Opener That Uses Fuming Nitric Acid as Dissolving Liquid **88:137**

Chemical Decapsulation Revisited **87:113**

Expedited Failure Analysis of GaAs FET Transistors, Si IMPATT Diodes and Schottky Diodes Using X-ray, SEM and EBIC Techniques **87:53**

Failure Analysis Applications of Plasma, **87:35**

Failure Analysis of Metallization Corrosion, **ASM Conference on Electronic Packaging, 1987, p 275.**

Techniques and New Etch Block Design to Enhance the Jet Etch Decapsulation **85:134**

An Improved Decapsulation Technique for Plastic Encapsulated Opto Coupler Devices **85:114**

Micro-Surgery as Used in Epoxy Laminate Failure Analysis **84:130**

Mechanical Decap Method for Plastic Devices **84:95**

Opening Techniques for IC Ceramic Packages **83:190**

Decapsulation of Silicone-Epoxy Copolymer Packages **82:73**

Three Decapsulation Methods for Epoxy Novolac Type Packages, **IRPS-1980, p 107.**

Improved Technique for Decapsulation of Epoxy-Packaged Semiconductor Devices and Microcircuits, by B. Wensink, **Solid State Technology, 1979, p 107.**

Tough Analysis Problems That Have a Solution Mike Jacques **78:124**

Nondestructive Decapsulation of Epoxy B/Novolac Packages **78:90**

defects, crystallographic

Crystallographic Imperfections and Their Effect on Micro-Electronic Performance **76:52**

defects, layout-dependent

Characterization of Layout-Dependent Defects on the 16-Mb DRAM **92:117**

defects, polysilicon

Failure Due to Pinholes in Polysilicon Conductors on CMOS Memory Devices **77:60**

defects, surface, oxide

Advanced Methods for Imaging Gate Oxide Defects With the Atomic Force Microscope **92:267**

Failure Analysis Techniques in Thin Gate Oxide Defects **92:181**

Failure Analysis of Double Polysilicon Thick Interlevel Oxide Failures **82:104**

Failure Analysis Technique for Examination of SNOS Type Defects **82:20**

Silicon Dioxide Defect Location and Analysis on VLSI DPS Structures **81:216**

Analysis of Surface Defects on Hybrid Microcircuits **79:153**

delamination (metals)

Factors Resulting in Bond Delamination of Au Backmetal Si Die Eutectically Bonded to Ni/Au **83:263**

delamination (plastics)

Evaluation of Methods for Delamination Detection by Acoustic Microscopy in Plastic-Packaged Integrated Circuits **92:425**

Identification of Package Defects in Plastic-Packaged Surface Mount IC's by Scanning Acoustic Microscopy (SAM) **89:61**

Nondestructive Evaluation of Thermally Shocked Plastic Integrated Circuit Packages Using Acoustic Microscopy **88:211**

Nondestructive Inspection of Encapsulated Capacitors: Flaw Growth Detection of Parts Submitted to 85% Humidity, 85 °C **85:193**

A Novel Technique for Evaluating a Gap Between Lead and Resin for Plastic Encapsulated Devices **85:120**

Flourescent Penetration Testing for Failure Analysis on Plastic Encapsulated Integrated Circuits **82:125**

delineation (decoration, staining)

Optimizing Various Electrochemical Etching Parameters for Measuring Specific Electrical Junctions in Silicon **92:151**

A Nonchemical Method of Delineating Junctions in Selected Semiconductor Devices **83:88**

Extension of Carbon Defect Decoration and Applications to Failure Analysis **78:146**

die bonds

Cold Sample Preparation Technique for Soft Solder Die Attach Examination **92:135**

Case Histories of Microelectronic Device Analyses Using Auger Electron Spectroscopy (AES) **89:373**

Thermal Fatigue Failures of Large Scale Package Type Power Transistor Modules **89:309**

Non-Destructive Inspection of Voids in Silver Hard Solder Layer of Power Transistor **88:47**

Fatigue of 63Sn-37Pb Solder Used in Electronic Packaging **88:53**

An Overview of Infrared Thermal Imaging Techniques in the Reliability and Failure Analysis of Power Transistors **88:63**

Characterization of Die Attach Integrity Using Destructive and Nondestructive Techniques **88:77**

A Closer Look at Delta-V_{BE} Characteristics **87:123**

Solder Paste Factors Affecting Quality and Reliability **85:59**

Scanning Laser Acoustic Microscopic (SLAM) Study of Die Attach Integrity **85:202**

Die Bonding on Non-Gold Plated Piece Parts **85:22**

Stress Analysis of Poor Gold-Silicon Die-Attachment for LSIs **84:180**

Die Attach Screen on A.T.E. **84:147**
Die Attach/Die Shear and Metal Migration **83:317**
A High Speed Approach to Die Bond Integrity Testing **83:254**
Reliability of CCD Video Line Store **82:182**
Failure Analysis of Die Attach Bonds **77:223**
A New Ultrasonic Technique for the Inspection of Semiconductor Package Diebonds **76:86**

die coating

An Analysis of Power Transistors with a Polyimide Die Coat **87:213**
Optimization of Polyimide Die Coat Cure Cycle via Chemical Analysis and Electrical Characterization **87:41**
Characterization of Polyimide Defects Utilizing a Backside Etch/Delamination Technique **86:223**

die cracking

Influence of Cracks on GaAs Solar Cells **87:59**
Acoustical Microscopy as a Tool for Nondestructive Testing of Finished Devices **87:21**
Scanning Laser Acoustic Microscopic (SLAM) Study of Die Attach Integrity **85:202**
The Use of Fractography in the Failure Analysis of Die Cracking **84:258**
Stress Analysis of Poor Gold-Silicon Die-Attachment for LSI **84:180**

dielectric breakdown

Sub-Micron Process and TDDB Characteristics **88:119**
A Non-Destructive Overvoltage Mode in CMOS **88:103**
Development of Automated TDDB System and Novel Evaluation Method on Thin Oxide Film **87:185**
CMOS Process Development and Time Dependent Dielectric Breakdown (TDDB) **85:36**
The Vsub Problem: The Characterization of an Unusual Dielectric Breakdown Phenomenon **85:72**

DPA (destructive physical analysis)

The Viscissitudes of Destructive Physical Analysis **81:294**
Destructive Physical Analysis (DPA) Experiences **80:125**
DPA Rejects—Are They Really Failures? **77:213**
A Refined Part Dissection Analysis System Continues to Assure Part Integrity **77:209**
Construction Analysis of Electronic Parts **77:198**

DRAM dynamic random access memory

Characterization of Layout-Dependent Defects on the 16-Mb DRAM **92:117**
Application of Multiple Failure Analysis Techniques to Resolve Moisture Related DRAM Failures **92:125**
Failure Analysis of Trench Capacitor CMOS DRAMs: The Backside Approach **91:225**
Comparison of the Quality of Foreign and Domestic Integrated Circuits by Transmission Electron Microscopy (TEM) **89:75**
Failure Analysis of DRAMs **88:31**
An Expert System for Help to Fault Diagnosis on VLSI Memories **88:153**
Degradation Mechanism Due to Hot Electron Trapping in High Density CMOS DRAM **88:89**
Failure Analysis and Failure Mechanisms of Triple Polysilicon VLSI Devices **86:201**

A Liquid Crystal Failure Analysis Method for Identification of Row Failure Leakage Sites on 256k DRAMs **86:169**
E-Beam Testing: Image & Signal Processing for the Failure Analysis of VLSI Components **85:89**
The Vsub problem: The Characterization of an Unusual Dielectric Breakdown Phenomenon **85:72**
Discovery of a New VLSI Failure Mechanism Through the Application of a Particular Failure Analysis Methodology **84:190**
Logical and Systematic Failure Analysis of 64k DRAM Using FTA and Reliability Analysis Techniques **84:59**
Substrate Voltage Bump Test for Dynamic RAM Devices **83:281**
A New Breakdown Phenomenon in High Density RAM **83:270**
Microcircuit Failure Analysis Using a Scanning Auger Multiprobe (SAM) **83:53**
Silicon Inclusion Failures in Aluminum Interconnects **83:30**
Isolation of Polysilicon for Failure Analysis **82:101**
Reliability Evaluation of 16K Dynamic RAMs in Plastic Packages **82:31**
RAM Cell Defect Localization and Analysis on VLSI Structures **82:13**
Development of the Method to Determine the Critical Energy for the Soft Error in Dynamic RAM **79:191**

drift

Advanced Technique for Analysis of Drift Failures in PN Junction Devices **78:69**

E-BEAM Testing

FACE: An Approach for Automatic IC Diagnosis by E-Beam Testing **92:15**
New Developments in the Failure Analysis of Passivated CMOS ICs by Electron Beam Testing **91:389**
Electron Beam Testing for Verification of Voltage Distribution on VLSI Circuits **82:149**

EBIC—electron beam induced current

Microprobing and EBIC for VLSI Technology **92:43**
EBIC and TEM Analyses of Piped Multi-Emitter Transistors **92:55**
CMOS Characterization Using EBIC: Applications and Performance **91:199**
Electron Beam Analysis of Laser Diodes and LEDS Using EBIC **89:87**
EBIC Observation of Subsurface Damage **88:225**
Failure Analysis of Inversion Mechanisms: Beyond Bake Recoverable **88:217**
Expedited Failure Analysis of GaAs FET Transistors, Si IMPATT Diodes and Schottky Diodes Using x-ray, SEM and EBIC Techniques **87:53**
Quantitative Measurement of Metallization Integrity Using EBIC **78:175**
Investigation of Gate Shorts in CMOS Devices by Electron Beam Induced Current and Internal SEM Probes **81:5**
Application of Scanning Electron Microscope, EBIC Mode to Semiconductor Evaluation and Failure Analysis **79:136**

EDXA, EDS energy dispersive X-ray spectroscopy

Quantitative Analysis of Thin Layers Using Microanalytical Techniques **86:5**
Analysis of On-Line Organic Microcontaminants in Semiconductor Assembly Plants **86:1**

Failure Analysis of ESD Damage in MOSFET Power Devices **86:83**
Materials Characterization by Analytical Electron Microscopy **84:46**
Failure Analysis of Electroplated Nickel on Ceramic IC Packages Using Electron Beam Techniques **83:182**
Reliability of SEM/EDX Analysis of P in SiO_2 **82:23**
A Non-Destructive Method to Approximate SiO_2 Passivation Thickness on Fabricated Semiconductor Devices Using WDXA for Oxygen **82:6**
A Comparison of Minimum Detection limits using Wavelength and Energy Dispersive Spectrometers **77:49**
EDXRF energy dispersive x-ray fluorescence Characterization of Multi-Layer Backmetal Systems via Multi-Technique Instrumental Analysis **87:137**

EGA (evolved gas analysis)

Analytical Pyrolysis and Evolved Gas Analysis of Electronic Polymers **86:11**

electrical testing and characterization

Failure Analysis of CMOS Devices with Anomalous I_{DD} Currents **91:381**
Automated Digital Integrated Circuit Tester (ADICT) **89:297**
Failure Analysis of Complex and High Pin Count Devices Using Computer Aided Electrical Characterization **89:261**
Failure Mode Analysis Methodology on High Density CMOS SRAMS **89:389**
Locating High Resistance Shorts in CMOS Circuits by Analyzing Supply Current Measurement Vectors **89:231**
Application of Quiescent Supply Current Signature Analysis to Failure Analysis of Integrated Circuits **89:27**
Nodal Waveform Analysis of Recoverable Gate Array Functional Failures **88:41**
Analysis of CMOS Microprocessor Failures Using a New Electrical Signature Analysis Technique **87:1**
Failure Analysis of VLSI Logic Functional Fails **86:179**
A Dual Trace, Multi-Pin Microcircuit Test Adaptor for Standard Curve Tracers **86:99**
A Diagnostic Technique for CMOS ICs Based on Changes in the Quiescent Power Supply Current Caused by Toggling the Inputs **86:185**
Data Retention Test Method of EPROMS **85:104**
Analysis of Failures in Large Arrays **85:85**
A Signature Analysis Technique for Defect Characterization of CMOS Static RAM Single Cell Fails **84:141**
A High Speed Approach to Die Bond Integrity Testing **83:254**
Software Model for Parallel Test Systems **83:250**
Testing of a Microprocessor with Pseudo-Random Vectors **83:247**
A Microprocessor Test Experience **83:244**
A Novel VLSI SRAM Failure Analysis Technique **83:40**
Fault Isolation in LSI Circuits Using Logic Simulator Driven Computer Guided Diagnosis **83:1**
Laser Die Probing for Complex CMOS **82:178**
Electron Irradiation Tests on Small Signal and Power GaAs FETs **82:168**
An Automated Device Characterizer Adds the Edge to Successful Failure Analysis **82:162**
Internal Node Testing by Tester Aided Voltage Contrast **82:156**

Electron Beam Testing for Verification of Voltage Distribution on VLSI Circuits 82:149

Problems Encountered During Automated Testing of Microelectronic Devices 82:140

An Electrical Test to Detect Mechanical Damage in Junction Transistors 81:39

Results of Testing and Programming of Fuseable Link PROMs 81:50

Simulation of Stuck-Open Faults in CMOS Integrated Circuits 81:53

Electrical Testing for Failure Analysis of Memory Devices—The Role of Memory Checker 81:57

Test Pattern Generation for Large Digital Hybrid Microcircuits 80:262

Problems Associated with Testing CMOS RAMS Using ATE 80:266

Fault Detection Efficiency Measurement Via Hardware Fault Simulation 80:255

Hybrid Die Acceptance Testing 80:250

Parallel Testing for CMOS Microcircuits 80:247

Test Equipment—The VLSI Challenge 80:245

ATE, the Device Specification and Part Failure 78:63

Automated Long Term, Uninterrupted, LED Test Complex 78:59

Automated Test Design for Analog Circuits 77:17

electromechanical devices

Failure Analysis of Coaxial Switch 81:182

The Relationship Between Corrosion and Physical Properties in Electrical Connections—An Investigation of Gold Alloy Palladium-Nickel and Tin-Lead Plating on Connections 81:173

Failure Analysis in Modern Disk Technology 81:159

Failure Prognosis Techniques for Rotating Machinery 78:44

Analytical Techniques for Studying Small Electrical Components 77:170

Failure Analysis of Electrical Contact Surfaces 77:159

Practical Approaches to Reed Capsules Failure Analysis 77:155

Field Performance Analysis and Corrective Action Methods for Electronic Equipment in the Telephone Industry 77:140

electromigration

Electromigration in Aluminum Bond Wires Subjected to Current Pulsing 91:183

The Validity of a Standard Electromigration Model at High Stress Conditions 89:115

Effect of the TiN/Ti Diffusion Barrier on the Al Electromigration 89:343

Electromigration-Accelerated Life Test of Al-Conductors Encapsulated in Plastic and Ceramic 87:219

Analysis of VLSI Metallization Microstructure by High Resolution Mechanical Cross Sectioning and Auger Analysis 85:98

ESD-Induced Electromigration: A New VLSI Failure Mechanism 84:214

Effect of Electromigration on GaAs FET Reliability 84:163

Die Attach/Die Shear and Metal Migration 83:317

Reliability of CCD Video Line Store 82:182

Characterization and Analysis of Certain Types of Defects in Channel Electron Multiplier Arrays (CEMAs) 77:280

electroplating

Failure Analysis of Electroplated Nickel on Ceramic IC Packages Using Electron Beam Techniques 83:182

emission microscopy (EMMI)

Emission Microscopy: A Powerful Tool for IS's Design Validation and Failure Analysis 92:289

An Economical Approach to Correlation and Calibration of Light Emission Microscopes 91:315

Characterization of Device Structure by Spectral Analysis of Photoemission 91:325

Emission Microscopy Applied to Optoelectronic Emitter Failure Analysis 91:353

Evaluation of Gate Defects in GaAs MESFETs by Emission Microscopy 91:363

Functional Failure Analysis Using Photoemission Microscopy 91:389

Failure Analysis of a Degraded Voltage Regulator Using Light Emission Microscopy 91:343

Case History: Wafer-Level Failure Analysis of Functional ICs with Elevated I_{DDQ} Caused by Resistive Gate Oxide Shorts 91:138

Emission Microscope Identification of Nonuniform Submicron Transistors with Hot Carrier Characteristics 91:113

A New Electrostatic Discharge Test Method for Charged Device Model 89:177

Second Generation Emission Microscopy and Its Applications 89:277

Leakage Detection Techniques: A Comparative Study 89:5

Localization of Defects in Gate Oxides by Means of Tunneling Current Microscopy IRPS-1986, p 95

encapsulant material

Characterization of Molding Compounds via Polymer Reconstruction Investigative Chromatopyrography 84:98

Partial Discharge Degradation of Insulating Materials for High Voltage Parts in Space 83:311

EOS (electrical overstress)

Inversion leakage on N-Channel DMOS FETs by Electrical Stress 91:581

Interpretation of Severe EOS Damage on Logic Circuitry 91:288

Electromigration in Aluminum Bond Wires Subjected to Current Pulsing 91:183

Failure Analysis of ESD/EOS Damage to HCMOS Gate Arrays 89:201

Photoemission Testing for EOS/ESD Failures in VLSI Devices: Advantages and Limitations 89:183

Failure Simulations Allow Estimation of Overvoltage Condition in TTL Electronics 89:209

A Non-Destructive Overvoltage Mode in CMOS 88:103

Simulation and Characterization of EOS and ESD Damage in MOSFET Transistors and Integrated Circuits 88:95

Limiting Phenomena in Power Transistors and the Interpretation of EOS Damage 87:169

Tuning-Induced Permanent Damage to GaAs IMPATT Diodes 86:125

The Interpretation of EOS Damage in Power SCRs 85:42

Observed Physical Effects and Failure Analysis of EOS/ESD on MOS Devices 84:205

SEM Study of the Dynamics of Electrical Overstress (EOS) in Power Transistors 84:221

The Interpretation of EOS Damage in Power Transistors 83:157

Transient Damage in Power Transistors 79:130

EPROM (erasable programmable read only memory)

Data Retention Test Method of EPROMS 85:104

Failure Modes of EPROM 84:79

Accelerated Retentivity Tests of 16K EPROMS 80:180

epoxy

Case History of Epoxy Contaminated Wire Bond Failures on Space Shuttle Hybrids 92:61

equipping the FA lab

Equipment and Techniques for Failure Analysis of LSI Digital Devices 81:277

ESCA (electron spectroscopy for chemical analysis)

Environmental Corrosion Studies of Cobalt Thin Metal Film for Magnetic Recording Media 85:6

Using ESCA in Failure Analysis: Some Recent Developments 84:35

Surface Analysis by Auger Spectroscopy and ESCA 76:9

ESD (electrostatic discharge)

Bilateral Latch-Up Failure in CMOS ESD Protection Circuits 92:223

Case Study of ESD Damage and Long-Term Implications on SOS RAMs 92:213

Specific Area TEM Analysis of ESD-Induced Silicon Melt Filaments on Output Transistors 92:27

Charged Device Model ESD Failure Mechanisms and Design Considerations 91:297

Interpretation of Failure Analysis Results on ESD-Damaged InP Laser Diodes 91:305

Failure Analysis of ESD/EOS Damage to HCMOS Gate Arrays 89:201

Breakdown of Thin Gate-Oxide by Application of Nanosecond Pulse as ESD Test 89:193

Photoemission Testing for EOS/ESD Failures in VLSI Devices: Advantages and Limitations 89:183

A New Electrostatic Discharge Test Method for Charged Device Model 89:177

Failure Analysis of Semiconductor Devices with the Aid of an Infrared (IR) Microscope 88:243

ESD Induced Damage in Vertical Power MOSFETs 88:111

Simulation and Characterization of EOS and ESD Damage in MOSFET Transistors and Integrated Circuits 88:95

Extended-Voltage ESD Testing 87:193

Latent ESD Metallization Damage 87:177

New ESD Test Method 86:91

ESD-Induced Electromigration: a New VLSI Failure Mechanism 84:214

Observed Physical Effects and Failure Analysis of EOS/ESD on MOS Devices 84:205

ESD Failure Analysis of NMOS VLSI Chips 83:152

Reliability of CCD Video Line Store 82:182

A Technique to Test for Electrostatic Discharge Damage to Thick Film Resistors 78:35

excimer lasers

Advanced Micromachining Techniques for Failure Analysis 91:35

expert systems

An Expert System for Help to Fault Diagnosis on VLSI Memories 88:153

failure analysis methodology

Failure Analysis Process 87:131

Discovery of a New VLSI Failure Mechanism through the Application of a Particular Failure Analysis Methodology **84:190**

failure mechanisms

Low-Voltage Failure Mechanisms for Ceramic Capacitors **81:101**

Incremental Failure Mechanism in Subminiature Hermetically Sealed Fuses **81:26**

Microwave Power GaAS FET Failure Mechanisms **80:83**

Failure Causes and Reliability Improvements in Metallized Polycarbonate Capacitors **79:77**

Advanced Technique for Analysis of Drift Failures in PN Junction Devices **78:69**

Failure Mechanisms in Discrete Semiconductor Devices **76:35**

Failure Mechanism in MOS Devices Versus HCl Gettering **76:104**

fault detection, location, isolation

An Automatic Fault Locating Method for Logic VLSIs **92:345**

Novel Electron-Beam Image-Based LSI Fault Technique Without Using CAD Database: Development and Its Application to Actual Devices **92:49**

A Failure Analysis Approach to Fault Isolation on Complex Microprocessors and Microcomputers **91:145**

Failure Site Location in Power MOSFET Devices by F.A.L.T. Probing **86:107**

Thermal Analysis of Microelectronic Devices Using a Hot Spot Detection System Based upon Vapor Bubble Technology **82:43**

Device Leakage Investigations Using Fluorescent Microthermography **IRPS-1986, p 157**

fault detection theory

A Brief Assessment of Failure Detection Methods **87:107**

ferroelectric liquid crystals

VLSI Chip Testing Method Using Ferroelectric Liquid Crystals **92:107**

field crystallization

Life Estimation for Slug Tantalum Capacitor **86:213**

field failures, FFRP

Selected Case Histories from the DoD Field Failure Return Program **92:67**

filler particles

Filler Particle induced Data Retention Failures in Plastic Encapsulated EPROM's **92:391**

fits

'FITS'—An Analytical Method of Choosing Burn-in Processes **82:53**

flip-chip bonds

Failure Analysis of Flip-Chip Bond Separation **87:239**

fluorescent die testing

Fluorescent Penetration Testing for Failure Analysis on Plastic Encapsulated Integrated Circuits **82:125**

focused ion beam (FIB)

Contrast Mechanisms in Focused Ion Beam Imaging **92:373**

Focused Ion Beam Applications for Design and Product Analysis **91:397**

Endpoint Detection and Microanalysis With Ga FIB SIMS **91:401**

The Study of Failure Mechanisms in IC Devices by Using Focused Ion Beam Technology **91:409**

A Technique for Preparing TEM Cross Sections to a Specific Area Using the FIB **91:417**

Imaging Dielectric Breakdown in MOS Transistors Using a Combination Transmission Electron Microscope and Focused Ion Beam **91:429**

Integration of a Focused Ion Beam System in a Failure Analysis Environment **91:85**

Advanced Micromachining Techniques for Failure Analysis **91:35**

Failure Analysis of Micron Technology VLSI Using Focused Ion Beams **89:249**

A New VLSI Diagnosis Technique: Focused Ion Beam Assisted Multi- Level Circuit Probing **IRPS-1987 p 111**.

FTIR (Fourier transform infrared spectroscopy)

Application of FTIR Microspectroscopy in Contamination Analysis **92:175**

An Analysis of Power Transistors with a Polyimide Die Coat **87:213**

Analysis of On-Line Organic Microcontaminants in Semiconductor Assembly Plants **88:1**

Analysis of On-line Organic Microcontaminants in Semiconductor Assembly Plants, **86:1**

Microscopic FTIR for Failure Analysis **85:177**

Fourier Transform Infrared Spectroscopy: Its Applications to Failure Analysis **84:53**

FTIR Microspectrophotometry—Advances and Applications **84:43**

fuses

Failure Analysis Techniques of TiW Fuse Failures on PLDs for Improved Programming Yield **91:271**

Reliability of Polysilicon Fuses in Plastic Molded Devices **84:112**

Study to Establish Limits for Reliable Fuse Application in Transient Environment **82:111**

Incremental Failure Mechanism in Subminiature Hermetically Sealed Fuses **81:26**

gallium arsenide (GaAs)

Interpretation of Life Test Results on DHBC In-GaAsP/InP Laser Diodes **92:253**

Storage Tests and Failure Analyses on AlGaAs/GaAs HEMTs **92:73**

High Reliability of Low-Noise Self-Aligned Gate GaAs MESFET with Noise Figure of 0.85db at 12GHz **91:191**

Reliability Evaluation on GaAs MMIC Components **89:335**

Reliability Evaluation of GaAs MESFETs and ICs **89:329**

Degradation Mechanism Due to Hot Electron Trapping in High Density CMOS DRAM **88:89**

The Effects of the Passivation Film on the Reliability of High Power GaAs MESFETS **83:302**

A Well Breakdown Phenomenon in High Density RAM **83:270**

Reliability Study of (AlGa)As Laser Diodes for Long Life Space Applications **82:25**

Reliability and Failure Mechanisms of GaAs-FETs **81:69**

Microwave Power GaAS FET Failure Mechanisms **80:83**

Voids in Aluminum Gate Power GaAsFETs Under Microwave Testing **78:155**

A Novel Electron Beam Technique for Studying GaAs FETs **77:114**

gas analysis

Analysis of Organic Gases Generated Inside a Magnetic Disk Drive **91:519**

gate arrays

Failure Analysis of ESD/EOS Damage to HCMOS Gate Arrays **89:201**

Photoemission Testing for EOS/ESD Failures in VLSI Devices: Advantages and Limitations **89:183**

Nodal Waveform Analysis of Recoverable Gate Array Functional Failures **88:41**

Fault Isolation in LSI Circuits Using Logic Simulator Driven Computer Guided Diagnosis **83:1**

gate oxide

Advanced Methods for Imaging Gate Oxide Defects With the Atomic Force Microscope **92:267**

Failure Analysis Techniques in Thin Gate Oxide Defects **92:181**

glass deposition

TEOS Glass Deposition by PECVD, A New Failure Analysis Technique **92:141**

glass voiding

Novel Microscopic Techniques in the Failure Analysis of Trimmed Ceramic Capacitors **86:131**

gold metallization

Reliability investigation on Gold Metallized RF Transistors **83:180**

HAST (highly accelerated stress testing)

Highly Accelerated Stress Testing on VLSI Plastic Components **88:167**

heat pumps

Thermoelectric Heat Pump Failures **91:167**

HEMTs

Storage Tests and Failure Analyses on AlGaAs/GaAs HEMTs **92:73**

hillocks (Al)

Submicron Precision Die Cross-Sectioning **89:243**

Die Attach/Die Shear and Metal Migration **83:317**

Application of Infrared Microscopy in the Design and Production of High Power Semiconductor Devices **82:188**

holography

A Holographic Technique for Failure Evaluation of Piece Part Leads Due to Vibration **76:75**

hybrid circuits

Ionic Contamination Induced Failure of a Hybrid Microcircuit **91:173**

Hybrid Microcircuit Failures Caused by Nickel Plating **85:1**

Capacitor End-Cap Dissolution Testing by Solder Reflow **83:122**

An Investigation of the Effect of Electric Field Equivalency (EFE) Dice Screening on Hybrid Yield **81:62**

ICP-AES (inductively coupled plasma-atomic emission spectroscopy)

Characterization of Multi-Layer Backmetal Systems via Multi-Technique Instrumental Analysis **87:137**

Analysis of Alloys and Plating Processes via Inductively Coupled Plasma Atomic Emission Spectroscopy (ICP-AES) **86:17**

Inductively Coupled Plasma Atomic Emission Spectroscopy and its Application in Testing and Failure Analysis **85:152**

image analyzers
New 3-D Image Analyzer of Radiograph and its Application for Electron Devices **79:154**

image processing
A Loosely Coupled Image Management System for Failure Analysis **91:205**

Practical Applications of Dynamic Fault Imaging **89:269**

Digital Processing and Color Coding of Voltage Contrast Images **85:63**

IMMA—ion microprobe mass analyzer
Ion Microprobe Mass Analyzer **76:10**

IMPATT diodes
Expedited Failure Analysis of GaAs FET Transistors, Si IMPATT Diodes and Schottky Diodes Using X-ray, SEM and EBIC Techniques **87:53**

Tuning-Induced Permanent Damage to GaAs IMPATT Diodes **86:125**

inclusions
Silicon Inclusion Failures in Aluminum Interconnects **83:30**

indium phosphide (InP)
Interpretation of Life Test Results on DHBC InGaAsP/InP Laser Diodes **92:253**

inductance testing
The Use of Infrared Imaging and Measurements of Inductance and Capacitance for Location of Faults in Multi-Layer Ceramic Circuit Boards **92:201**

infrared microscopy
Infrared Laser Microscopy of Structures on Heavily Doped Silicon **92:1**

Au-Al Bond Degradation Mechanism and the Analysis with Infrared Microscope **89:15**

Failure Analysis of Semiconductor Devices with the Aid of an Infrared Microscope **88:243**

Infrared Microscopy as Applied to Failure Analysis of P-DIP Devices **IRPS-1986, p 99**

A Study of Gold Ball Bond Intermetallic Formation in PEDs Using Infrared Microscopy **IRPS-1986, p 102**

Pulsed Infrared Microscopy for Debugging Latch-Up on CMOS Products **IRPS-1984, p 122**

An IR Microscopy Technique for Failure Analysis of Suspected Metallization Corrosion, **IRPS-1980, p 318.**

A Portable, X-Y Translating Infrared Microscope for Remote Inspection of Photovoltaic Solar Arrays **80:21**

infrared thermal imaging
The Use of Infrared Imaging and Measurements of Inductance and Capacitance for Location of Faults in Multi-Layer Ceramic Circuit Boards **92:201**

Diagnostic Test and Numerical Simulation of Solar Cells Subjected to High Reverse Current **89:323**

Thermal Fatigue Failures of Large Scale Package Type Power Transistor Modules **89:309**

An Overview of Infrared Thermal Imaging Techniques in the Reliability and Failure Analysis of Power Transistors **88:63**

A Closer Look at Delta-V_{BE} Characteristics **87:123**

Thermography Testing of P.C. Boards **83:219**

Diagnostic Test and Numerical Simulation of Solar Cells Subjected to High Reverse Current **89:323**

Degradation Mechanisms of GaAs MESFET Devices in High Humidity Conditions **89:141**

Diagnostic Techniques for Probing and Contacting Small Geometry and High Frequency Semiconductor Devices **89:239**

Reliability Evaluation of GaAs MESFETs and ICs **89:329**

Back-Etch: An Effective Tool for Characterization and Failure Analysis of MESFET Devices **88:235**

Influence of Cracks on GaAs Solar Cells **87:59**

Expedited Failure Analysis of GaAs FET Transistors, Si IMPATT Diodes and Schottky Diodes Using X-ray, SEM and EBIC Techniques **87:53**

Tuning-Induced Permanent Damage to GaAs IMPATT Diodes **86:125**

Effect of Electromigration on GaAs FET Reliability **84:163**

The Effects of the Passivation Film on the Reliability of High Power GaAs MESFETs **83:302**

Thermography Testing of P.C. Boards **83:219**

Application of infrared Microscopy in the Design and Production of High Power Semiconductor Devices **82:188**

Electron Irradiation Tests on Small Signal and Power GaAs FETs **82:168**

IC Thermal Response to Transient Current Overloads **77:28**

Short Pulse Thermal Measurements Using IR for Reliability Studies **77:24**

Noncontacting Infrared Maasurements of Operating Temperatures in Semiconducting Devices **77:21**

interconnects (contacts)
Silicon Inclusion Failures in Aluminum Interconnects **83:30**

ionic contamination
Model for Ionic Contamination Induced Failure of a Semiconductor **92:237**

ion beams
VLSI Failure Analysis and Characterization Applications of Photon and Ion Beams **91:1**

ion migration
Possible Causes for Ion-Migration in the Space-Use Electronic Equipments **92:93**

IR detectors
Solder Seal Failure Analysis of an IR Detector **82:359**

isolation
Fault Isolation in LSI Circuits Using Logic Simulator Driven Computer Guided Diagnosis **83:1**

Isolation of Polysilicon for Failure Analysis **82:101**

Failure Analysis Techniques Used to Physically Isolate LSI Circuits **79:177**

Low-Cost LSI Device Failure isolation Equipment **76:92**

isolation, photoresist
Application of Image Reversal Photoresist for Failure Analysis, **87:17**

isolation, probe
Methods of Trace Cutting for Diagnostic Probing of Semiconductor Devices **84:136**

A Laser Cutter for Failure Analysis **83:69**

Microprobing **IRPS 1980, p 117.**

Failure Analysis Techniques Used to Physically Isolate LSI Circuits **79:177**

ISS (ion scattering spectroscopy)
Applications of ISS/SIMS for Surface Analysis **76:13**

lasers
Advanced Micromachining Techniques for Failure Analysis **91:35**

Failure Analysis of Laser Optical Components **83:178**

A Laser Cutter for Failure Analysis **83:69**

laser diodes
Interpretation of Life Test Results on DHBC InGaAsP/InP Laser Diodes **92:253**

laser microscopy
Latchup Testing of Integrated Circuits **89:285**

A Laser Scanner For Solar Cell Evaluation and Failure Analysis **78:16**

A Holographic Technique for Failure Evaluation of Piece Part Leads Due to Vibration **76:75**

latch-up
Bilateral Latch-Up Failure in CMOS ESD Protection Circuits **92:223**

Advanced Photoemission Technique for Distinguishing Latch-up from Logic Failures on CMOS Devices **91:335**

Latchup Testing of Integrated Circuits **89:285**

lead crystals
Electrical Failures Caused by Lead Crystal Growth on Isolation Glass-A New Version of an Old Problem **92:397**

LED (light-emitting diode)
Automated Long Term, Uninterrupted, LED Test Complex **78:59**

leads
Failure Analysis of Electronic Component Leads **79:28**

A Holographic Technique for Failure Evaluation of Piece Part Leads Due to vibration **76:75**

leak testing
Failure Mechanisms for Seals on Large Packages **88:181**

Helium Leak Test for Small Components **86:195**

Zyglo® Penetrant Testing of Plastic Package Integrity **85:126**

A Novel Technique for Evaluating a Gap between Lead and Resin for Plastic Encapsulated Devices **85:120**

Fluorescent Penetration Testing for Failure Analysis on Plastic Encapsulated Integrated Circuits **82:125**

leakage current
A Technique to locate leakage current on Semiconductor Devices **77:90**

LIMS (laser ionization mass spectrometry)
Recent Advances in the Use of Laser Ionization Mass Spectrometry (LIMS) for Failure Analysis **84:22**

liquid crystal techniques
A Novel, Proactive Failure Analysis Technique Using Liquid Crystals **92:219**

VLSI Chip Testing Method Using Ferroelectric Liquid Crystals **92:107**

Advanced Liquid Crystal for Improved Hot Spot Detection Sensitivity **92:341**

Leakage Detection Techniques: A Comparative Study **89:5**

The Role of Failure Analysis in Problem Solving: Gate Oxide Integrity Problems **87:11**

A Liquid Crystal Failure Analysis Method for Identification of Row Failure Leakage Sites on 256k DRAMs **86:169**

Hot Spot Detection System Using Liquid Crystals with Precise Temperature Controller **86:139**

Improved Sensitivity for Hot Spot Detection Using Liquid Crystal **IRPS-1984, p 119**

Practical Liquid Crystal Applications in Failure Analysis **83:73**

Simple Liquid Crystal Techniques for Fault Site Isolation on Microcircuits **81:254**

Liquid Crystal Display Techniques for Analyzing Microprocessors **79:194**

LMMA, LAMMA (laser microprobe mass analysis)

Microanalytical Characterization of Semiconductor Device Contamination Using the Laser Microprobe Mass Analysis Technique **83:63**

low temperature testing

Low Temperature Test Aids **84:152**

LSM (laser scanning microscope; confocal LSM, etc)

Failure Analysis Techniques with the Confocal Laser Scanning Microscope **92:351**

magnetic decoration

Magnetic Decoration: A Technique to Locate Shorts by Current Path Imaging **84:254**

magnetic media

Pin-Hole Failure Analysis of Plastic-Based Magneto-Optical Disk **89:223**

Environmental Corrosion Studies of Cobalt Thin Metal Film for Magnetic Recording Media **85:6**

marking

Analyzing Marking Permanency Failures **91:283**

memories

Failure Analysis and Testing of Memory Devices for Improved Reliability and Manufacturing **91:277**

Analysis of Redundacy Related Failure Mechanisms VLSI Memory Devices **91:249**

Electrical Testing for Failure Analysis of Memory Devices—The Role of Memory Checker **81:57**

SEM Techniques for the Isolation of Failures in Memory Circuits **80:1**

Memory Array Failure Pattern Recognition **78:52**

MESFET (metal semiconductor field effect transistor)

High Reliability of Low-Noise Self-Aligned Gate GaAs MESFET with Noise Figure of 0.85db at l2GHz **91:191**

Degradation Mechanisms of GaAs MESFET Devices in High Humidity Conditions **89:141**

Reliability Evaluation of GaAs MESFETs and ICs **89:329**

Back-Etch: An Effective Tool for Characterization and Failure Analysis of MESFET Devices **88:235**

The Effects of the Passivation Film on the Reliability of High Power GaAs MESFETs **83:302**

metallization

Silicon Precipitate Nodule-Induced Failures of MOSFETs **91:161**

Practical Reliability and Failure Analysis for the Introduction of MDCl000A, A New MOSFET Turn-off Device **91:83**

Stress Voids, Creep Voids, and Silicon Nodules in Custom LSI Circuits: A Case History **89:355**

Identification of Nodules in Dual-Layer Aluminum-Copper Metallization on a Programmable Read Only Memory (PROM) **87:251**

Analysis of VLSI Metallization Microstructure by High Resolution Mechanical Cross Sectioning and Auger Analysis **85:98**

Novel Techniques for Analyzing Device High via Resistance **83:44**

Silicon Inclusion Failures in Aluminum Interconnects **83:30**

metallization, stress relief in

A New IC Failure Mechanism Due to Process-Induced Metal Stress Relief **92:413**

metallography (cross-sectioning)

Precision Cross Sectioning of Integrated Circuits **81:253**

Cost Savings with Unencapsulated Sectioning of Semiconductor Devices **80:136**

Precision Metallography of Microelectronic Devices **78:123**

In-Depth Evaluation of Microelectronic Devices by Cross Sectioning **77:203**

Rapid Evaluation of Processing Geometries by the Examination of Cleaved Samples in the SEM **77:132**

microanalysis—overviews

What Combination of Analytical Alphabet Technique is Best? **78:1**

Microanalysis—Past, Present and Future **76:1**

microprocessors

A Failure Analysis Approach to Fault Isolation on Complex Microprocessors and Microcomputers **91:145**

Testing of a Microprocessor with Pseudo-Random Vectors **83:247**

A Microprocessor Test Experience **83:244**

Limitations and Attributes of Microprocessor Testing Techniques **77:13**

microsectioning

Evaluation of the Planar Grinding Stage to Optimize Electrical Component Polishing **92:363**

Microstructural Preparation of Microelectronic Components and Devices Having Ceramic Substrates **91:483**

Microstructural Examination for Failures in Microelectronic Components **91:471**

New Cross Sectioning Techniques Using the Reactive Ion Etcher **91:88**

Precise Polishing Techniques for Technology Development and Failure Analysis **91:55**

Failure Analysis of Micron Technology VLSI Using Focused Ion Beams (FIB) **89:249**

Failure Analysis of DRAMs **88:31**

A Microfinishing Technique for Semiconductor Failure Analysis **88:161**

Single Stain Delineation of Semiconductor Devices **86:95**

Analysis of VLSI Metallization Microstructure by High Resolution Mechanical Cross Sectioning and Auger Analysis **85:98**

Improved Microsectioning Materials and Their Applications **83:93**

A Nonchemical Method of Delineating Junctions in Selected Semiconductor Devices **83:88**

Novel Techniques for Analyzing Device High via Resistance **83:44**

Non-encapsulated Microsectioning as a Construction and Failure Analysis Tool **IRPS-1982, p 221**.

Precision Diffusion Delineation in Silicon Microcircuits **81:1**

An Improved Scribe and Break Method to Obtain a Semiconductor Device Cross-Section **79:143**

microwave devices

Defect Analysis of MIC Substrates **83:218**

Failure Mechanisms in Microwave Solid State Devices **80:193**

Microwave Power GaAS FET Failure Mechanisms **80:83**

MLCC (multilayer ceramic capacitors)

Advanced Failure Analysis Techniques for Multilayer Ceramic Chip Capacitors **87:25**

Novel Microscopic Techniques in the Failure Analysis of Trimmed Ceramic Capacitors **86:131**

Nondestructive Inspection of Encapsulated Capacitors: Flaw Growth Detection of Parts Submitted to 85% Humidity, 85 °C **85:193**

Improved Microsectioning Materials and Their Applications **83:93**

MLCs Outperform Metallized-Film Capacitors under Actual Operating Conditions **82;209**

Extended Electrical Characterization of Ceramic Capacitors Failing under Low-Voltage Conditions **82:194**

Identification of Ceramic Capacitor Shorts by Voltage Contrast in Scanning Electron Microscope **82:203**

Acoustic Visualization of Stressed Ceramic Chip Capacitors **82:215**

MMIC (monolithic microwave integrated circuit)

Reliability Evaluation on GaAs MMIC Components **89:335**

moisture resistance

Acceleration Factor at a Pressure Cooker Test for the Surface Mount Device (SMD) **87:283**

Moisture-Resistance Test Using Unsaturated Pressure Cooker Equipment **86:189**

A Novel Method of Evaluating Moisture Resistance of Soldered Plastic Rncapsulated LSI by a New Ultrasonic Inspection System **86:173**

Humidity Resistance of (SOP-ICs) Small Outline Package ICs **86:235**

A Novel Technique for Evaluating a Gap between Lead and Resin for Plastic Encapsulated Devices **85:120**

Fluorescent Penetration Testing for Failure Analysis on Plastic Encapsulated Integrated Circuits **82:125**

A New Accelerated Humidity Test for Plastic Encapsulated ICs **82:65**

moisture sensors

Recent Evaluations of Al_2O_3 in Situ Moisture Sensors for Industrial Electronic Hermetic Components Application **83:300**

MOS capacitor

Investigation of Trapped Oxide Charge in MOS Devices **86:219**

MOSFET (metal oxide semiconductor field effect transistor)

ESD-Induced Damage in Vertical Power MOSFETs **88:111**

ISTFA Subject Index

Simulation and Characterization of EOS and ESD Damage in MOSFET Transistors and Integrated Circuits **88:95**

Failure Analysis of ESD Damage in MOSFET Power Devices **86:83**

Determination of Particulate Contamination in Gate Oxide **85:48**

Observed Physical Effects and Failure Analysis of EOS/ESD on MOS Devices **84:205**

nodules, in silicon

Silicon Precipitate Nodule-Induced Failures of MOSFETs **91:161**

noise

Noise Analysis & Potential Failure Mechanisms in Hybrid Microcircuits and Components **77:64**

optical beam-induced current (OBIC)

Logic Failure Analysis of CMOS Using a Laser Probe **IRPS-1984, p 69**

optical microscopy

Microlenses **85:165**

Innovative Dark Field Optical Techniques for VLSI Evaluation **83:8**

Application of the Optical Scanner to Semiconductor Failure Analysis **78:13**

Optical Techniques for Semiconductor Material and Circuit inspection **76:31**

Optical Microscopic Methods for Accentuating Surface Contrast **76:18**

optoelectronics

Typical Failure Mode and Mechanism of Opto-Electronic Devices **92:433**

Evaluation of Optical Fibers for Space Use **87:275**

An Improved Decapsulation Technique for Plastic Encapsulated Opto Coupler Devices **85:114**

Facet Degradation in High Power (AlGa)As Laser Diodes from −20 °C and 50 °C Life Tests **85:31**

Failure Analysis of Laser Optical Components **83:178**

Screening for Opto-Isolator Failure Modes to Improve Long Life Performance **77:294**

Characterization and Analysis of Certain Types of Defects in Channel Electron Multiplier Arrays (CEMAs) **77:280**

oxide defects

Gate Oxide Reliability and Defect Analysis of a High Performance CMOS Technology **89:109**

Application of Quiescent Supply Current Signature Analysis to Failure Analysis of Integrated Circuits **89:27**

Leakage Detection Techniques: A Comparative Study **89:5**

The Role of Failure Analysis in Problem Solving: Gate Oxide Integrity Problems **87:11**

Bubbles Induced under Thermally Grown SiO_2 by the Interaction of Phosphorus and Moisture in an Oxygen Plasma **86:121**

Failure Analysis and Failure Mechanisms of Triple Polysilicon VLSI Devices **86:201**

Determination of Particulate Contamination in Gate Oxide **85:48**

Failure Analysis of Double Polysilicon Thick Interlevel Oxide Failures **82:104**

oxide trapped charge

Recent Failure Analysis Applications of Secondary Ion Mass Spectrometry, SIMS **89:45**

Failure Analysis of Inversion Mechanisms: Beyond Bake Recoverable **88:217**

Investigation of Trapped Oxide Charge in MOS Devices **86:219**

packages, IC, ceramic

Failure Analysis of Electroplated Nickel on Ceramic IC Packages Using Electron Beam Techniques **83:182**

Factors Influencing Microcircuit Package Reliability **78:108**

particles

Circuit Failure Due to Fine Mode Particulate Air Pollution **92:329**

Techniques for Analysis of Small Particles **78:151**

Coverseal Particle Detection Using X-Rays **77:88**

Techniques in Post PIND Test Examination of Particle Contamination in Semiconductor Devices **77:56**

passivation cracking

Stress Analysis for Passivation and Interlevel-Insulation Film Cracks in Multilayer Aluminum Structures for Plastic-Packaged LSI **87:75**

Stress Analysis of Passivation Film Crack for Plastic Molded LSI Caused by Thermal Stress **83:275**

pattern recognition

GRAPHPATT/GSI—A Paperless Approach to VLSI RAM Failure Analysis with Automatic Wafer-Stager Control from Graphic Bit-Fail Maps **89:37**

Pattern Recognition Analysis: An Aid to Component Problem Solving **85:138**

Memory Array Failure Pattern Recognition **78:52**

Peltier devices

Thermoelectric Heat Pump Failures **91:167**

phos-glass PSG (phosphosilicate glass)

Comparative Study of Various Methods for Measuring Phosphorus Concentration in Phosphosilicate Glass **83:174**

Controlled Selective Etching of Silicon Dioxide and Phossil Glass **82:80**

Reliability of SEM/EDX Analysis of P in SiO_2 **82:23**

Geometry Differences In PSG Films After Etching with HF Allows the Determination of Phosphorous Concentration **78:137**

photoacoustic spectroscopy

Photoacoustic Spectroscopy as a Tool for the Nondestructive Evaluation of Materials and Devices **86:25**

photoemission

Advanced Photoemission Technique for Distinguishing Latch-up from Logic Failures on CMOS Devices **91:335**

photoresist

A Study for insoluble Layer Formation of Chemically Amplified Positive Resist Using Time-of-Flight Secondary Ion Mass Spectrometer **92:295**

photoresist; masking of specimens

Micro-Control of Photoresist Deposition for Failure Analysis of Microelectronic Circuits **92:335**

PIND (particle impact noise detection)

Methods of Particle Retrieval from Metal Packages **80:217**

PIND Testing and Its Problems **80:203**

Comparative PIND Tests of Hybrid Microcircuits **80:200**

Loose-Particle Detection in Micro-electronic Devices—A Review of an 1155 Task **79:1**

Development of the Method to Detect Foreign Materials In IC Packages **79:20**

PIND's Role as a Failure Analysis Tool **79:23**

The Effectivity of PIND Testing **79:27**

Techniques in Post PIND Test Examination of Particle Contamination in Semiconductor Devices **77:56**

pinholes

Failure Due to Pinholes in Polysilicon Conductors on CMOS Memory Devices **77:60**

planar cell capacitor

Sub-Micron Process and TDDB Characteristics **88:119**

plasma-enhanced chemical vapor deposition (PECVD)

TEOS Glass Deposition by PECVD, A New Failure Analysis Technique **92:141**

plasma etching and reactive ion etching (RIE)

Plasma Etching of Die Coatings in Tape Automated Bonded Devices **91:231**

Anisotropic Etching for Failure Analysis Applications **89:161**

Non-Destructive Passivation Deprocessing Using the RIE **89:257**

RIE for Failure Analysis **89:21**

Failure Analysis Applications of Plasma **87:35**

A Procedure for the Nondestructive Removal of Glassivation from Integrated Circuits **86:113**

Backside Etching Techniques **85:147**

Integrated Circuit Overlay Removal by Planar Plasma Etching **84:267**

Plasma Etching Techniques for Failure Analysis **81:288**

plating

Case Histories of Microelectronic Device Analysis Using Auger Electron Spectroscopy (AES) **89:373**

Analysis of Alloys and Plating Processes via Inductively Coupled Plasma Atomic Emission Spectroscopy (ICP/AES) **86:17**

Die Bonding on Non-Gold Plated Piece Parts **85:22**

Hybrid Microcircuit Failures Caused by Nickel Plating **85:1**

Comparative Analytical Techniques for the Determination of Pb in Solder Plated Leads **84:247**

Factors Resulting in Bond Delamination of Au Backmetal Si Die Eutectically Bonded to Ni/Au **83:263**

Failure Analysis of Electroplated Nickel on Ceramic IC Packages Using Electron Beam Techniques **83:p182**

PLCC (plastic leaded chip carrier)

Identification of Package Defects in Plastic-Packaged Surface Mount ICs by Scanning Acoustic Microscopy **89:61**

Acoustical Microscopy as a Tool for Nondestructive Testing of Finished Devices **87:21**

Failure Analysis of Surface Mounted Interconnection **86:73**

Microcircuit Packages Incorporating Radiation Shields for Use in Spacecraft Electronics **85:111**

polysilicon

Failure Analysis of Double Polysilicon Thick Interlevel Oxide Failures **82:104**

Isolation of Polysilicon for Failure Analysis **82:101**

printed wiring (circuit) boards— PWB or PCB

Thermal Cycling Induced Cracks in PWB Solder Connection **83:204**

Printed Wiring Board Test Coupons—How Reliable Are They? **81:91**

Automated Laser Inspection of Solder Joints **81:85**

Identifying "Internal Shorts" in Multilayer Printed Wiring Board **81:81**

Preparation of Printed Circuit Boards for Metallurgical Evaluation **79:101**

Development of Polymide-Glass Multi-layered Printed Wiring Boards (MLBS) for High Reliability Interconnection **79:95**

A New Failure Mechanism in Electro-Deposited Copper **77:258**

Failure Analysis Techniques for Epoxy Encapsulated Multilayer Board, Cordwood Constructed Modules **77:107**

Detection and Isolation of Sources of Ionic Contamination on PC boards **77:38**

New Techniques for SEM and Optical Microscopy to Examine Plated-Through Holes in PWB's **76:68**

power transistors

SEM Study of the Dynamics of Electrical Overstress (EOS) in Power Transistors **84:221**

The Interpretation of EOS Damage in Power Transistors **83:157**

probe and isolation

Failure Analysis Applications of an Excimer Laser **89:153**

Diagnostic Techniques for Probing and Contacting Small Geometry and High Frequency Semiconductor Devices **89:239**

Electron Beam Probing of VLSI Circuits through IEEE-488 Interface Control **89:81**

VLSI Failure Analysis Applications of the Excimer "MICROMILLING" Laser System **88:229**

Non-Contact Voltage Measurement on VLSI Circuits by Photoelectron Stimulation and Detection **88:133**

Procedures for Probing Underlying Structures in Multi-Layered VLSI Devices While Maintaining Electrical Integrity **88:127**

An Inexpensive Technique to Produce High Quality Probe Tips **87:149**

Application of Image Reversal Photoresist for Failure Analysis **87:17**

Failure Analysis of VLSI Logic Functional Fails **86:179**

Selective Submicron Trace Cutting and Thin Film Removal Using an Ultraviolet Laser **85:160**

Methods of Trace Cutting for Diagnostic Probing of Semiconductor Devices **84:136**

Dynamic Probing—An Essential Tool for Failure Analysis of VLSI Chips **84:90**

Laser Probe Technology for VLSI Failure Analysis **84:70**

Micromanipulator Acid Probe Isolation **84:67**

A Laser Cutter for Failure Analysis **83:69**

Failure Analysis of Double Polysilicon Thick Interlevel Oxide Failures **82:104**

Isolation of Polysilicon for Failure Analysis **82:101**

Circuit Surgery Using Xenon and YAG Lasers **82:86**

Controlled Selective Etching of Silicon Dioxide and Phossil Glass **82:80**

Laser Die Probing for Complex CMOS **82:178**

Failure Analysis Technique for Examination of SNOS Type Defects **82:20**

RAM Cell Defect Localization and Analysis on VLSI Structures **82:13**

A Micromechanical Probe in the SEM **79:145**

Procedural Guide

RADC Microelectronics Failure Analysis Techniques Procedural Guide—A Reality **81:265**

RADC Microelectronic Failure Analysis Procedural Guide—An Overview **79:170**

process improvement

1.0 μm CMOS Process improvement Case Study **92:101**

profiling, depth

Sputter Ion Depth Profiling Limitations of Microelectronic Materials **82:8**

Impurity Profiling in Semiconductor Devices **76:48**

PROM (programmable read only memory)

Identification of Nodules in Dual-Layer Aluminum-Copper Metallization on a Programmable Read Only Memory (PROM) **87:251**

Reliability of 16K-EE PROM **82:42**

Analysis of a Unique Failure Mode in a CMOS PROM **81:234**

Results of Testing and Programming of Fuseable Link PROMs **81:50**

radiation effects

Identification of Semiconductor Device Failures Caused by Entrapped Kr$_{85}$ **92:383**

The Degradation Study of MOS LSI Characteristics During X-Ray Inspection **92:419**

Evaluation of Optical Fibers for Space Use **87:275**

Microcircuit Packages Incorporating Radiation Shields for Use in Spacecraft Electronics **85:111**

Single Event Upset Mechanisms and Predictions **83:128**

Radiation Hardness Assurance **83:147**

Measurement of Single Event Upsets in the Laboratory **83:137**

Single Event Upset Mechanisms and Predictions **83:128**

Ground Testing of Devices for ionization Damage in Space **83:120**

Basic Mechanisms of Total Dose Effects **83:103**

The Radiation Background in Space: Definition, Evaluation, Application, and Effects **83:102**

Radiation Effects on Electronic Parts: An introduction **83:100**

Electron Irradiation Tests on Small Signal and Power GaAs FET **82:168**

Alpha Flux Measuring and Soft Error Evaluating Techniques **80:30**

Electron Beam Irradiation Damage and Contamination **77:121**

RAM (read-only memory)

Substrate Voltage Bump Test for Dynamic RAM Devices **83:281**

A New Breakdown Phenomenon in High Density RAM **83:270**

Reliability Evaluation of CMOS RAMs **82:133**

Reliability Evaluation of 16K Dynamic RAMs in Plastic Packages **82:31**

Ram Cell Defect Localization and Analysis on VLSI Structures **82:13**

Failure Analyais of 1K × 4 CMOS RAM **81:228**

A New Reliability Evaluation System Applicable to the Development of Dynamic RAMS **81:44**

RAMAN microprobe

Application of the RAMAN Microprobe to Quality Control in IC Manufacture and to Crystalline Analysis of Structural Ceramics **82:1**

Application of the Raman Microprobe to Quality Control in IC Manufacture and to Crystalline Analysis of Structural Ceramics **82:1**

RBS (rutherford backscattering spectrometry)

Auger Analysis of Refractory Metal Silicides **84:12**

Applications of Rutherford Backscattering Spectrometry for the Semiconductor Industry **84:27**

MeV Ion Beam analysis for Quality Control **82:2**

relays and switches

Analysis on a Cause of Rare Contact Failure in Sealed Contacts **86:153**

The Failure Analysis of Reed Relays by Scanning Auger Microscopy (SAM) **86:29**

reliability evaluation

Reliability Evaluation on GaAs MMIC Components **89:335**

Gate Oxide Reliability and Defect Analysis of a High Performance CMOS Technology **89:109**

Reliability Evaluation of GaAs MESFETs and ICs **89:329**

Life Estimation for Slug Tantalum Capacitor **86:213**

Reliability Improvement and Evaluation of TWTs for Communications Satellites **86:47**

Logical and Systematic Failure Analysis of 64K DRAM Using FTA and Reliability Analysis Techniques **84:59**

Reliability of a CCD Video line Store **82:182**

Reliability Evaluation of CMOS RAMs **82:133**

'FITS'—An Analytical Method of Choosing Burn-in Processes **82:53**

Reliability of 16K-EE PROM **82:42**

Reliability Evaluation of 16K Dynamic RAMs in Plastic Packages **82:31**

Reliability Study of (AlGa)As Laser Diodes for Long Life Space Applications **82:25**

Reliability Evaluation of Plastic Encapsulated Hybrid Microcircuit **81:208**

Memory Hybrid for Long Term High Reliability Spec Application **81:200**

Field Data on Electronic Parts of Color Television Sets **81:188**

Reliability of High-Power Pulsed Impatt Diodes **81:187**

Problems and Considerations for the Procurement of a 64K Bit CMOS RAM Reliability Evaluation of Plastic Encapsulated ICs Using a New Pressure Cooker Test **81:75**

Reliability and Failure Mechanisms of GaAs-FETs **81:69**

A New Reliability Evaluation System Applicable to the Development of Dynamic RAMS **81:44**

The Human Element in Reliability **80:77**

Factors Influencing Microcircuit Package Reliability **78:108**

Reliability Evaluation of imaging CCD's **78:70**

Reliability of Thermistor Bolometers **77:295**

The Management of Reliability Information **76:58**

resistors, thick-film

A Technique to Test for Electrostatic Discharge Damage to Thick Film Resistors **78:35**

resistors, thin-film

Electrochemical Voiding of Silicon-Chromium Thin Film Resistors: A Case History **91:157**

RF transistors

Reliability Investigation on Gold Metallized RF Transistors **83:180**

RGA (residual gas analysis)

Failure Mechanisms for Seals on Large Packages **88:181**

An Analysis of Power Transistors with a Polyimide Die Coat **87:213**

Optimization of Polyimide Die Coat Cure Cycle via Chemical Analysis and Electrical Characterization **87:41**

RIE (reactive ion etching)

Reactive Ion Etch Process Optimization for Failure Analysis Applications **92:33**

New Cross Sectioning Techniques Using the Reactive Ion Etcher **91:88**

ROM (read only memory)

Filler Particle induced Data Retention Failures in Plastic Encapsulated EPROM's **92:391**

Analysis of Failures in Large Arrays **85:85**

SAM (scanning Auger multiprobe)

Microcircuit Failure Analysis Using a Scanning Auger Multiprobe **83:53**

SAM (scanning acoustic microscopy)

Scanning Acoustic Microscopy: Reliability in Processing High Power Semiconductor Devices **91:119**

sample preparation

Cold Sample Preparation Technique for Soft Solder Die Attach Examination **92:135**

Sample Preparation Techniques for Optical and Electron Beam Analysis **76:62**

Schottky diodes

Progressive Deterioration of Glass-to-Metal Seals in Glass Body Diode Packages **89:361**

Expedited Failure Analysis of GaAs FET Transistors, Si IMPATT Diodes and Schottky Diodes Using X-ray, SEM and EBIC Techniques **87:53**

An Automated Device Characterizer Adds the Edge to Successful Failure Analysis **82:162**

SCR (silicon controlled rectifier)

The Interpretation of EOS Damage in Power SCRs **85:42**

scratch test

Adhesion and Cohesion Testing of Hard Coatings **86:87**

screening and testing

Screening Techniques for Detecting Latent Defects in Adhesive Bonds Associated with Microwave Hybrid Packages **88:173**

Screening Design System Generated VLSI Logic Chips for Performance **84:157**

Method of Estimating the Benefits of FRU Burn-in **83:290**

A New Screening Method for Metal-Nitride Oxide-Silicon (MNOS) Non-Volatile Memory IC **83:285**

Substrate Voltage Bump Test for Dynamic RAM Devices **83:281**

Failure Mechanism and Assurance Technique **79:105**

Results of a Screening Program for Distributor—Procured Microcircuits **79:111**

Screening and Testing Techniques for GaAs power MESFETs **79:114**

Study of Reliability of GaAs MESFET's **79:123**

ATE, the Device Specification and Part Failure **78:63**

Automated Test Design for Analog Circuits **77:17**

Limitations and Attributes of Microprocessor Testing Techniques **77:13**

SEM (scanning electron microscope)

A New Fast Signal Processing Devoted to Real Time Differential Imaging in Scanning Electron Microscopy **91:126**

Utilization of High Resolution SEM in the Failure analysis of Bipolar Chips **91:7**

Multibeam Wafer Inspection SEM **81:22**

SEM Stroboscopic Techniques—Their Application to Failure Analysis of LSI's **80:9**

SEM Techniques for the Isolation of Failures in Memory Circuits **80:1**

A Micromechanical Probe in the SEM **79:145**

SEM X-Section Investigation of IC Structures **78:160**

Methods to Augment SEM Metallization Inspection **78:170**

Applications of the Scanning Electron Microscope in the Development of Microtechnology **77:323**

Computer Controlled SEM for Image and Microtopography Analysis **77:108**

sensors, in situ

Recent Evaluations of Al, 0, In Situ Moisture Sensors for Industrial Electronic Hermetic Components Application **83:300**

shear test (die)

Characterization of Die Attach Integrity Using Destructive and Nondestructive Techniques **88:77**

Factors Resulting in Bond Delamination of Au Backmetal Si Die Eutectically Bonded to Ni/Au **83:263**

shear test (wire bond)

Thermalsonic Ball Bond Evaluation by a Bond Pluck Test **84:237**

shift register

Fault Isolation of Dynamic Shift Registers Using Unorthodox Electronics or SEM Voltage Contrast **77:111**

sichrome (SiCr)

Electrochemical Voiding of Silicon-Chromium Thin Film Resistors: A Case History **91:157**

SIDP (sputter ion depth profiling)

Sputter Ion Depth Profiling Limitations of Microelectronic Materials **82:8**

silicides

Transmission Electron Microscopy (TEM) in Silicon Device Technology **87:83**

Auger Analysis of Refractory Metal Silicides **84:12**

MeV Ion Beam Analysis for Quality Control **82:2**

silicon

Silicon Precipitate Nodule-Induced Failures of MOSFETs **91:161**

silicon nitride

Low Temperature Silicon Nitride Protection of Plastic Encapsulated Integrated Circuits **79:185**

silicon-germanium

Analysis of Epitaxial SiGe Structures via SIMS **91:17**

silver migration

Silver Migration in Lead Borosilicate Glass on Soda Lime Glass Substrate **80:153**

SIMS (secondary ion mass spectrometry)

A Study for insoluble Layer Formation of Chemically Amplified Positive Resist Using Time-of-Flight Secondary Ion Mass Spectrometer **92:295**

Analysis of Epitaxial SiGe Structures via SIMS **91:17**

Recent Failure Analysis Applications of Secondary Ion Mass Spectrometry, SIMS **89:45**

Surface Analysis of Contamination in Thin Film Coatings **87:155**

Role of Surface Chemistry in Packaging Failures **86:27**

Modern Analytical Usage of Secondary Ion Mass Spectrometry **84:1**

Investigation of Integrated Circuit Failures by Liquid Metal (SIMS) Imaging and Multitechnique Surface Analysis **84:3**

Microanalytical Characterization of Semiconductor Device Contamination Using Laser Microprobe Mass Analysis Technique **83:63**

Direct Lateral and In-Depth Distributional Analysis for Ionic Contaminants in Semiconductor Devices Using Secondary ion Mass Spectrometry **79:148**

single bit failures

Single Bit Failure Analysis: Once Straight Forward is Now Challenging-A Case Study of an Advanced Microprocessor SRAM Failure **92:79**

single-event upset (upset)

Measurement of Single Event Upsets in the Laboratory **83:137**

Single Event Upset Mechanisms and Predictions **83:128**

SLAM (scanning laser acoustic microscope)

Scanning Laser Acoustic Microscopic (SLAM) Study of Die Attach Integrity **85:202**

Nondestructive Evaluation of Ceramic Chip Capacitors by Means of Scanning Laser Acoustic Microscope (SLAM) **81:259**

Scanning Laser Acoustic Microscope Applied to Failure Analysis **78:25**

solar cells

Diagnostic Test and Numerical Simulation of Solar Cells Subjected to High Reverse Current **89:323**

Influence of Cracks on GaAs Solar Cells **87:59**

Testing and Analysis of Photovoltaic Modules for Electrochemical Corrosion **86:39**

A Portable, X-Y Translating Infrared Microscope for Remote Inspection of Photovoltaic Solar Arrays **80:21**

solder, indium

Ultrasonic Aluminum Wire Bonding and Lead-Indium Soldering to Gold Alloy Thick Film Conductors—Performance and Failure Mechanism **77:227**

solder failure (non die-bond)

Investigation of Solder Joint Failure **84:201**

Capacitor End-Cap Dissolution Testing by Solder Reflow **83:222**

Solder Evaluation Test **83:232**

Thermal Cycling Induced Cracks in PWB Solder Connections **83:204**

Solder Coating of Ceramic Capacitors; Wettabilty Problems **81:111**

solder heat testing

An Advanced Evaluation Method of Soldering Heat Resistance for Ultra Thin Plastic Encapsulated LSIs **91:213**

solderability

Reduced Solderability Effectiveness Due to Base Metal Oxide Formation **91:103**

Solderability Problem with Solder Coated Package Leads **87:201**

Investigation of Problems Resulting from Surface Mounted Device Board Attachment **85:56**

Comparative Analytical Techniques for the Determination of Pb in Solder Plated Leads **84:247**

solder joints

Cold Sample Preparation Technique for Soft Solder Die Attach Examination **92:135**

Correlation of Accelerated Test Failure Modes with Solder Joint Integrity and Performance **81:84**

SOS (silicon-on-sapphire)

Case Study of ESD Damage and Long-Term Implications on SOS RAMs **92:213**

space-level parts

Parts Technology Support to European Space Programmes **78:83**

specimen preparation

Evaluation of the Planar Grinding Stage to Optimize Electrical Component Polishing **92:363**

spin-on glass (SOG)

Applications of Spin on Glass to Precision Metallography of Semiconductors **92:157**

SRAM (static random access memory)

A Microprobing Technique for DC Characterization of Defects and Instabilities in Static RAMs **92:229**

Single Bit Failure Analysis: Once Straight Forward is Now Challenging-A Case Study of an Advanced Microprocessor SRAM Failure **92:79**

A Structured Approach for Failure Analysis of a 256K BiCMOS SRAM **89:167**

Anisotropic Etching for Failure Analysis Applications **89:161**

Gate Oxide Reliability and Defect Analysis of a High Performance CMOS Technology **89:109**

Failure Mode Analysis Methodology on High Density CMOS SRAMS **89:389**

Application of Quiescent Supply Current Signature Analysis to Failure Analysis of Integrated Circuits **89:27**

Screening for Faults in High Density SRAMs **88:195**

Failure Analysis by Dynamic Voltage Contrast Developement of a Semi-Automatic System **87:67**

Analysis of CMOS Microprocessor Failures Using a New Electrical Signature Analysis Technique **87:1**

A Signature Analysis Technique for Defect Characterization of CMOS Static RAM Single Cell Fails **84:141**

Investigation of Bipolar Microcircuit Cold Start Problem and its Impact on Reliability **84:85**

A Novel VLSI SRAM Failure Analysis Technique **83:40**

Reliability Evaluation of CMOS RAMs **82:133**

step coverage

Analytical Techniques and Procedures to Successfully Analyze Metal Step Coverage on Semiconductors **84:124**

STM (scanning tunneling microscope)

Applications of Scanning Kelvin Probe Force Microscope (SKPFM) for Failure Analysis **92:9**

strain

Use of Birefringent Patterns to Determine the Strain Induced in Silicon Chips by Bonding **77:36**

stress induced voiding

Stress Voids, Creep Voids, and Silicon Nodules in Custom LSI Circuits: A Case History **89:355**

Causes, Failure Analysis, and Accelerated Testing of Creep Voids in Aluminum Interconnect on Integrated Circuits **87:225**

surface mount devices (SOT, SOIC)

Identification of Package Defects in Plastic-Packaged Surface Mount IC's by Scanning Acoustic Microscopy **89:61**

Acceleration Factor at a Pressure Cooker Test for the Surface- Mount-Device **87:283**

Failure Analysis of Surface Mounted Interconnection **86:73**

A Novel Method of Evaluating Moisture Resistance of Soldered Plastic Encapsulated LSI by a New Ultrasonic Inspection System **86:173**

Investigation of Problems Resulting from Surface Mounted Device Board Attachment **85:56**

Capacitor End-Cap Dissolution Testing by Solder Reflow **83:222**

TAB (tape automated bonding)

Acoustic Microscopy: A Nondestructive Tool for Bond Evaluation on Tab Interconnections and Die Attach **84:243**

tantalum capacitors

Reliability Evaluation and Failure Diagnosis of Tantalum Capacitors Using Computer-Aided Frequential Analysis and Design of Experiments **92:189**

Life Estimation for Slug Tantalum Capacitor **86:213**

TEM (transmission electron microscopy)

EBIC and TEM Analyses of Piped Multi-Emitter Transistors **92:55**

Specific Area TEM Analysis of ESD-Induced Silicon Melt Filaments on Output Transistors **92:27**

Sample Preparation Methods for TEM Analysis of Semiconductor Devices **91:41**

Comparison of the Quality of Foreign and Domestic Integrated Circuits by Transmission Electron Microscopy **89:75**

A Technique for Producing Precise Large Area XTEM Samples of VLSI Devices **89:1**

TEM Studies of VLSI Metallization **88:145**

Identification of Nodules in Dual-Layer Aluminum-Copper Metallization on a Programmable Read Only Memory (PROM) **87:251**

TEM Sample Preparation Methods to Inspect Integrated Circuit Structures **87:103**

Applications of Transmission Electron Microscopy in LSI Process Developement **87:95**

Transmission Electron Microscopy in Silicon Device Technology **87:83**

IC Process Flow Analysis Using Transmission Electron Microscopy **87:91**

Backside Etching Techniques **85:147**

Sample Preparation for Ion Milling Sections of Semiconductor Structures for TEM Inspection **84:119**

Transmission Electron Microscope Techniques Applied to Microelectronics and Failure Analysis **78:30**

temperature sensors

Failure Analysis of a Resistance Temperature Detector (RTD) **89:385**

The Failure Analysis of a Bimetallic Thermostat **89:369**

Failure of Platinum Resistance Temperature Sensors **86:53**

TEOS (tetraethyl orthosilicate)

TEOS Glass Deposition by PECVD, A New Failure Analysis Technique **92:141**

testing

ReTAB for Test and Failure Analysis **91:221**

Test Pattern Generation for Large Digital Hybrid Microcircuits **80:262**

Hybrid Die Acceptance Testing **80:250**

Automated Test Program Generation and Fault Isolation for Boards and I.C's **78:185**

Limitations and Attributes of Microprocessor Testing Techniques **77:13**

testing, high temperature

Electric Field Equivalent—High Temperature Test (EFE-HTT) for Nonpackaged Dice **78:39**

test chips

A Microprocessor Test Experience **83:244**

TGA (thermogravimetric analysis)

An Analysis of Power Transistors with a Polyimide Die Coat **87:213**

thermal balancer

Outgassing Method Using Thermal Balancer **84:118**

thermal wave imaging

Failure Analysis of HgCdTe Infrared Detector Arrays **84:170**

Nondestructive Evaluation with Thermal Waves **83:80**

Thermal-Wave Microscopy **82:92**

thermoelectric devices

Thermoelectric Heat Pump Failures **91:167**

thick films

The Effect of Edge Preparation on the Failure of Thin Film Network Alumina Components **91:507**

Techniques for the Evaluation of Thick Film Materials **78:34**

thin film resistors

A Safe-Operating Area (SOA) Concept for Pulse-Operated Thin Film Resistors **83:294**

threshold shift

Degradation Mechanism Due to Hot Electron Trapping in High Density CMOS DRAM **88:89**

tin whiskers

Tin Whisker Induced Failure in Vacuum **92:407**

titanium-tungsten (TiW)
Failure Analysis Techniques of TiW Fuse Failures on PLDs for Improved Programming Yield **91:271**

traveling-wave tubes (TWTs)
Failure Mode Analysis and Life Time Estimation of Traveling Wave Tubes **80:196**

tunneling current microscopy
Case History: Wafer-Level Failure Analysis of Functional ICs with Elevated I_{DDQ} Caused by Resistive Gate Oxide Shorts **91:138**

ultrasonic inspection
A New Ultrasonic Technique for the Inspection of Semiconductor Package Diebonds **76:86**

vapor bubble test
Thermal Analysis of Microelectronic Devices Using a Hot Spot Detection System Based Upon Vapor Bubble Technology **82:43**

vibration effects
Screening Techniques for Detecting Latent Defects in Adhesive Bonds Associated with Microwave Hybrid Packages **88:173**

voltage contrast
Failure Analysis of Vendor-Produced CMOS Modules **89:99**

Practical Applications of Dynamic Fault Imaging **89:269**

Voltage Contrast Imaging of Passivated Devices **88:9**

Dynamic Voltage Contrast SEM Failure Analysis of a 32K ROM **88:15**

Electron Beam Testing of VLSI Circuits **88:21**

Overview of Voltage Contrast Techniques and Applications **88:1**

Failure Analysis by Dynamic Voltage Contrast Development of a Semi-Automatic System **87:67**

Electron-Beam Testing and Its Application to VLSI Technology, SPIE Conference on Characterization of Very High Speed Semiconductor Devices and ICs: Critical Review of Technology, March, 1987.

A Practical VLSI Characterization and Failure Analysis System for the IC User, **IRPS-1986, p 87**.

Fault Contrast: A New Voltage Contrast VLSI Diagnosis Technique, **IRPS-1986, p 109**.

E-Beam Testing: Image & Signal Processing for the Failure Analysis of VLSI Components **85:89**

LSI Failure Analysis Using an Electron-Beam Tester Directly Combined to an LSI Tester **85:78**

Digital Processing and Color Coding of Voltage Contrast Images **85:63**

Observation of Latch-Up Phenomena in CMOS ICs by Means of Digital Differential Voltage Contrast **84:265**

Fundamentals of Electron Beam Testing of Integrated Circuits, **Scanning, Vol 5, 1983, p 103**

Electron Beam Testing Methods and Applications, Scanning, Vol 5, 1983, p 14

Internal Node Testing by Tester Aided Voltage Contrast **82:156**

Identification of Ceramic Capacitor Shorts by Voltage Contrast in Scanning Electron Microscope **82:203**

Electron Beam Testing for Verification of Voltage Distribution on VLSI Circuits **82:149**

Automated Contactless SEM Testing for VLSI Development and Failure Analysis, **IRPS-1982, p 163**

Failure Analysis and Fault isolation of a 1 Kbit Schottky RAM by SEM Voltage Contrast **80:131**

The Practical Implementation of Voltage Contrast as a Diagnostic Tool, **IRPS-1983, P 167**.

Application of SEM EBIC Mode to Semiconductor Evaluation and Failure Analysis, **79:136**

Fault Isolation of Dynamic Shift Registers Using Unorthodox Electronics or SEM Voltage Contrast **77:111**

WDXA (wavelength-dispersive x-ray analysis)
A Non-Destructive Method to Approximate SiO_2 Passivation Thickness on Fabricated Semiconductor Devices Using Wavelength Dispersive X-Ray Analysis (WDXA) for Oxygen **82:6**

A Comparison of Minimum Detection limits using Wavelength and Energy Dispersive Spectrometers **77:49**

welds
Weld Reliability Data on Plastic Encapsulated Solid-State Devices **80:111**

Failure Analysis of EB Welded Metal Hybrid Packages **76:80**

whiskers
Degradation Mechanisms of GaAs MESFET Devices in High Humidity Conditions **89:141**

Tin Whiskers on Flat Pack Lead Plating between Solder Dip and Sealing Glass **85:16**

Crystalline Growth on RF Tuning Screws **85:10**

Comparative Analytical Techniques for the Determination of Pb in Solder Plated Leads **84:247**

wire bond
Case History of Epoxy Contaminated Wire Bond Failures on Space Shuttle Hybrids **92:61**

Wire Bonding—A Closer Look **91:525**

Failure of an Embedded Bond Wire in a CERDIP: Reliability Implications **89:133**

Thermal Fatigue Failures of Large Scale Package Type Power Transistor Modules **89:309**

Au-Al Bond Degradation Mechanism and the Analysis with Infrared Microscope **89:15**

Recognition and Morphology of Ultrasonic Vibration-Induced Damage in Gold Wire Bonds **86:227**

High-Temperature Long-Term Reliability Evaluation of Plastic Encapsulated LSI **85:129**

Thermalsonic Ball Bond Evaluation by a Bond Pluck Test **84:237**

Investigation of the Effect of Thallium on Gold/Aluminum Wire Bond Reliability **84:227**

wire fracture
Failure Analysis of Copper-Nickel Alloy Wire Fracture in Electrical Relay Components **87:259**

wire fusing current
Electromigration in Aluminum Bond Wires Subjected to Current Pulsing **91:183**

The DC Fusing Current and Safe Operating Current of Microelectronic Bonding Wires **89:121**

wire pull
A Preview of 100% Prestress Wire Bond Testing (Nondestructive and Pull Techniques Hazards and Economics) **78:106**

x-ray diffraction
Failure Analysis of Shorts in Color TV Tubes **86:163**

x-ray radiography
Transmitted Microfocus X-Ray Techniques **92:279**

Non-Destructive Failure Analysis of ICs Using Scanning Acoustic Tomography (SCAT) and High Resolution X-Ray Microscopy (HRXM) **89:69**

Characterization of Die Attach Integrity Using Destructive and Nondestructive techniques **88:77**

A High Speed X-Ray Topographic Camera For Wafer Defect Analysis **78:9**

Analysis of Electronic components by Micro X-Radiography **77:217**

Applications of Stereo X-Radiography to Failure Analysis **77:105**

Techniques for Improvement of Radiography for Electronic Parts **77:104**

XPS (x-ray photoelectron spectroscopy)
Surface Analysis of Contamination in Thin Film Coatings **87:155**

Non-Destructive Multitechnique Surface Analysis— The Role of XPS and ISS **84:40**

Using ESCA in Failure Analysis: Some Recent Applications, **84:35**

Zener diodes
Degradation of Zener Diodes Caused by Alloying Irregularities **86:145**

RESOURCES FOR FAILURE ANALYSIS

by Thomas W. Lee

This list is intended to be a base upon which you can build a personalized file of vendors for handy general reference, for yourself and your staff. Frequent updates are necessary, because the FA market is dynamic. Companies change locations; and have divestitures, acquisitions and reorganizations; all with consequent name or address changes. New FA equipment is invented quite frequently. It is recommended that this list be supplemented with a file of the vendor's equipment brochures.

Most of the items in this list should be relevant for your lab's mission. Some types may be relevant to FA labs only in certain situations. Microhardness testers are an example of such equipment. New equipment types are quickly becoming sophisticated. Defect detection, and in-line SEM tools are examples.

There are other databases available for additional information; magazines such as Evaluation Engineering, American Laboratory, etc, regularly produce vendor data issues.

This is not a preferred vendor list. The names in this list are not necessarily members of SEMI/SEMATECH. Appearance of a manufacturer's name is not intended to be a recommendation of suitability for a specific purpose. Inclusion should not be interpreted as representative of endorsement by SEMATECH, ISTFA/ASM, or the author, of a particular equipment manufacturer over others. Conversely, inadvertent omission of a manufacturer should not be interpreted as representative of disapproval or dissatisfaction.

Disclaimer

Major Subject Areas:
- Analytical Equipment
- Analytical Services
- Chemical Reagents
- Computer Equipment For F/A
- Confocal Microscopes
- Curve Tracers
- Decapping and Decapsulation
- E-Beam Probers
- EBIC Equipment
- Educational Services
- Electronic Equipment
- ESD Testers
- Environmental Test Equipment
- Focused Ion Beam (FIB) Systems
- Furnaces and Ovens
- Glass Wheels, Frosted
- Handtools, Tweezers, Etc
- Image Analyzers
- Information
- Infrared Microscopes and Imaging
- Lab Equipment
- Lapping Film
- Lasers
- Laser Confocal Microscopes
- Liquid Crystals
- Metallurgical Supplies
- Microscopes, Optical
- Microscopes, Measuring
- Office and Lab Furniture
- Photoemission Instruments
- Photographic Equipment
- Plasma and RIE Systems
- Probes and Associated Equipment
- Safety Equipment For Laboratories
- Scanning Electron Microscopes (SEM)
- Scanning Tunneling Microscopy and AFM
- SEM Supplies
- Seminars and Training For F/A
- Standards
- Testers, Electronic
- Testers, Hardness
- Testers, Temperature
- Tools and Renewable Lab Supplies
- Transmission Electron Microscopes
- Video Cameras and Imaging Systems
- X-Ray Analysis Equipment
- X-Ray Imaging Equipment

Analytical Instrumentation

Analect Instruments
17819 Gillette Ave
Irvine, California 92714
714-660-8801
FTIR systems

E G & G Ortec
100 Midland Road
Oak Ridge, Tennessee 37831-0895
615-482-4411
x-ray analytical systems

Leybold-Heraeus Vacuum Products Inc
LAS Group
5700 Mellon Road
Export, Pennsylvania 15632
412-327-5700
vacuum system components, XPS systems

Kevex Corporation
1101 Chess Drive
P.O. Box 4050
Foster City, California 94404
415-573-5866
x-ray spectrometers, XRF systems

Perkin-Elmer, Physical Electronics Division (PHI)
6509 Flying Cloud Drive
Eden Prarie, Minnesota 55344
612-828-6300
SAM, AES, ESCA, XPS, UPS systems and components

Pernicka Corporation
112 Racquette Drive
Fort Collins, Colorado 80524
303-224-0220
mass spec, vacuum science, and instrumentation

Physics International Company
2700 Merced Street
P.O. Box 1538
San Leandro, California 94577
415-577-7112
Ionscan ion implant monitors

Princeton Gamma-Tech
Solid State Measurements, Inc
600 Seco Rd
Monroeville, Pennsylvania 15146
412-373-3050
spreading resistance profilers

Varian
Eimac Salt Lake Division
1678 South Pioneer Road
Salt Lake City, Utah 84104
801-972-5000
replacement x-ray tubes for analytical systems

Analytical Services

Accurel Systems International
Sunnyvale, California

Advanced Materials Engineering Research
Sunnyvale, California

Alliance Analytical
10510 Research, S.E.
Albuquerque, New Mexico 87123
505-299-1967

Analytical Answers, Inc
4 Arrow Drive
Woburn, Massachusetts 01801
617-938-0300

Analytical Service Center
2132 Michelson Drive
Irvine, California 92715
714-975-0141

Alcatel Vacuum Products, Inc
40 Pond Park Road
Higham, Massachusetts 02043
617-938-0300 (chem)

Associated Testing Laboratories
53 Second Ane
Burlington, Massachusetts 01803
617-272-9050

Balzas Analytical Labs
1380 Borregas Avenue
Sunnyvale, California 94069-1004
408-745-0600 (chem)

Continental Testing Laboratories, Inc
8385 South U.S. Highway 17-92
Fern Park, Florida 32730
305-831-2700

DPA Labs, Inc
22512 Ward Ave
Simi Valley, California 93065-1860
805-581-9200

Charles Evans and Associates
1670 South Amphlett Boulevard
Suite 120
San Mateo, California 94402
415-572-1601

Dunegan-Endevco
Rancho Viejo Road
San Juan Capistrano, California 92675
714-831-9131

Eastman Kodak Company
Kodak Apparatus Division
Special Product Sales
901 Elmgrove Road
Rochester, New York 14650
716-726-9309

Failure Analysis Associates
149 Commonwealth Dr.
Menlo Park, California
415-326-9400

Geller Microanalytical Laboratory
1 Intercontinental Way
Peabody, Massachusetts 01960
508-535-5595

Grayfox Component Technology Center
(Control Data Corp)
1160 Grayfox Road
Arden Hills, Minnesota 55112-6994
612-482-2026, 6603

Hi-Rel Laboratories, Inc
911 S. Mountain Ave
Monrovia, California 91016
818-357-6083

Hughes Aircraft
El Segundo, California

IBM Analytical Services
Hopewell Junction, New York

Integrated Circuit Engineering (ICE)
15022 N. 75th Street
Scottsdale, Arizona 85260
602-998-9780

R. J. Lee Group, Inc
Berkeley, California
materials characterization

Martin-Marietta
Failure Analysis Laboratory
Denver, Colorado

Materials Analytical Services
2418 Blue Ridge Road
Suite 105
Raleigh, North Carolina 27607
919-881-7708

Measurement and Analysis Laboratory
General Electric Company
Aerospace Electronic Systems Department
M/D 747 Broad Street
Utica, New York 13503
315-797-1000, Ext 5125

Metrum Viking Laboratories
440 Bernardo Avenue
Mountain View, California 94043
415-969-5500

Northern Analytical Laboratory
23 Depot Street
Merrimack, New Hampshire 03054
603-429-9500 (chem)

Onieda Research Services
1 Halsey Road
Whitesboro, New York 13492
315-736-5480
508-670-2383

Perkin-Elmer, Physical Electronics Div
6509 Flying Cloud Drive
Eden Prarie, Minnesota 55344
612-828-6100

Pernicka Corporation
112 Racquette Drive
Fort Collins, Colorado 80524
303-224-0220

Photometrics
15151 Springdale St.
Huntington Beach, California 92649

QRL Analysis Corporation
77 Auriga Drive
Napean, Ontario, Canada K2E-7Z7
613-226-1885

Reliability Analysis Center
IIT Research Institute
P.O. Box 4700
Rome, New York 13440-8200

RIGA Analytical Lab, Inc
3375 Cott Blvd, Suite 132
Santa Clara, California 95051
408-496-6944

Semiconductor INSIGHTS, Inc
Head Office
P.O. Box 130
Stittsville, Ontario
Canada K2S 1A2
613-831-6116
fax 613-831-5001

Scanning Electron Analysis Laboratories (SEAL)
250 North Nash Street
El Segundo, California 90245
213-322-2011

Solecon Laboratories, Inc
1190-P Mountain View-Alviso Rd
Sunnyvale, California 94089
408-734-5045

Sonoscan, Incorporated
530 Green Street
Bensenville, Illinois 60106
708-766-7068

Structure Probe Inc, SPI Supplies
Box 656
West Chester, Pennsylvania 19381-0656
215-436-5400

Surface Science Laboratories
1206 Charleston Road
Mountain View, California 94043
415-962-8767

Technology Associates
51 Hillbrook Drive
Portola Valley, California 94025-7933
415-941-8272

TFI - Telemark (Cunningham Associates)
170 Nortech Parkway
San Jose, California 95134
408-945-8700

Trace Laboratories
Chatsworth, California

UTI Microlabs (Ultratest International)
Microlabs Division
142 Charcot Avenue
San Jose, California 95131
408-433-2244

Walter C. McCrone Associates, Inc
2820 South Michigan Avenue
Chicago, Illinois 60616
312-842-7100

Zenith Failure Analysis Services
1000 N. Milwaukee Avenue
Glenview, Illinois 60025
312-391-7000

Chemical Reagents and Lab Equipment

Aldrich Chemical Company
1001 West Saint Paul Avenue
Milwaukee, Wisconsin 53233
800-558-9160

Alfa Products
Johnson Matthey
P.O. Box 8247
Ward Hill, Massachusetts 01835-0747

Allied-Signal, Inc.
Planarization and Diffusion Products
1090 South Milpitas Blvd
Milpitas, California 95035
408-946-2411
fax 408-980-1430
ACCUGLASS spin-on glass (SOG)

J.T. Baker Chemical Company
222 Red School Lane
Phillipsburg, New Jersey 08865

Baxter Scientific Products
526 W. 21st Street
P.O. Box 27568
Tempe. Arizona 85282
602-968-3152

Capitol Scientific, Inc.
2500 Rutland Drive
P.O. Box 9250
Austin, Texas 78766
512-836-1167

Chemcyclopedia
American Chemical Society
1156 16th Street N.W.
Washington, D.C. 20036
1-202-872-4600

Dynaloy, Inc (Uresolve®)
7 Great Meadow Lane
Hanover, New Jersey 07936
201-887-9270

Resources for Failure Analysis

E.I. du Pont de Nemours & Co. Inc.
Chemicals and Pigments Department
Customer Service Center
Wilmington, Delaware 19898

Eastman Organic Chemicals
Eastman Kodak Company
343 State Street
Rochester, New York 14650

Fisher-Allied Scientific
International Headquarters
50 Fadhem Road
Springfield, New Jersey 07081
201-379-1400

Fluoroware, Inc.
102 Jonathan Blvd., North
Chaska, Minnesota 55318
612-448-3131
inert plastic labware

General Chemical Company
(formerly Allied Chemical)
Morristown, New Jersey
07960
(fuming nitric acid)

Puritan Products, Inc
635-711 Mill Street
Allentown, Pennsylvania 18103
215-435-4431
fuming nitric and fuming sulfuric acids

Sargent-Welch
Western Regional Office
1617 East Ball Road
Anaheim, California 92803
714-772-3550
laboratory apparatus and chemicals

Spectrum Chemical Mfg. Corp.
14422 South San Pedro Street
Gardena, California 90248-9985
310-516-8000
(f) 310-516-9843

VWR Scientific
P.O. Box 29027
Phoenix, Arizona 85038
(602)-269-7511
laboratory apparatus. Offices in all major cities

Computer Equipment for F/A

Special Note:

Visit local software and hardware stores and marts in your geographic area for current prices and fresh products. Customer service and technical assistance varies widely in quality from store to store.

Businessland
1001 Ridder Park Drive
San Jose, California 95131-9787
800-551-2468

Eastman Kodak
Rochester, New York
14650
800—242-2424

Hewlett-Packard Corporation
Palo Alto, California
ScanJet Plus® 9190A

Logitech, Inc
ScanMan® hand-held scanner

Marstek
17795 Skypark Circle, Suite F
Irvine, California 92714
714-833-7740
Scanners also available from Xerox, Kurzweil, and HP

PERX
1730 S. Amphlett Blvd.
Suite 222
San Mateo, California 94402
415-573-0834
oscilloscope, generator, spectrum analysis, and logic analyzers for PC, PS/2, and MAC

Sony Corporation
color cameras, imaging systems, color video printers

Truevision, Inc
7340 Shadeland Station
Indianapolis, Indiana 46256
Targa® and Tips® imaging software; systems

WordPerfect Corporation
1555 N. Technology Way
Orem, Utah 84057
801-225-5000
word processor WordPerfect 5.1® handles images and text files together, and is in worldwide use

World Videophone Teleconferencing Technologies, Ltd
2646 South Loop West
Suite 530
Houston, Texas 77054
713-665-2393

Special Note:

Refer also to magazines such as *Computer Shopper*, *BYTE*, and *PC*.

Curve Tracers

Hewlett-Packard Test and Measurement Sector
Santa Clara, California 41458
800452-4844
Parameter analyzer

Huntron Instruments, Inc
15123 Highway 99 North
Lynwood, Washington 98037
206-743-3171
Tracker® economy curve-tracers

Tektronix, Inc
P.O. Box 1700
Beaverton, Oregon 97075
800-547-1512
model 576, 370, 371, etc

Ultratest International, Inc (UTI)
San Jose, California

Decapping and Decapsulation

Aldrich Chemical Company
1001 West Saint Paul Avenue
Milwaukee, Wisconsin 53233
800-558-9160
fuming acids

Baxter Healthcare Corporation
210 Great Southwest Parkway
Grand Prarie, Texas 75050
red fuming nitric acid

B&G Enterprises
94 Hangar Way
Watsonville, Ca. 95076
408-728-3638
H_2SO_4 and HNO_3 decap machines

CEM Corporation
P.O. Box 200
Matthews, North Carolina 28106
704-821-7015
800-334-6317
microwave digestion systems

Dremel Manufacturing
Division of Emerson Electric Company
4915 21st Street
Racine, Wisconsin 53406
414-554-1390
Moto-tools®, carbide grinding bits

Mercator Control Systems, Inc
3211 S. Shannon Street
Santa Ana, California 92704
714-754-0703
fax 714-754-0905
plasma decapping

Wahl Clipper Corporation
2900 Locust Street
Sterling, Illinois 61081
815-625-6525
Sterling Minidrill®

ICOM Corporation
734 Silver Spur Road, Suite 201
Rolling Hills Estates, California 90274
213-544-0393
nitric acid decapsulation system

Puritan Products
635 Mill Street
Allentown, Pennsylvania 18103
215-435-4431
red fuming nitric acid

E-Beam Probers

Avantest America, Inc
300 Knightbridge Pkwy
Lincolnshire, Ill 60069

Schlumberger Technologies ATE
1601 Technology Drive
San Jose, California 95110
408-437-5000

Educational Services

Accelerated Testing
Convention Services
Attn. Wayne Nelson Seminars
P.O. Box 1216
Clifton Park, New York 12065-1216
f 518-383-6063

EBIC Equipment

Spectrum Sciences
3050 Oakmead Village Drive
Santa Clara, California 95051
408-727-1567
fax 408-727-1322
EBIC speciman current amplifiers
voltage contrast equipment for SEMs

Electronic Equipment

B&W Engineering Services
Commerce Park (H-10)
3303 Harbor Blvd
Coata Mesa, California 92626
714-540-9975
PIND testers

GENRAD
Component Test Division
37 Main Street
Bolton, Massachusetts 01740
digibridge® RLC testers, benchtop testers

Hewlett-Packard
P. O. Box 10301
Palo Alto, California 94303-0890
4145 B Semiconductor Parameter Analyzer; many other products

ITT/Pomona Electronics
1500 East Ninth Street
P.O. Box 2767
Pomona City, California 91769
714-629-3317
wide line of test leads, connectors

Keithley Instruments, Inc
30500 Bainbridge Road
Cleveland, Ohio 44139-2216
216-498-3100
216-248-6806

Sage Enterprise, Inc
254 Polaris Avenue
Mountain View, California 94043
415-969-5111
die bond evaluation testers

Tektronix, Inc
P.O. Box 1700
Beaverton, Oregon 97075
800-547-1512
oscilloscopes, test instruments, and computer systems

Tucker Electronics Company
1717 Reserve Street
Garland, Texas 75042-7621
new, used, and reconditioned electronics equipment

ESD Testers

IMCS
1300 Spacepark
Mountain View, California 94043
408-980-0660
ESD and latchup testers

Keytek Instrument Corporation
12 Cambridge Street
Burlington, Massachusetts 01803
617-272-5170
ESD and surge testers

Schaffner EMC, Inc
377 Route 17
Hasbrouck Heights, New Jersey 07604
201-288-6860
ESD and interference simulators

Environmental Test Equipment

Blue M Electric Company
Corporate Headquarters
Blue Island, Illinois 60646
312-385-9000
ovens

Dunegan/Endevco
Rancho Viejo Road
San Juan Capistrano, California 92675
714-831-9131
PIND testers

Express Test Corporation
1161-D San Antonio Road
Mountain View, California 99043
415-968-8818
autoclave accelerated test systems (HAST)

Helium Leak Testing, Inc
19348 Londelius Street
Northridge, California 91324
He leak detectors, calibrated leaks, etc

Magnaflux Corporation
7300 West Lawrence Avenue
Chicago, Illinois 60656
Spotcheck (R) dye penetrant, Zyglo (R)
mag. particle test supplies, fluoroscein die

Ransco
1400 E. Statham Parkway
Oxnard, California 93033-3918
805-487-7777
thermal stress screening systems

Spectrum Sciences
3050 Oakmead Village Drive
Santa Clara, California 95051
408-727-1567
nuclear instrumentation systems

Focused Ion Beam (FIB) Systems

FEI Company
19500 N.W. Gibbs Drive
Suite 100
Beaverton, Oregon 97006-6907
503-690-1500

Micrion Corporation
One Corporation Way
Centennial Park
Peabody, Massachusetts 01960-7990
508-531-6464

Microbeam, Inc.
1125 Business Center Drive
Newbury Park, California 91320
805-499-4629

Oxford Applied Research
Crawley Mill
Witney, Oxfordshire OX8 5TJ
0993-773 575

Schlumberger Technologies
ATE Division
1601 Technology Drive
San Jose, California 95110
408-437-9031

Seiko Instruments USA, Inc
1150 Ringwood Court
San Jose, California 95131
408-933-5821

Furnaces and Ovens

Barnstead/Thermolyne
(Thermolyne/Sybron)
2555 Kerper Boulevard
Dubuque, Iowa 52001

Blue M Corporation
Blue Island, Illinois

Gases, Compressed

Air Products and Chemicals
Specialty Gas Department
Box 538
Allentown, Pennsylvania 18105
215-481-8257

Alphagaz/Liquid Air Corporation
2121 North California Blvd.
Walnut Creek, California 94596
415-977-6500
800-248-1427

Linde/Union Carbide
Linde National Specialty Gas Office
P.O. Box 444
100 Davidson Avenue
Somerset, New Jersey 08873

Matheson Gas Products
Main Office
932 Paterson Plank Road
P.O. Box 85
East Rutherford, New Jersey 07073

Spectra Gases, Inc.
277 Coit Street
Irvington, New Jersey 07111
201-372-2060
800-932-0624

Handtools, Tweezers, Etc

Jensen Tools, Inc.
7815 South 46th Street
Phoenix, Arizona 85044-0020
602-968-6231

Specialized Products Company
3131 Premier Drive
Irving, Texas 75063-9968
800-527-5018

Techni-Tool
5 Apollo Road
P.O. Box 368
Plymouth Meeting, Pennsylvania 19462
215-825-4990
wide range of general purpose tools, etc

Terra Universal, Inc
700 N. Harbor Blvd
Anaheim California 92805-9949
714-526-0100
fax 714-992-2179
wide range of mfg products; glove boxes

Utica-Swiss Tool Company
Subsidiary, The Triangle Corporation
Orangeburg, South Carolina 29115
803-534-7010
quality hand tools

Image Analyzers

Buehler

Leco Corporation
St, Joseph, Michigan

Information, General

Joint Electron Device Engineering Council (JEDEC)
2001 Pennsylvania Avenue N.W.
Washington, D.C. 20006
202-457-4971
To order documents, call 800-854-7179 (standards and publications are printed by Global Engineering Documents Co.)

Microelectronic Manufacturing and Testing Desk Manual
Lake Publishing Corporation
Box 159
17730 West Peterson Road
Libertyville, Illinois 60048-9989
312-362-8711
fax 312-362-3484

Government-Industry Data Exchange Program (GIDEP)
U.S. Naval Warfare Assessment Center
Corona, California 91720
714-736-4677

Infrared Microscopes and Imaging

Compix, Inc
P.O. Box 885
Tualatin, Oregon 97062-0885
economy infrared equipment

EDO Corporation / Barnes Engineering Company
44 Commerce Road P. O. Box 53
Stamford, Connecticut 06904-0053
203-348-5381
infrared thermal imaging systems

FSI
FLIR Systems®
16505 SW 72nd Avenue
Portland, Oregon 97224-9901
503-684-3731

Hughes Aircraft Company
5250 West Century Boulevard
P.O. Box 90515
Los Angeles, California 90009
213-391-0711

Inframetrics
12 Oak Park Drive
Bedford, Massachusetts 01730
617-275-8990
infrared imagers

Research Devices, Inc
616 Springfield Ave
Berkeley Heights, New Jersey 07922
201-464-0668
short-wave infrared microscopes

UTI Instruments Company
325 N. Mathilda Avenue
P.O. Box 519
Sunnyvale, California 94088-3519
408-738-3301
thermal imaging systems

Resources for Failure Analysis

Lab Equipment

Alcatel Vacuum Products, Inc.
40 Pond Park Road
Higham, Massachusetts 02043
617-749-8710
fax 617-749-8660
vacuum pumps

Bid Service
640 Cokeman Avenue
Asbury Park, New Jersey 07712
908-775-8300
908-774-1443
Used lab and fab equipment

CuSaco Semiconductor Equipment Sales
P.O. Box 1136
Asbury Park, New Jersey 07712
908-502-0246
Used semiconductor manufacturing equipment

LABX System, Inc
One Elm Square
Route 114
North Andover, Massachusetts 01845
508-975-3336
Captair® freestanding ductless fume hoods

Lapping Film

Allied High Tech Products, Inc
2376 East Pacifica Place
Rancho Dominguez, California 90220
213-635-2466

Minnesota Mining and Manufacturing (3M)
Industrial Abrasives Division
Microfinishing Systems
3M Center
St. Paul, Minnesota 55144
OEM of Mylar® abrasive lapping film

Moyco Industries, Inc
S.E. Corner 21st and Clearfield Streets
Philadelphia, Pennsylvania 19132
215-229-0470

South Bay Technology, Inc
1120 Via Callejon
San Celmente, California 92672
714-492-2600

Laser Confocal Microscopes

Nikon, Inc. / Lasertec Corporation
Instrument Group
1300 Walt Whitman Road
Melville, New York 11747-3064
516-547-8500
fax 516-547-0306

Technical Instrument Company
San Francisco, California

Carl Zeiss, Inc
One Zeiss Drive
Thornwood, New Jersey 10594
914-747-1800

Lasers

Aegis Laser, Inc
4770 San Pablo Avenue
Emerviloe, California 94608

Florod Corporation
17360 So. Gramercy Place
Gardena, California 90247
213-532-2700
xenon and YAG laser cutters

Quantum Chromo Dynamics, Inc
2401 208th Street
Unit 9
Torrance, California 90501
laser cutters

Liquid Crystals

E-M Chemicals
5 Skyline Drive
Hawthorne, New York 10532
914-592-4660

Technology Associates
51 Hillbrook Drive
Portola Valley, California 94025
415-941-8272

Metallurgical Supplies

Allied High Tech Products, Inc
2376 East Pacifica Place
Rancho Dominguez, California 90220
213-635-2466
cross-sectioning equipment and supplies

Allied-Signal, Inc.
Planarization and Diffusion Products
1090 S. Milpitas Blvd
Milpitas, California 95035
408-946-2411
Accuglass® spin-on glass (SOG)

Buehler, Ltd.
41 Waukegan Road
Lake Bluff, Illinois 60044
312-295-6500
wide range of metallurgical products

Crystalite Corporation
13449 Beach Avenue
Marina Del Ray, California 90292
213-821-8206
diamond grinding wheels

Epoxy Technology, Incorporated
14 Fortune Drive
Billerica, Massachusetts 01821
508-667-3085
Epo-Tek®301-2 optically transparent epoxy

Excel Technologies, Inc
99 Phoenix Avenue
Enfield, Connecticut 06082
203-741-3435
fax 203-745-7212

Extec
90 Phoenix Avenue
Enfield, Connecticut 06082
203-741-3435

Fisher Scientific
Corporate Headquarters
711 Forbes Avenue
Pittsburgh, Pennsylvania 15219
412-562-8300
disposable supplies for Buehler, Leco, and Jarrett equipment

Grayco Optical Corporation
91-4 Colin Drive
Holbrook, New York 11741
516-472-3100
Attn: Ty Clinton
frosted glass wheels for die sectioning

Integrated Technologies, Inc.
70 Mill Road
Acushnet, Massachusetts 02743
508-998-3071
P-6204 spin-on coater

Leco Corporation
St, Joseph, Michigan

Minnesota Mining and Manufacturing (3M)
Industrial Abrasives Division
Microfinishing Systems
3M Center
St. Paul, Minnesota 55144
Mylar® abrasive lapping film

Moyco Industries, Inc
S.E. Corner 21st and Clearfield Streets
Philadelphia, Pennsylvania 19132
215-229-0470
abrasives, lapping film

South Bay Technology, Inc
1120 Via Callejon
San Clemente, California 92672
714-492-2600
slow saws, abrasive powder, slurry cutter, abrasive film

Struers, Inc
26100 First Street
Westlake, Ohio 44145
216-871-0071

23440 Hawthorne Blvd
Building Two, Suite 220
Torrance, Ca 90505
213-373-031

3111 Roberts Street
Suite L-1
East Hartford, Ct 06108
203-289-7427
abrasives, metallurgical equipment

T B W Industries, Inc
Forest Grove Road
Furlong, Pennsylvania 18925
215-794-8070
perforated flat diamond grinding wheels

Technology Associates
51 Hillbrook Drive
Portola Valley, California 94025
415-941-8272
frosted glass wheels, complete cross sectioning kits

Office and Lab Furniture

Herman Miller, Inc
Corporate Center
8500 Byron Road
Zeeland, Michigan 49464
616-772-3300
Actionlab® workstation and lab systems

Kalamazoo Technical Furniture, Inc
P.O. Box 1165
Kalamazoo, Michigan 49005-1165
616-385-2255
Teclab® work stations, lab furniture

Kewaunee
Technical Workstation Division
901 FM 20
Lockhart, Texas 78644-0930
800-824-6626
Sturdilite® workstations

Microscopes, Optical

American Optical
Scientific Instrument Division
Buffalo, New York 14215

Bausch and Lomb
Scientific Optical Products Division
Rochester, New York 14602
716-338-6000

Cambridge Instruments
Optical Systems Division
P.O. Box 123
Buffalo, New York 14240-0123
716-891-3000

Leica, Inc.
(Leica Mikroskopie und Systeme GmbH)
111 Deer Lake Road
Deerfield, Illinois 60015
(Leica, Leitz, Reichert and Wild Heerbrugg)

E. Leitz, Inc / Wild Heerbrugg
Link Drive
Rockleigh, New Jersey 07647
201-767-1000

Karl Suss America, Inc
P.O. Box 157
Suss Drive
Waterbury Center, Vermont 05677
802-244-5181

Nikon Incorporated
623 Stewart Avenue
Garden City, New York 11530
516-248-5200

Olympus Corporation/IFD of America
Precision Instrument Division
4 Nevada Drive
New Hyde Park, New York 11042
516-488-3880

Optische Werke AG
Hernalser Haupstr
219.A-1171
Wien, Austria 0222-461641

Reichert-Jung
Polyvar® microscopes

Unitron Instruments, Inc
101 Crossways Park West
Woodbury, New York 11797
measuring microscopes

Wild Heerbrugg Instruments, Inc.
465 Smith Street
Farmingdale, New York 11735
516-293-7400

Carl Zeiss, Inc
444 Fifth Avenue
New York, N.Y. 10018
212-730-4400
7082 Oberkochen
West Germany

Microscopes, Measuring

Unitron, Inc.
Instrument Group
170 Wilbur Place
P.O. Box 469
Bohemia, New York 11716
516-589-6666
fax 516-589-6975

Photoemission Instruments

Astromed Limited
Innovation Centre
Cambridge Science Park
Milton Road
Cambridge CB4 4GS
England 0223-420705
Represented in the USA by:
Microscience, Inc
41 Accord Park Drive
Norwell, Massachusetts 02061
617-871-0308
low-light CCD imagers

Hypervision
46560 Fremont Boulevard
Suite #414
Fremont, California 94538
415-651-7768
fax 415-651-1415

Varo Opto-Electronics
2201 West Walnut Street
P.O. Box 469015
Garland, Texas 75046-9015
214-487-4597
fax 214-487-4726
gated image intensifiers

Xybion Electronic Systems
8380 Miralani Drive
San Diego, California 92126
619-566-7850
fax 619-566-2032

Photographic Equipment

Kodak

Melgar Photographers
2971 Corvin Drive
Santa Clara, California 95051
408-733-4500
die photographs

Polaroid Corporation
Industrial Marketing Department
575 Technology Square
Cambridge, Massachusetts 02139
800-225-1618
film, photographic systems
CCD still video

Plasma and Rie Systems

Applied Materials, Inc.
3050 Bowers Avenue
Santa Clara, California 95054-3299
408-727-5555

Alcatel Comptech
San Jose, California

Ion and Plasma Equipment, Inc.
210 Hammond Avenue
Fremont, California 94539
415-490-4500
fax 415-490-4040
Giga-Etch® 100-E plasma decapper

ICOM Corporation
Rolling Hills Estates, California

IRIS (France)
San Jose, California
plasma decapping equipment, endpoint detection

LFE Plasma Systems
55 Green Street
Clinton, Massachusetts 01510
800-343-4813
barrel plasma reactors

March Instruments Incorporated
125 Mason Drive
Suite J
Concord, California 94520
415-827-1240
plasma etchers, reactive ion etchers

Mercator Control Systems, Inc
3211 S. Shannon Street
Santa Ana, California 92704
714-754-0703
fax 714-754-0905

Nextral
Santa Clara, California

Oxford Plasma Technology
130A Baker Ave Extension
Concord, Massachusetts 01742
508-369-7321
microetch components

Plasma Technology
1050-F East Duane Ave.
Sunnyvale, California 94086
408-732-3001

Plasma-Therm Industrial Products (IP), Inc.
9509 International Court
St. Petersburg, Florida 33716
813-577-4999

Samco International, Inc.
532 Weddell Drive
Suite 5
Sunnyvale, California 94089
408-734-0459
RIE machines and UV ozone cleaners, strippers

Technics, Inc.
6591 Sierra Lane
Dublin, California 94568
415-829-9000
Plasma, RIE, and ion beam systems

Tegal
11 Commerce Boulevard
Novato, California 94947
415-472-7500

Trion Technology
3019 Alvin Devine
Suite 240
Austin, Texas 78741
512-385-3616
fax 512-385-3328
RIE and CVD machines for FA

Probes and Associated Equipment

Alessi, Incorporated
35 Parker
Irvine, California 92714
714-830-0660

E-M Chemicals
5 Skyline Drive
Hawthorne, New York 10532
914-592-4660
liquid crystals

Infotronix Systems, Inc.
San Jose, California

Inter-Logic Systems
7201 E. Garden Grove Blvd
Garden Grove, California 92641
714-891-4456
programmable stage probers, probe cards

Karl Suss America, Inc
P.O. Box 157
Suss Drive
Waterbury Center, Vermont 05677
802-244-5181

The Micromanipulator Company, Inc
2801 Arrowhead Drive
Carson City, Nevada 89706
702-882-2400
probes, probers, C-V test stations

Lucas/Signatone Corporation
393-J Tomkins Court
Gilroy, California 95020
408-848-2851

Rucker and Kolls

Temptronic Corporation
55 Chapel Street
Newton, Massachusetts 02158
617-969-2501
hot chucks, spot heaters

Ultraprobe Corporation
125 Columbia Street
Suite A
Laguna Hills, California 92656
714-362-0300
non-contact ultraviolet probers

Wentworth Laboratories, Inc
500 Federal Road
Brookfield, Connecticut 06804
203-775-0448
probers

Safety Equipment for Laboratories

Aldrich Safety Products
Aldrich Chemical Company, Inc
1001 West Saint Paul Avenue
Milwaukee, Wisconsin 53233
800-558-9160

Direct Safety Company
7815 S. 46th Street
Phoenix, Arizona 85044-9974
800-462-3140

Eagle Manufacturing Company
Wellsburg, West Virginia 26070
lab safety disposal cans

Lab Safety Supply
P.O. Box 1368
Janesville, Wisconsin 53547-1368
608-754-2345
acid and hazardous material storage cabinets

Solkatronic Chemicals, Inc.
30 Two Bridges Road
Fairfield, New Jersey 07004-1530
800-521-3981
emergency gas bottle containment systems

Scanning Acoustical Microscopes (SAM)

Panametrics, Inc
221 Crescent Street
Waltham, Massachusetts 02254
617-899-2740
ultrasonic inspection systems

Sonoscan, Inc
530 E. Green Street
Bensenville, Illinois 60106
312-766-7088
scanning laser acoustic microscopes (SLAM)
C-mode scanning acoustic microscope (C-SAM)

Sonix, Incorporated (formerly Sonotek)
8700 Morrissette Drive
Springfield, Virginia 22152
703-440-0222
scanning ultrasonic microscope systems

Therma-Wave, Inc
47320 Mission Falls Court
Fremont, California 94539
415-490-3663
laser analytical systems

Scanning Electron Microscopes (SEM)

Amray, Inc.
160-T Middlesex Turnpike
Bedford, Massachusetts 01730
617-275-1400

Angstrom Technologies
1815 West Fiest Ave.
Suite 102
Mesa, Arizona 85202
602-649-1986
fax 602-649-1823

Cambridge Instrument Company, Inc
40 Robert Pitt Drive
Monsey, New York 10952
914-356-3331

Electro-Scan Corporation
66 Concord Street
Wilmington, Massachusetts 01887
508-988-0055
ESEM® environmental SEMs

Hitachi, Limited
Nissei Sangyo America, Ltd
460 E. Middlefield Road
Mountain View, California 94043
415-961-0461
SEMs and environmental SEMs (E-SEMs)

International Scientific Instruments (ISI)
Corporate Headquarters
3255-6C Scott Boulevard
Santa Clara, California 95051
408-727-9840

JEOL (USA), Inc
11 Dearborn Road
Peabody, Massachusetts 01960
508-535-5900

Ladd Research Industries, Inc
P.O. Box 1005
Burlington, Vermont 05402
802-658-4961

Leica, Inc
Deerfield, Illinois

Nanometrics

Perkin-Elmer/Etec
out of business

Philips Electronic Instruments Co.
85 McKee Drive
Mahwah, New Jersey 07430
201-529-3800

Topcon Technologies
Pleasanton, California

Tracor-Northern
2551 West Beltline Highway
Middleton, Wisconsin 53562
608-831-6511

Scanning Tunneling Microscopy (STM), SPM, AFM

Angstrom Technologies
1815 West Fiest Ave.
Suite 102
Mesa, Arizona 85202
602-649-1986
fax 602-649-1823

Burleigh Instruments, Inc
Burleigh Park
Fishers, New York 14453
716-924-9355
fax 716-924-9072

Digital Instruments, Inc
6780 Cortona Drive
Santa Barbara, California 93117
805-968-8116
fax 805-968-6627

LK Technologies
3910 Roll Ave.
Bloomington, Indiana 47403
812-332-4449
fax 812-332-4493

McAllister Technical Services
West 280 Prarie Ave
Cour d'Alene, Idaho 83814
208-772-9527
fax 208-772-3384

Park Scientific Instruments
476 Ellis Street
Mountain View, California 94043-2204
415-965-2976
fax 415-965-2946

RHK Technologies
1750 West Hamlin
Rochester Hills, Minnesota 48309
313-656-3116
fax 313-656-8347

Topometrix
1505 Wyatt Drive
Santa Clara. California 95054
408-982-9700
fax 408-982-9751

WYCO Corporation
2650 E. Elvira Road
Tucson, Arizona 85706
602-294-1297
fax 602-294-1799

SEM Supplies

Anatech, Ltd
(Technics, Inc)
5510 Vine Street
Alexandria, Virginia 22310
703-971-9200
"Hummer" sputter specimen coater
Nugget, Giga Etch depotting system

Balzers

Ernest F. Fullam, Inc
(EFFA)
P.O. Box 444
Schenectady, New York 12301
accessories for microscopy

FEI Company
19500 N.W. Gibbs Drive
Suite 100
Beaverton, Oregon 90006-6907
503-690-1500
products for ion and electron beam technology

Raith
70C Carolyn Road
Farmingdale, New York 11735
516-293-0870
fax 516-293-0187

Ted Pella, Inc
4595 Mountain Lakes Boulevard
Redding, California 96003
916-243-2200

Tousimis Research Corporation
P.O. Box 2189
Rockville, Maryland 20847-2189
301-881-2450
800-638-9558
supplies for TEM, SEM, microbeam analysis

SPI Supplies div Structure Probe, Inc
P.O. Box 342
West Chester, Pennsylvania 19380
215-436-5400
electron microscopy supplies

Seminars and Training For F/A

D M Data, Inc
6900 E. Camelback Road
Suite 1000
Scottsdale, Arizona 85251
602-945-9620

Integrated Circuit Engineering Corporation (ICE)
15022 N. 75th Street
Scottsdale, Arizona 85260
602-988-0780

Hi-Rel Laboratories, Inc
911 S. Mountain Ave
Monrovia, California 91016
818-357-6083
failure analysis seminars, workshops

Reliability Analysis Center
IIT Research Institute
P.O. Box 4700
Rome, New York 13440-8200
reliability data, seminars in TQM, testability, DFR, F/A, etc.

Scanning Electron Analysis Laboratories (SEAL), Inc
5301 Beethoven Street
Los Angeles, California 90066
213-306-4200

Technology Associates
51 Hillbrook Drive
Portola Valley, California 94025
415-941-8272

Standards

Geller Microanalytical Laboratory
One Intercontinental Way
Peabody, Massachusetts 01960
508-535-5595

Tools and Renewable Lab Supplies

Dwyer Instruments, Inc. (Magnehelic®)
P.O. Box 373
Junction Highways 12 and 212
Michigan City, Indiana 46360
219-879-8000

Testers, Electronic

Heuristic Physics Labs (HPL)
1649 So. Main Street
Milpitas, California 95035
408-263-1753

MOSAID Systems Incorporated
P.O. Box 13579
Kanata, Ontario, Canada K2K1X6
613-836-3977
bit-mapping RAM testers

Tektronix, Inc
P.O. Box 1700
Beaverton, Oregon 97075
800-547-1512

Teradyne, Inc.
Boston, Massachusetts

Testers, Hardness

Nano Instruments, Inc.
P.O. Box 14211
Knoxville, Tennessee 37914
615-927-0500
fax 615-927-3110

New Age Industries, Inc.
2300 Maryland Road
Willow Grove, Pennsylvania 19090
215-657-6040
fax 215-657-6594
handheld hardness testers

Krautkramer Branson
50 Industrial Park Road
Lewiston, Pennsylvania 17044-9990
717-242-0327
fax 717-242-2606
handheld hardness testers

Testers, Temperature

Omega Engineering, Inc.
One Omega Drive
Box 4047
Stamford, Connecticut 06907
203-322-1666
temperature measurement equipment

Temptronic Corporation
55 Chapel Street
Newton, Massachusetts 02158
617-969-2501

Transmission Electron Microscopes

Gatan, Incorporated
Pleasanton, California
TEMs, EELS, PEELS analyzers

Philips Electronic Instruments Company
Mahwah, New Jersey

Video Cameras and Imaging Systems

Digital Equipment Corporation
Continental Boulevard MK01/W83
P.O. Box 9501
Merrimack, New Hampshire 03054-9931

Knights Technology
Santa Clara, California

Kodak Image Acquisition Products
ESTEK Products Division
9625 Southern Pine Blvd
Charlotte, North Carolina 28273
800—423-2921

Polaroid Corporation
Industrial Marketing Department
575 Technology Square
Cambridge, Massachusetts 02139
800-225-1618

Semicaps, Inc.
Santa Clara, California
desktop imaging for microscopy

Sony Corporation
Communication Products
Professional Video Division
1600 Queen Anne Road
Teaneck, New Jersey 0766
201-833-5200

X-Ray Analysis Equipment

Fisions Instruments
Danvers, Massachusetts

Kevex Instruments/Fisons Instruments
355 Shoreway Road
San Carlos, California 94070
415-591-3600

Princeton Gamma-Tech Inc. (PGT)
Route 518 East of Rt 206
Rocky Hill, New Jersey 08540
609-924-7310

X-Ray Imaging Equipment

FeinFocus USA, Inc
5142 N. Clarendon Drive
Suite 160
Agoura Hills, California 91301

Hewlett-Packard
P. O. Box 10301
Palo Alto, California 94303-0890
Faxitron® Mod 43805N cabinet x-ray machines

Imaging Systems, Inc
DeForest, Wisconsin

IRT Corporation
3030 Callan Road
San Diego, California 92121
619-450-4343
RTX systems

Nicolet Mikrox
Imaging group
5225-5 Verona Road
Madison, Wisconsin 53711
608-273-5047
real-time x-ray imaging systems (RTX)

TEC
Knoxville, Tennessee
x-ray residual stress analyzers